Student Solutions Manual

Chemistry

ELEVENTH EDITION

Steven S. Zumdahl
University of Illinois

Susan A. Zumdahl
University of Illinois

Donald J. DeCoste
University of Illinois

Prepared by

Tom Hummel
University of Illinois

CENGAGE

Australia • Brazil • Canada • Mexico • Singapore • United Kingdom • United States

CENGAGE

ISBN: 978-0-357-85074-9

Cengage
200 Pier 4 Boulevard
Boston, MA 02210
USA

Cengage is a leading provider of customized learning solutions. Our employees reside in nearly 40 different countries and serve digital learners in 165 countries around the world. Find your local representative at **www.cengage.com**.

To learn more about Cengage platforms and services, register or access your online learning solution, or purchase materials for your course, visit **www.cengage.com.**

Printed at CLDPC, USA, 06-23

Table of Contents

CHAPTER 1

CHEMICAL FOUNDATIONS

Questions

19. A law summarizes what happens, e.g., law of conservation of mass in a chemical reaction or the ideal gas law, PV = nRT. A theory (model) is an attempt to explain why something happens. Dalton's atomic theory explains why mass is conserved in a chemical reaction. The kinetic molecular theory explains why pressure and volume are inversely related at constant temperature and moles of gas present, as well as explaining the other mathematical relationships summarized in PV = nRT.

21. No, it is useful whenever a systematic approach of observation and hypothesis testing can be used.

23. a. No b. Yes c. Yes

    Only statements b and c can be determined from experiment.

25. Many techniques of chemical analysis require relatively pure samples. Thus, a separation step often is necessary to remove materials that will interfere with the analytical measurement.

27. They are not equivalent. Doubling the temperature on the Kelvin scale is a larger increase than doubling the temperature on the Celsius scale. For example, doubling the Celsius temperature from 25°C to 50°C corresponds to taking the Kelvin temperature from 298 K to 333 K. However, doubling the Kelvin temperature from 298 K to 596 K corresponds with taking the Celsius temperature from 25°C to 323°C. Doubling the temperature are clearly different on the two scales.

29. The gas phase density is much smaller than the density of a solid or a liquid. The molecules in a solid and a liquid are very close together. In the gas phase, the molecules are very far apart from one another. In fact, the molecules are so far apart that a gas is considered to be mostly empty space. Because gases are mostly empty space, their density is very small.

31. The object that sinks has a greater density than water and the object that floats has a smaller density than water. Since both objects have the same mass, the sphere that sinks must have the smaller volume which makes it denser. Therefore, the object that floats has the larger volume along with the greater diameter.

Exercises

Significant Figures and Unit Conversions

33. a. exact b. inexact

c. exact

d. inexact (π has an infinite number of decimal places.)

35. a. $\underline{6.07} \times 10^{-15}$; 3 S.F.

b. $0.00\underline{3840}$; 4 S.F.

c. $\underline{17.00}$; 4 S.F.

d. $\underline{8} \times 10^{8}$; 1 S.F.

e. $\underline{463.8052}$; 7 S.F.

f. $\underline{3}00$; 1 S.F.

g. $\underline{301}$; 3 S.F.

h. $\underline{300.}$; 3 S.F.

37. When rounding, the last significant figure stays the same if the number after this significant figure is less than 5 and increases by one if the number is greater than or equal to 5.

a. 3.42×10^{-4}

b. 1.034×10^{4}

c. 1.7992×10^{1}

d. 3.37×10^{5}

39. Volume measurements are estimated to one place past the markings on the glassware. The first graduated cylinder is labeled to 0.2 mL volume increments, so we estimate volumes to the hundredths place. Realistically, the uncertainty in this graduated cylinder is ±0.05 mL. The second cylinder, with 0.02 mL volume increments, will have an uncertainty of ±0.005 mL. The approximate volume in the first graduated cylinder is 2.85 mL, and the volume in the other graduated cylinder is approximately 0.280 mL. The total volume would be:

$$\begin{array}{r} 2.85 \text{ mL} \\ \underline{+0.280 \text{ mL}} \\ 3.13 \text{ mL} \end{array}$$

We should report the total volume to the hundredths place because the volume from the first graduated cylinder is only read to the hundredths (read to two decimal places). The first graduated cylinder is the least precise volume measurement because the uncertainty of this instrument is in the hundredths place, while the uncertainty of the second graduated cylinder is to the thousandths place. It is always the lease precise measurement that limits the precision of a calculation.

41. For addition and/or subtraction, the result has the same number of decimal places as the number in the calculation with the fewest decimal places. When the result is rounded to the correct number of significant figures, the last significant figure stays the same if the number after this significant figure is less than 5 and increases by one if the number is greater than or equal to 5. The underline shows the last significant figure in the intermediate answers.

a. $53.5 + 5.612 + 6 = 6\underline{5}.112 = 65$

b. $10.67 - 9.5 - 0.634 = 0.\underline{5}36 = 0.5$

c. $3.25 \times 10^{3} + 6.174 \times 10^{2} = 32.5 \times 10^{2} + 6.174 \times 10^{2} = 38.\underline{6}74 \times 10^{2} = 3870$

d. $1.65 \times 10^{-2} - 9.73 \times 10^{-3} = 16.5 \times 10^{-3} - 9.73 \times 10^{-3} = 6.\underline{7}7 \times 10^{-3} = 0.0068$

When the exponents are different, it is easiest to apply the addition/subtraction rule when all numbers are based on the same power of 10.

43. a. For this problem, apply the multiplication/division rule first; then apply the addition/subtraction rule to arrive at the one-decimal-place answer. We will generally round off at intermediate steps in order to show the correct number of significant figures.

However, you should round off at the end of all the mathematical operations in order to avoid round-off error. The best way to do calculations is to keep track of the correct number of significant figures during intermediate steps, but round off at the end. For this problem, we underlined the last significant figure in the intermediate steps.

$$\frac{2.526}{3.1} + \frac{0.470}{0.623} + \frac{80.705}{0.4326} = 0.8\underline{1}48 + 0.75\underline{4}4 + 186.\underline{5}58 = 188.1$$

b. Here, the mathematical operation requires that we apply the addition/subtraction rule first, then apply the multiplication/division rule.

$$\frac{6.404 \times 2.91}{18.7 - 17.1} = \frac{6.404 \times 2.91}{1.\underline{6}} = 12$$

c. $6.071 \times 10^{-5} - 8.2 \times 10^{-6} - 0.521 \times 10^{-4} = 60.71 \times 10^{-6} - 8.2 \times 10^{-6} - 52.1 \times 10^{-6}$
$= 0.\underline{4}1 \times 10^{-6} = 4 \times 10^{-7}$

d. $$\frac{3.8 \times 10^{-12} + 4.0 \times 10^{-13}}{4 \times 10^{12} + 6.3 \times 10^{13}} = \frac{38 \times 10^{-13} + 4.0 \times 10^{-13}}{4 \times 10^{12} + 63 \times 10^{12}} = \frac{4\underline{2} \times 10^{-13}}{6\underline{7} \times 10^{12}} = 6.3 \times 10^{-26}$$

e. $$\frac{9.5 + 4.1 + 2.8 + 3.175}{4} = \frac{19.\underline{5}75}{4} = 4.89 = 4.9$$

Uncertainty appears in the first decimal place. The average of several numbers can only be as precise as the least precise number. Averages can be exceptions to the significant figure rules.

f. $$\frac{8.925 - 8.905}{8.925} \times 100 = \frac{0.02\underline{0}}{8.925} \times 100 = 0.22$$

45. a. $8.43 \text{ cm} \times \frac{1 \text{ m}}{100 \text{ cm}} \times \frac{1000 \text{ mm}}{\text{m}} = 84.3 \text{ mm}$ b. $2.41 \times 10^2 \text{ cm} \times \frac{1 \text{ m}}{100 \text{ cm}} = 2.41 \text{ m}$

c. $294.5 \text{ nm} \times \frac{1 \text{ m}}{1 \times 10^9 \text{ nm}} \times \frac{100 \text{ cm}}{\text{m}} = 2.945 \times 10^{-5} \text{ cm}$

d. $1.445 \times 10^4 \text{ m} \times \frac{1 \text{ km}}{1000 \text{ m}} = 14.45 \text{ km}$ e. $235.3 \text{ m} \times \frac{1000 \text{ mm}}{\text{m}} = 2.353 \times 10^5 \text{ mm}$

f. $903.3 \text{ nm} \times \frac{1 \text{ m}}{1 \times 10^9 \text{ nm}} \times \frac{1 \times 10^6 \text{ μm}}{\text{m}} = 0.9033 \text{ μm}$

47. a. $1.5 \text{ lb} \times \frac{453.6 \text{ g}}{1 \text{ lb}} = 680$ g; 715 g is the larger mass.

b. $1.90 \text{ L} \frac{1 \text{ qt}}{0.9463 \text{ L}} = 2.01$ qt; 1.90 L is the larger volume.

49. Conversion factors are found in Appendix 6. In general, the number of significant figures we

use in the conversion factors will be one more than the number of significant figures from the numbers given in the problem. This is usually sufficient to avoid round-off error.

$$3.91 \text{ kg} \times \frac{1 \text{ lb}}{0.4536 \text{ kg}} = 8.62 \text{ lb}; \ 0.62 \text{ lb} \times \frac{16 \text{ oz}}{\text{lb}} = 9.9 \text{ oz}$$

Baby's weight = 8 lb and 9.9 oz or, to the nearest ounce, 8 lb and 10. oz.

$$51.4 \text{ cm} \times \frac{1 \text{ in}}{2.54 \text{ cm}} = 20.2 \text{ in} \approx 20\ 1/4 \text{ in} = \text{baby's height}$$

51. a. $908 \text{ oz} \times \frac{1 \text{ lb}}{16 \text{ oz}} \times \frac{0.4536 \text{ kg}}{\text{lb}} = 25.7 \text{ kg}$

b. $12.8 \text{ L} \times \frac{1 \text{ qt}}{0.9463 \text{ L}} \times \frac{1 \text{ gal}}{4 \text{ qt}} = 3.38 \text{ gal}$

c. $125 \text{ mL} \times \frac{1 \text{ L}}{1000 \text{ mL}} \times \frac{1 \text{ qt}}{0.9463 \text{ L}} = 0.132 \text{ qt}$

53. a. $1.25 \text{ mi} \times \frac{8 \text{ furlongs}}{\text{mi}} = 10.0 \text{ furlongs}; \ 10.0 \text{ furlongs} \times \frac{40 \text{ rods}}{\text{furlong}} = 4.00 \times 10^2 \text{ rods}$

$$4.00 \times 10^2 \text{ rods} \times \frac{5.5 \text{ yd}}{\text{rod}} \times \frac{36 \text{ in}}{\text{yd}} \times \frac{2.54 \text{ cm}}{\text{in}} \times \frac{1 \text{ m}}{100 \text{ cm}} = 2.01 \times 10^3 \text{ m}$$

$$2.01 \times 10^3 \text{ m} \times \frac{1 \text{ km}}{1000 \text{ m}} = 2.01 \text{ km}$$

b. Let's assume we know this distance to ±1 yard. First, convert 26 miles to yards.

$$26 \text{ mi} \times \frac{5280 \text{ ft}}{\text{mi}} \times \frac{1 \text{ yd}}{3 \text{ ft}} = 45{,}760. \text{ yd}$$

26 mi + 385 yd = 45,760. yd + 385 yd = 46,145 yards

$$46{,}145 \text{ yard} \times \frac{1 \text{ rod}}{5.5 \text{ yd}} = 8390.0 \text{ rods}; \ 8390.0 \text{ rods} \times \frac{1 \text{ furlong}}{40 \text{ rods}} = 209.75 \text{ furlongs}$$

$$46{,}145 \text{ yard} \times \frac{36 \text{ in}}{\text{yd}} \times \frac{2.54 \text{ cm}}{\text{in}} \times \frac{1 \text{ m}}{100 \text{ cm}} = 42{,}195 \text{ m}; \ 42{,}195 \text{ m} \times \frac{1 \text{ km}}{1000 \text{ m}} = 42.195 \text{ km}$$

55. a. $1 \text{ troy lb} \times \frac{12 \text{ troy oz}}{\text{troy lb}} \times \frac{20 \text{ pw}}{\text{troy oz}} \times \frac{24 \text{ grains}}{\text{pw}} \times \frac{0.0648 \text{ g}}{\text{grain}} \times \frac{1 \text{ kg}}{1000 \text{ g}} = 0.373 \text{ kg}$

$$1 \text{ troy lb} = 0.373 \text{ kg} \times \frac{2.205 \text{ lb}}{\text{kg}} = 0.822 \text{ lb}$$

b. $1 \text{ troy oz} \times \frac{20 \text{ pw}}{\text{troy oz}} \times \frac{24 \text{ grains}}{\text{pw}} \times \frac{0.0648 \text{ g}}{\text{grain}} = 31.1 \text{ g}$

$$1 \text{ troy oz} = 31.1 \text{ g} \times \frac{1 \text{ carat}}{0.200 \text{ g}} = 156 \text{ carats}$$

c. 1 troy lb = 0.373 kg; $0.373 \text{ kg} \times \frac{1000 \text{ g}}{\text{kg}} \times \frac{1 \text{ cm}^3}{19.3 \text{ g}} = 19.3 \text{ cm}^3$

57. $15.6 \text{ g} \times \frac{1 \text{ capsule}}{0.65 \text{ g}} = 24$ capsules

59. $\text{warp } 1.71 = \left(5.00 \times \frac{3.00 \times 10^8 \text{ m}}{\text{s}}\right) \times \frac{1.094 \text{ yd}}{\text{m}} \times \frac{60 \text{ s}}{\text{min}} \times \frac{60 \text{ min}}{\text{h}} \times \frac{1 \text{ knot}}{2030 \text{ yd/h}}$

$= 2.91 \times 10^9$ knots

$$\left(5.00 \times \frac{3.00 \times 10^8 \text{ m}}{\text{s}}\right) \times \frac{1 \text{ km}}{1000 \text{ m}} \times \frac{1 \text{ mi}}{1.609 \text{ km}} \times \frac{60 \text{ s}}{\text{min}} \times \frac{60 \text{ min}}{\text{h}} = 3.36 \times 10^9 \text{ mi/h}$$

61. $1 \text{ s} \times \frac{1 \text{ min}}{60 \text{ s}} \times \frac{1 \text{ h}}{60 \text{ min}} \times \frac{65 \text{ mi}}{\text{h}} \times \frac{5280 \text{ ft}}{\text{mi}} = 95.3 \text{ ft} = 100 \text{ ft}$

If you take your eyes off the road for one second traveling at 65 mph, your car travels approximately 100 feet.

63. $180 \text{ lb} \times \frac{1 \text{ kg}}{2.205 \text{ lb}} \times \frac{8.0 \text{ mg}}{\text{kg}} = 650$ mg antibiotic/dose

$2 \text{ wk} \times \frac{7 \text{ days}}{\text{wk}} \times \frac{2 \text{ doses}}{\text{day}} \times \frac{650 \text{ mg}}{\text{dose}} = 18{,}000 \text{ mg} = 18$ g antibiotic in total

65. Volume of lake $= 100 \text{ mi}^2 \times \left(\frac{5280 \text{ ft}}{\text{mi}}\right)^2 \times 20 \text{ ft} = 6 \times 10^{10} \text{ ft}^3$

$$6 \times 10^{10} \text{ ft}^3 \times \left(\frac{12 \text{ in}}{\text{ft}} \times \frac{2.54 \text{ cm}}{\text{in}}\right)^3 \times \frac{1 \text{ mL}}{\text{cm}^3} \times \frac{0.4 \text{ µg}}{\text{mL}} = 7 \times 10^{14} \text{ µg mercury}$$

$$7 \times 10^{14} \text{ µg} \times \frac{1 \text{ g}}{1 \times 10^6 \text{ µg}} \times \frac{1 \text{ kg}}{1 \times 10^3 \text{ g}} = 7 \times 10^5 \text{ kg of mercury}$$

Temperature

67. $T_C = T_K - 273 = 292 - 273 = 19°C$; $T_C = \frac{5}{9}(T_F - 32) = \frac{5}{9}(72°F - 32) = 22.2°C = 22°C$

292 K is the coldest temperature, then 72°F, then 24°C is the hottest temperature.

69. a. $T_C = \frac{5}{9}(T_F - 32) = \frac{5}{9}(-459°F - 32) = -273°C$; $T_K = T_C + 273 = -273°C + 273 = 0 \text{ K}$

b. $T_C = \frac{5}{9}(-40.°F - 32) = -40.°C$; $T_K = -40.°C + 273 = 233 \text{ K}$

c. $T_C = \frac{5}{9}(68°F - 32) = 20.°C$; $T_K = 20.°C + 273 = 293$ K

d. $T_C = \frac{5}{9}(7 \times 10^7 °F - 32) = 4 \times 10^7 °C$; $T_K = 4 \times 10^7 °C + 273 = 4 \times 10^7$ K

71. a. $T_F = \frac{9}{5} \times T_C + 32 = \frac{9}{5} \times 39.2°C + 32 = 102.6°F$ (*Note*: 32 is exact.)

$T_K = T_C + 273.2 = 39.2 + 273.2 = 312.4$ K

b. $T_F = \frac{9}{5} \times (-25) + 32 = -13°F$; $T_K = -25 + 273 = 248$ K

c. $T_F = \frac{9}{5} \times (-273) + 32 = -459°F$; $T_K = -273 + 273 = 0$ K

d. $T_F = \frac{9}{5} \times 801 + 32 = 1470°F$; $T_K = 801 + 273 = 1074$ K

73. $T_F = \frac{9}{5} \times T_C + 32$; from the problem, we want the temperature where $T_F = 2T_C$.

Substituting:

$$2T_C = \frac{9}{5} \times T_C + 32,\ (0.2)T_C = 32,\ T_C = \frac{32}{0.2} = 160°C$$

$T_F = 2T_C$ when the temperature in Fahrenheit is 2(160) = 320°F. Because all numbers when solving the equation are exact numbers, the calculated temperatures are also exact numbers.

75. a. A change in temperature of 140°C is equal to 50°X. Therefore, $\frac{140\,°C}{50\,°X}$ is the unit conversion between a degree on the X scale to a degree on the Celsius scale. To account for the different zero points, −10° must be subtracted from the temperature on the X scale to get to the Celsius scale. The conversion between °X to °C is:

$$T_C = T_X \times \frac{140\,°C}{50\,°X} - 10°C,\ T_C = T_X \times \frac{14\,°C}{5\,°X} - 10°C$$

The conversion between °C to °X would be:

$$T_X = (T_C + 10°C)\,\frac{5\,°X}{14\,°C}$$

b. Assuming 10°C and $\frac{5\,°X}{14\,°C}$ are exact numbers:

$$T_X = (22.0°C + 10°C)\,\frac{5\,°X}{14\,°C} = 11.4°X$$

c. Assuming exact numbers in the temperature conversion formulas:

$$T_C = 58.0°X \times \frac{14°C}{5°X} - 10°C = 152°C$$

$$T_K = 152°C + 273 = 425\ K$$

$$T_F = \frac{9°F}{5°C} \times 152°C + 32°F = 306°F$$

Density

77. $\text{Density} = \frac{\text{mass}}{\text{volume}} = \frac{156\ g}{11.5\ cm^3} = 13.6\ g/cm^3$; from the table, the toxic liquid is mercury.

79. $\text{Mass} = 350\ lb \times \frac{453.6\ g}{lb} = 1.6 \times 10^5\ g;\ \ V = 1.2 \times 10^4\ in^3 \times \left(\frac{2.54\ cm}{in}\right)^3 = 2.0 \times 10^5\ cm^3$

$$\text{Density} = \frac{\text{mass}}{\text{volume}} = \frac{1 \times 10^5\ g}{2.0 \times 10^5\ cm^3} = 0.80\ g/cm^3$$

Because the material has a density less than water, it will float in water.

81. $V = \frac{4}{3}\pi r^3 = \frac{4}{3} \times 3.14 \times \left(7.0 \times 10^5\ km \times \frac{1000\ m}{km} \times \frac{100\ cm}{m}\right)^3 = 1.4 \times 10^{33}\ cm^3$

$$\text{Density} = \frac{\text{mass}}{\text{volume}} = \frac{2 \times 10^{36}\ kg \times \frac{1000\ g}{kg}}{1.4 \times 10^{33}\ cm^3} = 1.4 \times 10^6\ g/cm^3 = 1 \times 10^6\ g/cm^3$$

83. a. $5.0\ \text{carat} \times \frac{0.200\ g}{\text{carat}} \times \frac{1\ cm^3}{3.51\ g} = 0.28\ cm^3$

b. $2.8\ mL \times \frac{1\ cm^3}{mL} \times \frac{3.51\ g}{cm^3} \times \frac{1\ \text{carat}}{0.200\ g} = 49\ \text{carats}$

85. $V = 21.6\ mL - 12.7\ mL = 8.9\ mL;\ \ \text{density} = \frac{33.42\ g}{8.9\ mL} = 3.8\ g/mL = 3.8\ g/cm^3$

87. a. Both have the same mass of 1.0 kg.

b. 1.0 mL of mercury; mercury is more dense than water. *Note*: 1 mL = 1 cm^3.

$1.0\ mL \times \frac{13.6\ g}{mL} = 14$ g of mercury; $1.0\ mL \times \frac{0.998\ g}{mL} = 1.0$ g of water

c. Same; both represent 19.3 g of substance.

$19.3\ mL \times \frac{0.9982\ g}{mL} = 19.3$ g of water; $1.00\ mL \times \frac{19.32\ g}{mL} = 19.3$ g of gold

d. 1.0 L of benzene (880 g versus 670 g)

$75\ mL \times \frac{8.96\ g}{mL} = 670$ g of copper; $1.0\ L \times \frac{1000\ mL}{L} \times \frac{0.880\ g}{mL} = 880$ g of benzene

89. a. 1.0 kg feather; feathers are less dense than lead.

b. 100 g water; water is less dense than gold. c. Same; both volumes are 1.0 L.

91. $V = 1.00 \times 10^3 \text{ g} \times \frac{1 \text{ cm}^3}{22.57 \text{ g}} = 44.3 \text{ cm}^3$

$44.3 \text{ cm}^3 = l \times w \times h = 4.00 \text{ cm} \times 4.00 \text{ cm} \times h, \; h = 2.77 \text{ cm}$

Classification and Separation of Matter

93. A gas has molecules that are very far apart from each other, whereas a solid or liquid has molecules that are very close together. An element has the same type of atom, whereas a compound contains two or more different elements. Picture i represents an element that exists as two atoms bonded together (like H_2 or O_2 or N_2). Picture iv represents a compound (like CO, NO, or HF). Pictures iii and iv contain representations of elements that exist as individual atoms (like Ar, Ne, or He).

a. Picture iv represents a gaseous compound. Note that pictures ii and iii also contain a gaseous compound, but they also both have a gaseous element present.

b. Picture vi represents a mixture of two gaseous elements.

c. Picture v represents a solid element.

d. Pictures ii and iii both represent a mixture of a gaseous element and a gaseous compound.

95. Homogeneous: Having visibly indistinguishable parts (the same throughout).

Heterogeneous: Having visibly distinguishable parts (not uniform throughout).

a. heterogeneous (due to hinges, handles, locks, etc.)

b. homogeneous (hopefully; if you live in a heavily polluted area, air may be heterogeneous.)

c. homogeneous d. homogeneous (hopefully, if not polluted)

e. heterogeneous f. heterogeneous

97. a. pure b. mixture c. mixture d. pure e. mixture (copper and zinc)

f. pure g. mixture h. mixture i. mixture

Iron and uranium are elements. Water (H_2O) is a compound because it is made up of two or more different elements. Table salt is usually a homogeneous mixture composed mostly of sodium chloride (NaCl), but will usually contain other substances that help absorb water vapor (an anticaking agent).

99. Chalk is a compound because it loses mass when heated and appears to change into another substance with different physical properties (the hard chalk turns into a crumbly substance).

101. A physical change is a change in the state of a substance (solid, liquid, and gas are the three states of matter); a physical change does not change the chemical composition of the substance. A chemical change is a change in which a given substance is converted into another substance having a different formula (composition).

a. Vaporization refers to a liquid converting to a gas, so this is a physical change. The formula (composition) of the moth ball does not change.

b. This is a chemical change since hydrofluoric acid (HF) is reacting with glass (SiO_2) to form new compounds that wash away.

c. This is a physical change because all that is happening during the boiling process is the conversion of liquid alcohol to gaseous alcohol. The alcohol formula (C_2H_5OH) does not change.

d. This is a chemical change since the acid is reacting with cotton to form new compounds.

ChemWork Problems

103. The national debt is $2.0875 × 10^{13} and the wealthiest 1.00% represents 3.32 × 10^6 people.

$$\frac{2.0875 \times 10^{13}\text{ dollars}}{3.32 \times 10^{6}\text{ people}} = 6.29 \times 10^6 \text{ dollars/person} = \$6.29 \text{ million dollars per richest 1\%.}$$

105. $$4145\text{ mi} \times \frac{5280\text{ ft}}{\text{mi}} \times \frac{1\text{ fathom}}{6\text{ ft}} \times \frac{1\text{ cable length}}{100\text{ fathom}} = 3.648 \times 10^4 \text{ cable lengths}$$

$$4145\text{ mi} \times \frac{1\text{ km}}{0.62137\text{ mi}} \times \frac{1000\text{ m}}{\text{km}} = 6.671 \times 10^6\text{ m}$$

$$3.648 \times 10^4\text{ cable lengths} \times \frac{1\text{ nautical mile}}{10\text{ cable lengths}} = 3{,}648\text{ nautical miles}$$

107. Because each pill is 4.0% Lipitor by mass, for every 100.0 g of pills, there are 4.0 g of Lipitor present. Note that 100 pills is assumed to be an exact number.

$$100\text{ pills} \times \frac{2.5\text{ g}}{\text{pill}} \times \frac{4.0\text{ g Lipitor}}{100.0\text{ g pills}} \times \frac{1\text{ kg}}{1000\text{ g}} = 0.010\text{ kg Lipitor}$$

109. $T_C = \frac{5}{9}(T_F - 32) = \frac{5}{9}(134°F - 32) = 56.7°C$; phosphorus would be a liquid.

111. $$\text{Total volume} = \left(200.\text{ m} \times \frac{100\text{ cm}}{\text{m}}\right) \times \left(300.\text{ m} \times \frac{100\text{ cm}}{\text{m}}\right) \times 4.0\text{ cm} = 2.4 \times 10^9\text{ cm}^3$$

Volume of topsoil covered by 1 bag =

$$\left[10.\ \text{ft}^2 \times \left(\frac{12\ \text{in}}{\text{ft}}\right)^2 \times \left(\frac{2.54\ \text{cm}}{\text{in}}\right)^2\right] \times \left(1.0\ \text{in} \times \frac{2.54\ \text{cm}}{\text{in}}\right) = 2.4 \times 10^4\ \text{cm}^3$$

$$2.4 \times 10^9\ \text{cm}^3 \times \frac{1\ \text{bag}}{2.4 \times 10^4\ \text{cm}^3} = 1.0 \times 10^5\ \text{bags topsoil}$$

113. $1\ \text{light year} = 1\ \text{yr} \times \frac{365\ \text{day}}{\text{yr}} \times \frac{24\ \text{h}}{\text{day}} \times \frac{60\ \text{min}}{\text{h}} \times \frac{60\ \text{s}}{\text{min}} \times \frac{186{,}000\ \text{mi}}{\text{s}} = 5.87 \times 10^{12}\ \text{miles}$

$$9.6\ \text{parsecs} \times \frac{3.26\ \text{light yr}}{\text{parsec}} \times \frac{5.87 \times 10^{12}\ \text{mi}}{\text{light yr}} \times \frac{1.609\ \text{km}}{\text{mi}} \times \frac{1000\ \text{m}}{\text{km}} = 3.0 \times 10^{17}\ \text{m}$$

115. a. $0.25\ \text{lb} \times \frac{453.6\ \text{g}}{\text{lb}} \times \frac{1.0\ \text{g trytophan}}{100.0\ \text{g turkey}} = 1.1\ \text{g tryptophan}$

b. $0.25\ \text{qt} \times \frac{0.9463\ \text{L}}{\text{qt}} \times \frac{1.04\ \text{kg}}{\text{L}} \times \frac{1000\ \text{kg}}{\text{kg}} \times \frac{2.0\ \text{g tryptophan}}{100.0\ \text{g milk}} = 4.9\ \text{g tryptophan}$

117. a. False; sugar is generally considered to be the pure compound sucrose, $C_{12}H_{22}O_{11}$.

b. False; elements and compounds are pure substances.

c. True; air is a mixture of mostly nitrogen and oxygen gases.

d. False; gasoline has many additives, so it is a mixture.

e. True; compounds are broken down to elements by chemical change.

119. $5.4\ \text{L blood} \times \frac{1000\ \text{mL}}{\text{L}} \times \frac{250\ \text{mg cholesterol}}{100.0\ \text{mL blood}} \times \frac{1\ \text{g}}{1000\ \text{mg}} = 13.5\ \text{g} = 14\ \text{g cholesterol}$

121. $18.5\ \text{cm} \times \frac{10.0°\text{F}}{5.25\ \text{cm}} = 35.2°\text{F increase}$; $T_{final} = 98.6 + 35.2 = 133.8°\text{F}$

$T_C = 5/9\ (133.8 - 32) = 56.56°\text{C}$

123. $V = \frac{4}{3}\pi r^3 = \frac{4}{3} \times 3.14 \times \left(69\ \text{pm} \times \frac{1 \times 10^{-12}\ \text{m}}{\text{pm}} \times \frac{100\ \text{cm}}{\text{m}}\right)^3 = 1.4 \times 10^{-24}\ \text{cm}^3$

$$\text{Density} = \frac{\text{mass}}{\text{volume}} = \frac{3.35 \times 10^{-23}\ \text{g}}{1.4 \times 10^{-24}\ \text{cm}^3} = 24\ \text{g/cm}^3$$

125. a. Volume × density = mass; the orange block is more dense. Because mass (orange) > mass (blue) and because volume (orange) < volume (blue), the density of the orange block must be greater to account for the larger mass of the orange block.

b. Which block is more dense cannot be determined. Because mass (orange) > mass (blue) and because volume (orange) > volume (blue), the density of the orange block may or may not be larger than the blue block. If the blue block is more dense, its density cannot be so large that its mass is larger than the orange block's mass.

c. The blue block is more dense. Because mass (blue) = mass (orange) and because volume (blue) < volume (orange), the density of the blue block must be larger in order to equate the masses.

d. The blue block is denser. Because mass (blue) > mass (orange) and because the volumes are equal, the density of the blue block must be larger in order to give the blue block the larger mass.

127. $V = V_{final} - V_{initial}$; $d = \dfrac{28.90\text{ g}}{9.8\text{ cm}^3 - 6.4\text{ cm}^3} = \dfrac{28.90\text{ g}}{3.4\text{ cm}^3} = 8.5\text{ g/cm}^3$

$d_{max} = \dfrac{mass_{max}}{V_{min}}$; we get V_{min} from $9.7\text{ cm}^3 - 6.5\text{ cm}^3 = 3.2\text{ cm}^3$.

$d_{max} = \dfrac{28.93\text{ g}}{3.2\text{ cm}^3} = \dfrac{9.0\text{ g}}{\text{cm}^3}$; $d_{min} = \dfrac{mass_{min}}{V_{max}} = \dfrac{28.87\text{ g}}{9.9\text{ cm}^3 - 6.3\text{ cm}^3} = \dfrac{8.0\text{ g}}{\text{cm}^3}$

The density is $8.5 \pm 0.5\text{ g/cm}^3$.

Challenge Problems

129. In a subtraction, the result gets smaller, but the uncertainties add. If the two numbers are very close together, the uncertainty may be larger than the result. For example, let's assume we want to take the difference of the following two measured quantities, 999,999 ±2 and 999,996 ±2. The difference is 3 ±4. Because of the uncertainty, subtracting two similar numbers is poor practice.

131. a. $\dfrac{2.70 - 2.64}{2.70} \times 100 = 2\%$ b. $\dfrac{|16.12 - 16.48|}{16.12} \times 100 = 2.2\%$

c. $\dfrac{1.000 - 0.9981}{1.000} \times 100 = \dfrac{0.002}{1.000} \times 100 = 0.2\%$

133. Heavy pennies (old): mean mass = 3.08 ±0.05 g

Light pennies (new): mean mass $= \dfrac{(2.467 + 2.545 + 2.518)}{3} = 2.51 \pm 0.04\text{ g}$

Because we are assuming that volume is additive, let's calculate the volume of 100.0 g of each type of penny, then calculate the density of the alloy. For 100.0 g of the old pennies, 95 g will be Cu (copper) and 5 g will be Zn (zinc).

$V = 95\text{ g Cu} \times \dfrac{1\text{ cm}^3}{8.96\text{ g}} + 5\text{ g Zn} \times \dfrac{1\text{ cm}^3}{7.14\text{ g}} = 11.3\text{ cm}^3$ (carrying one extra sig. fig.)

Density of old pennies = $\frac{100.\ \text{g}}{11.3\ \text{cm}^3}$ = 8.8 g/cm^3

For 100.0 g of new pennies, 97.6 g will be Zn and 2.4 g will be Cu.

$$V = 2.4\ \text{g Cu} \times \frac{1\ \text{cm}^3}{8.96\ \text{g}} + 97.6\ \text{g Zn} \times \frac{1\ \text{cm}^3}{7.14\ \text{g}} = 13.94\ \text{cm}^3 \text{ (carrying one extra sig. fig.)}$$

Density of new pennies = $\frac{100.\ \text{g}}{13.94\ \text{cm}^3}$ = 7.17 g/cm^3

$d = \frac{\text{mass}}{\text{volume}}$; because the volume of both types of pennies are assumed equal, then:

$$\frac{d_{new}}{d_{old}} = \frac{\text{mass}_{new}}{\text{mass}_{old}} = \frac{7.17\ \text{g/cm}^3}{8.8\ \text{g/cm}^3} = 0.81$$

The calculated average mass ratio is: $\frac{\text{mass}_{new}}{\text{mass}_{old}} = \frac{2.51\ \text{g}}{3.08\ \text{g}} = 0.815$

To the first two decimal places, the ratios are the same. If the assumptions are correct, then we can reasonably conclude that the difference in mass is accounted for by the difference in alloy used.

135. Let x = mass of copper and y = mass of silver.

105.0 g = $x + y$ and 10.12 mL = $\frac{x}{8.96} + \frac{y}{10.5}$; solving and carrying 1 extra sig. fig.:

$$\left(10.12 = \frac{x}{8.96} + \frac{105.0 - x}{10.5}\right) \times 8.96 \times 10.5,\ 952.1 = (10.5)x + 940.8 - (8.96)x$$

$$11.3 = (1.54)x,\ x = 7.3\ \text{g};\ \text{mass \% Cu} = \frac{7.3\ \text{g}}{105.0\ \text{g}} \times 100 = 7.0\%\ \text{Cu}$$

137. a. One possibility is that rope B is not attached to anything and rope A and rope C are connected via a pair of pulleys and/or gears.

b. Try to pull rope B out of the box. Measure the distance moved by C for a given movement of A. Hold either A or C firmly while pulling on the other rope.

CHAPTER 2

ATOMS, MOLECULES, AND IONS

Questions

21. A compound will always contain the same numbers (and types) of atoms. A given amount of hydrogen will react only with a specific amount of oxygen. Any excess oxygen will remain unreacted.

23. Law of conservation of mass: Mass is neither created nor destroyed. The total mass before a chemical reaction always equals the total mass after a chemical reaction.

    Law of definite proportion: A given compound always contains exactly the same proportion of elements by mass. For example, water is always 1 g H for every 8 g oxygen.

    Law of multiple proportions: When two elements form a series of compounds, the ratios of the mass of the second element that combine with 1 g of the first element always can be reduced to small whole numbers: For CO_2 and CO discussed in Section 2.2, the mass ratios of oxygen that react with 1 g carbon in each compound are in a 2 : 1 ratio.

25. Mass is conserved in a chemical reaction because atoms are conserved. Chemical reactions involve the reorganization of atoms, so formulas change in a chemical reaction, but the number and types of atoms do not change. Because the atoms do not change in a chemical reaction, mass must not change. In this equation we have two oxygen atoms and four hydrogen atoms both before and after the reaction occurs.

27. J. J. Thomson's study of cathode-ray tubes led him to postulate the existence of negatively charged particles that we now call electrons. Thomson also postulated that atoms must contain positive charge for the atom to be electrically neutral. Ernest Rutherford and his alpha bombardment of metal foil experiments led him to postulate the nuclear atom–an atom with a tiny dense center of positive charge (the nucleus) with electrons moving about the nucleus at relatively large distances away; the distance is so large that an atom is mostly empty space.

29. The number and arrangement of electrons in an atom determine how the atom will react with other atoms, i.e., the electrons determine the chemical properties of an atom. The number of neutrons present determines the isotope identity and the mass number.

31. For lighter, stable isotopes, the number of protons in the nucleus is about equal to the number of neutrons. When the number of protons and neutrons is equal, the mass number (protons + neutrons) will be twice the atomic number (protons). Therefore, for lighter isotopes, the ratio of the mass number to the atomic number is close to 2. For example, consider ^{28}Si, which has 14 protons and (28 – 14 =) 14 neutrons. Here, the mass number to atomic number ratio is 28/14 = 2.0. For heavier isotopes, there are more neutrons than protons in the nucleus. Therefore, the ratio of the mass number to the atomic number increases steadily upward from 2 as the isotopes get heavier and heavier. For example, ^{238}U has 92 protons and (238 – 92 =) 146 neutrons. The ratio of the mass number to the atomic number for ^{238}U is 238/92 = 2.6.

33. Carbon is a nonmetal. Silicon and germanium are called metalloids because they exhibit both metallic and nonmetallic properties. Tin and lead are metals. Thus metallic character increases as one goes down a family in the periodic table. The metallic character decreases from left to right across the periodic table.

35. In the paste, sodium chloride will dissolve to form separate Na^+ and Cl^- ions. With the ions present and able to move about, electrical impulses will be conducted.

37. a. This represents ionic bonding. Ionic bonding is the electrostatic attraction between anions and cations.

    b. This represents covalent bonding where electrons are shared between two atoms. This could be the space-filling model for H_2O or SF_2 or NO_2, etc.

39. Statement d is true. For statement a, neutrons in the nucleus are neutral in charge. For statement b, the atom consists of a tiny dense nucleus with electrons around the nucleus at relatively large distances. An atom is mostly empty space. For statement c, the nucleus contains most of the mass of the nucleus. For statement e, the number of protons in a neutral atom is equal to the number of electrons.

41. Statements a and b are true. Element 118, Og, is a noble gas and will presumably be a nonmetal. For statement c, hydrogen has mostly nonmetallic properties. For statement d, a family of elements is also known as a group of elements. For statement e, two items are incorrect. When a metal reacts with a nonmetal, an ionic compound is produced, and the formula of the compound would be AX_2 (alkaline earth metals form 2+ ions and halo-gens form 1– ions in ionic compounds). The correct statement would be: When an alkaline earth metal, A, reacts with a halogen, X, the formula of the ionic compound formed should be AX_2.

43. Predicting formulas for covalent compounds is difficult, unlike ionic compound formulas where ion charges are used to determine formulas. Covalent compounds are not formed from ions. Also, sometimes many different formulas can form between two different nonmetals. For example, between nitrogen and oxygen, some compounds that form are NO, NO_2, N_2O, and N_2O_4. Because it is difficult to predict covalent compound formulas, prefixes are necessary to indicate the formula of the compound.

Exercises

Development of the Atomic Theory

45. a. The composition of a substance depends on the numbers of atoms of each element making up the compound (depends on the formula of the compound) and not on the composition of the mixture from which it was formed.

    b. Avogadro's hypothesis (law) implies that volume ratios are proportional to molecule ratios at constant temperature and pressure. $H_2(g) + Cl_2(g) \rightarrow 2\ HCl(g)$. From the balanced equation, the volume of HCl produced will be twice the volume of H_2 (or Cl_2) reacted.

47. From the law of definite proportions, a given compound always contains exactly the same proportion of elements by mass. The first sample of chloroform has a total mass of 12.0 g C + 106.4 g Cl + 1.01 g H = 119.41 g (carrying extra significant figures). The mass percent of carbon in this sample of chloroform is:

$$\frac{12.0 \text{ g C}}{119.41 \text{ g total}} \times 100 = 10.05\% \text{ C by mass}$$

From the law of definite proportions, the second sample of chloroform must also contain 10.05% C by mass. Let x = mass of chloroform in the second sample:

$$\frac{30.0 \text{ g C}}{x} \times 100 = 10.05, \quad x = 299 \text{ g chloroform}$$

49. Compound 1: 21.8 g C and 58.2 g O (80.0 – 21.8 = mass O)

Compound 2: 34.3 g C and 45.7 g O (80.0 – 34.3 = mass O)

The mass of carbon that combines with 1.0 g of oxygen is:

Compound 1: $\frac{21.8 \text{ g C}}{58.2 \text{ g O}} = 0.375 \text{ g C/g O}$

Compound 2: $\frac{34.3 \text{ g C}}{45.7 \text{ g O}} = 0.751 \text{ g C/g O}$

The ratio of the masses of carbon that combine with 1 g of oxygen is $\frac{0.751}{0.375} = \frac{2}{1}$; this supports the law of multiple proportions because this carbon ratio is a small whole number.

51. For CO and CO_2, it is easiest to concentrate on the mass of oxygen that combines with 1 g of carbon. From the formulas (two oxygen atoms per carbon atom in CO_2 versus one oxygen atom per carbon atom in CO), CO_2 will have twice the mass of oxygen that combines per gram of carbon as compared to CO. For CO_2 and C_3O_2, it is easiest to concentrate on the mass of carbon that combines with 1 g of oxygen. From the formulas (three carbon atoms per two oxygen atoms in C_3O_2 versus one carbon atom per two oxygen atoms in CO_2), C_3O_2 will have three times the mass of carbon that combines per gram of oxygen as compared to CO_2. As expected, the mass ratios are whole numbers as predicted by the law of multiple proportions.

53. Mass is conserved in a chemical reaction.

	ethanol	+ oxygen	→	water	+ carbon dioxide
Mass:	46.0 g	96.0 g		54.0 g	?

Mass of reactants = 46.0 + 96.0 = 142.0 g = mass of products

142.0 g = 54.0 g + mass of CO_2, mass of CO_2 = 142.0 – 54.0 = 88.0 g

55. To get the atomic mass of H to be 1.00, we divide the mass of hydrogen that reacts with 1.00 g of oxygen by 0.126; that is, $\frac{0.126}{0.126} = 1.00$. To get Na, Mg, and O on the same scale, we do the same division.

Na: $\frac{2.875}{0.126} = 22.8$; Mg: $\frac{1.500}{0.126} = 11.9$; O: $\frac{1.00}{0.126} = 7.94$

	H	O	Na	Mg
Relative value	1.00	7.94	22.8	11.9
Accepted value	1.008	16.00	22.99	24.31

For your information, the atomic masses of O and Mg are incorrect. The atomic masses of H and Na are close to the values given in the periodic table. Something must be wrong about the assumed formulas of the compounds. It turns out the correct formulas are H_2O, Na_2O, and MgO. The smaller discrepancies result from the error in the assumed atomic mass of H.

The Nature of the Atom

57. From section 2.5, the nucleus has "a diameter of about 10^{-13} cm" and the electrons "move about the nucleus at an average distance of about 10^{-8} cm from it." We will use these statements to help determine the densities. Density of hydrogen nucleus (contains one proton only):

$$V_{nucleus} = \frac{4}{3}\pi r^3 = \frac{4}{3}(3.14)(5 \times 10^{-14}\text{ cm})^3 = 5 \times 10^{-40}\text{ cm}^3$$

$$d = \text{density} = \frac{1.67 \times 10^{-24}\text{ g}}{5 \times 10^{-40}\text{ cm}^3} = 3 \times 10^{15}\text{ g/cm}^3$$

Density of H atom (contains one proton and one electron):

$$V_{atom} = \frac{4}{3}(3.14)(1 \times 10^{-8}\text{ cm})^3 = 4 \times 10^{-24}\text{ cm}^3$$

$$d = \frac{1.67 \times 10^{-24}\text{ g} + 9 \times 10^{-28}\text{ g}}{4 \times 10^{-24}\text{ cm}^3} = 0.4\text{ g/cm}^3$$

59. $5.93 \times 10^{-18}\text{ C} \times \frac{1\text{ electron charge}}{1.602 \times 10^{-19}\text{ C}}$ = 37 negative (electron) charges on the oil drop

61. Sn–tin; Pt–platinum; Hg–mercury; Mg–magnesium; K–potassium; Ag–silver

63. a. 6; the group 2A elements are Be, Mg, Ca, Sr, Ba, and Ra.

b. 6; the group 6A elements are O, S, Se, Te, Po, and Lv.

c. 4; the nickel family elements are Ni, Pd, Pt, amd Ds.

d. 7; the noble gas group 8A elements are He, Ne, Ar, Kr, Xe, Rn, and Og.

65. a. Metals: Mg, Ti, Au, Bi, Ge, Eu, and Am. Nonmetals: Si, B, At, Rn, and Br.

b. Si, Ge, B, and At. The elements at the boundary between the metals and the nonmetals are B, Si, Ge, As, Sb, Te, Po, and At. Aluminum has mostly properties of metals, so it is generally not classified as a metalloid.

67. a. transition metals b. alkaline earth metals c. alkali metals

d. noble gases e. halogens

69. The symbols are $^{24}_{13}Al$, $^{25}_{13}Al$, $^{26}_{13}Al$, $^{27}_{13}Al$, $^{28}_{13}Al$, $^{29}_{13}Al$, and $^{30}_{13}Al$.

71. a. Element 8 is oxygen. A = mass number = 9 + 8 = 17; $^{17}_{8}O$

b. Chlorine is element 17. $^{37}_{17}Cl$

c. Cobalt is element 27. $^{60}_{27}Co$

d. Z = 26; A = 26 + 31 = 57; $^{57}_{26}Fe$

e. Iodine is element 53. $^{131}_{53}I$

f. Lithium is element 3. $^{7}_{3}Li$

73. Z is the atomic number and is equal to the number of protons in the nucleus. A is the mass number and is equal to the number of protons plus neutrons in the nucleus. X is the symbol of the element. See the front cover of the text which has a listing of the symbols for the various elements and corresponding atomic number or see the periodic table on the cover to determine the identity of the various atoms. Because all of the atoms have equal numbers of protons and electrons, each atom is neutral in charge.

a. $^{23}_{11}Na$ b. $^{19}_{9}F$ c. $^{16}_{8}O$

75. a. $^{79}_{35}Br$: 35 protons, 79 – 35 = 44 neutrons. Because the charge of the atom is neutral, the number of protons = the number of electrons = 35.

b. $^{81}_{35}Br$: 35 protons, 46 neutrons, 35 electrons

c. $^{239}_{94}Pu$: 94 protons, 145 neutrons, 94 electrons

d. $^{133}_{55}Cs$: 55 protons, 78 neutrons, 55 electrons

e. $^{3}_{1}H$: 1 proton, 2 neutrons, 1 electron

f. $^{56}_{26}Fe$: 26 protons, 30 neutrons, 26 electrons

77. a. Ba is element 56. Ba^{2+} has 56 protons, so Ba^{2+} must have 54 electrons in order to have a net charge of 2+.

b. Zn is element 30. Zn^{2+} has 30 protons and 28 electrons.

c. N is element 7. N^{3-} has 7 protons and 10 electrons.

d. Rb is element 37. Rb^{+} has 37 protons and 36 electrons.

e. Co is element 27. Co^{3+} has 27 protons and 24 electrons.

f. Te is element 52. Te^{2-} has 52 protons and 54 electrons.

g. Br is element 35. Br^{-} has 35 protons and 36 electrons.

79. Atomic number = 63 (Eu); net charge = +63 – 60 = 3+; mass number = 63 + 88 = 151; symbol: $^{151}_{63}Eu^{3+}$

Atomic number = 50 (Sn); mass number = 50 + 68 = 118; net charge = +50 – 48 = 2+; symbol: $^{118}_{50}Sn^{2+}$

81.

Symbol	Number of protons in nucleus	Number of neutrons in nucleus	Number of electrons	Net charge
$^{238}_{92}U$	92	146	92	0
$^{40}_{20}Ca^{2+}$	20	20	18	2+
$^{51}_{23}V^{3+}$	23	28	20	3+
$^{89}_{39}Y$	39	50	39	0
$^{79}_{35}Br^{-}$	35	44	36	1–
$^{31}_{15}P^{3-}$	15	16	18	3–

83. In ionic compounds, metals lose electrons to form cations, and nonmetals gain electrons to form anions. Group 1A, 2A, and 3A metals form stable 1+, 2+, and 3+ charged cations, respectively. Group 5A, 6A, and 7A nonmetals form 3–, 2–, and 1– charged anions, respectively.

a. Lose 2 e^- to form Ra^{2+}. b. Lose 3 e^- to form In^{3+}. c. Gain 3 e^- to form P^{3-}.

d. Gain 2 e^- to form Te^{2-}. e. Gain 1 e^- to form Br^-. f. Lose 1 e^- to form Rb^+.

Nomenclature

85. a. sodium bromide b. rubidium oxide
c. calcium sulfide d. aluminum iodide
e. SrF_2 f. Al_2Se_3
g. K_3N h. Mg_3P_2

87. a. cesium fluoride b. lithium nitride
c. silver sulfide (Silver only forms stable 1+ ions in compounds, so no Roman numerals are needed.)
d. manganese(IV) oxide e. titanium(IV) oxide f. strontium phosphide

89. a. barium sulfite b. sodium nitrite
c. potassium permanganate d. potassium dichromate

91. a. dinitrogen tetroxide b. iodine trichloride
c. sulfur dioxide d. diphosphorus pentasulfide

93. a. copper(I) iodide b. copper(II) iodide c. cobalt(II) iodide
d. sodium carbonate e. sodium hydrogen carbonate or sodium bicarbonate
f. tetrasulfur tetranitride g. selenium tetrachloride h. sodium hypochlorite
i. barium chromate j. ammonium nitrate

95. In the case of sulfur, SO_4^{2-} is sulfate, and SO_3^{2-} is sulfite. By analogy:

SeO_4^{2-}: selenate; SeO_3^{2-}: selenite; TeO_4^{2-}: tellurate; TeO_3^{2-}: tellurite

97. a. SF_2 b. SF_6 c. NaH_2PO_4
d. Li_3N e. $Cr_2(CO_3)_3$ f. SnF_2
g. $NH_4C_2H_3O_2$ h. NH_4HSO_4 i. $Co(NO_3)_3$
j. Hg_2Cl_2; mercury(I) exists as Hg_2^{2+} ions. k. $KClO_3$ l. NaH

99. a. Na_2O b. Na_2O_2 c. KCN
d. $Cu(NO_3)_2$ e. $SeBr_4$ f. HIO_2
g. PbS_2 h. CuCl
i. GaAs (We would predict the stable ions to be Ga^{3+} and As^{3-}.)
j. CdSe (Cadmium only forms 2+ charged ions in compounds.)
k. ZnS (Zinc only forms 2+ charged ions in compounds.)
l. HNO_2 m. P_2O_5

101. a. nitric acid, HNO_3 b. perchloric acid, $HClO_4$ c. acetic acid, $HC_2H_3O_2$

d. sulfuric acid, H_2SO_4 e. phosphoric acid, H_3PO_4

103. a. $(NH_4)_2CrO_4$ is ammonium chromate.

b. HCl is hydrochloric acid. $HClO_4$ is perchloric acid.

c. P_2S_5 is diphosphorus pentasulfide.

d. Iron(II) nitrate is $Fe(NO_3)_2$.

ChemWork Problems

105. Answer b only contains alkaline earth metals combined with halogens to form alkaline halide compounds.

107. There should be no difference. The composition of insulin (the number and types of atoms) from both sources will be the same and therefore, it should have the some activity regardless of the source. As a practical note, trace contaminants in the two types of insulin may be different. These trace contaminants may be important towards the activity of insulin in the body.

109.

Number of protons in nucleus	Number of neutrons in nucleus	Symbol
9	10	$^{19}_{9}F$
13	14	$^{27}_{13}Al$
53	74	$^{127}_{53}I$
34	45	$^{79}_{34}Se$
16	16	$^{32}_{16}S$

111. a. $^{131}_{53}I$ has 53 protons and 131 – 53 = 78 neutrons.

b. $^{201}_{81}Tl$ has 81 protons and 201 – 81 = 120 neutrons.

113.

Symbol	Number of protons in nucleus	Number of neutrons in nucleus	Number of electrons
$^{120}_{50}Sn$	50	70	50
$^{25}_{12}Mg^{2+}$	12	13	10
$^{56}_{26}Fe^{2+}$	26	30	24
$^{79}_{34}Se$	34	45	34
$^{35}_{17}Cl$	17	18	17
$^{63}_{29}Cu$	29	34	29

115. $^{53}_{26}Fe^{2+}$ has 26 protons, 53 – 26 = 27 neutrons, and two fewer electrons than protons (24 electrons) to have a net charge of 2+.

117. From the Na_2X formula, X has a 2– charge. Because 36 electrons are present, X has 34 protons and 79 – 34 = 45 neutrons, and is selenium.

a. True. Nonmetals bond together using covalent bonds and are called covalent compounds.

b. False. The isotope has 34 protons.

c. False. The isotope has 45 neutrons.

d. False. The identity is selenium, Se.

119. a. $Pb(C_2H_3O_2)_2$: lead(II) acetate

b. $CuSO_4$: copper(II) sulfate

c. CaO: calcium oxide

d. $MgSO_4$: magnesium sulfate

e. $Mg(OH)_2$: magnesium hydroxide

f. $CaSO_4$: calcium sulfate

g. N_2O: dinitrogen monoxide or nitrous oxide (common name)

121. carbon tetrabromide, CBr_4; cobalt(II) phosphate, $Co_3(PO_4)_2$;

magnesium chloride, $MgCl_2$; nickel(II) acetate, $Ni(C_2H_3O_2)_2$;

calcium nitrate, $Ca(NO_3)_2$

123. K will lose 1 e^- to form K^+. Cs will lose 1 e^- to form Cs^+.

Br will gain 1 e^- to form Br^-. Sulfur will gain 2 e^- to form S^{2-}.

Se will gain 2 e^- to form Se^{2-}.

125. From the X_3P_2 formula, the charge on X is 2+. Since X^{2+} has 86 electrons, the ion must have 88 protons. From the periodic table, element 88 is radium, Ra. 230 – 88 = 142 neutrons.

127. a. Ca^{2+} and N^{3-}: Ca_3N_2, calcium nitride

b. K^+ and O^{2-}: K_2O, potassium oxide

c. Rb^+ and F^-: RbF, rubidium fluoride

d. Mg^{2+} and S^{2-}: MgS, magnesium sulfide

e. Ba^{2+} and I^-: BaI_2, barium iodide

f. Al^{3+} and Se^{2-}: Al_2Se_3, aluminum selenide

g. Cs^+ and P^{3-}: Cs_3P, cesium phosphide

h. In^{3+} and Br^-: $InBr_3$, indium(III) bromide. In also forms In^+ ions, but one would predict In^{3+} ions from its position in the periodic table.

129. a. Element 15 is phosphorus, P. This atom has 15 protons and 31 – 15 = 16 neutrons.

b. Element 53 is iodine, I. 53 protons; 74 neutrons

c. Element 19 is potassium, K. 19 protons; 20 neutrons

d. Element 70 is ytterbium, Yb. 70 protons; 103 neutrons

131. The law of multiple proportions does not involve looking at the ratio of the mass of one element with the total mass of the compounds. To illustrate the law of multiple proportions, we compare the mass of carbon that combines with 1.0 g of oxygen in each compound:

Compound 1: 27.2 g C and 72.8 g O (100.0 – 27.2 = mass O)

Compound 2: 42.9 g C and 57.1 g O (100.0 – 42.9 = mass O)

The mass of carbon that combines with 1.0 g of oxygen is:

Compound 1: $\frac{27.2 \text{ g C}}{72.8 \text{ g O}} = 0.374 \text{ g C/g O}$

Compound 2: $\frac{42.9 \text{ g C}}{57.1 \text{ g O}} = 0.751 \text{ g C/g O}$

$\frac{0.751}{0.374} = \frac{2}{1}$; because the ratio is a small whole number, this supports the law of multiple proportions.

133. The systematic name of Ta_2O_5 is tantalum(V) oxide. Tantalum is a transition metal and requires a Roman numeral. Sulfur is in the same group as oxygen, and its most common ion is S^{2-}. Therefore, the formula of the sulfur analogue would be Ta_2S_5.

Total number of protons in Ta_2O_5:

Ta, Z = 73, so 73 protons × 2 = 146 protons; O, Z = 8, so 8 protons × 5 = 40 protons

Total protons = 186 protons

Total number of protons in Ta_2S_5:

Ta, Z = 73, so 73 protons × 2 = 146 protons; S, Z = 16, so 16 protons × 5 = 80 protons

Total protons = 226 protons

Proton difference between Ta_2S_5 and Ta_2O_5: 226 protons – 186 protons = 40 protons

Challenge Problems

135. Copper (Cu), silver (Ag), and gold (Au) make up the coinage metals.

137. Avogadro proposed that equal volumes of gases (at constant temperature and pressure) contain equal numbers of molecules. In terms of balanced equations, Avogadro's hypothesis (law) implies that volume ratios will be identical to molecule ratios. Assuming one molecule of octane reacting, then 1 molecule of C_xH_y produces 8 molecules of CO_2 and 9 molecules of H_2O. $C_xH_y + n\ O_2 \rightarrow 8\ CO_2 + 9\ H_2O$. Because all the carbon in octane ends up as carbon in CO_2, octane must contain 8 atoms of C. Similarly, all hydrogen in octane ends up as hydrogen in H_2O, so one molecule of octane must contain 9 × 2 = 18 atoms of H. Octane formula = C_8H_{18}, and the ratio of C : H = 8 : 18 or 4 : 9.

139. The alchemists were incorrect. The solid residue must have come from the flask.

141. a. Both compounds have C_2H_6O as the formula. Because they have the same formula, their mass percent composition will be identical. However, these are different compounds with different properties because the atoms are bonded together differently. These compounds are called isomers of each other.

b. When wood burns, most of the solid material in wood is converted to gases, which escape. The gases produced are most likely CO_2 and H_2O.

c. The atom is not an indivisible particle but is instead composed of other smaller particles, called electrons, neutrons, and protons.

d. The two hydride samples contain different isotopes of either hydrogen and/or lithium. Although the compounds are composed of different isotopes, their properties are similar because different isotopes of the same element have similar properties (except, of course, their mass).

143. Most of the mass of the atom is due to the protons and the neutrons in the nucleus, and protons and neutrons have about the same mass (1.67×10^{-24} g). The ratio of the mass of the molecule to the mass of a nuclear particle will give a good approximation of the number of nuclear particles (protons and neutrons) present.

$$\frac{7.31 \times 10^{-23}\ \text{g}}{1.67 \times 10^{-24}\ \text{g}} = 43.8 \approx 44 \text{ nuclear particles}$$

Thus there are 44 protons and neutrons present. If the number of protons equals the number of neutrons, we have 22 protons in the molecule. One possibility would be the molecule CO_2 [6 + 2(8) = 22 protons].

CHAPTER 3

STOICHIOMETRY

Questions

27. The atomic mass of any isotope is a relative mass to a specific standard. The standard is one atom of the carbon-12 isotope weighing exactly $12.000\bar{0}$ u. One can determine from experiment how much heavier or lighter any specific isotope is than ^{12}C. From this information, we assign an atomic mass value to that isotope. For example, experiment tells one that ^{16}O is about 4/3 heavier than ^{12}C, so a mass of 4/3(12.00) = 16.00 u is assigned to ^{16}O.

    The atomic mass listed in the periodic table is also an average mass. Most elements in nature occur as a mixture of isotopes. The atomic mass of an element is the average mass of all the isotopes that make up a specific element, weighted by abundance.

29. Avogadro's number of dollars = 6.022×10^{23} dollars/mol dollars

    $$\frac{1 \text{ mol dollars} \times \dfrac{6.022 \times 10^{23} \text{ dollars}}{\text{mol dollars}}}{7.9 \times 10^{9} \text{ people}} = 7.6 \times 10^{13} \text{ dollars/person}$$

    1 trillion = 1,000,000,000,000 = 1×10^{12}; each person would have 76 trillion dollars.

31. Only in b are the empirical formulas the same for both compounds illustrated. In b, general formulas of X_2Y_4 and XY_2 are illustrated, and both have XY_2 for an empirical formula.

    For a, general formulas of X_2Y and X_2Y_2 are illustrated. The empirical formulas for these two compounds are the same as the molecular formulas. For c, general formulas of XY and XY_2 are illustrated; these general formulas are also the empirical formulas. For d, general formulas of XY_4 and X_2Y_6 are illustrated. XY_4 is also the molecular formula, but X_2Y_6 has the empirical formula of XY_3.

33. The mass percent of a compound is a constant no matter what amount of substance is present. Compounds always have constant composition.

35. The theoretical yield is the stoichiometric amount of product that should form if the limiting reactant is completely consumed, and the reaction has 100% yield.

Exercises

Atomic Masses and the Mass Spectrometer

37. Let A = average atomic mass; abundance of 70.95 u isotope = 100.00 – 60.16 = 39.84%

    A = 0.6016(68.95 u) + 0.3984(70.95 u) = 41.48 + 28.27 = 69.75 u

    Note: u is an abbreviation for amu (atomic mass units).

39. Let A = average atomic mass

$$A = 0.0140(203.973) + 0.2410(205.9745) + 0.2210(206.9759) + 0.5240(207.9766)$$

$A = 2.86 + 49.64 + 45.74 + 109.0 = 207.2$ u; from the periodic table, the element is Pb.

Note: u is an abbreviation for amu (atomic mass units).

41. Let A = mass of ^{185}Re:

$$186.207 = 0.6260(186.956) + 0.3740(A),\ 186.207 - 117.0 = 0.3740(A)$$

$$A = \frac{69.2}{0.3740} = 185 \text{ u } (A = 184.95 \text{ u without rounding to proper significant figures.})$$

43. Let x = % of ^{151}Eu and y = % of ^{153}Eu, then $x + y = 100$ and $y = 100 - x$.

$$151.96 = \frac{x(150.9196) + (100 - x)(152.9209)}{100}$$

$$15196 = (150.9196)x + 15292.09 - (152.9209)x,\quad -96 = -(2.0013)x$$

$x = 48\%$; 48% ^{151}Eu and 100 – 48 = 52% ^{153}Eu

45. There are three peaks in the mass spectrum, each 2 mass units apart. This is consistent with two isotopes differing in mass by two mass units. The peak at 157.84 corresponds to a Br_2 molecule composed of two atoms of the lighter isotope. This isotope has mass equal to 157.84/2 or 78.92. This corresponds to ^{79}Br. The second isotope is ^{81}Br with mass equal to 161.84/2 = 80.92. The peaks in the mass spectrum correspond to $^{79}Br_2$, $^{79}Br^{81}Br$, and $^{81}Br_2$ in order of increasing mass. The intensities of the highest and lowest masses tell us the two isotopes are present in about equal abundance. The actual abundance is 50.68% ^{79}Br and 49.32% ^{81}Br.

Moles and Molar Masses

47. When more than one conversion factor is necessary to determine the answer, we will usually put all the conversion factors into one calculation instead of determining intermediate answers. This method reduces round-off error and is a time saver.

$$500.\text{ atoms Fe} \times \frac{1 \text{ mol Fe}}{6.022 \times 10^{23} \text{ atoms Fe}} \times \frac{55.85 \text{ g Fe}}{\text{mol Fe}} = 4.64 \times 10^{-20} \text{ g Fe}$$

49. $$1.00 \text{ carat} \times \frac{0.200 \text{ g C}}{\text{carat}} \times \frac{1 \text{ mol C}}{12.01 \text{ g C}} \times \frac{6.022 \times 10^{23} \text{ atoms C}}{\text{mol C}} = 1.00 \times 10^{22} \text{ atoms C}$$

51. The formula of the metal phosphide compound would be M_3P_2. Let x = molar mass of M.

$238.0 = 3x + 2(30.97)$, $x = 58.69$ g/mol; from the periodic table, M = Ni.

53. $C_{17}H_{18}F_3NO$ (Prozac): 17(12.01) + 18(1.008) + 3(19.00) + 14.01 + 16.00 = 309.32 g/mol

$C_{17}H_{17}Cl_2N$ (Zoloft): 17(12.01) + 17(1.008) + 2(35.45) + 14.01 = 306.22 g/mol

55. a. The formula is NH_3. 14.01 g/mol + 3(1.008 g/mol) = 17.03 g/mol

b. The formula is N_2H_4. 2(14.01) + 4(1.008) = 32.05 g/mol

c. $(NH_4)_2Cr_2O_7$: 2(14.01) + 8(1.008) + 2(52.00) + 7(16.00) = 252.08 g/mol

57. a. $1.00 \text{ g } NH_3 \times \dfrac{1 \text{ mol } NH_3}{17.03 \text{ g } NH_3} = 0.0587 \text{ mol } NH_3$

b. $1.00 \text{ g } N_2H_4 \times \dfrac{1 \text{ mol } N_2H_4}{32.05 \text{ g } N_2H_4} = 0.0312 \text{ mol } N_2H_4$

c. $1.00 \text{ g } (NH_4)_2Cr_2O_7 \times \dfrac{1 \text{ mol } (NH_4)_2Cr_2O_7}{252.08 \text{ g } (NH_4)_2Cr_2O_7} = 3.97 \times 10^{-3} \text{ mol } (NH_4)_2Cr_2O_7$

59. a. $5.00 \text{ mol } NH_3 \times \dfrac{17.03 \text{ g } NH_3}{\text{mol } NH_3} = 85.2 \text{ g } NH_3$

b. $5.00 \text{ mol } N_2H_4 \times \dfrac{32.05 \text{ g } N_2H_4}{\text{mol } N_2H_4} = 160. \text{ g } N_2H_4$

c. $5.00 \text{ mol } (NH_4)_2Cr_2O_7 \times \dfrac{252.08 \text{ g } (NH_4)_2Cr_2O_7}{1 \text{ mol } (NH_4)_2Cr_2O_7} = 1260 \text{ g } (NH_4)_2Cr_2O_7$

61. Chemical formulas give atom ratios as well as mole ratios.

a. $5.00 \text{ mol } NH_3 \times \dfrac{1 \text{ mol N}}{\text{mol } NH_3} \times \dfrac{14.01 \text{ g N}}{\text{mol N}} = 70.1 \text{ g N}$

b. $5.00 \text{ mol } N_2H_4 \times \dfrac{2 \text{ mol N}}{\text{mol } N_2H_4} \times \dfrac{14.01 \text{ g N}}{\text{mol N}} = 140. \text{ g N}$

c. $5.00 \text{ mol } (NH_4)_2Cr_2O_7 \times \dfrac{2 \text{ mol N}}{\text{mol } (NH_4)_2Cr_2O_7} \times \dfrac{14.01 \text{ g N}}{\text{mol N}} = 140. \text{ g N}$

63. a. $1.00 \text{ g } NH_3 \times \dfrac{1 \text{ mol } NH_3}{17.03 \text{ g } NH_3} \times \dfrac{6.022 \times 10^{23} \text{ molecules } NH_3}{\text{mol } NH_3}$

$= 3.54 \times 10^{22} \text{ molecules } NH_3$

b. $1.00 \text{ g } N_2H_4 \times \dfrac{1 \text{ mol } N_2H_4}{32.05 \text{ g } N_2H_4} \times \dfrac{6.022 \times 10^{23} \text{ molecules } N_2H_4}{\text{mol } N_2H_4}$

$= 1.88 \times 10^{22} \text{ molecules } N_2H_4$

c. $1.00 \text{ g } (NH_4)_2Cr_2O_7 \times \dfrac{1 \text{ mol } (NH_4)_2Cr_2O_7}{252.08 \text{ g } (NH_4)_2Cr_2O_7}$

$\times \dfrac{6.022 \times 10^{23} \text{ formula units } (NH_4)_2Cr_2O_7}{\text{mol } (NH_4)_2Cr_2O_7} = 2.39 \times 10^{21} \text{ formula units } (NH_4)_2Cr_2O_7$

65. Using answers from Exercise 63:

a. 3.54×10^{22} molecules NH_3 $\times \dfrac{1 \text{ atom N}}{\text{molecule } NH_3} = 3.54 \times 10^{22}$ atoms N

b. 1.88×10^{22} molecules N_2H_4 $\times \dfrac{2 \text{ atoms N}}{\text{molecule } N_2H_4} = 3.76 \times 10^{22}$ atoms N

c. 2.39×10^{21} formula units $(NH_4)_2Cr_2O_7$ $\times \dfrac{2 \text{ atoms N}}{\text{formula unit } (NH_4)_2Cr_2O_7}$

$= 4.78 \times 10^{21}$ atoms N

67. $38 \text{ g } Al_2S_3 \times \dfrac{1 \text{ mol } Al_2S_3}{150.17 \text{ g}} \times \dfrac{3 \text{ mol S}}{\text{mol } Al_2S_3} \times \dfrac{6.02 \times 10^{23} \text{ atoms S}}{\text{mol S}} = 4.6 \times 10^{23}$ atoms S

69. Molar mass CH_4 = 12.0 + 4(1.0) = 16.0 g/mol; from the molar mass calculation, 16.0 g of methane is 1.00 mol of CH_4, contains 12.0 g C, and 4.0 g of H. So, 16.0 g CH_4 represents 6.02×10^{23} molecules of methane, not $16.0(6.02 \times 10^{23})$. Lastly, 16.0 g CH_4 contains 4 mol of H and $4(6.02 \times 10^{23})$ atoms of H. Answer d is the correct answer to the question.

71. $2.59 \times 10^{23} \text{ atoms H} \times \dfrac{1 \text{ mol H}}{6.022 \times 10^{23} \text{ atoms H}} \times \dfrac{1 \text{ mol } C_4H_{10}}{10 \text{ mol H}} \times \dfrac{58.12 \text{ g } C_4H_{10}}{\text{mol } C_4H_{10}}$

$= 2.50 \text{ g } C_4H_{10}$

73. a. $1.50 \text{ g } CO_2 \times \dfrac{1 \text{ mol}}{44.01 \text{ g}} = 3.41 \times 10^{-2} \text{ mol } CO_2$

b. 2.72×10^{21} molecules C_2H_5OH $\times \dfrac{1 \text{ mol}}{6.022 \times 10^{23} \text{ molecules}}$

$= 4.52 \times 10^{-3} \text{ mol } C_2H_5OH$

c. $20.0 \text{ mg } C_8H_{10}N_4O_2 \times \dfrac{1 \text{ g}}{1000 \text{ mg}} \times \dfrac{1 \text{ mol}}{194.20 \text{ g}} = 1.03 \times 10^{-4} \text{ mol } C_8H_{10}N_4O_2$

75. Molar mass of $C_6H_8O_6$ = 6(12.01) + 8(1.008) + 6(16.00) = 176.12 g/mol

$10 \text{ tablets} \times \dfrac{500.0 \text{ mg}}{\text{tablet}} \times \dfrac{1 \text{ g}}{1000 \text{ mg}} \times \dfrac{1 \text{ mol}}{176.12 \text{ g}} = 2.839 \times 10^{-2} \text{ mol } C_6H_8O_6$

$8 \text{ tablets} \times \dfrac{0.5000 \text{ g}}{\text{tablet}} \times \dfrac{1 \text{ mol}}{176.12 \text{ g}} \times \dfrac{6.022 \times 10^{23} \text{ molecules}}{\text{mol}} = 1.368 \times 10^{22}$ molecules

77. a. 2(12.01) + 3(1.008) + 3(35.45) + 2(16.00) = 165.39 g/mol

b. $500.0 \text{ g} \times \dfrac{1 \text{ mol}}{165.39 \text{ g}} = 3.023 \text{ mol } C_2H_3Cl_3O_2$

c. $2.0 \times 10^{-2} \text{ mol} \times \dfrac{165.39 \text{ g}}{\text{mol}} = 3.3 \text{ g } C_2H_3Cl_3O_2$

d. $5.0 \text{ g } C_2H_3Cl_3O_2 \times \frac{1 \text{ mol}}{165.39 \text{ g}} \times \frac{6.022 \times 10^{23} \text{ molecules}}{\text{mol}} \times \frac{3 \text{ atoms Cl}}{\text{molecule}}$

$= 5.5 \times 10^{22}$ atoms of chlorine

e. $1.0 \text{ g Cl} \times \frac{1 \text{ mol Cl}}{35.45 \text{ g}} \times \frac{1 \text{ mol } C_2H_3Cl_3O_2}{3 \text{ mol Cl}} \times \frac{165.39 \text{ g } C_2H_3Cl_3O_2}{\text{mol } C_2H_3Cl_3O_2} = 1.6 \text{ g } C_2H_3Cl_3O_2$

f. $500 \text{ molecules} \times \frac{1 \text{ mol}}{6.022 \times 10^{23} \text{ molecules}} \times \frac{165.39 \text{ g}}{\text{mol}} = 1.373 \times 10^{-19} \text{ g } C_2H_3Cl_3O_2$

Percent Composition

79. a. $C_3H_4O_2$: Molar mass = 3(12.01) + 4(1.008) + 2(16.00) = 36.03 + 4.032 + 32.00

= 72.06 g/mol

$\text{Mass \% C} = \frac{36.03 \text{ g C}}{72.06 \text{ g compound}} \times 100 = 50.00\% \text{ C}$

$\text{Mass \% H} = \frac{4.032 \text{ g H}}{72.06 \text{ g compound}} \times 100 = 5.595\% \text{ H}$

Mass % O = 100.00 – (50.00 + 5.595) = 44.41% O or:

$\% \text{ O} = \frac{32.00 \text{ g}}{72.06 \text{ g}} \times 100 = 44.41\% \text{ O}$

b. $C_4H_6O_2$: Molar mass = 4(12.01) + 6(1.008) + 2(16.00) = 48.04 + 6.048 + 32.00

= 86.09 g/mol

$\text{Mass \% C} = \frac{48.04 \text{ g}}{86.09 \text{ g}} \times 100 = 55.80\% \text{ C}$; $\text{mass \% H} = \frac{6.048 \text{ g}}{86.09 \text{ g}} \times 100 = 7.025\% \text{ H}$

Mass % O = 100.00 – (55.80 + 7.025) = 37.18% O

c. C_3H_3N: Molar mass = 3(12.01) + 3(1.008) + 1(14.01) = 36.03 + 3.024 + 14.01

= 53.06 g/mol

$\text{Mass \% C} = \frac{36.03 \text{ g}}{53.06 \text{ g}} \times 100 = 67.90\% \text{ C}$; $\text{mass \% H} = \frac{3.024 \text{ g}}{53.06 \text{ g}} \times 100 = 5.699\% \text{ H}$

$\text{Mass \% N} = \frac{14.01 \text{ g}}{53.06 \text{ g}} \times 100 = 26.40\% \text{ N}$ or % N = 100.00 − (67.90 + 5.699)

= 26.40% N

81. NO: $\text{Mass \% N} = \frac{14.01 \text{ g N}}{30.01 \text{ g NO}} \times 100 = 46.68\% \text{ N}$

NO_2: $\text{Mass \% N} = \frac{14.01 \text{ g N}}{46.01 \text{ g } NO_2} \times 100 = 30.45\% \text{ N}$

N_2O: $\text{Mass \% N} = \frac{2(14.01) \text{ g N}}{44.02 \text{ g } N_2O} \times 100 = 63.65\% \text{ N}$

From the calculated mass percent values, only NO is 46.7% N by mass, so NO could be this species. Any other compound having NO as an empirical formula could also be the compound.

83. Assuming 100.0 g of sarin, there are 22.8 g of O present. The moles of sarin in 100.0 g is:

$$22.8 \text{ g O} \times \frac{1 \text{ mol O}}{16.00 \text{ g}} \times \frac{1 \text{ mol sarin}}{2 \text{ mol O}} = 0.713 \text{ mol sarin}$$

Molar mass is a ratio between the mass of substance and moles of substance.

$$\text{Molar mass} = \frac{\text{mass}}{\text{mol}} = \frac{100.0 \text{ g}}{0.713 \text{ mol}} = 140. \text{ g/mol}$$

85. There are 0.390 g Cu for every 100.000 g of fungal laccase. Assuming 100.00 g fungal laccase:

$$\text{mol fungal laccase} = 0.390 \text{ g Cu} \times \frac{1 \text{ mol Cu}}{63.55 \text{ g Cu}} \times \frac{1 \text{ mol fungal laccase}}{4 \text{ mol Cu}} = 1.53 \times 10^{-3} \text{ mol}$$

$$\frac{x \text{ g fungal laccase}}{\text{mol fungal laccase}} = \frac{100.000 \text{ g}}{1.53 \times 10^{-3} \text{ mol}}, \; x = \text{molar mass} = 6.54 \times 10^{4} \text{ g/mol}$$

Empirical and Molecular Formulas

87. a. Molar mass of CH_2O = $1 \text{ mol C}\left(\frac{12.01 \text{ g C}}{\text{mol C}}\right) + 2 \text{ mol H}\left(\frac{1.008 \text{ g H}}{\text{mol H}}\right)$

$$+ 1 \text{ mol O}\left(\frac{16.00 \text{ g O}}{\text{mol O}}\right) = 30.03 \text{ g/mol}$$

$$\% \text{C} = \frac{12.01 \text{ g C}}{30.03 \text{ g } CH_2O} \times 100 = 39.99\% \text{ C}; \; \% \text{H} = \frac{2.016 \text{ g H}}{30.03 \text{ g } CH_2O} \times 100 = 6.713\% \text{ H}$$

$$\% \text{O} = \frac{16.00 \text{ g O}}{30.03 \text{ g } CH_2O} \times 100 = 53.28\% \text{ O or } \% \text{O} = 100.00 - (39.99 + 6.713) = 53.30\%$$

b. Molar mass of $C_6H_{12}O_6$ = 6(12.01) + 12(1.008) + 6(16.00) = 180.16 g/mol

$$\% \text{C} = \frac{6(12.01) \text{ g C}}{180.16 \text{ g } C_6H_{12}O_6} \times 100 = 40.00\%; \; \% \text{H} = \frac{12(1.008) \text{ g H}}{180.16 \text{ g } C_6H_{12}O_6} \times 100 = 6.714\%$$

% O = 100.00 – (40.00 + 6.714) = 53.29%

c. Molar mass of $HC_2H_3O_2$ = 2(12.01) + 4(1.008) + 2(16.00) = 60.05 g/mol

$$\% \text{C} = \frac{2(12.01) \text{ g C}}{60.05 \text{ g } HC_2H_3O_2} \times 100 = 40.00\%; \; \% \text{H} = \frac{4(1.008) \text{ g H}}{60.05 \text{ g } HC_2H_3O_2} \times 100 = 6.714\%$$

% O = 100.00 – (40.00 + 6.714) = 53.29%

89. a. The molecular formula is N_2O_4. The smallest whole number ratio of the atoms (the empirical formula) is NO_2.

b. Molecular formula: C_3H_6; empirical formula: CH_2

c. Molecular formula: P_4O_{10}; empirical formula: P_2O_5

d. Molecular formula: $C_6H_{12}O_6$; empirical formula: CH_2O

91. Mass O in compound = 6.91 g Fe_xO_y – 5.00 g Fe = 1.91 g O

$$5.00 \text{ g Fe} \times \frac{1 \text{ mol Fe}}{55.85 \text{ g Fe}} = 0.0895 \text{ mol Fe}; \; 1.91 \text{ g O} \times \frac{1 \text{ mol O}}{16.00 \text{ g O}} = 0.119 \text{ mol O}$$

The mol ratio between mol O to mol Fe is: $\frac{0.119 \text{ mol O}}{0.0895 \text{ mol Fe}} = 1.33$; this is a 4:3 mol ratio.

Fe_3O_4 is the empirical formula.

93. Out of 100.00 g of compound, there are:

$$62.58 \text{ g C} \times \frac{1 \text{ mol C}}{12.01 \text{ g C}} = 5.211 \text{ mol C}; \; 9.63 \text{ g H} \times \frac{1 \text{ mol H}}{1.008 \text{ g H}} = 9.55 \text{ mol H}$$

$$27.79 \text{ g O} \times \frac{1 \text{ mol O}}{16.00 \text{ g O}} = 1.737 \text{ mol O}$$

Dividing each mole value by the smallest number:

$$\frac{5.211}{1.737} = 3.000; \; \frac{9.55}{1.737} = 5.50; \; \frac{1.737}{1.737} = 1.000$$

The C : H : O mol ratio is 3 : 5.5 : 1. Because a whole number ratio is required, this is 6 : 11 : 2. So the empirical formula is $C_6H_{11}O_2$.

95. Compound I: Mass O = 0.6498 g Hg_xO_y – 0.6018 g Hg = 0.0480 g O

$$0.6018 \text{ g Hg} \times \frac{1 \text{ mol Hg}}{200.6 \text{ g Hg}} = 3.000 \times 10^{-3} \text{ mol Hg}$$

$$0.0480 \text{ g O} \times \frac{1 \text{ mol O}}{16.00 \text{ g O}} = 3.00 \times 10^{-3} \text{ mol O}$$

The mole ratio between Hg and O is 1 : 1, so the empirical formula of compound I is HgO.

Compound II: Mass Hg = 0.4172 g Hg_xO_y – 0.016 g O = 0.401 g Hg

$$0.401 \text{ g Hg} \times \frac{1 \text{ mol Hg}}{200.6 \text{ g Hg}} = 2.00 \times 10^{-3} \text{ mol Hg}; \; 0.016 \text{ g O} \times \frac{1 \text{ mol O}}{16.00 \text{ g O}} = 1.0 \times 10^{-3} \text{ mol O}$$

The mole ratio between Hg and O is 2 : 1, so the empirical formula is Hg_2O.

97. Out of 100.0 g, there are:

$$69.6 \text{ g S} \times \frac{1 \text{ mol S}}{32.07 \text{ g S}} = 2.17 \text{ mol S}; \;\; 30.4 \text{ g N} \times \frac{1 \text{ mol N}}{14.01 \text{ g N}} = 2.17 \text{ mol N}$$

The empirical formula is SN because the mole values are in a 1 : 1 mole ratio.

The empirical formula mass of SN is ~46 g/mol. Because 184/46 = 4.0, the molecular formula is S_4N_4.

99. Assuming 100.00 g of compound:

$$47.08 \text{ g C} \times \frac{1 \text{ mol C}}{12.01 \text{ g C}} = 3.920 \text{ mol C}; \;\; 6.59 \text{ g H} \times \frac{1 \text{ mol H}}{1.008 \text{ g H}} = 6.54 \text{ mol H}$$

$$46.33 \text{ g Cl} \times \frac{1 \text{ mol Cl}}{35.45 \text{ g Cl}} = 1.307 \text{ mol Cl}$$

Dividing all mole values by 1.307 gives:

$$\frac{3.920}{1.307} = 2.999; \;\; \frac{6.54}{1.307} = 5.00; \;\; \frac{1.307}{1.307} = 1.000$$

The empirical formula is C_3H_5Cl. The empirical formula mass is 3(12.01) + 5(1.008) + 1(35.45) = 76.52 g/mol.

$$\frac{\text{Molar mass}}{\text{Empirical formula mass}} = \frac{153}{76.52} = 2.00; \text{ the molecular formula is } (C_3H_5Cl)_2 = C_6H_{10}Cl_2.$$

101. When combustion data are given, it is assumed that all the carbon in the compound ends up as carbon in CO_2 and all the hydrogen in the compound ends up as hydrogen in H_2O. In the sample of fructose combusted, the masses of C and H are:

$$\text{mass C} = 2.20 \text{ g } CO_2 \times \frac{1 \text{ mol } CO_2}{44.01 \text{ g } CO_2} \times \frac{1 \text{ mol C}}{1 \text{ mol } CO_2} \times \frac{12.01 \text{ g C}}{\text{mol C}} = 0.600 \text{ g C}$$

$$\text{mass H} = 0.900 \text{ g } H_2O \times \frac{1 \text{ mol } H_2O}{18.02 \text{ g } H_2O} \times \frac{2 \text{ mol H}}{1 \text{ mol } H_2O} \times \frac{1.008 \text{ g H}}{\text{mol H}} = 0.101 \text{ g H}$$

Mass O = 1.50 g fructose – 0.600 g C – 0.101 g H = 0.799 g O

So, in 1.50 g of the fructose, we have:

$$0.600 \text{ g C} \times \frac{1 \text{ mol C}}{12.01 \text{ g C}} = 0.0500 \text{ mol C}; \;\; 0.101 \text{ g H} \times \frac{1 \text{ mol H}}{1.008 \text{ g H}} = 0.100 \text{ mol H}$$

$$0.799 \text{ g O} \times \frac{1 \text{ mol O}}{16.00 \text{ g O}} = 0.0499 \text{ mol O}$$

Dividing by the smallest number: $\frac{0.100}{0.0499} = 2.00$; the empirical formula is CH_2O.

103. a. $0.2833 \text{ g } CO_2 \times \frac{1 \text{ mol } CO_2}{44.01 \text{ g } CO_2} \times \frac{1 \text{ mol C}}{1 \text{ mol } CO_2} \times \frac{12.01 \text{ g C}}{\text{mol C}} = 0.07731 \text{ g C}$

$$0.1160 \text{ g } H_2O \times \frac{1 \text{ mol } H_2O}{18.02 \text{ g } H_2O} \times \frac{2 \text{ mol H}}{1 \text{ mol } H_2O} \times \frac{1.008 \text{ g H}}{\text{mol H}} = 0.01298 \text{ g H}$$

Mass O = 0.1006 g compound – 0.07731 g C – 0.01298 g H = 0.0103 g O

$$\text{Mass percent O} = \frac{0.0103 \text{ g O}}{0.1006 \text{ g compound}} \times 100 = 10.2\% \text{ O}$$

b. From the work in part a, 0.1006 g menthol contains 0.07731 g C, 0.01298 g H, and 0.0103 g O. Let’s use this mass breakdown to solve the empirical formula problem.

$$0.07731 \text{ g C} \times \frac{1 \text{ mol C}}{12.01 \text{ g C}} = 0.006437 \text{ mol C}; \ 0.01298 \text{ g H} \times \frac{1 \text{ mol H}}{1.008 \text{ g H}} = 0.01288 \text{ mol H}$$

$$0.0103 \text{ g O} \times \frac{1 \text{ mol O}}{16.00 \text{ g O}} = 6.44 \times 10^{-4} \text{ mol O}$$

Dividing all mole values by 6.44×10^{-4} gives:

$$\frac{0.006437}{6.44 \times 10^{-4}} = 10.0; \ \frac{0.01288}{6.44 \times 10^{-4}} = 20.0$$

The empirical formula for menthol is $C_{10}H_{20}O$.

105. The combustion data allow determination of the amount of hydrogen in cumene. One way to determine the amount of carbon in cumene is to determine the mass percent of hydrogen in the compound from the data in the problem; then determine the mass percent of carbon by difference (100.0 – mass % H = mass % C).

$42.8 \times 10^{-3} \text{ g } H_2O \times \frac{1 \text{ mol } H_2O}{18.02 \text{ g } H_2O} \times \frac{2 \text{ mol H}}{\text{mol } H_2O} \times \frac{1.008 \text{ g H}}{\text{mol H}} \times \frac{1000 \text{ mg}}{\text{g}} = 4.79$ mg H

$$\text{Mass \% H} = \frac{4.79 \text{ mg H}}{47.6 \text{ mg cumene}} \times 100 = 10.1\% \text{ H}; \ \text{mass \% C} = 100.0 - 10.1 = 89.9\% \text{ C}$$

Now solve the empirical formula problem. Out of 100.0 g cumene, we have:

$$89.9 \text{ g C} \times \frac{1 \text{ mol C}}{12.01 \text{ g C}} = 7.49 \text{ mol C}; \ 10.1 \text{ g H} \times \frac{1 \text{ mol H}}{1.008 \text{ g H}} = 10.0 \text{ mol H}$$

$\frac{10.0}{7.49} = 1.34 \approx \frac{4}{3}$; the mole H to mole C ratio is 4 : 3. The empirical formula is C_3H_4.

Empirical formula mass ≈ 3(12) + 4(1) = 40 g/mol.

The molecular formula must be $(C_3H_4)_3$ or C_9H_{12} because the molar mass of this formula will be between 115 and 125 g/mol (molar mass ≈ 3 × 40 g/mol = 120 g/mol).

Balancing Chemical Equations

107. When balancing reactions, start with elements that appear in only one of the reactants and one of the products, and then go on to balance the remaining elements.

a. $C_6H_{12}O_6(s) + O_2(g) \rightarrow CO_2(g) + H_2O(g)$

Balance C atoms: $C_6H_{12}O_6 + O_2 \rightarrow 6\ CO_2 + H_2O$

Balance H atoms: $C_6H_{12}O_6 + O_2 \rightarrow 6\ CO_2 + 6\ H_2O$

Lastly, balance O atoms: $C_6H_{12}O_6(s) + 6\ O_2(g) \rightarrow 6\ CO_2(g) + 6\ H_2O(g)$

b. $Fe_2S_3(s) + HCl(g) \rightarrow FeCl_3(s) + H_2S(g)$

Balance Fe atoms: $Fe_2S_3 + HCl \rightarrow 2\ FeCl_3 + H_2S$

Balance S atoms: $Fe_2S_3 + HCl \rightarrow 2\ FeCl_3 + 3\ H_2S$

There are 6 H and 6 Cl on right, so balance with 6 HCl on left:

$Fe_2S_3(s) + 6\ HCl(g) \rightarrow 2\ FeCl_3(s) + 3\ H_2S(g)$.

c. $CS_2(l) + NH_3(g) \rightarrow H_2S(g) + NH_4SCN(s)$

C and S balanced; balance N:

$CS_2 + 2\ NH_3 \rightarrow H_2S + NH_4SCN$

H is also balanced. $CS_2(l) + 2\ NH_3(g) \rightarrow H_2S(g) + NH_4SCN(s)$

109. $2\ H_2O_2(aq) \xrightarrow[\text{catalyst}]{MnO_2} 2\ H_2O(l) + O_2(g)$

111. a. $3\ Ca(OH)_2(aq) + 2\ H_3PO_4(aq) \rightarrow 6\ H_2O(l) + Ca_3(PO_4)_2(s)$

b. $Al(OH)_3(s) + 3\ HCl(aq) \rightarrow AlCl_3(aq) + 3\ H_2O(l)$

c. $2\ AgNO_3(aq) + H_2SO_4(aq) \rightarrow Ag_2SO_4(s) + 2\ HNO_3(aq)$

113. a. The formulas of the reactants and products are $C_6H_6(l) + O_2(g) \rightarrow CO_2(g) + H_2O(g)$. To balance this combustion reaction, notice that all of the carbon in C_6H_6 has to end up as carbon in CO_2 and all of the hydrogen in C_6H_6 has to end up as hydrogen in H_2O. To balance C and H, we need 6 CO_2 molecules and 3 H_2O molecules for every 1 molecule of C_6H_6. We do oxygen last. Because we have 15 oxygen atoms in 6 CO_2 molecules and 3 H_2O molecules, we need 15/2 O_2 molecules in order to have 15 oxygen atoms on the reactant side.

$C_6H_6(l) + \frac{15}{2} O_2(g) \rightarrow 6\ CO_2(g) + 3\ H_2O(g)$; multiply by two to give whole numbers.

$2\ C_6H_6(l) + 15\ O_2(g) \rightarrow 12\ CO_2(g) + 6\ H_2O(g)$

b. The formulas of the reactants and products are $C_4H_{10}(g) + O_2(g) \rightarrow CO_2(g) + H_2O(g)$.

$C_4H_{10}(g) + \frac{13}{2} O_2(g) \rightarrow 4\ CO_2(g) + 5\ H_2O(g)$; multiply by two to give whole numbers.

$2\ C_4H_{10}(g) + 13\ O_2(g) \rightarrow 8\ CO_2(g) + 10\ H_2O(g)$

c. $C_{12}H_{22}O_{11}(s) + 12\ O_2(g) \rightarrow 12\ CO_2(g) + 11\ H_2O(g)$

d. $2\ Fe(s) + \frac{3}{2} O_2(g) \rightarrow Fe_2O_3(s)$; for whole numbers: $4\ Fe(s) + 3\ O_2(g) \rightarrow 2\ Fe_2O_3(s)$

e. $2\ FeO(s) + \frac{1}{2} O_2(g) \rightarrow Fe_2O_3(s)$; for whole numbers, multiply by two.

$4\ FeO(s) + O_2(g) \rightarrow 2\ Fe_2O_3(s)$

115. a. $SiO_2(s) + C(s) \rightarrow Si(s) + CO(g)$; Si is balanced.

Balance oxygen atoms: $SiO_2 + C \rightarrow Si + 2\ CO$

Balance carbon atoms: $SiO_2(s) + 2\ C(s) \rightarrow Si(s) + 2\ CO(g)$

b. $SiCl_4(l) + Mg(s) \rightarrow Si(s) + MgCl_2(s)$; Si is balanced.

Balance Cl atoms: $SiCl_4 + Mg \rightarrow Si + 2\ MgCl_2$

Balance Mg atoms: $SiCl_4(l) + 2\ Mg(s) \rightarrow Si(s) + 2\ MgCl_2(s)$

c. $Na_2SiF_6(s) + Na(s) \rightarrow Si(s) + NaF(s)$; Si is balanced.

Balance F atoms: $Na_2SiF_6 + Na \rightarrow Si + 6\ NaF$

Balance Na atoms: $Na_2SiF_6(s) + 4\ Na(s) \rightarrow Si(s) + 6\ NaF(s)$

Reaction Stoichiometry

117. $6\ Li(s) + N_2(g) \rightarrow 2\ Li_3N(s)$

$$20.0\text{ g Li} \times \frac{1\text{ mol Li}}{6.941\text{ g Li}} \times \frac{2\text{ mol Li}_3\text{N}}{6\text{ mol Li}} \times \frac{34.83\text{ g Li}_3\text{N}}{\text{mol Li}_3\text{N}} = 33.5\text{ g Li}_3\text{N}$$

119. The stepwise method to solve stoichiometry problems is outlined in the text. Instead of calculating intermediate answers for each step, we will combine conversion factors into one calculation. This practice reduces round-off error and saves time.

$Fe_2O_3(s) + 2\ Al(s) \rightarrow 2\ Fe(l) + Al_2O_3(s)$

$$15.0\text{ g Fe} \times \frac{1\text{ mol Fe}}{55.85\text{ g Fe}} = 0.269\text{ mol Fe};\ 0.269\text{ mol Fe} \times \frac{2\text{ mol Al}}{2\text{ mol Fe}} \times \frac{26.98\text{ g Al}}{\text{mol Al}} = 7.26\text{ g Al}$$

$$0.269\text{ mol Fe} \times \frac{1\text{ mol Fe}_2\text{O}_3}{2\text{ mol Fe}} \times \frac{159.70\text{ g Fe}_2\text{O}_3}{\text{mol Fe}_2\text{O}_3} = 21.5\text{ g Fe}_2\text{O}_3$$

$$0.269\text{ mol Fe} \times \frac{1\text{ mol Al}_2\text{O}_3}{2\text{ mol Fe}} \times \frac{101.96\text{ g Al}_2\text{O}_3}{\text{mol Al}_2\text{O}_3} = 13.7\text{ g Al}_2\text{O}_3$$

121. $$1.000\text{ kg Al} \times \frac{1000\text{ g Al}}{\text{kg Al}} \times \frac{1\text{ mol Al}}{26.98\text{ g Al}} \times \frac{3\text{ mol NH}_4\text{ClO}_4}{3\text{ mol Al}} \times \frac{117.49\text{ g NH}_4\text{ClO}_4}{\text{mol NH}_4\text{ClO}_4} = 4355\text{ g} = 4.355\text{ kg NH}_4\text{ClO}_4$$

123. a. $$1.0 \times 10^2\text{ mg NaHCO}_3 \times \frac{1\text{ g}}{1000\text{ mg}} \times \frac{1\text{ mol NaHCO}_3}{84.01\text{ g NaHCO}_3} \times \frac{1\text{ mol C}_6\text{H}_8\text{O}_7}{3\text{ mol NaHCO}_3} \times \frac{192.12\text{ g C}_6\text{H}_8\text{O}_7}{\text{mol C}_6\text{H}_8\text{O}_7} = 0.076\text{ g or 76 mg C}_6\text{H}_8\text{O}_7$$

b. $$0.10\text{ g NaHCO}_3 \times \frac{1\text{ mol NaHCO}_3}{84.01\text{ g NaHCO}_3} \times \frac{3\text{ mol CO}_2}{3\text{ mol NaHCO}_3} \times \frac{44.01\text{ g CO}_2}{\text{mol CO}_2} = 0.052\text{ g or 52 mg CO}_2$$

125. $$1.0 \times 10^4\text{ kg waste} \times \frac{3.0\text{ kg NH}_4^+}{100\text{ kg waste}} \times \frac{1000\text{ g}}{\text{kg}} \times \frac{1\text{ mol NH}_4^+}{18.04\text{ g NH}_4^+} \times \frac{1\text{ mol C}_5\text{H}_7\text{O}_2\text{N}}{55\text{ mol NH}_4^+} \times \frac{113.12\text{ g C}_5\text{H}_7\text{O}_2\text{N}}{\text{mol C}_5\text{H}_7\text{O}_2\text{N}} = 3.4 \times 10^4\text{ g tissue if all NH}_4^+\text{ converted}$$

Because only 95% of the NH_4^+ ions react:

mass of tissue = $(0.95)(3.4 \times 10^4\text{ g}) = 3.2 \times 10^4$ g or 32 kg bacterial tissue

127. $$1.0\text{ ton CuO} \times \frac{907\text{ kg}}{\text{ton}} \times \frac{1000\text{ g}}{\text{kg}} \times \frac{1\text{ mol CuO}}{79.55\text{ g CuO}} \times \frac{1\text{ mol C}}{2\text{ mol CuO}} \times \frac{12.01\text{ g C}}{\text{mol C}} \times \frac{100.\text{ g coke}}{95\text{ g C}} = 7.2 \times 10^4\text{ g or 72 kg coke}$$

Limiting Reactants and Percent Yield

129. The product formed in the reaction is NO_2; the other species present in the product representation is excess O_2. Therefore, NO is the limiting reactant. In the pictures, 6 NO molecules react with 3 O_2 molecules to form 6 NO_2 molecules.

$6\ NO(g) + 3\ O_2(g) \rightarrow 6\ NO_2(g)$

For smallest whole numbers, the balanced reaction is:

$2\ NO(g) + O_2(g) \rightarrow 2\ NO_2(g)$

131. In the following table we have listed three rows of information. The "Initial" row is the number of molecules present initially, the "Change" row is the number of molecules that react to reach completion, and the "After" row is the number of molecules present after completion. To determine the limiting reactant, let's calculate how much of one reactant is necessary to react with the other.

$$10 \text{ molecules } O_2 \times \frac{4 \text{ molecules } NH_3}{5 \text{ molecules } O_2} = 8 \text{ molecules } NH_3 \text{ to react with all the } O_2$$

Because we have 10 molecules of NH_3 and only 8 molecules of NH_3 are necessary to react with all of the O_2, O_2 is limiting. Now use the 10 molecules of O_2 and the molecule relationships given in the balanced equation to determine the number of molecules of each product formed, then complete the table.

	$4\ NH_3(g)$	+	$5\ O_2(g)$	$\rightarrow$	$4\ NO(g)$	+	$6\ H_2O(g)$
Initial	10 molecules		10 molecules		0		0
Change	−8 molecules		−10 molecules		+8 molecules		+12 molecules
After	2 molecules		0		8 molecules		12 molecules

The total number of molecules present after completion = 2 molecules NH_3 + 0 molecules O_2 + 8 molecules NO + 12 molecules H_2O = 22 molecules.

133. a. The strategy we will generally use to solve limiting reactant problems is to assume each reactant is limiting, and then calculate the quantity of product each reactant could produce if it were limiting. The reactant that produces the smallest quantity of product is the limiting reactant (runs out first) and therefore determines the mass of product that can be produced.

Assuming N_2 is limiting:

$$1.00 \times 10^3 \text{ g } N_2 \times \frac{1 \text{ mol } N_2}{28.02 \text{ g } N_2} \times \frac{2 \text{ mol } NH_3}{\text{mol } N_2} \times \frac{17.03 \text{ g } NH_3}{\text{mol } NH_3} = 1.22 \times 10^3 \text{ g } NH_3$$

Assuming H_2 is limiting:

$$5.00 \times 10^2 \text{ g } H_2 \times \frac{1 \text{ mol } H_2}{2.016 \text{ g } H_2} \times \frac{2 \text{ mol } NH_3}{3 \text{ mol } H_2} \times \frac{17.03 \text{ g } NH_3}{\text{mol } NH_3} = 2.82 \times 10^3 \text{ g } NH_3$$

Because N_2 produces the smaller mass of product (1220 g vs. 2820 g NH_3), N_2 is limiting and 1220 g NH_3 can be produced. As soon as 1220 g of NH_3 is produced, all of the N_2 has run out. Even though we have enough H_2 to produce more product, there is no more N_2 present as soon as 1220 g of NH_3 have been produced.

b. $$1.00 \times 10^3 \text{ g } N_2 \times \frac{1 \text{ mol } N_2}{28.02 \text{ g } N_2} \times \frac{3 \text{ mol } H_2}{\text{mol } N_2} \times \frac{2.016 \text{ g } H_2}{\text{mol } H_2} = 216 \text{ g } H_2 \text{ reacted}$$

Excess H_2 = 500. g H_2 initially − 216 g H_2 reacted = 284 g H_2 in excess (unreacted)

135. Assuming BaO_2 is limiting:

$$1.50 \text{ g } BaO_2 \times \frac{1 \text{ mol } BaO_2}{169.3 \text{ g } BaO_2} \times \frac{1 \text{ mol } H_2O_2}{\text{mol } BaO_2} \times \frac{34.02 \text{ g } H_2O_2}{\text{mol } H_2O_2} = 0.301 \text{ g } H_2O_2$$

Assuming HCl is limiting:

$$88.0 \text{ mL} \times \frac{0.0272 \text{ g HCl}}{\text{mL}} \times \frac{1 \text{ mol HCl}}{36.46 \text{ g HCl}} \times \frac{1 \text{ mol } H_2O_2}{2 \text{ mol HCl}} \times \frac{34.02 \text{ g } H_2O_2}{\text{mol } H_2O_2} = 1.12 \text{ g } H_2O_2$$

BaO_2 produces the smaller amount of H_2O_2, so it is limiting and a mass of 0.301 g of H_2O_2 can be produced.

Initial mol HCl present: $88.0 \text{ mL} \times \frac{0.0272 \text{ g HCl}}{\text{mL}} \times \frac{1 \text{ mol HCl}}{36.46 \text{ g HCl}} = 6.57 \times 10^{-2} \text{ mol HCl}$

The amount of HCl reacted:

$$1.50 \text{ g } BaO_2 \times \frac{1 \text{ mol } BaO_2}{169.3 \text{ g } BaO_2} \times \frac{2 \text{ mol HCl}}{\text{mol } BaO_2} = 1.77 \times 10^{-2} \text{ mol HCl}$$

Excess mol HCl = 6.57×10^{-2} mol − 1.77×10^{-2} mol = 4.80×10^{-2} mol HCl

Mass of excess HCl = $4.80 \times 10^{-2} \text{ mol HCl} \times \frac{36.46 \text{ g HCl}}{\text{mol HCl}} = 1.75$ g HCl unreacted

137. To solve limiting-reagent problems, we will generally assume each reactant is limiting and then calculate how much product could be produced from each reactant. The reactant that produces the smallest amount of product will run out first and is the limiting reagent.

$$5.00 \times 10^6 \text{ g } NH_3 \times \frac{1 \text{ mol } NH_3}{17.03 \text{ g } NH_3} \times \frac{2 \text{ mol HCN}}{2 \text{ mol } NH_3} = 2.94 \times 10^5 \text{ mol HCN}$$

$$5.00 \times 10^6 \text{ g } O_2 \times \frac{1 \text{ mol } O_2}{32.00 \text{ g } O_2} \times \frac{2 \text{ mol HCN}}{3 \text{ mol } O_2} = 1.04 \times 10^5 \text{ mol HCN}$$

$$5.00 \times 10^6 \text{ g } CH_4 \times \frac{1 \text{ mol } CH_4}{16.04 \text{ g } CH_4} \times \frac{2 \text{ mol HCN}}{2 \text{ mol } CH_4} = 3.12 \times 10^5 \text{ mol HCN}$$

O_2 is limiting because it produces the smallest amount of HCN. Although more product could be produced from NH_3 and CH_4, only enough O_2 is present to produce 1.04×10^5 mol HCN. The mass of HCN produced is:

$$1.04 \times 10^5 \text{ mol HCN} \times \frac{27.03 \text{ g HCN}}{\text{mol HCN}} = 2.81 \times 10^6 \text{ g HCN}$$

$$5.00 \times 10^6 \text{ g } O_2 \times \frac{1 \text{ mol } O_2}{32.00 \text{ g } O_2} \times \frac{6 \text{ mol } H_2O}{3 \text{ mol } O_2} \times \frac{18.02 \text{ g } H_2O}{1 \text{ mol } H_2O} = 5.63 \times 10^6 \text{ g } H_2O$$

139. $$20.0 \text{ mL } Br_2 \times \frac{3.10 \text{ g } Br_2}{\text{mL}} \times \frac{1 \text{ mol } Br_2}{159.80 \text{ g } Br_2} \times \frac{2 \text{ mol } AlBr_3}{3 \text{ mol } Br_2} \times \frac{266.68 \text{ g } AlBr_3}{\text{mol } AlBr_3} = 69.0 \text{ g } AlBr_3$$

The theoretical yield of the reaction is 69.0 g $AlBr_3$. From the problem, the actual yield was 50.3 g.

$$\text{Percent yield} = \frac{\text{actual}}{\text{theoretical}} \times 100 = \frac{50.3 \text{ g}}{69.0 \text{ g}} \times 100 = 72.9\%$$

141. $C_2H_6(g) + Cl_2(g) \rightarrow C_2H_5Cl(g) + HCl(g)$

If C_2H_6 is limiting:

$$300.\ g\ C_2H_6 \times \frac{1\ mol\ C_2H_6}{30.07\ g\ C_2H_6} \times \frac{1\ mol\ C_2H_5Cl}{mol\ C_2H_6} \times \frac{64.51\ g\ C_2H_5Cl}{mol\ C_2H_5Cl} = 644\ g\ C_2H_5Cl$$

If Cl_2 is limiting:

$$650.\ g\ Cl_2 \times \frac{1\ mol\ Cl_2}{70.90\ g\ Cl_2} \times \frac{1\ mol\ C_2H_5Cl}{mol\ Cl_2} \times \frac{64.51\ g\ C_2H_5Cl}{mol\ C_2H_5Cl} = 591\ g\ C_2H_5Cl$$

Cl_2 is limiting because it produces the smaller quantity of product. Hence, the theoretical yield for this reaction is 591 g C_2H_5Cl. The percent yield is:

$$\text{percent yield} = \frac{\text{actual}}{\text{theoretical}} \times 100 = \frac{490.\ g}{591\ g} \times 100 = 82.9\%$$

143. $2\ Cu(s) + S(s) \rightarrow Cu_2S(s)$; the theoretical yield of Cu_2S is:

$$10.0\ g\ Cu_2S\ (actual) \times \frac{100.0\ g\ Cu_2S\ (theoretical)}{73.0\ g\ Cu_2S\ (actual)} = 13.7\ g\ Cu_2S\ (theoretical)$$

$$13.7\ g\ Cu_2S \times \frac{1\ mol\ Cu_2S}{159.17\ g\ Cu_2S} \times \frac{2\ mol\ Cu}{1\ mol\ Cu_2S} \times \frac{63.55\ g\ Cu}{mol\ Cu} = 10.9\ g\ Cu$$

If 10.9 g of Cu is reacted, 13.7 g of Cu_2S will be produced in theory. And with a 73.0% yield, 10.0 g of Cu_2S will be the actual yield.

ChemWork Problems

145. $^{12}C_2{}^1H_6$: 2(12.000000) + 6(1.007825) = 30.046950 u

$^{12}C^1H_2{}^{16}O$: 1(12.000000) + 2(1.007825) + 1(15.994915) = 30.010565 u

$^{14}N^{16}O$: 1(14.003074) + 1(15.994915) = 29.997989 u

The peak results from $^{12}C^1H_2{}^{16}O$.

147. a. $1.05 \times 10^{-20}\ g\ Fe \times \frac{1\ mol\ Fe}{55.85\ g\ Fe} \times \frac{6.022 \times 10^{23}\ atoms\ Fe}{mol\ Fe} = 113\ atoms\ Fe$

b. The total number of platinum atoms is 14 × 20 = 280 atoms (exact number). The mass of these atoms is:

$$280\ atoms\ Pt \times \frac{1\ mol\ Pt}{6.022 \times 10^{23}\ atoms\ Pt} \times \frac{195.1\ g\ Pt}{mol\ Pt} = 9.071 \times 10^{-20}\ g\ Pt$$

c. $9.071 \times 10^{-20}\ g\ Ru \times \frac{1\ mol\ Ru}{101.1\ g\ Ru} \times \frac{6.022 \times 10^{23}\ atoms\ Ru}{mol\ Ru} = 540.3 = 540.\ atoms\ Ru$

149. $15.00\text{ mL} \times \frac{3.97\text{ g Al}_2\text{O}_3}{\text{mL}} \times \frac{1\text{ mol Al}_2\text{O}_3}{101.96\text{ g Al}_2\text{O}_3} \times \frac{2\text{ mol Al}}{1\text{ mol Al}_2\text{O}_3} \times \frac{6.022 \times 10^{23}\text{ atoms Al}}{\text{mol Al}}$

$= 7.03 \times 10^{23}$ atoms Al

151. Mol compound = $1\text{ molecule} \times \frac{1\text{ mol}}{6.022 \times 10^{23}\text{ molecules}} = 1.661 \times 10^{-24}$ mol

Molar mass = $\frac{4.65 \times 10^{-23}\text{ g}}{1.661 \times 10^{-24}\text{ mol}}$ = 28.0 g/mol; N_2 and CO are possible identities.

153. Molar mass XeF_n = $\frac{0.368\text{ g XeF}_n}{9.03 \times 10^{20}\text{ molecules XeF}_n \times \frac{1\text{ mol XeF}_n}{6.022 \times 10^{23}\text{ molecules}}}$ = 245 g/mol

245 g = 131.3 g + n(19.00 g), n = 5.98; formula = XeF_6

155. Molar mass = 20(12.01) + 29(1.008) + 19.00 + 3(16.00) = 336.43 g/mol

Mass % C = $\frac{20(12.01)\text{ g C}}{336.43\text{ g compound}} \times 100 = 71.40\%$ C

Mass % H = $\frac{29(1.008)\text{ g H}}{336.43\text{ g compound}} \times 100 = 8.689\%$ H

Mass % F = $\frac{19.00\text{ g F}}{336.43\text{ g compound}} \times 100 = 5.648\%$ F

Mass % O = 100.00 – (71.40 + 8.689 + 5.648) = 14.26% O or:

% O = $\frac{3(16.00)\text{ g O}}{336.43\text{ g compound}} \times 100 = 14.27\%$ O

157. a. NO: Mass % N = $\frac{14.01\text{ g N}}{30.01\text{ g NO}} \times 100 = 46.68\%$ N

b. N_2O: Mass % N = $\frac{2(14.01)\text{ g N}}{44.02\text{ g N}_2\text{O}} \times 100 = 63.65\%$ N

c. NH_3: Mass % N = $\frac{14.01\text{ g N}}{17.03\text{ g NH}_3} \times 100 = 82.27\%$ N

d. SNH: Mass % N = $\frac{14.01\text{ g N}}{47.09\text{ g SNH}} \times 100 = 29.75\%$ N

From the calculations above, the order from smallest to largest mass percent N is SNH < NO < N_2O < NH_3.

159. Out of 100.00 g of adrenaline, there are:

$56.79\text{ g C} \times \frac{1\text{ mol C}}{12.01\text{ g C}} = 4.729$ mol C; $6.56\text{ g H} \times \frac{1\text{ mol H}}{1.008\text{ g H}} = 6.51$ mol H

$$28.37 \text{ g O} \times \frac{1 \text{ mol O}}{16.00 \text{ g O}} = 1.773 \text{ mol O}; \; 8.28 \text{ g N} \times \frac{1 \text{ mol N}}{14.01 \text{ g N}} = 0.591 \text{ mol N}$$

Dividing each mole value by the smallest number:

$$\frac{4.729}{0.591} = 8.00; \; \frac{6.51}{0.591} = 11.0; \; \frac{1.773}{0.591} = 3.00; \; \frac{0.591}{0.591} = 1.00$$

This gives adrenaline an empirical formula of $C_8H_{11}O_3N$.

161. There are many valid methods to solve this problem. We will assume 100.00 g of compound, and then determine from the information in the problem how many moles of compound equals 100.00 g of compound. From this information, we can determine the mass of one mole of compound (the molar mass) by setting up a ratio. Assuming 100.00 g cyanocobalamin:

$$\text{mol cyanocobalamin} = 4.34 \text{ g Co} \times \frac{1 \text{ mol Co}}{58.93 \text{ g Co}} \times \frac{1 \text{ mol cyanocobalamin}}{\text{mol Co}}$$

$$= 7.36 \times 10^{-2} \text{ mol cyanocobalamin}$$

$$\frac{x \text{ g cyanocobalamin}}{1 \text{ mol cyanocobalamin}} = \frac{100.00 \text{ g}}{7.36 \times 10^{-2} \text{ mol}}, \; x = \text{molar mass} = 1.36 \times 10^3 \text{ g/mol}$$

163. Empirical formula mass = 12.01 + 1.008 = 13.02 g/mol; because 104.14/13.02 = 7.998 ≈ 8, the molecular formula for styrene is $(CH)_8 = C_8H_8$.

$$2.00 \text{ g } C_8H_8 \times \frac{1 \text{ mol } C_8H_8}{104.14 \text{ g } C_8H_8} \times \frac{8 \text{ mol H}}{\text{mol } C_8H_8} \times \frac{6.022 \times 10^{23} \text{ atoms H}}{\text{mol H}} = 9.25 \times 10^{22} \text{ atoms H}$$

165. $17.3 \text{ g H} \times \frac{1 \text{ mol H}}{1.008 \text{ g H}} = 17.2 \text{ mol H}; \; 82.7 \text{ g C} \times \frac{1 \text{ mol C}}{12.01 \text{ g C}} = 6.89 \text{ mol C}$

$\frac{17.2}{6.89} = 2.50$; the empirical formula is C_2H_5.

The empirical formula mass is ~29 g/mol, so two times the empirical formula would put the compound in the correct molar mass range. Molecular formula = $(C_2H_5)_2 = C_4H_{10}$.

$$1.04 \times 10^{23} \text{ atoms C} \times \frac{1 \text{ mol C}}{6.022 \times 10^{23} \text{ atoms C}} \times \frac{1 \text{ mol } C_4H_{10}}{4 \text{ mol C}} \times \frac{58.12 \text{ g } C_4H_{10}}{\text{mol } C_4H_{10}}$$

$$= 2.51 \text{ g } C_4H_{10}$$

167. Mass of H_2O = 0.755 g $CuSO_4 \bullet xH_2O$ – 0.483 g $CuSO_4$ = 0.272 g H_2O

$$0.483 \text{ g } CuSO_4 \times \frac{1 \text{ mol } CuSO_4}{159.62 \text{ g } CuSO_4} = 0.00303 \text{ mol } CuSO_4$$

$$0.272 \text{ g } H_2O \times \frac{1 \text{ mol } H_2O}{18.02 \text{ g } H_2O} = 0.0151 \text{ mol } H_2O$$

$$\frac{0.0151 \text{ mol } H_2O}{0.00303 \text{ g } CuSO_4} = \frac{4.98 \text{ mol } H_2O}{1 \text{ mol } CuSO_4}; \text{ compound formula} = CuSO_4 \bullet 5H_2O, \; x = 5$$

169. $1.20 \text{ g } CO_2 \times \frac{1 \text{ mol } CO_2}{44.01 \text{ g}} \times \frac{1 \text{ mol C}}{\text{mol } CO_2} \times \frac{1 \text{ mol } C_{24}H_{30}N_3O}{24 \text{ mol C}} \times \frac{376.51 \text{ g}}{\text{mol } C_{24}H_{30}N_3O}$
$= 0.428 \text{ g } C_{24}H_{30}N_3O$

$\frac{0.428 \text{ g } C_{24}H_{30}N_3O}{1.00 \text{ g sample}} \times 100 = 42.8\% \ C_{24}H_{30}N_3O$ (LSD)

171. Assuming 100.00 g of tetrodotoxin:

$$41.38 \text{ g C} \times \frac{1 \text{ mol C}}{12.01 \text{ g C}} = 3.445 \text{ mol C}; \quad 13.16 \text{ g N} \times \frac{1 \text{ mol N}}{14.01 \text{ g N}} = 0.9393 \text{ mol N}$$

$$5.37 \text{ g H} \times \frac{1 \text{ mol H}}{1.008 \text{ g H}} = 5.33 \text{ mol H}; \quad 40.09 \text{ g O} \times \frac{1 \text{ mol O}}{16.00 \text{ g O}} = 2.506 \text{ mol O}$$

Divide by the smallest number:

$$\frac{3.445}{0.9393} = 3.668; \quad \frac{5.33}{0.9393} = 5.67; \quad \frac{2.506}{0.9393} = 2.668$$

To get whole numbers for each element, multiply through by 3.

Empirical formula: $(C_{3.668}H_{5.67}NO_{2.668})_3 = C_{11}H_{17}N_3O_8$; the mass of the empirical formula is 319.3 g/mol.

$$\text{Molar mass tetrodotoxin} = \frac{1.59 \times 10^{-21} \text{ g}}{3 \text{ molecules} \times \frac{1 \text{ mol}}{6.022 \times 10^{23} \text{ molecules}}} = 319 \text{ g/mol}$$

Because the empirical mass and molar mass are the same, the molecular formula is the same as the empirical formula, $C_{11}H_{17}N_3O_8$.

$$165 \text{ lb} \times \frac{1 \text{ kg}}{2.2046 \text{ lb}} \times \frac{10. \ \mu\text{g}}{\text{kg}} \times \frac{1 \times 10^{-6} \text{ g}}{\mu\text{g}} \times \frac{1 \text{ mol}}{319.3 \text{ g}} \times \frac{6.022 \times 10^{23} \text{ molecules}}{1 \text{ mol}}$$

$= 1.4 \times 10^{18}$ molecules tetrodotoxin is the LD_{50} dosage

173. $126 \text{ g } B_5H_9 \times \frac{1 \text{ mol } B_5H_9}{63.12 \text{ g } B_5H_9} \times \frac{9 \text{ mol } H_2O}{2 \text{ mol } B_5H_9} \times \frac{18.02 \text{ g } H_2O}{\text{mol } H_2O} = 162 \text{ g } H_2O$

$192 \text{ g } O_2 \times \frac{1 \text{ mol } O_2}{32.00 \text{ g } O_2} \times \frac{9 \text{ mol } H_2O}{12 \text{ mol } O_2} \times \frac{18.02 \text{ g } H_2O}{\text{mol } H_2O} = 81.1 \text{ g } H_2O$

Because O_2 produces the smallest quantity of product, O_2 is limiting and 81.1g H_2O can be produced.

175. a. False; this is what is done when equations are balanced.

b. False; the coefficients give molecule ratios as well as mole ratios between reactants and products.

c. False; the reactants are on the left, with the products on the right.

d. True

e. True; for mass to be conserved, there must be the same number of atoms as well as the same type of atoms on both sides.

177. $453 \text{ g Fe} \times \frac{1 \text{ mol Fe}}{55.85 \text{ g Fe}} \times \frac{1 \text{ mol } Fe_2O_3}{2 \text{ mol Fe}} \times \frac{159.70 \text{ g } Fe_2O_3}{\text{mol } Fe_2O_3} = 648 \text{ g } Fe_2O_3$

Mass percent $Fe_2O_3 = \frac{648 \text{ g } Fe_2O_3}{752 \text{ g ore}} \times 100 = 86.2\%$

179. Assuming 1 mole of vitamin A (286.4 g vitamin A):

$$\text{mol C} = 286.4 \text{ g vitamin A} \times \frac{0.8386 \text{ g C}}{\text{g vitamin A}} \times \frac{1 \text{ mol C}}{12.01 \text{ g C}} = 20.00 \text{ mol C}$$

$$\text{mol H} = 286.4 \text{ g vitamin A} \times \frac{0.1056 \text{ g H}}{\text{g vitamin A}} \times \frac{1 \text{ mol H}}{1.008 \text{ g H}} = 30.00 \text{ mol H}$$

Because 1 mole of vitamin A contains 20 mol C and 30 mol H, the molecular formula of vitamin A is $C_{20}H_{30}E$. To determine E, let's calculate the molar mass of E:

$$286.4 \text{ g} = 20(12.01) + 30(1.008) + \text{molar mass E}, \quad \text{molar mass E} = 16.0 \text{ g/mol}$$

From the periodic table, E = oxygen, and the molecular formula of vitamin A is $C_{20}H_{30}O$.

181. X_2Z: 40.0% X and 60.0% Z by mass; $\frac{\text{mol X}}{\text{mol Z}} = 2 = \frac{40.0/A_x}{60.0/A_z} = \frac{(40.0)A_z}{(60.0)A_x}$ or $A_z = 3A_x$,

where A = molar mass. For XZ_2, molar mass = $A_x + 2A_z = A_x + 2(3A_x) = 7A_x$.

Mass percent X = $\frac{A_x}{7A_x} \times 100 = 14.3\%$ X; % Z = 100.0 – 14.3 = 85.7% Z

Challenge Problems

183. The volume of a gas is proportional to the number of molecules of gas. Thus the formulas are:

I: NH_3; II: N_2H_4; III: HN_3

The mass ratios are:

I: $\frac{82.25 \text{ g N}}{17.75 \text{ g H}} = \frac{4.634 \text{ g N}}{\text{g H}}$; II: $\frac{6.949 \text{ g N}}{\text{g H}}$; III: $\frac{41.7 \text{ g N}}{\text{g H}}$

If we set the atomic mass of H equal to 1.008, then the atomic mass, A, for nitrogen is:

I: 14.01; II: 14.01; III. 14.0

For example, for compound I: $\frac{A}{3(1.008)} = \frac{4.634}{1}$, A = 14.01

185. First, we will determine composition in mass percent. We assume that all the carbon in the 0.213 g CO_2 came from the 0.157 g of the compound and that all the hydrogen in the 0.0310 g H_2O came from the 0.157 g of the compound.

$$0.213 \text{ g } CO_2 \times \frac{12.01 \text{ g C}}{44.01 \text{ g } CO_2} = 0.0581 \text{ g C}; \ \%\text{ C} = \frac{0.0581 \text{ g C}}{0.157 \text{ g compound}} \times 100 = 37.0\% \text{ C}$$

$$0.0310 \text{ g } H_2O \times \frac{2.016 \text{ g H}}{18.02 \text{ g } H_2O} = 3.47 \times 10^{-3} \text{ g H}; \ \%\text{ H} = \frac{3.47 \times 10^{-3} \text{ g}}{0.157 \text{g}} \times 100 = 2.21\% \text{ H}$$

We get the mass percent of N from the second experiment:

$$0.0230 \text{ g } NH_3 \times \frac{14.01 \text{ g N}}{17.03 \text{ g } NH_3} = 1.89 \times 10^{-2} \text{ g N}$$

$$\%\text{ N} = \frac{1.89 \times 10^{-2} \text{ g}}{0.103 \text{ g}} \times 100 = 18.3\% \text{ N}$$

The mass percent of oxygen is obtained by difference:

$$\%\text{ O} = 100.00 - (37.0 + 2.21 + 18.3) = 42.5\% \text{ O}$$

So, out of 100.00 g of compound, there are:

$$37.0 \text{ g C} \times \frac{1 \text{ mol C}}{12.01 \text{ g C}} = 3.08 \text{ mol C}; \ \ 2.21 \text{ g H} \times \frac{1 \text{ mol H}}{1.008 \text{ g H}} = 2.19 \text{ mol H}$$

$$18.3 \text{ g N} \times \frac{1 \text{ mol N}}{14.01 \text{ g N}} = 1.31 \text{ mol N}; \ \ 42.5 \text{ g O} \times \frac{1 \text{ mol O}}{16.00 \text{ g O}} = 2.66 \text{ mol O}$$

Lastly, and often the hardest part, we need to find simple whole-number ratios. Divide all mole values by the smallest number:

$$\frac{3.08}{1.31} = 2.35; \ \frac{2.19}{1.31} = 1.67; \ \frac{1.31}{1.31} = 1.00; \ \frac{2.66}{1.31} = 2.03$$

Multiplying all these ratios by 3 gives an empirical formula of $C_7H_5N_3O_6$.

187. $Fe(s) + \frac{1}{2} O_2(g) \rightarrow FeO(s)$; $2 Fe(s) + \frac{3}{2} O_2(g) \rightarrow Fe_2O_3(s)$

$$20.00 \text{ g Fe} \times \frac{1 \text{ mol Fe}}{55.85 \text{ g}} = 0.3581 \text{ mol}$$

$$(11.20 - 3.24) \text{ g } O_2 \times \frac{1 \text{ mol } O_2}{32.00 \text{ g}} = 0.2488 \text{ mol } O_2 \text{ consumed (1 extra sig. fig.)}$$

Let's assume x moles of Fe reacts to form x moles of FeO. Then $0.3581 - x$, the remaining moles of Fe, reacts to form Fe_2O_3. Balancing the two equations in terms of x:

$$x \text{ Fe} + \frac{1}{2} x \text{ } O_2 \rightarrow x \text{ FeO}$$

$$(0.3581 - x) \text{ mol Fe} + \frac{3}{2}\left(\frac{0.3581 - x}{2}\right) \text{mol } O_2 \rightarrow \left(\frac{0.3581 - x}{2}\right) \text{mol } Fe_2O_3$$

Setting up an equation for total moles of O_2 consumed:

$$\frac{1}{2}x + \frac{3}{4}(0.3581 - x) = 0.2488 \text{ mol } O_2, \quad x = 0.0791 = 0.079 \text{ mol FeO}$$

$$0.079 \text{ mol FeO} \times \frac{71.85 \text{ g FeO}}{\text{mol}} = 5.7 \text{ g FeO produced}$$

$$\text{Mol } Fe_2O_3 \text{ produced} = \frac{0.3581 - 0.079}{2} = 0.140 \text{ mol } Fe_2O_3$$

$$0.140 \text{ mol } Fe_2O_3 \times \frac{159.70 \text{ g } Fe_2O_3}{\text{mol}} = 22.4 \text{ g } Fe_2O_3 \text{ produced}$$

189. The two relevant equations are:

$$Zn(s) + 2\,HCl(aq) \rightarrow ZnCl_2(aq) + H_2(g) \text{ and } Mg(s) + 2\,HCl(aq) \rightarrow MgCl_2(aq) + H_2(g)$$

Let x = mass Mg, so $10.00 - x$ = mass Zn. From the balanced equations, moles H_2 produced = moles Zn reacted + moles Mg reacted.

$$\text{Mol } H_2 = 0.5171 \text{ g } H_2 \times \frac{1 \text{ mol } H_2}{2.016 \text{ g } H_2} = 0.2565 \text{ mol } H_2$$

$$0.2565 = \frac{x}{24.31} + \frac{10.00 - x}{65.38}; \text{ solving: } x = 4.008 \text{ g Mg}$$

$$\frac{4.008 \text{ g}}{10.00 \text{ g}} \times 100 = 40.08\% \text{ Mg}$$

191. We know that water is a product, so one of the elements in the compound is hydrogen.

$$X_aH_b + O_2 \rightarrow H_2O + ?$$

To balance the H atoms, the mole ratio between X_aH_b and $H_2O = \frac{2}{b}$.

$$\text{Mol compound} = \frac{1.39 \text{ g}}{62.09 \text{ g/mol}} = 0.0224 \text{ mol}; \text{ mol } H_2O = \frac{1.21 \text{ g}}{18.02 \text{ g/mol}} = 0.0671 \text{ mol}$$

$\frac{2}{b} = \frac{0.0224}{0.0671}$, $b = 6$; X_aH_6 has a molar mass of 62.09 g/mol.

$62.09 = a(\text{molar mass of X}) + 6(1.008)$, $a(\text{molar mass of X}) = 56.04$

Some possible identities for X could be Fe ($a = 1$), Si ($a = 2$), N ($a = 4$), and Li ($a = 8$). N fits the data best, so N_4H_6 is the most likely formula.

193. Total mass of copper used:

$$10{,}000 \text{ boards} \times \frac{(8.0 \text{ cm} \times 16.0 \text{ cm} \times 0.060 \text{ cm})}{\text{board}} \times \frac{8.96 \text{ g}}{\text{cm}^3} = 6.9 \times 10^5 \text{ g Cu}$$

Amount of Cu to be recovered = $0.80 \times (6.9 \times 10^5 \text{ g}) = 5.5 \times 10^5$ g Cu.

$$5.5 \times 10^5 \text{ g Cu} \times \frac{1 \text{ mol Cu}}{63.55 \text{ g Cu}} \times \frac{1 \text{ mol Cu(NH}_3)_4\text{Cl}_2}{\text{mol Cu}} \times \frac{202.59 \text{ g Cu(NH}_3)_4\text{Cl}_2}{\text{mol Cu(NH}_3)_4\text{Cl}_2}$$
$$= 1.8 \times 10^6 \text{ g Cu(NH}_3)_4\text{Cl}_2$$

$$5.5 \times 10^5 \text{ g Cu} \times \frac{1 \text{ mol Cu}}{63.55 \text{ g Cu}} \times \frac{4 \text{ mol NH}_3}{\text{mol Cu}} \times \frac{17.03 \text{ g NH}_3}{\text{mol NH}_3} = 5.9 \times 10^5 \text{ g NH}_3$$

195. 10.00 g XCl_2 + excess $Cl_2 \rightarrow$ 12.55 g XCl_4; 2.55 g Cl reacted with XCl_2 to form XCl_4. XCl_4 contains 2.55 g Cl and 10.00 g XCl_2. From the mole ratios, 10.00 g XCl_2 must also contain 2.55 g Cl; mass X in XCl_2 = 10.00 – 2.55 = 7.45 g X.

$$2.55 \text{ g Cl} \times \frac{1 \text{ mol Cl}}{35.45 \text{ g Cl}} \times \frac{1 \text{ mol XCl}_2}{2 \text{ mol Cl}} \times \frac{1 \text{ mol X}}{\text{mol XCl}_2} = 3.60 \times 10^{-2} \text{ mol X}$$

So 3.60×10^{-2} mol X has a mass equal to 7.45 g X. The molar mass of X is:

$$\frac{7.45 \text{ g X}}{3.60 \times 10^{-2} \text{ mol X}} = 207 \text{ g/mol X; atomic mass} = 207 \text{ u, so X is Pb.}$$

197. Consider the case of aluminum plus oxygen. Aluminum forms Al^{3+} ions; oxygen forms O^{2-} anions. The simplest compound between the two elements is Al_2O_3. Similarly, we would expect the formula of any Group 6A element with Al to be Al_2X_3. Assuming this, out of 100.00 g of compound, there are 12.35 g Al and 87.65 g of the unknown element, X. Let's use this information to determine the molar mass of X, which will allow us to identify X from the periodic table.

$$12.35 \text{ g Al} \times \frac{1 \text{ mol Al}}{26.98 \text{ g Al}} \times \frac{3 \text{ mol X}}{2 \text{ mol Al}} = 0.6866 \text{ mol X}$$

87.65 g of X must represent 0.6866 mol of X.

$$\text{Molar mass of X} = \frac{87.65 \text{ g X}}{0.6866 \text{ mol X}} = 127.7 \text{ g/mol X}$$

From the periodic table, the unknown element is tellurium, and the formula is Al_2Te_3.

199. $LaH_{2.90}$ is the formula. If only La^{3+} is present, LaH_3 would be the formula. If only La^{2+} is present, LaH_2 would be the formula. Let x = mol La^{2+} and y = mol La^{3+}:

$(La^{2+})_x(La^{3+})_yH_{(2x+3y)}$ where $x + y = 1.00$ and $2x + 3y = 2.90$

Solving by simultaneous equations:

$$\begin{aligned} 2x + 3y &= 2.90 \\ -2x - 2y &= -2.00 \\ \hline y &= 0.90 \text{ and } x = 0.10 \end{aligned}$$

$LaH_{2.90}$ contains $\frac{1}{10}$ La^{2+}, or 10.% La^{2+}, and $\frac{9}{10}$ La^{3+}, or 90.% La^{3+}.

201. The balanced equations are:

$$4\ NH_3(g) + 5\ O_2(g) \rightarrow 4\ NO(g) + 6\ H_2O(g) \text{ and } 4\ NH_3(g) + 7\ O_2(g) \rightarrow 4\ NO_2(g) + 6\ H_2O(g)$$

Let $4x$ = number of moles of NO formed, and let $4y$ = number of moles of NO_2 formed. Then:

$$4x\ NH_3 + 5x\ O_2 \rightarrow 4x\ NO + 6x\ H_2O \text{ and } 4y\ NH_3 + 7y\ O_2 \rightarrow 4y\ NO_2 + 6y\ H_2O$$

All the NH_3 reacted, so $4x + 4y = 2.00$. $10.00 - 6.75 = 3.25$ mol O_2 reacted, so $5x + 7y = 3.25$.

Solving by the method of simultaneous equations:

$$\begin{array}{r} 20x + 28y = 13.0 \\ \underline{-20x - 20y = -10.0} \\ 8y = 3.0, \end{array} \quad y = 0.38;\ \ 4x + 4 \times 0.38 = 2.00,\ \ x = 0.12$$

Mol NO = $4x = 4 \times 0.12 = 0.48$ mol NO formed

CHAPTER 4

TYPES OF CHEMICAL REACTIONS AND SOLUTION STOICHIOMETRY

Questions

19. a. Polarity is a term applied to covalent compounds. Polar covalent compounds have an unequal sharing of electrons in bonds that results in unequal charge distribution in the overall molecule. Polar molecules have a partial negative end and a partial positive end. These are not full charges as in ionic compounds but are charges much smaller in magnitude. Water is a polar molecule and dissolves other polar solutes readily. The oxygen end of water (the partial negative end of the polar water molecule) aligns with the partial positive end of the polar solute, whereas the hydrogens of water (the partial positive end of the polar water molecule) align with the partial negative end of the solute. These opposite charge attractions stabilize polar solutes in water. This process is called hydration. Nonpolar solutes do not have permanent partial negative and partial positive ends; nonpolar solutes are not stabilized in water and do not dissolve.

    b. KF is a soluble ionic compound, so it is a strong electrolyte. KF(aq) actually exists as separate hydrated K^+ ions and hydrated F^- ions in solution: $C_6H_{12}O_6$ is a polar covalent molecule that is a nonelectrolyte. $C_6H_{12}O_6$ is hydrated as described in part a.

    c. RbCl is a soluble ionic compound, so it exists as separate hydrated Rb^+ ions and hydrated Cl^- ions in solution. AgCl is an insoluble ionic compound, so the ions stay together in solution and fall to the bottom of the container as a precipitate.

    d. HNO_3 is a strong acid and exists as separate hydrated H^+ ions and hydrated NO_3^- ions in solution. CO is a polar covalent molecule and is hydrated as explained in part a.

21. During a dilution, only water is added. There is the same quantity of solute in a larger volume. Because of this, the concentration decreases. Answer d is correct.

23. Only statement b is true. A concentrated solution can also contain a nonelectrolyte dissolved in water, e.g., concentrated sugar water. Acids are either strong or weak electrolytes. Some ionic compounds are not soluble in water, so they are not labeled as a specific type of electrolyte.

25. Zinc Chloride is not one of the insoluble chlorides, so it is soluble. All nitrate salts are soluble, so iron(II) nitrate is soluble. The rest of the compounds, c-e, will be insoluble by the solubility rules. Pb^{2+} ions form an insoluble sulfate, and most sulfide and chromate salts are insoluble.

27. Use the solubility rules in Table 4.1. Some soluble bromides by Rule 2 would be NaBr, KBr, and NH_4Br (there are others). The insoluble bromides by Rule 3 would be AgBr, $PbBr_2$, and Hg_2Br_2. Similar reasoning is used for the other parts to this problem.

Sulfates: Na_2SO_4, K_2SO_4, and $(NH_4)_2SO_4$ (and others) would be soluble, and $BaSO_4$, $CaSO_4$, and $PbSO_4$ (or Hg_2SO_4) would be insoluble.

Hydroxides: NaOH, KOH, $Ca(OH)_2$ (and others) would be soluble, and $Al(OH)_3$, $Fe(OH)_3$, and $Cu(OH)_2$ (and others) would be insoluble.

Phosphates: Na_3PO_4, K_3PO_4, $(NH_4)_3PO_4$ (and others) would be soluble, and Ag_3PO_4, $Ca_3(PO_4)_2$, and $FePO_4$ (and others) would be insoluble.
Lead: $PbCl_2$, $PbBr_2$, PbI_2, $Pb(OH)_2$, $PbSO_4$, and PbS (and others) would be insoluble. $Pb(NO_3)_2$ would be a soluble Pb^{2+} salt.

29. The Brønsted-Lowry definitions are best for our purposes. An acid is a proton donor, and a base is a proton acceptor. A proton is an H^+ ion. Neutral hydrogen has 1 electron and 1 proton, so an H^+ ion is just a proton. An acid-base reaction is the transfer of an H^+ ion (a proton) from an acid to a base.

31. a. The species reduced is the element that gains electrons. The reducing agent causes reduction to occur by itself being oxidized. The reducing agent generally refers to the entire formula of the compound/ion that contains the element oxidized.

b. The species oxidized is the element that loses electrons. The oxidizing agent causes oxidation to occur by itself being reduced. The oxidizing agent generally refers to the entire formula of the compound/ion that contains the element reduced.

c. For simple binary ionic compounds, the actual charges on the ions are the same as the oxidation states. For covalent compounds, nonzero oxidation states are imaginary charges the elements would have if they were held together by ionic bonds (assuming the bond is between two different nonmetals). Nonzero oxidation states for elements in covalent compounds are not actual charges. Oxidation states for covalent compounds are a bookkeeping method to keep track of electrons in a reaction.

Exercises

Aqueous Solutions: Strong and Weak Electrolytes

33. a. $NaBr(s) \rightarrow Na^+(aq) + Br^-(aq)$

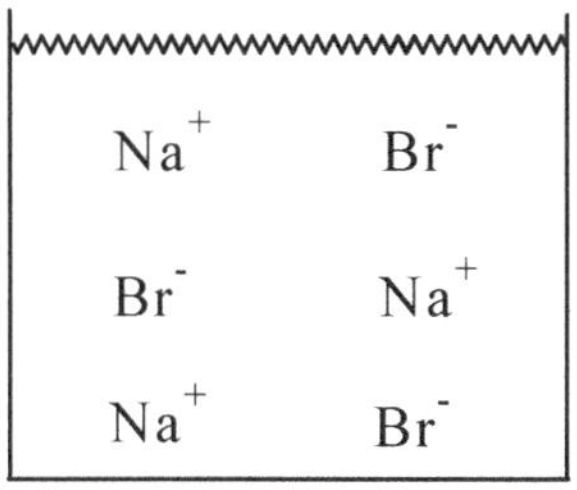

Your drawing should show equal number of Na^+ and Br^- ions.

b. $MgCl_2(s) \rightarrow Mg^{2+}(aq) + 2\ Cl^-(aq)$

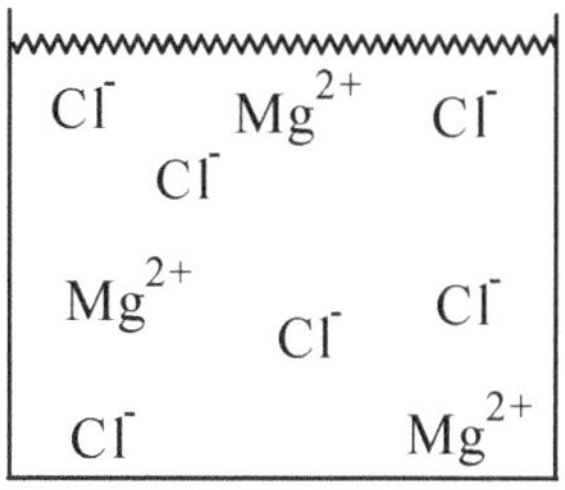

Your drawing should show twice the number of Cl^- ions as Mg^{2+} ions.

c. $Al(NO_3)_3(s) \rightarrow Al^{3+}(aq) + 3\ NO_3^-(aq)$

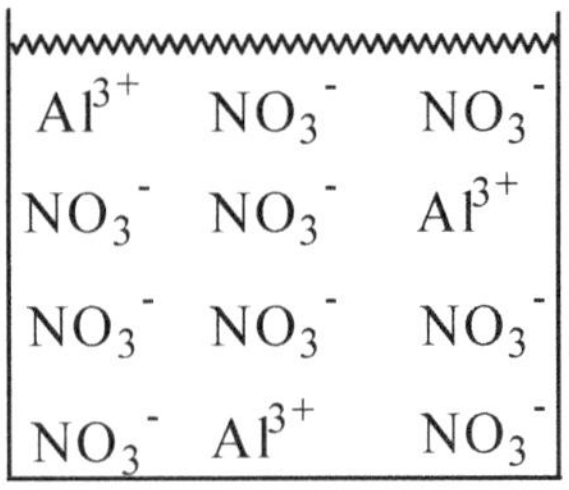

d. $(NH_4)_2SO_4(s) \rightarrow 2\ NH_4^+(aq) + SO_4^{2-}(aq)$

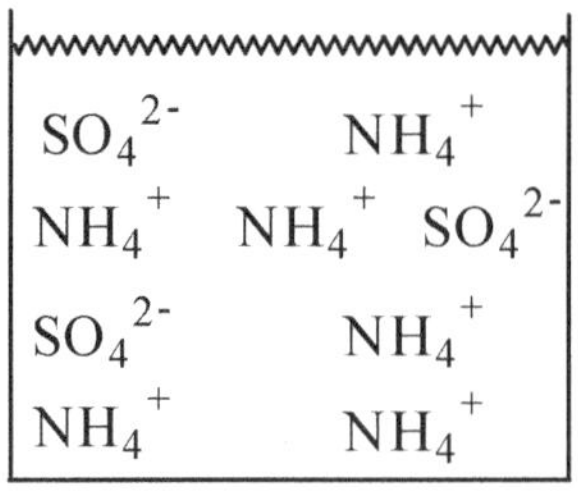

For e-i, your drawings should show equal numbers of the cations and anions present because each salt is a 1 : 1 salt. The ions present are listed in the following dissolution reactions.

e. $NaOH(s) \rightarrow Na^+(aq) + OH^-(aq)$

f. $FeSO_4(s) \rightarrow Fe^{2+}(aq) + SO_4^{2-}(aq)$

g. $KMnO_4(s) \rightarrow K^+(aq) + MnO_4^-\ (aq)$

h. $HClO_4(aq) \rightarrow H^+(aq) + ClO_4^-(aq)$

i. $NH_4C_2H_3O_2(s) \rightarrow NH_4^+(aq) + C_2H_3O_2^-(aq)$

35. By the sulfate solubility rule, $CuSO_4$ is soluble. It will exist in solution as hydrated Cu^{2+} and SO_4^{2-} ions moving independently of each other. Answer d is correct.

37. $CaCl_2(s) \rightarrow Ca^{2+}(aq) + 2\ Cl^-(aq)$

Solution Concentration: Molarity

39. a. $5.623\text{ g NaHCO}_3 \times \dfrac{1\text{ mol NaHCO}_3}{84.01\text{ g NaHCO}_3} = 6.693 \times 10^{-2}\text{ mol NaHCO}_3$

$$M = \frac{6.693 \times 10^{-2}\text{ mol}}{250.0\text{ mL}} \times \frac{1000\text{ mL}}{\text{L}} = 0.2677\ M\ \text{NaHCO}_3$$

b. $0.1846\text{ g K}_2\text{Cr}_2\text{O}_7 \times \dfrac{1\text{ mol K}_2\text{Cr}_2\text{O}_7}{294.20\text{ g K}_2\text{Cr}_2\text{O}_7} = 6.275 \times 10^{-4}\text{ mol K}_2\text{Cr}_2\text{O}_7$

$$M = \frac{6.275 \times 10^{-4}\text{ mol}}{500.0 \times 10^{-3}\text{ L}} = 1.255 \times 10^{-3}\ M\ \text{K}_2\text{Cr}_2\text{O}_7$$

c. $0.1025\text{ g Cu} \times \dfrac{1\text{ mol Cu}}{63.55\text{ g Cu}} = 1.613 \times 10^{-3}\text{ mol Cu} = 1.613 \times 10^{-3}\text{ mol Cu}^{2+}$

$$M = \frac{1.613 \times 10^{-3}\text{ mol Cu}^{2+}}{200.0\text{ mL}} \times \frac{1000\text{ mL}}{\text{L}} = 8.065 \times 10^{-3}\ M\ \text{Cu}^{2+}$$

41. a. $M_{Ca(NO_3)_2} = \dfrac{0.100\text{ mol Ca(NO}_3)_2}{0.100\text{ L}} = 1.00\ M$

$Ca(NO_3)_2(s) \rightarrow Ca^{2+}(aq) + 2\ NO_3^-(aq)$; $M_{Ca^{2+}} = 1.00\ M$; $M_{NO_3^-} = 2(1.00) = 2.00\ M$

b. $M_{Na_2SO_4} = \dfrac{2.5 \text{ mol } Na_2SO_4}{1.25 \text{ L}} = 2.0\ M$

$Na_2SO_4(s) \rightarrow 2\ Na^+(aq) + SO_4^{2-}(aq)$; $M_{Na^+} = 2(2.0) = 4.0\ M$; $M_{SO_4^{2-}} = 2.0\ M$

c. $5.00 \text{ g } NH_4Cl \times \dfrac{1 \text{ mol } NH_4Cl}{53.49 \text{ g } NH_4Cl} = 0.0935 \text{ mol } NH_4Cl$

$M_{NH_4Cl} = \dfrac{0.0935 \text{ mol } NH_4Cl}{0.5000 \text{ L}} = 0.187\ M$

$NH_4Cl(s) \rightarrow NH_4^+(aq) + Cl^-(aq)$; $M_{NH_4^+} = M_{Cl^-} = 0.187\ M$

d. $1.00 \text{ g } K_3PO_4 \times \dfrac{1 \text{ mol } K_3PO_4}{212.27 \text{ g}} = 4.71 \times 10^{-3} \text{ mol } K_3PO_4$

$M_{K_3PO_4} = \dfrac{4.71 \times 10^{-3} \text{ mol}}{0.2500 \text{ L}} = 0.0188\ M$

$K_3PO_4(s) \rightarrow 3\ K^+(aq) + PO_4^{3-}(aq)$; $M_{K^+} = 3(0.0188) = 0.0564\ M$; $M_{PO_4^{3-}} = 0.0188\ M$

43. Mol solute = volume (L) × molarity $\left(\dfrac{\text{mol}}{\text{L}}\right)$; $AlCl_3(s) \rightarrow Al^{3+}(aq) + 3\ Cl^-(aq)$

Mol Cl^- = $0.1000 \text{ L} \times \dfrac{0.30 \text{ mol } AlCl_3}{\text{L}} \times \dfrac{3 \text{ mol } Cl^-}{\text{mol } AlCl_3} = 9.0 \times 10^{-2} \text{ mol } Cl^-$

$MgCl_2(s) \rightarrow Mg^{2+}(aq) + 2\ Cl^-(aq)$

Mol Cl^- = $0.0500 \text{ L} \times \dfrac{0.60 \text{ mol } MgCl_2}{\text{L}} \times \dfrac{2 \text{ mol } Cl^-}{\text{mol } MgCl_2} = 6.0 \times 10^{-2} \text{ mol } Cl^-$

$NaCl(s) \rightarrow Na^+(aq) + Cl^-(aq)$

Mol Cl^- = $0.2000 \text{ L} \times \dfrac{0.40 \text{ mol NaCl}}{\text{L}} \times \dfrac{1 \text{ mol } Cl^-}{\text{mol NaCl}} = 8.0 \times 10^{-2} \text{ mol } Cl^-$

100.0 mL of 0.30 M $AlCl_3$ contains the most moles of Cl^- ions.

45. Molar mass of NaOH = 22.99 + 16.00 + 1.008 = 40.00 g/mol

Mass NaOH = $0.2500 \text{ L} \times \dfrac{0.400 \text{ mol NaOH}}{\text{L}} \times \dfrac{40.00 \text{ g NaOH}}{\text{mol NaOH}} = 4.00 \text{ g NaOH}$

47. $0.0150 \text{ L} \times \dfrac{137 \text{ mmol Na}}{\text{L}} \times \dfrac{1 \text{ mol}}{1000 \text{ mmol}} \times \dfrac{22.99 \text{ g Na}}{\text{mol Na}} = 0.0472 \text{ g Na}$

49. Mol $CaCl_2$ present = $0.230 \text{ L } C_{12}H_{22}O_{11} \times \dfrac{0.275 \text{ mol } C_{12}H_{22}O_{11}}{\text{L } C_{12}H_{22}O_{11}} = 6.33 \times 10^{-2} \text{ mol } C_{12}H_{22}O_{11}$

The volume of $C_{12}H_{22}O_{11}$ solution after evaporation is:

$$6.33 \times 10^{-2} \text{ mol } C_{12}H_{22}O_{11} \times \frac{1 \text{ L } C_{12}H_{22}O_{11}}{1.10 \text{ mol } C_{12}H_{22}O_{11}} = 5.75 \times 10^{-2} \text{ L} = 57.5 \text{ mL } C_{12}H_{22}O_{11}$$

Volume H_2O evaporated = 230. mL − 57.5 mL = 173 mL H_2O evaporated

51. a. $2.00 \text{ L} \times \dfrac{0.250 \text{ mol NaOH}}{\text{L}} \times \dfrac{40.00 \text{ g NaOH}}{\text{mol NaOH}} = 20.0 \text{ g NaOH}$

Place 20.0 g NaOH in a 2-L volumetric flask; add water to dissolve the NaOH, and fill to the mark with water, mixing several times along the way.

b. $2.00 \text{ L} \times \dfrac{0.250 \text{ mol NaOH}}{\text{L}} \times \dfrac{1 \text{ L stock}}{1.00 \text{ mol NaOH}} = 0.500 \text{ L}$

Add 500. mL of 1.00 *M* NaOH stock solution to a 2-L volumetric flask; fill to the mark with water, mixing several times along the way.

c. $2.00 \text{ L} \times \dfrac{0.100 \text{ mol } K_2CrO_4}{\text{L}} \times \dfrac{194.20 \text{ g } K_2CrO_4}{\text{mol } K_2CrO_4} = 38.8 \text{ g } K_2CrO_4$

Like the solution made in part a, instead using 38.8 g K_2CrO_4.

d. $2.00 \text{ L} \times \dfrac{0.100 \text{ mol } K_2CrO_4}{\text{L}} \times \dfrac{1 \text{ L stock}}{1.75 \text{ mol } K_2CrO_4} = 0.114 \text{ L}$

Like the solution made in part b, instead using 114 mL of the 1.75 *M* K_2CrO_4 stock solution.

53. $10.8 \text{ g } (NH_4)_2SO_4 \times \dfrac{1 \text{ mol}}{132.15 \text{ g}} = 8.17 \times 10^{-2} \text{ mol } (NH_4)_2SO_4$

$$\text{Molarity} = \frac{8.17 \times 10^{-2} \text{ mol}}{100.0 \text{ mL}} \times \frac{1000 \text{ mL}}{\text{L}} = 0.817\ M\ (NH_4)_2SO_4$$

Moles of $(NH_4)_2SO_4$ in final solution:

$$10.00 \times 10^{-3} \text{ L} \times \frac{0.817 \text{ mol}}{\text{L}} = 8.17 \times 10^{-3} \text{ mol}$$

$$\text{Molarity of final solution} = \frac{8.17 \times 10^{-3} \text{ mol}}{(10.00 + 50.00) \text{ mL}} \times \frac{1000 \text{ mL}}{\text{L}} = 0.136\ M\ (NH_4)_2SO_4$$

$(NH_4)_2SO_4(s) \rightarrow 2\ NH_4^+(aq) + SO_4^{2-}(aq)$; $M_{NH_4^+} = 2(0.136) = 0.272\ M$; $M_{SO_4^{2-}} = 0.136\ M$

55. $\text{Mol } Na_2CO_3 = 0.0700 \text{ L} \times \dfrac{3.0 \text{ mol } Na_2CO_3}{\text{L}} = 0.21 \text{ mol } Na_2CO_3$

$Na_2CO_3(s) \rightarrow 2\ Na^+(aq) + CO_3^{2-}(aq)$; mol Na^+ = 2(0.21 mol) = 0.42 mol

$$\text{Mol NaHCO}_3 = 0.0300\ \text{L} \times \frac{1.0\ \text{mol NaHCO}_3}{\text{L}} = 0.030\ \text{mol NaHCO}_3$$

$NaHCO_3(s) \rightarrow Na^+(aq) + HCO_3^-(aq)$; mol Na^+ = 0.030 mol

$$M_{\text{Na}^+} = \frac{\text{total mol Na}^+}{\text{total volume}} = \frac{0.42\ \text{mol} + 0.030\ \text{mol}}{0.0700\ \text{L} + 0.0300\ \text{L}} = \frac{0.45\ \text{mol}}{0.1000\ \text{L}} = 4.5\ M\ \text{Na}^+$$

57. $$\text{Stock solution} = \frac{10.0\ \text{mg}}{500.0\ \text{mL}} = \frac{10.0 \times 10^{-3}\ \text{g}}{500.0\ \text{mL}} = \frac{2.00 \times 10^{-5}\ \text{g steroid}}{\text{mL}}$$

$$100.0 \times 10^{-6}\ \text{L stock} \times \frac{1000\ \text{mL}}{\text{L}} \times \frac{2.00 \times 10^{-5}\ \text{g steroid}}{\text{mL}} = 2.00 \times 10^{-6}\ \text{g steroid}$$

This is diluted to a final volume of 100.0 mL.

$$\frac{2.00 \times 10^{-6}\ \text{g steroid}}{100.0\ \text{mL}} \times \frac{1000\ \text{mL}}{\text{L}} \times \frac{1\ \text{mol steroid}}{336.43\ \text{g steroid}} = 5.94 \times 10^{-8}\ M\ \text{steroid}$$

Precipitation Reactions

59. The solubility rules referenced in the following answers are outlined in Table 4.1 of the text.

a. Soluble: Most nitrate salts are soluble (Rule 1).
b. Soluble: Most chloride salts are soluble except for Ag^+, Pb^{2+}, and Hg_2^{2+} (Rule 3).
c. Soluble: Most sulfate salts are soluble except for $BaSO_4$, $PbSO_4$, Hg_2SO_4, and $CaSO_4$ (Rule 4.)
d. Insoluble: Most hydroxide salts are only slightly soluble (Rule 5).

Note: We will interpret the phrase "slightly soluble" as meaning insoluble and the phrase "marginally soluble" as meaning soluble. So the marginally soluble hydroxides $Ba(OH)_2$, $Sr(OH)_2$, and $Ca(OH)_2$ will be assumed soluble unless noted otherwise.

e. Insoluble: Most sulfide salts are only slightly soluble (Rule 6). Again, "slightly soluble" is interpreted as "insoluble" in problems like these.
f. Insoluble: Rule 5 (see answer d).
g. Insoluble: Most phosphate salts are only slightly soluble (Rule 6).

61. In these reactions, soluble ionic compounds are mixed together. To predict the precipitate, switch the anions and cations in the two reactant compounds to predict possible products; then use the solubility rules in Table 4.1 to predict if any of these possible products are insoluble (are the precipitate). Note that the phrase "slightly soluble" in Table 4.1 is interpreted to mean insoluble, and the phrase "marginally soluble" is interpreted to mean soluble.

a. Possible products = $FeCl_2$ and K_2SO_4; both salts are soluble, so no precipitate forms.
b. Possible products = $Al(OH)_3$ and $Ba(NO_3)_2$; precipitate = $Al(OH)_3(s)$
c. Possible products = $CaSO_4$ and NaCl; precipitate = $CaSO_4(s)$
d. Possible products = KNO_3 and NiS; precipitate = NiS(s)

63. Precipitates will form in all four beakers. In beaker 1, $PbCl_2(s)$ forms, in beaker 2, $Pb(OH)_2(s)$ forms, in beaker 3, $Pb_3(PO_4)_2(s)$ forms, and in beaker 4, $PbSO_4(s)$ forms.

65. For the following answers, the balanced formula equation is first, followed by the complete ionic equation, then the net ionic equation.

a. No reaction occurs since all possible products are soluble salts.

b. $2\ Al(NO_3)_3(aq) + 3\ Ba(OH)_2(aq) \rightarrow 2\ Al(OH)_3(s) + 3\ Ba(NO_3)_2(aq)$

$2\ Al^{3+}(aq) + 6\ NO_3^-(aq) + 3\ Ba^{2+}(aq) + 6\ OH^-(aq) \rightarrow$
$2\ Al(OH)_3(s) + 3\ Ba^{2+}(aq) + 6\ NO_3^-(aq)$

$Al^{3+}(aq) + 3\ OH^-(aq) \rightarrow Al(OH)_3(s)$

c. $CaCl_2(aq) + Na_2SO_4(aq) \rightarrow CaSO_4(s) + 2\ NaCl(aq)$

$Ca^{2+}(aq) + 2\ Cl^-(aq) + 2\ Na^+(aq) + SO_4^{2-}(aq) \rightarrow CaSO_4(s) + 2\ Na^+(aq) + 2\ Cl^-(aq)$

$Ca^{2+}(aq) + SO_4^{2-}(aq) \rightarrow CaSO_4(s)$

d. $K_2S(aq) + Ni(NO_3)_2(aq) \rightarrow 2\ KNO_3(aq) + NiS(s)$

$2\ K^+(aq) + S^{2-}(aq) + Ni^{2+}(aq) + 2\ NO_3^-(aq) \rightarrow 2\ K^+(aq) + 2\ NO_3^-(aq) + NiS(s)$

$Ni^{2+}(aq) + S^{2-}(aq) \rightarrow NiS(s)$

67. a. When $CuSO_4(aq)$ is added to $Na_2S(aq)$, the precipitate that forms is $CuS(s)$. Therefore, Na^+ (the gray spheres) and SO_4^{2-} (the bluish green spheres) are the spectator ions.

$CuSO_4(aq) + Na_2S(aq) \rightarrow CuS(s) + Na_2SO_4(aq)$; $Cu^{2+}(aq) + S^{2-}(aq) \rightarrow CuS(s)$

b. When $CoCl_2(aq)$ is added to $NaOH(aq)$, the precipitate that forms is $Co(OH)_2(s)$. Therefore, Na^+ (the gray spheres) and Cl^- (the green spheres) are the spectator ions.

$CoCl_2(aq) + 2\ NaOH(aq) \rightarrow Co(OH)_2(s) + 2\ NaCl(aq)$

$Co^{2+}(aq) + 2\ OH^-(aq) \rightarrow Co(OH)_2(s)$

c. When $AgNO_3(aq)$ is added to $KI(aq)$, the precipitate that forms is $AgI(s)$. Therefore, K^+ (the red spheres) and NO_3^- (the blue spheres) are the spectator ions.

$AgNO_3(aq) + KI(aq) \rightarrow AgI(s) + KNO_3(aq)$; $Ag^+(aq) + I^-(aq) \rightarrow AgI(s)$

69. a. $(NH_4)_2SO_4(aq) + Ba(NO_3)_2(aq) \rightarrow 2\ NH_4NO_3(aq) + BaSO_4(s)$

$Ba^{2+}(aq) + SO_4^{2-}(aq) \rightarrow BaSO_4(s)$

b. $Pb(NO_3)_2(aq) + 2\ NaCl(aq) \rightarrow PbCl_2(s) + 2\ NaNO_3(aq)$

$Pb^{2+}(aq) + 2\ Cl^-(aq) \rightarrow PbCl_2(s)$

c. Potassium phosphate and sodium nitrate are both soluble in water. No reaction occurs.

d. No reaction occurs because all possible products are soluble.

e. $CuCl_2(aq) + 2\ NaOH(aq) \rightarrow Cu(OH)_2(s) + 2\ NaCl(aq)$

$Cu^{2+}(aq) + 2\ OH^-(aq) \rightarrow Cu(OH)_2(s)$

71. For effective separation of the ions, only one precipitate can be filtered off at a time. The NaCl solution cannot be added first since AgCl(s) and $PbCl_2(s)$ both form, and the Na_2S solution cannot be added first since all three cations form an insoluble sulfide. So, the Na_2SO_4 solution is added first removing the Pb^{2+} ion as $PbSO_4(s)$. After the $PbSO_4$ precipitate is removed, the NaCl solution is added next removing the Ag^+ ion as AgCl(s). After the AgCl precipitate has been removed, the Na_2S solution is added last to precipitate out Ni^{2+} as NiS(s)

73. $2\ AgNO_3(aq) + Na_2CrO_4(aq) \rightarrow Ag_2CrO_4(s) + 2\ NaNO_3(aq)$

$$0.0750\ L \times \frac{0.100\ mol\ AgNO_3}{L} \times \frac{1\ mol\ Na_2CrO_4}{2\ mol\ AgNO_3} \times \frac{161.98\ g\ Na_2CrO_4}{mol\ Na_2CrO_4} = 0.607\ g\ Na_2CrO_4$$

75. $Fe(NO_3)_3(aq) + 3\ NaOH(aq) \rightarrow Fe(OH)_3(s) + 3\ NaNO_3(aq)$

Assuming $Fe(NO_3)_3$ is limiting:

$$0.0750\ L \times \frac{0.105\ mol\ Fe(NO_3)_3}{L} \times \frac{1\ mol\ Fe(OH)_3}{mol\ Fe(NO_3)_3} \times \frac{106.87\ g\ Fe(OH)_3}{mol\ Fe(OH)_3} = 0.842\ g\ Fe(OH)_3$$

Assuming NaOH is limiting:

$$0.125\ L \times \frac{0.150\ mol\ NaOH}{L} \times \frac{1\ mol\ Fe(OH)_3}{3\ mol\ NaOH} \times \frac{106.87\ g\ Fe(OH)_3}{mol\ Fe(OH)_3} = 0.668\ g\ Fe(OH)_3$$

Because NaOH produces the smaller mass of the $Fe(OH)_3$ precipitate, NaOH is the limiting reagent and 0.668 g $Fe(OH)_3$ can form.

77. a. $3\ AgNO_3(aq) + K_3PO_4(aq) \rightarrow Ag_3PO_4(s) + 3\ KNO_3(aq)$

Assuming $AgNO_3$ is limiting:

$$50.00\ mL\ AgNO_3 \times \frac{1\ L}{1000\ mL} \times \frac{1.00\ mol\ AgNO_3}{L\ AgNO_3} \times \frac{1\ mol\ Ag_3PO_4}{3\ mol\ AgNO_3} = 0.0167\ mol\ Ag_3PO_4$$

Assuming K_3PO_4 is limiting:

$$25.0\ mL\ K_3PO_4 \times \frac{1\ L}{1000\ mL} \times \frac{1.00\ mol\ K_3PO_4}{L\ K_3PO_4} \times \frac{1\ mol\ Ag_3PO_4}{mol\ K_3PO_4} = 0.0250\ mol\ Ag_3PO_4$$

The $AgNO_3$ reagent produces the smaller quantity of Ag_3PO_4, so $AgNO_3$ is limiting and 0.0167 mol Ag_3PO_4 can form.

b. initial mol of PO_4^{3-} = $0.0250 \text{ L} \times \frac{1.00 \text{ mol } K_3PO_4}{\text{L } K_3PO_4} \times \frac{1 \text{ mol } PO_4^{3-}}{\text{mol } K_3PO_4} = 0.0250 \text{ mol } PO_4^{3-}$

mol PO_4^{3-} in the precipitate = $0.0167 \text{ mol } Ag_3PO_4 \times \frac{1 \text{ mol } PO_4^{3-}}{\text{mol } Ag_3PO_4} = 0.0167 \text{ mol } PO_4^{3-}$

mol PO_4^{3-} remaining in solution = 0.0250 mol – 0.0167 mol = 0.0083 mol PO_4^{3-}

PO_4^{3-} concentration after reaction = $\frac{0.0083 \text{ mol } PO_4^{3-}}{0.05000 \text{ L} + 0.02500 \text{ L}} = 0.11 \ M \ PO_4^{3-}$

79. a. The balanced reaction is $2 \text{ KOH(aq)} + Mg(NO_3)_2\text{(aq)} \rightarrow Mg(OH)_2\text{(s)} + 2 \text{ KNO}_3\text{(aq)}$.

b. The precipitate is magnesium hydroxide.

c. Assuming KOH is limiting:

$$0.1000 \text{ L KOH} \times \frac{0.200 \text{ mol KOH}}{\text{L KOH}} \times \frac{1 \text{ mol } Mg(OH)_2}{2 \text{ mol KOH}} \times \frac{58.33 \text{ g } Mg(OH)_2}{\text{mol } Mg(OH)_2} = 0.583 \text{ g } Mg(OH)_2$$

Assuming $Mg(NO_3)_2$ is limiting:

$$0.1000 \text{ L } Mg(NO_3)_2 \times \frac{0.200 \text{ mol } Mg(NO_3)_2}{\text{L } Mg(NO_3)_2} \times \frac{1 \text{ mol } Mg(OH)_2}{\text{mol } Mg(NO_3)_2} \times \frac{58.33 \text{ g } Mg(OH)_2}{\text{mol } Mg(OH)_2} = 1.17 \text{ g } Mg(OH)_2$$

The KOH reagent is limiting because it produces the smaller quantity of the $Mg(OH)_2$ precipitate. So 0.583 g $Mg(OH)_2$ can form.

d. The net ionic equation for this reaction is $Mg^{2+}\text{(aq)} + 2 \text{ OH}^-\text{(aq)} \rightarrow Mg(OH)_2\text{(s)}$.

Because KOH is the limiting reagent, all of the OH^- is used up in the reaction. So M_{OH^-} = 0 M. Note that K^+ is a spectator ion, so it is still present in solution after precipitation was complete. Also present will be the excess Mg^{2+} and NO_3^- (the other spectator ion).

$$\text{Total } Mg^{2+} = 0.1000 \text{ L } Mg(NO_3)_2 \times \frac{0.200 \text{ mol } Mg(NO_3)_2}{\text{L } Mg(NO_3)_2} \times \frac{1 \text{ mol } Mg^{2+}}{\text{mol } Mg(NO_3)_2} = 0.0200 \text{ mol } Mg^{2+}$$

$$\text{Mol } Mg^{2+} \text{ reacted} = 0.1000 \text{ L KOH} \times \frac{0.200 \text{ mol KOH}}{\text{L KOH}} \times \frac{1 \text{ mol } Mg(NO_3)_2}{2 \text{ mol KOH}} \times \frac{1 \text{ mol } Mg^{2+}}{\text{mol } Mg(NO_3)_2} = 0.0100 \text{ mol } Mg^{2+}$$

$$M_{Mg^{2+}} = \frac{\text{mol excess } Mg^{2+}}{\text{total volume}} = \frac{(0.0200 - 0.0100) \text{ mol } Mg^{2+}}{0.1000 \text{ L} + 0.1000 \text{ L}} = 5.00 \times 10^{-2} \ M \ Mg^{2+}$$

The spectator ions are K^+ and NO_3^-. The moles of each are:

$$\text{mol K}^+ = 0.1000 \text{ L KOH} \times \frac{0.200 \text{ mol KOH}}{\text{L KOH}} \times \frac{1 \text{ mol K}^+}{\text{mol KOH}} = 0.0200 \text{ mol K}^+$$

$$\text{mol NO}_3^- = 0.1000 \text{ L Mg(NO}_3)_2 \times \frac{0.200 \text{ mol Mg(NO}_3)_2}{\text{L Mg(NO}_3)_2} \times \frac{2 \text{ mol NO}_3^-}{\text{mol Mg(NO}_3)_2} = 0.0400 \text{ mol NO}_3^-$$

The concentrations are:

$$\frac{0.0200 \text{ mol K}^+}{0.2000 \text{ L}} = 0.100\ M\ \text{K}^+; \quad \frac{0.0400 \text{ mol NO}_3^-}{0.2000 \text{ L}} = 0.200\ M\ \text{NO}_3^-$$

81. $M_2SO_4(aq) + CaCl_2(aq) \rightarrow CaSO_4(s) + 2\ MCl(aq)$

$$1.36 \text{ g CaSO}_4 \times \frac{1 \text{ mol CaSO}_4}{136.15 \text{ g CaSO}_4} \times \frac{1 \text{ mol M}_2\text{SO}_4}{\text{mol CaSO}_4} = 9.99 \times 10^{-3} \text{ mol M}_2\text{SO}_4$$

From the problem, 1.42 g M_2SO_4 was reacted, so:

$$\text{molar mass} = \frac{1.42 \text{ g M}_2\text{SO}_4}{9.99 \times 10^{-3} \text{ mol M}_2\text{SO}_4} = 142 \text{ g/mol}$$

142 u = 2(atomic mass M) + 32.07 + 4(16.00), atomic mass M = 23 u

From periodic table, M = Na (sodium).

Acid-Base Reactions

83. All the bases in this problem are ionic compounds containing OH^-. The acids are either strong or weak electrolytes. The best way to determine if an acid is a strong or weak electrolyte is to memorize all the strong electrolytes (strong acids). Any other acid you encounter that is not a strong acid will be a weak electrolyte (a weak acid), and the formula should be left unaltered in the complete ionic and net ionic equations. The strong acids to recognize are HCl, HBr, HI, HNO_3, $HClO_4$, and H_2SO_4. For the following answers, the order of the equations are formula, complete ionic, and net ionic.

a. $2\ HClO_4(aq) + Mg(OH)_2(s) \rightarrow 2\ H_2O(l) + Mg(ClO_4)_2(aq)$

$2\ H^+(aq) + 2\ ClO_4^-(aq) + Mg(OH)_2(s) \rightarrow 2\ H_2O(l) + Mg^{2+}(aq) + 2\ ClO_4^-(aq)$

$2\ H^+(aq) + Mg(OH)_2(s) \rightarrow 2\ H_2O(l) + Mg^{2+}(aq)$

b. $HCN(aq) + NaOH(aq) \rightarrow H_2O(l) + NaCN(aq)$

$HCN(aq) + Na^+(aq) + OH^-(aq) \rightarrow H_2O(l) + Na^+(aq) + CN^-(aq)$

$HCN(aq) + OH^-(aq) \rightarrow H_2O(l) + CN^-(aq)$

c. $HCl(aq) + NaOH(aq) \rightarrow H_2O(l) + NaCl(aq)$

$H^+(aq) + Cl^-(aq) + Na^+(aq) + OH^-(aq) \rightarrow H_2O(l) + Na^+(aq) + Cl^-(aq)$

$H^+(aq) + OH^-(aq) \rightarrow H_2O(l)$

85. All the acids in this problem are strong electrolytes (strong acids). The acids to recognize as strong electrolytes are HCl, HBr, HI, HNO_3, $HClO_4$, and H_2SO_4.

a. $KOH(aq) + HNO_3(aq) \rightarrow H_2O(l) + KNO_3(aq)$

b. $Ba(OH)_2(aq) + 2\ HCl(aq) \rightarrow 2\ H_2O(l) + BaCl_2(aq)$

c. $3\ HClO_4(aq) + Fe(OH)_3(s) \rightarrow 3\ H_2O(l) + Fe(ClO_4)_3(aq)$

d. $AgOH(s) + HBr(aq) \rightarrow AgBr(s) + H_2O(l)$

e. $Sr(OH)_2(aq) + 2\ HI(aq) \rightarrow 2\ H_2O(l) + SrI_2(aq)$

87. If we begin with 50.00 mL of 0.200 *M* NaOH, then:

$$50.00 \times 10^{-3}\ \text{L} \times \frac{0.200\ \text{mol}}{\text{L}} = 1.00 \times 10^{-2}\ \text{mol NaOH is to be neutralized}$$

a. $NaOH(aq) + HCl(aq) \rightarrow NaCl(aq) + H_2O(l)$

$$1.00 \times 10^{-2}\ \text{mol NaOH} \times \frac{1\ \text{mol HCl}}{\text{mol NaOH}} \times \frac{1\ \text{L}}{0.100\ \text{mol}} = 0.100\ \text{L or 100. mL}$$

b. $HNO_3(aq) + NaOH(aq) \rightarrow H_2O(l) + NaNO_3(aq)$

$$1.00 \times 10^{-2}\ \text{mol NaOH} \times \frac{1\ \text{mol HNO}_3}{\text{mol NaOH}} \times \frac{1\ \text{L}}{0.150\ \text{mol HNO}_3} = 6.67 \times 10^{-2}\ \text{L or 66.7 mL}$$

c. $HC_2H_3O_2(aq) + NaOH(aq) \rightarrow H_2O(l) + NaC_2H_3O_2(aq)$

$$1.00 \times 10^{-2}\ \text{mol NaOH} \times \frac{1\ \text{mol HC}_2\text{H}_3\text{O}_2}{\text{mol NaOH}} \times \frac{1\ \text{L}}{0.200\ \text{mol HC}_2\text{H}_3\text{O}_2} = 5.00 \times 10^{-2}\ \text{L}$$

$$= 50.0\ \text{mL}$$

89. $Ba(OH)_2(aq) + 2\ HCl(aq) \rightarrow BaCl_2(aq) + 2\ H_2O(l)$; $H^+(aq) + OH^-(aq) \rightarrow H_2O(l)$

$$75.0 \times 10^{-3}\ \text{L} \times \frac{0.250\ \text{mol HCl}}{\text{L}} = 1.88 \times 10^{-2}\ \text{mol HCl} = 1.88 \times 10^{-2}\ \text{mol H}^+$$

$$+ 1.88 \times 10^{-2}\ \text{mol Cl}^-$$

$$225.0 \times 10^{-3}\ \text{L} \times \frac{0.0550\ \text{mol Ba(OH)}_2}{\text{L}} = 1.24 \times 10^{-2}\ \text{mol Ba(OH)}_2$$

$$= 1.24 \times 10^{-2}\ \text{mol Ba}^{2+} + 2.48 \times 10^{-2}\ \text{mol OH}^-$$

The net ionic equation requires a 1 : 1 mole ratio between OH^- and H^+. The actual mole OH^- to mole H^+ ratio is greater than 1 : 1, so OH^- is in excess. Because 1.88×10^{-2} mol OH^- will be neutralized by the H^+, we have $(2.48 - 1.88) \times 10^{-2} = 0.60 \times 10^{-2}$ mol OH^- in excess.

$$M_{\text{OH}^-} = \frac{\text{mol OH}^-\ \text{excess}}{\text{total volume}} = \frac{6.0 \times 10^{-3}\ \text{mol OH}^-}{0.0750\ \text{L} + 0.2250\ \text{L}} = 2.0 \times 10^{-2}\ M\ \text{OH}^-$$

91. $HCl(aq) + NaOH(aq) \rightarrow H_2O(l) + NaCl(aq)$

$$24.16 \times 10^{-3}\ \text{L NaOH} \times \frac{0.106\ \text{mol NaOH}}{\text{L NaOH}} \times \frac{1\ \text{mol HCl}}{\text{mol NaOH}} = 2.56 \times 10^{-3}\ \text{mol HCl}$$

$$\text{Molarity of HCl} = \frac{2.56 \times 10^{-3}\ \text{mol}}{25.00 \times 10^{-3}\ \text{L}} = 0.102\ M\ \text{HCl}$$

93. $2\ HNO_3(aq) + Ca(OH)_2(aq) \rightarrow 2\ H_2O(l) + Ca(NO_3)_2(aq)$

$$35.00 \times 10^{-3}\ \text{L HNO}_3 \times \frac{0.0500\ \text{mol HNO}_3}{\text{L HNO}_3} \times \frac{1\ \text{mol Ca(OH)}_2}{2\ \text{mol HNO}_3} \times \frac{1\ \text{L Ca(OH)}_2}{0.0200\ \text{mol Ca(OH)}_2}$$
$$= 0.0438\ \text{L} = 43.8\ \text{mL Ca(OH)}_2$$

95. Because KHP is a monoprotic acid, the reaction is (KHP is an abbreviation for potassium hydrogen phthalate):

$$NaOH(aq) + KHP(aq) \rightarrow NaKP(aq) + H_2O(l)$$

$$0.1082\ \text{g KHP} \times \frac{1\ \text{mol KHP}}{204.22\ \text{g KHP}} \times \frac{1\ \text{mol NaOH}}{\text{mol KHP}} = 5.298 \times 10^{-4}\ \text{mol NaOH}$$

There are 5.298×10^{-4} mol of sodium hydroxide in 34.67 mL of solution. Therefore, the concentration of sodium hydroxide is:

$$\frac{5.298 \times 10^{-4}\ \text{mol}}{34.67 \times 10^{-3}\ \text{L}} = 1.528 \times 10^{-2}\ M\ \text{NaOH}$$

Oxidation-Reduction Reactions

97. Apply the rules in Table 4.2.

a. $KMnO_4$ is composed of K^+ and MnO_4^- ions. Assign oxygen an oxidation state of −2, which gives manganese a +7 oxidation state because the sum of oxidation states for all atoms in MnO_4^- must equal the 1− charge on MnO_4^-. K, +1; O, −2; Mn, +7.

b. Assign O a −2 oxidation state, which gives nickel a +4 oxidation state. Ni, +4; O, −2.

c. $Na_4Fe(OH)_6$ is composed of Na^+ cations and $Fe(OH)_6^{4-}$ anions. $Fe(OH)_6^{4-}$ is composed of an iron cation and 6 OH^- anions. For an overall anion charge of 4−, iron must have a +2 oxidation state. As is usually the case in compounds, assign O a −2 oxidation state and H a +1 oxidation state. Na, +1; Fe, +2; O, −2; H, +1.

d. $(NH_4)_2HPO_4$ is made of NH_4^+ cations and HPO_4^{2-} anions. Assign +1 as the oxidation state of H and −2 as the oxidation state of O. In NH_4^+, $x + 4(+1) = +1$, $x = -3$ = oxidation state of N. In HPO_4^{2-}, $+1 + y + 4(-2) = -2$, $y = +5$ = oxidation state of P.

e. O, −2; P, +3

f. O, −2; $3x + 4(-2) = 0$, $x = +8/3$ = oxidation state of Fe; this is the average oxidation state of the three iron ions in Fe_3O_4. In the actual formula unit, there are two Fe^{3+} ions and one Fe^{2+} ion.

g. O, −2; F, −1; Xe, +6

h. F, −1; S, +4

i. O, −2; C, +2

j. H, +1; O, −2; C, 0

99. a. −3 b. −3 c. $2(x) + 4(+1) = 0,\ x = -2$

d. +2 e. +1 f. +4

g. +3 h. +5 i. 0

101. All the statements are false. The oxygen oxidation state is −2 on both the reactant and product sides of the equation. Oxygen is neither oxidized nor reduced. Carbon goes from the +2 oxidation state in CO to the +4 oxidation state in CO_2. With the increase in oxidation states, carbon is oxidized, so CO is the reducing agent. Note that nitrogen goes from the +2 oxidation state in NO to the 0 oxidation state in N_2. Nitrogen is reduced and NO is the oxidizing agent.

103. To determine if the reaction is an oxidation-reduction reaction, assign oxidation states. If the oxidation states change for some elements, then the reaction is a redox reaction. If the oxidation states do not change, then the reaction is not a redox reaction. In redox reactions, the species oxidized (called the reducing agent) shows an increase in oxidation states, and the species reduced (called the oxidizing agent) shows a decrease in oxidation states.

	Redox?	Oxidizing Agent	Reducing Agent	Substance Oxidized	Substance Reduced
a.	Yes	Ag^+	Cu	Cu	Ag^+
b.	No	–	–	–	–
c.	No	–	–	–	–
d.	Yes	$SiCl_4$	Mg	Mg	$SiCl_4$ (Si)
e.	No	–	–	–	–

In b, c, and e, no oxidation numbers change.

105. Use the method of half-reactions described in Section 4.10 of the text to balance these redox reactions. The first step always is to separate the reaction into the two half-reactions, and then to balance each half-reaction separately.

a. $3\ I^- \rightarrow I_3^- + 2e^-$

$ClO^- \rightarrow Cl^-$

$2e^- + 2H^+ + ClO^- \rightarrow Cl^- + H_2O$

Adding the two balanced half-reactions so electrons cancel:

$$3\ I^-(aq) + 2\ H^+(aq) + ClO^-(aq) \rightarrow I_3^-(aq) + Cl^-(aq) + H_2O(l)$$

b. $As_2O_3 \rightarrow H_3AsO_4$

$As_2O_3 \rightarrow 2\ H_3AsO_4$

Left 3 – O; right 8 – O

$NO_3^- \rightarrow NO + 2\ H_2O$

$4\ H^+ + NO_3^- \rightarrow NO + 2\ H_2O$

$(3\ e^- + 4\ H^+ + NO_3^- \rightarrow NO + 2\ H_2O) \times 4$

Right hand side has 5 extra O. Balance the oxygen atoms first using H_2O, then balance H using H^+, and finally, balance charge using electrons. This gives:

$$(5\ H_2O + As_2O_3 \rightarrow 2\ H_3AsO_4 + 4\ H^+ + 4\ e^-) \times 3$$

Common factor is a transfer of 12 e^-. Add half-reactions so that electrons cancel.

$$12\ e^- + 16\ H^+ + 4\ NO_3^- \rightarrow 4\ NO + 8\ H_2O$$
$$15\ H_2O + 3\ As_2O_3 \rightarrow 6\ H_3AsO_4 + 12\ H^+ + 12\ e^-$$

$$7\ H_2O(l) + 4\ H^+(aq) + 3\ As_2O_3(s) + 4\ NO_3^-(aq) \rightarrow 4\ NO(g) + 6\ H_3AsO_4(aq)$$

c. $(2\ Br^- \rightarrow Br_2 + 2\ e^-) \times 5$

$MnO_4^- \rightarrow Mn^{2+} + 4\ H_2O$

$(5\ e^- + 8\ H^+ + MnO_4^- \rightarrow Mn^{2+} + 4\ H_2O) \times 2$

Common factor is a transfer of 10 e^-.

$$10\ Br^- \rightarrow 5\ Br_2 + 10\ e^-$$
$$10\ e^- + 16\ H^+ + 2\ MnO_4^- \rightarrow 2\ Mn^{2+} + 8\ H_2O$$

$$16\ H^+(aq) + 2\ MnO_4^-(aq) + 10\ Br^-(aq) \rightarrow 5\ Br_2(l) + 2\ Mn^{2+}(aq) + 8\ H_2O(l)$$

d. $CH_3OH \rightarrow CH_2O$

$(CH_3OH \rightarrow CH_2O + 2\ H^+ + 2\ e^-) \times 3$

$Cr_2O_7^{2-} \rightarrow 2\ Cr^{3+}$

$14\ H^+ + Cr_2O_7^{2-} \rightarrow 2\ Cr^{3+} + 7\ H_2O$

$6\ e^- + 14\ H^+ + Cr_2O_7^{2-} \rightarrow 2\ Cr^{3+} + 7\ H_2O$

Common factor is a transfer of 6 e^-.

$$3\ CH_3OH \rightarrow 3\ CH_2O + 6\ H^+ + 6\ e^-$$
$$6\ e^- + 14\ H^+ + Cr_2O_7^{2-} \rightarrow 2\ Cr^{3+} + 7\ H_2O$$

$$8\ H^+(aq) + 3\ CH_3OH(aq) + Cr_2O_7^{2-}(aq) \rightarrow 2\ Cr^{3+}(aq) + 3\ CH_2O(aq) + 7\ H_2O(l)$$

107. Use the same method as with acidic solutions. After the final balanced equation, convert H^+ to OH^- as described in Section 4.10 of the text. The extra step involves converting H^+ into H_2O by adding equal moles of OH^- to each side of the reaction. This converts the reaction to a basic solution while still keeping it balanced.

a. $Al \rightarrow Al(OH)_4^-$

$4\ H_2O + Al \rightarrow Al(OH)_4^- + 4\ H^+$

$4\ H_2O + Al \rightarrow Al(OH)_4^- + 4\ H^+ + 3\ e^-$

$MnO_4^- \rightarrow MnO_2$

$3\ e^- + 4\ H^+ + MnO_4^- \rightarrow MnO_2 + 2\ H_2O$

$$4\ H_2O + Al \rightarrow Al(OH)_4^- + 4\ H^+ + 3\ e^-$$
$$3\ e^- + 4\ H^+ + MnO_4^- \rightarrow MnO_2 + 2\ H_2O$$

$$2\ H_2O(l) + Al(s) + MnO_4^-(aq) \rightarrow Al(OH)_4^-(aq) + MnO_2(s)$$

H^+ doesn't appear in the final balanced reaction, so we are done.

b. $Cl_2 \rightarrow Cl^-$ $\quad$ $Cl_2 \rightarrow OCl^-$

$2\ e^- + Cl_2 \rightarrow 2\ Cl^-$ $\quad$ $2\ H_2O + Cl_2 \rightarrow 2\ OCl^- + 4\ H^+ + 2\ e^-$

$$\begin{array}{c} 2\ e^- + Cl_2 \rightarrow 2\ Cl^- \\ 2\ H_2O + Cl_2 \rightarrow 2\ OCl^- + 4\ H^+ + 2\ e^- \\ \hline 2\ H_2O + 2\ Cl_2 \rightarrow 2\ Cl^- + 2\ OCl^- + 4\ H^+ \end{array}$$

Now convert to a basic solution. Add 4 OH^- to both sides of the equation. The 4 OH^- will react with the 4 H^+ on the product side to give 4 H_2O. After this step, cancel identical species on both sides (2 H_2O). Applying these steps gives: $4\ OH^- + 2\ Cl_2 \rightarrow 2\ Cl^- + 2\ OCl^- + 2\ H_2O$, which can be further simplified to:

$$2\ OH^-(aq) + Cl_2(g) \rightarrow Cl^-(aq) + OCl^-(aq) + H_2O(l)$$

c. $NO_2^- \rightarrow NH_3$ $\quad$ $Al \rightarrow AlO_2^-$

$6\ e^- + 7\ H^+ + NO_2^- \rightarrow NH_3 + 2\ H_2O$ $\quad$ $(2\ H_2O + Al \rightarrow AlO_2^- + 4\ H^+ + 3\ e^-) \times 2$

Common factor is a transfer of 6 e^-.

$$\begin{array}{c} 6e^- + 7\ H^+ + NO_2^- \rightarrow NH_3 + 2\ H_2O \\ 4\ H_2O + 2\ Al \rightarrow 2\ AlO_2^- + 8\ H^+ + 6\ e^- \\ \hline OH^- + 2\ H_2O + NO_2^- + 2\ Al \rightarrow NH_3 + 2\ AlO_2^- + H^+ + OH^- \end{array}$$

Reducing gives $OH^-(aq) + H_2O(l) + NO_2^-(aq) + 2\ Al(s) \rightarrow NH_3(g) + 2\ AlO_2^-(aq)$.

109. $NaCl + H_2SO_4 + MnO_2 \rightarrow Na_2SO_4 + MnCl_2 + Cl_2 + H_2O$

We could balance this reaction by the half-reaction method, which is generally the preferred method. However, sometimes a redox reaction is not so complicated and thus balancing by inspection is a possibility. Let's try inspection here. To balance Cl^-, we need 4 NaCl:

$$4\ NaCl + H_2SO_4 + MnO_2 \rightarrow Na_2SO_4 + MnCl_2 + Cl_2 + H_2O$$

Balance the Na^+ and SO_4^{2-} ions next:

$$4\ NaCl + 2\ H_2SO_4 + MnO_2 \rightarrow 2\ Na_2SO_4 + MnCl_2 + Cl_2 + H_2O$$

On the left side: 4 H and 10 O; on the right side: 8 O not counting H_2O

We need 2 H_2O on the right side to balance H and O:

$$4\ NaCl(aq) + 2\ H_2SO_4(aq) + MnO_2(s) \rightarrow 2\ Na_2SO_4(aq) + MnCl_2(aq) + Cl_2(g) + 2\ H_2O(l)$$

111. $(H_2C_2O_4 \rightarrow 2\ CO_2 + 2\ H^+ + 2\ e^-) \times 5$ $\quad$ $(5\ e^- + 8\ H^+ + MnO_4^- \rightarrow Mn^{2+} + 4\ H_2O) \times 2$

$$\begin{array}{c} 5\ H_2C_2O_4 \rightarrow 10\ CO_2 + 10\ H^+ + 10\ e^- \\ 10\ e^- + 16\ H^+ + 2\ MnO_4^- \rightarrow 2\ Mn^{2+} + 8\ H_2O \\ \hline 6\ H^+(aq) + 5\ H_2C_2O_4(aq) + 2\ MnO_4^-(aq) \rightarrow 10\ CO_2(g) + 2\ Mn^{2+}(aq) + 8\ H_2O(l) \end{array}$$

$$0.1058 \text{ g } H_2C_2O_4 \times \frac{1 \text{ mol } H_2C_2O_4}{90.04 \text{ g}} \times \frac{2 \text{ mol } MnO_4^-}{5 \text{ mol } H_2C_2O_4} = 4.700 \times 10^{-4} \text{ mol } MnO_4^-$$

$$\text{Molarity} = \frac{4.700 \times 10^{-4} \text{ mol } MnO_4^-}{28.97 \text{ mL}} \times \frac{1000 \text{ mL}}{\text{L}} = 1.622 \times 10^{-2}\ M\ MnO_4^-$$

113. a. $(Fe^{2+} \rightarrow Fe^{3+} + e^-) \times 5$ $5\ e^- + 8\ H^+ + MnO_4^- \rightarrow Mn^{2+} + 4\ H_2O$

The balanced equation is:

$$8\ H^+(aq) + MnO_4^-(aq) + 5\ Fe^{2+}(aq) \rightarrow 5\ Fe^{3+}(aq) + Mn^{2+}(aq) + 4\ H_2O(l)$$

$$20.62 \times 10^{-3} \text{ L soln} \times \frac{0.0216 \text{ mol } MnO_4^-}{\text{L soln}} \times \frac{5 \text{ mol } Fe^{2+}}{\text{mol } MnO_4^-} = 2.23 \times 10^{-3} \text{ mol } Fe^{2+}$$

$$\text{Molarity} = \frac{2.23 \times 10^{-3} \text{ mol } Fe^{2+}}{50.00 \times 10^{-3} \text{ L}} = 4.46 \times 10^{-2}\ M\ Fe^{2+}$$

b. $(Fe^{2+} \rightarrow Fe^{3+} + e^-) \times 6$ $6\ e^- + 14\ H^+ + Cr_2O_7^{2-} \rightarrow 2\ Cr^{3+} + 7\ H_2O$

The balanced equation is:

$$14\ H^+(aq) + Cr_2O_7^{2-}(aq) + 6\ Fe^{2+}(aq) \rightarrow 6\ Fe^{3+}(aq) + 2\ Cr^{3+}(aq) + 7\ H_2O(l)$$

$$50.00 \times 10^{-3} \text{ L} \times \frac{4.46 \times 10^{-2} \text{ mol } Fe^{2+}}{\text{L}} \times \frac{1 \text{ mol } Cr_2O_7^{2-}}{6 \text{ mol } Fe^{2+}} \times \frac{1 \text{ L}}{0.0150 \text{ mol } Cr_2O_7^{2-}}$$
$$= 2.48 \times 10^{-2} \text{ L or } 24.8 \text{ mL}$$

ChemWork Problems

115. Desired uncertainty is 1% of 0.02, or ±0.0002. So we want the solution to be 0.0200 ± 0.0002 *M,* or the concentration should be between 0.0198 and 0.0202 *M.* We should use a 1-L volumetric flask to make the solution. They are good to ±0.1%. We want to weigh out between 0.0198 mol and 0.0202 mol of KIO_3.

Molar mass of KIO_3 = 39.10 + 126.9 + 3(16.00) = 214.0 g/mol

$$0.0198 \text{ mol} \times \frac{214.0 \text{ g}}{\text{mol}} = 4.237 \text{ g}; \quad 0.0202 \text{ mol} \times \frac{214.0 \text{ g}}{\text{mol}} = 4.323 \text{ g (carrying extra sig. figs.)}$$

We should weigh out between 4.24 and 4.32 g of KIO_3. We should weigh it to the nearest milligram, or nearest 0.1 mg. Dissolve the KIO_3 in water, and dilute (with mixing along the way) to the mark in a 1-L volumetric flask. This will produce a solution whose concentration is within the limits and is known to at least the fourth decimal place.

117. $$32.0 \text{ g } C_{12}H_{22}O_{11} \times \frac{1 \text{ mol } C_{12}H_{22}O_{11}}{342.30 \text{ g}} = 0.0935 \text{ mol } C_{12}H_{22}O_{11} \text{ added to blood}$$

The blood sugar level would increase by:

$$\frac{0.0935 \text{ mol } C_{12}H_{22}O_{11}}{5.0 \text{ L}} = 0.019 \text{ mol/L}$$

119. Stock solution = $\frac{0.6706\text{ g}}{100.0\text{ mL}} = \frac{6.706 \times 10^{-3}\text{ g oxalic acid}}{\text{mL}}$

$$10.00\text{ mL stock} \times \frac{6.706 \times 10^{-3}\text{ g oxalic acid}}{\text{mL}} = 6.706 \times 10^{-2}\text{ g oxalic acid}$$

This is diluted to a final volume of 250.0 mL.

$$\frac{6.706 \times 10^{-2}\text{ g }H_2C_2O_4}{250.0\text{ mL}} \times \frac{1000\text{ mL}}{\text{L}} \times \frac{1\text{ mol }H_2C_2O_4}{90.04\text{ g}} = 2.979 \times 10^{-3}\ M\ H_2C_2O_4$$

121. There are other possible correct choices for most of the following answers. We have listed only three possible reactants in each case.

a. $AgNO_3$, $Pb(NO_3)_2$, and $Hg_2(NO_3)_2$ would form precipitates with the Cl^- ion.

$Ag^+(aq) + Cl^-(aq) \rightarrow AgCl(s)$; $Pb^{2+}(aq) + 2\ Cl^-(aq) \rightarrow PbCl_2(s)$

$Hg_2^{2+}(aq) + 2\ Cl^-(aq) \rightarrow Hg_2Cl_2(s)$

b. Na_2SO_4, Na_2CO_3, and Na_3PO_4 would form precipitates with the Ca^{2+} ion.

$Ca^{2+}(aq) + SO_4^{2-}(aq) \rightarrow CaSO_4(s)$; $Ca^{2+}(aq) + CO_3^{2-}(aq) \rightarrow CaCO_3(s)$

$3\ Ca^{2+}(aq) + 2\ PO_4^{3-}(aq) \rightarrow Ca_3(PO_4)_2(s)$

c. NaOH, Na_2S, and Na_2CO_3 would form precipitates with the Fe^{3+} ion.

$Fe^{3+}(aq) + 3\ OH^-(aq) \rightarrow Fe(OH)_3(s)$; $2\ Fe^{3+}(aq) + 3\ S^{2-}(aq) \rightarrow Fe_2S_3(s)$

$2\ Fe^{3+}(aq) + 3\ CO_3^{2-}(aq) \rightarrow Fe_2(CO_3)_3(s)$

d. $BaCl_2$, $Pb(NO_3)_2$, and $Ca(NO_3)_2$ would form precipitates with the SO_4^{2-} ion.

$Ba^{2+}(aq) + SO_4^{2-}(aq) \rightarrow BaSO_4(s)$; $Pb^{2+}(aq) + SO_4^{2-}(aq) \rightarrow PbSO_4(s)$

$Ca^{2+}(aq) + SO_4^{2-}(aq) \rightarrow CaSO_4(s)$

e. Na_2SO_4, NaCl, and NaI would form precipitates with the Hg_2^{2+} ion.

$Hg_2^{2+}(aq) + SO_4^{2-}(aq) \rightarrow Hg_2SO_4(s)$; $Hg_2^{2+}(aq) + 2\ Cl^-(aq) \rightarrow Hg_2Cl_2(s)$

$Hg_2^{2+}(aq) + 2\ I^-(aq) \rightarrow Hg_2I_2(s)$

f. NaBr, Na_2CrO_4, and Na_3PO_4 would form precipitates with the Ag^+ ion.

$Ag^+(aq) + Br^-(aq) \rightarrow AgBr(s)$; $2\ Ag^+(aq) + CrO_4^{2-}(aq) \rightarrow Ag_2CrO_4(s)$

$3\ Ag^+(aq) + PO_4^{3-}(aq) \rightarrow Ag_3PO_4(s)$

123. $2\ NaOH(aq) + Ni(NO_3)_2(aq) \rightarrow Ni(OH)_2(s) + 2\ NaNO_3(aq)$

$$0.1500\text{ L} \times \frac{0.249\text{ mol }Ni(NO_3)_2}{\text{L}} \times \frac{2\text{ mol NaOH}}{1\text{ mol }Ni(NO_3)_2} \times \frac{1\text{ L NaOH}}{0.100\text{ mol NaOH}} = 0.747\text{ L}$$

$$= 747\text{ mL NaOH}$$

125. $3\ (NH_4)_2CrO_4(aq) + 2\ Cr(NO_2)_3(aq) \rightarrow 6\ NH_4NO_2(aq) + Cr_2(CrO_4)_3(s)$

$$0.203\text{ L} \times \frac{0.307\text{ mol }(NH_4)_2CrO_4}{\text{L}} \times \frac{1\text{ mol }Cr_2(CrO_4)_3}{3\text{ mol }(NH_4)_2CrO_4} \times \frac{452.00\text{ g }Cr_2(CrO_4)_3}{\text{mol }Cr_2(CrO_4)_3} = 9.39\text{ g }Cr_2(CrO_4)_3$$

$$0.137\text{ L} \times \frac{0.269\text{ mol }Cr(NO_2)_3}{\text{L}} \times \frac{1\text{ mol }Cr_2(CrO_4)_3}{2\text{ mol }Cr(NO_2)_3} \times \frac{452.00\text{ g }Cr_2(CrO_4)_3}{\text{mol }Cr_2(CrO_4)_3} = 8.33\text{ g }Cr_2(CrO_4)_3$$

The $Cr(NO_2)_3$ reagent produces the smaller amount of product, so $Cr(NO_2)_3$ is limiting and the theoretical yield of $Cr_2(CrO_4)_3$ is 8.33 g.

$$0.880 = \frac{\text{actual yield}}{8.33\text{ g}},\ \text{actual yield} = (8.33\text{ g})(0.880) = 7.33\text{ g }Cr_2(CrO_4)_3\text{ isolated}$$

127. $XCl_2(aq) + 2\ AgNO_3(aq) \rightarrow 2\ AgCl(s) + X(NO_3)_2(aq)$

$$1.38\text{ g AgCl} \times \frac{1\text{ mol AgCl}}{143.4\text{ g}} \times \frac{1\text{ mol }XCl_2}{2\text{ mol AgCl}} = 4.81 \times 10^{-3}\text{ mol }XCl_2$$

$$\frac{1.00\text{ g }XCl_2}{4.91 \times 10^{-3}\text{ mol }XCl_2} = 208\text{ g/mol};\ x + 2(35.45) = 208,\ x = 137\text{ g/mol}$$

From the periodic table, the metal X is barium (Ba).

129. $Zn_2P_2O_7$: 2(65.38) + 2(30.97) + 7(16.00) = 304.70 g/mol

All of the zinc in $Zn_2P_2O_7$ came from the zinc in the foot powder.

$$0.4089\text{ g }Zn_2P_2O_7 \times \frac{1\text{ mol }Zn_2P_2O_7}{304.70\text{ g }Zn_2P_2O_7} \times \frac{2\text{ mol Zn}}{\text{mol }Zn_2P_2O_7} \times \frac{65.38\text{ g Zn}}{\text{mol}} = 0.1755\text{ g Zn}$$

$$\text{Mass \% Zn} = \frac{0.1755\text{ g Zn}}{1.343\text{ g foot powder}} \times 100 = 13.07\%\text{ Zn}$$

131. All the Tl in TlI came from Tl in Tl_2SO_4. The conversion from TlI to Tl_2SO_4 uses the molar masses and formulas of each compound.

$$0.1824\text{ g TlI} \times \frac{204.4\text{ g Tl}}{331.3\text{ g TlI}} \times \frac{504.9\text{ g }Tl_2SO_4}{408.8\text{ g Tl}} = 0.1390\text{ g }Tl_2SO_4$$

$$\text{Mass \% }Tl_2SO_4 = \frac{0.1390\text{ g }Tl_2SO_4}{9.486\text{ g pesticide}} \times 100 = 1.465\%\ Tl_2SO_4$$

133. With the ions present, the only possible precipitate is $Cr(OH)_3$.

$Cr(NO_3)_3(aq) + 3\ NaOH(aq) \rightarrow Cr(OH)_3(s) + 3\ NaNO_3(aq)$

Mol NaOH used to form precipitate =

$$2.06\text{ g }Cr(OH)_3 \times \frac{1\text{ mol }Cr(OH)_3}{103.02\text{ g}} \times \frac{3\text{ mol NaOH}}{\text{mol }Cr(OH)_3} = 6.00 \times 10^{-2}\text{ mol}$$

$NaOH(aq) + HCl(aq) \rightarrow NaCl(aq) + H_2O(l)$

Mol NaOH used to react with HCl =

$$0.1000 \text{ L} \times \frac{0.400 \text{ mol HCl}}{\text{L}} \times \frac{1 \text{ mol NaOH}}{\text{mol HCl}} = 4.00 \times 10^{-2} \text{ mol}$$

$$M_{\text{NaOH}} = \frac{\text{total mol NaOH}}{\text{volume}} = \frac{6.00 \times 10^{-2} \text{ mol} + 4.00 \times 10^{-2} \text{ mol}}{0.0500 \text{ L}} = 2.00\ M \text{ NaOH}$$

135. Using HA as an abbreviation for the monoprotic acid acetylsalicylic acid:

$HA(aq) + NaOH(aq) \rightarrow H_2O(l) + NaA(aq)$

$$\text{Mol HA} = 0.03517 \text{ L NaOH} \times \frac{0.5065 \text{ mol NaOH}}{\text{L NaOH}} \times \frac{1 \text{ mol HA}}{\text{mol NaOH}} = 1.781 \times 10^{-2} \text{ mol HA}$$

From the problem, 3.210 g HA was reacted, so:

$$\text{molar mass} = \frac{3.210 \text{ g HA}}{1.781 \times 10^{-2} \text{ mol HA}} = 180.2 \text{ g/mol}$$

137. Let HA = unknown monoprotic acid; $HA(aq) + NaOH(aq) \rightarrow NaA(aq) + H_2O(l)$

$$\text{Mol HA present} = 0.0250 \text{ L} \times \frac{0.500 \text{ mol NaOH}}{\text{L}} \times \frac{1 \text{ mol HA}}{1 \text{ mol NaOH}} = 0.0125 \text{ mol HA}$$

$$\frac{x \text{ g HA}}{\text{mol HA}} = \frac{2.20 \text{ g HA}}{0.0125 \text{ mol HA}}, \quad x = \text{molar mass of HA} = 176 \text{ g/mol}$$

Empirical formula weight ≈ 3(12) + 4(1) + 3(16) = 88 g/mol.

Because 176/88 = 2.0, the molecular formula is $(C_3H_4O_3)_2 = C_6H_8O_6$.

139. $2\ HNO_3(aq) + Ca(OH)_2(aq) \rightarrow 2\ H_2O(l) + Ca(NO_3)_2(aq)$

$$34.66 \times 10^{-3} \text{ L } HNO_3 \times \frac{0.944 \text{ mol } HNO_3}{\text{L } HNO_3} \times \frac{1 \text{ mol } Ca(OH)_2}{2 \text{ mol } HNO_3} = 1.64 \times 10^{-2} \text{ mol } Ca(OH)_2$$

$$\text{Molarity of } Ca(OH)_2 = \frac{1.64 \times 10^{-2} \text{ mol}}{50.00 \times 10^{-3} \text{ L}} = 0.328\ M\ Ca(OH)_2$$

141. $$0.104 \text{ g AgCl} \times \frac{1 \text{ mol AgCl}}{143.4 \text{ g AgCl}} \times \frac{1 \text{ mol } Cl^-}{\text{mol AgCl}} \times \frac{35.45 \text{ g } Cl^-}{\text{mol } Cl^-} = 2.57 \times 10^{-2} \text{ g } Cl^-$$

All of the Cl^- in the AgCl precipitate came from the chlorisondamine chloride compound in the medication. So we need to calculate the quantity of $C_{14}H_{20}Cl_6N_2$ which contains 2.57×10^{-2} g Cl^-.

Molar mass of $C_{14}H_{20}Cl_6N_2$ = 14(12.01) + 20(1.008) + 6(35.45) + 2(14.01) = 429.02 g/mol

There are 6(35.45) = 212.70 g chlorine for every mole (429.02 g) of $C_{14}H_{20}Cl_6N_2$.

$$2.57 \times 10^{-2}\text{ g Cl}^- \times \frac{429.02\text{ g }C_{14}H_{20}Cl_6N_2}{212.70\text{ g Cl}^-} = 5.18 \times 10^{-2}\text{ g }C_{14}H_{20}Cl_6N_2$$

$$\text{Mass \% chlorisondamine chloride} = \frac{5.18 \times 10^{-2}\text{ g}}{1.28\text{ g}} \times 100 = 4.05\%$$

143. $Pb^{2+} + 2\ Cl^- \rightarrow PbCl_2(s)$

$$0.0292\text{ L} \times \frac{0.100\text{ mol }Pb^{2+}}{\text{L}} \times \frac{2\text{ mol Cl}^-}{\text{mol }Pb^{2+}} \times \frac{35.45\text{ g Cl}^-}{\text{mol Cl}^-} = 0.207\text{ g Cl}^-$$

Mass % Cl = 0.207 g Cl^-/0.455 g douglasite = 45.5% Cl^-

145. $Mn \rightarrow Mn^{2+} + 2\ e^-$ $\quad$ $HNO_3 \rightarrow NO_2$

$HNO_3 \rightarrow NO_2 + H_2O$

$(e^- + H^+ + HNO_3 \rightarrow NO_2 + H_2O) \times 2$

$$Mn \rightarrow Mn^{2+} + 2\ e^-$$
$$2\ e^- + 2\ H^+ + 2\ HNO_3 \rightarrow 2\ NO_2 + 2\ H_2O$$

$$2\ H^+(aq) + Mn(s) + 2\ HNO_3(aq) \rightarrow Mn^{2+}(aq) + 2\ NO_2(g) + 2\ H_2O(l)$$ or

$$4\ H^+(aq) + Mn(s) + 2\ NO_3^-(aq) \rightarrow Mn^{2+}(aq) + 2\ NO_2(g) + 2\ H_2O(l)$$ (HNO₃ is a strong acid.)

$(4\ H_2O + Mn^{2+} \rightarrow MnO_4^- + 8\ H^+ + 5\ e^-) \times 2$ $\quad$ $(2\ e^- + 2\ H^+ + IO_4^- \rightarrow IO_3^- + H_2O) \times 5$

$$8\ H_2O + 2\ Mn^{2+} \rightarrow 2\ MnO_4^- + 16\ H^+ + 10\ e^-$$
$$10\ e^- + 10\ H^+ + 5\ IO_4^- \rightarrow 5\ IO_3^- + 5\ H_2O$$

$$3\ H_2O(l) + 2\ Mn^{2+}(aq) + 5\ IO_4^-(aq) \rightarrow 2\ MnO_4^-(aq) + 5\ IO_3^-(aq) + 6\ H^+(aq)$$

Challenge Problems

147. a. $$5.0\text{ ppb Hg in water} = \frac{5.0\text{ ng Hg}}{\text{g soln}} = \frac{5.0 \times 10^{-9}\text{ g Hg}}{\text{mL soln}}$$

$$\frac{5.0 \times 10^{-9}\text{ g Hg}}{\text{mL}} \times \frac{1\text{ mol Hg}}{200.6\text{ g Hg}} \times \frac{1000\text{ mL}}{\text{L}} = 2.5 \times 10^{-8}\ M\text{ Hg}$$

b. $$\frac{1.0 \times 10^{-9}\text{ g }CHCl_3}{\text{mL}} \times \frac{1\text{ mol }CHCl_3}{119.37\text{ g }CHCl_3} \times \frac{1000\text{ mL}}{\text{L}} = 8.4 \times 10^{-9}\ M\ CHCl_3$$

c. $$10.0\text{ ppm As} = \frac{10.0\ \mu\text{g As}}{\text{g soln}} = \frac{10.0 \times 10^{-6}\text{ g As}}{\text{mL soln}}$$

$$\frac{10.0 \times 10^{-6}\text{ g As}}{\text{mL}} \times \frac{1\text{ mol As}}{74.92\text{ g As}} \times \frac{1000\text{ mL}}{\text{L}} = 1.33 \times 10^{-4}\ M\text{ As}$$

d. $$\frac{0.10 \times 10^{-6}\text{ g DDT}}{\text{mL}} \times \frac{1\text{ mol DDT}}{354.46\text{ g DDT}} \times \frac{1000\text{ mL}}{\text{L}} = 2.8 \times 10^{-7}\ M\text{ DDT}$$

149. a. $0.308 \text{ g AgCl} \times \dfrac{35.45 \text{ g Cl}}{143.4 \text{ g AgCl}} = 0.0761 \text{ g Cl}$; $\% \text{ Cl} = \dfrac{0.0761 \text{ g}}{0.256 \text{ g}} \times 100 = 29.7\% \text{ Cl}$

Cobalt(III) oxide, Co_2O_3: 2(58.93) + 3(16.00) = 165.86 g/mol

$0.145 \text{ g } Co_2O_3 \times \dfrac{117.86 \text{ g Co}}{165.86 \text{ g } Co_2O_3} = 0.103 \text{ g Co}$; $\% \text{ Co} = \dfrac{0.103 \text{ g}}{0.416 \text{ g}} \times 100 = 24.8\% \text{ Co}$

The remainder, 100.0 – (29.7 + 24.8) = 45.5%, is water.

Assuming 100.0 g of compound:

$$45.5 \text{ g } H_2O \times \frac{2.016 \text{ g H}}{18.02 \text{ g } H_2O} = 5.09 \text{ g H}; \quad \% \text{ H} = \frac{5.09 \text{ g H}}{100.0 \text{ g compound}} \times 100 = 5.09\% \text{ H}$$

$$45.5 \text{ g } H_2O \times \frac{16.00 \text{ g O}}{18.02 \text{ g } H_2O} = 40.4 \text{ g O}; \quad \% \text{ O} = \frac{40.4 \text{ g O}}{100.0 \text{ g compound}} \times 100 = 40.4\% \text{ O}$$

The mass percent composition is 24.8% Co, 29.7% Cl, 5.09% H, and 40.4% O.

b. Out of 100.0 g of compound, there are:

$$24.8 \text{ g Co} \times \frac{1 \text{ mol}}{58.93 \text{ g Co}} = 0.421 \text{ mol Co}; \quad 29.7 \text{ g Cl} \times \frac{1 \text{ mol}}{35.45 \text{ g Cl}} = 0.838 \text{ mol Cl}$$

$$5.09 \text{ g H} \times \frac{1 \text{ mol}}{1.008 \text{ g H}} = 5.05 \text{ mol H}; \quad 40.4 \text{ g O} \times \frac{1 \text{ mol}}{16.00 \text{ g O}} = 2.53 \text{ mol O}$$

Dividing all results by 0.421, we get $CoCl_2{\bullet}6H_2O$ for the empirical formula, which is also the actual formula given the information in the problem. The $\bullet 6H_2O$ represent six waters of hydration in the chemical formula.

c. $CoCl_2{\bullet}6H_2O(aq) + 2\ AgNO_3(aq) \rightarrow 2\ AgCl(s) + Co(NO_3)_2(aq) + 6\ H_2O(l)$

$CoCl_2{\bullet}6H_2O(aq) + 2\ NaOH(aq) \rightarrow Co(OH)_2(s) + 2\ NaCl(aq) + 6\ H_2O(l)$

$Co(OH)_2 \rightarrow Co_2O_3$ This is an oxidation-reduction reaction. Thus we also need to include an oxidizing agent. The obvious choice is O_2.

$4\ Co(OH)_2(s) + O_2(g) \rightarrow 2\ Co_2O_3(s) + 4\ H_2O(l)$

151. $Zn(s) + 2\ AgNO_2(aq) \rightarrow 2\ Ag(s) + Zn(NO_2)_2(aq)$

Let x = mass of Ag and y = mass of Zn after the reaction has stopped. Then $x + y = 29.0$ g. Because the moles of Ag produced will equal two times the moles of Zn reacted:

$$(19.0 - y) \text{ g Zn} \times \frac{1 \text{ mol Zn}}{65.38 \text{ g Zn}} \times \frac{2 \text{ mol Ag}}{1 \text{ mol Zn}} = x \text{ g Ag} \times \frac{1 \text{ mol Ag}}{107.9 \text{ g Ag}}$$

Simplifying:

$$3.059 \times 10^{-2}(19.0 - y) = (9.268 \times 10^{-3})x$$

Substituting $x = 29.0 - y$ into the equation gives:

$$3.059 \times 10^{-2}(19.0 - y) = 9.268 \times 10^{-3}(29.0 - y)$$

Solving:

$$0.581 - (3.059 \times 10^{-2})y = 0.269 - (9.268 \times 10^{-3})y,\ (2.132 \times 10^{-2})y = 0.312,\ y = 14.6 \text{ g Zn}$$

14.6 g Zn is present, and 29.0 – 14.6 = 14.4 g Ag is also present after the reaction is stopped.

153. $0.298 \text{ g } BaSO_4 \times \dfrac{96.07 \text{ g } SO_4^{2-}}{233.4 \text{ g } BaSO_4} = 0.123 \text{ g } SO_4^{2-}$; $\% \text{ sulfate} = \dfrac{0.123 \text{ g } SO_4^{2-}}{0.205 \text{ g}} = 60.0\%$

Assume we have 100.0 g of the mixture of Na_2SO_4 and K_2SO_4. There are:

$$60.0 \text{ g } SO_4^{2-} \times \frac{1 \text{ mol}}{96.07 \text{ g}} = 0.625 \text{ mol } SO_4^{2-}$$

There must be 2 × 0.625 = 1.25 mol of 1+ cations to balance the 2− charge of SO_4^{2-}.

Let x = number of moles of K^+ and y = number of moles of Na^+; then $x + y = 1.25$.

The total mass of Na^+ and K^+ must be 40.0 g in the assumed 100.0 g of mixture. Setting up an equation:

$$x \text{ mol } K^+ \times \frac{39.10 \text{ g}}{\text{mol}} + y \text{ mol } Na^+ \times \frac{22.99 \text{ g}}{\text{mol}} = 40.0 \text{ g}$$

So we have two equations with two unknowns: $x + y = 1.25$ and $(39.10)x + (22.99)y = 40.0$

$x = 1.25 - y$, so $39.10(1.25 - y) + (22.99)y = 40.0$

$48.9 - (39.10)y + (22.99)y = 40.0,\ -(16.11)y = -8.9$

$y = 0.55$ mol Na^+ and $x = 1.25 - 0.55 = 0.70$ mol K^+

Therefore:

$$0.70 \text{ mol } K^+ \times \frac{1 \text{ mol } K_2SO_4}{2 \text{ mol } K^+} = 0.35 \text{ mol } K_2SO_4;\ 0.35 \text{ mol } K_2SO_4 \times \frac{174.27 \text{ g}}{\text{mol}} = 61 \text{ g } K_2SO_4$$

We assumed 100.0 g; therefore, the mixture is 61% K_2SO_4 and 39% Na_2SO_4.

155. $Pb^{2+}(aq) + 2\ Cl^-(aq) \rightarrow PbCl_2(s)$

$$3.407 \text{ g } PbCl_2 \times \frac{1 \text{ mol } PbCl_2}{278.1 \text{ g } PbCl_2} \times \frac{1 \text{ mol } Pb^{2+}}{\text{mol } PbCl_2} = 0.01225 \text{ mol } Pb^{2+}$$

$$\frac{0.01225 \text{ mol}}{2.00 \times 10^{-3} \text{ L}} = 6.13\ M\ Pb^{2+} = 6.13\ M\ Pb(NO_3)_2$$

This is also the $Pb(NO_3)_2$ concentration in the 80.0 mL of evaporated solution.

$$\text{Original concentration} = \frac{\text{moles } Pb(NO_3)_2}{\text{original volume}} = \frac{0.0800 \text{ L} \times 6.13 \text{ mol/L}}{0.1000 \text{ L}} = 4.90\ M\ Pb(NO_3)_2$$

157. 0.2750 L × 0.300 mol/L = 0.0825 mol H^+; let y = volume (L) delivered by Y and z = volume (L) delivered by Z.

$$H^+(aq) + OH^-(aq) \rightarrow H_2O(l);\quad \underbrace{y(0.150 \text{ mol/L}) + z(0.250 \text{ mol/L})}_{\text{mol } OH^-} = 0.0825 \text{ mol } H^+$$

$$0.2750 \text{ L} + y + z = 0.655 \text{ L},\quad y + z = 0.380,\quad z = 0.380 - y$$

$$y(0.150) + (0.380 - y)(0.250) = 0.0825,\ \text{solving: } y = 0.125 \text{ L},\ z = 0.255 \text{ L}$$

$$\text{Flow rate for Y} = \frac{125 \text{ mL}}{60.65 \text{ min}} = 2.06 \text{ mL/min};\ \text{flow rate for Z} = \frac{255 \text{ mL}}{60.65 \text{ min}} = 4.20 \text{ mL/min}$$

159. $2\ H_3PO_4(aq) + 3\ Ba(OH)_2(aq) \rightarrow 6\ H_2O(l) + Ba_3(PO_4)_2(s)$

$$0.01420 \text{ L} \times \frac{0.141 \text{ mol } H_3PO_4}{\text{L}} \times \frac{3 \text{ mol } Ba(OH)_2}{2 \text{ mol } H_3PO_4} \times \frac{1 \text{ L } Ba(OH)_2}{0.0521 \text{ mol } Ba(OH)_2} = 0.0576 \text{ L}$$

$$= 57.6 \text{ mL } Ba(OH)_2$$

161. $$\text{Mol } OH^- \text{ added} = 0.0500 \text{ L} \times \frac{0.20 \text{ mol } Ba(OH)_2}{\text{L}} \times \frac{2 \text{ mol } OH^-}{\text{mol } Ba(OH)_2} = 0.020 \text{ mol } OH^-$$

If a monoprotic acid: $HA(aq) + OH^-(aq) \rightarrow A^-(aq) + H_2O(l)$

$$\text{Molarity of HA} = \frac{0.020 \text{ mol } OH^- \times \frac{1 \text{ mol HA}}{\text{mol } OH^-}}{0.1000 \text{ L}} = 0.20\ M;$$ answers a-c are incorrect.

If a diprotic acid: $H_2A(aq) + 2\ OH^-(aq) \rightarrow A^{2-}(aq) + 2\ H_2O(l)$

$$\text{Molarity of } H_2A = \frac{0.020 \text{ mol } OH^- \times \frac{1 \text{ mol } H_2A}{2 \text{ mol } OH^-}}{0.1000 \text{ L}} = 0.10\ M$$

Answer e is correct.

163. The pertinent equations are:

$$2\ NaOH(aq) + H_2SO_4(aq) \rightarrow Na_2SO_4(aq) + 2\ H_2O(l)$$

$$HCl(aq) + NaOH(aq) \rightarrow NaCl(aq) + H_2O(l)$$

$$\text{Amount of NaOH added} = 0.0500 \text{ L} \times \frac{0.213 \text{ mol}}{\text{L}} = 1.07 \times 10^{-2} \text{ mol NaOH}$$

Amount of NaOH neutralized by HCl:

$$0.01321 \text{ L HCl} \times \frac{0.103 \text{ mol HCl}}{\text{L HCl}} \times \frac{1 \text{ mol NaOH}}{\text{mol HCl}} = 1.36 \times 10^{-3} \text{ mol NaOH}$$

The difference, 9.3×10^{-3} mol, is the amount of NaOH neutralized by the sulfuric acid.

$$9.3 \times 10^{-3} \text{ mol NaOH} \times \frac{1 \text{ mol } H_2SO_4}{2 \text{ mol NaOH}} = 4.7 \times 10^{-3} \text{ mol } H_2SO_4$$

$$\text{Concentration of } H_2SO_4 = \frac{4.7 \times 10^{-3} \text{ mol}}{0.1000 \text{ L}} = 4.7 \times 10^{-2}\ M\ H_2SO_4$$

165. $\text{Mol } C_6H_8O_7 = 0.250 \text{ g } C_6H_8O_7 \times \frac{1 \text{ mol } C_6H_8O_7}{192.12 \text{ g } C_6H_8O_7} = 1.30 \times 10^{-3} \text{ mol } C_6H_8O_7$

Let H_xA represent citric acid, where x is the number of acidic hydrogens. The balanced neutralization reaction is:

$$H_xA(aq) + x\ OH^-(aq) \rightarrow x\ H_2O(l) + A^{x-}(aq)$$

$$\text{Mol } OH^- \text{ reacted} = 0.0372 \text{ L} \times \frac{0.105 \text{ mol } OH^-}{\text{L}} = 3.91 \times 10^{-3} \text{ mol } OH^-$$

$$x = \frac{\text{mol } OH^-}{\text{mol citric acid}} = \frac{3.91 \times 10^{-3} \text{ mol}}{1.30 \times 10^{-3} \text{ mol}} = 3.01$$

Therefore, the general acid formula for citric acid is H_3A, meaning that citric acid has three acidic hydrogens per citric acid molecule (citric acid is a triprotic acid).

167. X^{2-} contains 36 electrons, so X^{2-} has 34 protons, which identifies X as selenium (Se). The name of H_2Se would be hydroselenic acid following the conventions described in Chapter 2.

$$H_2Se(aq) + 2\ OH^-(aq) \rightarrow Se^{2-}(aq) + 2\ H_2O(l)$$

$$0.0356 \text{ L} \times \frac{0.175 \text{ mol } OH^-}{\text{L}} \times \frac{1 \text{ mol } H_2Se}{2 \text{ mol } OH^-} \times \frac{80.98 \text{ g } H_2Se}{\text{mol } H_2Se} = 0.252 \text{ g } H_2Se$$

169. $\text{Mol KHP used} = 0.4016 \text{ g} \times \frac{1 \text{ mol}}{204.22 \text{ g}} = 1.967 \times 10^{-3} \text{ mol KHP}$

Because 1 mole of NaOH reacts completely with 1 mole of KHP, the NaOH solution contains 1.967×10^{-3} mol NaOH.

$$\text{Molarity of NaOH} = \frac{1.967 \times 10^{-3} \text{ mol}}{25.06 \times 10^{-3} \text{ L}} = \frac{7.849 \times 10^{-2} \text{ mol}}{\text{L}}$$

$$\text{Maximum molarity} = \frac{1.967 \times 10^{-3} \text{ mol}}{25.01 \times 10^{-3} \text{ L}} = \frac{7.865 \times 10^{-2} \text{ mol}}{\text{L}}$$

$$\text{Minimum molarity} = \frac{1.967 \times 10^{-3} \text{ mol}}{25.11 \times 10^{-3} \text{ L}} = \frac{7.834 \times 10^{-2} \text{ mol}}{\text{L}}$$

We can express this as $0.07849 \pm 0.00016\ M$. An alternative way is to express the molarity as $0.0785 \pm 0.0002\ M$. This second way shows the actual number of significant figures in the molarity. The advantage of the first method is that it shows that we made all our individual measurements to four significant figures.

171. a.

$$7\ H_2O + 2\ Cr^{3+} \rightarrow Cr_2O_7^{2-} + 14\ H^+ + 6\ e^-$$
$$(2\ e^- + S_2O_8^{2-} \rightarrow 2\ SO_4^{2-}) \times 3$$

$$7\ H_2O(l) + 2\ Cr^{3+}(aq) + 3\ S_2O_8^{2-}(aq) \rightarrow Cr_2O_7^{2-}(aq) + 14\ H^+(aq) + 6\ SO_4^{2-}(aq)$$

$$(Fe^{2+} \rightarrow Fe^{3+} + e^-) \times 6$$
$$6\ e^- + 14\ H^+ + Cr_2O_7^{2-} \rightarrow 2\ Cr^{3+} + 7\ H_2O$$

$$14\ H^+(aq) + 6\ Fe^{2+}(aq) + Cr_2O_7^{2-}(aq) \rightarrow 2\ Cr^{3+}(aq) + 6\ Fe^{3+}(aq) + 7\ H_2O(l)$$

b. $8.58 \times 10^{-3}\ L \times \dfrac{0.0520\ mol\ Cr_2O_7^{2-}}{L} \times \dfrac{6\ mol\ Fe^{2+}}{mol\ Cr_2O_7^{2-}} = 2.68 \times 10^{-3}$ mol of excess Fe^{2+}

Fe^{2+} (total) = 3.000 g $Fe(NH_4)_2(SO_4)_2{\bullet}6H_2O \times \dfrac{1\ mol}{392.17\ g} = 7.650 \times 10^{-3}$ mol Fe^{2+}

$7.650 \times 10^{-3} - 2.68 \times 10^{-3} = 4.97 \times 10^{-3}$ mol Fe^{2+} reacted with $Cr_2O_7^{2-}$ generated from the Cr plating.

The Cr plating contained:

$$4.97 \times 10^{-3}\ mol\ Fe^{2+} \times \frac{1\ mol\ Cr_2O_7^{2-}}{6\ mol\ Fe^{2+}} \times \frac{2\ mol\ Cr^{3+}}{mol\ Cr_2O_7^{2-}} = 1.66 \times 10^{-3}\ mol\ Cr^{3+}$$
$$= 1.66 \times 10^{-3}\ mol\ Cr$$

$$1.66 \times 10^{-3}\ mol\ Cr \times \frac{52.00\ g\ Cr}{mol\ Cr} = 8.63 \times 10^{-2}\ g\ Cr$$

Volume of Cr plating = $8.63 \times 10^{-2}\ g \times \dfrac{1\ cm^3}{7.19\ g} = 1.20 \times 10^{-2}\ cm^3$ = area × thickness

Thickness of Cr plating = $\dfrac{1.20 \times 10^{-2}\ cm^3}{40.0\ cm^2} = 3.00 \times 10^{-4}\ cm = 300.\ \mu m$

CHAPTER 5

GASES

Questions

25. The column of water would have to be 13.6 times taller than a column of mercury. When the pressure of the column of liquid standing on the surface of the liquid is equal to the pressure of air on the rest of the surface of the liquid, then the height of the column of liquid is a measure of atmospheric pressure. Because water is 13.6 times less dense than mercury, the column of water must be 13.6 times longer than that of mercury to match the force exerted by the columns of liquid standing on the surface.

27. The P versus 1/V plot is incorrect. The plot should be linear with positive slope and a *y*-intercept of zero. PV = k, so P = k(1/V). This is in the form of the straight-line equation $y = mx + b$. The *y*-axis is pressure, the *x*-axis is 1/V, and the *y*-intercept is the origin.

29. Pressure is directly related to the mole fraction of the various gases present in a mixture. If the molar masses of the three gases X, Y, and Z are equal, then the mole fraction of each gas would be 1/3, and the partial pressure of each gas in the mixture equals 1/3 of the total pressure. If the molar mass of gas X is greater than the other molar masses, then fewer moles of gas X are present compared to moles of gas Y and Z. This results in a mole fraction of X less than 1/3, and a partial pressure of X less than 1/3 the total pressure (answer b).

31. At STP (T = 273.2 K and P = 1.000 atm), the volume of 1.000 mol of gas is:

$$V = \frac{nRT}{P} = \frac{1.000\ \text{mol} \times \dfrac{0.08206\ \text{L atm}}{\text{K mol}} \times 273.2\ \text{K}}{1.000\ \text{atm}} = 22.42\ \text{L}$$

At STP, the volume of 1.000 mole of any gas is 22.42 L, assuming the gas behaves ideally. Therefore, the molar volume of He(g) and N_2(g) at STP both equal 22.42 L/mol. If the temperature increases to 25.0°C (298.2 K), the volume of 1.000 mole of a gas will be larger than 22.42 L/mole because molar volume is directly related to the temperature at constant pressure. If 1.000 mole of a gas is collected over water at a total pressure of 1.000 atm, the partial pressure of the collected gas will be less than 1.000 atm because water vapor is present ($P_{total} = P_{gas} + P_{H_2O}$). At some partial pressure below 1.000 atm, the volume of 1.000 mole of a gas will be larger than 22.42 L/mol because molar volume is inversely related to the pressure at constant temperature.

33. a. For an ideal gas, KE_{avg} = (3/2)RT. So as temperature increases, the average kinetic energy will increase.

b. $\mu_{avg} \propto \mu_{rms} \propto (T)^{1/2}$; as temperature increases, the average velocity of the gas molecules increase.

c. At constant temperature, the lighter the gas molecules (the smaller the molar mass), the faster the average velocity. This must be true for the average kinetic energies to be the same at constant T.

35. No; at any nonzero Kelvin temperature, there is a distribution of kinetic energies. Similarly, there is a distribution of velocities at any nonzero Kelvin temperature. The reason there is a distribution of kinetic energies at any specific temperature is because there is a distribution of velocities for any gas sample at any specific temperature.

37. $2\ NO_2(g) \rightarrow N_2(g) + 2\ O_2(g)$; at constant V and T, pressure is directly proportional to the moles of gas present. As NO_2 is converted to products, the moles of gas increase from 2 total moles to 3 total moles (a 50% increase). Because of this, the total pressure will increase by 50% as reactants are converted into products.

From the stoichiometry in the balanced equation, the moles of O_2 produced will equal the moles of NO reacted. Because pressure is directly related to moles of gas present, the partial pressure of O_2 will equal the pressure of NO_2 reacted ($P_{O2} = 4.0$ atm). Also from the stoichiometry, there are twice the number of moles of O_2 produced as moles of N_2. This will result in a N_2 partial pressure which is half the O_2 partial pressure ($P_{N2} = 2.0$ atm). Note that the total pressure of gases present is 2.0 + 4.0 = 6.0 atm as stated in the problem.

39. The values of a are: H_2, $\dfrac{0.244\ \text{atm L}^2}{\text{mol}^2}$; CO_2, 3.59; N_2, 1.39; CH_4, 2.25

Because a is a measure of intermolecular attractions, the attractions are greatest for CO_2.

41. PV = nRT; Figure 5.6 is illustrating how well Boyle's law works. Boyle's law studies the pressure-volume relationship for a gas at constant moles of gas (n) and constant temperature (T). At constant n and T, the PV product for an ideal gas equals a constant value of nRT, no matter what the pressure of the gas. Figure 5.6 plots the PV product versus P for three different gases. The ideal value for the PV product is shown with a dotted line at about a value of 22.41 L atm. From the plot, it looks like the plot for Ne is closest to the dotted line, so we can conclude that of the three gases in the plot, Ne behaves most ideally. The O_2 plot is also fairly close to the dotted line, so O_2 also behaves fairly ideally. CO_2, on the other hand, has a plot farthest from the ideal plot; hence CO_2 behaves least ideally.

Exercises

Pressure

43. a. $4.8\ \text{atm} \times \dfrac{760\ \text{mm Hg}}{\text{atm}} = 3.6 \times 10^3\ \text{mm Hg}$

b. $3.6 \times 10^3\ \text{mm Hg} \times \dfrac{1\ \text{torr}}{\text{mm Hg}} = 3.6 \times 10^3\ \text{torr}$

c. $4.8\ \text{atm} \times \dfrac{1.013 \times 10^5\ \text{Pa}}{\text{atm}} = 4.9 \times 10^5\ \text{Pa}$

d. $4.8\ \text{atm} \times \dfrac{14.7\ \text{psi}}{\text{atm}} = 71\ \text{psi}$

45. $6.5\text{ cm} \times \dfrac{10\text{ mm}}{\text{cm}} = 65\text{ mm Hg} = 65\text{ torr};\ 65\text{ torr} \times \dfrac{1\text{ atm}}{760\text{ torr}} = 8.6 \times 10^{-2}\text{ atm}$

$8.6 \times 10^{-2}\text{ atm} \times \dfrac{1.013 \times 10^5\text{ Pa}}{\text{atm}} = 8.7 \times 10^3\text{ Pa}$

47. If the levels of mercury in each arm of the manometer are equal, then the pressure in the flask is equal to atmospheric pressure. When they are unequal, the difference in height in millimeters will be equal to the difference in pressure in millimeters of mercury between the flask and the atmosphere. Which level is higher will tell us whether the pressure in the flask is less than or greater than atmospheric.

a. $P_{flask} < P_{atm};\ P_{flask} = 760. - 118 = 642\text{ torr}$

$642\text{ torr} \times \dfrac{1\text{ atm}}{760\text{ torr}} = 0.845\text{ atm}$

$0.845\text{ atm} \times \dfrac{1.013 \times 10^5\text{ Pa}}{\text{atm}} = 8.56 \times 10^4\text{ Pa}$

b. $P_{flask} > P_{atm};\ P_{flask} = 760.\text{ torr} + 215\text{ torr} = 975\text{ torr}$

$975\text{ torr} \times \dfrac{1\text{ atm}}{760\text{ torr}} = 1.28\text{ atm}$

$1.28\text{ atm} \times \dfrac{1.013 \times 10^5\text{ Pa}}{\text{atm}} = 1.30 \times 10^5\text{ Pa}$

c. $P_{flask} = 635 - 118 = 517\text{ torr};\ P_{flask} = 635 + 215 = 850.\text{ torr}$

Gas Laws

49. At constant n and T, PV = nRT = constant, so $P_1V_1 = P_2V_2$. Solving for P_2:

$$P_2 = \frac{P_1V_1}{V_2} = \frac{5.20\text{ atm} \times 0.400\text{ L}}{2.14\text{ L}} = 0.972\text{ atm}$$

As predicted by Boyle's law, as the volume of a gas increases, pressure decreases.

51. $n_{Ar} = 27.1\text{ g Ar} \times \dfrac{1\text{ mol Ar}}{39.95\text{ g}} = 0.678\text{ mol}$; at constant T and P, Avogadro's law holds ($V \propto n$).

$$\frac{V_{Ar}}{n_{Ar}} = \frac{V_{Ne}}{n_{Ne}},\ V_{Ne} = \frac{V_{Ar}n_{Ne}}{n_{Ar}} = \frac{4.21\text{ L} \times 1.29\text{ mol}}{0.678\text{ mol}} = 8.01\text{ L}$$

As expected, as n increases, V increases.

53. a. $PV = nRT,\ V = \dfrac{nRT}{P} = \dfrac{2.00\text{ mol} \times \dfrac{0.08206\text{ L atm}}{\text{K mol}} \times (155 + 273)\text{ K}}{5.00\text{ atm}} = 14.0\text{ L}$

b. $PV = nRT, \quad n = \frac{PV}{RT} = \frac{0.300 \text{ atm} \times 2.00 \text{ L}}{\frac{0.08206 \text{ L atm}}{\text{K mol}} \times 155 \text{ K}} = 4.72 \times 10^{-2} \text{ mol}$

c. $PV = nRT, \quad T = \frac{PV}{nR} = \frac{4.47 \text{ atm} \times 25.0 \text{ L}}{2.01 \text{ mol} \times \frac{0.08206 \text{ L atm}}{\text{K mol}}} = 678 \text{ K} = 405°\text{C}$

d. $PV = nRT, \quad P = \frac{nRT}{V} = \frac{10.5 \text{ mol} \times \frac{0.08206 \text{ L atm}}{\text{K mol}} \times (273 + 75) \text{ K}}{2.25 \text{ L}} = 133 \text{ atm}$

55. $n = \frac{PV}{RT} = \frac{2.70 \text{ atm} \times 200.0 \text{ L}}{\frac{0.08206 \text{ L atm}}{\text{K mol}} \times (273 + 24) \text{ K}} = 22.2 \text{ mol}$

For He: $22.2 \text{ mol} \times \frac{4.003 \text{ g He}}{\text{mol}} = 88.9 \text{ g He}$

For H_2: $22.2 \text{ mol} \times \frac{2.016 \text{ g } H_2}{\text{mol}} = 44.8 \text{ g } H_2$

57. $PV = nRT, \quad n = \frac{PV}{RT} = \frac{14.5 \text{ atm} \times (75.0 \times 10^{-3} \text{ L})}{\frac{0.08206 \text{ L atm}}{\text{K mol}} \times 295 \text{ K}} = 0.0449 \text{ mol } O_2$

$0.0449 \text{ mol } O_2 \times \frac{32.00 \text{ g } O_2}{\text{mol } O_2} = 1.44 \text{ g } O_2$

59. $PV = nRT; \quad T = \frac{PV}{nR} = \frac{76.0 \text{ torr} \times \frac{1 \text{ atm}}{760 \text{ torr}} \times 37.0 \text{ L}}{0.200 \text{ mol} \times \frac{0.08206 \text{ L atm}}{\text{K mol}}} = 225 \text{ K}; \; 225 - 273 = -48°\text{C}$

61. $n = \frac{PV}{RT} = \frac{745 \text{ torr} \times \frac{1 \text{ atm}}{760 \text{ torr}} \times 0.45 \text{ L}}{\frac{0.08206 \text{ L atm}}{\text{K mol}} \times 295 \text{ K}} = 0.018 \text{ mol air}$

$0.018 \text{ mol} \times \frac{6.022 \times 10^{23} \text{ particles}}{\text{mol}} = 1.1 \times 10^{22} \text{ air particles}$

63. For a gas at two conditions: $\frac{P_1V_1}{n_1T_1} = \frac{P_2V_2}{n_2T_2}$

Because V is constant: $\frac{P_1}{n_1T_1} = \frac{P_2}{n_2T_2}, \quad n_2 = \frac{n_1P_2T_1}{P_1T_2}$

$$n_2 = \frac{0.50 \text{ mol} \times 2.5 \text{ atm} \times 298 \text{ K}}{0.75 \text{ atm} \times 323 \text{ K}} = 1.5 \text{ mol}$$

Moles of gas added = $n_2 - n_1 = 1.5 - 0.50 = 1.0$ mol

65. At two conditions: $\frac{P_1V_1}{n_1T_1} = \frac{P_2V_2}{n_2T_2}$; all gases are assumed to follow the ideal gas law. The identity of the gas in container B is unimportant as long as we know the moles of gas present.

$$\frac{P_B}{P_A} = \frac{V_A n_B T_B}{V_B n_A T_A} = \frac{1.0 \text{ L} \times 2.0 \text{ mol} \times 560. \text{ K}}{2.0 \text{ L} \times 1.0 \text{ mol} \times 280. \text{ K}} = 2.0$$

The pressure of the gas in container B is twice the pressure of the gas in container A.

67. At constant n and V, $\frac{P_1}{T_1} = \frac{P_2}{T_2}$, $T_2 = \frac{T_1P_2}{P_1} = (21 + 273 \text{ K}) \times \frac{2500 \text{ torr}}{758 \text{ torr}} = 970 \text{ K} = 7.0 \times 10^2 °\text{C}$

The flask will burst when the temperature reaches 7.0×10^2°C.

69. $\frac{PV}{T} = nR = \text{constant}, \quad \frac{P_1V_1}{T_1} = \frac{P_2V_2}{T_2}$

$$P_2 = \frac{P_1V_1T_2}{V_2T_1} = 710. \text{ torr} \times \frac{5.0 \times 10^2 \text{ mL}}{25 \text{ mL}} \times \frac{(273 + 820.) \text{ K}}{(273 + 30.) \text{ K}} = 5.1 \times 10^4 \text{ torr}$$

71. PV = nRT, n is constant. $\frac{PV}{T} = nR = \text{constant}, \quad \frac{P_1V_1}{T_1} = \frac{P_2V_2}{T_2}, \quad V_2 = \frac{V_1P_1T_2}{P_2T_1}$

$$V_2 = 1.00 \text{ L} \times \frac{760 .\text{torr}}{220 . \text{ torr}} \times \frac{(273 - 31) \text{ K}}{(273 + 23) \text{ K}} = 2.82 \text{ L}; \ \Delta V = 2.82 - 1.00 = 1.82 \text{ L}$$

Gas Density, Molar Mass, and Reaction Stoichiometry

73. STP: T = 273 K and P = 1.00 atm; at STP, the molar volume of a gas is 22.42 L.

$$2.00 \text{ L } O_2 \times \frac{1 \text{ mol } O_2}{22.42 \text{ L}} \times \frac{4 \text{ mol Al}}{3 \text{ mol } O_2} \times \frac{26.98 \text{ g Al}}{\text{mol Al}} = 3.21 \text{ g Al}$$

Note: We could also solve this problem using PV = nRT, where n_{O_2} = PV/RT. You don't have to memorize 22.42 L/mol at STP.

75. $2 \text{ NaN}_3(s) \rightarrow 2 \text{ Na}(s) + 3 \text{ N}_2(g)$

$$n_{N_2} = \frac{PV}{RT} = \frac{1.00 \text{ atm} \times 70.0 \text{ L}}{\frac{0.08206 \text{ L atm}}{\text{K mol}} \times 273 \text{ K}} = 3.12 \text{ mol } N_2 \text{ needed to fill air bag.}$$

$$\text{Mass NaN}_3 \text{ reacted} = 3.12 \text{ mol } N_2 \times \frac{2 \text{ mol NaN}_3}{3 \text{ mol } N_2} \times \frac{65.02 \text{ g NaN}_3}{\text{mol NaN}_3} = 135 \text{ g NaN}_3$$

77. $$n_{H_2} = \frac{PV}{RT} = \frac{1.0\text{ atm} \times \left[4800\text{ m}^3 \times \left(\frac{100\text{ cm}}{\text{m}}\right)^3 \times \frac{1\text{ L}}{1000\text{ cm}^3}\right]}{\frac{0.08206\text{ L atm}}{\text{K mol}} \times 273\text{ K}} = 2.1 \times 10^5\text{ mol}$$

2.1×10^5 mol H_2 is in the balloon. This is 80.% of the total amount of H_2 that had to be generated:

$$0.80(\text{total mol } H_2) = 2.1 \times 10^5,\ \text{total mol } H_2 = 2.6 \times 10^5\text{ mol}$$

$$2.6 \times 10^5\text{ mol } H_2 \times \frac{1\text{ mol Fe}}{\text{mol } H_2} \times \frac{55.85\text{ g Fe}}{\text{mol Fe}} = 1.5 \times 10^7\text{ g Fe}$$

$$2.6 \times 10^5\text{ mol } H_2 \times \frac{1\text{ mol } H_2SO_4}{\text{mol } H_2} \times \frac{98.09\text{ g } H_2SO_4}{\text{mol } H_2SO_4} \times \frac{100\text{ g reagent}}{98\text{ g } H_2SO_4}$$
$$= 2.6 \times 10^7\text{ g of 98\% sulfuric acid}$$

79. $Kr(g) + 2\ Cl_2(g) \rightarrow KrCl_4(s)$; $n_{Kr} = \frac{PV}{RT} = \frac{0.500\text{ atm} \times 15.0\text{ L}}{\frac{0.08206\text{ L atm}}{\text{K mol}} \times 623\text{ K}} = 0.147\text{ mol Kr}$

We could do the same calculation for Cl_2. However, the only variable that changed is the pressure. Because the partial pressure of Cl_2 is triple that of Kr, moles of Cl_2 = 3(0.147) = 0.441 mol Cl_2. The balanced equation requires 2 moles of Cl_2 to react with every mole of Kr. However, we actually have three times as many moles of Cl_2 as we have of Kr. So Cl_2 is in excess and Kr is the limiting reagent.

$$0.147\text{ mol Kr} \times \frac{1\text{ mol } KrCl_4}{\text{mol Kr}} \times \frac{225.60\text{ g } KrCl_4}{\text{mol Kr}} = 33.2\text{ g } KrCl_4$$

81. $CH_3OH + 3/2\ O_2 \rightarrow CO_2 + 2\ H_2O$ or $2\ CH_3OH(l) + 3\ O_2(g) \rightarrow 2\ CO_2(g) + 4\ H_2O(g)$

$$50.0\text{ mL} \times \frac{0.850\text{ g}}{\text{mL}} \times \frac{1\text{ mol}}{32.04\text{ g}} = 1.33\text{ mol } CH_3OH(l)\text{ available}$$

$$n_{O_2} = \frac{PV}{RT} = \frac{2.00\text{ atm} \times 22.8\text{ L}}{\frac{0.08206\text{ L atm}}{\text{K mol}} \times 300.\text{ K}} = 1.85\text{ mol } O_2\text{ available}$$

Assuming CH_3OH is limiting:

$$1.33\text{ mol } CH_3OH \times \frac{4\text{ mol } H_2O}{2\text{ mol } CH_3OH} = 2.66\text{ mol } H_2O$$

Assuming O_2 is limiting:

$$1.85\text{ mol } O_2 \times \frac{4\text{ mol } H_2O}{3\text{ mol } O_2} = 2.47\text{ mol } H_2O$$

Because the O_2 reactant produces the smaller quantity of H_2O, O_2 is limiting and 2.47 mol of H_2O can be produced.

83. a. $CH_4(g) + NH_3(g) + O_2(g) \rightarrow HCN(g) + H_2O(g)$; balancing H first, then O, gives:

$$CH_4 + NH_3 + \frac{3}{2}O_2 \rightarrow HCN + 3\ H_2O \text{ or } 2\ CH_4(g) + 2\ NH_3(g) + 3\ O_2(g) \rightarrow 2\ HCN(g) + 6\ H_2O(g)$$

b. PV = nRT, T and P constant; $\frac{V_1}{n_1} = \frac{V_2}{n_2}$, $\frac{V_1}{V_2} = \frac{n_1}{n_2}$

The volumes are all measured at constant T and P, so the volumes of gas present are directly proportional to the moles of gas present (Avogadro's law). Because Avogadro's law applies, the balanced reaction gives mole relationships as well as volume relationships.

If CH_4 is limiting: $20.0\text{ L } CH_4 \times \frac{2\text{ L HCN}}{2\text{ L } CH_4} = 20.0\text{ L HCN}$

If NH_3 is limiting: $20.0\text{ L } NH_3 \times \frac{2\text{ L HCN}}{2\text{ L } NH_3} = 20.0\text{ L HCN}$

If O_2 is limiting: $20.0\text{ L } O_2 \times \frac{2\text{ L HCN}}{3\text{ L } O_2} = 13.3\text{ L HCN}$

O_2 produces the smallest quantity of product, so O_2 is limiting and 13.3 L HCN can be produced.

85. d = (molar mass)P/RT; density is directly proportional to the molar mass of a gas. Helium, with the smallest molar mass of all the noble gases, will have the smallest density at STP.

87. Molar mass = $\frac{dRT}{P}$, where d = density of gas in units of g/L.

$$\text{Molar mass} = \frac{3.164\text{ g/L} \times \frac{0.08206\text{ L atm}}{\text{K mol}} \times 273.2\text{ K}}{1.000\text{ atm}} = 70.98\text{ g/mol}$$

The gas is diatomic, so the average atomic mass = 70.93/2 = 35.47 u. From the periodic table, this is chlorine, and the identity of the gas is Cl_2.

89. $$d_{UF_6} = \frac{P \times (\text{molar mass})}{RT} = \frac{\left(745\text{ torr} \times \frac{1\text{ atm}}{760\text{ torr}}\right) \times 352.0\text{ g/mol}}{\frac{0.08206\text{ L atm}}{\text{K mol}} \times 333\text{ K}} = 12.6\text{ g/L}$$

Partial Pressure

91. The container has 5 He atoms, 3 Ne atoms, and 2 Ar atoms for a total of 10 atoms. The mole fractions of the various gases will be equal to the molecule fractions.

$\chi_{He} = \frac{5\text{ He atoms}}{10\text{ total atoms}} = 0.50$; $\chi_{Ne} = \frac{3\text{ Ne atoms}}{10\text{ total atoms}} = 0.30$

$\chi_{Ar} = 1.00 - 0.50 - 0.30 = 0.20$

$P_{He} = \chi_{He} \times P_{total} = 0.50(1.00 \text{ atm}) = 0.50 \text{ atm}$

$P_{Ne} = \chi_{Ne} \times P_{Total} = 0.30(1.00 \text{atm}) = 0.30 \text{ atm}$

$P_{Ar} = 1.00 \text{ atm} - 0.50 \text{ atm} - 0.30 \text{ atm} = 0.20 \text{ atm}$

93. $n_{He} = 15.2 \text{ g He} \times \dfrac{1 \text{ mol He}}{4.003 \text{ g He}} = 3.80 \text{ mol } H_2;\ n_{O_2} = 30.6 \text{ g } O_2 \times \dfrac{1 \text{ mol } O_2}{32.00 \text{ g } O_2}$

$= 0.956 \text{ mol } O_2$

$$P_{He} = \frac{n_{He} \times RT}{V} = \frac{3.80 \text{ mol} \times \dfrac{0.08206 \text{ L atm}}{\text{K mol}} \times (273 + 22) \text{ K}}{5.00 \text{ L}} = 18.4 \text{ atm}$$

$$P_{O_2} = \frac{n_{O_2} \times RT}{V} = 4.63 \text{ atm};\ P_{total} = P_{He} + P_{O_2} = 18.4 \text{ atm} + 4.63 \text{ atm} = 23.0 \text{ atm}$$

95. Treat each gas separately and determine how the partial pressure of each gas changes when the container volume increases. Once the partial pressures of H_2 and N_2 are determined, the total pressure will be the sum of these two partial pressures. At constant n and T, the relationship $P_1V_1 = P_2V_2$ holds for each gas.

For H_2: $P_2 = \dfrac{P_1V_1}{V_2} = 475 \text{ torr} \times \dfrac{2.00 \text{ L}}{3.00 \text{ L}} = 317 \text{ torr}$

For N_2: $P_2 = 0.200 \text{ atm} \times \dfrac{1.00 \text{ L}}{3.00 \text{ L}} = 0.0667 \text{ atm};\ 0.0667 \text{ atm} \times \dfrac{760 \text{ torr}}{\text{atm}} = 50.7 \text{ torr}$

$P_{total} = P_{H_2} + P_{N_2} = 317 + 50.7 = 368 \text{ torr}$

97. $P_1V_1 = P_2V_2$; the total volume is 1.00 L + 1.00 L + 2.00 L = 4.00 L.

For He: $P_2 = \dfrac{P_1V_1}{V_2} = 200. \text{ torr} \times \dfrac{1.00 \text{ L}}{4.00 \text{ L}} = 50.0 \text{ torr He}$

For Ne: $P_2 = 0.400 \text{ atm} \times \dfrac{1.00 \text{ L}}{4.00 \text{ L}} = 0.100 \text{ atm};\ 0.100 \text{ atm} \times \dfrac{760 \text{ torr}}{\text{atm}} = 76.0 \text{ torr Ne}$

For Ar: $P_2 = 24.0 \text{ kPa} \times \dfrac{2.00 \text{ L}}{4.00 \text{ L}} = 12.0 \text{ kPa};\ 12.0 \text{ kPa} \times \dfrac{1 \text{ atm}}{101.3 \text{ kPa}} \times \dfrac{760 \text{ torr}}{\text{atm}}$

$= 90.0 \text{ torr Ar}$

$P_{total} = 50.0 + 76.0 + 90.0 = 216.0 \text{ torr}$

99. Mole fraction cyclopropane $= \chi = \dfrac{P_{cyclopropane}}{P_{total}} = \dfrac{170. \text{ torr}}{170. \text{ torr} + 570. \text{ torr}} = 0.230$

101. a. Mole fraction $CH_4 = \chi_{CH_4} = \dfrac{P_{CH_4}}{P_{total}} = \dfrac{0.175\ \text{atm}}{0.175\ \text{atm} + 0.250\ \text{atm}} = 0.412$

$\chi_{O_2} = 1.000 - 0.412 = 0.588$

b. $PV = nRT$, $n_{total} = \dfrac{P_{total} \times V}{RT} = \dfrac{0.425\ \text{atm} \times 10.5\ \text{L}}{\dfrac{0.08206\ \text{L atm}}{\text{K mol}} \times 338\ \text{K}} = 0.161\ \text{mol}$

c. $\chi_{CH_4} = \dfrac{n_{CH_4}}{n_{total}}$, $n_{CH_4} = \chi_{CH_4} \times n_{total} = 0.412 \times 0.161\ \text{mol} = 6.63 \times 10^{-2}\ \text{mol}\ CH_4$

$6.63 \times 10^{-2}\ \text{mol}\ CH_4 \times \dfrac{16.04\ \text{g}\ CH_4}{\text{mol}\ CH_4} = 1.06\ \text{g}\ CH_4$

$n_{O_2} = 0.588 \times 0.161\ \text{mol} = 9.47 \times 10^{-2}\ \text{mol}\ O_2$; $9.47 \times \text{mol}\ O_2 \times \dfrac{32.00\ \text{g}\ O_2}{\text{mol}\ O_2}$

$= 3.03\ \text{g}\ O_2$

103. To calculate the volume of gas, we can use P_{total} and n_{total} ($V = n_{total}RT/P_{total}$), or we can use P_{He} and n_{He} ($V = n_{He}RT/P_{He}$). Because n_{H_2O} is unknown, we will use P_{He} and n_{He}.

$P_{He} + P_{H_2O} = 1.00\ \text{atm} = 760.\ \text{torr}$, $P_{He} + 17.5\ \text{torr} = 760.\ \text{torr}$, $P_{He} = 743\ \text{torr}$

$n_{He} = 0.586\ \text{g} \times \dfrac{1\ \text{mol}}{4.003\ \text{g}} = 0.146\ \text{mol He}$

$$V = \frac{n_{He}RT}{P_{He}} = \frac{0.146\ \text{mol} \times \dfrac{0.08206\ \text{L atm}}{\text{K mol}} \times 293\ \text{K}}{743\ \text{torr} \times \dfrac{1\ \text{atm}}{760\ \text{torr}}} = 3.59\ \text{L}$$

105. $P_{total} = P_{H_2} + P_{H_2O}$, $1.032\ \text{atm} = P_{H_2} + 32\ \text{torr} \times \dfrac{1\ \text{atm}}{760\ \text{torr}}$, $P_{H_2} = 1.032 - 0.042 = 0.990\ \text{atm}$

$$n_{H_2} = \frac{P_{H_2}V}{RT} = \frac{0.990\ \text{atm} \times 0.240\ \text{L}}{\dfrac{0.08206\ \text{L atm}}{\text{K mol}} \times 303\ \text{K}} = 9.56 \times 10^{-3}\ \text{mol}\ H_2$$

$9.56 \times 10^{-3}\ \text{mol}\ H_2 \times \dfrac{1\ \text{mol Zn}}{\text{mol}\ H_2} \times \dfrac{65.38\ \text{g Zn}}{\text{mol Zn}} = 0.625\ \text{g Zn}$

107. $10.10\ \text{atm} - 7.62\ \text{atm} = 2.48\ \text{atm}$ is the pressure of the amount of F_2 reacted.

$PV = nRT$, V and T are constant. $\dfrac{P}{n} = \text{constant}$, $\dfrac{P_1}{n_1} = \dfrac{P_2}{n_2}$ or $\dfrac{P_1}{P_2} = \dfrac{n_1}{n_2}$

$\dfrac{\text{Moles } F_2 \text{ reacted}}{\text{Moles Xe reacted}} = \dfrac{2.48\ \text{atm}}{1.24\ \text{atm}} = 2.00$; so $Xe + 2\ F_2 \rightarrow XeF_4$

109. $2\ HN_3(g) \rightarrow 3\ N_2(g) + H_2(g)$; at constant V and T, P is directly proportional to n. In the reaction, we go from 2 moles of gaseous reactants to 4 moles of gaseous products. Because moles doubled, the final pressure will double (P_{total} = 6.0 atm). Similarly, from the 2 : 1 mole ratio between HN_3 and H_2, the partial pressure of H_2 will be 3.0/2 = 1.5 atm. The partial pressure of N_2 will be (3/2)3.0 atm = 4.5 atm. This is from the 2 : 3 mole ratio between HN_3 and N_2.

111. $2\ SO_2(g) + O_2(g) \rightarrow 2\ SO_3(g)$; because P and T are constant, volume ratios will equal mole ratios ($V_f/V_i = n_f/n_i$). Let x = mol SO_2 = mol O_2 present initially. From the balanced equation, 2 mol of SO_2 react for every 1 mol of O_2 that reacts. Because we have equal moles of SO_2 and O_2 present initially, and because SO_2 is used up twice as fast as O_2, SO_2 is the limiting reagent. Therefore, no SO_2 will be present after the reaction goes to completion. However, excess $O_2(g)$ will be present as well as the $SO_3(g)$ produced.

$$\text{Mol } O_2 \text{ reacted} = x \text{ mol } SO_2 \times \frac{1 \text{ mol } O_2}{2 \text{ mol } SO_2} = x/2 \text{ mol } O_2$$

Mol O_2 remaining = x mol O_2 initially – $x/2$ mol O_2 reacted = $x/2$ mol O_2

$$\text{Mol } SO_3 \text{ produced} = x \text{ mol } SO_2 \times \frac{2 \text{ mol } SO_3}{2 \text{ mol } SO_2} = x \text{ mol } SO_3$$

Total moles gas initially = x mol SO_2 + x mol O_2 = $2x$

Total moles gas after reaction = $x/2$ mol O_2 + x mol SO_3 = $(3/2)x = (1.5)x$

$$\frac{n_f}{n_i} = \frac{V_f}{V_i} = \frac{(1.5)\,x}{2x} = \frac{1.5}{2} = 0.75; \quad V_f/V_i = 0.75 : 1 \text{ or } 3 : 4$$

The volume of the reaction container shrinks to 75% of the initial volume. This makes sense because the total mol of gas decreased as the reaction occurs at constant P and T.

113. $$150 \text{ g } (CH_3)_2N_2H_2 \times \frac{1 \text{ mol } (CH_3)_2N_2H_2}{60.10 \text{ g}} \times \frac{3 \text{ mol } N_2}{\text{mol } (CH_3)_2N_2H_2} = 7.5 \text{ mol } N_2 \text{ produced}$$

$$P_{N_2} = \frac{nRT}{V} = \frac{7.5 \text{ mol} \times \frac{0.08206 \text{ L atm}}{\text{K mol}} \times 400.\text{ K}}{250 \text{ L}} = 0.98 \text{ atm}$$

We could do a similar calculation for P_{H_2O} and P_{CO_2} and then calculate P_{total} (= $P_{N_2} + P_{H_2O} + P_{CO_2}$). Or we can recognize that 9 total moles of gaseous products form for every mole of $(CH_3)_2N_2H_2$ reacted (from the balanced equation given in the problem). This is three times the moles of N_2 produced. Therefore, P_{total} will be three times larger than P_{N_2}.

$P_{total} = 3 \times P_{N_2} = 3 \times 0.98$ atm = 2.9 atm.

Kinetic Molecular Theory and Real Gases

115. KE_{avg} = (3/2)RT; the average kinetic energy depends only on temperature. At each temperature, CH_4 and N_2 will have the same average KE. For energy units of joules (J), use R =

8.3145 J/K•mol. To determine average KE per molecule, divide the molar KE_{avg} by Avogadro's number, 6.022×10^{23} molecules/mol.

At 273 K: $KE_{avg} = \frac{3}{2} \times \frac{8.3145\ J}{K\ mol} \times 273\ K = 3.40 \times 10^3\ J/mol = 5.65 \times 10^{-21}$ J/molecule

At 546 K: $KE_{avg} = \frac{3}{2} \times \frac{8.3145\ J}{K\ mol} \times 546\ K = 6.81 \times 10^3\ J/mol = 1.13 \times 10^{-20}$ J/molecule

117. $\mu_{rms} = \left(\frac{3RT}{M}\right)^{1/2}$, where $R = \frac{8.3145\ J}{K\ mol}$ and M = molar mass in kg.

For CH_4, $M = 1.604 \times 10^{-2}$ kg, and for N_2, $M = 2.802 \times 10^{-2}$ kg.

$$\text{For } CH_4 \text{ at 273 K: } \mu_{rms} = \left(\frac{3 \times \frac{8.3145\ J}{K\ mol} \times 273\ K}{1.604 \times 10^{-2}\ kg/mol}\right)^{1/2} = 652\ m/s$$

Similarly, μ_{rms} for CH_4 at 546 K is 921 m/s.

$$\text{For } N_2 \text{ at 273 K: } \mu_{rms} = \left(\frac{3 \times \frac{8.3145\ J}{K\ mol} \times 273\ K}{2.802 \times 10^{-2}\ kg/mol}\right)^{1/2} = 493\ m/s$$

Similarly, for N_2 at 546 K, $\mu_{rms} = 697$ m/s.

119. The number of gas particles is constant, so at constant moles of gas, either a temperature change or a pressure change results in the smaller volume. If the temperature is constant, an increase in the external pressure would cause the volume to decrease. Gases are mostly empty space so gases are easily compressible.

If the pressure is constant, a decrease in temperature would cause the volume to decrease. As the temperature is lowered, the gas particles move with a slower average velocity and don't collide with the container walls as frequently and as forcefully. As a result, the internal pressure decreases. To keep the pressure constant, the volume of the container must decrease in order to increase the gas particle collisions per unit area.

121.

	a	b	c	d
Avg. KE	increase	decrease	same ($KE \propto T$)	same
Avg. velocity	increase	decrease	same ($\frac{1}{2}mv^2 = KE \propto T$)	same
Wall coll. freq	increase	decrease	increase	increase

Average kinetic energy and average velocity depend on T. As T increases, both average kinetic energy and average velocity increase. At constant T, both average kinetic energy and average velocity are constant. The collision frequency is proportional to the average velocity (as velocity increases, it takes less time to move to the next collision) and to the quantity n/V (as molecules per volume increase, collision frequency increases).

123. a. They will all have the same average kinetic energy because they are all at the same temperature [$KE_{avg} = (3/2)RT$].

b. Flask C; H_2 has the smallest molar mass. At constant T, the lighter molecules have the faster average velocity. This must be true for the average kinetic energies to be the same.

125. Graham's law of effusion: $\frac{\text{Rate}_A}{\text{Rate}_B} = \left(\frac{M_B}{M_A}\right)^{1/2}$, $9 = \left(\frac{M_B}{M_A}\right)^{1/2}$; squaring both sides:

$81 = \frac{M_B}{M_A}$, $M_B = 81M_A$; answer e is correct

127. Graham's law of effusion: $\frac{\text{Rate}_1}{\text{Rate}_2} = \left(\frac{M_2}{M_1}\right)^{1/2}$

Let the unknown = gas 1 and O_2 = gas 2:

$$\frac{31.50}{30.50} = \left(\frac{32.00}{M_1}\right)^{1/2},\quad 1.067 = \frac{32.00}{M_1},\quad M_1 = 29.99 \text{ g/mol}$$

Of the choices, NO with a molar mass of 30.01 g/mol, best fits the data. So the unknown gas is nitrogen monoxide (NO).

129. $\frac{\text{Rate}_A}{\text{Rate}_B} = \left(\frac{M_B}{M_A}\right)^{1/2}$, $\frac{1.0 \text{ L}/3.0 \text{ min}}{1.0 \text{ mL}/12.0 \text{ min}} = \left(\frac{M_B}{M_A}\right)^{1/2}$, $4.0 = \left(\frac{M_B}{M_A}\right)^{1/2}$; squaring both sides:

$16 = \frac{M_B}{M_A}$, $M_B = 16M_A$; the molar mass of B is 16 times the molar mass of A.

Answer e is correct. The molar mass of O_2 (gas B) is 32.0 g/mol and the molar mass of H_2 (gas A) is 2.0 g/mol.

131. A gas behaves most ideally at high temperatures and low pressure. Answer b combines the highest temperature and lowest pressure, so answer b is correct.

133. a. PV = nRT

$$P = \frac{nRT}{V} = \frac{0.5000 \text{ mol} \times \frac{0.08206 \text{ L atm}}{\text{K mol}} \times (25.0 + 273.2) \text{ K}}{1.0000 \text{ L}} = 12.24 \text{ atm}$$

b. $\left[P + a\left(\frac{n}{V}\right)^2\right](V - nb) = nRT$; for N_2: $a = 1.39$ atm L^2/mol^2 and $b = 0.0391$ L/mol

$$\left[P + 1.39\left(\frac{0.5000}{1.0000}\right)^2 \text{ atm}\right](1.0000 \text{ L} - 0.5000 \times 0.0391 \text{ L}) = 12.24 \text{ L atm}$$

$(P + 0.348 \text{ atm})(0.9805 \text{ L}) = 12.24 \text{ L atm}$

$$P = \frac{12.24 \text{ L atm}}{0.9805 \text{ L}} - 0.348 \text{ atm} = 12.48 - 0.348 = 12.13 \text{ atm}$$

c. The ideal gas law is high by 0.11 atm, or $\frac{0.11}{12.13} \times 100 = 0.91\%$.

Atmospheric Chemistry

135. $\chi_{He} = 5.24 \times 10^{-6}$ from Table 5.4. $P_{He} = \chi_{He} \times P_{total} = 5.24 \times 10^{-6} \times 1.0 \text{ atm} = 5.2 \times 10^{-6} \text{ atm}$

$$\frac{n}{V} = \frac{P}{RT} = \frac{5.2 \times 10^{-6} \text{ atm}}{\frac{0.08206 \text{ L atm}}{\text{K mol}} \times 298 \text{ K}} = 2.1 \times 10^{-7} \text{ mol He/L}$$

$$\frac{2.1 \times 10^{-7} \text{ mol}}{\text{L}} \times \frac{1 \text{ L}}{1000 \text{ cm}^3} \times \frac{6.022 \times 10^{23} \text{ atoms}}{\text{mol}} = 1.3 \times 10^{14} \text{ atoms He/cm}^3$$

137. $S(s) + O_2(g) \rightarrow SO_2(g)$, combustion of coal

$2\ SO_2(g) + O_2(g) \rightarrow 2\ SO_3(g)$, reaction with atmospheric O_2

$SO_3(g) + H_2O(l) \rightarrow H_2SO_4(aq)$, reaction with atmospheric H_2O

139. a. If we have 1.0×10^6 L of air, then there are 3.0×10^2 L of CO.

$$P_{CO} = \chi_{CO}P_{total};\ \chi_{CO} = \frac{V_{CO}}{V_{total}} \text{ because } V \propto n;\ P_{CO} = \frac{3.0 \times 10^2}{1.0 \times 10^6} \times 628 \text{ torr} = 0.19 \text{ torr}$$

b. $n_{CO} = \frac{P_{CO}V}{RT}$; assuming 1.0 m^3 air, 1 m^3 = 1000 L:

$$n_{CO} = \frac{\frac{0.19}{760} \text{ atm} \times (1.0 \times 10^3 \text{ L})}{\frac{0.08206 \text{ L atm}}{\text{K mol}} \times 273 \text{ K}} = 1.1 \times 10^{-2} \text{ mol CO}$$

$$1.1 \times 10^{-2} \text{ mol} \times \frac{6.02 \times 10^{23} \text{ molecules}}{\text{mol}} = 6.6 \times 10^{21} \text{ CO molecules in } 1.0 \text{ m}^3 \text{ of air}$$

c. $$\frac{6.6 \times 10^{21} \text{ molecules}}{\text{m}^3} \times \left(\frac{1 \text{ m}}{100 \text{ cm}}\right)^3 = \frac{6.6 \times 10^{15} \text{ molecules CO}}{\text{cm}^3}$$

ChemWork Problems

141. Processes b, c, and d will all result in a doubling of the pressure. Process b doubles the pressure because the absolute temperature is doubled (from 200. K to 400. K). Process c has

the effect of halving the volume, which would double the pressure by Boyle's law. Process d doubles the pressure because the moles of gas are doubled (28 g N_2 is 1 mol of N_2 and 32 g O_2 is 1 mol of O_2). Process a won't double the pressure because 28 g O_2 is less than one mol of gas and process e won't double the temperature since the absolute temperature is not doubled (goes from 303 K to 333 K).

143. At constant P and T, volume is directly proportional to the moles of gas present.

$\frac{V_{He}}{V_{Ne}} = 4 = \frac{n_{He}}{n_{Ne}}$; let's assume we have 1 mol of Ne present. For the volume of the helium balloon to be 4 times larger, we will need 4 mol of He. The mass of 1 mol of Ne is 20.18 g and the mass of 4 moles of He is 4(4.003) = 16.01 g. The He:Ne mass ratio is:

$\frac{16.01}{20.18} = 0.7934$; this is approximately a 4:5 mass ratio between He and Ne.

145. a. PV = nRT

PV = constant

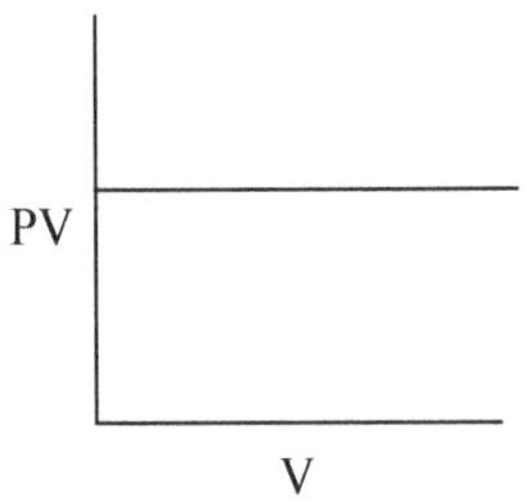

b. PV = nRT

$$P = \left(\frac{nR}{V}\right) \times T = \text{const} \times T$$

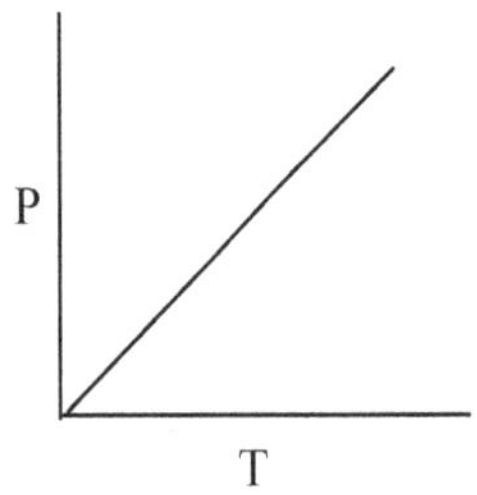

c. PV = nRT

$$T = \left(\frac{P}{nR}\right) \times V = \text{const} \times V$$

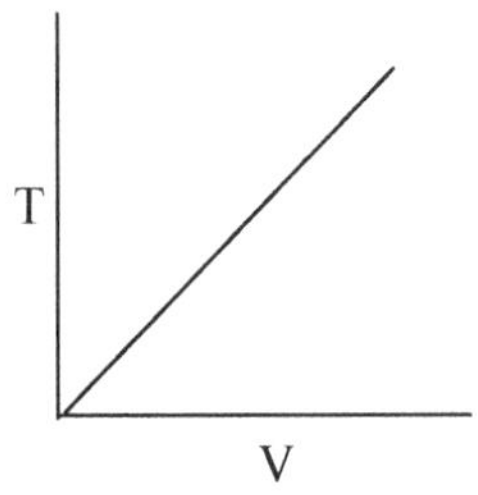

d. PV = nRT

PV = constant

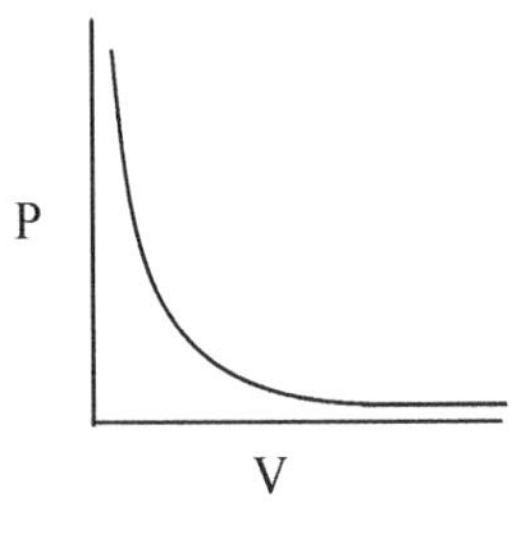

e. $P = \frac{nR}{V} = \frac{\text{constant}}{V}$

$$P = \text{constant} \times \frac{1}{V}$$

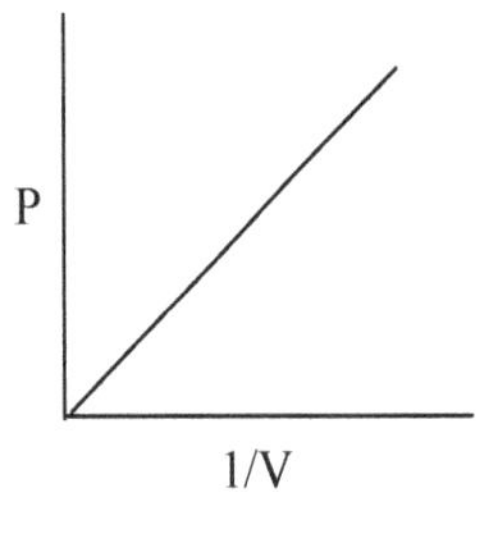

f. PV = nRT

$$\frac{PV}{T} = nR = \text{constant}$$

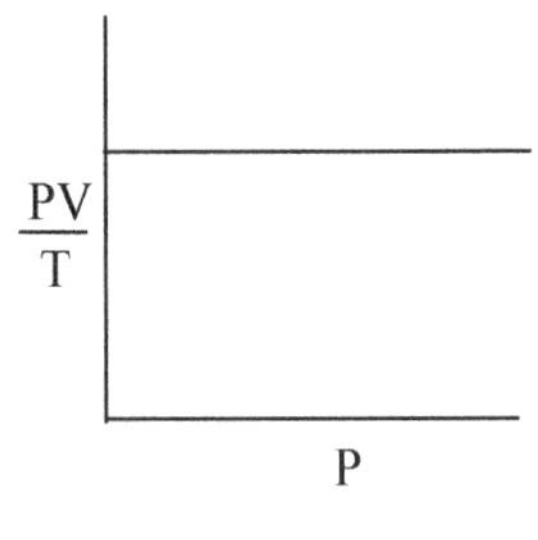

Note: The equation for a straight line is $y = mx + b$, where y is the y-axis and x is the x-axis. Any equation that has this form will produce a straight line with slope equal to m and a y intercept equal to b. Plots b, c, and e have this straight-line form.

147. $$14.1 \times 10^2 \text{ in Hg•in}^3 \times \frac{2.54 \text{ cm}}{\text{in}} \times \frac{10 \text{ mm}}{1 \text{ cm}} \times \frac{1 \text{ atm}}{760 \text{ mm}} \times \left(\frac{2.54 \text{ cm}}{\text{in}}\right)^3 \times \frac{1 \text{ L}}{1000 \text{ cm}^3} = 0.772 \text{ atm•L}$$

Boyle's law: PV = k, where k = nRT; from Example 5.3, the k values are around 22 atm•L. Because k = nRT, we can assume that Boyle's data and the Example 5.3 data were taken at different temperatures and/or had different sample sizes (different moles).

149. PV = nRT, n is constant. $\frac{PV}{T} = nR = \text{constant}$, $\frac{P_1V_1}{T_1} = \frac{P_2V_2}{T_2}$, $V_2 = \frac{V_1P_1T_2}{P_2T_1}$

$$V_2 = 855 \text{ L} \times \frac{730. \text{ torr}}{605 \text{ torr}} \times \frac{(273 + 15) \text{ K}}{(273 + 25) \text{ K}} = 997 \text{ L}; \ \Delta V = 997 - 855 = 142 \text{ L}$$

151. Assume some mass of the mixture. If we had 100.0 g of the gas, we would have 50.0 g He and 50.0 g Xe.

$$\chi_{He} = \frac{n_{He}}{n_{He} + n_{Xe}} = \frac{\frac{50.0 \text{ g}}{4.003 \text{ g/mol}}}{\frac{50.0 \text{ g}}{4.003 \text{ g/mol}} + \frac{50.0 \text{ g}}{131.3 \text{ g/mol}}} = \frac{12.5 \text{ mol He}}{12.5 \text{ mol He} + 0.381 \text{ mol Xe}} = 0.970$$

No matter what the initial mass of mixture is assumed, the mole fraction of helium will always be 0.970.

$P_{He} = \chi_{He}P_{total} = 0.970 \times 600. \text{ torr} = 582 \text{ torr}$; $P_{Xe} = 600. - 582 = 18 \text{ torr}$

153. P × (molar mass) = dRT, $d = \frac{\text{mass}}{\text{volume}}$, P × (molar mass) $= \frac{\text{mass}}{V} \times RT$

Mass of gas = 135.87 g – 134.66 g = 1.21 g

$$\text{Molar mass} = \frac{\text{mass} \times RT}{PV} = \frac{1.21 \text{ g} \times \frac{0.08206 \text{ L atm}}{\text{K mol}} \times 304 \text{ K}}{0.967 \text{ atm} \times 0.936 \text{ L}} = 33.3 \text{ g/mol}$$

155. $P_{total} = P_{N_2} + P_{H_2O}$, $P_{N_2} = 726 \text{ torr} - 23.8 \text{ torr} = 702 \text{ torr} \times \frac{1 \text{ atm}}{760 \text{ torr}} = 0.924 \text{ atm}$

$$n_{N_2} = \frac{P_{N_2} \times V}{RT} = \frac{0.924 \text{ atm} \times 31.8 \times 10^{-3} \text{ L}}{\frac{0.08206 \text{ L atm}}{\text{K mol}} \times 298 \text{ K}} = 1.20 \times 10^{-3} \text{ mol } N_2$$

$$\text{Mass of N in compound} = 1.20 \times 10^{-3} \text{ mol } N_2 \times \frac{28.02 \text{ g } N_2}{\text{mol}} = 3.36 \times 10^{-2} \text{ g nitrogen}$$

$$\text{Mass \% N} = \frac{3.36 \times 10^{-2} \text{ g}}{0.253 \text{ g}} \times 100 = 13.3\% \text{ N}$$

157. We will apply Boyle's law to solve. $PV = nRT = \text{constant}$, $P_1V_1 = P_2V_2$

Let condition (1) correspond to He from the tank that can be used to fill balloons. We must leave 1.0 atm of He in the tank, so $P_1 = 200. - 1.00 = 199$ atm and $V_1 = 15.0$ L. Condition (2) will correspond to the filled balloons with $P_2 = 1.00$ atm and $V_2 = N(2.00\text{ L})$, where N is the number of filled balloons, each at a volume of 2.00 L.

199 atm × 15.0 L = 1.00 atm × N(2.00 L), N = 1492.5; we can't fill 0.5 of a balloon, so N = 1492 balloons or, to 3 significant figures, 1490 balloons.

159. $$1.00 \times 10^3 \text{ kg Mo} \times \frac{1000\text{ g}}{\text{kg}} \times \frac{1\text{ mol Mo}}{95.94\text{ g Mo}} = 1.04 \times 10^4 \text{ mol Mo}$$

$$1.04 \times 10^4 \text{ mol Mo} \times \frac{1\text{ mol MoO}_3}{\text{mol Mo}} \times \frac{7/2\text{ mol O}_2}{\text{mol MoO}_3} = 3.64 \times 10^4 \text{ mol O}_2$$

$$V_{O_2} = \frac{n_{O_2}RT}{P} = \frac{3.64 \times 10^4 \text{ mol} \times \frac{0.08206\text{ L atm}}{\text{K mol}} \times 290.\text{ K}}{1.00\text{ atm}} = 8.66 \times 10^5 \text{ L of O}_2$$

$$8.66 \times 10^5 \text{ L O}_2 \times \frac{100\text{ L air}}{21\text{ L O}_2} = 4.1 \times 10^6 \text{ L air}$$

$$1.04 \times 10^4 \text{ mol Mo} \times \frac{3\text{ mol H}_2}{\text{mol Mo}} = 3.12 \times 10^4 \text{ mol H}_2$$

$$V_{H_2} = \frac{3.12 \times 10^4 \text{ mol} \times \frac{0.08206\text{ L atm}}{\text{K mol}} \times 290.\text{ K}}{1.00\text{ atm}} = 7.42 \times 10^5 \text{ L of H}_2$$

161. The partial pressures can be determined by using the mole fractions.

$P_{methane} = P_{total} \times \chi_{methane} = 1.44\text{ atm} \times 0.915 = 1.32\text{ atm}$; $P_{ethane} = 1.44 - 1.32 = 0.12\text{ atm}$

Determining the number of moles of natural gas combusted:

$$n_{natural\ gas} = \frac{PV}{RT} = \frac{1.44\text{ atm} \times 15.00\text{ L}}{\frac{0.08206\text{ L atm}}{\text{K mol}} \times 293\text{ K}} = 0.898 \text{ mol natural gas}$$

$$n_{methane} = n_{natural\ gas} \times \chi_{methane} = 0.898\text{ mol} \times 0.915 = 0.822\text{ mol methane}$$

$$n_{ethane} = 0.898 - 0.822 = 0.076\text{ mol ethane}$$

$$CH_4(g) + 2\ O_2(g) \rightarrow CO_2(g) + 2\ H_2O(l); \quad 2\ C_2H_6 + 7\ O_2(g) \rightarrow 4\ CO_2(g) + 6\ H_2O(l)$$

$$0.822\text{ mol CH}_4 \times \frac{2\text{ mol H}_2\text{O}}{1\text{ mol CH}_4} \times \frac{18.02\text{ g H}_2\text{O}}{\text{mol H}_2\text{O}} = 29.6\text{ g H}_2\text{O}$$

$$0.076\text{ mol C}_2\text{H}_6 \times \frac{6\text{ mol H}_2\text{O}}{2\text{ mol C}_2\text{H}_6} \times \frac{18.02\text{ g H}_2\text{O}}{\text{mol H}_2\text{O}} = 4.1\text{ g H}_2\text{O}$$

The total mass of H_2O produced = 29.6 g + 4.1 g = 33.7 g H_2O.

163. Out of 100.00 g of compound there are:

$$58.51 \text{ g C} \times \frac{1 \text{ mol C}}{12.01 \text{ g C}} = 4.872 \text{ mol C}; \quad \frac{4.872}{2.435} = 2.001$$

$$7.37 \text{ g H} \times \frac{1 \text{ mol H}}{1.008 \text{ g H}} = 7.31 \text{ mol H}; \quad \frac{7.31}{2.435} = 3.00$$

$$34.12 \text{ g N} \times \frac{1 \text{ mol N}}{14.01 \text{ g N}} = 2.435 \text{ mol N}; \quad \frac{2.435}{2.435} = 1.000$$

The empirical formula is C_2H_3N.

$$\frac{\text{Rate}_1}{\text{Rate}_2} = \left(\frac{M_2}{M_1}\right)^{1/2}; \text{ let gas (1) = He; } 3.20 = \left(\frac{M_2}{4.003}\right)^{1/2}, \; M_2 = 41.0 \text{ g/mol}$$

The empirical formula mass of $C_2H_3N \approx 2(12.0) + 3(1.0) + 1(14.0) = 41.0$ g/mol. So the molecular formula is also C_2H_3N.

165. $$0.2766 \text{ g } CO_2 \times \frac{12.01 \text{ g C}}{44.01 \text{ g } CO_2} = 7.548 \times 10^{-2} \text{ g C}; \% \text{C} = \frac{7.548 \times 10^{-2} \text{ g}}{0.1023 \text{ g}} \times 100 = 73.78\% \text{ C}$$

$$0.0991 \text{ g } H_2O \times \frac{2.016 \text{ g II}}{18.02 \text{ g } H_2O} = 1.11 \times 10^{-2} \text{ g H}; \; \% \text{H} = \frac{1.11 \times 10^{-2} \text{ g}}{0.1023 \text{ g}} \times 100 = 10.9\% \text{ H}$$

$$PV = nRT, \quad n_{N_2} = \frac{PV}{RT} = \frac{1.00 \text{ atm} \times 27.6 \times 10^{-3} \text{ L}}{\frac{0.08206 \text{ L atm}}{\text{K mol}} \times 273 \text{ K}} = 1.23 \times 10^{-3} \text{ mol } N_2$$

$$1.23 \times 10^{-3} \text{ mol } N_2 \times \frac{28.02 \text{ g } N_2}{\text{mol } N_2} = 3.45 \times 10^{-2} \text{ g nitrogen}$$

$$\text{Mass \% N} = \frac{3.45 \times 10^{-2} \text{ g}}{0.4831 \text{ g}} \times 100 = 7.14\% \text{ N}$$

Mass % O = 100.00 – (73.78 + 10.9 + 7.14) = 8.2% O

Out of 100.00 g of compound, there are:

$$73.78 \text{ g C} \times \frac{1 \text{ mol}}{12.01 \text{ g}} = 6.143 \text{ mol C}; \; 7.14 \text{ g N} \times \frac{1 \text{ mol}}{14.01 \text{ g}} = 0.510 \text{ mol N}$$

$$10.9 \text{ g H} \times \frac{1 \text{ mol}}{1.008 \text{ g}} = 10.8 \text{ mol H}; \; 8.2 \text{ g O} \times \frac{1 \text{ mol}}{16.00 \text{ g}} = 0.51 \text{ mol O}$$

Dividing all values by 0.51 gives an empirical formula of $C_{12}H_{21}NO$.

$$\text{Molar mass} = \frac{dRT}{P} = \frac{\frac{4.02 \text{ g}}{\text{L}} \times \frac{0.08206 \text{ L atm}}{\text{K mol}} \times 400. \text{ K}}{256 \text{ torr} \times \frac{1 \text{ atm}}{760 \text{ torr}}} = 392 \text{ g/mol}$$

Empirical formula mass of $C_{12}H_{21}NO \approx 195$ g/mol; $\frac{392}{195} \approx 2$

Thus the molecular formula is $C_{24}H_{42}N_2O_2$.

167. All the gases have the same average kinetic energy since they are all at the same temperature [$KE_{avg} = (3/2)RT$]. At constant T, the lighter the gas molecule, the faster the average velocity [$\mu_{avg} \propto \mu_{rms} \propto (1/M)^{1/2}$]. The average velocity order is:

F_2 (38.00 g/mol) < N_2 (28.02 g/mol) < He (4.003 g/mol)
slowest fastest

169. Statements a-c are false. For statement a, F_2, with the smaller molar mass, will have the faster average velocity, so the F_2 molecules will collide with the container walls more frequently. For statement b, both gas samples are at the same P, V, and T, so n (the moles of gas present) must be equal. For statement c, because both gas samples are at the same T, they will have the same average kinetic energy. Statement d is true because the force of collisions is proportional to the mass of the gas particles. The more massive SO_2 molecules will collide with the container walls more forcefully, but with less frequency than the F_2 collisions.

Challenge Problems

171. $BaO(s) + CO_2(g) \rightarrow BaCO_3(s)$; $CaO(s) + CO_2(g) \rightarrow CaCO_3(s)$

$$n_i = \frac{P_i V}{RT} = \text{initial moles of } CO_2 = \frac{\frac{750.}{760}\text{ atm} \times 1.50\text{ L}}{\frac{0.08206\text{ L atm}}{\text{K mol}} \times 303.2\text{ K}} = 0.0595 \text{ mol } CO_2$$

$$n_f = \frac{P_f V}{RT} = \text{final moles of } CO_2 = \frac{\frac{230.}{760}\text{ atm} \times 1.50\text{ L}}{\frac{0.08206\text{ L atm}}{\text{K mol}} \times 303.2\text{ K}} = 0.0182 \text{ mol } CO_2$$

$0.0595 - 0.0182 = 0.0413$ mol CO_2 reacted

Because each metal reacts 1 : 1 with CO_2, the mixture contains a total of 0.0413 mol of BaO and CaO. The molar masses of BaO and CaO are 153.3 and 56.08 g/mol, respectively.

Let x = mass of BaO and y = mass of CaO, so:

$$x + y = 5.14 \text{ g} \quad \text{and} \quad \frac{x}{153.3} + \frac{y}{56.08} = 0.0413 \text{ mol} \quad \text{or} \quad x + (2.734)y = 6.33$$

Solving by simultaneous equations:

$$\begin{array}{rcl} x + (2.734)y &=& 6.33 \\ \underline{-x \qquad\quad -y} &=& \underline{-5.14} \\ (1.734)y &=& 1.19, \quad y - 1.19/1.734 = 0.686 \end{array}$$

$y = 0.686$ g CaO and $5.14 - y = x = 4.45$ g BaO

$$\text{Mass \% BaO} = \frac{4.45 \text{ g BaO}}{5.14 \text{ g}} \times 100 = 86.6\% \text{ BaO}; \ \%\text{CaO} = 100.0 - 86.6 = 13.4\% \text{ CaO}$$

173. Assuming 1.000 L of the hydrocarbon (C_xH_y), then the volume of products will be 4.000 L, and the mass of products ($H_2O + CO_2$) will be:

$$1.391 \text{ g/L} \times 4.000 \text{ L} = 5.564 \text{ g products}$$

$$\text{Mol } C_xH_y = n_{C_xH_y} = \frac{PV}{RT} = \frac{0.959 \text{ atm} \times 1.000 \text{ L}}{\frac{0.08206 \text{ L atm}}{\text{K mol}} \times 298 \text{ K}} = 0.0392 \text{ mol}$$

$$\text{Mol products} = n_p = \frac{PV}{RT} = \frac{1.51 \text{ atm} \times 4.000 \text{ L}}{\frac{0.08206 \text{ L atm}}{\text{K mol}} \times 375 \text{ K}} = 0.196 \text{ mol}$$

$$C_xH_y + \text{oxygen} \rightarrow x\ CO_2 + y/2\ H_2O$$

Setting up two equations:

$(0.0392)x + 0.0392(y/2) = 0.196$ (moles of products)

$(0.0392)x(44.01 \text{ g/mol}) + 0.0392(y/2)(18.02 \text{ g/mol}) = 5.564 \text{ g}$ (mass of products)

Solving: $x = 2$ and $y = 6$, so the formula of the hydrocarbon is C_2H_6.

175. a. The reaction is $CH_4(g) + 2\ O_2(g) \rightarrow CO_2(g) + 2\ H_2O(g)$.

$$PV = nRT, \ \frac{PV}{n} = RT = \text{constant}, \ \frac{P_{CH_4}V_{CH_4}}{n_{CH_4}} = \frac{P_{air}V_{air}}{n_{air}}$$

The balanced equation requires 2 mol O_2 for every mole of CH_4 that reacts. For three times as much oxygen, we would need 6 mol O_2 per mole of CH_4 reacted ($n_{O_2} = 6n_{CH_4}$). Air is 21% mole percent O_2, so $n_{O_2} = (0.21)n_{air}$. Therefore, the moles of air we would need to deliver the excess O_2 are:

$$n_{O_2} = (0.21)n_{air} = 6n_{CH_4}, \ n_{air} = 29n_{CH_4}, \ \frac{n_{air}}{n_{CH_4}} = 29$$

In 1 minute:

$$V_{air} = V_{CH_4} \times \frac{n_{air}}{n_{CH_4}} \times \frac{P_{CH_4}}{P_{air}} = 200. \text{ L} \times 29 \times \frac{1.50 \text{ atm}}{1.00 \text{ atm}} = 8.7 \times 10^3 \text{ L air/min}$$

b. If x mol of CH_4 were reacted, then $6x$ mol O_2 were added, producing $(0.950)x$ mol CO_2 and $(0.050)x$ mol of CO. In addition, $2x$ mol H_2O must be produced to balance the hydrogens.

$$CH_4(g) + 2\ O_2(g) \rightarrow CO_2(g) + 2\ H_2O(g); \ CH_4(g) + 3/2\ O_2(g) \rightarrow CO(g) + 2\ H_2O(g)$$

Amount O_2 reacted:

$$(0.950)x \text{ mol } CO_2 \times \frac{2 \text{ mol } O_2}{\text{mol } CO_2} = (1.90)x \text{ mol } O_2$$

$$(0.050)x \text{ mol CO} \times \frac{1.5 \text{ mol } O_2}{\text{mol CO}} = (0.075)x \text{ mol } O_2$$

Amount of O_2 left in reaction mixture = $(6.00)x - (1.90)x - (0.075)x = (4.03)x$ mol O_2

Amount of $N_2 = (6.00)x \text{ mol } O_2 \times \frac{79 \text{ mol } N_2}{21 \text{ mol } O_2} = (22.6)x \approx 23x \text{ mol } N_2$

The reaction mixture contains:

$$(0.950)x \text{ mol } CO_2 + (0.050)x \text{ mol CO} + (4.03)x \text{ mol } O_2 + (2.00)x \text{ mol } H_2O + 23x \text{ mol } N_2 = (30.)x \text{ mol of gas total}$$

$$\chi_{CO} = \frac{(0.050)x}{(30.)x} = 0.0017; \quad \chi_{CO_2} = \frac{(0.950)x}{(30.)x} = 0.032; \quad \chi_{O_2} = \frac{(4.03)x}{(30.)x} = 0.13$$

$$\chi_{H_2O} = \frac{(2.00)x}{(30.)x} = 0.067; \quad \chi_{N_2} = \frac{23x}{(30.)x} = 0.77$$

177. a. Volume of hot air: $V = \frac{4}{3}\pi r^3 = \frac{4}{3}\pi(2.50 \text{ m})^3 = 65.4 \text{ m}^3$

(*Note*: Radius = diameter/2 = 5.00/2 = 2.50 m)

$$65.4 \text{ m}^3 \times \left(\frac{10 \text{ dm}}{\text{m}}\right)^3 \times \frac{1 \text{ L}}{\text{dm}^3} = 6.54 \times 10^4 \text{ L}$$

$$n = \frac{PV}{RT} = \frac{\left(745 \text{ torr} \times \frac{1 \text{ atm}}{760 \text{ torr}}\right) \times 6.54 \times 10^4 \text{ L}}{\frac{0.08206 \text{ L atm}}{\text{K mol}} \times (273 + 65) \text{ K}} = 2.31 \times 10^3 \text{ mol air}$$

Mass of hot air = $2.31 \times 10^3 \text{ mol} \times \frac{29.0 \text{ g}}{\text{mol}} = 6.70 \times 10^4 \text{ g}$

Air displaced: $n = \frac{PV}{RT} = \frac{\frac{745}{760} \text{ atm} \times 6.54 \times 10^4 \text{ L}}{\frac{0.08206 \text{ L atm}}{\text{K mol}} \times (273 + 21) \text{ K}} = 2.66 \times 10^3 \text{ mol air}$

Mass of air displaced = $2.66 \times 10^3 \text{ mol} \times \frac{29.0 \text{ g}}{\text{mol}} = 7.71 \times 10^4 \text{ g}$

Lift = $7.71 \times 10^4 \text{ g} - 6.70 \times 10^4 \text{ g} = 1.01 \times 10^4 \text{ g}$

b. Mass of air displaced is the same, 7.71×10^4 g. Moles of He in balloon will be the same as moles of air displaced, 2.66×10^3 mol, because P, V, and T are the same.

$$\text{Mass of He} = 2.66 \times 10^3 \text{ mol} \times \frac{4.003 \text{ g}}{\text{mol}} = 1.06 \times 10^4 \text{ g}$$

$$\text{Lift} = 7.71 \times 10^4 \text{ g} - 1.06 \times 10^4 \text{ g} = 6.65 \times 10^4 \text{ g}$$

c. Hot air: $n = \dfrac{PV}{RT} = \dfrac{\frac{630.}{760} \text{ atm} \times (6.54 \times 10^4 \text{ L})}{\frac{0.08206 \text{ L atm}}{\text{K mol}} \times 338 \text{ K}} = 1.95 \times 10^3 \text{ mol air}$

$$1.95 \times 10^3 \text{ mol} \times \frac{29.0 \text{ g}}{\text{mol}} = 5.66 \times 10^4 \text{ g of hot air}$$

$$\text{Air displaced:} \quad n = \frac{PV}{RT} = \frac{\frac{630.}{760} \text{ atm} \times (6.54 \times 10^4 \text{ L})}{\frac{0.08206 \text{ L atm}}{\text{K mol}} \times 294 \text{ K}} = 2.25 \times 10^3 \text{ mol air}$$

$$2.25 \times 10^3 \text{ mol} \times \frac{29.0 \text{ g}}{\text{mol}} = 6.53 \times 10^4 \text{ g of air displaced}$$

$$\text{Lift} = 6.53 \times 10^4 \text{ g} - 5.66 \times 10^4 \text{ g} = 8.7 \times 10^3 \text{ g}$$

179. a. Average molar mass of air = 0.790 × 28.02 g/mol + 0.210 × 32.00 g/mol = 28.9 g/mol

Molar mass of helium = 4.003 g/mol

A given volume of air at a given set of conditions has a larger density than helium at those conditions due to the larger average molar mass of air. We need to heat the air to a temperature greater than 25°C in order to lower the air density (by driving air molecules out of the hot air balloon) until the density is the same as that for helium (at 25°C and 1.00 atm).

b. To provide the same lift as the helium balloon (assume V = 1.00 L), the mass of air in the hot air balloon (V = 1.00 L) must be the same as that in the helium balloon. Let MM = molar mass:

$$P \bullet MM = dRT, \quad \text{mass} = \frac{MM \bullet PV}{RT}; \quad \text{solving: mass He} = 0.164 \text{ g}$$

$$\text{Mass air} = 0.164 \text{ g} = \frac{28.9 \text{ g/mol} \times 1.00 \text{ atm} \times 1.00 \text{ L}}{\frac{0.08206 \text{ L atm}}{\text{K mol}} \times T}$$

T = 2150 K (a very high temperature)

181. d = molar mass(P/RT); at constant P and T, the density of gas is directly proportional to the molar mass of the gas. Thus the molar mass of the gas has a value which is 1.38 times that of the molar mass of O_2.

Molar mass = 1.38(32.00 g/mol) = 44.2 g/mol

Because H_2O is produced when the unknown binary compound is combusted, the unknown must contain hydrogen. Let A_xH_y be the formula for unknown compound.

$$\text{Mol } A_xH_y = 10.0 \text{ g } A_xH_y \times \frac{1 \text{ mol } A_xH_y}{44.2 \text{ g}} = 0.226 \text{ mol } A_xH_y$$

$$\text{Mol H} = 16.3 \text{ g } H_2O \times \frac{1 \text{ mol } H_2O}{18.02 \text{ g}} \times \frac{2 \text{ mol H}}{\text{mol } H_2O} = 1.81 \text{ mol H}$$

$$\frac{1.81 \text{ mol H}}{0.226 \text{ mol } A_xH_y} = 8 \text{ mol H/mol } A_xH_y;\ A_xH_y = A_xH_8$$

The mass of the x moles of A in the A_xH_8 formula is:

$$44.2 \text{ g} - 8(1.008 \text{ g}) = 36.1 \text{ g}$$

From the periodic table and by trial and error, some possibilities for A_xH_8 are ClH_8, F_2H_8, C_3H_8, and Be_4H_8. C_3H_8 and Be_4H_8 fit the data best, and because C_3H_8 (propane) is a known substance, C_3H_8 is the best possible identity from the data in this problem.

183. $\dfrac{\text{Rate}_1}{\text{Rate}_2} = \left(\dfrac{M_2}{M_1}\right)^{1/2}$; let N_2O = gas 1 and the lachrymator (tear gas) = gas 2:

$$\frac{\text{Rate}_1}{\text{Rate}_2} = \left(\frac{176}{44.02}\right)^{1/2} = (4.00)^{1/2} = 2.00$$

The rate of effusion of N_2O is twice the rate of effusion of the tear gas. So in a given amount of time, one would expect the N_2O gas to travel twice as far as the tear gas. There are 9 rows difference between row 1 and row 10. Let x = rows N_2O travels and y = rows the tear gas travels. Setting up two equations:

$$\frac{x}{y} = 2.00 \text{ and } x + y = 9 \text{ rows; solving: } x = 6 \text{ and } y = 3.$$

So N_2O travels from row 1 to row 7 and the tear gas travels from row 10 to row 7 where the two gases meet causing the row 7 students to simultaneously laugh and cry.

CHAPTER 6

THERMOCHEMISTRY

Questions

17. Path-dependent functions for a trip from Chicago to Denver are those quantities that depend on the route taken. One can fly directly from Chicago to Denver, or one could fly from Chicago to Atlanta to Los Angeles and then to Denver. Some path-dependent quantities are miles traveled, fuel consumption of the airplane, time traveling, airplane snacks eaten, etc. State functions are path-independent; they only depend on the initial and final states. Some state functions for an airplane trip from Chicago to Denver would be longitude change, latitude change, elevation change, and overall time zone change.

19. $2\ C_8H_{18}(l) + 25\ O_2(g) \rightarrow 16\ CO_2(g) + 18\ H_2O(g)$; the combustion of gasoline is exothermic (as is typical of combustion reactions). For exothermic reactions, heat is released into the surroundings giving a negative q value. To determine the sign of w, concentrate on the moles of gaseous reactants versus the moles of gaseous products. In this reaction, we go from 25 moles of reactant gas molecules to 16 + 18 = 34 moles of product gas molecules. As reactants are converted to products, an expansion will occur because the moles of gas increase. When a gas expands, the system does work on the surroundings, and w is a negative value.

21. $\Delta E = q + w$; only statement d is false. When a gas expands, the system is doing work on the surroundings. For statement a, when work is done by the system, this has a negative sign, and when heat is flowing into a system, this has a positive sign. So, when more work is done (–) than heat flowing in (+), ΔE will decrease. For statement b, at constant volume where only PV work is allowed, w = 0, so $\Delta E = q$. For statement c, ΔE will increase when heat and work are both added to the system.

23. Water has a relatively large heat capacity, so it takes a lot of energy to increase the temperature of a large body of water. Because of this, the temperature fluctuations of a large body of water (oceans) are small compared to the temperatures fluctuations of air. Hence, oceans act as a heat reservoir for areas close to them which results in smaller temperature changes as compared to areas farther away from the oceans.

25. Given:

$$CH_4(g) + 2\ O_2(g) \rightarrow CO_2(g) + 2\ H_2O(l) \qquad \Delta H = -891\text{ kJ}$$
$$CH_4(g) + 2\ O_2(g) \rightarrow CO_2(g) + 2\ H_2O(g) \qquad \Delta H = -803\text{ kJ}$$

Using Hess's law:

$$H_2O(l) + 1/2\ CO_2(g) \rightarrow 1/2\ CH_4(g) + O_2(g) \qquad \Delta H_1 = -1/2(-891\text{ kJ})$$
$$1/2\ CH_4(g) + O_2(g) \rightarrow 1/2\ CO_2(g) + H_2O(g) \qquad \Delta H_2 = 1/2(-803\text{ kJ})$$
$$\overline{\qquad H_2O(l) \rightarrow H_2O(g) \qquad \Delta H = \Delta H_1 + \Delta H_2 = 44\text{ kJ}}$$

The enthalpy of vaporization of water is 44 kJ/mol.

Note: When an equation is reversed, the sign on ΔH is reversed. When the coefficients in a balanced equation are multiplied by an integer, then the value of ΔH is multiplied by the same integer.

27. The zero point for ΔH_f° values are elements in their standard state. All substances are measured in relationship to this zero point.

29. No matter how insulated your thermos bottle, some heat will always escape into the surroundings. If the temperature of the thermos bottle (the surroundings) is high, less heat initially will escape from the coffee (the system); this results in your coffee staying hotter for a longer period of time.

31. Fossil fuels contain carbon; the incomplete combustion of fossil fuels produces CO(g) instead of CO_2(g). This occurs when the amount of oxygen reacting is not sufficient to convert all the carbon to CO_2. Carbon monoxide is a poisonous gas to humans.

Exercises

Potential and Kinetic Energy

33. $KE = \frac{1}{2}mv^2$; convert mass and velocity to SI units. $1\ J = \frac{1\ kg\ m^2}{s^2}$

$$\text{Mass} = 5.25\ oz \times \frac{1\ lb}{16\ oz} \times \frac{1\ kg}{2.205\ lb} = 0.149\ kg$$

$$\text{Velocity} = \frac{1.0 \times 10^2\ mi}{h} \times \frac{1\ h}{60\ min} \times \frac{1\ min}{60\ s} \times \frac{1760\ yd}{mi} \times \frac{1\ m}{1.094\ yd} = \frac{45\ m}{s}$$

$$KE = \frac{1}{2}mv^2 = \frac{1}{2} \times 0.149\ kg \times \left(\frac{45\ m}{s}\right)^2 = 150\ J$$

35. a. Potential energy is energy due to position. Initially, ball A has a higher potential energy than ball B because the position of ball A is higher than the position of ball B. In the final position, ball B has the higher position so ball B has the higher potential energy.

b. As ball A rolled down the hill, some of the potential energy lost by A has been converted to random motion of the components of the hill (frictional heating). The remainder of the lost potential energy was added to B to initially increase its kinetic energy and then to increase its potential energy.

Heat and Work

37. ΔE = q + w = 45 kJ + (−29 kJ) = 16 kJ

39. Step 1: $\Delta E_1 = q + w = 72\ J + 35\ J = 107\ J$; step 2: $\Delta E_2 = 35\ J - 72\ J = -37\ J$

$\Delta E_{overall} = \Delta E_1 + \Delta E_2 = 107\ J - 37\ J = 70.\ J$

41. For step 1, heat is removed from the system and work is done on the system for a compression. $\Delta E_{step\ 1} = -20 + 15 = -5$ J. For step 2, heat is added to the system and work is done on the surroundings for an expansion. $\Delta E_{step\ 2} = 25 + w_2$. For the overall process, $\Delta E_{overall} = \Delta E_1 + \Delta E_2$. Because the initial and final states are the same in the overall process (a cyclic process), $\Delta E_{overall}$ = 0. Note: this is true for any state function in a cyclic process.

$0 = \Delta E_1 + \Delta E_2 = -5 + (25 + w_2),\ w_2 = -20\ J$

Statements a and c are true. For statement b, $\Delta E_2 = -\ \Delta E_1 = 5\ J$

43. $w = -P\Delta V$; we need the final volume of the gas. Because T and n are constant, $P_1V_1 = P_2V_2$.

$$V_2 = \frac{V_1\ P_1}{P_2} = \frac{10.0\ L(15.0\ atm)}{2.00\ atm} = 75.0\ L$$

$$w = -P\Delta V = -2.00\ atm(75.0\ L - 10.0\ L) = -130.\ L\ atm \times \frac{101.3\ J}{L\ atm} \times \frac{1\ kJ}{1000\ J} = -13.2\ kJ = \text{work}$$

45. In this problem, $q = w = -950.$ J.

$$-950.\ J \times \frac{1\ L\ atm}{101.3\ J} = -9.38\ L\ atm \text{ of work done by the gases}$$

$$w = -P\Delta V,\ -9.38\ L\ atm = \frac{-650.}{760} atm \times (V_f - 0.040\ L),\ V_f - 0.040 = 11.0\ L,\ V_f = 11.0\ L$$

47. $$q = \text{molar heat capacity} \times \text{mol} \times \Delta T = \frac{20.8\ J}{^{o}C\ mol} \times 39.1\ mol \times (38.0 - 0.0)^{\circ}C = 30{,}900\ J = 30.9\ kJ$$

$$w = -P\Delta V = -1.00\ atm \times (998\ L - 876\ L) = -122\ L\ atm \times \frac{101.3\ J}{L\ atm} = -12{,}400\ J = -12.4\ kJ$$

$\Delta E = q + w = 30.9\ kJ + (-12.4\ kJ) = 18.5\ kJ$

Properties of Enthalpy

49. This is an endothermic reaction, so heat must be absorbed in order to convert reactants into products. The high-temperature environment of internal combustion engines provides the heat.

51 a. Heat is absorbed from the water (it gets colder) as KBr dissolves, so this is an endothermic process.

b. Heat is released as CH_4 is burned, so this is an exothermic process.

c. Heat is released to the water (it gets hot) as H_2SO_4 is added, so this is an exothermic process.

d. Heat must be added (absorbed) to boil water, so this is an endothermic process.

53. When a gas is converted to a liquid, energy is released, so q is negative. The volume occupied by 1 mol of a liquid is very small compared to the volume of 1 mol of a gas. So, when a gas is converted into a liquid, we have a compression, which makes w positive.

55. $4\ Fe(s) + 3\ O_2(g) \rightarrow 2\ Fe_2O_3(s)$ $\Delta H = -1652$ kJ; note that 1652 kJ of heat is released when 4 mol Fe reacts with 3 mol O_2 to produce 2 mol Fe_2O_3.

a. $4.00 \text{ mol Fe} \times \dfrac{-1652 \text{ kJ}}{4 \text{ mol Fe}} = -1650 \text{ kJ}$; 1650 kJ of heat released

b. $1.00 \text{ mol } Fe_2O_3 \times \dfrac{-1652 \text{ kJ}}{2 \text{ mol } Fe_2O_3} = -826 \text{ kJ}$; 826 kJ of heat released

c. $1.00 \text{ g Fe} \times \dfrac{1 \text{ mol Fe}}{55.85 \text{ g}} \times \dfrac{-1652 \text{ kJ}}{4 \text{ mol Fe}} = -7.39 \text{ kJ}$; 7.39 kJ of heat released

d. $10.0 \text{ g Fe} \times \dfrac{1 \text{ mol Fe}}{55.85 \text{ g Fe}} \times \dfrac{-1652 \text{ kJ}}{4 \text{ mol Fe}} = -73.9 \text{ kJ}$

$$2.00 \text{ g } O_2 \times \frac{1 \text{ mol } O_2}{32.00 \text{ g } O_2} \times \frac{-1652 \text{ kJ}}{3 \text{ mol } O_2} = -34.4 \text{ kJ}$$

Because 2.00 g O_2 releases the smaller quantity of heat, O_2 is the limiting reactant and 34.4 kJ of heat can be released from this mixture.

57. From Example 6.3, $q = 1.3 \times 10^8$ J. Because the heat transfer process is only 60.% efficient, the total energy required is $1.3 \times 10^8 \text{ J} \times \dfrac{100. \text{ J}}{60. \text{ J}} = 2.2 \times 10^8$ J.

$$\text{Mass } C_3H_8 = 2.2 \times 10^8 \text{ J} \times \frac{1 \text{ mol } C_3H_8}{2221 \times 10^3 \text{ J}} \times \frac{44.09 \text{ g } C_3H_8}{\text{mol } C_3H_8} = 4.4 \times 10^3 \text{ g } C_3H_8$$

59. When a liquid is converted into gas, there is an increase in volume. The 2.5 kJ/mol quantity is the work done by the vaporization process in pushing back the atmosphere.

Calorimetry and Heat Capacity

61. Specific heat capacity is defined as the amount of heat necessary to raise the temperature of one gram of substance by one degree Celsius. Therefore, $H_2O(l)$ with the largest heat capacity value requires the largest amount of heat for this process. The amount of heat for $H_2O(l)$ is:

$$\text{energy} = s \times m \times \Delta T = \frac{4.18 \text{ J}}{^\circ\text{C g}} \times 25.0 \text{ g} \times (37.0^\circ\text{C} - 15.0^\circ\text{C}) = 2.30 \times 10^3 \text{ J}$$

The largest temperature change when a certain amount of energy is added to a certain mass of substance will occur for the substance with the smallest specific heat capacity. This is Hg(l), and the temperature change for this process is:

$$\Delta T = \frac{\text{energy}}{s \times m} = \frac{10.7\ \text{kJ} \times \frac{1000\ \text{J}}{\text{kJ}}}{\frac{0.14\ \text{J}}{^\circ\text{C g}} \times 550.\ \text{g}} = 140^\circ\text{C}$$

63. $s = \text{specific heat capacity} = \frac{q}{m \times \Delta T} = \frac{133\ \text{J}}{5.00\ \text{g} \times (55.1 - 25.2)^\circ\text{C}} = 0.890\ \text{J}/^\circ\text{C}\bullet\text{g}$

From Table 6.1, the substance is solid aluminum.

65. | Heat loss by hot water | = | heat gain by cooler water |

The magnitudes of heat loss and heat gain are equal in calorimetry problems. The only difference is the sign (positive or negative). To avoid sign errors, keep all quantities positive and, if necessary, deduce the correct signs at the end of the problem. This means for ΔT, put the higher temperature first. Water has a specific heat capacity = s = 4.18 J/°C•g = 4.18 J/K•g (ΔT in °C = ΔT in K).

Heat loss by hot water $= s \times m \times \Delta T = \frac{4.18\ \text{J}}{\text{K g}} \times 50.0\ \text{g} \times (330.\ \text{K} - T_f)$

Heat gain by cooler water $= \frac{4.18\ \text{J}}{\text{K g}} \times 30.0\ \text{g} \times (T_f - 280.\ \text{K})$; heat loss = heat gain, so:

$$\frac{209\ \text{J}}{\text{K}} \times (330.\ \text{K} - T_f) = \frac{125\ \text{J}}{\text{K}} \times (T_f - 280.\ \text{K})$$

$$6.90 \times 10^4 - 209T_f = 125T_f - 3.50 \times 10^4,\ 334T_f = 1.040 \times 10^5,\ T_f = 311\ \text{K}$$

Note that the final temperature is closer to the temperature of the more massive hot water, which is as it should be.

67. |Heat gained by water | = |heat lost by steel | = s × m × ΔT, where s = specific heat capacity.

Heat gain $= \frac{4.18\text{J}}{^\circ\text{C g}} \times 75.0\ \text{g} \times (T_f - 20.0^\circ\text{C})$

A common error in calorimetry problems is sign errors. Keeping all quantities positive helps to eliminate sign errors. This means for ΔT, put the higher temperature first.

Heat loss $= \frac{0.490\ \text{J}}{^\circ\text{C g}} \times 25.0 \times (85.0^\circ\text{C} - T_f)$

$$\frac{4.18\ \text{J}}{^\circ\text{C g}} \times 75.0\ \text{g} \times (T_f - 20.0^\circ\text{C}) = \frac{0.490\ \text{J}}{^\circ\text{C g}} \times 25.0 \times (85.0^\circ\text{C} - T_f)$$

$$314T_f - 6270 = 1040 - 12.3T_f,\ 326T_f = 7310,\ T_f = 22.4^\circ\text{C}$$

69. | Heat gained by water | = | heat lost by nickel | = s × m × ΔT, where s = specific heat capacity.

$$\text{Heat gain} = \frac{4.18\ \text{J}}{^\circ\text{C g}} \times 150.0\ \text{g} \times (25.0^\circ\text{C} - 23.5^\circ\text{C}) = 940\ \text{J}$$

A common error in calorimetry problems is sign errors. Keeping all quantities positive, specifically ΔT, helps to eliminate sign errors.

$$\text{Heat loss} = 940\ \text{J} = \frac{0.444\ \text{J}}{^\circ\text{C g}} \times \text{mass} \times (99.8 - 25.0)\ ^\circ\text{C},\ \ \text{mass} = \frac{940}{0.444 \times 74.8} = 28\ \text{g}$$

71. | Heat loss by Al + heat loss by Fe | = | heat gain by water |; keeping all quantities positive to avoid sign error:

$$\frac{0.89\ \text{J}}{^\circ\text{C g}} \times 5.00\ \text{g Al} \times (100.0^\circ\text{C} - T_f) + \frac{0.45\ \text{J}}{^\circ\text{C g}} \times 10.00\ \text{g Fe} \times (100.0 - T_f)$$

$$= \frac{4.18\ \text{J}}{^\circ\text{C g}} \times 97.3\ \text{g}\ H_2O \times (T_f - 22.0^\circ\text{C})$$

$$4.5(100.0 - T_f) + 4.5(100.0 - T_f) = 407(T_f - 22.0),\ \ 450 - (4.5)T_f + 450 - (4.5)T_f = 407T_f - 8950$$

$$416T_f = 9850,\ \ T_f = 23.7^\circ\text{C}$$

73. 50.0×10^{-3} L × 0.100 mol/L = 5.00×10^{-3} mol of both $AgNO_3$ and HCl are reacted. Thus 5.00×10^{-3} mol of AgCl will be produced because there is a 1 : 1 mole ratio between reactants.

| Heat lost by chemicals | = | heat gained by solution |

$$\text{Heat gain} = \frac{4.18\ \text{J}}{^\circ\text{C g}} \times 100.0\ \text{g} \times (23.40 - 22.60)^\circ\text{C} = 330\ \text{J}$$

Heat loss = 330 J; this is the heat evolved (exothermic reaction) when 5.00×10^{-3} mol of AgCl is produced. So q = −330 J and ΔH (heat per mol AgCl formed) is negative with a value of:

$$\Delta H = \frac{-330\ \text{J}}{5.00 \times 10^{-3}\ \text{mol}} \times \frac{1\ \text{kJ}}{1000\ \text{J}} = -66\ \text{kJ/mol}$$

Note: Sign errors are common with calorimetry problems. However, the correct sign for ΔH can be determined easily from the ΔT data; i.e., if ΔT of the solution increases, then the reaction is exothermic because heat was released, and if ΔT of the solution decreases, then the reaction is endothermic because the reaction absorbed heat from the water. For calorimetry problems, keep all quantities positive until the end of the calculation and then decide the sign for ΔH. This will help eliminate sign errors.

75. $NH_4NO_3(s) \rightarrow NH_4^+(aq) + NO_3^-(aq)$ ΔH = ?; mass of solution = 60.00 g + 4.25 g = 64.25 g

| Heat lost by solution | = | heat gained as NH_4NO_3 dissolves |; to help eliminate sign errors, we will keep all quantities positive (q and ΔT) and then deduce the correct sign for ΔH at the

end of the problem. Here, because temperature decreases as NH_4NO_3 dissolves, heat is absorbed as NH_4NO_3 dissolves, so this is an endothermic process (ΔH is positive).

$$\text{Heat lost by solution} = \frac{4.18\ \text{J}}{^\circ\text{C g}} \times 64.25\ \text{g} \times (22.0 - 16.9)^\circ\text{C} = 1400\ \text{J}$$

Heat gained as NH_4NO_3 dissolves = 1400 J

$$\Delta H = \frac{1400\ \text{J}}{4.25\ \text{g NH}_4\text{NO}_3} \times \frac{80.05\ \text{g NH}_4\text{NO}_3}{\text{mol NH}_4\text{NO}_3} \times \frac{1\ \text{kJ}}{1000\ \text{J}} = 26\ \text{kJ/mol NH}_4\text{NO}_3\ \text{dissolving}$$

77. Because ΔH is exothermic, the temperature of the solution will increase as $CaCl_2(s)$ dissolves. Keeping all quantities positive:

$$\text{heat loss as CaCl}_2\ \text{dissolves} = 11.0\ \text{g CaCl}_2 \times \frac{1\ \text{mol CaCl}_2}{110.98\ \text{g CaCl}_2} \times \frac{81.5\ \text{kJ}}{\text{mol CaCl}_2} = 8.08\ \text{kJ}$$

$$\text{heat gained by solution} = 8.08 \times 10^3\ \text{J} = \frac{4.18\ \text{J}}{^\circ\text{C g}} \times (125 + 11.0)\ \text{g} \times (T_f - 25.0^\circ\text{C})$$

$$T_f - 25.0^\circ\text{C} = \frac{8.08 \times 10^3}{4.18 \times 136} = 14.2^\circ\text{C},\ T_f = 14.2^\circ\text{C} + 25.0^\circ\text{C} = 39.2^\circ\text{C}$$

79. $\text{Heat gain by calorimeter} = \frac{1.56\ \text{kJ}}{^\circ\text{C}} \times 3.2^\circ\text{C} = 5.0\ \text{kJ} = \text{heat loss by quinine}$

Heat loss = 5.0 kJ, which is the heat evolved (exothermic reaction) by the combustion of 0.1964 g of quinone. Because we are at constant volume, $q_V = \Delta E$.

$$\Delta E_{comb} = \frac{-5.0\ \text{kJ}}{0.1964\ \text{g}} = -25\ \text{kJ/g}; \qquad \Delta E_{comb} = \frac{-25\ \text{kJ}}{\text{g}} \times \frac{108.09\ \text{g}}{\text{mol}} = -2700\ \text{kJ/mol}$$

81. a. $\text{Heat gain by calorimeter} = \text{heat loss by CH}_4 = 6.79\ \text{g CH}_4 \times \frac{1\ \text{mol CH}_4}{16.04\ \text{g}} \times \frac{802\ \text{kJ}}{\text{mol}}$

$= 340.\ \text{kJ}$

$$\text{Heat capacity of calorimeter} = \frac{340.\ \text{kJ}}{10.8\ ^\circ\text{C}} = 31.5\ \text{kJ/}^\circ\text{C}$$

b. $\text{Heat loss by C}_2\text{H}_2 = \text{heat gain by calorimeter} = 16.9^\circ\text{C} \times \frac{31.5\ \text{kJ}}{^\circ\text{C}} = 532\ \text{kJ}$

A bomb calorimeter is at constant volume, so the heat released/gained = $q_V = \Delta E$:

$$\Delta E_{comb} = \frac{-532\ \text{kJ}}{12.6\ \text{g C}_2\text{H}_2} \times \frac{26.04\ \text{g}}{\text{mol C}_2\text{H}_2} = -1.10 \times 10^3\ \text{kJ/mol}$$

Hess's Law

83. Information given:

$$C(s) + O_2(g) \rightarrow CO_2(g) \qquad \Delta H = -393.7\ \text{kJ}$$
$$CO(g) + 1/2\ O_2(g) \rightarrow CO_2(g) \qquad \Delta H = -283.3\ \text{kJ}$$

Using Hess's law:

$$2\ C(s) + 2\ O_2(g) \rightarrow 2\ CO_2(g) \qquad \Delta H_1 = 2(-393.7\ kJ)$$
$$2\ CO_2(g) \rightarrow 2\ CO(g) + O_2(g) \qquad \Delta H_2 = -2(-283.3\ kJ)$$
$$\overline{2\ C(s) + O_2(g) \rightarrow 2\ CO(g) \qquad \Delta H = \Delta H_1 + \Delta H_2 = -220.8\ kJ}$$

Note: When an equation is reversed, the sign on ΔH is reversed. When the coefficients in a balanced equation are multiplied by an integer, then the value of ΔH is multiplied by the same integer.

85.

$$2\ N_2(g) + 6\ H_2(g) \rightarrow 4\ NH_3(g) \qquad \Delta H = -2(92\ kJ)$$
$$6\ H_2O(g) \rightarrow 6\ H_2(g) + 3\ O_2(g) \qquad \Delta H = -3(-484\ kJ)$$
$$\overline{2\ N_2(g) + 6\ H_2O(g) \rightarrow 3\ O_2(g) + 4\ NH_3(g) \qquad \Delta H = 1268\ kJ}$$

No, because the reaction is very endothermic (requires a lot of heat to react), it would not be a practical way of making ammonia because of the high energy costs required.

87.

$$C_6H_4(OH)_2 \rightarrow C_6H_4O_2 + H_2 \qquad \Delta H = 177.4\ kJ$$
$$H_2O_2 \rightarrow H_2 + O_2 \qquad \Delta H = -(-191.2\ kJ)$$
$$2\ H_2 + O_2 \rightarrow 2\ H_2O(g) \qquad \Delta H = 2(-241.8\ kJ)$$
$$2\ H_2O(g) \rightarrow 2\ H_2O(l) \qquad \Delta H = 2(-43.8\ kJ)$$
$$\overline{C_6H_4(OH)_2(aq) + H_2O_2(aq) \rightarrow C_6H_4O_2(aq) + 2\ H_2O(l) \qquad \Delta H = -202.6\ kJ}$$

89.

$$CaC_2 \rightarrow Ca + 2\ C \qquad \Delta H = -(-62.8\ kJ)$$
$$CaO + H_2O \rightarrow Ca(OH)_2 \qquad \Delta H = -653.1\ kJ$$
$$2\ CO_2 + H_2O \rightarrow C_2H_2 + 5/2\ O_2 \qquad \Delta H = -(-1300.\ kJ)$$
$$Ca + 1/2\ O_2 \rightarrow CaO \qquad \Delta H = -635.5\ kJ$$
$$2\ C + 2\ O_2 \rightarrow 2\ CO_2 \qquad \Delta H = 2(-393.5\ kJ)$$
$$\overline{CaC_2(s) + 2\ H_2O(l) \rightarrow Ca(OH)_2(aq) + C_2H_2(g) \qquad \Delta H = -713\ kJ}$$

Standard Enthalpies of Formation

91. The change in enthalpy that accompanies the formation of 1 mole of a compound from its elements, with all substances in their standard states, is the standard enthalpy of formation for a compound. The reactions that refer to ΔH_f° are:

$$Na(s) + 1/2\ Cl_2(g) \rightarrow NaCl(s);\quad H_2(g) + 1/2\ O_2(g) \rightarrow H_2O(l)$$

$$6\ C(graphite, s) + 6\ H_2(g) + 3\ O_2(g) \rightarrow C_6H_{12}O_6(s)$$

$$Pb(s) + S(rhombic, s) + 2\ O_2(g) \rightarrow PbSO_4(s)$$

93. The standard enthalpy of formation for a compound is the change in enthalpy that accompanies the formation of 1 mole of a compound from its elements, with all substances in their standard states. For CuO, the enthalpy of formation equation is $Cu(s) + ½\ O_2(g) \rightarrow CuO(s)$. Now,

rearrange the given equations to add up to the enthalpy of formation equation. Applying Hess's law will allow determination of $\Delta H^{o}_{f,\,CuO}$.

$$\begin{array}{ll} CuO(s) + Cu(s) \rightarrow Cu_2O(s) & \Delta H^\circ = -(11\ kJ) \\ Cu_2O(s) + \tfrac{1}{2}\,O_2(g) \rightarrow 2\ CuO(s)] & \Delta H^\circ = \tfrac{1}{2}\,(-288\ kJ) \\ \hline Cu(s) + \tfrac{1}{2}\,O_2(g) \rightarrow CuO(s) & \Delta H^{o}_{f,\,CuO} = -\ 155\ kJ/mol \end{array}$$

95. $\Delta H^\circ = \sum n_p \Delta H^\circ_{f,\ products} - \sum n_r \Delta H^\circ_{f,\ reactants}$; elements in their standard state have $\Delta H^\circ_f = 0$ by definition.

$P_4O_{10}(s)\ +\ 6\ H_2O(l)\ \rightarrow\ 4\ H_3PO_4(s)$

$\Delta H^\circ = (4\ mol\ H_3PO_4 \times \Delta H^{o}_{f,\,H_3PO_4}) - (1\ mol\ P_4O_{10} \times \Delta H^{o}_{f,\,P_4O_{10}} + 6\ mol\ H_2O \times \Delta H^{o}_{f,\,H_2O})$

$\Delta H^\circ = [4(-1279\ kJ)] - [-2984\ kJ + 6(-286\ kJ)] = -416.\ kJ/mol$

$2.00\ mol\ H_2O \times \dfrac{-416\ kJ}{6\ mol\ H_2O} = -139\ kJ$; 139 kJ of heat is released for the reaction

97. In general, $\Delta H^\circ = \sum n_p \Delta H^\circ_{f,\ products} - \sum n_r \Delta H^\circ_{f,\ reactants}$, and all elements in their standard state have $\Delta H^\circ_f = 0$ by definition.

a. The balanced equation is $2\ NH_3(g) + 3\ O_2(g) + 2\ CH_4(g) \rightarrow 2\ HCN(g) + 6\ H_2O(g)$.

$$\Delta H^\circ = (2\ mol\ HCN \times \Delta H^\circ_{f,\,HCN} + 6\ mol\ H_2O(g) \times \Delta H^\circ_{f,\,H_2O}) - (2\ mol\ NH_3 \times \Delta H^\circ_{f,\,NH_3} + 2\ mol\ CH_4 \times \Delta H^\circ_{f,\,CH_4})$$

$\Delta H^\circ = [2(135.1) + 6(-242)] - [2(-46) + 2(-75)] = -940.\ kJ$

b. $Ca_3(PO_4)_2(s) + 3\ H_2SO_4(l) \rightarrow 3\ CaSO_4(s) + 2\ H_3PO_4(l)$

$$\Delta H^\circ = \left[3\ mol\ CaSO_4(s)\left(\frac{-1433\ kJ}{mol}\right) + 2\ mol\ H_3PO_4(l)\left(\frac{-1267\ kJ}{mol}\right)\right] - \left[1\ mol\ Ca_3(PO_4)_2(s)\left(\frac{-4126\ kJ}{mol}\right) + 3\ mol\ H_2SO_4(l)\left(\frac{-814\ kJ}{mol}\right)\right]$$

$\Delta H^\circ = -6833\ kJ - (-6568\ kJ) = -265\ kJ$

c. $NH_3(g) + HCl(g) \rightarrow NH_4Cl(s)$

$\Delta H^\circ = (1\ mol\ NH_4Cl \times \Delta H^\circ_{f,\,NH_4Cl}) - (1\ mol\ NH_3 \times \Delta H^\circ_{f,\,NH_3} + 1\ mol\ HCl \times \Delta H^\circ_{f,\,HCl})$

$$\Delta H° = \left[1\text{ mol}\left(\frac{-314\text{ kJ}}{\text{mol}}\right)\right] - \left[1\text{ mol}\left(\frac{-46\text{ kJ}}{\text{mol}}\right) + 1\text{ mol}\left(\frac{-92\text{ kJ}}{\text{mol}}\right)\right]$$

$\Delta H° = -314\text{ kJ} + 138\text{ kJ} = -176\text{ kJ}$

99. a. $4\ NH_3(g) + 5\ O_2(g) \rightarrow 4\ NO(g) + 6\ H_2O(g)$; $\Delta H° = \sum n_p \Delta H^\circ_{f,\ products} - \sum n_r \Delta H^\circ_{f,\ reactants}$

$$\Delta H° = \left[4\text{ mol}\left(\frac{90.\text{ kJ}}{\text{mol}}\right) + 6\text{ mol}\left(\frac{-242\text{ kJ}}{\text{mol}}\right)\right] - \left[4\text{ mol}\left(\frac{-46\text{ kJ}}{\text{mol}}\right)\right] = -908\text{ kJ}$$

$2\ NO(g) + O_2(g) \rightarrow 2\ NO_2(g)$

$$\Delta H° = \left[2\text{ mol}\left(\frac{34\text{ kJ}}{\text{mol}}\right)\right] - \left[2\text{ mol}\left(\frac{90.\text{ kJ}}{\text{mol}}\right)\right] = -112\text{ kJ}$$

$3\ NO_2(g) + H_2O(l) \rightarrow 2\ HNO_3(aq) + NO(g)$

$$\Delta H° = \left[2\text{ mol}\left(\frac{-207\text{ kJ}}{\text{mol}}\right) + 1\text{ mol}\left(\frac{90.\text{ kJ}}{\text{mol}}\right)\right] - \left[3\text{ mol}\left(\frac{34\text{ kJ}}{\text{mol}}\right) + 1\text{ mol}\left(\frac{-286\text{ kJ}}{\text{mol}}\right)\right]$$

$-140.\text{ kJ}$

Note: All ΔH°_f values are assumed ±1 kJ.

b. $12\ NH_3(g) + 15\ O_2(g) \rightarrow 12\ NO(g) + 18\ H_2O(g)$
$12\ NO(g) + 6\ O_2(g) \rightarrow 12\ NO_2(g)$
$12\ NO_2(g) + 4\ H_2O(l) \rightarrow 8\ HNO_3(aq) + 4\ NO(g)$
$4\ H_2O(g) \rightarrow 4\ H_2O(l)$

$12\ NH_3(g) + 21\ O_2(g) \rightarrow 8\ HNO_3(aq) + 4\ NO(g) + 14\ H_2O(g)$

The overall reaction is exothermic because each step is exothermic.

101. $3\ Al(s) + 3\ NH_4ClO_4(s) \rightarrow Al_2O_3(s) + AlCl_3(s) + 3\ NO(g) + 6\ H_2O(g)$

$$\Delta H° = \left[6\text{ mol}\left(\frac{-242\text{ kJ}}{\text{mol}}\right) + 3\text{ mol}\left(\frac{90.\text{ kJ}}{\text{mol}}\right) + 1\text{ mol}\left(\frac{-704\text{ kJ}}{\text{mol}}\right) + 1\text{ mol}\left(\frac{-1676\text{ kJ}}{\text{mol}}\right)\right]$$

$$- \left[3\text{ mol}\left(\frac{-295\text{ kJ}}{\text{mol}}\right)\right] = -2677\text{ kJ}$$

103. $2\ ClF_3(g) + 2\ NH_3(g) \rightarrow N_2(g) + 6\ HF(g) + Cl_2(g)$ $\Delta H° = -1196\text{ kJ}$

$\Delta H° = (6\ \Delta H^\circ_{f,\ HF}) - (2\ \Delta H^\circ_{f,\ ClF_3} + 3\ \Delta H^\circ_{f,\ NH_3})$

$$-1196\text{ kJ} = 6\text{ mol}\left(\frac{-271\text{ kJ}}{\text{mol}}\right) - 2\ \Delta H^\circ_{f,\ ClF_3} - 2\text{ mol}\left(\frac{-46\text{ kJ}}{\text{mol}}\right)$$

$$-1196\text{ kJ} = -1626\text{ kJ} - 2\ \Delta H^\circ_{f,\ ClF_3} + 92\text{ kJ},\quad \Delta H^\circ_{f,\ ClF_3} = \frac{(-1626 + 92 + 1196)\text{ kJ}}{2\text{ mol}} = \frac{-169\text{ kJ}}{\text{mol}}$$

Energy Consumption and Sources

105. $C(s) + H_2O(g) \rightarrow H_2(g) + CO(g)$ $\Delta H° = -110.5 \text{ kJ} - (-242 \text{ kJ}) = 132 \text{ kJ}$

107. $C_2H_5OH(l) + 3\ O_2(g) \rightarrow 2\ CO_2(g) + 3\ H_2O(l)$

$\Delta H° = [2(-393.5 \text{ kJ}) + 3(-286 \text{ kJ})] - (-278 \text{ kJ}) = -1367 \text{ kJ/mol ethanol}$

$$\frac{-1367 \text{ kJ}}{\text{mol}} \times \frac{1 \text{ mol}}{46.07 \text{ g}} = -29.67 \text{ kJ/g}$$

109. $C_3H_8(g) + 5\ O_2(g) \rightarrow 3\ CO_2(g) + 4\ H_2O(l)$

$\Delta H° = [3(-393.5 \text{ kJ}) + 4(-286 \text{ kJ})] - (-104 \text{ kJ}) = -2221 \text{ kJ/mol } C_3H_8$

$$\frac{-2221 \text{ kJ}}{\text{mol}} \times \frac{1 \text{ mol}}{44.09 \text{ g}} = \frac{-50.37 \text{ kJ}}{\text{g}} \quad \text{versus } -47.7 \text{ kJ/g for octane (Example 6.11)}$$

The fuel values are very close. An advantage of propane is that it burns more cleanly. The boiling point of propane is −42°C. Thus it is more difficult to store propane, and there are extra safety hazards associated with using high-pressure compressed-gas tanks.

111. The molar volume of a gas at STP is 22.42 L (from Chapter 5).

$$4.19 \times 10^6 \text{ kJ} \times \frac{1 \text{ mol } CH_4}{891 \text{ kJ}} \times \frac{22.42 \text{ L } CH_4}{\text{mol } CH_4} = 1.05 \times 10^5 \text{ L } CH_4$$

ChemWork Problems

113. $$2.0 \text{ h} \times \frac{5500 \text{ kJ}}{\text{h}} \times \frac{1 \text{ mol } H_2O}{40.6 \text{ kJ}} \times \frac{18.02 \text{ g } H_2O}{\text{mol}} = 4900 \text{ g} = 4.9 \text{ kg } H_2O$$

115. $313 \text{ g He} \times \dfrac{1 \text{ mol He}}{4.003 \text{ g He}} = 78.2 \text{ mol}$; $q = \text{molar heat capacity} \times \text{mol} \times \Delta T$

$$q = \frac{20.8 \text{ J}}{°\text{C mol}} \times 78.2 \text{ mol} \times (-41.6°\text{C}) = -67{,}700 \text{ J} = -67.7 \text{ kJ}$$

$$w = -P\Delta V = -1.00 \text{ atm} \times (1643 \text{ L} - 1910. \text{ L}) = 267 \text{ L atm} \times \frac{101.3 \text{ J}}{\text{L atm}} = 2.70 \times 10^4 \text{ J} = 27.0 \text{ kJ}$$

$\Delta E = q + w = -67.7 \text{ kJ kJ} + 27.0 \text{ kJ} = -40.7 \text{ kJ}$

117. a. $2\ SO_2(g) + O_2(g) \rightarrow 2\ SO_3(g)$; $w = -P\Delta V$; because the volume of the piston apparatus decreased as reactants were converted to products ($\Delta V < 0$), w is positive ($w > 0$).

b. $COCl_2(g) \rightarrow CO(g) + Cl_2(g)$; because the volume increased ($\Delta V > 0$), w is negative ($w < 0$).

c. $N_2(g) + O_2(g) \rightarrow 2\ NO(g)$; because the volume did not change ($\Delta V = 0$), no PV work is done ($w = 0$).

To predict the sign of w for a reaction, compare the coefficients of all the product gases in the balanced equation to the coefficients of all the reactant gases. When a balanced reaction has more moles of product gases than moles of reactant gases (as in b), the reaction will expand in volume (ΔV positive), and the system does work on the surroundings. When a balanced reaction has a decrease in the moles of gas from reactants to products (as in a), the reaction will contract in volume (ΔV negative), and the surroundings will do compression work on the system. When there is no change in the moles of gas from reactants to products (as in c), $\Delta V = 0$ and $w = 0$.

119. $w = -P\Delta V$; Δn = moles of gaseous products – moles of gaseous reactants. Only gases can do PV work (we ignore solids and liquids). When a balanced reaction has more moles of product gases than moles of reactant gases (Δn positive), the reaction will expand in volume (ΔV positive), and the system will do work on the surroundings. For example, in reaction c, $\Delta n = 2 - 0 = 2$ moles, and this reaction would do expansion work against the surroundings. When a balanced reaction has a decrease in the moles of gas from reactants to products (Δn negative), the reaction will contract in volume (ΔV negative), and the surroundings will do compression work on the system, e.g., reaction a, where $\Delta n = 0 - 1 = -1$. When there is no change in the moles of gas from reactants to products, $\Delta V = 0$ and $w = 0$, e.g., reaction b, where $\Delta n = 2 - 2 = 0$.

When $\Delta V > 0$ ($\Delta n > 0$), then $w < 0$, and the system does work on the surroundings (c and e).

When $\Delta V < 0$ ($\Delta n < 0$), then $w > 0$, and the surroundings do work on the system (a and d).

When $\Delta V = 0$ ($\Delta n = 0$), then $w = 0$ (b).

121. $$54.0\text{ g }B_2H_6 \times \frac{1\text{ mol }B_2H_6}{27.67\text{ g }B_2H_6} \times \frac{2035\text{ kJ heat released}}{\text{mol }B_2H_6} = 3.97 \times 10^3\text{ kJ heat released}$$

123. The kcals in one serving size of cookies is:

$$4\text{ g fat} \times \frac{8\text{ kcal}}{\text{g fat}} + 20\text{ g carbs} \times \frac{4\text{ kcal}}{\text{g carb}} + 2\text{ g protein} \times \frac{4\text{ kcal}}{\text{g protein}} = 120\text{ kcal}$$

$$120\text{ kcal} \times \frac{4.184\text{ kJ}}{\text{kcal}} \times \frac{1\text{ mile}}{170\text{ kJ}} = 3.0\text{ miles}$$

125. $HNO_3(aq) + KOH(aq) \rightarrow H_2O(l) + KNO_3(aq) \qquad \Delta H = -56\text{ kJ}$

$$0.2000\text{ L} \times \frac{0.400\text{ mol }HNO_3}{\text{L}} \times \frac{56\text{ kJ heat released}}{\text{mol }HNO_3} = 4.5\text{ kJ heat released if }HNO_3\text{ limiting}$$

$$0.1500\text{ L} \times \frac{0.500\text{ mol KOH}}{\text{L}} \times \frac{56\text{ kJ heat released}}{\text{mol KOH}} = 4.2\text{ kJ heat released if KOH limiting}$$

Because the KOH reagent produces the smaller quantity of heat released, KOH is limiting and 4.2 kJ of heat released.

127. $|q_{surr}| = |q_{solution} + q_{cal}|$; we normally assume that q_{cal} is zero (no heat gain/loss by the calorimeter). However, if the calorimeter has a nonzero heat capacity, then some of the heat absorbed

by the endothermic reaction came from the calorimeter. If we ignore q_{cal}, then q_{surr} is too small, giving a calculated ΔH value that is less positive (smaller) than it should be.

129. $HCl(aq) + NaOH(aq) \rightarrow H_2O(l) + NaCl(aq) \quad \Delta H = -56 \text{ kJ}$

$$0.1500 \text{ L} \times \frac{0.50 \text{ mol HCl}}{\text{L}} \times \frac{56 \text{ kJ heat released}}{\text{mol HCl}} = 4.2 \text{ kJ heat released if HCl limiting}$$

$$0.0500 \text{ L} \times \frac{1.00 \text{ mol NaOH}}{\text{L}} \times \frac{56 \text{ kJ heat released}}{\text{mol NaOH}} = 2.8 \text{ kJ heat released if NaOH limiting}$$

Because the NaOH reagent produces the smaller quantity of heat released, NaOH is limiting and 4.2 kJ of heat released.

$$q = \text{specific heat capacity} \times \text{mass} \times \Delta T, \ 2800 \text{ J} = \frac{4.184 \text{ J}}{^\circ\text{C g}} \times 200.0 \text{ g} \times \Delta T, \ \Delta T = 3.3^\circ\text{C}$$

This is an exothermic reaction so the temperature will increase by 3.3°C. T_{final} = 48.2°C + 3.3°C = 51.5°C

131. Heat released = 1.056 g × 26.42 kJ/g = 27.90 kJ = heat gain by water and calorimeter

$$\text{Heat gain} = 27.90 \text{ kJ} = \left(\frac{4.18 \text{ kJ}}{^\circ\text{C kg}} \times 0.987 \text{ kg} \times \Delta T\right) + \left(\frac{6.66 \text{ kJ}}{^\circ\text{C}} \times \Delta T\right)$$

$$27.90 = (4.13 + 6.66)\Delta T = (10.79)\Delta T, \ \Delta T = 2.586^\circ\text{C}$$

$$2.586^\circ\text{C} = T_f - 23.32^\circ\text{C}, \ T_f = 25.91^\circ\text{C}$$

133.

$2\,H_2(g) + 2\,F_2(g) \rightarrow 4\,HF(g)$	$\Delta H = 2\Delta H_2$
$2\,C(s) + 4\,F_2(g) \rightarrow 2\,CF_4(g)$	$\Delta H = 2\Delta H_3$
$C_2H_4(g) \rightarrow 2\,C(s) + 2\,H_2(g)$	$\Delta H = -\Delta H_4$
$C_2H_4(g) + 6\,F_2(g) \rightarrow 2\,CF_4(g) + 4\,HF(g)$	$\Delta H_1 = 2\,\Delta H_2 + 2\,\Delta H_3 - \Delta H_4$

Answer e is correct.

135. We want ΔH for $N_2H_4(l) + O_2(g) \rightarrow N_2(g) + 2\,H_2O(l)$. It will be easier to calculate ΔH for the combustion of four moles of N_2H_4 because we will avoid fractions.

$9\,H_2 + 9/2\,O_2 \rightarrow 9\,H_2O$	$\Delta H = 9(-286 \text{ kJ})$
$3\,N_2H_4 + 3\,H_2O \rightarrow 3\,N_2O + 9\,H_2$	$\Delta H = -3(-317 \text{ kJ})$
$2\,NH_3 + 3\,N_2O \rightarrow 4\,N_2 + 3\,H_2O$	$\Delta H = -1010. \text{ kJ}$
$N_2H_4 + H_2O \rightarrow 2\,NH_3 + 1/2\,O_2$	$\Delta H = -(-143 \text{ kJ})$
$4\,N_2H_4(l) + 4\,O_2(g) \rightarrow 4\,N_2(g) + 8\,H_2O(l)$	$\Delta H = -2490. \text{ kJ}$

$$\text{For } N_2H_4(l) + O_2(g) \rightarrow N_2(g) + 2\,H_2O(l) \quad \Delta H = \frac{-2490. \text{ kJ}}{4} = -623 \text{ kJ}$$

Note: By the significant figure rules, we could report this answer to four significant figures. However, because the ΔH values given in the problem are only known to ±1 kJ, our final answer will at best be ±1 kJ.

137. $Na_2SO_4(aq) + Ba(NO_3)_2(aq) \rightarrow BaSO_4(s) + 2\ NaNO_3(aq) \qquad \Delta H = ?$

$$1.00\ L \times \frac{2.00\ mol\ Na_2SO_4}{L} \times \frac{1\ mol\ BaSO_4}{mol\ Na_2SO_4} = 2.00\ mol\ BaSO_4 \text{ if } Na_2SO_4 \text{ limiting}$$

$$2.00\ L \times \frac{0.750\ mol\ Ba(NO_3)_2}{L} \times \frac{1\ mol\ BaSO_4}{mol\ Ba(NO_3)_2} = 1.50\ mol\ BaSO_4 \text{ if } Ba(NO_3)_2 \text{ limiting}$$

The $Ba(NO_3)_2$ reagent produces the smaller quantity of product, so $Ba(NO_3)_2$ is limiting and 1.50 mol $BaSO_4$ can form. Heat gain by solution = heat loss by reaction.

$$\text{Mass of solution} = 3.00\ L \times \frac{1000\ mL}{L} \times \frac{2.00\ g}{mL} = 6.00 \times 10^3\ g$$

$$\text{Heat gain by solution} = \frac{6.37\ J}{^\circ C\ g} \times 6.00 \times 10^3\ g \times (42.0 - 30.0)^\circ C = 4.59 \times 10^5\ J$$

Because the solution gained heat, the reaction is exothermic; $q = -4.59 \times 10^5$ J for the reaction.

$$\Delta H = \frac{-4.59 \times 10^5\ J}{1.50\ mol\ BaSO_4} = -3.06 \times 10^5\ J/mol = -306\ kJ/mol$$

139. |Heat gain by water| = |heat loss by Al|; we will keep all quantities positive avoid sign errors.

$$\frac{4.18\ J}{^\circ C\ g} \times \text{mass} \times (24.9^\circ C - 22.3^\circ C) = 25.0\ g\ Al \times \frac{1\ mol\ Al}{26.98\ g\ Al} \times \frac{24.03\ J}{^\circ C\ mol} \times (82.4^\circ C - 24.9^\circ C)$$

11(mass) = 1280, mass = 116 g = 120 g H_2O

141. a. $C_2H_4(g) + O_3(g) \rightarrow CH_3CHO(g) + O_2(g) \quad \Delta H^\circ = -166\ kJ - [143\ kJ + 52\ kJ] = -361\ kJ$

b. $O_3(g) + NO(g) \rightarrow NO_2(g) + O_2(g) \quad \Delta H^\circ = 34\ kJ - [90.\ kJ + 143\ kJ] = -199\ kJ$

c. $SO_3(g) + H_2O(l) \rightarrow H_2SO_4(aq) \quad \Delta H^\circ = -909\ kJ - [-396\ kJ + (-286\ kJ)] = -227\ kJ$

d. $2\ NO(g) + O_2(g) \rightarrow 2\ NO_2(g) \quad \Delta H^\circ = 2(34)\ kJ - 2(90.)\ kJ = -112\ kJ$

143. $N_2(g) + 2\ O_2(g) \rightarrow 2\ NO_2(g) \qquad \Delta H = 67.7\ kJ$

$$n_{N_2} = \frac{PV}{RT} = \frac{3.50\ atm \times 0.250\ L}{\frac{0.08206\ L\ atm}{K\ mol} \times 373\ K} = 2.86 \times 10^{-2}\ mol\ N_2$$

$$n_{O_2} = \frac{PV}{RT} = \frac{3.50\ atm \times 0.450\ L}{\frac{0.08206\ L\ atm}{K\ mol} \times 373\ K} = 5.15 \times 10^{-2}\ mol\ O_2$$

$$2.86 \times 10^{-2}\ mol\ N_2 \times \frac{2\ mol\ NO_2}{1\ mol\ N_2} = 5.72 \times 10^{-2}\ mol\ NO_2 \text{ produced if } N_2 \text{ is limiting.}$$

$$5.15 \times 10^{-2}\ mol\ O_2 \times \frac{2\ mol\ NO_2}{2\ mol\ O_2} = 5.15 \times 10^{-2}\ mol\ NO_2 \text{ produced if } O_2 \text{ is limiting.}$$

O_2 is limiting because it produces the smaller quantity of product. The heat required is:

$$5.15 \times 10^{-2} \text{ mol } NO_2 \times \frac{67.7 \text{ kJ}}{2 \text{ mol } NO_2} = 1.74 \text{ kJ}$$

145. Heat loss by U = heat gain by heavy water; volume of cube = (cube edge)^3

Mass of heavy water = $1.00 \times 10^3 \text{ mL} \times \frac{1.11 \text{ g}}{\text{mL}} = 1110 \text{ g}$

Heat gain by heavy water = $\frac{4.211 \text{ J}}{^\circ\text{C g}} \times 1110 \text{ g} \times (28.5 - 25.5)^\circ\text{C} = 1.4 \times 10^4 \text{ J}$

Heat loss by U = $1.4 \times 10^4 \text{ J} = \frac{0.117 \text{ J}}{^\circ\text{C g}} \times \text{mass} \times (200.0 - 28.5)^\circ\text{C}$, mass = 7.0×10^2 g U

$7.0 \times 10^2 \text{ g U} \times \frac{1 \text{ cm}^3}{19.05 \text{ g}} = 37 \text{ cm}^3$; cube edge = $(37 \text{ cm}^3)^{1/3} = 3.3$ cm

Challenge Problems

147. $A(l) \rightarrow A(g)$ $\Delta H_{vap} = 30.7$ kJ; at constant pressure, $\Delta H = q_p = 30.7$ kJ

Because PV = nRT, at constant pressure and temperature: $w = -P\Delta V = -RT\Delta n$, where:

Δn = moles of gaseous products − moles of gaseous reactants = 1 − 0 = 1

$w = -RT\Delta n = -8.3145 \text{ J/K•mol}(80. + 273 \text{ K})(1 \text{ mol}) = -2940 \text{ J} = -2.94 \text{ kJ}$

$\Delta E = q + w = 30.7 \text{ kJ} + (-2.94 \text{ kJ}) = 27.8 \text{ kJ}$

149. Because PV = nRT, at constant pressure and temperature, $P\Delta V = RT\Delta n$, where Δn = moles of gaseous products − moles of gaseous reactants. For units of joules, use R = 8.3145 J/K·mol.

$\Delta H = \Delta E + RT\Delta n$; $RT\Delta n = 8.3145 \text{ J/K·mol}(273 + 110. \text{ K})(25 - 0) = 7.96 \times 10^4 \text{ J} = 79.6 \text{ kJ}$

$\Delta H = \Delta E + RT\Delta n = -4674 \text{ kJ} + 79.6 \text{ kJ} = -4594 \text{ kJ}$

151. Energy used in 8.0 hours = 40. kWh = $\frac{40. \text{ kJ h}}{\text{s}} \times \frac{3600 \text{ s}}{\text{h}} = 1.4 \times 10^5 \text{ kJ}$

Energy from the sun in 8.0 hours = $\frac{10. \text{ kJ}}{\text{s m}^2} \times \frac{60 \text{ s}}{\text{min}} \times \frac{60 \text{ min}}{\text{h}} \times 8.0 \text{ h} = 2.9 \times 10^4 \text{ kJ/m}^2$

Only 22% of the sunlight is converted into electricity:

$0.22 \times (2.9 \times 10^4 \text{ kJ/m}^2) \times \text{area} = 1.4 \times 10^5 \text{ kJ}$, area = 22 m^2

153. $400 \text{ kcal} \times \frac{4.18 \text{ kJ}}{\text{kcal}} = 1.7 \times 10^3 \text{ kJ} \approx 2 \times 10^3 \text{ kJ}$

$$PE = mgz = \left(180 \text{ lb} \times \frac{1 \text{ kg}}{2.205 \text{ lb}}\right) \times \frac{9.81 \text{ m}}{\text{s}^2} \times \left(8 \text{ in} \times \frac{2.54 \text{ cm}}{\text{in}} \times \frac{1 \text{ m}}{100 \text{ cm}}\right) = 160 \text{ J} \approx 200 \text{ J}$$

200 J of energy is needed to climb one step. The total number of steps to climb are:

$$2 \times 10^6 \text{ J} \times \frac{1 \text{ step}}{200 \text{ J}} = 1 \times 10^4 \text{ steps}$$

155. There are five parts to this problem. We need to calculate:

(1) q required to heat $H_2O(s)$ from –30. °C to 0°C; use the specific heat capacity of $H_2O(s)$

(2) q required to convert 1 mol $H_2O(s)$ at 0°C into 1 mol $H_2O(l)$ at 0°C; use ΔH_{fusion}

(3) q required to heat $H_2O(l)$ from 0°C to 100.°C; use the specific heat capacity of $H_2O(l)$

(4) q required to convert 1 mol $H_2O(l)$ at 100.°C into 1 mol $H_2O(g)$ at 100.°C; use $\Delta H_{vaporization}$

(5) q required to heat $H_2O(g)$ from 100.°C to 140.°C; use the specific heat capacity of $H_2O(g)$

We will sum up the heat required for all five parts, and this will be the total amount of heat required to convert 1.00 mol of $H_2O(s)$ at –30.°C to $H_2O(g)$ at 140.°C.

$q_1 = 2.03 \text{ J/°C•g} \times 18.02 \text{ g} \times [0 - (-30.)]\text{°C} = 1.1 \times 10^3 \text{ J}$

$q_2 = 1.00 \text{ mol} \times 6.02 \times 10^3 \text{ J/mol} = 6.02 \times 10^3 \text{ J}$

$q_3 = 4.18 \text{ J/°C•g} \times 18.02 \text{ g} \times (100. - 0)\text{°C} = 7.53 \times 10^3 \text{ J}$

$q_4 = 1.00 \text{ mol} \times 40.7 \times 10^3 \text{ J/mol} = 4.07 \times 10^4 \text{ J}$

$q_5 = 2.02 \text{ J/°C•g} \times 18.02 \text{ g} \times (140. - 100.)\text{°C} = 1.5 \times 10^3 \text{ J}$

$q_{total} = q_1 + q_2 + q_3 + q_4 + q_5 = 5.69 \times 10^4 \text{ J} = 56.9 \text{ kJ}$

157. $88.0 \text{ g } N_2O \times \frac{1 \text{ mol } N_2O}{44.02 \text{ g } N_2O} = 2.00 \text{ mol } N_2O$

At constant pressure, $q_p = \Delta H$.

$\Delta H = (2.00 \text{ mol})(38.7 \text{ J/°C•mol})(55\text{°C} - 165\text{°C}) = -8510 \text{ J} = -8.51 \text{ kJ} = q_p$

$w = -P\Delta V = -nR\Delta T = -(2.00 \text{ mol})(8.3145 \text{ J/K•mol})(-110. \text{ K}) = 1830 \text{ J} = 1.83 \text{ kJ}$

$\Delta E = q + w = -8.51 \text{ kJ} + 1.83 \text{ kJ} = -6.68 \text{ kJ}$

CHAPTER 7

ATOMIC STRUCTURE AND PERIODICITY

Questions

27. The equations relating the terms are $\nu\lambda = c$, $E = h\nu$, and $E = hc/\lambda$. Frequency is defined as the number of wave cycles that pass a certain part per second, so statement a is true. Statement c is true because wavelength and frequency are inversely related. Statements b and c are false. All electromagnetic radiation travels at the same speed, the speed of light, c. And frequency and photon energy are directly related. As the frequency is halved, photon energy is halved.

29. The photoelectric effect refers to the phenomenon in which electrons are emitted from the surface of a metal when light strikes it. The light must have a certain minimum frequency (energy) to remove electrons from the surface of a metal. Light having a frequency below the minimum value results in no electrons being emitted, whereas light at or higher than the minimum frequency does cause electrons to be emitted. For light having a frequency higher than the minimum frequency, the excess energy is transferred into kinetic energy for the emitted electron. Albert Einstein explained the photoelectric effect by applying quantum theory.

31. Example 7.3 calculates the deBroglie wavelength of a ball and of an electron. The ball has a wavelength on the order of 10^{-34} m. This is incredibly short and, as far as the wave- particle duality is concerned, the wave properties of large objects are insignificant. The electron, with its tiny mass, also has a short wavelength; on the order of 10^{-10} m. However, this wavelength is significant because it is on the same order as the spacing between atoms in a typical crystal. For very tiny objects like electrons, the wave properties are important. The wave properties must be considered, along with the particle properties, when hypothesizing about the electron motion in an atom.

33. Reference the energy-level diagram in given in question 32. Notice that the energy levels get closer and closer together as n increases. For hydrogen (Z = 1), the energy levels in units of joules are given by the equation $E_n = -2.178 \times 10^{-18}(1/n^2)$. As n increases, the differences between $1/n^2$ for consecutive energy levels becomes smaller and smaller.

    If you draw in the electronic transitions in this problem on an energy level diagram, you can see that the $3 \rightarrow 1$ transition (answer e) will have the largest ΔE value, which will correspond to the emission having the shortest wavelength transition ($E = hc/\lambda$). The other transitions that differ by 2 energy levels will have much smaller ΔE values and longer wavelength transitions. The $2 \rightarrow 1$ transition will have the next shortest wavelength transition. The largest differences in allowed energy levels is between allowed energy states with the lowest n value.

35. The Bohr model was an important step in the development of the current quantum mechanical model of the atom. The idea that electrons can only occupy certain, allowed energy levels is illustrated nicely (and relatively easily). We talk about the Bohr model to present the idea of quantized energy levels.

37. When the p and d orbital functions are evaluated at various points in space, the results sometimes have positive values and sometimes have negative values. The term phase is often associated with the + and – signs. For example, a sine wave has alternating positive and negative phases. This is analogous to the positive and negative values (phases) in the p and d orbitals.

39. a. The magnetic quantum number (m_ℓ) is related to the orientation in space of an orbital in relation to the other orbitals in the atom.

    b. The principal quantum number (n) is related to the size and the energy of the orbital.

    c. The electron spin quantum number (m_s) is related to the two oppositely aligned magnetic moments produced by an electron.

    d. The angular momentum quantum number (ℓ) is related to the shape of an atomic orbital.

41. If one more electron is added to a half-filled subshell, electron-electron repulsions will increase because two electrons must now occupy the same atomic orbital. This may slightly decrease the stability of the atom.

43. The valence electrons are strongly attracted to the nucleus for elements with large ionization energies. One would expect these species to readily accept another electron and have very exothermic electron affinities. The noble gases are an exception; they have a large ionization energy but have an endothermic electron affinity. Noble gases have a filled valence shell of electrons. The added electron in a noble gas must go into a higher n value atomic orbital, having a significantly higher energy, and this is very unfavorable.

45. For hydrogen and hydrogen-like (one-electron ions), all atomic orbitals with the same n value have the same energy. For polyatomic atoms/ions, the energy of the atomic orbitals also depends on ℓ. Because there are more nondegenerate energy levels for polyatomic atoms/ions as compared to hydrogen, there are many more possible electronic transitions resulting in more complicated line spectra.

47. Yes, the maximum number of unpaired electrons in any configuration corresponds to a minimum in electron-electron repulsions.

49. a. The energy levels for a one electron atom or ion only depend on n. For example, the 3s, 3p, and 3d orbitals have the same energy for hydrogen or any other one electron species.

    b. The energy levels for an atom or ion with more than one electron depend on n and ℓ. For example, the 3s, 3p, and 3d orbitals have different energies for atoms or ions with more than one electron.

Exercises

Light and Matter

51. $$\nu = \frac{c}{\lambda} = \frac{2.998 \times 10^8 \text{ m/s}}{660 \text{ nm} \times \dfrac{1 \text{ m}}{1 \times 10^9 \text{ nm}}} = 4.5 \times 10^{14} \text{ s}^{-1}$$

53. $\nu = \dfrac{c}{\lambda} = \dfrac{3.00 \times 10^8 \text{ m/s}}{1.0 \times 10^{-2} \text{ m}} = 3.0 \times 10^{10} \text{ s}^{-1}$

$E = h\nu = 6.63 \times 10^{-34} \text{ J s} \times 3.0 \times 10^{10} \text{ s}^{-1} = 2.0 \times 10^{-23} \text{ J/photon}$

$\dfrac{2.0 \times 10^{-23} \text{ J}}{\text{photon}} \times \dfrac{6.02 \times 10^{23} \text{ photons}}{\text{mol}} = 12 \text{ J/mol}$

55. 280 nm: $\nu = \dfrac{c}{\lambda} = \dfrac{3.00 \times 10^8 \text{ m/s}}{280 \text{ nm} \times \dfrac{1 \text{ m}}{1 \times 10^9 \text{ nm}}} = 1.1 \times 10^{15} \text{ s}^{-1}$

320 nm: $\nu = \dfrac{3.00 \times 10^8 \text{ m/s}}{320 \times 10^{-9} \text{ nm}} = 9.4 \times 10^{14} \text{ s}^{-1}$

The compounds in the sunscreen absorb ultraviolet B (UVB) electromagnetic radiation having a frequency from $9.4 \times 10^{14} \text{ s}^{-1}$ to $1.1 \times 10^{15} \text{ s}^{-1}$.

57. The wavelength is the distance between consecutive wave peaks. Wave *a* shows 4 wavelengths, and wave *b* shows 8 wavelengths.

Wave *a*: $\lambda = \dfrac{1.6 \times 10^{-3} \text{ m}}{4} = 4.0 \times 10^{-4} \text{ m}$

Wave *b*: $\lambda = \dfrac{1.6 \times 10^{-3} \text{ m}}{8} = 2.0 \times 10^{-4} \text{ m}$

Wave *a* has the longer wavelength. Because frequency and photon energy are both inversely proportional to wavelength, wave *b* will have the higher frequency and larger photon energy since it has the shorter wavelength.

$\nu = \dfrac{c}{\lambda} = \dfrac{2.998 \times 10^8 \text{ m/s}}{2.0 \times 10^{-4} \text{ m}} = 1.5 \times 10^{12} \text{ s}^{-1}$

$E = \dfrac{hc}{\lambda} = \dfrac{6.626 \times 10^{-34} \text{ J s} \times 2.998 \times 10^8 \text{ m/s}}{2.0 \times 10^{-4} \text{ m}} = 9.9 \times 10^{-22} \text{ J}$

Because both waves are examples of electromagnetic radiation, both waves travel at the same speed, c, the speed of light. From Figure 7.2 of the text, both of these waves represent infrared electromagnetic radiation.

59. a. $\lambda = \dfrac{c}{\nu} = \dfrac{3.00 \times 10^8 \text{ m/s}}{6.0 \times 10^{13} \text{ s}^{-1}} = 5.0 \times 10^{-6} \text{ m}$

b. From Figure 7.2, this is infrared electromagnetic radiation.

c. $E = h\nu = 6.63 \times 10^{-34} \text{ J s} \times 6.0 \times 10^{13} \text{ s}^{-1} = 4.0 \times 10^{-20} \text{ J/photon}$

$\dfrac{4.0 \times 10^{-20} \text{ J}}{\text{photon}} \times \dfrac{6.022 \times 10^{23} \text{ photons}}{\text{mol}} = 2.4 \times 10^4 \text{ J/mol}$

d. Frequency and photon energy are directly related (E = hν). Because $5.4 \times 10^{13}\ s^{-1}$ electromagnetic radiation (EMR) has a lower frequency than $6.0 \times 10^{13}\ s^{-1}$ EMR, the $5.4 \times 10^{13}\ s^{-1}$ EMR will have less energetic photons.

61. The energy needed to remove a single electron is:

$$\frac{279.7\ \text{kJ}}{\text{mol}} \times \frac{1\ \text{mol}}{6.0221 \times 10^{23}} = 4.645 \times 10^{-22}\ \text{kJ} = 4.645 \times 10^{-19}\ \text{J}$$

$$E = \frac{hc}{\lambda},\ \lambda = \frac{hc}{E} = \frac{6.6261 \times 10^{-34}\ \text{J s} \times 2.9979 \times 10^{8}\ \text{m/s}}{4.645 \times 10^{-19}\ \text{J}} = 4.277 \times 10^{-7}\ \text{m} = 427.7\ \text{nm}$$

63. Ionization energy = energy to remove an electron; the energy required for the removal of one electron from a sulfur atom is:

$$\frac{1005\ \text{kJ}}{\text{mol}} \times \frac{1000\ \text{J}}{\text{kJ}} \times \frac{1\ \text{mol}}{6.0221 \times 10^{23}\ \text{atoms}} = 1.669 \times 10^{-18}\ \text{J/atom}$$

To ionize an electron from one sulfur atom requires $E_{photon} = 1.669 \times 10^{-18}$ J.

$$\lambda = \frac{hc}{E_{photon}} = \frac{6.6261 \times 10^{-34}\ \text{J s} \times 2.9979 \times 10^{8}\ \text{m/s}}{1.669 \times 10^{-18}\ \text{J}} = 1.190 \times 10^{-7}\ \text{m} = 119.0\ \text{nm}$$

65. a. 10.% of speed of light = $0.10 \times 3.00 \times 10^{8}$ m/s = 3.0×10^{7} m/s

$$\lambda = \frac{h}{mv},\ \lambda = \frac{6.63 \times 10^{-34}\ \text{J s}}{9.11 \times 10^{-31}\ \text{kg} \times 3.0 \times 10^{7}\ \text{m/s}} = 2.4 \times 10^{-11}\ \text{m} = 2.4 \times 10^{-2}\ \text{nm}$$

Note: For units to come out, the mass must be in kg because $1\ \text{J} = \frac{1\ \text{kg m}^2}{\text{s}^2}$.

b. $$\lambda = \frac{h}{mv} = \frac{6.63 \times 10^{-34}\ \text{J s}}{0.055\ \text{kg} \times 35\ \text{m/s}} = 3.4 \times 10^{-34}\ \text{m} = 3.4 \times 10^{-25}\ \text{nm}$$

This number is so small that it is insignificant. We cannot detect a wavelength this small. The meaning of this number is that we do not have to worry about the wave properties of large objects.

67. $$\lambda = \frac{h}{mv},\ m = \frac{h}{\lambda v} = \frac{6.63 \times 10^{-34}\ \text{J s}}{1.5 \times 10^{-15}\ \text{m} \times (0.90 \times 3.00 \times 10^{8}\ \text{m/s})} = 1.6 \times 10^{-27}\ \text{kg}$$

This particle is probably a proton or a neutron.

Hydrogen Atom: The Bohr Model

69. For the H atom (Z = 1): $E_n = -2.178 \times 10^{-18}\ \text{J}/n^2$; for a spectral transition, $\Delta E = E_f - E_i$:

$$\Delta E = -2.178 \times 10^{-18}\ \text{J} \left(\frac{1}{n_f^2} - \frac{1}{n_i^2} \right)$$

where n_i and n_f are the levels of the initial and final states, respectively. A positive value of ΔE always corresponds to an absorption of light, and a negative value of ΔE always corresponds to an emission of light.

a. $\Delta E = -2.178 \times 10^{-18}\ J\left(\frac{1}{2^2} - \frac{1}{3^2}\right) = -2.178 \times 10^{-18}\ J\left(\frac{1}{4} - \frac{1}{9}\right)$

$\Delta E = -2.178 \times 10^{-18} J \times (0.2500 - 0.1111) = -3.025 \times 10^{-19} J$

The photon of light must have precisely this energy (3.025×10^{-19} J).

$$|\Delta E| = E_{photon} = h\nu = \frac{hc}{\lambda},\ \lambda = \frac{hc}{|\Delta E|} = \frac{6.6261 \times 10^{-34}\ J\ s \times 2.9979 \times 10^{8}\ m/s}{3.025 \times 10^{-19}\ J}$$

$$= 6.567 \times 10^{-7}\ m = 656.7\ nm$$

From Figure 7.2, this is visible electromagnetic radiation (red light).

b. $\Delta E = -2.178 \times 10^{-18}\ J\left(\frac{1}{2^2} - \frac{1}{4^2}\right) = -4.084 \times 10^{-19} J$

$$\lambda = \frac{hc}{|\Delta E|} = \frac{6.6261 \times 10^{-34}\ J\ s \times 2.9979 \times 10^{8}\ m/s}{4.084 \times 10^{-19}\ J} = 4.864 \times 10^{-7}\ m = 486.4\ nm$$

This is visible electromagnetic radiation (green-blue light).

c. $\Delta E = -2.178 \times 10^{-18} J\left(\frac{1}{1^2} - \frac{1}{2^2}\right) = -1.634 \times 10^{-18} J$

$$\lambda = \frac{6.6261 \times 10^{-34}\ J\ s \times 2.9979 \times 10^{8}\ m/s}{1.634 \times 10^{-18}\ J} = 1.216 \times 10^{-7}\ m = 121.6\ nm$$

This is ultraviolet electromagnetic radiation.

The transitions in an energy level diagram are:

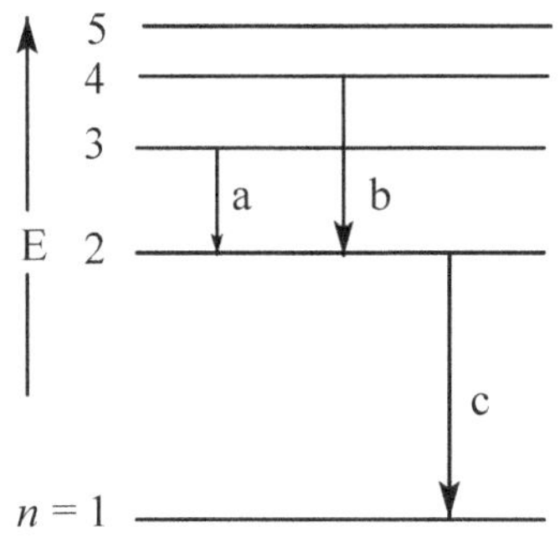

a. 3 → 2

b. 4 → 2

c. 2 → 1

Energy levels are not to scale.

71. a. There are three possible transitions for an electron in the $n = 4$ level (4 → 3, 4 → 2, and 4 → 1). If an electron drops to the $n = 3$ level, two additional transitions can occur (3 → 2, and 3 → 1), and one more transition can occur from the $n = 2$ level (2 → 1). This gives a total of 6 possible transitions for an electron in the $n = 4$ level. Because each transition corresponds to a different ΔE, there will be a total of 6 different wavelength emissions.

b. The lowest energy transition (smallest ΔE) is the $n = 4$ to $n = 3$ electronic transition.

c. Because energy and wavelength are inversely related, the shortest wavelength emission will correspond to the transition with the largest energy change (largest ΔE). This is the transition from $n = 4$ to $n = 1$.

73. $$\Delta E = -2.178 \times 10^{-18}\ \text{J}\left(\frac{1}{n_f^2} - \frac{1}{n_i^2}\right) = -2.178 \times 10^{-18}\ \text{J}\left(\frac{1}{5^2} - \frac{1}{1^2}\right) = 2.091 \times 10^{-18}\ \text{J} = E_{photon}$$

$$\lambda = \frac{hc}{E} = \frac{6.6261 \times 10^{-34}\ \text{J s} \times 2.9979 \times 10^8\ \text{m/s}}{2.091 \times 10^{-18}\ \text{J}} = 9.500 \times 10^{-8}\ \text{m} = 95.00\ \text{nm}$$

Because wavelength and energy are inversely related, visible light ($\lambda \approx$ 400–700 nm) is not energetic enough to excite an electron in hydrogen from $n = 1$ to $n = 5$.

$$\Delta E = -2.178 \times 10^{-18}\ \text{J}\left(\frac{1}{6^2} - \frac{1}{2^2}\right) = 4.840 \times 10^{-19}\ \text{J}$$

$$\lambda = \frac{hc}{E} = \frac{6.6261 \times 10^{-34}\ \text{J s} \times 2.9979 \times 10^8\ \text{m/s}}{4.840 \times 10^{-19}\ \text{J}} = 4.104 \times 10^{-7}\ \text{m} = 410.4\ \text{nm}$$

Visible light with $\lambda = 410.4$ nm will excite an electron from the $n = 2$ to the $n = 6$ energy level.

75. Ionization from $n = 1$ corresponds to the transition $n_i = 1 \rightarrow n_f = \infty$, where $E_\infty = 0$.

$$\Delta E = E_\infty - E_1 = -E_1 = 2.178 \times 10^{-18}\left(\frac{1}{1^2}\right) = 2.178 \times 10^{-18}\ \text{J} = E_{photon}$$

$$\lambda = \frac{hc}{E} = \frac{6.6261 \times 10^{-34}\ \text{J s} \times 2.9979 \times 10^8\ \text{m/s}}{2.178 \times 10^{-18}\ \text{J}} = 9.120 \times 10^{-8}\ \text{m} = 91.20\ \text{nm}$$

To ionize from $n = 2$, $\Delta E = E_\infty - E_2 = -E_2 = 2.178 \times 10^{-18}\left(\frac{1}{2^2}\right) = 5.445 \times 10^{-19}\ \text{J}$

$$\lambda = \frac{6.6261 \times 10^{-34}\ \text{J s} \times 2.9979 \times 10^8\ \text{m/s}}{5.445 \times 10^{-19}\ \text{J}} = 3.648 \times 10^{-7}\ \text{m} = 364.8\ \text{nm}$$

77. $$|\Delta E| = E_{photon} = h\nu = \frac{hc}{\lambda},\ E_{photon} = \frac{6.626 \times 10^{-34}\ \text{J s} \times 2.998 \times 10^8\ \text{m/s}}{434 \times 10^{-9}\ \text{m}} = 4.58 \times 10^{-19}\ \text{J}$$

$\Delta E = -4.58 \times 10^{-19}$ J because we have an emission.

$$-4.58 \times 10^{-19}\ \text{J} = E_n - E_5 = -2.178 \times 10^{-18}\ \text{J}\left(\frac{1}{n^2} - \frac{1}{5^2}\right)$$

$$\frac{1}{n^2} - \frac{1}{25} = 0.210,\quad \frac{1}{n^2} = 0.250,\quad n^2 = 4,\ n = 2$$

The electronic transition is from $n = 5$ to $n = 2$.

Quantum Mechanics, Quantum Numbers, and Orbitals

79. a. $\Delta p = m\Delta v = 9.11 \times 10^{-31}\ \text{kg} \times 0.100\ \text{m/s} = \dfrac{9.11 \times 10^{-32}\ \text{kg m}}{\text{s}}$

$$\Delta p \Delta x \geq \frac{h}{4\pi},\ \Delta x = \frac{h}{4\pi \Delta p} = \frac{6.626 \times 10^{-34}\ \text{J s}}{4 \times 3.142 \times (9.11 \times 10^{-32}\ \text{kg m/s})} = 5.79 \times 10^{-4}\ \text{m}$$

b. $$\Delta x = \frac{h}{4\pi \Delta p} = \frac{6.626 \times 10^{-34}\ \text{J s}}{4 \times 3.142 \times 0.145\ \text{kg} \times 0.100\ \text{m/s}} = 3.64 \times 10^{-33}\ \text{m}$$

c. The diameter of an H atom is roughly $\sim 10^{-8}$ cm. The uncertainty in position is much larger than the size of the atom.

d. The uncertainty is insignificant compared to the size of a baseball.

81. $n = 1, 2, 3, \ldots$; $\ell = 0, 1, 2, \ldots (n - 1)$; $m_\ell = -\ell \ldots -2, -1, 0, 1, 2, \ldots +\ell$

83. 2s: $n = 2$, $\ell = 0$, and $m_\ell = 0$; 4p: $n = 4$, $\ell = 1$, and $m_\ell = -1, 0,$ or 1; $6d_{xy}$: $n = 6$, $\ell = 2$, and m_ℓ has one of the values of –-2 or –1 or 0 or 1 or 2. Which specific m_ℓ designates the d_{xy} orbital cannot be predicted. For an f orbital, $\ell = 3$. Since $\ell = 0, 1, 2, \ldots (n - 1)$, the first possible f orbital is when $n = 4$. Hence a 3f orbital is a forbidden orbital.

85. a. allowed

b. For $\ell = 3$, m_ℓ can range from –3 to +3; thus +4 is not allowed.

c. n cannot equal zero.

d. ℓ cannot be a negative number.

87. ψ^2 gives the probability of finding the electron at that point.

Polyelectronic Atoms

89. He: $1s^2$; Ne: $1s^2 2s^2 2p^6$; Ar: $1s^2 2s^2 2p^6 3s^2 3p^6$; each peak in the diagram corresponds to a subshell with different values of n. Corresponding subshells are closer to the nucleus for heavier elements because of the increased nuclear charge.

91. 5p: three orbitals $3d_{z^2}$: one orbital 4d: five orbitals

$n = 5$: $\ell = 0$ (1 orbital), $\ell = 1$ (3 orbitals), $\ell = 2$ (5 orbitals), $\ell = 3$ (7 orbitals), $\ell = 4$ (9 orbitals); total for $n = 5$ is 25 orbitals.

$n = 4$: $\ell = 0$ (1), $\ell = 1$ (3), $\ell = 2$ (5), $\ell = 3$ (7); total for $n = 4$ is 16 orbitals.

93. a. $n = 4$: ℓ can be 0, 1, 2, or 3. Thus we have s ($2\ e^-$), p ($6\ e^-$), d ($10\ e^-$), and f ($14\ e^-$) orbitals present. Total number of electrons to fill these orbitals is 32.

b. $n = 5$, $m_\ell = +1$: For $n = 5$, $\ell = 0, 1, 2, 3, 4$. For $\ell = 1, 2, 3, 4$, all can have $m_\ell = +1$. There are four distinct orbitals having these quantum numbers, which can hold 8 electrons.

c. $n = 5$, $m_s = +1/2$: For $n = 5$, $\ell = 0, 1, 2, 3, 4$. Number of orbitals = 1, 3, 5, 7, 9 for each value of ℓ, respectively. There are 25 orbitals with $n = 5$. They can hold 50 electrons, and 25 of these electrons can have $m_s = +1/2$.

d. $n = 3$, $\ell = 2$: These quantum numbers define a set of 3d orbitals. There are 5 degenerate 3d orbitals that can hold a total of 10 electrons.

e. $n = 2$, $\ell = 1$: These define a set of 2p orbitals. There are 3 degenerate 2p orbitals that can hold a total of 6 electrons.

95. a. Na: $1s^22s^22p^63s^1$; Na has 1 unpaired electron.

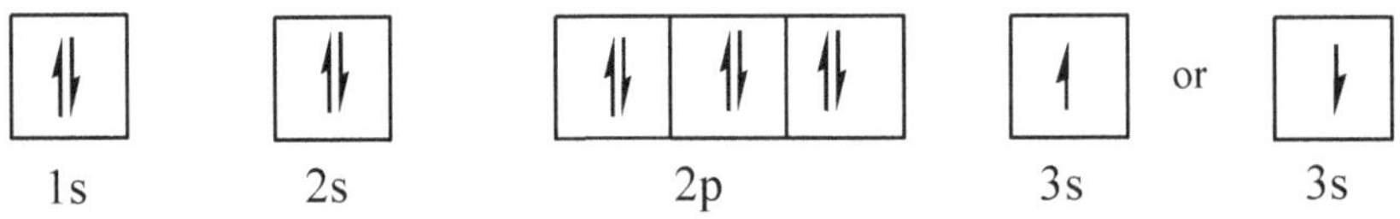

b. Co: $1s^22s^22p^63s^23p^64s^23d^7$; Co has 3 unpaired electrons.

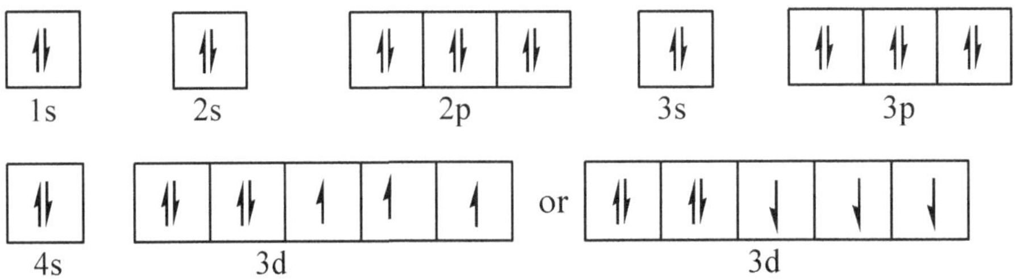

c. Kr: $1s^22s^22p^63s^23p^64s^23d^{10}4p^6$; Kr has 0 unpaired electrons.

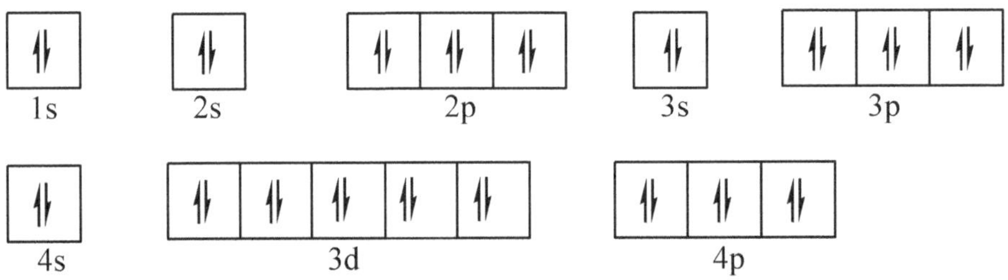

97. Si: $1s^22s^22p^63s^23p^2$ or $[Ne]3s^23p^2$; Ga: $1s^22s^22p^63s^23p^64s^23d^{10}4p^1$ or $[Ar]4s^23d^{10}4p^1$

As: $[Ar]4s^23d^{10}4p^3$; Ge: $[Ar]4s^23d^{10}4p^2$; Al: $[Ne]3s^23p^1$; Cd: $[Kr]5s^24d^{10}$

S: $[Ne]3s^23p^4$; Se: $[Ar]4s^23d^{10}4p^4$

99. a. Zr: $[Kr]5s^24d^2$; two 4d electrons are present.

b. Cd: $[Kr]5s^24d^{10}$; ten 4d electrons are present.

c. Ir: $[Kr]5s^24d^{10}5p^66s^24f^{14}5d^7$; ten 4d electrons are present along with seven 5d electrons.

d. Fe: $[Ar]4s^23d^6$; zero 4d electrons are present, but there are six 3d electrons.

101. For an element to have only 12 p electrons, the element can be Ar through Zn. The elements from Ar through Zn that 0 unpaired electrons are Ar ($3p^6$, ↿⇂ ↿⇂ ↿⇂), Ca ($4s^2$, ↿⇂), and Zn ($3d^{10}$, ↿⇂ ↿⇂ ↿⇂ ↿⇂ ↿⇂).

103. a. Both In and I have one unpaired 5p electron, but only the nonmetal I would be expected to form a covalent compound with the nonmetal F. One would predict an ionic compound to form between the metal In and the nonmetal F.

I: $[Kr]5s^24d^{10}5p^5$ ↿⇂ ↿⇂ ↿
5p

b. From the periodic table, this will be element 120. Element 120: $[Rn]7s^25f^{14}6d^{10}7p^68s^2$

c. Rn: $[Xe]6s^24f^{14}5d^{10}6p^6$; note that the next discovered noble gas will also have 4f electrons (as well as 5f electrons).

d. This is chromium, which is an exception to the predicted filling order. Cr has 6 unpaired electrons, and the next most is 5 unpaired electrons for Mn.

Cr: $[Ar]4s^13d^5$ ↿ ↿ ↿ ↿ ↿ ↿
4s 3d

105. a. The complete ground state electron for this neutral atom is $1s^22s^22p^63s^23p^4$. This atom has 2 + 2 + 6 + 2 + 4 = 16 electrons. Because the atom is neutral, it also has 16 protons, making the atom sulfur, S.

b. Complete excited state electron configuration: $1s^22s^12p^4$; this neutral atom has 2 + 1 + 4 = 7 electrons, which means it has 7 protons, which identifies it as nitrogen, N.

c. Complete ground state electron configuration: $1s^22s^22p^63s^23p^64s^23d^{10}4p^5$; this 1– charged ion has 35 electrons. Because the overall charge is 1–, this ion has 34 protons which identifies it as selenium. The ion is Se^-.

107. Hg: $1s^22s^22p^63s^23p^64s^23d^{10}4p^65s^24d^{10}5p^66s^24f^{14}5d^{10}$

a. From the electron configuration for Hg, we have $3s^2$, $3p^6$, and $3d^{10}$ electrons; 18 total electrons with $n = 3$.

b. $3d^{10}$, $4d^{10}$, $5d^{10}$; 30 electrons are in the d atomic orbitals.

c. $2p^6$, $3p^6$, $4p^6$, $5p^6$; each set of np orbitals contain one p_z atomic orbital. Because we have 4 sets of np orbitals and two electrons can occupy the p_z orbital, there are 4(2) = 8 electrons in p_z atomic orbitals.

d. All the electrons are paired in Hg, so one-half of the electrons are spin up ($m_s = +1/2$) and the other half are spin down ($m_s = -1/2$). 40 electrons have spin up.

109. B: $1s^22s^22p^1$

	n	ℓ	m_ℓ	m_s
1s	1	0	0	+1/2
1s	1	0	0	−1/2
2s	2	0	0	+1/2
2s	2	0	0	−1/2
2p*	2	1	−1	+1/2

*This is only one of several possibilities for the 2p electron. The 2p electron in B could have $m_\ell = -1$, 0 or +1 and $m_s = +1/2$ or $-1/2$ for a total of six possibilities.

N: $1s^22s^22p^3$

	n	ℓ	m_ℓ	m_s	
1s	1	0	0	+1/2	
1s	1	0	0	−1/2	
2s	2	0	0	+1/2	
2s	2	0	0	−1/2	
2p	2	1	−1	+1/2	
2p	2	1	0	+1/2	(Or all 2p electrons could have $m_s = -1/2$.)
2p	2	1	+1	+1/2	

111. Group 1A: 1 valence electron; ns^1; Li: $[He]2s^1$; $2s^1$ is the valence electron configuration for Li.

Group 2A: 2 valence electrons; ns^2; Ra: $[Rn]7s^2$; $7s^2$ is the valence electron configuration for Ra.

Group 3A: 3 valence electrons; ns^2np^1; Ga: $[Ar]4s^23d^{10}4p^1$; $4s^24p^1$ is the valence electron configuration for Ga. Note that valence electrons for the representative elements of Groups 1A-8A are considered those electrons in the highest n value, which for Ga is $n = 4$. We do not include the 3d electrons as valence electrons because they are not in $n = 4$ level.

Group 4A: 4 valence electrons; ns^2np^2; Si: $[Ne]3s^23p^2$; $3s^23p^2$ is the valence electron configuration for Si.

Group 5A: 5 valence electrons; ns^2np^3; Sb: $[Kr]5s^24d^{10}5p^3$; $5s^25p^3$ is the valence electron configuration for Sb.

Group 6A: 6 valence electrons; ns^2np^4; Po: $[Xe]6s^24f^{14}5d^{10}6p^4$; $6s^26p^4$ is the valence electron configuration for Po.

Group 7A: 7 valence electrons; ns^2np^5; Ts (element 117): $[Rn]7s^25f^{14}6d^{10}7p^5$; $7s^27p^5$ is the valence electron configuration for Ts.

Group 8A: 8 valence electrons; ns^2np^6; Ne: $[He]2s^22p^6$; $2s^22p^6$ is the valence electron configuration for Ne.

113. O: $1s^22s^22p_x^22p_y^2$ (↑↓ ↑↓ __); there are no unpaired electrons in this oxygen atom. This configuration would be an excited state, and in going to the more stable ground state (↑↓ ↑ ↑), energy would be released.

115. None of the s block elements have 2 unpaired electrons. In the p block, the elements with either ns^2np^2 or ns^2np^4 valence electron configurations have 2 unpaired electrons. For elements 1-36, these are elements C, Si, and Ge (with ns^2np^2) and elements O, S, and Se (with ns^2np^4). For the d block, the elements with configurations nd^2 or nd^8 have two unpaired electrons. For elements 1-36, these are Ti ($3d^2$) and Ni ($3d^8$). A total of 8 elements from the first 36 elements have two unpaired electrons in the ground state.

117. We get the number of unpaired electrons by examining the incompletely filled subshells. The paramagnetic substances have unpaired electrons, and the ones with no unpaired electrons are not paramagnetic (they are called diamagnetic).

Li: $1s^22s^1$ ↑ ; paramagnetic with 1 unpaired electron.
2s

N: $1s^22s^22p^3$ ↑ ↑ ↑ ; paramagnetic with 3 unpaired electrons.
2p

Ni: $[Ar]4s^23d^8$ ↑↓ ↑↓ ↑↓ ↑ ↑ ; paramagnetic with 2 unpaired electrons.
3d

Te: $[Kr]5s^24d^{10}5p^4$ ↑↓ ↑ ↑ ; paramagnetic with 2 unpaired electrons.
5p

Ba: $[Xe]6s^2$ ↑↓ ; not paramagnetic because no unpaired electrons are present.
6s

Hg: $[Xe]6s^24f^{14}5d^{10}$ ↑↓ ↑↓ ↑↓ ↑↓ ↑↓ ; not paramagnetic because no unpaired electrons.
5d

The Periodic Table and Periodic Properties

119. Size (radius) decreases left to right across the periodic table, and size increases from top to bottom of the periodic table.

a. S < Se < Te b. Br < Ni < K c. F < Si < Ba

All follow the general radius trend.

121. The ionization energy trend is the opposite of the radius trend; ionization energy (IE), in general, increases left to right across the periodic table and decreases from top to bottom of the periodic table.

a. Te < Se < S b. K < Ni < Br c. Ba < Si < F

All follow the general ionization energy trend.

123. a. He (From the general radius trend.) b. Cl

c. Element 116 (Lv) is under Po, element 119 is the next alkali metal to be discovered (under Fr), and element 120 is the next alkaline earth metal to be discovered (under Ra). From the general radius trend, element 116 (Lv) will be the smallest.

d. Si

e. Na^+; this ion has the fewest electrons as compared to the other sodium species present. Na^+ has the smallest number of electron-electron repulsions, which makes it the smallest ion with the largest ionization energy.

125. The false statement is c. Element 118 (Og) has a ground state electron configuration of $[Rn]7s^25f^{14}6d^{10}7p^6$. Element 116 is Lv; its electron configuration is $[Rn]7s^25f^{14}6d^{10}7p^4$. A $7p^4$ configuration has 2 unpaired electrons. Element 117 (Ts) is a halogen, and all halogens have 7 valence electrons. For example, Ts has a valence electron configuration of $7s^27p^5$. Element 119 will be an alkali metal under Fr. Alkali metals form 1+ charged ions, so X_2S is correct since sulfur forms 2– charged anions when in ionic compounds. Element 120 will be an alkaline earth metal under Ra. Alkaline earth metals form 2+ charged ions when in ionic compounds.

127. The elements are Si (a), P (b), O (c), N (d), and As (e). From the ionization energy trend, O is closest to the upper righthand corner of the periodic table where elements with the largest ionization energies exist. However, there is an exception to the ionization energy trend between groups 5A and 6A for row 2 elements. Because of this exception, N (answer d) will have the largest ionization energy.

129. As: $[Ar]4s^23d^{10}4p^3$; Se: $[Ar]4s^23d^{10}4p^4$; the general ionization energy trend predicts that Se should have a higher ionization energy than As. Se is an exception to the general ionization energy trend. There are extra electron-electron repulsions in Se because two electrons are in the same 4p orbital, resulting in a lower ionization energy for Se than predicted.

131. a. As we remove succeeding electrons, the electron being removed is closer to the nucleus, and there are fewer electrons left repelling it. The remaining electrons are more strongly attracted to the nucleus, and it takes more energy to remove these electrons.

b. Al: $1s^22s^22p^63s^23p^1$; for I_4, we begin removing an electron with $n = 2$. For I_3, we remove an electron with $n = 3$ (the last valence electron). In going from $n = 3$ to $n = 2$, there is a big jump in ionization energy because the $n = 2$ electrons are closer to the nucleus on average than the $n = 3$ electrons. Since the $n = 2$ electrons are closer, on average, to the nucleus, they are held more tightly and require a much larger amount of energy to remove compared to the $n = 3$ electrons. In general, valence electrons are much easier to remove than inner-core electrons.

133. a. More favorable electron affinity: C and Br; the electron affinity trend is very erratic. Both N and Ar have positive electron affinity values (unfavorable) due to their electron configurations (see text for detailed explanation).

b. Higher ionization energy: N and Ar (follows the ionization energy trend)

c. Larger size: C and Br (follows the radius trend)

135. Al(−44), Si(−120), P(−74), S(−200.4), Cl(−348.7); based on the increasing nuclear charge, we would expect the electron affinity values to become more exothermic as we go from left to right in the period. Phosphorus is out of line. The reaction for the electron affinity of P is:

$$P(g) + e^- \rightarrow P^-(g)$$
$$[Ne]3s^23p^3 \qquad [Ne]3s^23p^4$$

The additional electron in P^- will have to go into an orbital that already has one electron. There will be greater repulsions between the paired electrons in P^-, causing the electron affinity of P to be less favorable than predicted based solely on attractions to the nucleus.

137. The electron affinity trend is very erratic. In general, electron affinity becomes more positive down the periodic table, and becomes more negative from left to right across the periodic table. But there are many exceptions.

a. Se < S; S is most exothermic.

b. I < Br < F < Cl; Cl is most exothermic. (F is an exception).

139. a. $Se^{3+}(g) \rightarrow Se^{4+}(g) + e^-$

b. $S^-(g) + e^- \rightarrow S^{2-}(g)$

c. $Fe^{3+}(g) + e^- \rightarrow Fe^{2+}(g)$

d. $Mg(g) \rightarrow Mg^+(g) + e^-$

141. C: $1s^22s^22p^2$; in the ground state, carbon has three different orbitals that hold electrons. The electrons in each orbital will have a separate peak on the PES spectrum, so we should have 3 different peaks in the spectrum. The peak at the lowest binding energy corresponds to the peak with the easiest electrons to remove. This will be the 2p electrons. The peak at the highest energy corresponds to the 1s orbital electrons since they will have the largest binding energy, and the middle peak corresponds to the 2s electrons. The area under each peak is directly related to the number of electrons in each orbital. Since all orbitals in carbon have 2 electrons, each peak area will be equal. This is shown on the PES spectrum for carbon; 3 peaks with equal area.

Alkali Metals

143. It should be potassium peroxide (K_2O_2) because K^+ ions are stable in ionic compounds. K^{2+} ions are not stable; the second ionization energy of K is very large compared to the first.

145. $$\nu = \frac{c}{\lambda} = \frac{2.9979 \times 10^8 \text{ m/s}}{455.5 \times 10^{-9} \text{ m}} = 6.582 \times 10^{14} \text{s}^{-1}$$

$$E = h\nu = 6.6261 \times 10^{-34} \text{ J s} \times 6.582 \times 10^{14} \text{s}^{-1} = 4.361 \times 10^{-19} \text{ J}$$

147. Yes; the ionization energy general trend is to decrease down a group, and the atomic radius trend is to increase down a group. The data in Table 7.7 confirm both general trends.

149. No; lithium metal is very reactive. It will react somewhat violently with water, making it completely unsuitable for human consumption. Lithium has a low first ionization energy, so it is more likely that the lithium prescribed will be in the form of a soluble lithium salt (a soluble ionic compound with Li^+ as the cation).

151. a. $6\ Li(s) + N_2(g) \rightarrow 2\ Li_3N(s)$ b. $2\ Rb(s) + S(s) \rightarrow Rb_2S(s)$

ChemWork Problems

153. $\lambda = \frac{c}{\nu} = \frac{2.998 \times 10^8\ m/s}{580. \times 10^3\ s^{-1}} = 516.9\ m$ (1 extra sig fig)

Minimum antenna height = ½ λ = ½(516.9 m) = 258 m

155. All oxygen family elements have ns^2np^4 valence electron configurations, so this nonmetal is from the oxygen family.

a. 2 + 4 = 6 valence electrons.

b. O, S, Se, and Te are the nonmetals from the oxygen family (Po is a metal).

c. Because oxygen family nonmetals form 2− charged ions in ionic compounds, K_2X would be the predicted formula, where X is the unknown nonmetal.

d. From the size trend, this element would have a smaller radius than barium.

e. From the ionization energy trend, this element would have a smaller ionization energy than fluorine.

157. $E = \frac{310\ kJ}{mol} \times \frac{1\ mol}{6.022 \times 10^{23}} = 5.15 \times 10^{-22}\ kJ = 5.15 \times 10^{-19}\ J$

$$E = \frac{hc}{\lambda},\ \lambda = \frac{hc}{E} = \frac{6.626 \times 10^{-34}\ J\ s \times 2.998 \times 10^8\ m/s}{5.15 \times 10^{-19}\ J} = 3.86 \times 10^{-7}\ m = 386\ nm$$

159. $60 \times 10^6\ km \times \frac{1000\ m}{km} \times \frac{1\ s}{3.00 \times 10^8\ m} = 200\ s$ (about 3 minutes)

161. $\Delta E = -R_H \left(\frac{1}{n_f^2} - \frac{1}{n_i^2}\right) = -2.178 \times 10^{-18}\ J \left(\frac{1}{2^2} - \frac{1}{6^2}\right) = -4.840 \times 10^{-19}\ J$

$$\lambda = \frac{hc}{|\Delta E|} = \frac{6.6261 \times 10^{-34}\ J\ s \times 2.9979 \times 10^8\ m/s}{4.840 \times 10^{-19}\ J} = 4.104 \times 10^{-7}\ m \times \frac{100\ cm}{m}$$

$$= 4.104 \times 10^{-5}\ cm$$

From the spectrum, $\lambda = 4.104 \times 10^{-5}$ cm is violet light, so the $n = 6$ to $n = 2$ visible spectrum line is violet.

163. Ionization from $n = 1$ corresponds to the transition $n_i = 1 \rightarrow n_f = \infty$, where $E_\infty = 0$.

$$\Delta E = E_\infty - E_1 = -E_1 = 2.178 \times 10^{-18}\left(\frac{1}{1^2}\right) = 2.178 \times 10^{-18}\ J$$

To ionize from $n = 3$, $\Delta E = E_\infty - E_3 = -E_3 = 2.178 \times 10^{-18}\left(\frac{1}{3^2}\right) = \frac{2.178 \times 10^{-18}\text{ J}}{9}$

The ionization energy for an H atom is 9 times larger from the ground state as compared to the ionization energy from $n = 3$.

165. a. False; an electron in a 2s orbital would be on average, closer to the nucleus than an electron in a 3s orbital.

b. True; the Bohr model is fundamentally incorrect.

c. True; d. False; these two terms come from different atomic structure models.

e. True; when $n = 3$, ℓ can equal 0 (s orbital), 1 (p orbitals), and 2 (d orbitals).

167. a. True for H only. b. True for all atoms. c. True for all atoms.

169. As: $1s^2 2s^2 2p^6 3s^2 3p^6 4s^2 3d^{10} 4p^3$

$\ell = 1$ are p orbitals. $2p^6$, $3p^6$, and $4p^3$ are the p orbitals used. So 15 electrons have $\ell = 1$.

The s, p, and d orbitals all have one of the orbitals with $m_\ell = 0$. There are four orbitals with $m_\ell = 0$ from the various s orbitals used, there are three orbitals with $m_\ell = 0$ from the various p orbitals used, and there is one orbital with $m_\ell = 0$ from the 3d orbitals used. We have a total of $4 + 3 + 1 = 8$ orbitals with $m_\ell = 0$. Seven of these orbitals are filled with 2 electrons, and the 4p orbitals are only half-filled. The number of electrons with $m_\ell = 0$ is $7 \times (2\ e^-) + 1 \times (1\ e^-) = 15$ electrons.

The p and d orbitals all have one of the degenerate orbitals with $m_\ell = 1$. There are 3 orbitals with $m_\ell = 1$ from the various p orbitals used and there is one orbital with $m_\ell = 1$ from the 3d orbitals used. We have a total of $3 + 1 = 4$ orbitals with $m_\ell = 1$. Three of these orbitals are filled with 2 electrons, and the 4p orbitals are only half-filled. The number of electrons with $m_\ell = 1$ is $3 \times (2\ e^-) + 1 \times (1\ e^-) = 7$ electrons.

171. The X^+ ion has $2 + 2 + 6 + 2 + 6 + 2 + 8 + 2 + 2 + 4 = 36$ electron. Rubidium has 37 electrons; the ion is Rb^+ which has 36 electrons.

173. From the radii trend, the smallest-size element (excluding hydrogen) would be the one in the most upper right corner of the periodic table. This would be O. The largest-size element would be the one in the most lower left of the periodic table. Thus K would be the largest. The ionization energy trend is the exact opposite of the radii trend. So K, with the largest size, would have the smallest ionization energy. From the general ionization energy trend, O should have the largest ionization energy. However, there is an exception to the general ionization energy trend between N and O. Due to this exception, N would have the largest ionization energy of the elements examined.

175. More favorable electron affinity: K, Br, and Se (follows general electron affinity trend). Higher ionization energy: K, Br, and Se (follows general ionization energy trend). Larger size: Cs, Te, and Ge (follows general atomic radius trend).

177. $E = hc/\lambda = h\nu$; photon energy and wavelength are inversely related, while photon energy and frequency are directly related. Statement b is true. From the ionization energy trend, S has a larger ionization energy than Se. Therefore, S will need the more energetic EMR photons to ionize it; the more energetic photon has the shorter wavelength. For answer a, Cl has a larger ionization energy than Cs, so Cl will require shorter wavelength EMR. For c, Mg requires the more energetic photon to ionize it. For d, P has the larger ionization energy, so it needs the higher frequency EMR. For e, all EMR travels at the same speed, so this statement is false.

179. For consecutive ionization energies, valence electrons are much easier to remove than inner core electrons. For element X, there is a huge jump in ionization energies between I_5 and I_6. This leads to the conclusion that element X has 5 valence electrons. When the sixth electron is removed, an inner core electron is removed. The big jump in ionization energies for element Y is between 4 and 5. Element Y has 4 valence electrons. Since the X and Y are from row 3, X = P and Y = Si.

181. Cd: $1s^2 2s^2 2p^6 3s^2 3p^6 4s^2 3d^{10} 4p^6 5s^2 4d^{10}$

a. Electrons with $\ell = 2$ are the d electrons. Since the 3d and 4d orbitals are filled, there are 20 electrons with $\ell = 2$ in cadmium.

b. The n = 4 electrons are the $4s^2$, $4p^6$, and $4d^{10}$ electrons, for a total of 18 electrons.

c. The various p and d orbitals in cadmium can have one of their orbitals with $m_\ell = -1$. So each of the 2p, 3p, 3d, 4p, and 4d orbitals have a $m_\ell = -1$ orbital. All of these various p and d orbitals are filled, so there are 2(5) = 10 electrons with $m_\ell = -1$.

d. All orbitals in cadmium are filled and one-half of the electrons in these orbitals contain an electron with $m_s = -1/2$. So there are ½(48) = 24 electrons with $m_s = -1/2$.

183. a.

$Na(g) \rightarrow Na^+(g) + e^-$	$I_1 = 495$ kJ
$Cl(g) + e^- \rightarrow Cl^-(g)$	EA = −348.7 kJ
$Na(g) + Cl(g) \rightarrow Na^+(g) + Cl^-g)$	ΔH = 146 kJ

b.

$Mg(g) \rightarrow Mg^+(g) + e^-$	$I_1 = 735$ kJ
$F(g) + e^- \rightarrow F^-(g)$	EA = −327.8 kJ
$Mg(g) + F(g) \rightarrow Mg^+(g) + F^-(g)$	ΔH = 407 kJ

c.

$Mg^+(g) \rightarrow Mg^{2+}(g) + e^-$	$I_2 = 1445$ kJ
$F(g) + e^- \rightarrow F^-(g)$	EA = −327.8 kJ
$Mg^+(g) + F(g) \rightarrow Mg^{2+}(g) + F^-(g)$	ΔH = 1117 kJ

d. Using parts b and c, we get:

$Mg(g) + F(g) \rightarrow Mg^+(g) + F^-(g)$	ΔH = 407 kJ
$Mg^+(g) + F(g) \rightarrow Mg^{2+}(g) + F^-(g)$	ΔH = 1117 kJ
$Mg(g) + 2\,F(g) \rightarrow Mg^{2+}(g) + 2\,F^-(g)$	ΔH = 1524 kJ

Challenge Problems

185. $\lambda = \dfrac{h}{mv}$, where m = mass and v = velocity; $v_{rms} = \sqrt{\dfrac{3RT}{m}}$, $\lambda = \dfrac{h}{m\sqrt{\dfrac{3RT}{m}}} = \dfrac{h}{\sqrt{3RTm}}$

For one atom, $R = \dfrac{8.3145 \text{ J}}{\text{K mol}} \times \dfrac{1 \text{ mol}}{6.022 \times 10^{23} \text{ atoms}} = 1.381 \times 10^{-23}$ J/K•atom

$2.31 \times 10^{-11} \text{ m} = \dfrac{6.626 \times 10^{-34} \text{ J s}}{\sqrt{m}\sqrt{3(1.381 \times 10^{-23})(373 \text{ K})}}$, $m = 5.32 \times 10^{-26}$ kg $= 5.32 \times 10^{-23}$ g

$\text{Molar mass} = \dfrac{5.32 \times 10^{-23} \text{ g}}{\text{atom}} \times \dfrac{6.022 \times 10^{23} \text{ atoms}}{\text{mol}} = 32.0$ g/mol

The atom is sulfur (S).

187. a. Because wavelength is inversely proportional to energy, the spectral line to the right of B (at a larger wavelength) represents the lowest possible energy transition; this is $n = 4$ to $n = 3$. The B line represents the next lowest energy transition, which is $n = 5$ to $n = 3$, and the A line corresponds to the $n = 6$ to $n = 3$ electronic transition.

b. Because this spectrum is for a one-electron ion, $E_n = -2.178 \times 10^{-18} \text{ J } (Z^2/n^2)$. To determine ΔE and, in turn, the wavelength of spectral line A, we must determine Z, the atomic number of the one electron species. Use spectral line B data to determine Z.

$$\Delta E_{5 \to 3} = -2.178 \times 10^{-18} \text{ J}\left(\frac{Z^2}{3^2} - \frac{Z^2}{5^2}\right) = -2.178 \times 10^{-18}\left(\frac{16 Z^2}{9 \times 25}\right)$$

$$E = \frac{hc}{\lambda} = \frac{6.6261 \times 10^{-34} \text{ J s}(2.9979 \times 10^8 \text{ m/s})}{142.5 \times 10^{-9} \text{ m}} = 1.394 \times 10^{-18} \text{ J}$$

Because an emission occurs, $\Delta E_{5 \to 3} = -1.394 \times 10^{-18}$ J.

$$\Delta E = -1.394 \times 10^{-18} \text{ J} = -2.178 \times 10^{-18} \text{ J}\left(\frac{16 Z^2}{9 \times 25}\right), \quad Z^2 = 9.001, \quad Z = 3; \text{ the ion is } Li^{2+}.$$

Solving for the wavelength of line A:

$$\Delta E_{6 \to 3} = -2.178 \times 10^{-18}(3)^2\left(\frac{1}{3^2} - \frac{1}{6^2}\right) = -1.634 \times 10^{-18} \text{ J}$$

$$\lambda = \frac{hc}{|\Delta E|} = \frac{6.6261 \times 10^{-34} \text{ J s}(2.9979 \times 10^8 \text{ m/s})}{1.634 \times 10^{-18} \text{ J}} = 1.216 \times 10^{-7} \text{ m} = 121.6 \text{ nm}$$

189. For one-electron species, $E_n = -R_H Z^2/n^2$. The ground state ionization energy is the energy change for the $n = 1 \to n = \infty$ transition. So:

ionization energy $= E_\infty - E_1 = -E_1 = R_H Z^2/n^2 = R_H Z^2$

$$\frac{4.72 \times 10^4 \text{ kJ}}{\text{mol}} \times \frac{1 \text{ mol}}{6.022 \times 10^{23}} \times \frac{1000 \text{ J}}{\text{kJ}} = 2.178 \times 10^{-18} \text{ J } (Z^2); \text{ solving: } Z = 6$$

Element 6 is carbon (X = carbon), and the charge for a one-electron carbon ion is 5+ (m = 5). The one-electron ion is C^{5+}.

191. For r = a_o and $\theta = 0°$ (Z = 1 for H):

$$\psi_{2p_z} = \frac{1}{4(2\pi)^{1/2}}\left(\frac{1}{5.29 \times 10^{-11}}\right)^{3/2} (1)\ e^{-1/2} \cos 0 = 1.57 \times 10^{14};\ \psi^2 = 2.46 \times 10^{28}$$

For r = a_o and $\theta = 90°$, $\psi_{2p_z} = 0$ since cos 90° = 0; $\psi^2 = 0$; there is no probability of finding an electron in the $2p_z$ orbital with $\theta = 0°$. As expected, the xy plane, which corresponds to $\theta = 0°$, is a node for the $2p_z$ atomic orbital.

193. a. 1st period: $p = 1,\ q = 1,\ r = 0,\ s = \pm1/2$ (2 elements)

2nd period: $p = 2,\ q = 1,\ r = 0,\ s = \pm1/2$ (2 elements)

3rd period: $p = 3,\ q = 1,\ r = 0,\ s = \pm1/2$ (2 elements)

$p = 3,\ q = 3,\ r = -2,\ s = \pm1/2$ (2 elements)
$p = 3,\ q = 3,\ r = 0,\ s = \pm1/2$ (2 elements)
$p = 3,\ q = 3,\ r = +2,\ s = \pm1/2$ (2 elements)

4th period: $p = 4$; q and r values are the same as with $p = 3$ (8 total elements)

1							2
3							4
5	6	7	8	9	10	11	12
13	14	15	16	17	18	19	20

b. Elements 2, 4, 12, and 20 all have filled shells and will be least reactive.

c. Draw similarities to the modern periodic table.

XY could be X^+Y^-, $X^{2+}Y^{2-}$, or $X^{3+}Y^{3-}$. Possible ions for each are:

X^+ could be elements 1, 3, 5, or 13; Y^- could be 11 or 19.

X^{2+} could be 6 or 14; Y^{2-} could be 10 or 18.

X^{3+} could be 7 or 15; Y^{3-} could be 9 or 17.

Note: X^{4+} and Y^{4-} ions probably won't form.

XY_2 will be $X^{2+}(Y^-)_2$; See above for possible ions.

X_2Y will be $(X^+)_2Y^{2-}$ See above for possible ions.

XY_3 will be $X^{3+}(Y^-)_3$; See above for possible ions.

X_2Y_3 will be $(X^{3+})_2(Y^{2-})_3$; See above for possible ions.

d. $p = 4,\ q = 3,\ r = -2,\ s = \pm 1/2$ (2 electrons)
$p = 4,\ q = 3,\ r = 0,\ s = \pm 1/2$ (2 electrons)
$p = 4,\ q = 3,\ r = +2,\ s = \pm 1/2$ (2 electrons)

A total of 6 electrons can have $p = 4$ and $q = 3$.

e. $p = 3,\ q = 0,\ r = 0$; this is not allowed; q must be odd. Zero electrons can have these quantum numbers.

f. $p = 6,\ q = 1,\ r = 0,\ s = \pm 1/2$ (2 electrons)

$p = 6,\ q = 3,\ r = -2, 0, +2;\ s = \pm 1/2$ (6 electrons)

$p = 6,\ q = 5,\ r = -4, -2, 0, +2, +4;\ s = \pm 1/2$ (10 electrons)

Eighteen electrons can have $p = 6$.

195. The ratios for Mg, Si, P, Cl, and Ar are about the same. However, the ratios for Na, Al, and S are higher. For Na, the second ionization energy is extremely high because the electron is taken from $n = 2$ (the first electron is taken from $n = 3$). For Al, the first electron requires a bit less energy than expected by the trend due to the fact it is a 3p electron versus a 3s electron. For S, the first electron requires a bit less energy than expected by the trend due to electrons being paired in one of the p orbitals.

197. a. Assuming the Bohr model applies to the 1s electron, $E_{1s} = -R_H Z^2/n^2 = -R_H (Z_{eff})^2$, where $n = 1$. Ionization energy $= E_\infty - E_{1s} = 0 - E_{1s} = R_H (Z_{eff})^2$.

$$\frac{2.462 \times 10^6\ \text{kJ}}{\text{mol}} \times \frac{1\ \text{mol}}{6.0221 \times 10^{23}} \times \frac{1000\ \text{J}}{\text{kJ}} = 2.178 \times 10^{-18}\ \text{J}\ (Z_{eff})^2,\ Z_{eff} = 43.33$$

b. Silver is element 47, so $Z = 47$ for silver. Our calculated Z_{eff} value is less than 47. Electrons in other orbitals can penetrate the 1s orbital. Thus a 1s electron can be slightly shielded from the nucleus by these penetrating electrons, giving a Z_{eff} close to but less than Z.

CHAPTER 8

BONDING: GENERAL CONCEPTS

Questions

21. a. This diagram represents a polar covalent bond as in HCl. In a polar covalent bond, there is an electron rich region (indicated by the red color) and an electron poor region (indicated by the blue color). In HCl, the more electronegative Cl atom (on the red side of the diagram) has a slightly greater ability to attract the bonding electrons than does H (on the blue side of the diagram), which in turn produces a dipole moment.

    b. This diagram represents an ionic bond as in NaCl. Here, the electronegativity differences between the Na and Cl are so great that the valence electron of sodium is transferred to the chlorine atom. This results in the formation of a cation, an anion, and an ionic bond.

    c. This diagram represents a pure covalent bond as in H_2. Both atoms attract the bonding electrons equally, so there is no bond dipole formed. This is illustrated in the electrostatic potential diagram as the various red and blue colors are equally distributed about the molecule. The diagram shows no one region that is red nor one region that is blue (there is no specific partial negative end and no specific partial positive end), so the molecule is nonpolar.

23. Of the compounds listed, P_2O_5 is the only compound containing only covalent bonds. $(NH_4)_2SO_4$, $Ca_3(PO_4)_2$, K_2O, and KCl are all compounds composed of ions, so they exhibit ionic bonding. The polyatomic ions in $(NH_4)_2SO_4$ are NH_4^+ and SO_4^{2-}. Covalent bonds exist between the N and H atoms in NH_4^+ and between the S and O atoms in SO_4^{2-}. Therefore, $(NH_4)_2SO_4$ contains both ionic and covalent bonds. The same is true for $Ca_3(PO_4)_2$. The bonding is ionic between the Ca^{2+} and PO_4^{3-} ions and covalent between the P and O atoms in PO_4^{3-}. Therefore, $(NH_4)_2SO_4$ and $Ca_3(PO_4)_2$ are the compounds with both ionic and covalent bonds.

25. Electronegativity increases left to right across the periodic table and decreases from top to bottom. Hydrogen has an electronegativity value between B and C in the second row and identical to P in the third row. Going further down the periodic table, H has an electronegativity value between As and Se (row 4) and identical to Te (row 5). It is important to know where hydrogen fits into the electronegativity trend, especially for rows 2 and 3. If you know where H fits into the trend, then you can predict bond dipole directions for nonmetals bonded to hydrogen.

27. For ions, concentrate on the number of protons and the number of electrons present. The species whose nucleus holds the electrons most tightly will be smallest. For example, anions are larger than the neutral atom. The anion has more electrons held by the same number of protons in the nucleus. These electrons will not be held as tightly, resulting in a bigger size for the anion as

compared to the neutral atom. For isoelectronic ions, the same number of electrons are held by different numbers of protons in the various ions. The ion with the most protons holds the electrons tightest and is smallest in size.

29. Fossil fuels contain a lot of carbon and hydrogen atoms. Combustion of fossil fuels (reaction with O_2) produces CO_2 and H_2O. Both these compounds have very strong bonds. Because stronger product bonds are formed than reactant bonds broken, combustion reactions are very exothermic.

31. $:C{\equiv}O:$ Carbon: FC = 4 – 2 – 1/2(6) = –1; oxygen: FC = 6 – 2 – 1/2(6) = +1

Electronegativity predicts the opposite polarization. The two opposing effects seem to partially cancel to give a much less polar molecule than expected.

Exercises

Chemical Bonds and Electronegativity

33. Using the periodic table, the general trend for electronegativity is:

(1) Increase as we go from left to right across a period

(2) Decrease as we go down a group

Using these trends, the expected orders are:

a. C < N < O b. Se < S < Cl c. Sn < Ge < Si d. Tl < Ge < S

35. The most polar bond will have the greatest difference in electronegativity between the two atoms. From positions in the periodic table, we would predict:

a. Ge–F b. P–Cl c. S–F d. Ti–Cl

37. The general trends in electronegativity used in Exercises 33 and 35 are only rules of thumb. In this exercise, we use experimental values of electronegativities and can begin to see several exceptions. The order of EN from Figure 8.3 is:

a. C (2.5) < N (3.0) < O (3.5) same as predicted

b. Se (2.4) < S (2.5) < Cl (3.0) same

c. Si = Ge = Sn (1.8) different

d. Tl (1.8) = Ge (1.8) < S (2.5) different

Most polar bonds using actual EN values:

a. Si–F and Ge–F have equal polarity (Ge–F predicted).

b. P–Cl (same as predicted)

c. S–F (same as predicted) d. Ti–Cl (same as predicted)

39. Use the electronegativity trend to predict the partial negative end and the partial positive end of the bond dipole (if there is one). To do this, you need to remember that H has electronegativity between B and C and identical to P. Answers b, d, and e are incorrect. For d (Br_2), the bond between two Br atoms will be a pure covalent bond, where there is equal sharing of the bonding electrons, and no dipole moment exists. For b and e, the bond polarities are reversed. In Cl–I, the more electronegative Cl atom will be the partial negative end of the bond dipole, with I having the partial positive end. In O–P, the more electronegative oxygen will be the partial negative end of the bond dipole, with P having the partial positive end. In the following, we used arrows to indicate the bond dipole. The arrow always points to the partial negative end of a bond dipole (which always is the most electronegative atom in the bond).

41. Bonding between a metal and a nonmetal is generally ionic. Bonding between two nonmetals is covalent, and in general, the bonding between two different nonmetals is usually polar covalent. When two different nonmetals have very similar electronegativities, the bonding is pure covalent or just covalent.

a. ionic b. covalent c. polar covalent

d. ionic e. polar covalent f. covalent

43. In general bonds between a metal and a nonmetal are usually ionic and bonds between two nonmetals are covalent. MgS is ionic, and the others are covalent. The electronegativities of C and H are close to each other, so C-H bonds are usually considered nonpolar (like F_2). Of the remaining covalent bonds, the electronegativity difference between As and F is larger than the electronegativity difference between Se-O (from the electronegativity trend). So As-F is the most polar covalent bond of the bonds listed.

45. Electronegativity values increase from left to right across the periodic table. The order of electronegativities for the atoms from smallest to largest electronegativity will be H = P < C < N < O < F. The most polar bond will be F–H since it will have the largest difference in electronegativities, and the least polar bond will be P–H since it will have the smallest difference in electronegativities ($\Delta EN = 0$). The order of the bonds in decreasing polarity will be F–H > O–H > N–H > C–H > P–H.

47. A permanent dipole moment exists in a molecule if the molecule has one specific area with a partial negative end (a red end in an electrostatic potential diagram) and a different specific region with a partial positive end (a blue end in an electrostatic potential diagram). If the blue and red colors are equally distributed in the electrostatic potential diagrams, then no permanent dipole exists.

a. Has a permanent dipole. b. Has no permanent dipole.

c. Has no permanent dipole. d. Has a permanent dipole.

e. Has no permanent dipole. f. Has no permanent dipole.

Ions and Ionic Compounds

49. Al^{3+}: $[He]2s^22p^6$; Ba^{2+}: $[Kr]5s^24d^{10}5p^6$; Se^{2-}: $[Ar]4s^23d^{10}4p^6$; I^-: $[Kr]5s^24d^{10}5p^6$

51. a. Li^+ and N^{3-} are the expected ions. The formula of the compound would be Li_3N (lithium nitride).

    b. Ga^{3+} and O^{2-}; Ga_2O_3, gallium(III) oxide or gallium oxide

    c. Rb^+ and Cl^-; RbCl, rubidium chloride d. Ba^{2+} and S^{2-}; BaS, barium sulfide

53. a. Mg^{2+}: $1s^22s^22p^6$; K^+: $1s^22s^22p^63s^23p^6$; Al^{3+}: $1s^22s^22p^6$

    b. N^{3-}, O^{2-}, and F^-: $1s^22s^22p^6$; Te^{2-}: $[Kr]5s^24d^{10}5p^6$

55. a. Sc^{3+}: [Ar] b. Te^{2-}: [Xe] c. Ce^{4+}: [Xe] and Ti^{4+}: [Ar] d. Ba^{2+}: [Xe]

    All these ions have the noble gas electron configuration shown in brackets.

57. a. Na^+ has 10 electrons. F^-, O^{2-}, and N^{3-} are some possible anions also having 10 electrons.

    b. Ca^{2+} has 18 electrons. Cl^-, S^{2-}, and P^{3-} also have 18 electrons.

    c. Al^{3+} has 10 electrons. F^-, O^{2-}, and N^{3-} also have 10 electrons.

    d. Rb^+ has 36 electrons. Br^-, Se^{2-}, and As^{3-} also have 36 electrons.

59. Se^{2-}, Br^-, Rb^+, Sr^{2+}, Y^{3+}, and Zr^{4+} are some ions that are isoelectronic with Kr (36 electrons). In terms of size, the ion with the most protons will hold the electrons tightest and will be the smallest. The size trend is:

$$Zr^{4+} < Y^{3+} < Sr^{2+} < Rb^+ < Br^- < Se^{2-}$$

smallest largest

61. In each group, the smallest species is O^-, Mg^{2+}, and Mg^{2+}. As electrons are added to a species, size increases. Conversely, as electrons are removed from a species, size decreases. Between Mg^{2+} and O^{2-}, both have 10 electrons. Mg^{2+} has 12 protons compared to 8 protons for O^{2-}. Mg^{2+} with the most protons will be smallest.

63. a. $Fe > Fe^{2+} > Fe^{3+}$ b. $Au^+ > Ag^+ > Cu^+$ c. $S^{2-} > S^- > S$

    d. $Br^- > Cl^- > F^-$ e. $Se^{2-} > Rb^+ > Yb^{3+}$

    For answer a, as electrons are removed from an atom, size decreases. Answers b and d follow the radius trend. For answer c, as electrons are added to an atom, size increases. Answer e follows the trend for an isoelectronic series; i.e., the smallest ion has the most protons.

65. Lattice energy is proportional to $-Q_1Q_2/r$, where Q is the charge of the ions and r is the distance between the centers of the ions. The more negative the lattice energy, the more stable the ionic

compound. So greater charged ions as well as smaller sized ions lead to more negative lattice energy values and more stable ionic compounds.

a. NaCl; Na^+ is smaller than K^+.

b. LiF; F^- is smaller than Cl^-.

c. MgO; O^{2-} has a greater charge than OH^-.

d. $Fe(OH)_3$; Fe^{3+} has a greater charge than Fe^{2+}.

e. Na_2O; O^{2-} has a greater charge than Cl^-.

f. MgO; both ions are smaller in MgO.

67.

$K(s) \rightarrow K(g)$	$\Delta H = 90.$ kJ (sublimation)
$K(g) \rightarrow K^+(g) + e^-$	$\Delta H = 419$ kJ (ionization energy)
$1/2\ Cl_2(g) \rightarrow Cl(g)$	$\Delta H = 239/2$ kJ (bond energy)
$Cl(g) + e^- \rightarrow Cl^-(g)$	$\Delta H = -349$ kJ (electron affinity)
$K^+(g) + Cl^-(g) \rightarrow KCl(s)$	$\Delta H = -690.$ kJ (lattice energy)
$K(s) + 1/2\ Cl_2(g) \rightarrow KCl(s)$	$\Delta H_f^\circ = -411$ kJ/mol

69. From the data given, it takes less energy to produce $Mg^+(g) + O^-(g)$ than to produce $Mg^{2+}(g) + O^{2-}(g)$. However, the lattice energy for $Mg^{2+}O^{2-}$ will be much more exothermic than that for Mg^+O^- due to the greater charges in $Mg^{2+}O^{2-}$. The favorable lattice energy term dominates, and $Mg^{2+}O^{2-}$ forms.

71. Use Figure 8.11 as a template for this problem.

$Li(s) \rightarrow Li(g)$	$\Delta H_{sub} = ?$
$Li(g) \rightarrow Li^+(g) + e^-$	$\Delta H = 520.$ kJ
$1/2\ I_2(g) \rightarrow I(g)$	$\Delta H = 151/2$ kJ
$I(g) + e^- \rightarrow I^-(g)$	$\Delta H = -295$ kJ
$Li^+(g) + I^-(g) \rightarrow LiI(s)$	$\Delta H = -753$ kJ
$Li(s) + 1/2\ I_2(g) \rightarrow LiI(s)$	$\Delta H = -292$ kJ

$\Delta H_{sub} + 520. + 151/2 - 295 - 753 = -292,\ \ \Delta H_{sub} = 161$ kJ

73. Ca^{2+} has a greater charge than Na^+, and Se^{2-} is smaller than Te^{2-}. The effect of charge on the lattice energy is greater than the effect of size. We expect the trend from most exothermic lattice energy to least exothermic to be:

CaSe	>	CaTe	>	Na_2Se	>	Na_2Te	
(−2862)		(−2721)		(−2130)		(−2095)	This is what we observe.

Bond Energies

75. When the overall reactant bond strengths are stronger than the product bond strengths, the reaction will be endothermic. This is what we have with this reaction; the bond strengths of the N_2 and O_2 bonds are overall stronger than the bond strengths of two NO bonds.

77. a. H—H + Cl—Cl ⟶ 2 H—Cl

Bonds broken:

1 H–H (432 kJ/mol)
1 Cl–Cl (239 kJ/mol)

Bonds formed:

2 H–Cl (427 kJ/mol)

$\Delta H = \Sigma D_{\text{broken}} - \Sigma D_{\text{formed}}$, $\Delta H = 432 \text{ kJ} + 239 \text{ kJ} - 2(427) \text{ kJ} = -183 \text{ kJ}$

b. 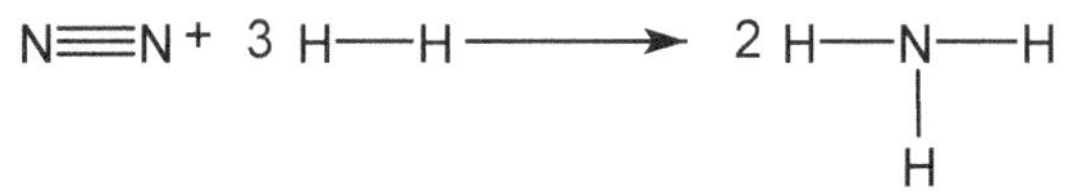

Bonds broken:

1 N≡N (941 kJ/mol)
3 H–H (432 kJ/mol)

Bonds formed:

6 N–H (391 kJ/mol)

$\Delta H = 941 \text{ kJ} + 3(432) \text{ kJ} - 6(391) \text{ kJ} = -109 \text{ kJ}$

79. 

Bonds broken: 1 C–N (305 kJ/mol) Bonds formed: 1 C–C (347 kJ/mol)

$\Delta H = \Sigma D_{\text{broken}} - \Sigma D_{\text{formed}}$, $\Delta H = 305 - 347 = -42 \text{ kJ}$

Note: Sometimes some of the bonds remain the same between reactants and products. To save time, only break and form bonds that are involved in the reaction.

81. 2 H—O—H + 2 F—F ⟶ 4 H—F + O=O

Bonds broken:

4 O–H (467 kJ/mol)
2 F–F (154 kJ/mol)

Bonds formed:

4 H–F (565 kJ/mol)
1 O=O (495 kJ/mol)

$\Delta H = 4(467) + 2(154) - [4(565) + 495] = -579 \text{ kJ}$

83. H–C≡C–H + 5/2 O=O → 2 O=C=O + H–O–H

Bonds broken:

2 C–H (413 kJ/mol)
1 C≡C (839 kJ/mol)
5/2 O=O (495 kJ/mol)

Bonds formed:

2 × 2 C=O (799 kJ/mol)
2 O–H (467 kJ/mol)

$\Delta H = 2(413 \text{ kJ}) + 839 \text{ kJ} + 5/2\ (495 \text{ kJ}) - [4(799 \text{ kJ}) + 2(467 \text{ kJ})] = -1228 \text{ kJ}$

85.

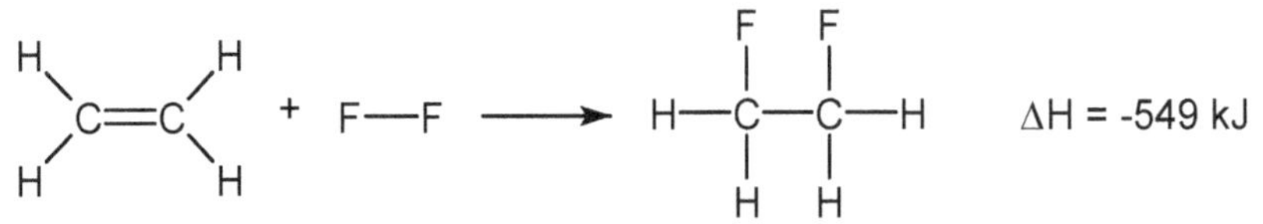

Bonds broken:

1 C=C (614 kJ/mol)
1 F–F (154 kJ/mol)

Bonds formed:

1 C–C (347 kJ/mol)
2 C–F (D_{CF} = C–F bond energy)

$\Delta H = -549\text{ kJ} = 614\text{ kJ} + 154\text{ kJ} - [347\text{ kJ} + 2D_{CF}]$, $2D_{CF} = 970.$, $D_{CF} = 485$ kJ/mol

87. a. $\Delta H° = 2\Delta H^{o}_{f,\,HCl} = 2\text{ mol}(-92\text{ kJ/mol}) = -184\text{ kJ}$ (–183 kJ from bond energies)

b. $\Delta H° = 2\Delta H^{o}_{f,\,NH_3} = 2\text{ mol}(-46\text{ kJ/mol}) = -92\text{ kJ}$ (–109 kJ from bond energies)

Comparing the values for each reaction, bond energies seem to give a reasonably good estimate for the enthalpy change of a reaction. The estimate is especially good for gas phase reactions.

89. a. Using SF_4 data: $SF_4(g) \rightarrow S(g) + 4\,F(g)$

$$\Delta H° = 4D_{SF} = 278.8 + 4\,(79.0) - (-775) = 1370.\text{ kJ}$$

$$D_{SF} = \frac{1370.\text{ kJ}}{4\text{ mol SF bonds}} = 342.5\text{ kJ/mol} = \text{S–F bond energy}$$

Using SF_6 data: $SF_6(g) \rightarrow S(g) + 6\,F(g)$

$$\Delta H° = 6D_{SF} = 278.8 + 6\,(79.0) - (-1209) = 1962\text{ kJ}$$

$$D_{SF} = \frac{1962\text{ kJ}}{6\text{ mol}} = 327.0\text{ kJ/mol} = \text{S–F bond energy}$$

b. The S–F bond energy in Table 8.5 is 327 kJ/mol. The value in the table was based on the S–F bond in SF_6.

c. S(g) and F(g) are not the most stable forms of the elements at 25°C and 1 atm. The most stable forms are $S_8(s)$ and $F_2(g)$; $\Delta H^{o}_{f} = 0$ for these two species.

91. H–H(g) + O=O(g) → H–O–O–H(g) $\Delta H_{rxn} = \Delta H^{o}_{f,\,H_2O_2}$

Bonds broken:

1 H–H (432 kJ/mol)
1 O=O (495 kJ/mol)

Bonds formed:

2 H–O (467 kJ/mol)
1 O–O (146 kJ/mol)

$\Delta H^{o}_{f,\,H_2O_2} = 432 + 495 - [2(467) + 146] = -153$ kJ/mol

Lewis Structures and Resonance

93. Drawing Lewis structures is mostly trial and error. However, the first two steps are always the same. These steps are (1) count the valence electrons available in the molecule/ion, and (2) attach all atoms to each other with single bonds (called the skeletal structure). Unless noted otherwise, the atom listed first is assumed to be the atom in the middle, called the central atom, and all other atoms in the formula are attached to this atom. The most notable exceptions to the rule are formulas that begin with H, e.g., H_2O, H_2CO, etc. Hydrogen can never be a central atom since this would require H to have more than two electrons. In these compounds, the atom listed second is assumed to be the central atom.

After counting valence electrons and drawing the skeletal structure, the rest is trial and error. We place the remaining electrons around the various atoms to satisfy the octet rule (or duet rule for H). Keep in mind that practice makes perfect. After practicing, you can (and will) become very adept at drawing Lewis structures.

a. F_2 has 2(7) = 14 valence electrons.

Skeletal structure Lewis structure

b. O_2 has 2(6) = 12 valence electrons.

Skeletal structure Lewis structure

c. CO has 4 + 6 = 10 valence electrons.

Skeletal structure Lewis structure

d. CH_4 has 4 + 4(1) = 8 valence electrons.

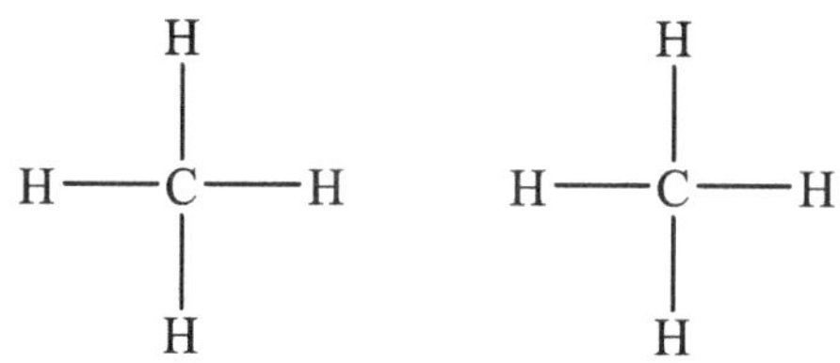

Skeletal structure Lewis structure

e. NH_3 has 5 + 3(1) = 8 valence electrons.

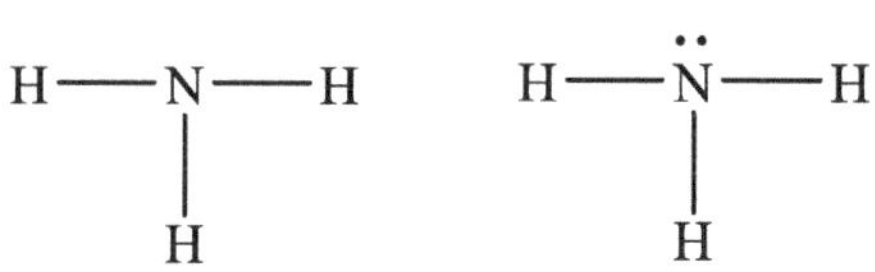

Skeletal structure Lewis structure

f. H_2O has 2(1) + 6 = 8 valence electrons.

Skeletal structure Lewis structure

g. HF has 1 + 7 = 8 valence electrons.

Skeletal structure

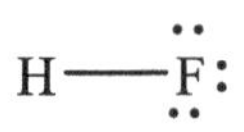

Lewis structure

95. The Lewis structure uses 14 e^- valence electrons and the formula of the ion is X_2^{2-}. Let's set-up an equation for determining the total number of valence electrons in X_2^{2-}:

$$14 = 2(x) + 2, x = 6 \text{ number of valence electrons of X}$$

Element X is a group 6A nonmetal. Possible identities are O, S, Se, and Te. O_2^{2-} is most likely.

97. Drawing Lewis structures is mostly trial and error. However, the first two steps are always the same. These steps are (1) count the valence electrons available in the molecule/ion, and (2) attach all atoms to each other with single bonds (called the skeletal structure). Unless noted otherwise, the atom listed first is assumed to be the atom in the middle, called the central atom, and all other atoms in the formula are attached to this atom. The most notable exceptions to the rule are formulas that begin with H, e.g., H_2O, H_2CO, etc. Hydrogen can never be a central atom since this would require H to have more than two electrons. In these compounds, the atom listed second is assumed to be the central atom.

After counting valence electrons and drawing the skeletal structure, the rest is trial and error. We place the remaining electrons around the various atoms in an attempt to satisfy the octet rule (or duet rule for H).

a. CCl_4 has 4 + 4(7) = 32 valence electrons.

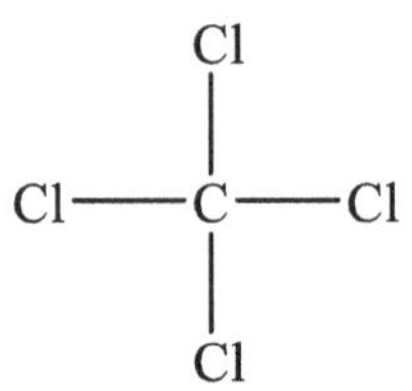

Skeletal structure

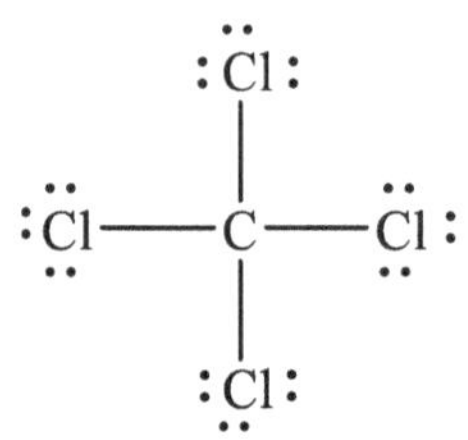

Lewis structure

b. NCl_3 has 5 + 3(7) = 26 valence electrons.

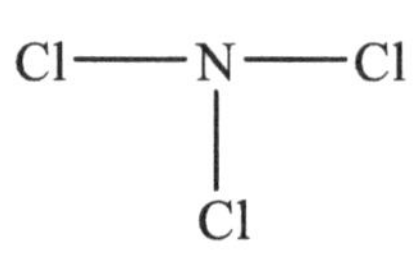

Skeletal structure

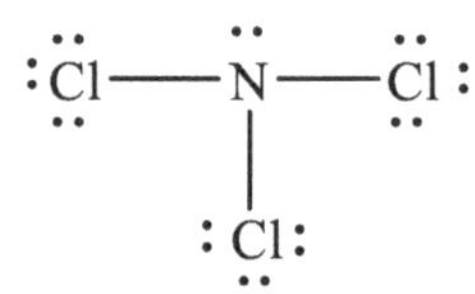

Lewis structure

c. $SeCl_2$ has 6 + 2(7) = 20 valence electrons.

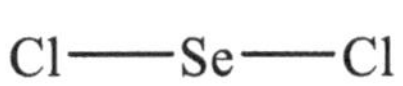

Skeletal structure

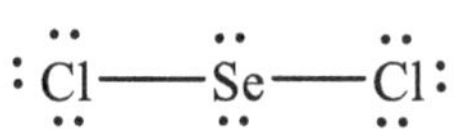

Lewis structure

d. ICl has 7 + 7 = 14 valence electrons.

Skeletal structure

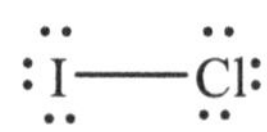

Lewis structure

99. BeH_2, $2 + 2(1) = 4$ valence electrons

BH_3, $3 + 3(1) = 6$ valence electrons

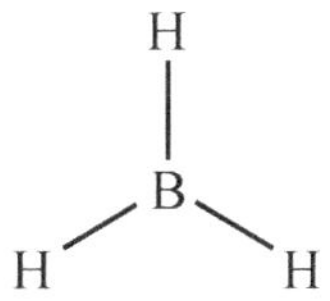

101. PF_5, $5 + 5(7) = 40$ valence electrons

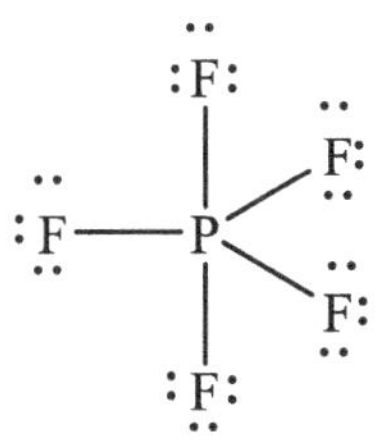

SF_4, $6 + 4(7) = 34\ e^-$

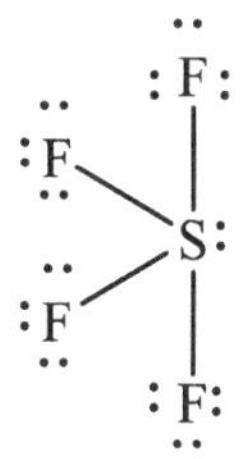

ClF_3, $7 + 3(7) = 28\ e^-$

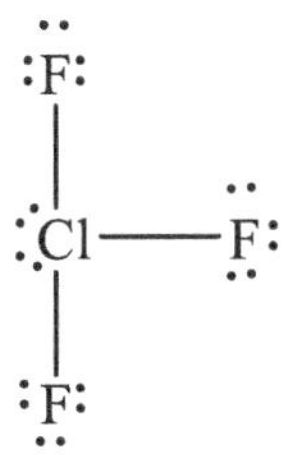

Br_3^-, $3(7) + 1 = 22\ e^-$

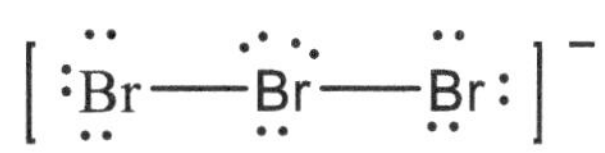

Row 3 and heavier nonmetals can have more than 8 electrons around them when they have to. Row 3 and heavier elements have empty d orbitals that are close in energy to valence s and p orbitals. These empty d orbitals can accept extra electrons.

For example, P in PF_5 has its five valence electrons in the 3s and 3p orbitals. These s and p orbitals have room for three more electrons, and if it must, P can use the empty 3d orbitals for any electrons above 8.

103. BH_3 only has 6 valence electrons, so it must be an exception to the octet rule for the central B atom. NS_2 has an odd number of valence electrons ($17\ e^-$), so it must be an exception to the octet rule for one of the atoms in the compound. For the other two compounds, here are the Lewis structures:

BrF_3, $7 + 3(7) = 28\ e^-$

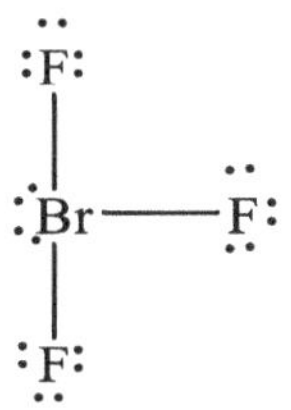

OCl_2, $6 + 2(7) = 20$

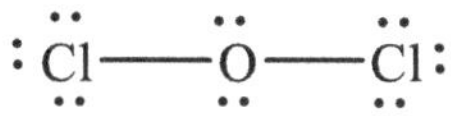

BrF_3 must also be an exception to the octet rule. With 28 valence electrons, it is impossible to satisfy the octet rule for all atom in the compound. OCl_2 satisfies the octet rule.

105. The Lewis structure has 34 valence electrons, and the formula of the ion is $EF_2O_2^-$. Let x = the number of valence electrons of E:

$34 = x + 2(7) + 2(6) + 1, x = 7$

E is a halogen. In addition, E must be a row 3 or heavier element since this ion has more than eight electrons around the central E atom (row 2 elements never have more than eight electrons around them). Some possible halogens for E are Cl, Br, I, and At.

107. a. NO_2^- has 5 + 2(6) + 1 = 18 valence electrons. The skeletal structure is O–N–O.

To get an octet about the nitrogen and only use 18 e^-, we must form a double bond to one of the oxygen atoms.

Because there is no reason to have the double bond to a particular oxygen atom, we can draw two resonance structures. Each Lewis structure uses the correct number of electrons and satisfies the octet rule, so each is a valid Lewis structure. Resonance structures occur when you have multiple bonds that can be in various positions. We say the actual structure is an average of these two resonance structures.

NO_3^- has 5 + 3(6) + 1 = 24 valence electrons. We can draw three resonance structures for NO_3^-, with the double bond rotating among the three oxygen atoms.

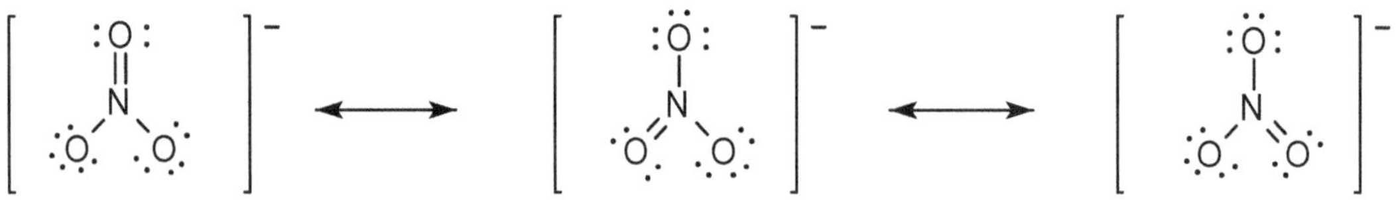

N_2O_4 has 2(5) + 4(6) = 34 valence electrons. We can draw four resonance structures for N_2O_4.

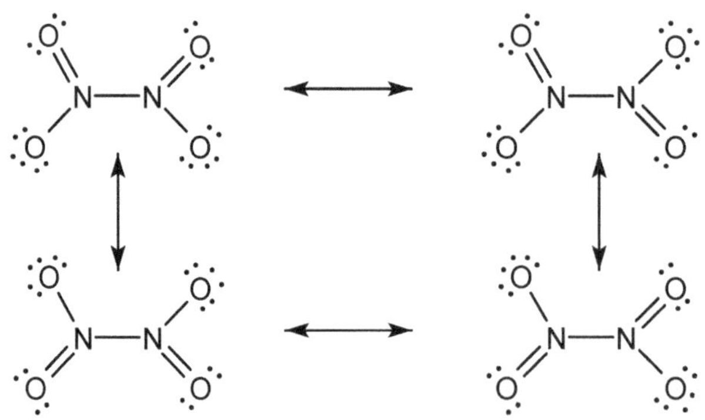

b. OCN^- has 6 + 4 + 5 + 1 = 16 valence electrons. We can draw three resonance structures for OCN^-.

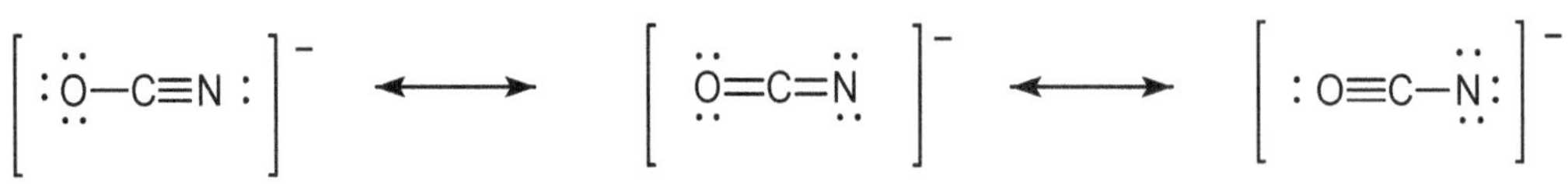

SCN^- has 6 + 4 + 5 + 1 = 16 valence electrons. Three resonance structures are possible.

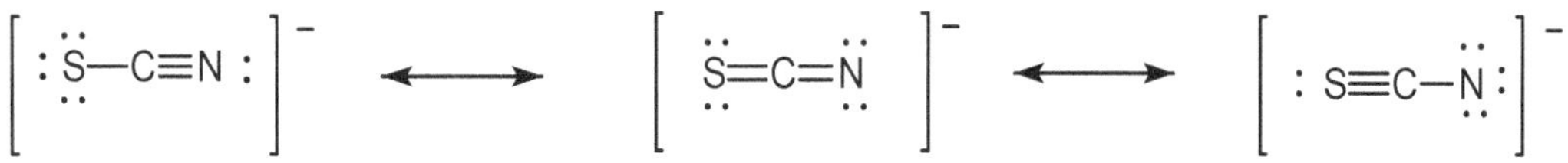

N_3^- has 3(5) + 1 = 16 valence electrons. As with OCN^- and SCN^-, three different resonance structures can be drawn.

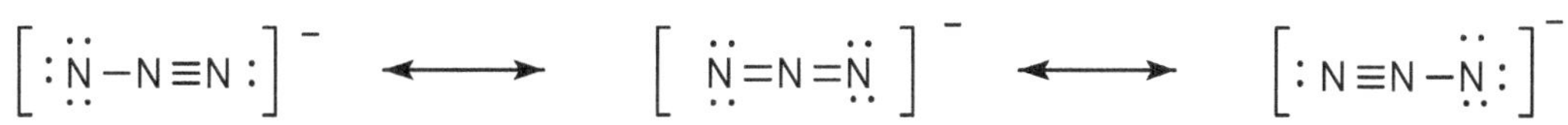

109. Benzene has 6(4) + 6(1) = 30 valence electrons. Two resonance structures can be drawn for benzene. The actual structure of benzene is an average of these two resonance structures; i.e., all carbon-carbon bonds are equivalent with a bond length and bond strength somewhere between a single and a double bond.

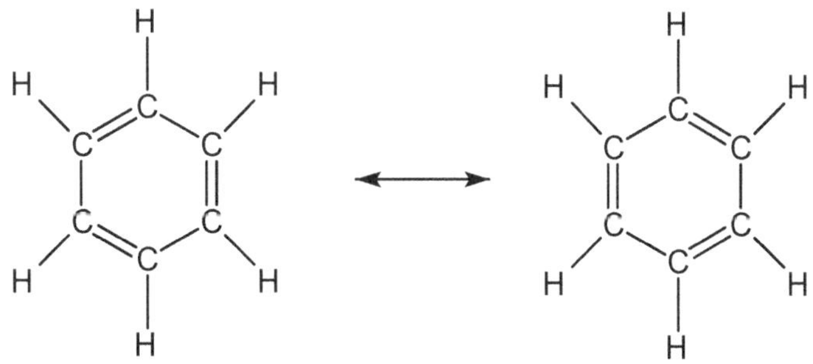

111. We will use a hexagon to represent the six-member carbon ring, and we will omit the four hydrogen atoms and the three lone pairs of electrons on each chlorine. If no resonance existed, we could draw four different molecules:

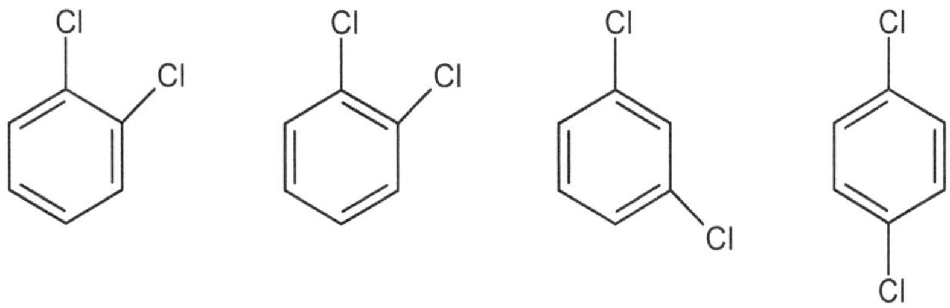

If the double bonds in the benzene ring exhibit resonance, then we can draw only three different dichlorobenzenes. The circle in the hexagon represents the delocalization of the three double bonds in the benzene ring (see Exercise 109).

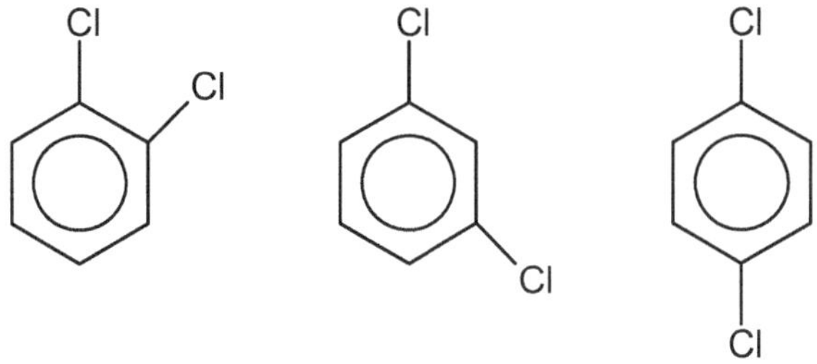

With resonance, all carbon-carbon bonds are equivalent. We can't distinguish between a single and double bond between adjacent carbons that have a chlorine attached. That only three isomers are observed supports the concept of resonance.

113. CH_3NCO has 4 + 3(1) + 5 + 4 + 6 = 22 valence electrons. Three resonance structures can be drawn for methyl isocyanate.

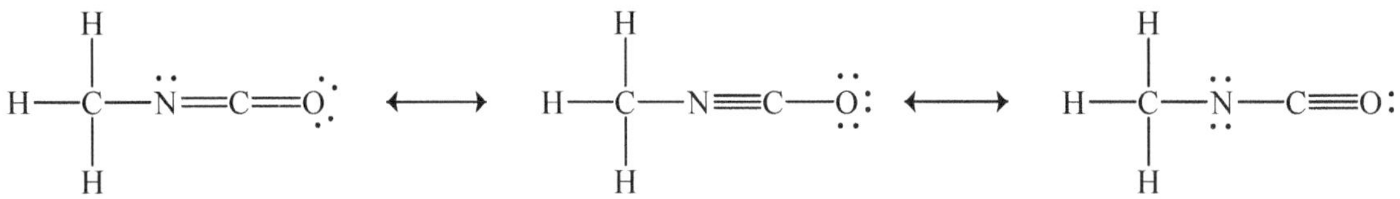

115. The Lewis structures for the various species are:

CO (10 e^-):  Triple bond between C and O.

CO_2 (16 e^-): 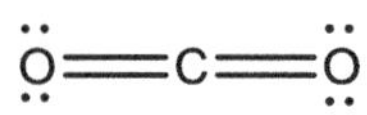 Double bond between C and O.

CO_3^{2-} (24 e^-): 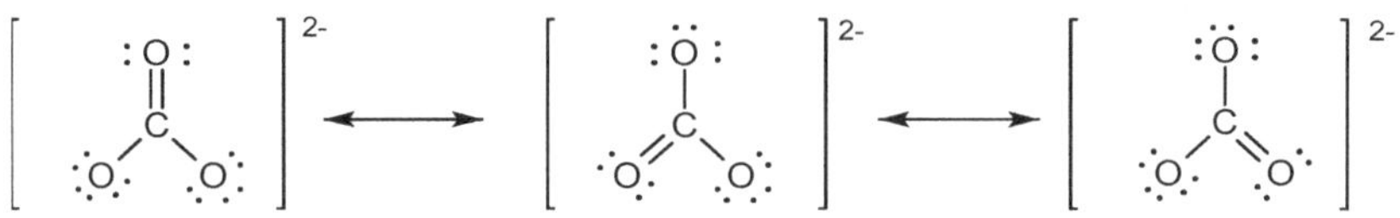

Average of 1 1/3 bond between C and O in CO_3^{2-}.

CH_3OH (14 e^-): 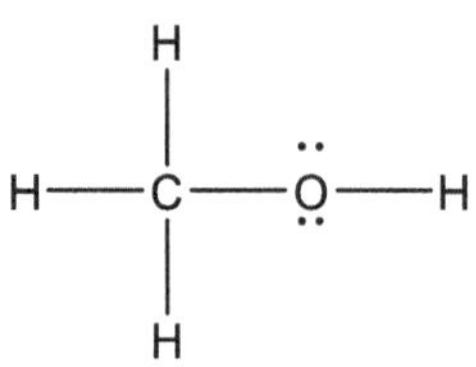 Single bond between C and O.

As the number of bonds increases between two atoms, bond strength increases, and bond length decreases. Then:

Longest → shortest C – O bond: $CH_3OH > CO_3^{2-} > CO_2 > CO$

Weakest → strongest C – O bond: $CH_3OH < CO_3^{2-} < CO_2 < CO$

Formal Charge

117. BF_3 has 3 + 3(7) = 24 valence electrons. The two Lewis structures to consider are:

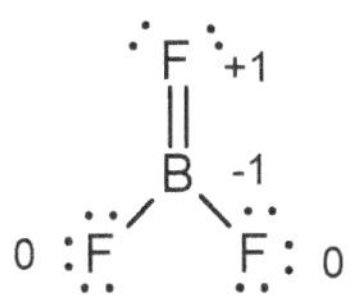

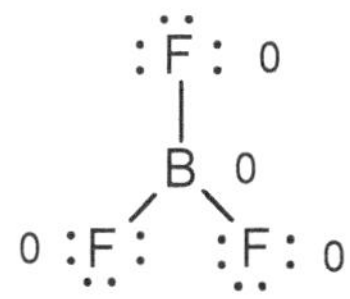

The formal charges for the various atoms are assigned in the Lewis structures. Formal charge = number of valence electrons on free atom – number of lone pair electrons on atoms – 1/2 (number of shared electrons of atom). For B in the first Lewis structure, formal charge (FC) = 3 – 0 – 1/2(8) = –1. For F in the first structure with the double bond, FC = 7 – 4 – 1/2(4) = +1. The others all have a formal charge equal to zero [FC = 7 – 6 – 1/2(2) = 0].

The first Lewis structure obeys the octet rule but has a +1 formal charge on the most electronegative element there is, fluorine, and a negative formal charge on a much less electronegative element, boron. This is just the opposite of what we expect: negative formal charge on F and positive formal charge on B. The other Lewis structure does not obey the octet rule for B but has a zero formal charge on each element in BF_3. Because structures generally want to minimize formal charge, then BF_3 with only single bonds is best from a formal charge point of view.

119. See Exercise 98 for the Lewis structures of $POCl_3$, SO_4^{2-}, ClO_4^- and PO_4^{3-}. All these compounds/ions have similar Lewis structures to those of SO_2Cl_2 and XeO_4 shown below. Formal charge = [number of valence electrons on free atom] – [number of lone pair electrons on atom + 1/2(number of shared electrons of atom)].

a. $POCl_3$: P, FC = 5 – 1/2(8) = +1

b. SO_4^{2-}: S, FC = 6 – 1/2(8) = +2

c. ClO_4^-: Cl, FC = 7 – 1/2(8) = +3

d. PO_4^{3-}: P, FC = 5 – 1/2(8) = +1

e. SO_2Cl_2, 6 + 2(6) + 2(7) = 32 e^-

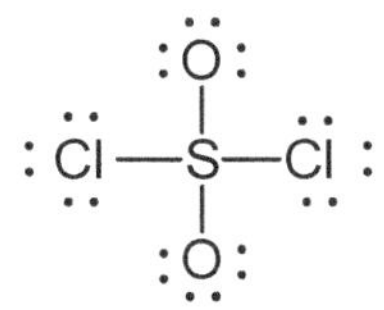

S, FC = 6 – 1/2(8) = +2

f. XeO_4, 8 + 4(6) = 32 e^-

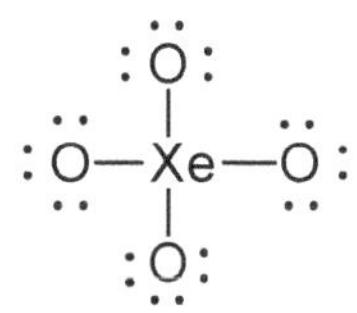

Xe, FC = 8 – 1/2(8) = +4

g. ClO_3^-, 7 + 3(6) + 1 = 26 e^-

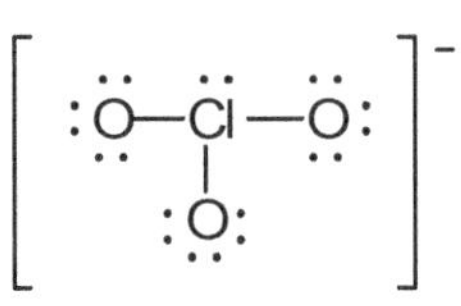

Cl, FC = 7 – 2 – 1/2(6) = +2

h. NO_4^{3-}, 5 + 4(6) + 3 = 32 e^-

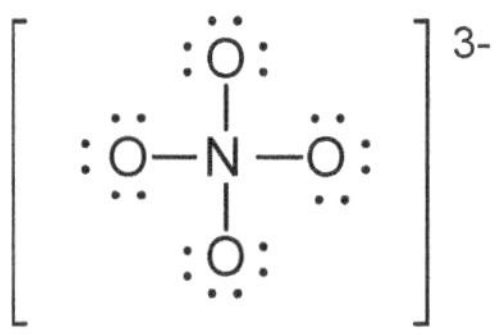

N, FC = 5 – 1/2(8) = +1

121. O_2F_2 has 2(6) + 2(7) = 26 valence e^-. The formal charge and oxidation number (state) of each atom is below the Lewis structure of O_2F_2.

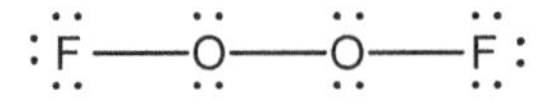

	F	O	O	F
Formal Charge	0	0	0	0
Oxid. Number	-1	+1	+1	-1

Oxidation states are more useful when accounting for the reactivity of O_2F_2. We are forced to assign +1 as the oxidation state for oxygen due to the bonding to fluorine. Oxygen is very electronegative, and +1 is not a stable oxidation state for this element.

123. SCl, 6 + 7 = 13; the formula could be SCl (13 valence electrons), S_2Cl_2 (26 valence electrons), S_3Cl_3 (39 valence electrons), etc. For a formal charge of zero on S, we will need each sulfur in the Lewis structure to have two bonds to it and two lone pairs [FC = 6 – 4 – 1/2(4) = 0]. Cl will need one bond and three lone pairs for a formal charge of zero [FC = 7 – 6 – 1/2(2) = 0]. Since chlorine wants only one bond to it, it will not be a central atom here. Only S_2Cl_2 can have a Lewis structure with a formal charge of zero on all atoms. The structure is:

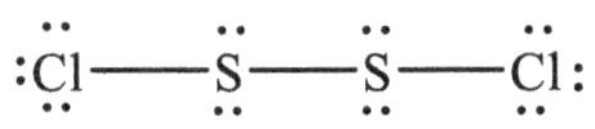

125. For formal charge values of zero:

(1) each carbon in the structure has 4 bonding pairs of electrons and no lone pairs;

(2) each N has 3 bonding pairs of electrons and 1 lone pair of electrons;

(3) each O has 2 bonding pairs of electrons and 2 lone pairs of electrons;

(4) each H is attached by only a single bond (1 bonding pair of electrons).

Following these guidelines, the Lewis structure is:

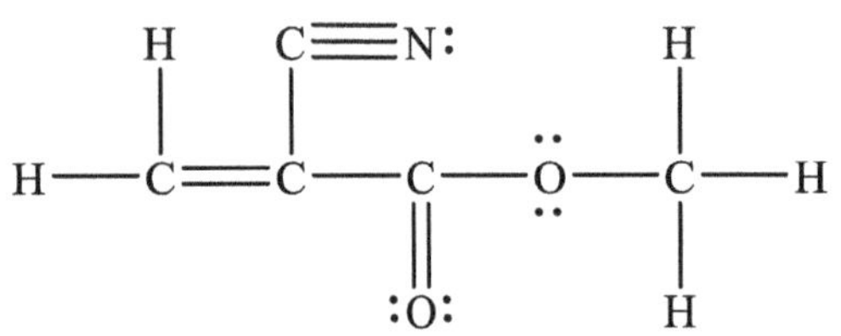

Molecular Structure and Polarity

127. The first step always is to draw a valid Lewis structure when predicting molecular structure. When resonance is possible, only one of the possible resonance structures is necessary to predict the correct structure because all resonance structures give the same structure. The Lewis structures are in Exercises 97 and 107. The structures and bond angles for each follow.

97: a. CCl_4: tetrahedral, 109.5°

b. NCl_3: trigonal pyramid, <109.5°

c. $SeCl_2$: V-shaped or bent, <109.5°

d. ICl: linear, but there is no bond angle present

Note: NCl_3 and $SeCl_2$ both have lone pairs of electrons on the central atom that result in bond angles that are something less than predicted from a tetrahedral arrangement (109.5°). However, we cannot predict the exact number. For the solutions manual, we will insert a less than sign to indicate this phenomenon. For bond angles equal to 120°, the lone pair phenomenon isn't as significant as compared to smaller bond angles. For these molecules, for example, NO_2^-, we will insert an approximate sign in front of the 120° to note that there may be a slight distortion from the VSEPR predicted bond angle.

107: a. NO_2^-: V-shaped, ≈120°; NO_3^-: trigonal planar, 120°

N_2O_4: trigonal planar, 120° about both N atoms

b. OCN^-, SCN^-, and N_3^- are all linear with 180° bond angles.

129. From the Lewis structures (see Exercise 101), Br_3^- would have a linear molecular structure, ClF_3 would have a T-shaped molecular structure, and SF_4 would have a see-saw molecular structure. For example, consider ClF_3 (28 valence electrons):

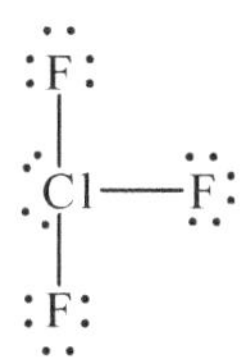

The central Cl atom is surrounded by five electron pairs, which requires a trigonal bipyramid geometry. Since there are three bonded atoms and two lone pairs of electrons about Cl, we describe the molecular structure of ClF_3 as T-shaped with predicted bond angles of about 90°. The actual bond angles will be slightly less than 90° due to the stronger repulsive effect of the lone-pair electrons as compared to the bonding electrons.

131. a. V-shaped or bent b. see-saw c. trigonal pyramid

d. trigonal bipyramid e. tetrahedral

133. a. SeO_3, $6 + 3(6) = 24\ e^-$

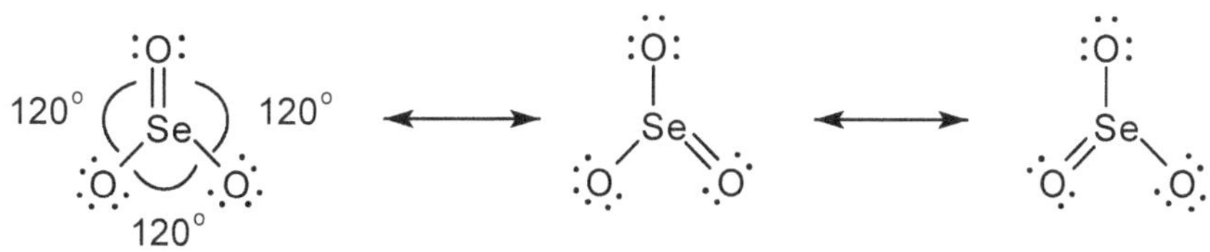

SeO_3 has a trigonal planar molecular structure with all bond angles equal to 120°. Note that any one of the resonance structures could be used to predict molecular structure and bond angles.

b. SeO_2, $6 + 2(6) = 18\ e^-$

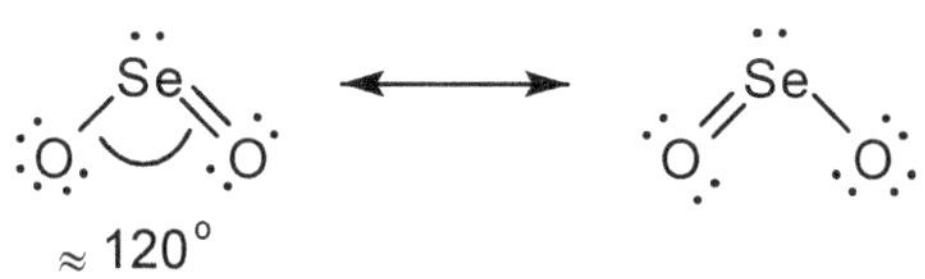

SeO_2 has a V-shaped molecular structure. We would expect the bond angle to be approximately 120° as expected for trigonal planar geometry.

Note: Both SeO_3 and SeO_2 structures have three effective pairs of electrons about the central atom. All the structures are based on a trigonal planar geometry, but only SeO_3 is described as having a trigonal planar structure. Molecular structure always describes the relative positions of the atoms.

135. a. $XeCl_2$ has 8 + 2(7) = 22 valence electrons.

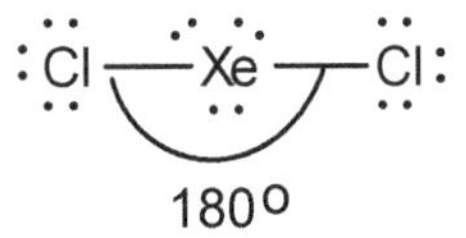

There are five pairs of electrons about the central Xe atom. The structure will be based on a trigonal bipyramid geometry. The most stable arrangement of the atoms in $XeCl_2$ is a linear molecular structure with a 180° bond angle.

b. ICl_3 has 7 + 3(7) = 28 valence electrons.

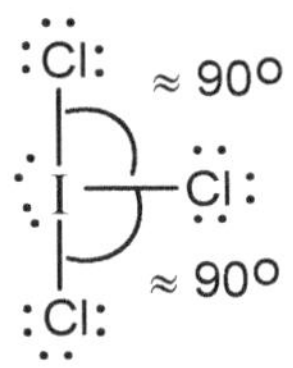

T-shaped; the ClICl angles are ≈90°. Since the lone pairs will take up more space, the ClICl bond angles will probably be slightly less than 90°.

c. TeF_4 has 6 + 4(7) = 34 valence electrons.

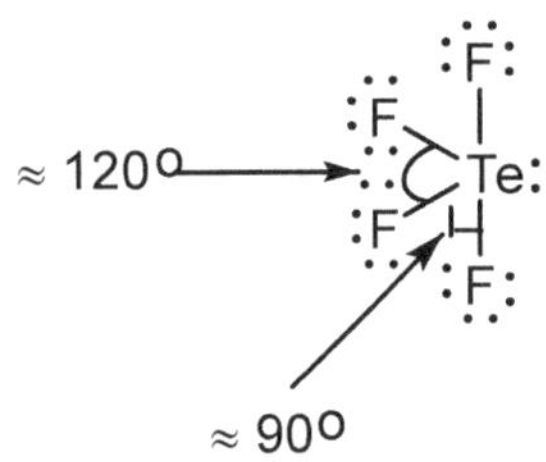

See-saw or teeter-totter or distorted tetrahedron

d. PCl_5 has 5 + 5(7) = 40 valence electrons.

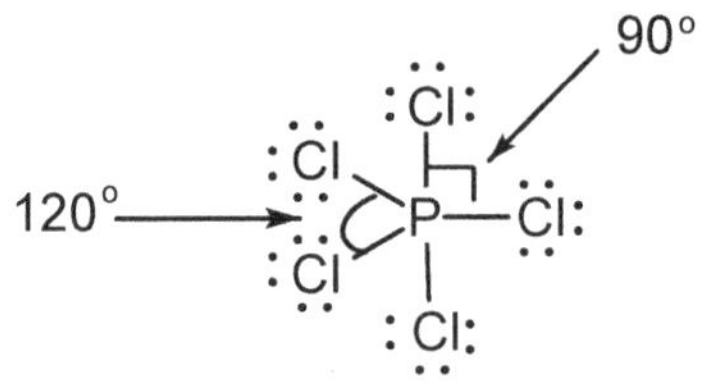

Trigonal bipyramid

All the species in this exercise have five pairs of electrons around the central atom. All the structures are based on a trigonal bipyramid geometry, but only in PCl_5 are all the pairs, bonding pairs. Thus PCl_5 is the only one for which we describe the molecular structure as trigonal bipyramid. Still, we had to begin with the trigonal bipyramid geometry to get to the structures (and bond angles) of the others.

137. To have the trigonal pyramid structure, the Lewis structure must be:

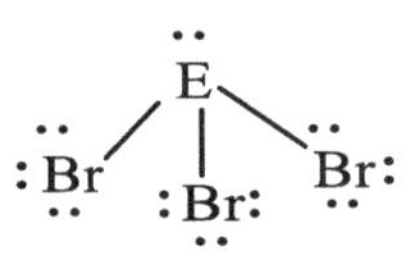

This structure requires 26 valence electrons and has a formula of EBr_3 formula.

$26 = x + 3(7),\ x = 5$ valence electrons

E has 5 valence electrons, so it is from Group 5A. Since the compound is covalent, some possible nonmetals in Group 5A for E are N, P, and As.

139. SeO_3 and SeO_2 both have polar bonds, but only SeO_2 has a dipole moment. The three bond dipoles from the three polar Se–O bonds in SeO_3 will all cancel when summed together. Hence SeO_3 is nonpolar since the overall molecule has no resulting dipole moment. In SeO_2, the two Se–O bond dipoles do not cancel when summed together; hence SeO_2 has a net dipole moment (is polar). Since O is more electronegative than Se, the negative end of the dipole moment is between the two O atoms, and the positive end is around the Se atom. The arrow in the following illustration represents the overall dipole moment in SeO_2. Note that to predict polarity for SeO_2, either of the two resonance structures can be used.

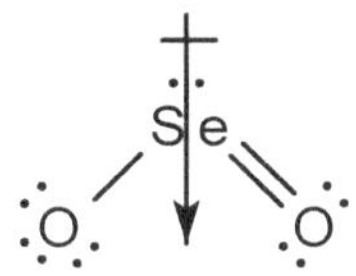

141. All have polar bonds, but only TeF_4 and ICl_3 have dipole moments. The bond dipoles from the five P–Cl bonds in PCl_5 cancel each other when summed together, so PCl_5 has no net dipole moment. The bond dipoles in $XeCl_2$ also cancel:

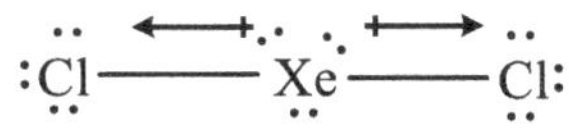

Because the bond dipoles from the two Xe–Cl bonds are equal in magnitude but point in opposite directions, they cancel each other, and $XeCl_2$ has no net dipole moment (is nonpolar). For TeF_4 and ICl_3, the arrangement of these molecules is such that the individual bond dipoles do *not* all cancel, so each has an overall net dipole moment (is polar).

143. Molecules that have an overall dipole moment are called polar molecules, and molecules that do not have an overall dipole moment are called nonpolar molecules.

a. OCl_2, $6 + 2(7) = 20\ e^-$

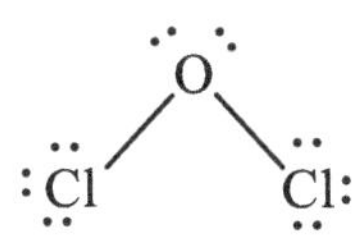

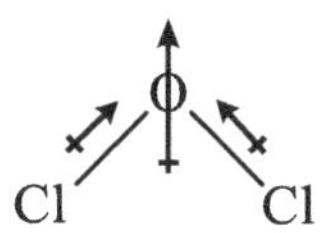

V-shaped, polar; OCl_2 is polar because the two O–Cl bond dipoles don't cancel each other. The resulting dipole moment is shown in the drawing.

KrF_2, $8 + 2(7) = 22\ e^-$

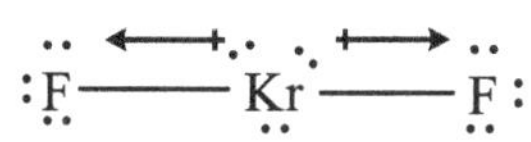

Linear, nonpolar; the molecule is nonpolar because the two Kr–F bond dipoles cancel each other.

BeH_2, $2 + 2(1) = 4\ e^-$

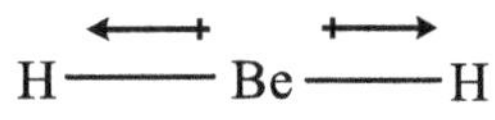

Linear, nonpolar; Be–H bond dipoles are equal and point in opposite directions. They cancel each other. BeH_2 is nonpolar.

SO_2, $6 + 2(6) = 18\ e^-$

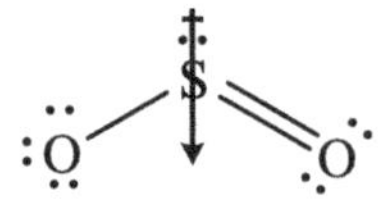

V-shaped, polar; the S–O bond dipoles do not cancel, so SO_2 is polar (has a net dipole moment). Only one resonance structure is shown.

Note: All four species contain three atoms. They have different structures because the number of lone pairs of electrons around the central atom are different in each case.

b. SO_3, $6 + 3(6) = 24\ e^-$

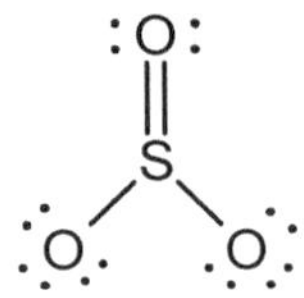

Trigonal planar, nonpolar; bond dipoles cancel. Only one resonance structure is shown.

NF_3, $5 + 3(7) = 26\ e^-$

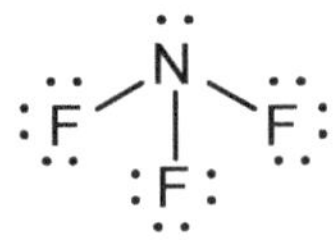

Trigonal pyramid, polar; bond dipoles do not cancel.

IF_3 has $7 + 3(7) = 28$ valence electrons.

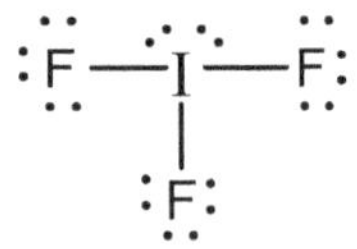

T-shaped, polar; bond dipoles do not cancel.

Note: Each molecule has the same number of atoms but different structures because of differing numbers of lone pairs around each central atom.

c. CF_4, $4 + 4(7) = 32\ e^-$

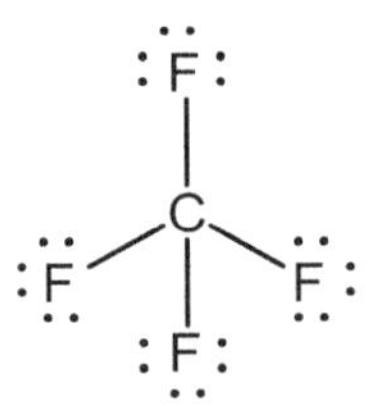

Tetrahedral, nonpolar; bond dipoles cancel.

SeF_4, $6 + 4(7) = 34\ e^-$

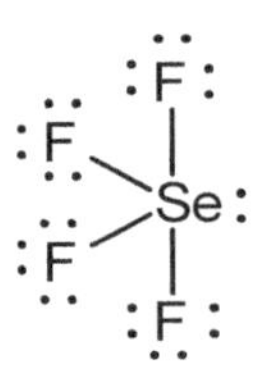

See-saw, polar; bond dipoles do not cancel.

KrF_4, 8 + 4(7) = 36 valence electrons

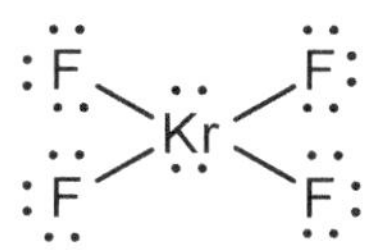

Square planar, nonpolar; bond dipoles cancel.

Note: Again, each molecule has the same number of atoms but different structures because of differing numbers of lone pairs around the central atom.

d. IF_5, 7 + 5(7) = 42 e^-

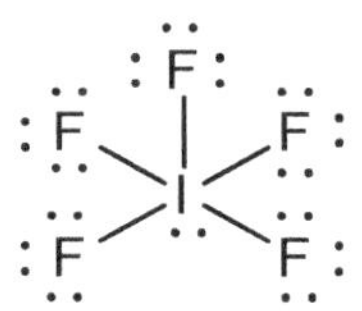

Square pyramid, polar; bond dipoles do not cancel.

AsF_5, 5 + 5(7) = 40 e^-

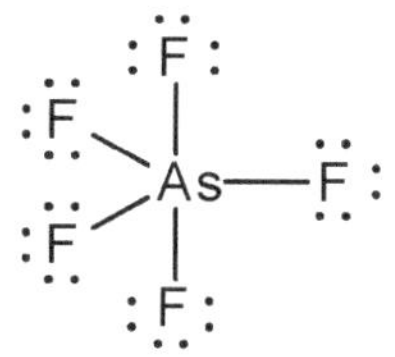

Trigonal bipyramid, nonpolar; bond dipoles cancel.

Note: Yet again, the molecules have the same number of atoms but different structures because of the presence of differing numbers of lone pairs.

145. EO_3^- is the formula of the ion. The Lewis structure has 26 valence electrons. Let x = number of valence electrons of element E.

$26 = x + 3(6) + 1$, $x = 7$ valence electrons

Element E is a halogen because halogens have seven valence electrons. Some possible identities are F, Cl, Br, and I. The EO_3^- ion has a trigonal pyramid molecular structure with bond angles of less than 109.5° (<109.5°).

147. N_2F_2: 2(5) + 2(7) = 24 e^-; in the Lewis structure, the two central nitrogen atoms exhibit trigonal planar geometry with ~120° bond angles. The two possible arrangements for the atoms in the N_2F_2 Lewis structures are:

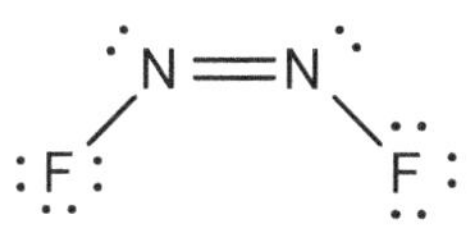

Polar

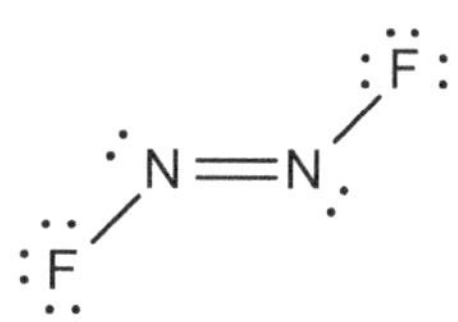

Nonpolar

In the first structure, the N–F bond dipoles are both pointing at angles downward somewhat, so they will add together to make this structure of N_2F_2 polar. In the second structure, the N_2F_2 bond dipoles point in opposite directions from one another; they will cancel each other out making this structure nonpolar.

ChemWork Problems

149. Bonding between a metal and a nonmetal is generally ionic. Bonding between two nonmetals is covalent, and in general, the bonding between two different nonmetals is usually polar covalent. When two different nonmetals have very similar electronegativities, the bonding is pure covalent or nonpolar covalent.

a. nonpolar covalent b. ionic c. ionic d. polar covalent

e. polar covalent f. polar covalent g. polar covalent h. ionic

151. The elements are identified by their electron configurations:

$[Ar]4s^1 3d^5$ = Cr; $[Ne]3s^2 3p^3$ = P; $[Ar]4s^2 3d^{10} 4p^3$ = As; $[Ne]3s^2 3p^5$ = Cl

Following the electronegativity trend, the order is Cr < As < P < Cl.

153. a. Radius: $N^+ < N < N^-$; IE: $N^- < N < N^+$

N^+ has the fewest electrons held by the seven protons in the nucleus, whereas N^- has the most electrons held by the seven protons. The seven protons in the nucleus will hold the electrons most tightly in N^+ and least tightly in N^-. Therefore, N^+ has the smallest radius with the largest ionization energy (IE), and N^- is the largest species with the smallest IE.

b. Radius: $Cl^+ < Cl < Se < Se^-$; IE: $Se^- < Se < Cl < Cl^+$

The general trends tell us that Cl has a smaller radius than Se and a larger IE than Se. Cl^+, with fewer electron-electron repulsions than Cl, will be smaller than Cl and have a larger IE. Se^-, with more electron-electron repulsions than Se, will be larger than Se and have a smaller IE.

c. Radius: $Ca^{2+} < K^+ < Cl^-$; IE: $Cl^- < K^+ < Ca^{2+}$

These ions are isoelectronic. The species with the most protons (Ca^{2+}) will hold the electrons most tightly and will have the smallest radius and largest IE. The ion with the fewest protons (Cl^-) will hold the electrons least tightly and will have the largest radius and smallest IE.

155. a.

$HF(g) \rightarrow H(g) + F(g)$	$\Delta H = 565$ kJ
$H(g) \rightarrow H^+(g) + e^-$	$\Delta H = 1312$ kJ
$F(g) + e^- \rightarrow F^-(g)$	$\Delta H = -327.8$ kJ
$HF(g) \rightarrow H^+(g) + F^-(g)$	$\Delta H = 1549$ kJ

b.

$HCl(g) \rightarrow H(g) + Cl(g)$	$\Delta H = 427$ kJ
$H(g) \rightarrow H^+(g) + e^-$	$\Delta H = 1312$ kJ
$Cl(g) + e^- \rightarrow Cl^-(g)$	$\Delta H = -348.7$ kJ
$HCl(g) \rightarrow H^+(g) + Cl^-(g)$	$\Delta H = 1390.$ kJ

c.

$$\begin{array}{ll} HI(g) \rightarrow H(g) + I(g) & \Delta H = 295 \text{ kJ} \\ H(g) \rightarrow H^+(g) + e^- & \Delta H = 1312 \text{ kJ} \\ I(g) + e^- \rightarrow I^-(g) & \Delta H = -295.2 \text{ kJ} \\ \hline HI(g) \rightarrow H^+(g) + I^-(g) & \Delta H = 1312 \text{ kJ} \end{array}$$

d.

$$\begin{array}{ll} H_2O(g) \rightarrow OH(g) + H(g) & \Delta H = 467 \text{ kJ} \\ H(g) \rightarrow H^+(g) + e^- & \Delta H = 1312 \text{ kJ} \\ OH(g) + e^- \rightarrow OH^-(g) & \Delta H = -180. \text{ kJ} \\ \hline H_2O(g) \rightarrow H^+(g) + OH^-(g) & \Delta H = 1599 \text{ kJ} \end{array}$$

157. The stable species are:

a. NaBr: In $NaBr_2$, the sodium ion would have a 2+ charge, assuming that each bromine has a 1− charge. Sodium doesn't form stable Na^{2+} ionic compounds.

b. ClO_4^-: ClO_4 has 31 valence electrons, so it is impossible to satisfy the octet rule for all atoms in ClO_4. The extra electron from the 1− charge in ClO_4^- allows for complete octets for all atoms.

c. XeO_4: We can't draw a Lewis structure that obeys the octet rule for SO_4 (30 electrons), unlike XeO_4 (32 electrons).

d. SeF_4: Both compounds require the central atom to expand its octet. O is too small and doesn't have low-energy d orbitals to expand its octet (which is true for all row 2 elements).

159. a. $XeCl_4$, $8 + 4(7) = 36\ e^-$

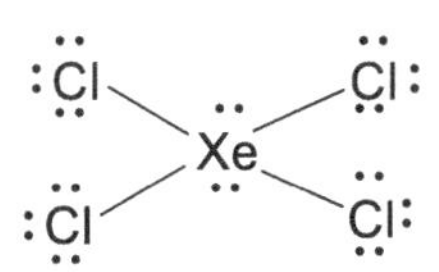

Square planar, 90°, nonpolar

$XeCl_2$, $8 + 2(7) = 22\ e^-$

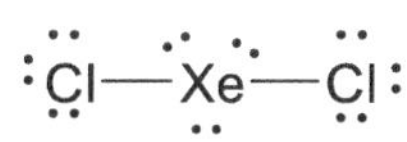

Linear, 180°, nonpolar

Both compounds have a central Xe atom with lone pairs and terminal Cl atoms, and both compounds do not satisfy the octet rule. In addition, both are nonpolar because the Xe–Cl bond dipoles and lone pairs around Xe are arranged in such a manner that they cancel each other out. The last item in common is that both have 180° bond angles. Although we haven't emphasized this, the bond angles between the Cl atoms on the diagonal in $XeCl_4$ are 180° apart from each other.

b. All of these are polar covalent compounds. The bond dipoles do not cancel out each other when summed together. The reason the bond dipoles are not symmetrically arranged in these compounds is that they all have at least one lone pair of electrons on the central atom, which disrupts the symmetry. Note that there are molecules that have lone pairs and are nonpolar, e.g., $XeCl_4$ and $XeCl_2$ in the preceding problem. A lone pair on a central atom does not guarantee a polar molecule.

161. a. $SeCl_4$ (34 e^-)

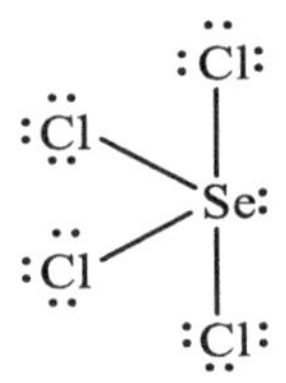

see-saw, ≈90° and ≈120°, polar

b. SO_2 (18 e^-)

V-shaped, ≈120°, polar

c. KrF_4 (36 e^-)

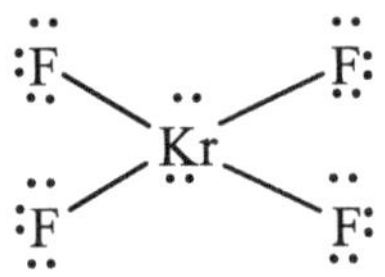

square planar, 90°, nonpolar

d. CBr_4 (32 e^-)

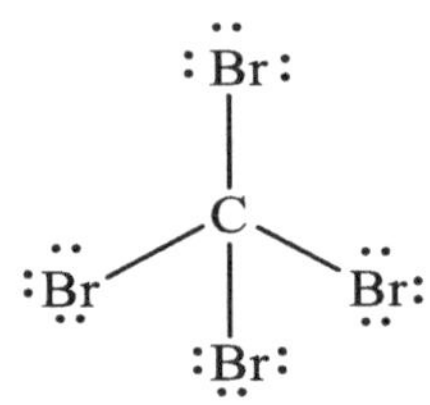

tetrahedral, 109.5°, nonpolar

e. IF_3 (28 e^-)

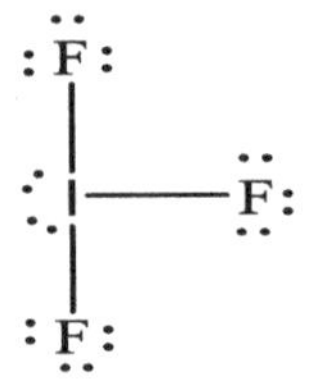

T-shaped, ≈90°, polar

f. ClF_5 (42 e^-)

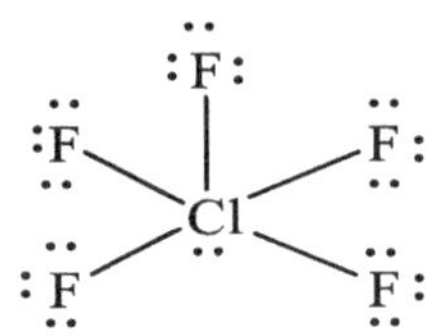

square pyramid, ≈90°, polar

163. NO_3^- has three resonance structures with the double bond rotated to each of the three oxygen atoms. Whenever resonance structures can be drawn for a substance, the actual bonding between the atoms is an average of all the resonance structures. In NO_3^-, the average bond is 1 1/3 bonds (a single in two resonance structures and a double in the third structure). The actual nitrogen-oxygen bond will all be equivalent with a bond length shorter than a single bond but longer than a double bond. With an average bond of 1 1/3, the bond length will be closer to the N-O single bond length with an estimate of 1.67 units long.

165. SiF_4, 4 + 4(7) = 32 e^- SeF_4, 6 + 4(7) = 34 e^- XeF_4, 8 + 4(7) = 36 e^-

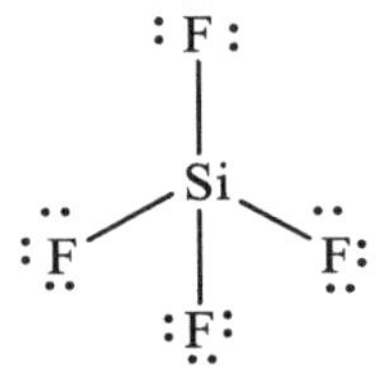

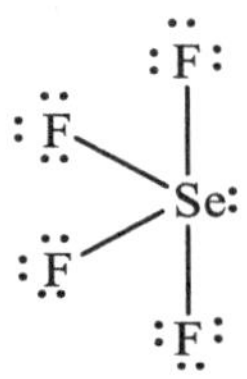

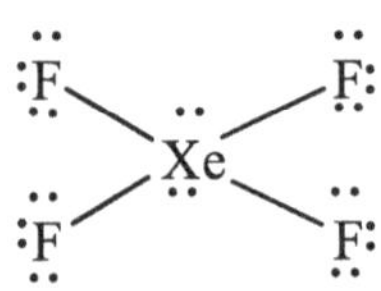

SiF_4 is tetrahedral and SeF_4 is polar (the bond dipoles cancel each other in SiF_4 and XeF_4).

167. The bonds are H–H, F–F, O=O, N≡N, and Cl–Cl. A triple bond is the strongest bond, so N_2 should have the largest bond energy value.

169.

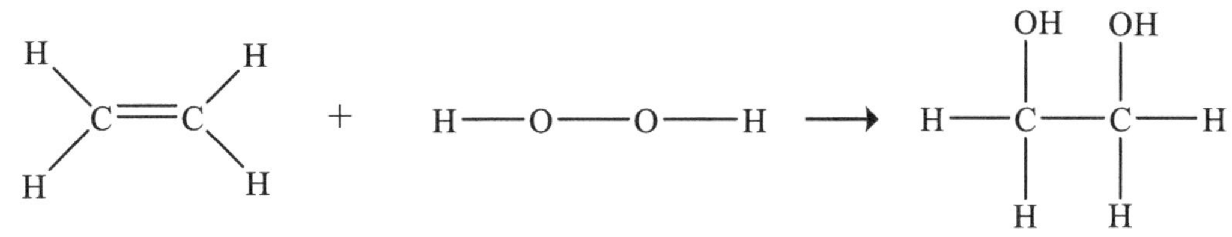

Bonds broken:

1 C=C (614 kJ/mol)
1 O–O (146 kJ/mol)

Bonds formed:

1 C–C (347 kJ/mol)
2 C–O (358 kJ/mol)

$\Delta H = 614\text{ kJ} + 146\text{ kJ} - [347\text{ kJ} + 2(358\text{ kJ})] = -303\text{ kJ}$

Note: Sometimes some of the bonds remain the same between reactants and products. To save time, only break and form bonds that are involved in the reaction.

171.

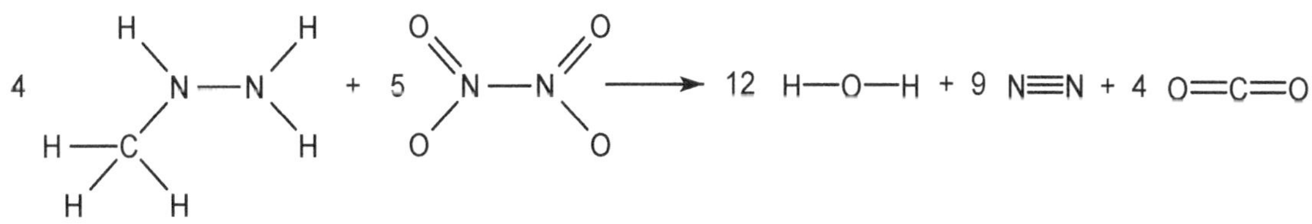

Bonds broken:

9 N–N (160. kJ/mol)
4 N–C (305 kJ/mol)
12 C–H (413 kJ/mol)
12 N–H (391 kJ/mol)
10 N=O (607 kJ/mol)
10 N–O (201 kJ/mol)

Bonds formed:

24 O–H (467 kJ/mol)
9 N≡N (941 kJ/mol)
8 C=O (799 kJ/mol)

$\Delta H = [9(160.) + 4(305) + 12(413) + 12(391) + 10(607) + 10(201)]$

$- [24(467) + 9(941) + 8(799)]$

$\Delta H = 20{,}388\text{ kJ} - 26{,}069\text{ kJ} = -5681\text{ kJ}$

173. Yes, each structure has the same number of effective pairs around the central atom, giving the same predicted molecular structure for each compound/ion. (A multiple bond is counted as a single group of electrons.)

175. TeF_5^- has 6 + 5(7) + 1 = 42 valence electrons.

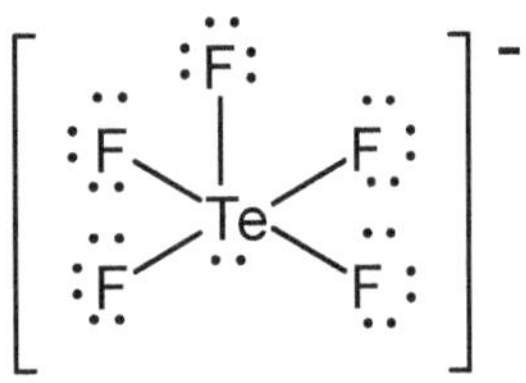

The lone pair of electrons around Te exerts a stronger repulsion than the bonding pairs of electrons. This pushes the four square-planar F atoms away from the lone pair and reduces the bond angles between the axial F atom and the square-planar F atoms.

177. Assuming 100.00 g of compound: $42.81 \text{ g F} = \dfrac{1 \text{ mol X}}{19.00 \text{ g F}} = 2.253 \text{ mol F}$

The number of moles of X in XF_5 is: $2.53 \text{ mol F} \times \dfrac{1 \text{ mol X}}{5 \text{ mol F}} = 0.4506 \text{ mol X}$

This number of moles of X has a mass of 57.19 g (= 100.00 g – 42.81 g). The molar mass of X is:

$\dfrac{57.19 \text{ g X}}{0.4506 \text{ mol X}} = 126.9$ g/mol; this is element I and the compound is IF_5.

IF_5, 7 + 5(7) = 42 e^-

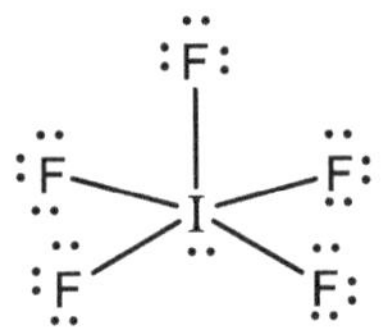

The molecular structure is square pyramid.

Challenge Problems

179. a. There are two attractions of the form $\dfrac{(+1)(-1)}{r}$, where r = 1 × 10^{-10} m = 0.1 nm.

$$V = 2 \times (2.31 \times 10^{-19} \text{ J nm})\left[\frac{(+1)(-1)}{0.1 \text{ nm}}\right] = -4.62 \times 10^{-18} \text{ J} = -5 \times 10^{-18} \text{ J}$$

b. There are four attractions of +1 and –1 charges at a distance of 0.1 nm from each other. The two negative charges and the two positive charges repel each other across the diagonal of the square. This is at a distance of $\sqrt{2} \times 0.1$ nm.

$$V = 4 \times (2.31 \times 10^{-19})\left[\frac{(+1)(-1)}{0.1}\right] + 2.31 \times 10^{-19}\left[\frac{(+1)(+1)}{\sqrt{2}\,(0.1)}\right] + 2.31 \times 10^{-19}\left[\frac{(-1)(-1)}{\sqrt{2}\,(0.1)}\right]$$

$V = -9.24 \times 10^{-18}\ J + 1.63 \times 10^{-18}\ J + 1.63 \times 10^{-18}\ J = -5.98 \times 10^{-18}\ J = -6 \times 10^{-18}\ J$

Note: There is a greater net attraction in arrangement b than in a.

181. The reaction is:

$1/2\ I_2(g) + 1/2\ Cl_2(g) \rightarrow ICl(g) \qquad \Delta H_f^\circ = ?$

Using Hess's law:

$1/2\ I_2(s) \rightarrow 1/2\ I_2(g)$	$\Delta H = 1/2(62\ kJ)$	(Appendix 4)
$1/2\ I_2(g) \rightarrow I(g)$	$\Delta H = 1/2(149\ kJ)$	(Table 8.5)
$1/2\ Cl_2(g) \rightarrow Cl(g)$	$\Delta H = 1/2(239\ kJ)$	(Table 8.5)
$I(g) + Cl(g) \rightarrow ICl(g)$	$\Delta H = -208\ kJ$	(Table 8.5)
$1/2\ I_2(s) + 1/2\ Cl_2(g) \rightarrow ICl(g)$	$\Delta H = 17\ kJ$ so $\Delta H_f^\circ = 17\ kJ/mol$	

183. See Figure 8.11 to see the data supporting MgO as an ionic compound. Note that the lattice energy is large enough to overcome all of the other processes (removing two electrons from Mg, etc.). The bond energy for O_2 (247 kJ/mol) and electron affinity (737 kJ/mol) are the same when making CO. However, ionizing carbon to form a C^{2+} ion must be too large. See Figure 7.30 to see that the first ionization energy for carbon is about 350 kJ/mol greater than the first ionization energy for magnesium. If all other numbers were equal, the overall energy change would be down to ~250 kJ/mol (see Figure 8.11). It is not unreasonable to assume that the second ionization energy for carbon is more than 250 kJ/mol greater than the second ionization energy of magnesium. This would result in a positive ΔH value for the formation of CO as an ionic compound. One wouldn't expect CO to be ionic if the energetics were unfavorable.

185. As the halogen atoms get larger, it becomes more difficult to fit three halogen atoms around the small nitrogen atom, and the NX_3 molecule becomes less stable.

187. a. i. $C_6H_6N_{12}O_{12} \rightarrow 6\ CO + 6\ N_2 + 3\ H_2O + 3/2\ O_2$

The NO_2 groups are assumed to have one N–O single bond and one N=O double bond, and each carbon atom has one C–H single bond. We must break and form all bonds.

Bonds broken:	Bonds formed:
3 C–C (347 kJ/mol)	6 C≡O (1072 kJ/mol)
6 C–H (413 kJ/mol)	6 N≡N (941 kJ/mol)
12 C–N (305 kJ/mol)	6 H–O (467 kJ/mol)
6 N–N (160. kJ/mol)	3/2 O=O (495 kJ/mol)
6 N–O (201 kJ/mol)	$\Sigma D_{formed} = 15{,}623\ kJ$
6 N=O (607 kJ/mol)	
$\Sigma D_{broken} = 12{,}987\ kJ$	

$\Delta H = \Sigma D_{broken} - \Sigma D_{formed} = 12{,}987\ kJ - 15{,}623\ kJ = -2636\ kJ$

ii. $C_6H_6N_{12}O_{12} \rightarrow 3\ CO + 3\ CO_2 + 6\ N_2 + 3\ H_2O$

Note: The bonds broken will be the same for all three reactions.

Bonds formed:

3 C≡O (1072 kJ/mol)
6 C=O (799 kJ/mol)
6 N≡N (941 kJ/mol)
6 H–O (467 kJ/mol)

$\Sigma D_{formed} = 16{,}458$ kJ

$\Delta H = 12{,}987\text{ kJ} - 16{,}458\text{ kJ} = -3471\text{ kJ}$

iii. $C_6H_6N_{12}O_{12} \rightarrow 6\ CO_2 + 6\ N_2 + 3\ H_2$

Bonds formed:

12 C=O (799 kJ/mol)
6 N≡N (941 kJ/mol)
3 H–H (432 kJ/mol)

$\Sigma D_{formed} = 16{,}530.$ kJ

$\Delta H = 12{,}987\text{ kJ} - 16{,}530.\text{ kJ} = -3543\text{ kJ}$

b. Reaction iii yields the most energy per mole of CL-20, so it will yield the most energy per kilogram.

$$\frac{-3543\text{ kJ}}{\text{mol}} \times \frac{1\text{ mol}}{438.23\text{ g}} \times \frac{1000\text{ g}}{\text{kg}} = -8085\text{ kJ/kg}$$

189. For carbon atoms to have a formal charge of zero, each C atom must satisfy the octet rule by forming four bonds (with no lone pairs). For nitrogen atoms to have a formal charge of zero, each N atom must satisfy the octet rule by forming three bonds and have one lone pair of electrons. For oxygen atoms to have a formal charge of zero, each O atom must satisfy the octet rule by forming two bonds and have two lone pairs of electrons. With these bonding requirements in mind, then the Lewis structure of histidine, where all atoms have a formal charge of zero, is:

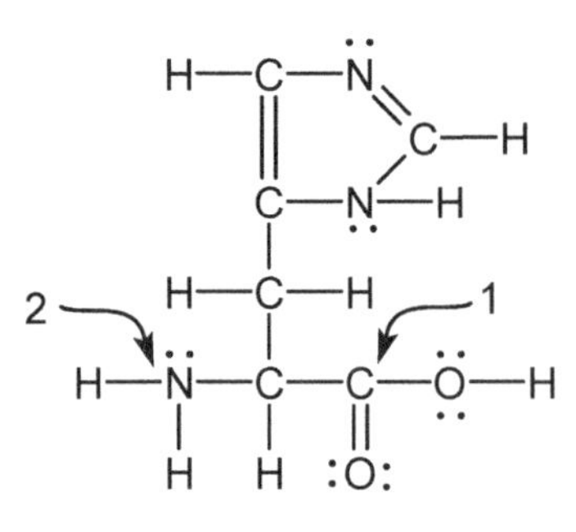

We would expect 120° bond angles about the carbon atom labeled 1 and ≈109.5° bond angles about the nitrogen atom labeled 2. The nitrogen bond angles should be slightly smaller than 109.5° due to the lone pair of electrons on nitrogen.

191. a. $BrFI_2$, 7 +7 + 2(7) = 28 e^-; two possible structures exist with Br as the central atom; each has a T-shaped molecular structure.

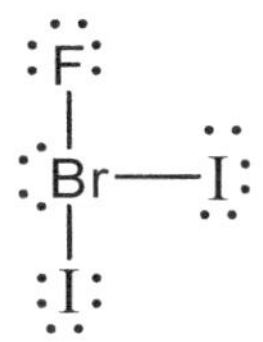

90° bond angles between I atoms

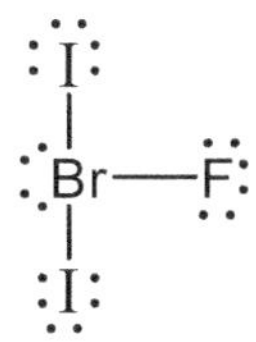

180° bond angles between I atoms

b. XeO_2F_2, 8 + 2(6) + 2(7) = 34 e^-; three possible structures exist with Xe as the central atom; each has a see-saw molecular structure.

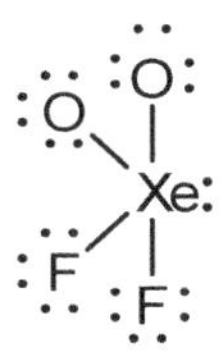

90° bond angle between O atoms

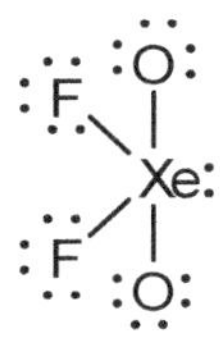

180° bond angle between O atoms

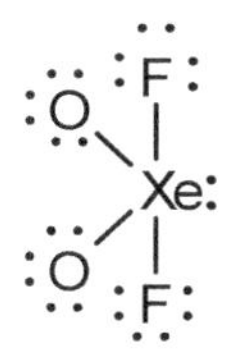

120° bond angle between O atoms

c. $TeF_2Cl_3^-$; 6 + 2(7) + 3(7) + 1 = 42 e^-; three possible structures exist with Te as the central atom; each has a square pyramid molecular structure.

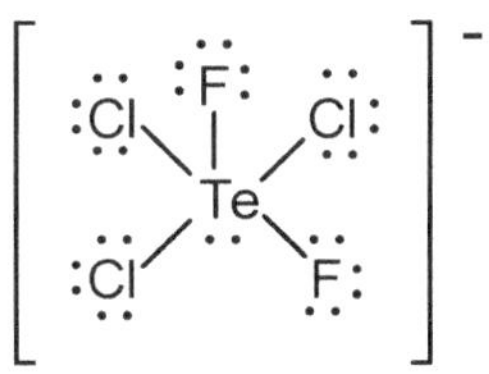

One F is 180° from the lone pair.

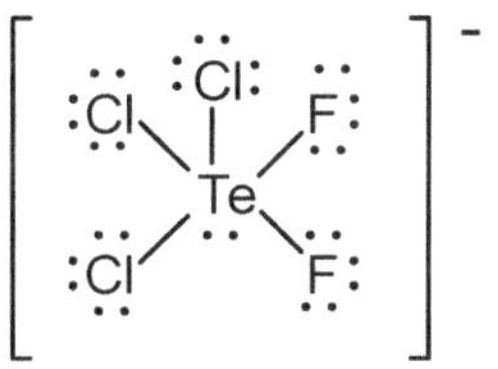

Both F atoms are 90° from the lone pair and 90° from each other.

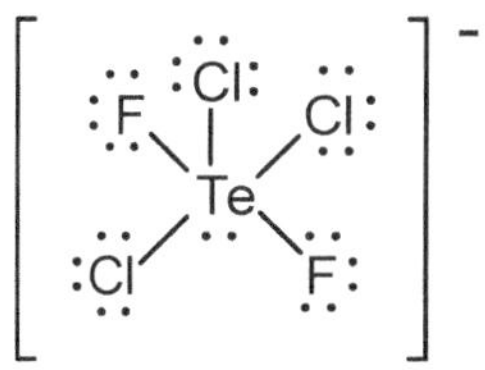

Both F atoms are 90° from the lone pair and 180° from each other.

193. The complete Lewis structure follows. All but two of the carbon atoms exhibit 109.5° bond angles. The two carbon atoms that contain the double bond exhibit 120° bond angles (see * in the following Lewis structure).

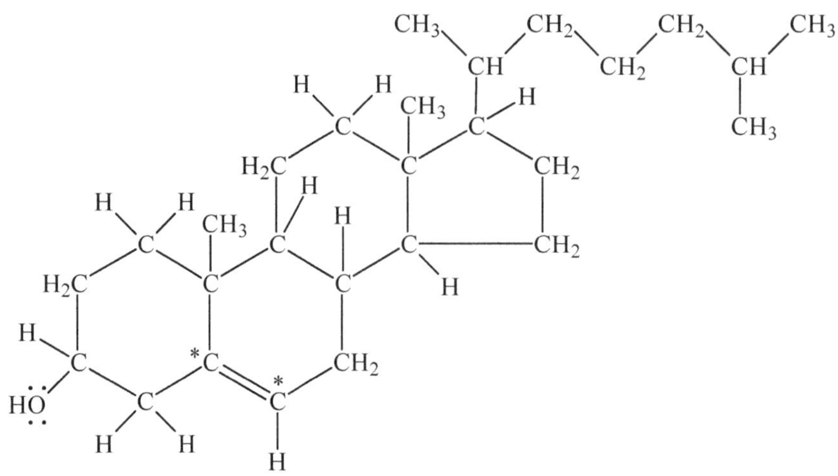

No; most of the carbons are not in the same plane since a majority of carbon atoms exhibit a tetrahedral structure (109.5° bond angles). *Note*: HO, CH, CH_2, H_2C, and CH_3 are shorthand for oxygen and carbon atoms singly bonded to hydrogen atoms.

CHAPTER 9

COVALENT BONDING: ORBITALS

Questions

13. In hybrid orbital theory, some or all of the valence atomic orbitals of the central atom in a molecule are mixed together to form hybrid orbitals; these hybrid orbitals point to where the bonded atoms and lone pairs are oriented. The sigma bonds are formed from the hybrid orbitals overlapping head to head with an appropriate orbital from the bonded atom. The π bonds, in hybrid orbital theory, are formed from unhybridized p atomic orbitals. The p orbitals overlap side to side to form the π bond, where the π electrons occupy the space above and below a line joining the atoms (the internuclear axis). Assuming the z-axis is the internuclear axis, then the p_z atomic orbital will always be hybridized whether the hybridization is sp, sp^2, sp^3, dsp^3 or d^2sp^3. For sp hybridization, the p_x and p_y atomic orbitals are unhybridized; they are used to form two π bonds to the bonded atom(s). For sp^2 hybridization, either the p_x or the p_y atomic orbital is hybridized (along with the s and p_z orbitals); the other p orbital is used to form a π bond to a bonded atom. For sp^3 hybridization, the s and all the p orbitals are hybridized; no unhybridized p atomic orbitals are present, so no π bonds form with sp^3 hybridization. For dsp^3 and d^2sp^3 hybridization, we just mix in one or two d orbitals into the hybridization process. Which specific d orbitals are used is not important to our discussion.

15. We use d orbitals when we have to, i.e., we use d orbitals when the central atom on a molecule has more than eight electrons around it. In hybrid orbital theory, the d orbitals are used to accommodate the electrons over eight. Row 2 elements never have more than eight electrons around them, so they never hybridize d orbitals. We rationalize this by saying there are no d orbitals close in energy to the valence 2s and 2p orbitals (2d orbitals are forbidden energy levels). However, for row 3 and heavier elements, there are 3d, 4d, 5d, etc. orbitals that will be close in energy to the valence s and p orbitals. It is row 3 and heavier nonmetals that hybridize d orbitals when they have to.

    For sulfur, the valence electrons are in 3s and 3p orbitals. Therefore, 3d orbitals are closest in energy and are available for hybridization. Arsenic would hybridize 4d orbitals to go with the valence 4s and 4p orbitals, whereas iodine would hybridize 5d orbitals since the valence electrons are in $n = 5$.

17. 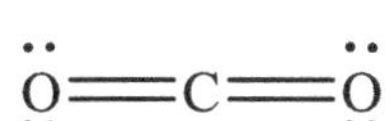

    The darker green orbitals about carbon are sp hybrid orbitals. The lighter green orbitals about each oxygen are sp^2 hybrid orbitals, and the gold orbitals about all of the atoms are unhybridized p atomic orbitals. In each double bond in CO_2, one sigma and one π bond exists. The two carbon-oxygen sigma bonds are formed from overlap of sp hybrid orbitals from carbon with a sp^2 hybrid orbital from each oxygen. The two carbon-oxygen π bonds are formed from side-to-side overlap of the unhybridized p atomic orbitals from carbon with an unhybridized p atomic orbital from each oxygen. These two π bonds are oriented perpendicular to each other as illustrated in the figure.

19. Bonding and antibonding molecular orbitals are both solutions to the quantum mechanical treatment of the molecule. Bonding orbitals form when in-phase orbitals combine to give constructive interference. This results in enhanced electron probability located between the two nuclei. The end result is that a bonding MO is lower in energy than the atomic orbitals from which it is composed. Antibonding orbitals form when out-of-phase orbitals combine. The mismatched phases produce destructive interference leading to a node of electron probability between the two nuclei. With electron distribution pushed to the outside, the energy of an antibonding orbital is higher than the energy of the atomic orbitals from which it is composed.

21. H_2: $(\sigma_{1s})^2$ Bond Order = (2–0)/2 = 1

H_2^-: $(\sigma_{1s})^2(\sigma_{1s}*)^1$ Bond Order = (2–1)/2 = ½

The calculated bond order for H_2 is two times larger than the bond order for H_2^-. From the simple bond order calculation, the H_2 bond is predicted to be twice as strong as the H_2^- bond.

23. The localized electron model does not deal effectively with molecules containing unpaired electrons. We can draw all of the possible structures for NO with its odd number of valence electrons but still not have a good feel for whether the bond in NO is weaker or stronger than the bond in NO^-. MO theory can handle odd electron species without any modifications. From the MO electron configurations, the bond order is 2.5 for NO and 2 for NO^-. Therefore, NO should have the stronger bond (and it does). In addition, hybrid orbital theory does not predict that NO^- is paramagnetic. The MO theory correctly makes this prediction.

25. A localized bond has the electron probability centered between two nuclei or centered above and below two nuclei. A delocalized bond has electron probability centered over multiple nuclei; it occurs when unhybridized p atomic orbitals on 3 or more atoms in a molecule/ion all combine. We use delocalize π bonding to explain why molecules that exhibit resonance have equivalent bonds in length and strength.

Exercises

The Localized Electron Model and Hybrid Orbitals

27. H_2O has 2(1) + 6 = 8 valence electrons.

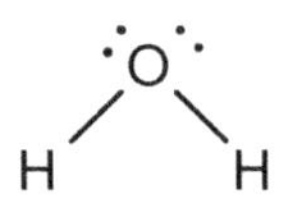

H_2O has a tetrahedral arrangement of the electron pairs about the O atom that requires sp^3 hybridization. Two of the four sp^3 hybrid orbitals are used to form bonds to the two hydrogen atoms, and the other two sp^3 hybrid orbitals hold the two lone pairs on oxygen. The two O–H bonds are formed from overlap of the sp^3 hybrid orbitals from oxygen with the 1s atomic orbitals from the hydrogen atoms. Each O–H covalent bond is called a sigma (σ) bond since the shared electron pair in each bond is centered in an area on a line running between the two atoms.

29. H_2CO has 2(1) + 4 + 6 = 12 valence electrons.

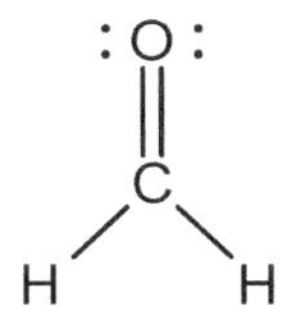

The central carbon atom has a trigonal planar arrangement of the electron pairs that requires sp^2 hybridization. The two C–H sigma bonds are formed from overlap of the sp^2 hybrid orbitals from carbon with the hydrogen 1s atomic orbitals. The double bond between carbon and oxygen consists of one σ and one π bond. The oxygen atom, like the carbon atom, also has a trigonal planar arrangement of the electrons that requires sp^2 hybridization. The σ bond in the double bond is formed from overlap of a carbon sp^2 hybrid orbital with an oxygen sp^2 hybrid orbital. The π bond in the double bond is formed from overlap of the unhybridized p atomic orbitals. Carbon and oxygen each has one unhybridized p atomic orbital that is parallel with the other. When two parallel p atomic orbitals overlap, a π bond results where the shared electron pair occupies the space above and below a line joining the atoms in the bond.

31. Ethane, C_2H_6, has 2(4) + 6(1) = 14 valence electrons.

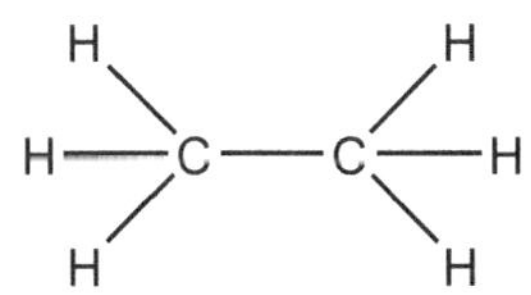

The carbon atoms are sp^3 hybridized. The six C–H sigma bonds are formed from overlap of the sp^3 hybrid orbitals from C with the 1s atomic orbitals from the hydrogen atoms. The carbon-carbon sigma bond is formed from overlap of an sp^3 hybrid orbital from each C atom.

Ethanol, C_2H_6O has 2(4) + 6(1) + 6 = 20 e^-

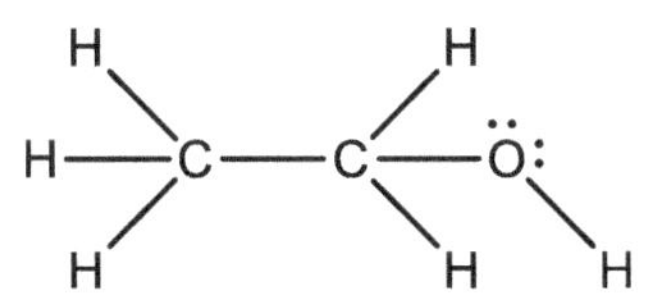

The two C atoms and the O atom are sp^3 hybridized. All bonds are formed from overlap with these sp^3 hybrid orbitals. The C–H and O–H sigma bonds are formed from overlap of sp^3 hybrid orbitals with hydrogen 1s atomic orbitals. The C–C and C–O sigma bonds are formed from overlap of the sp^3 hybrid orbitals from each atom.

33. See Exercises 8.97 and 8.107 for the Lewis structures. To predict the hybridization, first determine the arrangement of electron pairs about each central atom using the VSEPR model; then use the information in Figure 9.24 of the text to deduce the hybridization required for that arrangement of electron pairs.

8.97 a. CCl_4: C is sp^3 hybridized. b. NCl_3: N is sp^3 hybridized.

c. $SeCl_2$: Se is sp^3 hybridized. d. ICl: Both I and Cl are sp^3 hybridized.

8.107 a. The central N atom is sp^2 hybridized in NO_2^- and NO_3^-. In N_2O_4, both central N atoms are sp^2 hybridized.

b. In OCN^- and SCN^-, the central carbon atoms in each ion are sp hybridized, and in N_3^-, the central N atom is also sp hybridized.

35. All exhibit dsp^3 hybridization. All of these molecules/ions have a trigonal bipyramid arrangement of electron pairs about the central atom; all have central atoms with dsp^3 hybridization. See Exercise 8.101 for the Lewis structures.

37. The molecules in Exercise 8.133 all have a trigonal planar arrangement of electron pairs about the central atom, so all have central atoms with sp^2 hybridization. The molecules in Exercise 8.134 all have a tetrahedral arrangement of electron pairs about the central atom, so all have central atoms with sp^3 hybridization. See Exercises 8.133 and 8.134 for the Lewis structures.

39. a.

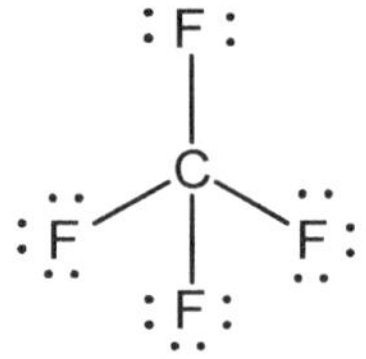

tetrahedral sp^3
109.5° nonpolar

b.

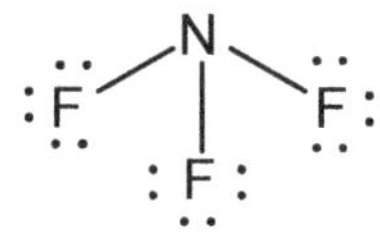

trigonal pyramid sp^3
<109.5° polar

The angles in NF_3 should be slightly less than 109.5° because the lone pair requires more space than the bonding pairs.

c.

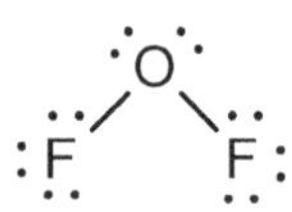

V-shaped sp^3
<109°.5 polar

d.

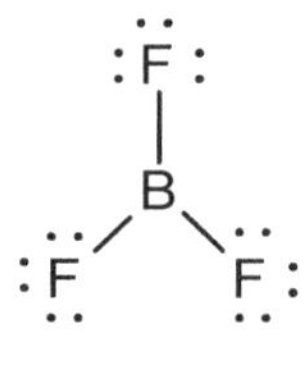

trigonal planar sp^2
120° nonpolar

e. H–Be–H linear, 180°, sp, nonpolar

41. a. The molecular structure of TeF_4 is see-saw. Bond angle a is ≈120° and angle b is ≈90°. The hybridization of the central Te atom is dsp^3. The individual bond dipoles in TeF_4 will not all cancel each other, so it is a polar molecule.

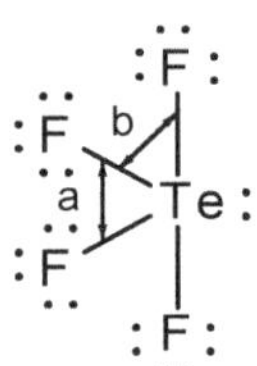

b.

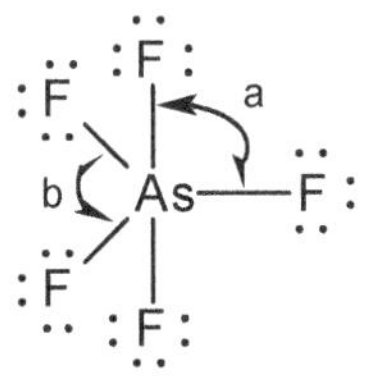

trigonal bipyramid — dsp^3

a) 90°, b) 120° — nonpolar

c.

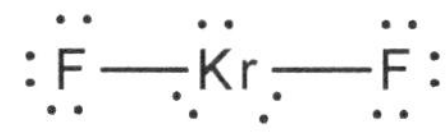

linear — dsp^3

180° — nonpolar

d.

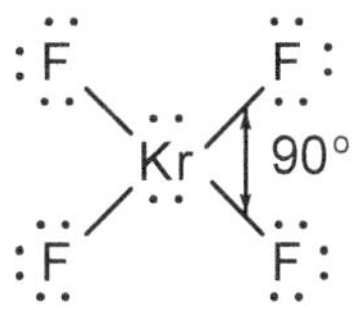

square planar — d^2sp^3

90° — nonpolar

e.

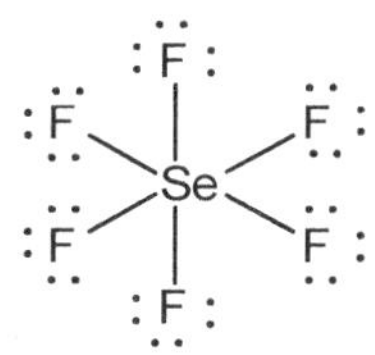

octahedral — d^2sp^3

90° — nonpolar

f.

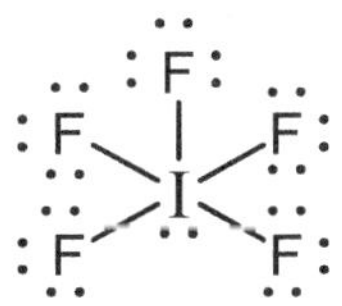

square pyramid — d^2sp^3

≈90° — polar

g.

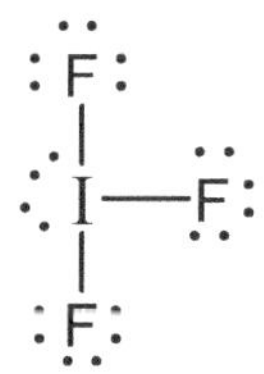

T-shaped — dsp^3

≈90° — polar

43. With d^2sp^3 hybridization, we have six pairs of electrons about the central atom. There are three molecular structures that have six pairs of electrons about the central atom. They are octahedral, square pyramid, and square planar. Only the square pyramid molecular structure has 5 bonded atoms, so QF_5 has a square pyramid molecular structure. IF_5 is an example of a compound with a square pyramid molecular structure:

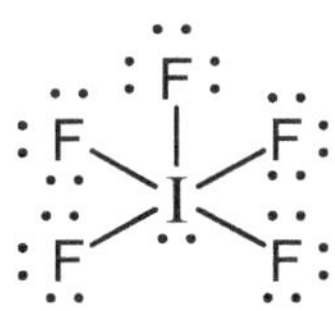

This Lewis structure has 42 valence electrons. Since the five F atoms have 35 valence electrons, the central Q atom must have 7 valence electrons. So, Q is a halogen. But since the central atom exceeds the octet rule, Q can't be F. Some possible identities for Q are Cl, Br, I, and At.

45. ICl_5 has $7 + 5(7) = 42$ valence electrons.

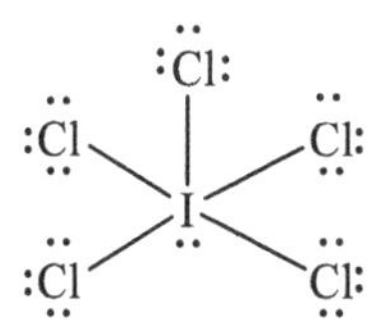

The only statement that is false is answer e. The central atom is d^2sp^3 hybridized. The bond dipoles will not all cancel when added together, hence ICl_5 is polar.

47.

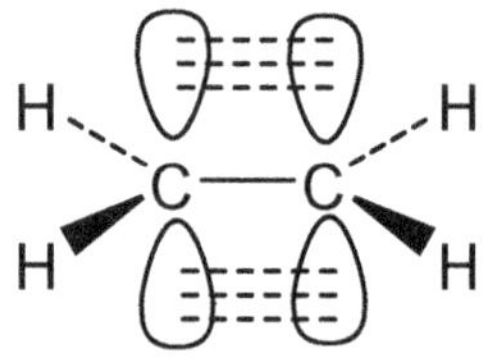

For the p orbitals to properly line up to form the π bond, all six atoms are forced into the same plane. If the atoms are not in the same plane, then the π bond could not form since the p orbitals would no longer be parallel to each other.

49. Statement e is false. The triple bond consists of one σ bond formed from overlap of *sp* hybrid orbitals from each C and two π bonds from overlap of unhybridized p atomic orbitals from each carbon.

51. The two nitrogen atoms in urea both have a tetrahedral arrangement of electron pairs, so both atoms are sp^3 hybridized. The carbon atom has a trigonal planar arrangement of electron pairs, so C is sp^2 hybridized. O is also sp^2 hybridized because it also has a trigonal planar arrangement of electron pairs.

Each of the four N–H sigma bonds are formed from overlap of an sp^3 hybrid orbital from nitrogen with a 1s orbital from hydrogen. Each of the two N–C sigma bonds are formed from an sp^3 hybrid orbital from N with an sp^2 hybrid orbital from carbon. The double bond between carbon and oxygen consists of one σ and one π bond. The σ bond in the double bond is formed from overlap of a carbon sp^2 hybrid orbital with an oxygen sp^2 hybrid orbital. The π bond in the double bond is formed from overlap of the unhybridized p atomic orbitals. Carbon and oxygen each have one unhybridized p atomic orbital, and they are assumed to be parallel to each other. When two parallel p atomic orbitals overlap side to side, a π bond results.

53. To complete the Lewis structures, just add lone pairs of electrons to satisfy the octet rule for the atoms with fewer than eight electrons.

Biacetyl ($C_4H_6O_2$) has 4(4) + 6(1) + 2(6) = 34 valence electrons.

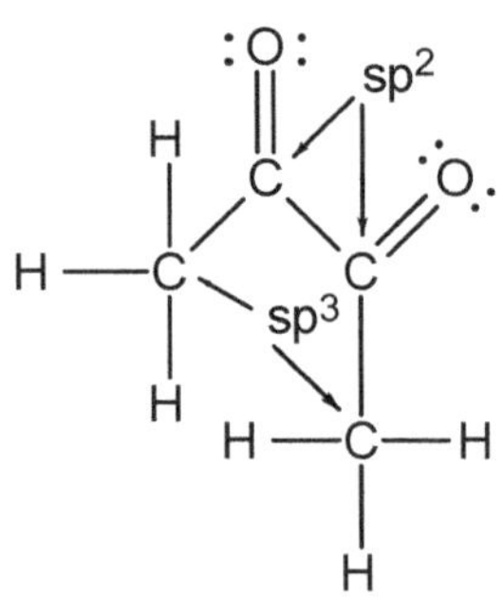

All CCO angles are 120°. The six atoms are not forced to lie in the same plane because of free rotation about the carbon-carbon single (sigma) bonds. There are 11 σ and 2 π bonds in biacetyl.

Acetoin ($C_4H_8O_2$) has 4(4) + 8(1) + 2(6) = 36 valence electrons.

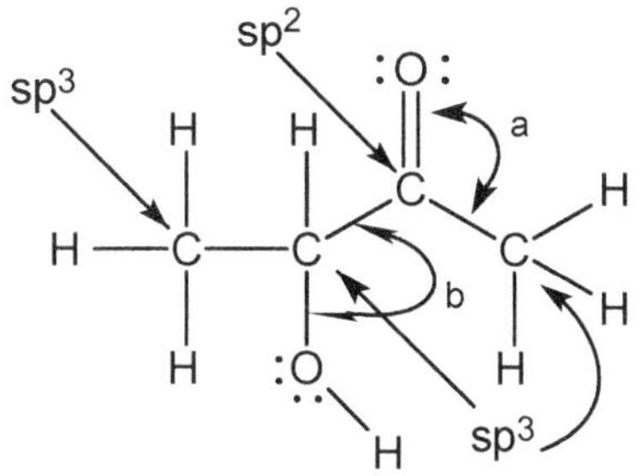

The carbon with the doubly bonded O is sp^2 hybridized. The other three C atoms are sp^3 hybridized. Angle a = 120° and angle b = 109.5°. There are 13 σ and 1 π bonds in acetoin.

Note: All single bonds are σ bonds, all double bonds are one σ and one π bond, and all triple bonds are one σ and two π bonds.

55. a. Add lone pairs to complete octets for each O and N.

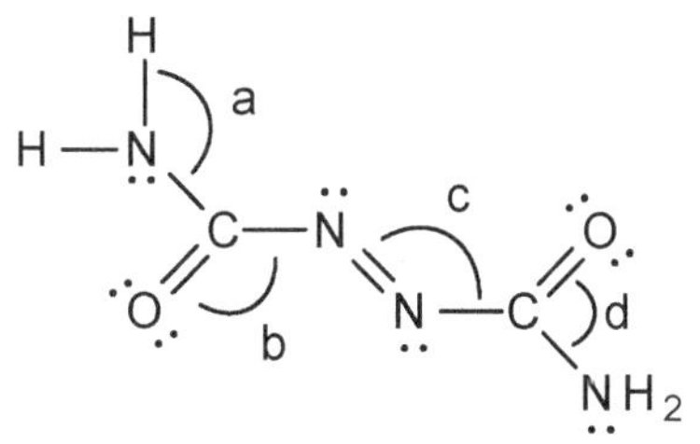

Azodicarbonamide

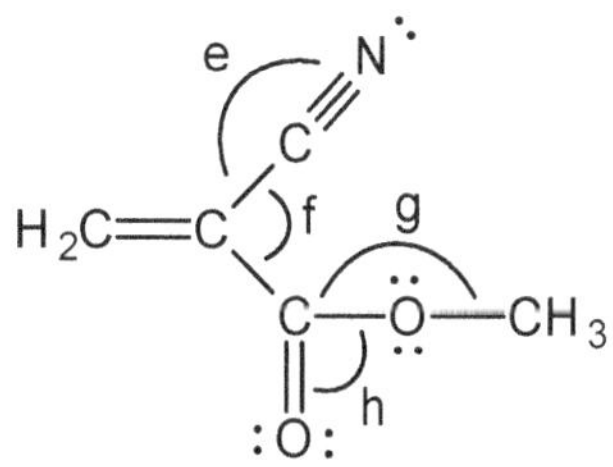

methyl cyanoacrylate

Note: NH_2, CH_2 (H_2C), and CH_3 are shorthand for nitrogen or carbon atoms singly bonded to hydrogen atoms.

b. In azodicarbonamide, the two carbon atoms are sp^2 hybridized, the two nitrogen atoms with hydrogens attached are sp^3 hybridized, and the other two nitrogen atoms are sp^2 hybridized. In methyl cyanoacrylate, the CH_3 carbon is sp^3 hybridized, the carbon with the triple bond is sp hybridized, and the other three carbons are sp^2 hybridized.

c. Azodicarbonamide contains three π bonds and methyl cyanoacrylate contains four π bonds.

d. a) ≈109.5° b) 120° c) ≈120° d) 120° e) 180°

f) 120° g) ≈109.5° h) 120°

57. To complete the Lewis structure, just add lone pairs of electrons to satisfy the octet rule for the atoms that have fewer than eight electrons.

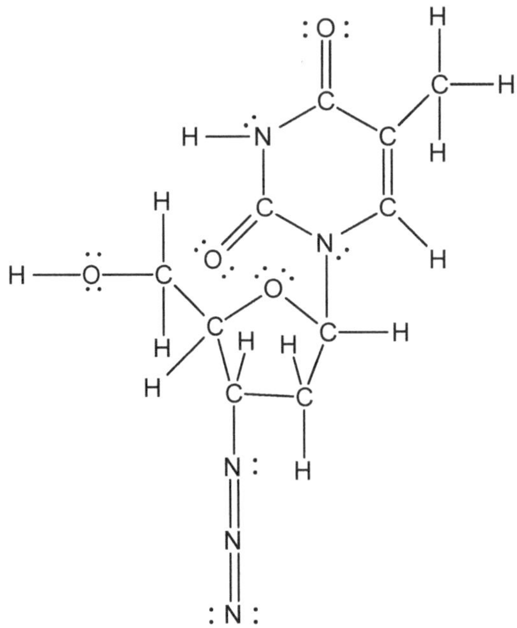

a. 6 b. 4 c. The center N in –N=N=N group

d. 33 σ e. 5 π bonds f. 180°

g. ≈109.5° h. sp^3

The Molecular Orbital Model

59. a. The bonding molecular orbital is on the right and the antibonding molecular orbital is on the left. The bonding MO has the greatest electron probability between the nuclei, while the antibonding MO has greatest electron probability outside the space between the two nuclei.

b. The bonding MO is lower in energy. Because the electrons in the bonding MO have the greatest probability of lying between the two nuclei, these electrons are attracted to two different nuclei, resulting in a lower energy.

61. If we calculate a nonzero bond order for a molecule, then we predict that it can exist (is stable).

H_2^+: $(\sigma_{1s})^1$ B.O. = bond order = (1–0)/2 = 1/2, stable

H_2: $(\sigma_{1s})^2$ B.O. = (2–0)/2 = 1, stable

H_2^-: $(\sigma_{1s})^2(\sigma_{1s}{*})^1$ B.O. = (2–1)/2 = 1/2, stable

H_2^{2-}: $(\sigma_{1s})^2(\sigma_{1s}{*})^2$ B.O. = (2–2)/2 = 0, not stable

63. If we calculate a nonzero bond order for a molecule, then we predict that it can exist (is stable).

N_2^{2-}: $(\sigma_{2s})^2(\sigma_{2s}*)^2(\pi_{2p})^4(\sigma_{2p})^2(\pi_{2p}*)^2$ B.O. = bond order = (8–4)/2 = 2, stable
O_2^{2-}: $(\sigma_{2s})^2(\sigma_{2s}*)^2(\sigma_{2p})^2(\pi_{2p})^4(\pi_{2p}*)^4$ B.O. = (8–6)/2 = 1, stable
F_2^{2-}: $(\sigma_{2s})^2(\sigma_{2s}*)^2(\sigma_{2p})^2(\pi_{2p})^4(\pi_{2p}*)^4(\sigma_{2p}*)^2$ B.O. = (8–8)/2 = 0, not stable

65. The electron configurations are:

a. Li_2: $(\sigma_{2s})^2$ B.O. = (2–0)/2 = 1, diamagnetic (0 unpaired e^-)

b. C_2: $(\sigma_{2s})^2(\sigma_{2s}*)^2(\pi_{2p})^4$ B.O. = (6–2)/2 = 2, diamagnetic (0 unpaired e^-)

c. S_2: $(\sigma_{3s})^2(\sigma_{3s}*)^2(\sigma_{3p})^2(\pi_{3p})^4(\pi_{3p}*)^2$ B.O. = (8–4)/2 = 2, paramagnetic (2 unpaired e^-)

67. N_2^+ and N_2^- each have a bond order of 2.5.

N_2^+: $(\sigma_{2s})^2(\sigma_{2s}*)^2(\pi_{2p})^4(\sigma_{2p})^1$ B.O. = bond order = (7–2)/2 = 2.5

N_2^-: $(\sigma_{2s})^2(\sigma_{2s}*)^2(\pi_{2p})^4(\sigma_{2p})^2(\pi_{2p}*)^1$ B.O. = bond order = (8–3)/2 = 2.5

69. O_2^{2-}: $(\sigma_{2s})^2(\sigma_{2s}*)^2(\sigma_{2p})^2(\pi_{2p})^4(\pi_{2p}*)^4$ B.O. = (8–6)/2 = 1; 0 unpaired e^-

O_2^-: $(\sigma_{2s})^2(\sigma_{2s}*)^2(\sigma_{2p})^2(\pi_{2p})^4(\pi_{2p}*)^3$ B.O. = (8–5)/2 = 1.5; 0 unpaired e^-

Because O_2^- has a larger bond order than O_2^{2-}, the superoxide bond length is expected to be shorter than the peroxide bond length.

71. The electron configurations are (assuming the same orbital order as that for N_2):

a. CO: $(\sigma_{2s})^2(\sigma_{2s}*)^2(\pi_{2p})^4(\sigma_{2p})^2$ B.O. = (8-2)/2 = 3, diamagnetic

b. CO^+: $(\sigma_{2s})^2(\sigma_{2s}*)^2(\pi_{2p})^4(\sigma_{2p})^1$ B.O. = (7-2)/2 = 2.5, paramagnetic

c. CO^{2+}: $(\sigma_{2s})^2(\sigma_{2s}*)^2(\pi_{2p})^4$ B.O. = (6-2)/2 = 2, diamagnetic

Because bond order is directly proportional to bond energy and inversely proportional to bond length:

Shortest → longest bond length: $CO < CO^+ < CO^{2+}$

Smallest → largest bond energy: $CO^{2+} < CO^+ < CO$

73. a. H_2: $(\sigma_{1s})^2$

b. B_2: $(\sigma_{2s})^2(\sigma_{2s}*)^2(\pi_{2p})^2$

c. C_2^{2-}: $(\sigma_{2s})^2(\sigma_{2s}*)^2(\pi_{2p})^4(\sigma_{2p})^2$

d. OF: $(\sigma_{2s})^2(\sigma_{2s}*)^2(\sigma_{2p})^2(\pi_{2p})^4(\pi_{2p}*)^3$

The bond strength will weaken if the electron removed comes from a bonding orbital. Of the molecules listed, H_2, B_2, and C_2^{2-} would be expected to have their bond strength weaken as an electron is removed. OF has the electron removed from an antibonding orbital, so its bond strength increases.

75. The two types of overlap that result in bond formation for p orbitals are in-phase side-to-side overlap (π bond) and in-phase head-to-head overlap (σ bond).

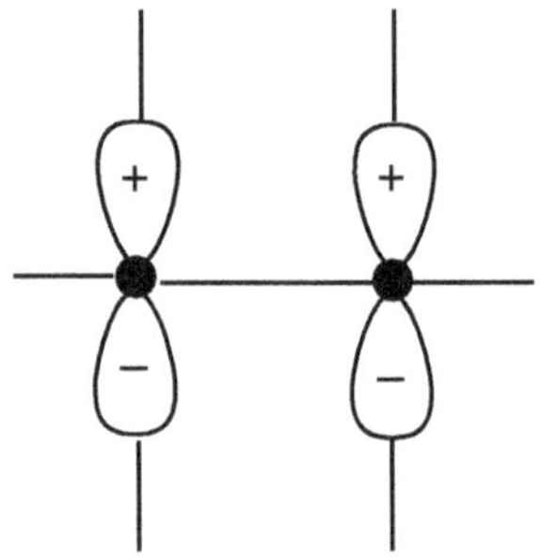

π_{2p} (in-phase; the signs match up)

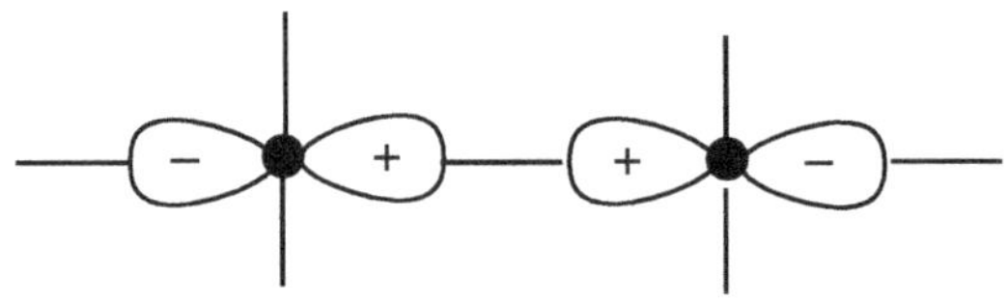

σ_{2p} (in-phase; the signs match up)

77. a. The electron density would be closer to F on average. The F atom is more electronegative than the H atom, and the 2p orbital of F is lower in energy than the 1s orbital of H.

b. The bonding MO would have more fluorine 2p character since it is closer in energy to the fluorine 2p atomic orbital.

c. The antibonding MO would place more electron density closer to H and would have a greater contribution from the higher-energy hydrogen 1s atomic orbital.

79. C_2^{2-} has 10 valence electrons. The Lewis structure predicts sp hybridization for each carbon with two unhybridized p orbitals on each carbon.

$[:C\equiv C:]^{2-}$ sp hybrid orbitals form the σ bond and the two unhybridized p atomic orbitals from each carbon form the two π bonds.

MO: $(\sigma_{2s})^2(\sigma_{2s}^*)^2(\pi_{2p})^4(\sigma_{2p})^2$, B.O. = (8 – 2)/2 = 3

Both give the same picture, a triple bond composed of one σ and two π bonds. Both predict the ion will be diamagnetic. Lewis structures deal well with diamagnetic (all electrons paired) species. The Lewis model cannot really predict magnetic properties.

81. Statement b is false. All carbon–carbon bonds in benzene are equivalent. Each carbon is sp^2 hybridized, leaving 1 unhybridized p atomic orbital that is perpendicular to the plane of the molecule. All six of the unhybridized p orbitals combine to form a delocalized π bonding system resulting in equivalent carbon–carbon bonds in length and strength.

83. O_3 and NO_2^- are isoelectronic, so we only need consider one of them since the same bonding ideas apply to both. The Lewis structures for O_3 are:

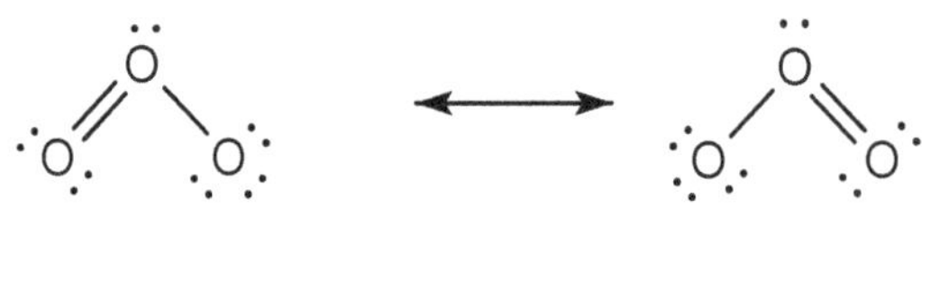

For each of the two resonance forms, the central O atom is sp^2 hybridized with one unhybridized p atomic orbital. The sp^2 hybrid orbitals are used to form the two sigma bonds to the central atom and hold the lone pair of electrons on the central O atom. The localized electron view of the π bond uses unhybridized p atomic orbitals. The π bond resonates between the two positions in the Lewis structures; the actual structure of O_3 is an average of the two resonance structures:

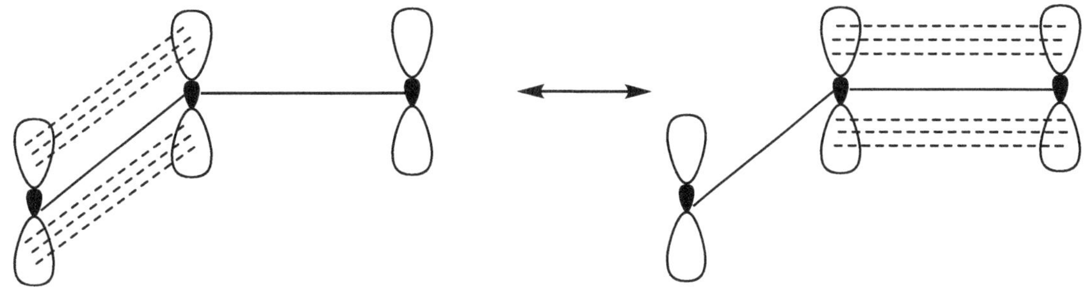

In the MO picture of the π bond, all three unhybridized p orbitals overlap at the same time, resulting in π electrons that are delocalized over the entire surface of the molecule. This is represented as:

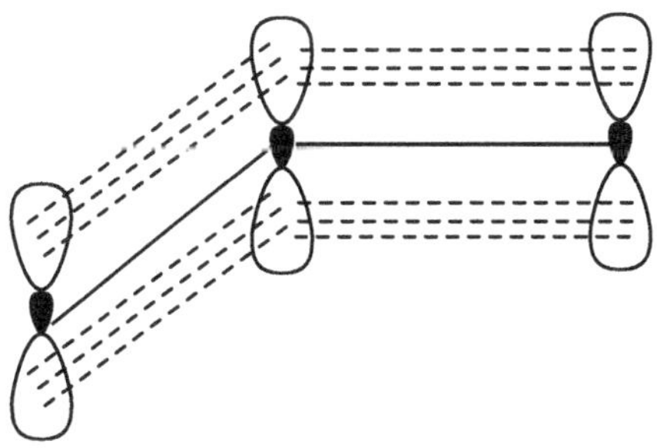

or

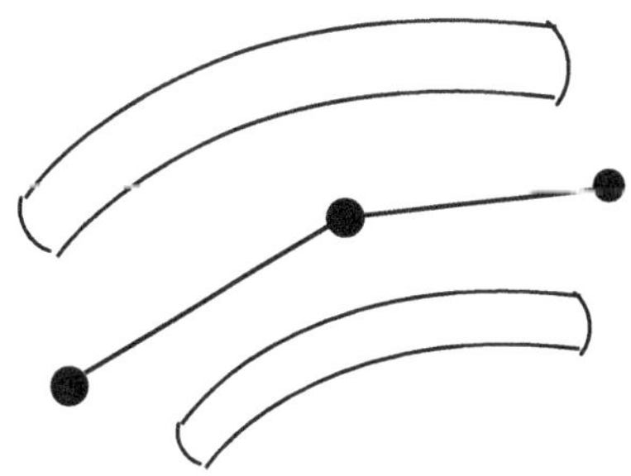

ChemWork Problems

85. Only statement c is true. Here an atom with a trigonal planar arrangement of electron pairs which is sp^2 hybridized is bonded to an atom with a tetrahedral arrangement of electron pairs which is sp^3 hybridized. These hybrid orbitals would combine to form a σ bond.

For statement a, only valence orbitals combine to form hybrid orbitals. For statement b, the hybrid orbitals always point to where the electron pairs are located. Hybrid orbitals are used to explain the bond angles about a central atom. For statement d, hybrid orbitals always combine head-to-head resulting in σ bonds. And for statement e, the unhybridized p atomic orbital go to form the π bonds. With sp^2 hybridization, only one p atomic orbital is unhybridized, so only 1 π bond can form.

87. SeO_2 (18 e^-)

V-shaped, ≈120°, polar, sp^2 hybridized

PCl_3 (26 e^-)

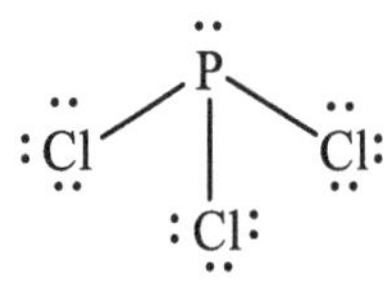

trigonal pyramid, < 109.5°

polar, sp^3 hybridized

NNO (16 e^-)

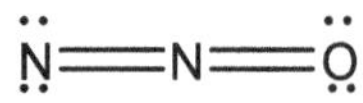

+2 others

linear, 180°

polar, sp hybridized

COS (16 e^-)

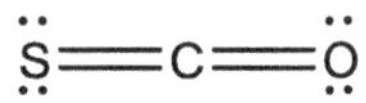

+2 others

linear, 180°

polar, sp hybridized

PF_3 (26 e^-)

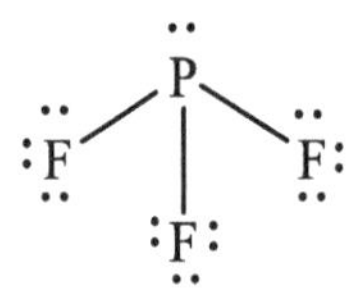

trigonal pyramid, < 109.5°

polar, sp^3 hybridized

All these compounds are polar. In each compound, the bond dipoles do not cancel each other out, leading to a polar compound. SeO_2 is the only compound exhibiting ~120° bond angles. PCl_3 and PF_3 both have a tetrahedral arrangement of electron pairs, so these are the compounds that have central atoms that are sp^3 hybridized. NNO and COS both have linear molecular structure.

89. a. XeO_3, $8 + 3(6) = 26\ e^-$

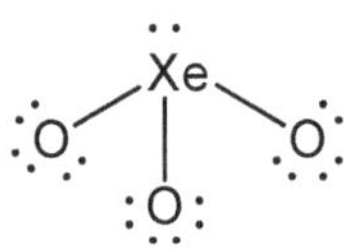

trigonal pyramid; sp^3

b. XeO_4, $8 + 4(6) = 32\ e^-$

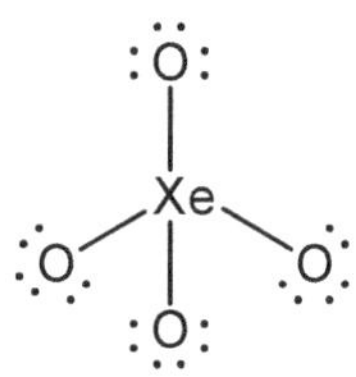

tetrahedral; sp^3

c. $XeOF_4$, $8 + 6 + 4(7) = 42\ e^-$

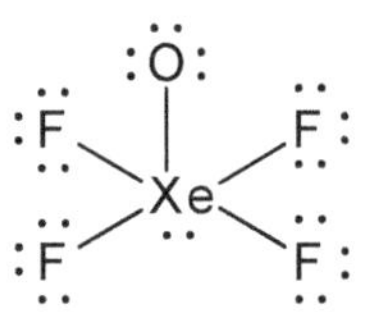

or

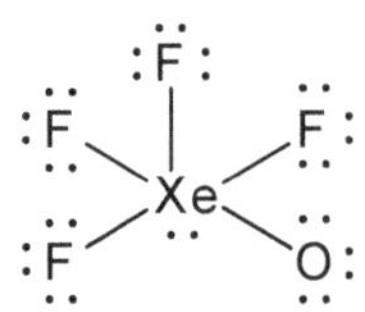

square pyramid; d^2sp^3

d. $XeOF_2$, $8 + 6 + 2(7) = 28\ e^-$

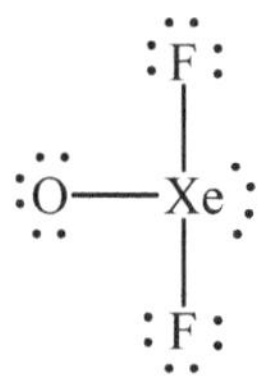

or

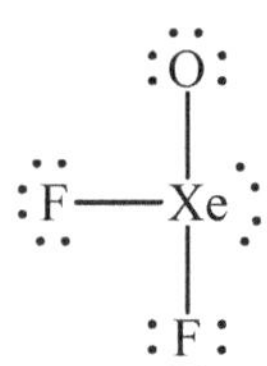

T-shaped; dsp^3

e. XeO_3F_2 has 8 + 3(6) + 2(7) = 40 valence electrons.

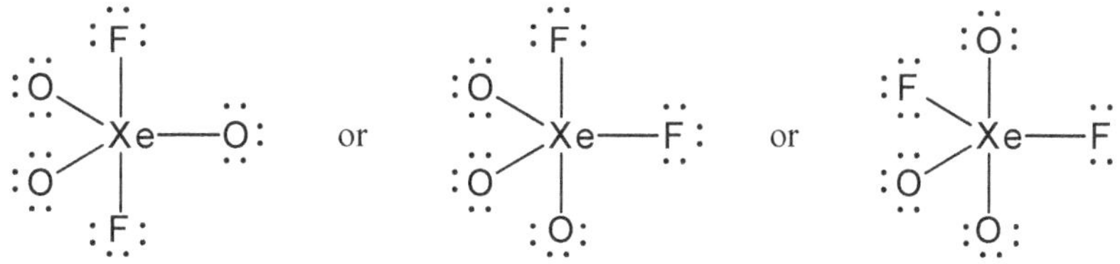

trigonal bipyramid; dsp^3

91. The Lewis structure for N_2 is: $:N{\equiv}N:$

Each nitrogen has two effective electron pairs, so each N is sp hybridized. The sp hybrid orbitals combine to form the σ bond between the two N atoms. The two unhybridized p atomic orbitals on each N combine to form two π bonds in the triple bond. Answer c is correct.

93. Molecule A has a tetrahedral arrangement of electron pairs because it is sp^3 hybridized. Molecule B has 6 electron pairs about the central atom, so it is d^2sp^3 hybridized. Molecule C has two σ and two π bonds to the central atom, so it either has two double bonds to the central atom (as in CO_2) or one triple bond and one single bond (as in HCN). Molecule C is consistent with a linear arrangement of electron pairs exhibiting sp hybridization. There are many correct possibilities for each molecule; an example of each is:

Molecule A: CH_4

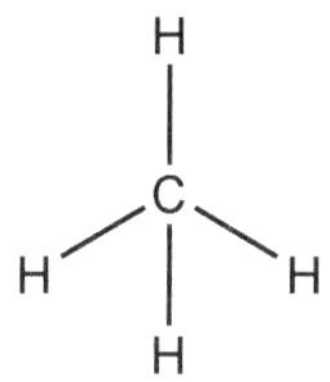

tetrahedral; 109.5°; sp^3

Molecule B: XeF_4

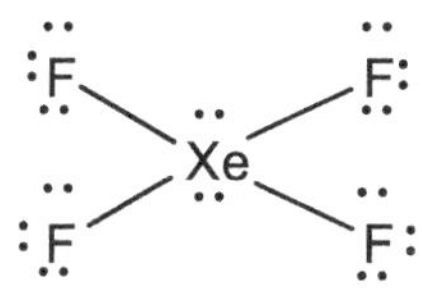

square planar; 90°; d^2sp^3

Molecule C: CO_2 or HCN

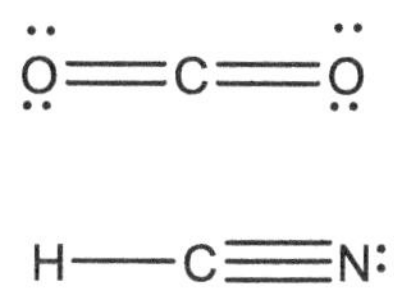

linear; 180°; sp

95. a. No, some atoms are in different places. Thus these are not resonance structures; they are different compounds.

b. For the first Lewis structure, all nitrogen atoms are sp^3 hybridized and all carbon atoms are sp^2 hybridized. In the second Lewis structure, all nitrogen atoms and carbon atoms are sp^2 hybridized.

c. For the reaction:

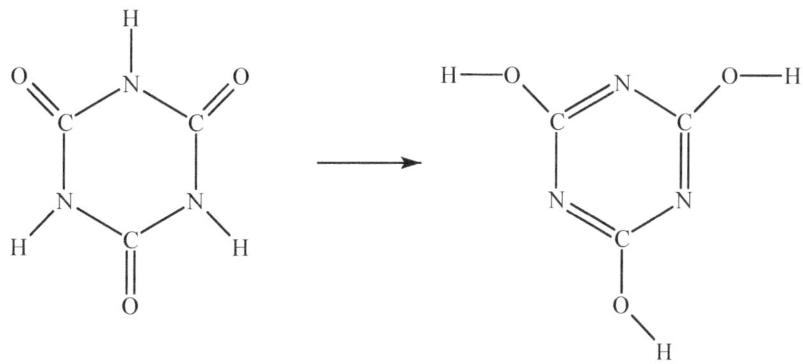

Bonds broken:	Bonds formed:
3 C=O (745 kJ/mol)	3 C=N (615 kJ/mol)
3 C–N (305 kJ/mol)	3 C–O (358 kJ/mol)
3 N–H (391 kJ/mol)	3 O–H (467 kJ/mol)

$\Delta H = 3(745) + 3(305) + 3(391) - [3(615) + 3(358) + 3(467)]$

$\Delta H = 4323 \text{ kJ} - 4320 \text{ kJ} = 3 \text{ kJ}$

The bonds are slightly stronger in the first structure with the carbon-oxygen double bonds since ΔH for the reaction is positive. However, the value of ΔH is so small that the best conclusion is that the bond strengths are comparable in the two structures.

97. For carbon, nitrogen, and oxygen atoms to have formal charge values of zero, each C atom will form four bonds to other atoms and have no lone pairs of electrons, each N atom will form three bonds to other atoms and have one lone pair of electrons, and each O atom will form two bonds to other atoms and have two lone pairs of electrons. Using these bonding requirements gives the following two resonance structures for vitamin B_6:

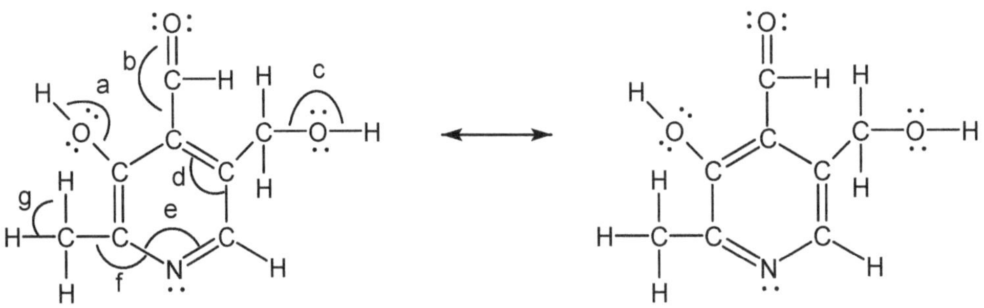

a. 21 σ bonds; 4 π bonds (The electrons in the three π bonds in the ring are delocalized.)

b. Angles a), c), and g): ≈109.5°; angles b), d), e), and f): ≈120°

c. 6 sp^2 carbons; the five carbon atoms in the ring are sp^2 hybridized, as is the carbon with the double bond to oxygen.

d. 4 sp^3 atoms; the two carbons that are not sp^2 hybridized are sp^3 hybridized, and the oxygens marked with angles a and c are sp^3 hybridized.

e. Yes, the π electrons in the ring are delocalized. The atoms in the ring are all sp^2 hybridized. This leaves a p orbital perpendicular to the plane of the ring from each atom. Overlap of all six of these p orbitals results in a π molecular orbital system where the electrons are delocalized above and below the plane of the ring (similar to benzene in Figure 9.48 of the text).

99. 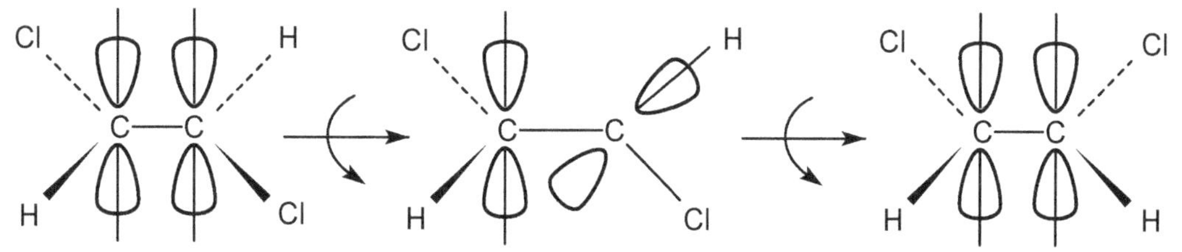

To rotate about the double bond, the molecule must go through an intermediate stage where the π bond is broken and the sigma bond remains intact. Bond energies are 347 kJ/mol for C–C and 614 kJ/mol for C=C. If we take the single bond as the strength of the σ bond, then the strength of the π bond is (614 – 347 =) 267 kJ/mol. In theory, 267 kJ/mol must be supplied to rotate about a carbon-carbon double bond.

101. a. N_2F_2 has 2(5) + 2(7) = 24 valence electrons.

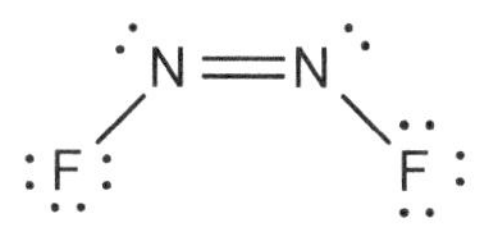

V-shaped about both N's;
≈120° about both N's;
both N atoms: sp^2

polar

Can also be:

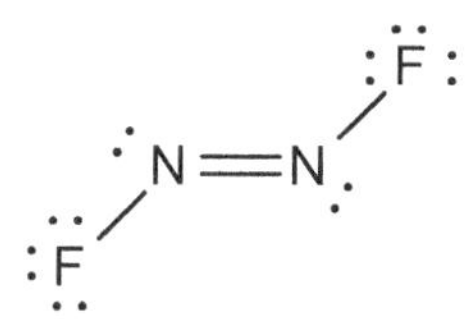

nonpolar

These are distinctly different molecules.

b. C_4H_6 has 4(4) + 6(1) = 22 valence electrons.

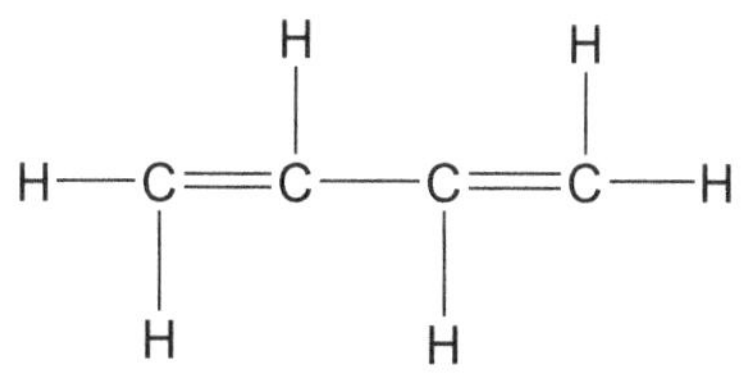

All C atoms are trigonal planar with 120° bond angles and sp^2 hybridization. Because C and H have similar electronegativity values, the C–H bonds are essentially nonpolar, so the molecule is nonpolar. All neutral compounds composed of only C and H atoms are nonpolar.

103. a. The Lewis structures for NNO and NON are:

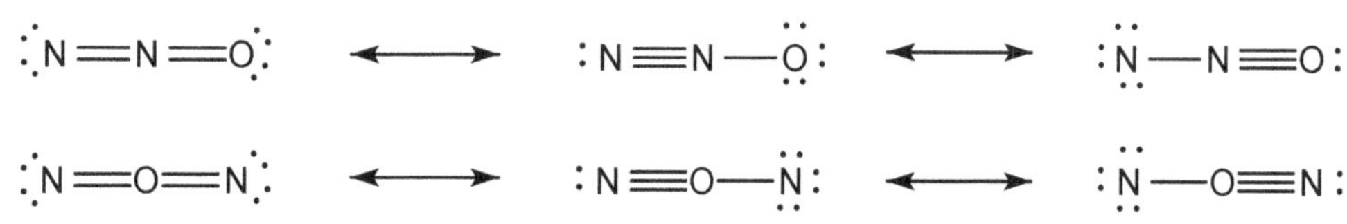

The NNO structure is correct. From the Lewis structures, we would predict both NNO and NON to be linear. However, we would predict NNO to be polar and NON to be nonpolar. Since experiments show N_2O to be polar, NNO is the correct structure.

b. Formal charge = number of valence electrons of atoms – [(number of lone pair electrons) + 1/2(number of shared electrons)].

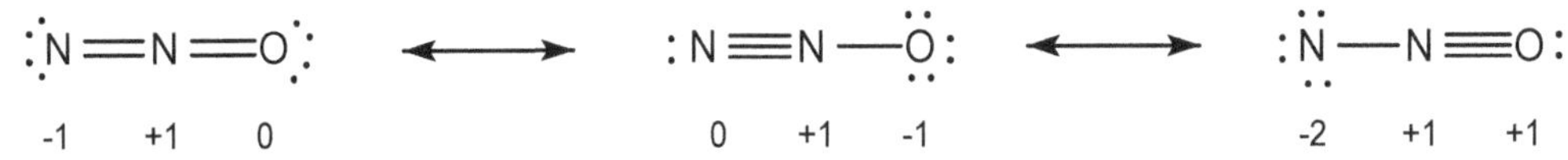

The formal charges for the atoms in the various resonance structures are below each atom. The central N is sp hybridized in all the resonance structures. We can probably ignore the third resonance structure on the basis of the relatively large formal charges as compared to the first two resonance structures.

c. The sp hybrid orbitals from the center N overlap with atomic orbitals (or appropriate hybrid orbitals) from the other two atoms to form the two sigma bonds. The remaining two unhybridized p orbitals from the center N overlap with two p orbitals from the peripheral N to form the two π bonds.

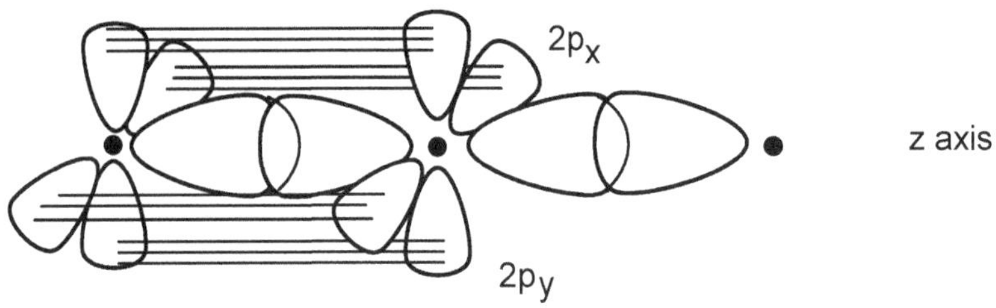

105. N_2: $(\sigma_{2s})^2(\sigma_{2s}^*)^2(\pi_{2p})^4(\sigma_{2p})^2$; N_2^+: $(\sigma_{2s})^2(\sigma_{2s}^*)^2(\pi_{2p})^4(\sigma_{2p})^1$

N_2^-: $(\sigma_{2s})^2(\sigma_{2s}^*)^2(\pi_{2p})^4(\sigma_{2p})^2(\pi_{2p}^*)^1$

107. O_2: $(\sigma_{2s})^2(\sigma_{2s}^*)^2(\sigma_{2p})^2(\pi_{2p})^4(\pi_{2p}^*)^2$ B.O. = bond order = (8 – 4)/2 = 2

N_2: $(\sigma_{2s})^2(\sigma_{2s}^*)^2(\pi_{2p})^4(\sigma_{2p})^2$ B.O. = (8 – 2)/2 = 3

In O_2, an antibonding electron is removed, which will increase the bond order to 2.5 [= (8 – 3)/2]. The bond order increases as an electron is removed, so the bond strengthens. In N_2, a bonding electron is removed, which decreases the bond order to 2.5 = [(7 – 2)/2]. So the bond strength weakens as an electron is removed from N_2.

109. F_2: $(\sigma_{2s})^2(\sigma_{2s}^*)^2(\sigma_{2p})^2(\pi_{2p})^4(\pi_{2p}^*)^4$; F_2 should have a lower ionization energy than F. The electron removed from F_2 is in a π_{2p}^* antibonding molecular orbital that is higher in energy than the 2p atomic orbitals from which the electron in atomic fluorine is removed. Because the electron removed from F_2 is higher in energy than the electron removed from F, it should be easier to remove an electron from F_2 than from F.

111. Side-to-side in-phase overlap of these d orbitals would produce a π bonding molecular orbital. There would be no probability of finding an electron on the axis joining the two nuclei, which is characteristic of π MOs.

113. Element X has 36 protons, which identifies it as Kr. Element Y has one less electron than Y^-, so the electron configuration of Y is $1s^22s^22p^5$. This is F.

KrF_3^+, $8 + 3(7) - 1 = 28$ e$^-$

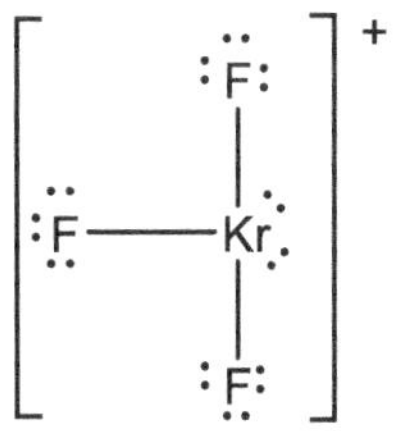

T-shaped, dsp^3

115. a. Li_2: $(\sigma_{2s})^2$ B.O. = (2 – 0)/2 = 1

B_2: $(\sigma_{2s})^2(\sigma_{2s}^*)^2(\pi_{2p})^2$ B.O. = (4 – 2)/2 = 1; both have a bond order of 1.

b. B_2 has four more electrons than Li_2, so four electrons must be removed from B_2 to make it isoelectronic with Li_2. The isoelectronic ion is B_2^{4+}.

c. To form B_2^{4+}, it takes 6455 kJ of energy to remove 4 mol of electrons from 1 mol of B_2.

$$1.5 \text{ kg } B_2 \times \frac{1000 \text{ g}}{1 \text{ kg}} \times \frac{1 \text{ mol } B_2}{21.62 \text{ g } B_2} \times \frac{6455 \text{ kJ}}{\text{mol } B_2} = 4.5 \times 10^5 \text{ kJ}$$

117. SO_2 (18 e$^-$)

a. True; the central sulfur has a trigonal planar arrangement of electron pairs, hence the bond angle is ~120° and the central sulfur atom is sp^2 hybridized.

b. False; all the S−O bonds are equivalent in strength with a bond length value somewhere between a single and a double bond.

c. True; d. True; e. False; the two resonance structures are shown above.

Challenge Problems

119. The following Lewis structure has a formal charge of zero for all of the atoms in the molecule.

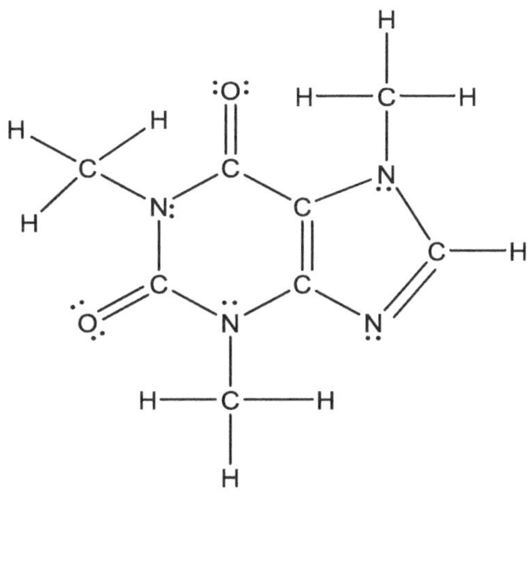

The three C atoms each bonded to three H atoms are sp^3 hybridized (tetrahedral geometry); the other five C atoms with trigonal planar geometry are sp^2 hybridized. The one N atom with the double bond is sp^2 hybridized, and the other three N atoms are sp^3 hybridized. The answers to the questions are:

- 6 total C and N atoms exhibit 120° bond angles
- 6 total C and N atoms are sp^3 hybridized
- 0 C and N atoms are sp hybridized (linear geometry)
- 25 σ bonds and 4 π bonds

121. a. NCN^{2-} has 5 + 4 + 5 + 2 = 16 valence electrons.

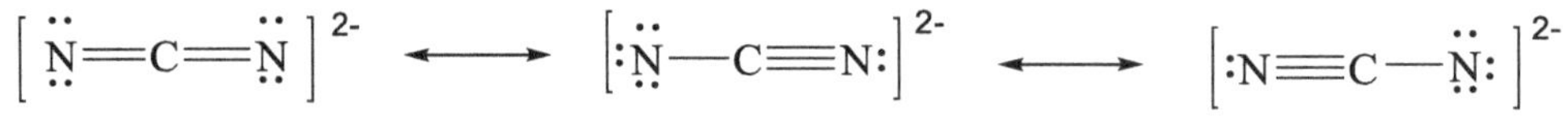

H_2NCN has 2(1) + 5 + 4 + 5 = 16 valence electrons.

favored by formal charge

$NCNC(NH_2)_2$ has 5 + 4 + 5 + 4 + 2(5) + 4(1) = 32 valence electrons.

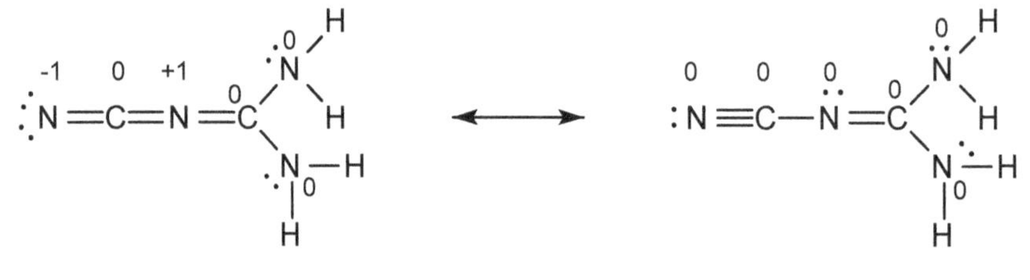

favored by formal charge

Melamine ($C_3N_6H_6$) has 3(4) + 6(5) + 6(1) = 48 valence electrons.

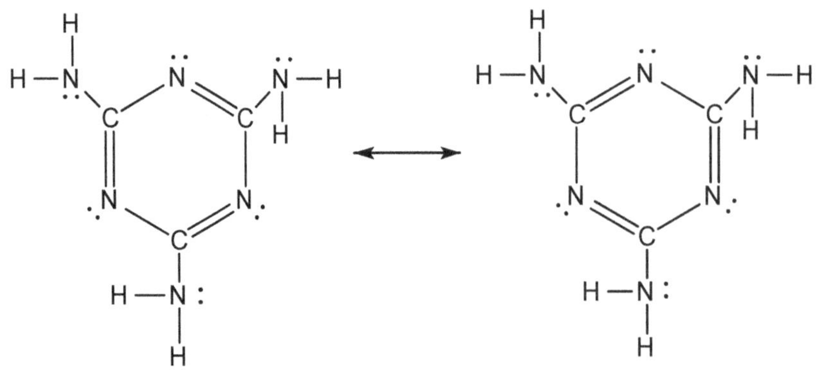

b. NCN^{2-}: C is sp hybridized. Each resonance structure predicts a different hybridization for the N atom. Depending on the resonance form, N is predicted to be sp, sp^2, or sp^3 hybridized. For the remaining compounds, we will give hybrids for the favored resonance structures as predicted from formal charge considerations.

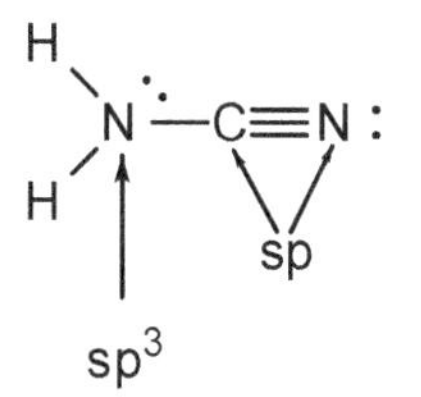

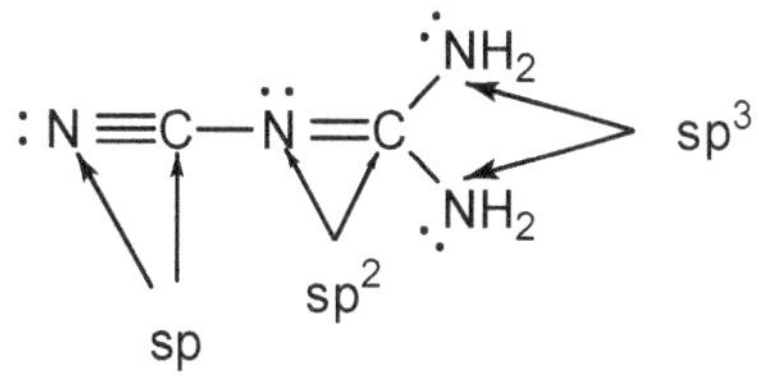

Melamine: N in NH_2 groups are all sp^3 hybridized; atoms in ring are all sp^2 hybridized.

c. NCN^{2-}: 2 σ and 2 π bonds; H_2NCN: 4 σ and 2 π bonds; dicyandiamide: 9 σ and 3 π bonds; melamine: 15 σ and 3 π bonds

d. The π-system forces the ring to be planar, just as the benzene ring is planar (see Figure 9.48 of the text).

e. The structure:

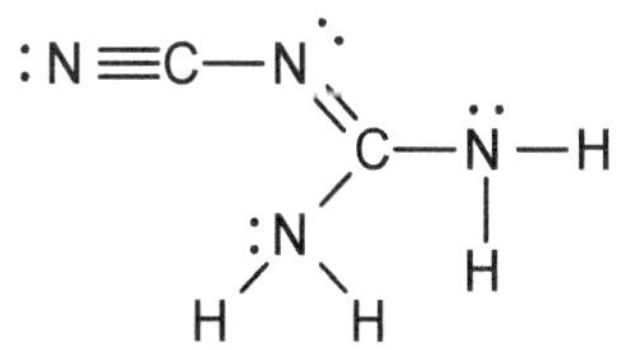

best agrees with experiments because it has three different CN bonds. This structure is also favored based on formal charge.

123. a. $E = \dfrac{hc}{\lambda} = \dfrac{(6.626 \times 10^{-34}\ \text{J s})(2.998 \times 10^{8}\ \text{m/s})}{25 \times 10^{-9}\ \text{m}} = 7.9 \times 10^{-18}\ \text{J}$

$$7.9 \times 10^{-18}\ \text{J} \times \frac{6.022 \times 10^{23}}{\text{mol}} \times \frac{1\ \text{kJ}}{1000\ \text{J}} = 4800\ \text{kJ/mol}$$

Using ΔH values from the various reactions, 25-nm light has sufficient energy to ionize N_2 and N and to break the triple bond. Thus N_2, N_2^+, N, and N^+ will all be present, assuming excess N_2.

b. To produce atomic nitrogen but no ions, the range of energies of the light must be from 941 kJ/mol to just below 1402 kJ/mol.

$$\frac{941\ \text{kJ}}{\text{mol}} \times \frac{1\ \text{mol}}{6.022 \times 10^{23}} \times \frac{1000\ \text{J}}{1\ \text{kJ}} = 1.56 \times 10^{-18}\ \text{J/photon}$$

$$\lambda = \frac{hc}{E} = \frac{(6.6261 \times 10^{-34}\ \text{J s})(2.998 \times 10^{8}\ \text{m/s})}{1.56 \times 10^{-18}\ \text{J}} = 1.27 \times 10^{-7}\ \text{m} = 127\ \text{nm}$$

$$\frac{1402\ \text{kJ}}{\text{mol}} \times \frac{1\ \text{mol}}{6.0221 \times 10^{23}} \times \frac{1000\ \text{J}}{\text{kJ}} = 2.328 \times 10^{-18}\ \text{J/photon}$$

$$\lambda = \frac{hc}{E} = \frac{(6.6261 \times 10^{-34}\ \text{J s})(2.9979 \times 10^{8}\ \text{m/s})}{2.328 \times 10^{-18}\ \text{J}} = 8.533 \times 10^{-8}\ \text{m} = 85.33\ \text{nm}$$

Light with wavelengths in the range of 85.33 nm $< \lambda \le$ 127 nm will produce N but no ions.

c. N_2: $(\sigma_{2s})^2(\sigma_{2s}^*)^2(\pi_{2p})^4(\sigma_{2p})^2$; the electron removed from N_2 is in the σ_{2p} molecular orbital, which is lower in energy than the 2p atomic orbital from which the electron in atomic nitrogen is removed. Because the electron removed from N_2 is lower in energy than the electron removed from N, the ionization energy of N_2 is greater than that for N.

125. O=N–Cl: The bond order of the NO bond in NOCl is 2 (a double bond).

NO: From molecular orbital theory, the bond order of this NO bond is 2.5. (See Figure 9.40 of the text.)

Both reactions apparently involve only the breaking of the N–Cl bond. However, in the reaction ONCl $\rightarrow$ NO + Cl, some energy is released in forming the stronger NO bond, lowering the value of ΔH. Therefore, the apparent N–Cl bond energy is artificially low for this reaction. The first reaction involves only the breaking of the N–Cl bond.

127. The ground state MO electron configuration for He_2 is $(\sigma_{1s})^2(\sigma_{1s}^*)^2$, giving a bond order of 0. Therefore, He_2 molecules are not predicted to be stable (and are not stable) in the lowest- energy ground state. However, in a high-energy environment, electron(s) from the anti-bonding orbitals in He_2 can be promoted into higher-energy bonding orbitals, thus giving a nonzero bond order and a "reason" to form. For example, a possible excited-state MO electron configuration for He_2 would be $(\sigma_{1s})^2(\sigma_{1s}^*)^1(\sigma_{2s})^1$, giving a bond order of (3 – 1)/2 = 1. Thus excited He_2 molecules can form, but they spontaneously break apart as the electron(s) fall back to the ground state, where the bond order equals zero.

129. The electron configurations are:

N_2: $(\sigma_{2s})^2(\sigma_{2s}^*)^2(\pi_{2p})^4(\sigma_{2p})^2$

O_2: $(\sigma_{2s})^2(\sigma_{2s}^*)^2(\sigma_{2p})^2(\pi_{2p})^4(\pi_{2p}^*)^2$

N_2^{2-}: $(\sigma_{2s})^2(\sigma_{2s}^*)^2(\pi_{2p})^4(\sigma_{2p})^2(\pi_{2p}^*)^2$

N_2^-: $(\sigma_{2s})^2(\sigma_{2s}^*)^2(\pi_{2p})^4(\sigma_{2p})^2(\pi_{2p}^*)^1$

O_2^+: $(\sigma_{2s})^2(\sigma_{2s}^*)^2(\sigma_{2p})^2(\pi_{2p})^4(\pi_{2p}^*)^1$

Note: The ordering of the σ_{2p} and π_{2p} orbitals is not important to this question.

The species with the smallest ionization energy has the electron that is easiest to remove. From the MO electron configurations, O_2, N_2^{2-}, N_2^-, and O_2^+ all contain electrons in the same higher-energy antibonding orbitals (π_{2p}^*), so they should have electrons that are easier to remove as compared to N_2, which has no π_{2p}^* electrons. To differentiate which has the easiest π_{2p}^* to remove, concentrate on the number of electrons in the orbitals attracted to the number of protons in the nucleus.

N_2^{2-} and N_2^- both have 14 protons in the two nuclei combined. Because N_2^{2-} has more electrons, one would expect N_2^{2-} to have more electron repulsions, which translates into having an easier electron to remove. Between O_2 and O_2^+, the electron in O_2 should be easier to remove. O_2 has one more electron than O_2^+, and one would expect the fewer electrons in O_2^+ to be better attracted to the nuclei (and harder to remove). Between N_2^{2-} and O_2, both have 16 electrons; the difference is the number of protons in the nucleus. Because N_2^{2-} has two fewer protons than O_2, one would expect the N_2^{2-} to have the easiest electron to remove, which translates into the smallest ionization energy.

131. a. The CO bond is polar with the negative end at the more electronegative oxygen atom. We would expect metal cations to be attracted to and bond to the oxygen end of CO on the basis of electronegativity.

b.  FC (carbon) = 4 – 2 – 1/2(6) = –1

FC (oxygen) = 6 – 2 – 1/2(6) = +1

From formal charge, we would expect metal cations to bond to the carbon (with the negative formal charge).

c. In molecular orbital theory, only orbitals with proper symmetry overlap to form bonding orbitals. The metals that form bonds to CO are usually transition metals, all of which have outer electrons in the d orbitals. The only molecular orbitals of CO that have proper symmetry to overlap with d orbitals are the π_{2p}* orbitals, whose shape is similar to the d orbitals. Because the antibonding molecular orbitals have more carbon character (carbon is less electronegative than oxygen), one would expect the bond to form through carbon.

CHAPTER 10

LIQUIDS AND SOLIDS

Questions

19. Answer a is correct. Intermolecular forces are the forces between molecules that hold the substances together in the solid and liquid phases. Hydrogen bonding is a specific type of intermolecular forces. In this figure, the dotted lines represent the hydrogen bonding interactions that hold individual H_2O molecules together in the solid and liquid phases. The solid lines represent the O–H covalent bonds.

21. The halogens are all nonpolar substances, and nonpolar substances only exhibit London dispersion forces. In general, the larger the molar mass, the stronger the LD forces. I2 has the largest molar mass, so it has the highest boiling point.

23. Ideal gas molecules are assumed to exhibit no intermolecular forces and are assumed to be volumeless. Real gases deviate from ideal gas behavior because real gases do exhibit intermolecular forces and real gas molecules do have a volume. Between two gases, the gas with the weaker intermolecular forces and the smaller size should behave more ideally at some conditions. Both N_2 and CO have similar size (molar mass), but the nonpolar N_2 molecules only exhibit London dispersion forces while the polar CO molecules exhibit additional dipole force. N_2 will have the weaker intermolecular forces and should behave more ideally than CO.

25. Atoms have an approximately spherical shape (on average). It is impossible to pack spheres together without some empty space among the spheres.

27. Atoms touch along the edge in a simple cubic unit cell, so Po has a simple cubic unit cell. Atoms touch along the face diagonal in a face-centered unit cell, so Al has a face-centered unit cell. Atoms touch along the body-diagonal in a body-centered unit cell, so Na has a body-centered unit cell.

29. An alloy is a substance that contains a mixture of elements and has metallic properties. In a substitutional alloy, some of the host metal atoms are replaced by other metal atoms of similar size, e.g., brass, pewter, plumber's solder. An interstitial alloy is formed when some of the interstices (holes) in the closest packed metal structure are occupied by smaller atoms, e.g., carbon steels.

31. a. As the strength of the intermolecular forces increase, the rate of evaporation decreases.

    b. As temperature increases, the rate of evaporation increases.

    c. As surface area increases, the rate of evaporation increases.

33. $C_2H_5OH(l) \rightarrow C_2H_5OH(g)$ is an endothermic process. Heat is absorbed when liquid ethanol vaporizes; the internal heat from the body provides this heat, which results in the cooling of the body.

35. Sublimation will occur, allowing water to escape as $H_2O(g)$.

37. The strength of intermolecular forces determines relative boiling points. The types of intermolecular forces for covalent compounds are London dispersion forces, dipole forces, and hydrogen bonding. Because the three compounds are assumed to have similar molar mass and shape, the strength of the London dispersion forces will be about equal among the three compounds. One of the compounds will be nonpolar, so it only has London dispersion forces. The other two compounds will be polar, so they have additional dipole forces and will boil at a higher temperature than the nonpolar compound. One of the polar compounds will have an H covalently bonded to either N, O, or F. This gives rise to the strongest type of covalent intermolecular forces, hydrogen bonding. The compound that hydrogen bonds will have the highest boiling point, whereas the polar compound with no hydrogen bonding will boil at a temperature in the middle of the other compounds.

39. a. Both CO_2 and H_2O are molecular solids. Both have an ordered array of the individual molecules, with the molecular units occupying the lattice points. A difference within each solid lattice is the strength of the intermolecular forces. CO_2 is nonpolar and only exhibits London dispersion forces. H_2O exhibits the relatively strong hydrogen-bonding interactions. The differences in strength is evidenced by the solid-phase changes that occur at 1 atm. $CO_2(s)$ sublimes at a relatively low temperature of $-78°C$. In sublimation, all of the intermolecular forces are broken. However, $H_2O(s)$ doesn't have a phase change until $0°C$, and in this phase change from ice to water, only a fraction of the intermolecular forces are broken. The higher temperature and the fact that only a portion of the intermolecular forces are broken are attributed to the strength of the intermolecular forces in $H_2O(s)$ as compared to $CO_2(s)$.

    Related to the intermolecular forces are the relative densities of the solid and liquid phases for these two compounds. $CO_2(s)$ is denser than $CO_2(l)$, whereas $H_2O(s)$ is less dense than $H_2O(l)$. For $CO_2(s)$ and for most solids, the molecules pack together as close as possible; hence solids are usually more dense than the liquid phase. H_2O is an exception to this. Water molecules are particularly well suited for hydrogen bonding interaction with each other because each molecule has two polar O–H bonds and two lone pairs on the oxygen. This can lead to the association of four hydrogen atoms with each oxygen atom: two by covalent bonds and two by dipoles. To keep this arrangement (which maximizes the hydrogen-bonding interactions), the $H_2O(s)$ molecules occupy positions that create empty space in the lattice. This translates into a smaller density for $H_2O(s)$ as compared to $H_2O(l)$.

    b. Both NaCl and CsCl are ionic compounds with the anions at the lattice points of the unit cells and the cations occupying the empty spaces created by anions (called holes). In NaCl, the Cl^- anions occupy the lattice points of a face-centered unit cell, with the Na^+ cations occupying the octahedral holes. Octahedral holes are the empty spaces created by six Cl^- ions. CsCl has the Cl^- ions at the lattice points of a simple cubic unit cell, with the Cs^+ cations occupying the middle of the cube.

41. Chalk is composed of the ionic compound calcium carbonate ($CaCO_3$). The electrostatic forces in ionic compounds are much stronger than the intermolecular forces in covalent compounds. Therefore, $CaCO_3$ should have a much higher boiling point than the covalent compounds found in motor oil and in H_2O. Motor oil is composed of nonpolar C–C and C–H bonds. The intermolecular forces in motor oil are therefore London dispersion forces. We generally consider these forces to be weak. However, with compounds that have large molar

masses, these London dispersion forces add up significantly and can overtake the relatively strong hydrogen-bonding interactions in water.

43. The mathematical equation that relates the vapor pressure of a substance to temperature is:

$$\ln P_{vap} = -\frac{\Delta H_{vap}}{R}\left(\frac{1}{T}\right) + C$$

$$y \qquad m \qquad x \quad + b$$

This equation is in the form of the straight-line equation ($y = mx + b$) If one plots ln P_{vap} versus 1/T with temperature in Kelvin, the slope (m) of the straight line is $-\Delta H_{vap}/R$. Because ΔH_{vap} is always positive, the slope of the straight line will be negative.

Exercises

Intermolecular Forces and Physical Properties

45. Ionic compounds have ionic forces. Covalent compounds all have London dispersion (LD) forces, whereas polar covalent compounds have dipole forces and/or hydrogen bonding forces. For hydrogen-bonding (H-bonding) forces, the covalent compound must have either a N–H, O–H, or F–H bond in the molecule.

a. LD only

b. dipole, LD

c. H-bonding, LD

d. ionic

e. LD only (CH_4 is a nonpolar covalent compound.)

f. dipole, LD

g. ionic

47. NaF, CaF_2, and CrF_3 are ionic compounds that exhibit extremely strong ionic forces. HF is a covalent compound which can form hydrogen-bonding interactions. This leaves CF_4. The Lewis structure is:

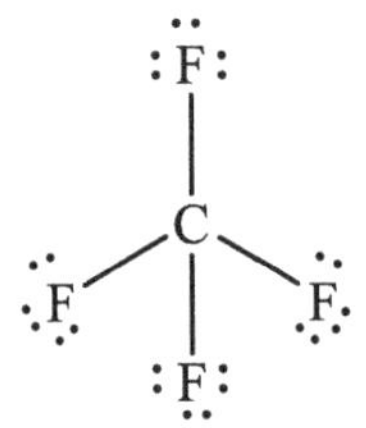

CF_4 is a nonpolar covalent compound and nonpolar compounds only exhibit London dispersion forces. Just one of the compounds (CF_4) only exhibits LD forces.

49. a. OCS; OCS is polar and has dipole-dipole forces in addition to London dispersion (LD) forces. All polar molecules have dipole forces. CO_2 is nonpolar and only has LD forces. To predict polarity, draw the Lewis structure and deduce whether the individual bond dipoles cancel.

b. SeO_2; both SeO_2 and SO_2 are polar compounds, so they both have dipole forces as well as LD forces. However, SeO_2 is a larger molecule, so it would have stronger LD forces.

c. $H_2NCH_2CH_2NH_2$; more extensive hydrogen bonding (H-bonding) is possible because two NH_2 groups are present.

d. H_2CO; H_2CO is polar, whereas CH_3CH_3 is nonpolar. H_2CO has dipole forces in addition to LD forces. CH_3CH_3 only has LD forces.

e. CH_3OH; CH_3OH can form relatively strong H-bonding interactions, unlike H_2CO.

51. Ar exists as individual atoms that are held together in the condensed phases by London dispersion forces. The molecule that will have a boiling point closest to Ar will be a nonpolar substance with about the same molar mass as Ar (39.95 g/mol); this same size nonpolar substance will have about equivalent strength of London dispersion forces. Of the choices, only Cl_2 (70.90 g/mol) and F_2 (38.00 g/mol) are nonpolar. Because F_2 has a molar mass closest to that of Ar, one would expect the boiling point of F_2 to be close to that of Ar.

53. The Lewis structure for CF_2H_2 is:

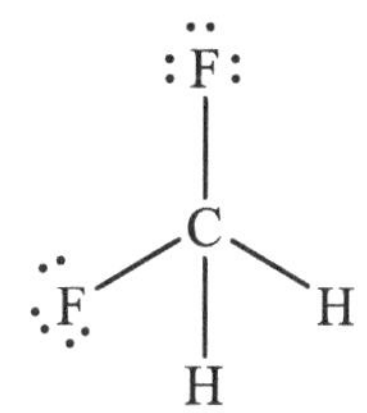

To form hydrogen-bonding intermolecular forces, you must have either a N–H, an O–H, and/or a H–F covalent bond in the compound. CF_2H_2 has a hydrogen and a fluorine in the formula, but the compound does not have a H-F bond. So CF_2H_2 does not exhibit H-bonding (statement a is false). All the other statements are true.

55. a. Kr and HBr have similar molar masses, so the London dispersion forces in each substance should be about the same. However, HBr is a polar compound while Kr is nonpolar. HBr has additional dipole forces so it boils at a higher temperature.

b. HF is capable of H-bonding; HBr is not.

c. LiCl is ionic, and HF is a molecular solid with hydrogen bonding and LD forces. Ionic forces are much stronger than the forces for molecular solids.

d. n-Hexane is a larger molecule, so it has stronger LD forces.

57. Boiling points and freezing points are assumed directly related to the strength of the intermolecular forces, whereas vapor pressure is inversely related to the strength of the intermolecular forces.

a. HBr; HBr is polar, whereas Kr and Cl_2 are nonpolar. HBr has dipole forces unlike Kr and Cl_2. So HBr has the stronger intermolecular forces and the highest boiling point.

b. NaCl; the ionic forces in NaCl are much stronger than the intermolecular forces for molecular substances, so NaCl has the highest melting point.

c. I_2; all are nonpolar, so the largest molecule (I_2) will have the strongest LD (London Dispersion) forces and the lowest vapor pressure.

d. N_2; nonpolar and smallest, so it has the weakest intermolecular forces.

e. CH_4; smallest, nonpolar molecule, so it has the weakest LD forces.

f. HF; HF can form relatively strong H-bonding interactions, unlike the others.

g. $CH_3CH_2CH_2OH$; H-bonding, unlike the others, so it has strongest intermolecular forces.

Properties of Liquids

59. The attraction of H_2O for glass is stronger than the $H_2O–H_2O$ attraction. The meniscus is concave to increase the area of contact between glass and H_2O. The Hg–Hg attraction is greater than the Hg–glass attraction. The meniscus is convex to minimize the Hg–glass contact.

61. The structure of H_2O_2 is H–O–O–H, which produces greater hydrogen bonding than in water. Thus the intermolecular forces are stronger in H_2O_2 than in H_2O, resulting in a higher normal boiling point for H_2O_2 and a lower vapor pressure.

Structures and Properties of Solids

63. $n\lambda = 2d\sin\theta,\ d = \dfrac{n\lambda}{2\sin\theta} = \dfrac{1 \times 154\text{ pm}}{2 \times \sin 14.22°} = 313\text{ pm} = 3.13 \times 10^{-10}\text{m}$

65. $\lambda = \dfrac{2d\sin\theta}{n} = \dfrac{2 \times 1.36 \times 10^{-10}\text{ m} \times \sin 15.0°}{1} = 7.04 \times 10^{-11}\text{m} = 0.704\text{ Å} = 70.4\text{ pm}$

67. A cubic closest packed structure has a face-centered cubic unit cell. In a face-centered cubic unit, there are:

$$8\text{ corners} \times \frac{1/8\text{ atom}}{\text{corner}} + 6\text{ faces} \times \frac{1/2\text{ atom}}{\text{face}} = 4\text{ atoms}$$

The atoms in a face-centered cubic unit cell touch along the face diagonal of the cubic unit cell. Using the Pythagorean formula, where l = length of the face diagonal and r = radius of the atom:

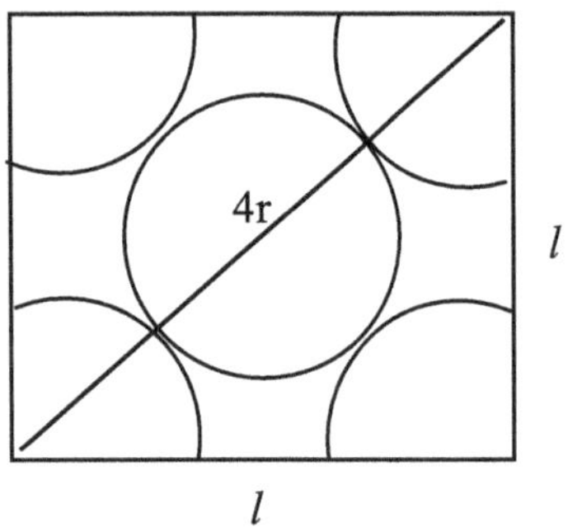

$$l^2 + l^2 = (4r)^2$$

$$2l^2 = 16r^2$$

$$l = r\sqrt{8}$$

$l = r\sqrt{8} = 197 \times 10^{-12}\text{m} \times \sqrt{8} = 5.57 \times 10^{-10}\text{m} = 5.57 \times 10^{-8}\text{cm}$

Volume of a unit cell $= l^3 = (5.57 \times 10^{-8}\text{cm})^3 = 1.73 \times 10^{-22}\text{cm}^3$

$$\text{Mass of a unit cell} = 4 \text{ Ca atoms} \times \frac{1 \text{ mol Ca}}{6.022 \times 10^{23} \text{ atoms}} \times \frac{40.08 \text{ g Ca}}{\text{mol Ca}} = 2.662 \times 10^{-22} \text{ g Ca}$$

$$\text{Density} = \frac{\text{mass}}{\text{volume}} = \frac{2.662 \times 10^{-22} \text{ g}}{1.73 \times 10^{-22} \text{ cm}^3} = 1.54 \text{ g/cm}^3$$

69. There are four Ni atoms in each unit cell. For a unit cell:

$$\text{density} = \frac{\text{mass}}{\text{volume}} = 6.84 \text{ g/cm}^3 = \frac{4 \text{ Ni atoms} \times \frac{1 \text{ mol Ni}}{6.022 \times 10^{23} \text{ atoms}} \times \frac{58.69 \text{ g Ni}}{\text{mol Ni}}}{l^3}$$

Solving: $l = 3.85 \times 10^{-8}$ cm = cube edge length

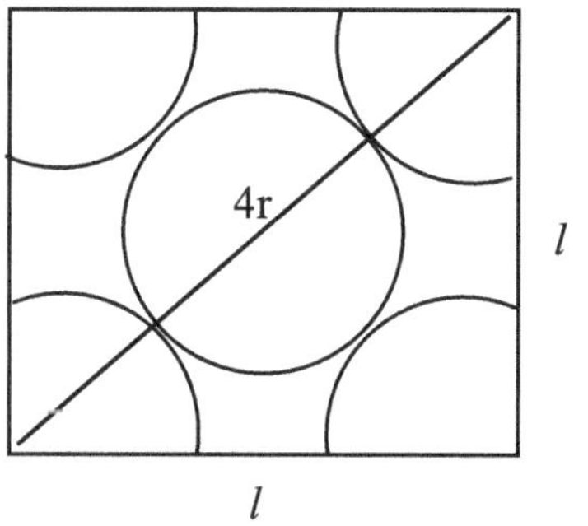

For a face-centered cube:

$(4r)^2 = l^2 + l^2 = 2l^2$

$r\sqrt{8} = l, \; r = l/\sqrt{8}$

$r = 3.85 \times 10^{-8} \text{ cm} / \sqrt{8}$

$r = 1.36 \times 10^{-8} \text{ cm} = 136 \text{ pm}$

71. A face-centered cubic unit cell contains four atoms. For a unit cell:

$$\text{mass of X} = \text{volume} \times \text{density} = (4.09 \times 10^{-8} \text{ cm})^3 \times 10.5 \text{ g/cm}^3 = 7.18 \times 10^{-22} \text{ g}$$

$$\text{mol X} = 4 \text{ atoms X} \times \frac{1 \text{ mol X}}{6.022 \times 10^{23} \text{ atoms}} = 6.642 \times 10^{-24} \text{ mol X}$$

$$\text{Molar mass} = \frac{7.18 \times 10^{-22} \text{ g X}}{6.642 \times 10^{-24} \text{ mol X}} = 108 \text{ g/mol; the metal is silver (Ag).}$$

73. For a body-centered unit cell, 8 corners $\times \dfrac{1/8 \text{ Ti}}{\text{corner}}$ + Ti at body center = 2 Ti atoms.

All body-centered unit cells have two atoms per unit cell. For a unit cell where l = cube edge length:

$$\text{density} = 4.50 \text{ g/cm}^3 = \frac{2 \text{ atoms Ti} \times \frac{1 \text{ mol Ti}}{6.022 \times 10^{23} \text{ atoms}} \times \frac{47.88 \text{ g Ti}}{\text{mol Ti}}}{l^3}$$

Solving: l = edge length of unit cell = 3.28×10^{-8} cm = 328 pm

Assume Ti atoms just touch along the body diagonal of the cube, so body diagonal = 4 × radius of atoms = 4r.

The triangle we need to solve is:

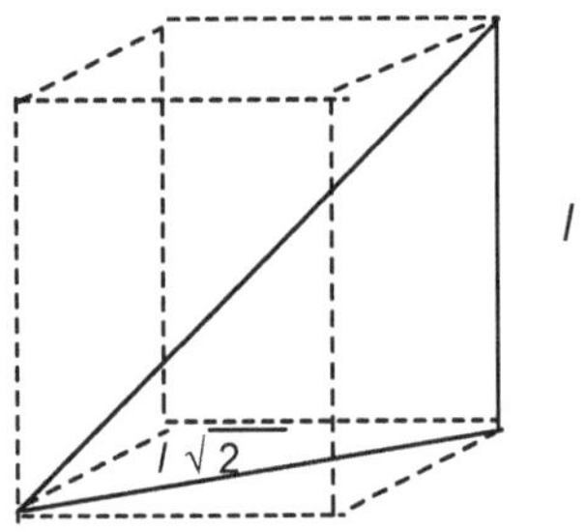

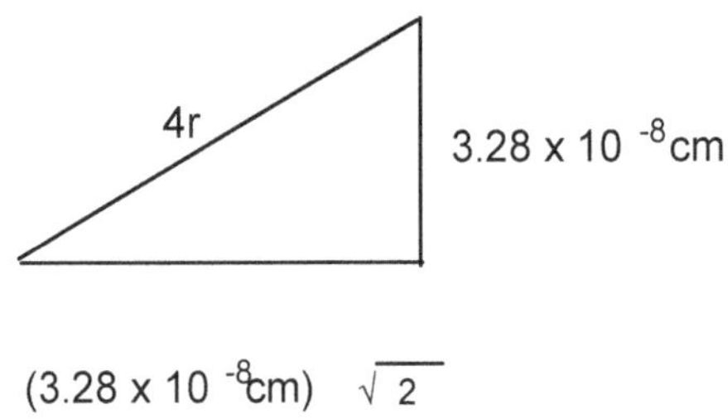

$$(4r)^2 = (3.28 \times 10^{-8}\,\text{cm})^2 + [(3.28 \times 10^{-8}\,\text{cm})\sqrt{2}\,]^2,\ \ r = 1.42 \times 10^{-8}\,\text{cm} = 142\ \text{pm}$$

For a body-centered unit cell (bcc), the radius of the atom is related to the cube edge length by: $4r = l\sqrt{3}$ or $l = 4r/\sqrt{3}$.

75. If gold has a face-centered cubic structure, then there are four atoms per unit cell, and from Exercise 67:

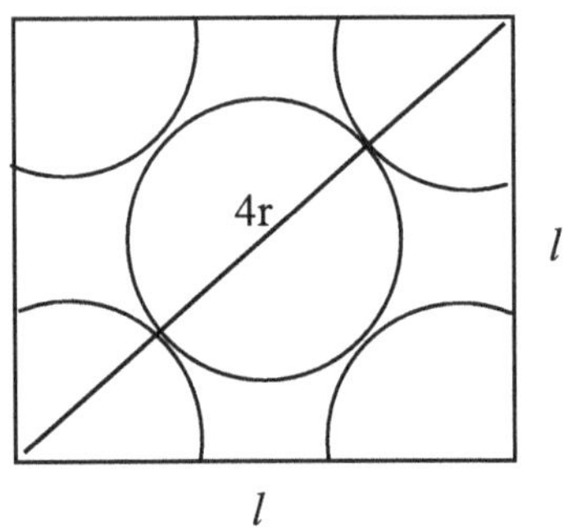

$2l^2 = 16r^2$

$l = r\sqrt{8} = (144\ \text{pm})\sqrt{8} = 407\ \text{pm}$

$l = 407 \times 10^{-12}\ \text{m} = 4.07 \times 10^{-8}\ \text{cm}$

$$\text{Density} = \frac{4\ \text{atoms Au} \times \dfrac{1\ \text{mol Au}}{6.022 \times 10^{23}\ \text{atoms}} \times \dfrac{197.0\ \text{g Au}}{\text{mol Au}}}{(4.07 \times 10^{-8}\ \text{cm})^3} = 19.4\ \text{g/cm}^3$$

If gold has a body-centered cubic structure, then there are two atoms per unit cell, and from Exercise 73:

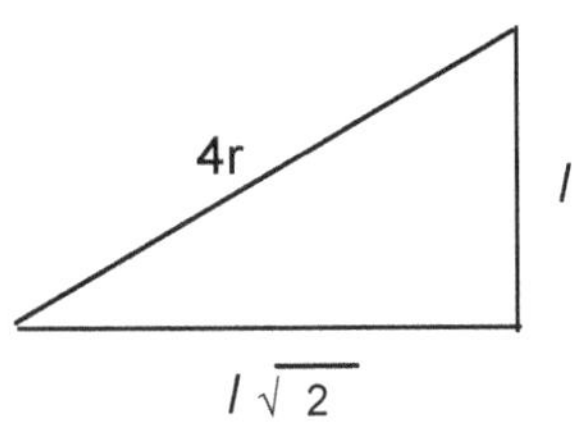

$16r^2 = l^2 + 2l^2$

$l = 4r/\sqrt{3} = 333\ \text{pm} = 333 \times 10^{-12}\ \text{m}$

$l = 333 \times 10^{-10}\ \text{cm} = 3.33 \times 10^{-8}\ \text{cm}$

$$\text{Density} = \frac{2\ \text{atoms Au} \times \dfrac{1\ \text{mol Au}}{6.022 \times 10^{23}\ \text{atoms}} \times \dfrac{197.0\ \text{g Au}}{\text{mol Au}}}{(3.33 \times 10^{-8}\ \text{cm})^3} = 17.7\ \text{g/cm}^3$$

The measured density of gold is consistent with a face-centered cubic unit cell.

77. In a face-centered unit cell (a cubic closest packed structure), the atoms touch along the face diagonal:

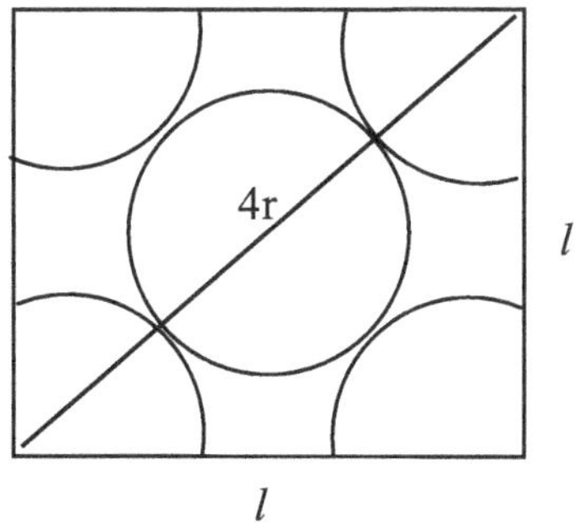

$(4r)^2 = l^2 + l^2$

$l = r\sqrt{8}$

$V_{cube} = l^3 = (r\sqrt{8})^3 = (22.63)r^3$

There are four atoms in a face-centered cubic cell (see Exercise 53). Each atom has a volume of $(4/3)\pi r^3$ = volume of a sphere.

$$V_{atoms} = 4 \times \frac{4}{3}\pi r^3 = (16.76)r^3$$

So $\frac{V_{atoms}}{V_{cube}} = \frac{(16.76)r^3}{(22.63)r^3}$ = 0.7406, or 74.06% of the volume of each unit cell is occupied by atoms.

In a simple cubic unit cell, the atoms touch along the cube edge l:

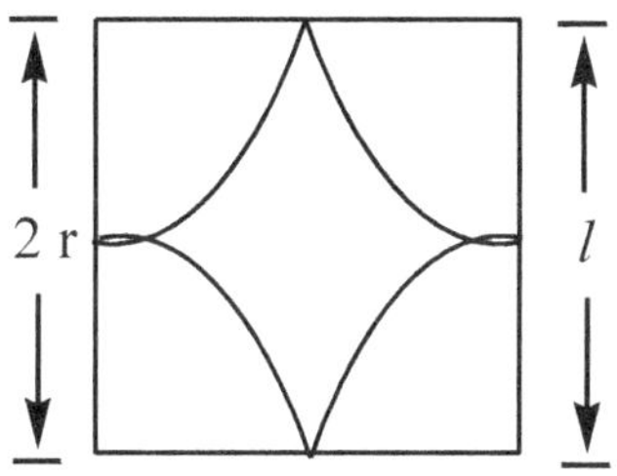

$2(\text{radius}) = 2r = l$

$V_{cube} = l^3 = (2r)^3 = 8r^3$

There is one atom per simple cubic cell (8 corner atoms × 1/8 atom per corner = 1 atom/unit cell). Each atom has an assumed volume of $(4/3)\pi r^3$ = volume of a sphere.

$$V_{atom} = \frac{4}{3}\pi r^3 = (4.189)r^3$$

So $\frac{V_{atom}}{V_{cube}} = \frac{(4.189)r^3}{8r^3}$ = 0.5236, or 52.36% of the volume of each unit cell is occupied by atoms.

A cubic closest packed structure (face-centered cubic unit cell) packs the atoms much more efficiently than a simple cubic structure.

79. Doping silicon with phosphorus produces an n-type semiconductor. The phosphorus adds electrons at energies near the conduction band of silicon. Electrons do not need as much energy to move from filled to unfilled energy levels, so conduction increases. Doping silicon with

gallium produces a p-type semiconductor. Because gallium has fewer valence electrons than silicon, holes (unfilled energy levels) at energies in the previously filled molecular orbitals are created, which induces greater electron movement (greater conductivity).

81. In has fewer valence electrons than Se. Thus Se doped with In would be a p-type semiconductor.

83. $E_{gap} = 2.5 \text{ eV} \times 1.6 \times 10^{-19} \text{ J/eV} = 4.0 \times 10^{-19} \text{ J}$; we want $E_{gap} = E_{light} = hc/\lambda$, so:

$$\lambda = \frac{hc}{E} = \frac{(6.63 \times 10^{-34} \text{ J s})(3.00 \times 10^{8} \text{ m/s})}{4.0 \times 10^{-19} \text{ J}} = 5.0 \times 10^{-7} \text{ m} = 5.0 \times 10^{2} \text{ nm}$$

85. Sodium chloride structure: $8 \text{ corners} \times \frac{1/8 \text{ Cl}^-}{\text{corner}} + 6 \text{ faces} \times \frac{1/2 \text{ Cl}^-}{\text{face}} = 4 \text{ Cl}^- \text{ ions}$

$12 \text{ edges} \times \frac{1/4 \text{ Na}^+}{\text{edge}} + 1 \text{ Na}^+$ at body center = 4 Na^+ ions; NaCl is the formula.

Cesium chloride structure: 1 Cs^+ ion at body center; $8 \text{ corners} \times \frac{1/8 \text{ Cl}^-}{\text{corner}} = 1 \text{ Cl}^- \text{ ion}$

CsCl is the formula.

Zinc sulfide structure: There are four Zn^{2+} ions inside the cube.

$8 \text{ corners} \times \frac{1/8 \text{ S}^{2-}}{\text{corner}} + 6 \text{ faces} \times \frac{1/2 \text{ S}^{2-}}{\text{face}} = 4 \text{ S}^{2-}$ ions; ZnS is the formula.

Titanium oxide structure: $8 \text{ corners} \times \frac{1/8 \text{ Ti}^{4+}}{\text{corner}} + 1 \text{ Ti}^{4+}$ at body center = 2 Ti^{4+} ions

$4 \text{ faces} \times \frac{1/2 \text{ O}^{2-}}{\text{face}} + 2 \text{ O}^{2-}$ inside cube = 4 O^{2-} ions; TiO_2 is the formula.

87. There is one octahedral hole per closest packed anion in a closest packed structure. If one-half of the octahedral holes are filled, then there is a 2 : 1 ratio of fluoride ions to cobalt ions in the crystal. The formula is CoF_2, which is composed of Co^{2+} and F^- ions.

89. In a cubic closest packed array of anions, there are twice the number of tetrahedral holes as anions present, and an equal number of octahedral holes as anions present. A cubic closest packed array of sulfide ions will have four S^{2-} ions, eight tetrahedral holes, and four octahedral holes. In this structure we have 1/8(8) = 1 Zn^{2+} ion and 1/2(4) = 2 Al^{3+} ions present, along with the 4 S^{2-} ions. The formula is $ZnAl_2S_4$.

91. 8 F^- ions at corners × 1/8 F^-/corner = 1 F^- ion per unit cell; because there is one cubic hole per cubic unit cell, there is a 2 : 1 ratio of F^- ions to metal ions in the crystal if only half of the body centers are filled with the metal ions. The formula is MF_2, where M^{2+} is the metal ion.

93. From Figure 10.36, MgO has the NaCl structure containing 4 Mg^{2+} ions and 4 O^{2-} ions per face-centered unit cell.

$$4 \text{ MgO formula units} \times \frac{1 \text{ mol MgO}}{6.022 \times 10^{23} \text{ atoms}} \times \frac{40.31 \text{ g MgO}}{1 \text{ mol MgO}} = 2.678 \times 10^{-22} \text{ g MgO}$$

$$\text{Volume of unit cell} = 2.678 \times 10^{-22} \text{ g MgO} \times \frac{1 \text{ cm}^3}{3.58 \text{ g}} = 7.48 \times 10^{-23} \text{ cm}^3$$

Volume of unit cell = l^3, l = cube edge length; $l = (7.48 \times 10^{-23} \text{ cm}^3)^{1/3} = 4.21 \times 10^{-8}$ cm

For a face-centered unit cell, the O^{2-} ions touch along the face diagonal:

$$\sqrt{2}\, l = 4 r_{O^{2-}}, \quad r_{O^{2-}} = \frac{\sqrt{2} \times 4.21 \times 10^{-8} \text{ cm}}{4} = 1.49 \times 10^{-8} \text{ cm}$$

The cube edge length goes through two radii of the O^{2-} anions and the diameter of the Mg^{2+} cation, so:

$$l = 2 r_{O^{2-}} + 2 r_{Mg^{2+}}, \quad 4.21 \times 10^{-8} \text{ cm} = 2(1.49 \times 10^{-8} \text{ cm}) + 2 r_{Mg^{2+}}, \quad r_{Mg^{2+}} = 6.15 \times 10^{-9} \text{ cm}$$

95. CsCl is a simple cubic array of Cl^- ions with Cs^+ in the middle of each unit cell. There is one Cs^+ and one Cl^- ion in each unit cell. Cs^+ and Cl^- ions touch along the body diagonal.

Body diagonal = $2 r_{Cs^+} + 2 r_{Cl^-} = \sqrt{3}\, l$, l = length of cube edge

In each unit cell:

$$\text{mass} = 1 \text{ CsCl formula unit} \times \frac{1 \text{ mol CsCl}}{6.022 \times 10^{23} \text{ formula units}} \times \frac{168.4 \text{ g CsCl}}{\text{mol CsCl}} = 2.796 \times 10^{-22} \text{ g}$$

$$\text{volume} = l^3 = 2.796 \times 10^{-22} \text{ g CsCl} \times \frac{1 \text{ cm}^3}{3.97 \text{ g CsCl}} = 7.04 \times 10^{-23} \text{ cm}^3$$

$l^3 = 7.04 \times 10^{-23}$ cm^3, $l = 4.13 \times 10^{-8}$ cm = 413 pm = length of cube edge

$$2 r_{Cs^+} + 2 r_{Cl^-} = \sqrt{3}\, l = \sqrt{3}(413 \text{ pm}) = 715 \text{ pm}$$

The distance between ion centers = $r_{Cs^+} + r_{Cl^-}$ = 715 pm/2 = 358 pm

From ionic radius: r_{Cs^+} = 169 pm and r_{Cl^-} = 181 pm; $r_{Cs^+} + r_{Cl^-}$ = 169 + 181 = 350. pm

The distance calculated from the density is 8 pm (2.3%) greater than that calculated from tables of ionic radii.

97. a. CO_2: molecular b. SiO_2: network c. Si: atomic, network

d. CH_4: molecular e. Ru: atomic, metallic f. I_2: molecular

g. KBr: ionic h. H_2O: molecular i. NaOH: ionic

j. U: atomic, metallic k. $CaCO_3$: ionic l. PH_3: molecular

99. a. The unit cell consists of Ni at the cube corners and Ti at the body center or Ti at the cube corners and Ni at the body center.

b. $8 \times 1/8 = 1$ atom from corners + 1 atom at body center; empirical formula = NiTi

c. Both have a coordination number of 8 (both are surrounded by 8 atoms).

101. Structure 1 (on left)

$$8 \text{ corners} \times \frac{1/8 \text{ Ca}}{\text{corner}} = 1 \text{ Ca atom}$$

$$6 \text{ faces} \times \frac{1/2 \text{ O}}{\text{face}} = 3 \text{ O atoms}$$

1 Ti at body center. Formula = $CaTiO_3$

Structure 2 (on right)

$$8 \text{ corners} \times \frac{1/8 \text{ Ti}}{\text{corner}} = 1 \text{ Ti atom}$$

$$12 \text{ edges} \times \frac{1/4 \text{ O}}{\text{corner}} = 3 \text{ O atoms}$$

1 Ca at body center. Formula = $CaTiO_3$

In the extended lattice of both structures, each Ti atom is surrounded by six O atoms.

103. a. Y: 1 Y in center; Ba: 2 Ba in center

$$\text{Cu: } 8 \text{ corners} \times \frac{1/8 \text{ Cu}}{\text{corner}} = 1 \text{ Cu}, \ 8 \text{ edges} \times \frac{1/4 \text{ Cu}}{\text{edge}} = 2 \text{ Cu}, \ \text{total} = 3 \text{ Cu atoms}$$

$$\text{O: } 20 \text{ edges} \times \frac{1/4 \text{ O}}{\text{edge}} = 5 \text{ O atoms}, \ 8 \text{ faces} \times \frac{1/2 \text{ O}}{\text{face}} = 4 \text{ O atoms}, \ \text{total} = 9 \text{ O atoms}$$

Formula: $YBa_2Cu_3O_9$

b. The structure of this superconductor material follows the second perovskite structure described in Exercise 101. The $YBa_2Cu_3O_9$ structure is three of these cubic perovskite unit cells stacked on top of each other. The oxygen atoms are in the same places, Cu takes the place of Ti, two of the calcium atoms are replaced by two barium atoms, and one Ca is replaced by Y.

c. Y, Ba, and Cu are the same. Some oxygen atoms are missing.

$$12 \text{ edges} \times \frac{1/4 \text{ O}}{\text{edge}} = 3 \text{ O}, \ 8 \text{ faces} \times \frac{1/2 \text{ O}}{\text{face}} = 4 \text{ O}, \ \text{total} = 7 \text{ O atoms}$$

Superconductor formula is $YBa_2Cu_3O_7$.

Phase Changes and Phase Diagrams

105. The compound with the weakest intermolecular forces will have the highest vapor pressure. Compounds d and e are nonpolar compounds consisting of only nonpolar C–C and C–H bonds. The only intermolecular force that nonpolar compounds have is London dispersion forces. Compound e with the smaller molar mass will have the weakest LD forces and the highest

vapor pressure. The compound with the lowest vapor pressure has the strongest intermolecular forces. Compound a-c all have dipole forces, but compound b can also form the relatively strong hydrogen-bonding forces. Compound b has the lowest vapor pressure at 20°C.

107. If we graph ln P_{vap} versus 1/T with temperature in Kelvin, the slope of the resulting straight line will be $-\Delta H_{vap}/R$.

P_{vap}	ln P_{vap}	T (Li)	1/T	T (Mg)	1/T
1 torr	0	1023 K	9.775×10^{-4} K^{-1}	893 K	11.2×10^{-4} K^{-1}
10.	2.3	1163	8.598×10^{-4}	1013	9.872×10^{-4}
100.	4.61	1353	7.391×10^{-4}	1173	8.525×10^{-4}
400.	5.99	1513	6.609×10^{-4}	1313	7.616×10^{-4}
760.	6.63	1583	6.317×10^{-4}	1383	7.231×10^{-4}

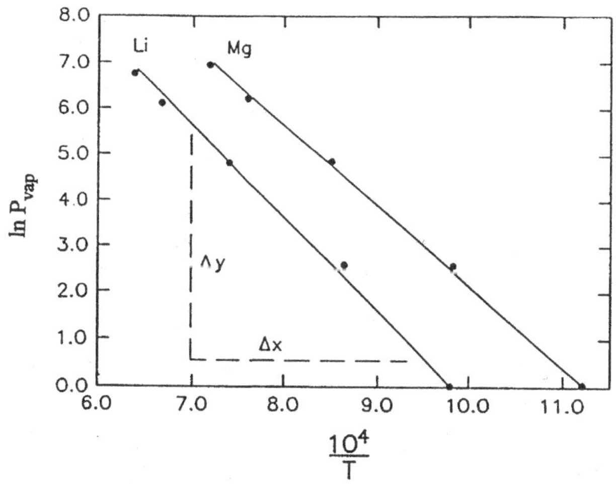

For Li:

We get the slope by taking two points (x, y) that are on the line we draw. For a line, slope = $\Delta y/\Delta x$, or we can determine the straight-line equation using a calculator. The general straight-line equation is $y = mx + b$, where m = slope and b = y intercept.

The equation of the Li line is: $\ln P_{vap} = -1.90 \times 10^4(1/T) + 18.6$, slope = -1.90×10^4 K

Slope = $-\Delta H_{vap}/R$, ΔH_{vap} = $-$slope $\times$ R = 1.90×10^4 K $\times$ 8.3145 J/K•mol

$\Delta H_{vap} = 1.58 \times 10^5$ J/mol = 158 kJ/mol

For Mg:

The equation of the line is: $\ln P_{vap} = -1.67 \times 10^4(1/T) + 18.7$, slope = -1.67×10^4 K

ΔH_{vap} = $-$slope $\times$ R = 1.67×10^4 K $\times$ 8.3145 J/K•mol

$\Delta H_{vap} = 1.39 \times 10^5$ J/mol = 139 kJ/mol

The bonding is stronger in Li because ΔH_{vap} is larger for Li.

109. At 100.°C (373 K), the vapor pressure of H_2O is 1.00 atm = 760. torr. For water, ΔH_{vap} = 40.7 kJ/mol.

$$\ln\left(\frac{P_1}{P_2}\right) = \frac{\Delta H_{vap}}{R}\left(\frac{1}{T_2} - \frac{1}{T_1}\right) \text{ or } \ln\left(\frac{P_2}{P_1}\right) = \frac{\Delta H_{vap}}{R}\left(\frac{1}{T_1} - \frac{1}{T_2}\right)$$

$$\ln\left(\frac{520.\text{ torr}}{760.\text{ torr}}\right) = \frac{40.7 \times 10^3 \text{ J/mol}}{8.3145 \text{ J/K}\bullet\text{mol}}\left(\frac{1}{373\text{ K}} - \frac{1}{T_2}\right),\ -7.75 \times 10^{-5} = \left(\frac{1}{373\text{ K}} - \frac{1}{T_2}\right)$$

$$-7.75 \times 10^{-5} = 2.68 \times 10^{-3} - \frac{1}{T_2},\ \frac{1}{T_2} = 2.76 \times 10^{-3},\ T_2 = \frac{1}{2.76 \times 10^{-3}} = 362 \text{ K or } 89°\text{C}$$

111. $$\ln\left(\frac{P_1}{P_2}\right) = \frac{\Delta H_{vap}}{R}\left(\frac{1}{T_2} - \frac{1}{T_1}\right),\ \ln\left(\frac{836\text{ torr}}{213\text{ torr}}\right) = \frac{\Delta H_{vap}}{8.3145 \text{ J/K}\bullet\text{mol}}\left(\frac{1}{313\text{ K}} - \frac{1}{353\text{ K}}\right)$$

Solving: $\Delta H_{vap} = 3.1 \times 10^4$ J/mol; for the normal boiling point, P = 1.00 atm = 760. torr.

$$\ln\left(\frac{760.\text{ torr}}{213\text{ torr}}\right) = \frac{3.1 \times 10^4 \text{ J/mol}}{8.3145 \text{ J/K}\bullet\text{mol}}\left(\frac{1}{313\text{ K}} - \frac{1}{T_1}\right),\ \frac{1}{313} - \frac{1}{T_1} = 3.4 \times 10^{-4}$$

T_1 = 350. K = 77°C; the normal boiling point of CCl_4 is 77°C.

113. Let P_1 = 760. torr, T_1 = 273.2 + 34.6 = 307.8 K, P_2 = 400. torr, T_2 = 273.2 + 17.9 = 291.1 K:

$$\ln\left(\frac{P_1}{P_2}\right) = \frac{\Delta H_{vap}}{R}\left(\frac{1}{T_2} - \frac{1}{T_1}\right),\ \ln\left(\frac{760\text{ torr}}{400.\text{ torr}}\right) = \frac{\Delta H_{vap}}{8.3145 \text{ J/K}\bullet\text{mol}}\left(\frac{1}{291.1\text{ K}} - \frac{1}{307.8\text{ K}}\right)$$

Solving: $\Delta H_{vap} = 2.86 \times 10^4$ J/mol = 28.6 kJ/mol

115.

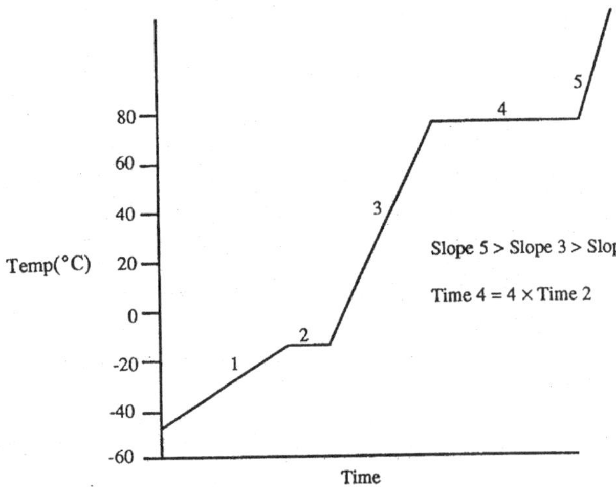

117. a. Many more intermolecular forces must be broken to convert a liquid to a gas as compared with converting a solid to a liquid. Because more intermolecular forces must be broken, much more energy is required to vaporize a liquid than is required to melt a solid. Therefore, ΔH_{vap} is much larger than ΔH_{fus}.

b. $1.00 \text{ g Na} \times \frac{1 \text{ mol Na}}{22.99 \text{ g}} \times \frac{2.60 \text{ kJ}}{\text{mol Na}} = 0.113 \text{ kJ} = 113 \text{ J}$ to melt 1.00 g Na

c. $1.00 \text{ g Na} \times \frac{1 \text{ mol Na}}{22.99 \text{ g}} \times \frac{97.0 \text{ kJ}}{\text{mol Na}} = 4.22 \text{ kJ} = 4220 \text{ J}$ to vaporize 1.00 g Na

d. This is the reverse process of that described in part c, so the energy change is the same quantity but opposite in sign. Therefore, $q = -4220$ J; i.e., 4220 of heat will be released.

119. Let's break up the process in a series of steps. For each step, we will calculate the energy released, then sum up the energy for all the steps to determine the total amount energy removed.

$X(g, 100.0°C) \rightarrow X(g, 75°C)$,

$$q_1 = s_g \times m \times \Delta T = \frac{1.0 \text{ J}}{\text{g °C}} \times 250. \text{ g} \times (-25°C) = -6.3 \times 10^3 \text{ J} = -6.3 \text{ kJ}$$

$H_2O(g, 75°C) \rightarrow H_2O(l, 75°C)$, $q_2 = 250. \text{ g} \times \frac{1 \text{ mol}}{75.0 \text{ g}} \times \frac{-20. \text{ kJ}}{\text{mol}} = -67 \text{ kJ}$

$H_2O(l, 75°C) \rightarrow H_2O(l, -15°C)$, $q_3 = \frac{2.5 \text{ J}}{\text{g °C}} \times 250. \text{ g} \times (-90.°C) = -5.6 \times 10^4 \text{ J} = -56 \text{ kJ}$

$H_2O(l, -15°C) \rightarrow H_2O(s, -15°C)$, $q_4 = 250 \times \frac{1 \text{ mol}}{75 \text{ g}} \times \frac{-5.0 \text{ kJ}}{\text{mol}} = -17 \text{ kJ}$

$H_2O(s, -15°C) \rightarrow H_2O(s, -50°C)$, $q_5 = \frac{3.0 \text{ J}}{\text{g °C}} \times 250. \text{ g} \times (-35°C) = -2.6 \times 10^4 \text{ J} = -26 \text{ kJ}$

$q_{total} = q_1 + q_2 + q_3 + q_4 + q_5 = -6.3 + -67 + -56 + -17 + -26 = -172 \text{ kJ}$

172 kJ of energy must be removed to do the overall process.

121. To calculate q_{total}, break up the heating process into five steps.

$H_2O(s, -20.0°C) \rightarrow H_2O(s, 0.0°C)$, $\Delta T = 20.0°C$; let s_{ice} = specific heat capacity of ice:

$$q_1 = s_s \times m \times \Delta T = \frac{2.03 \text{ J}}{\text{g °C}} \times 500. \text{ g} \times 20.0°C = 2.03 \times 10^4 \text{ J} = 20.3 \text{ kJ}$$

$H_2O(s, 0.0°C) \rightarrow H_2O(l, 0.0°C)$, $q_2 = 500. \text{ g } H_2O \times \frac{1 \text{ mol}}{18.02 \text{ g}} \times \frac{6.02 \text{ kJ}}{\text{mol}} = 167 \text{ kJ}$

$H_2O(l, 0.0°C) \rightarrow H_2O(l, 100.0°C)$, $q_3 = \frac{4.18 \text{ J}}{\text{g °C}} \times 500. \text{ g} \times 100.0°C = 2.09 \times 10^5 \text{J} = 209 \text{ kJ}$

$H_2O(l, 100.0°C) \rightarrow H_2O(g, 100.0°C)$, $q_4 = 500. \text{ g} \times \frac{1 \text{ mol}}{18.02 \text{ g}} \times \frac{40.7 \text{ kJ}}{\text{mol}} = 1130 \text{ kJ}$

$H_2O(g, 100.0°C) \rightarrow H_2O(g, 250.0°C),\ q_5 = \frac{2.02\ J}{g\ °C} \times 500.\ g \times 150.0°C = 1.52 \times 10^5\ J$

$= 152\ kJ$

$q_{total} = q_1 + q_2 + q_3 + q_4 + q_5 = 20.3 + 167 + 209 + 1130 + 152 = 1680\ kJ$

123. Let x = mass of water; break the process up into three steps, then set the total energy added to the energy required for each step. Step 1 is heating water from 50.0°C to 100.0°C, step 2 is boiling the water at 100.0°C, and step three is heating gaseous water at 100.0°C to 150.0 °C.

$q_1 = \frac{4.18\ J}{g\ °C} \times x \times 50.0°C = 209x;\ q_2 = x \times \frac{1\ mol}{18.02\ g} \times \frac{40.7 \times 10^3\ J}{mol} = 2260x$

$q_3 = \frac{2.02\ J}{g\ °C} \times x \times 50.0°C = 101x$

$385 \times 10^3 = q_1 + q_2 + q_3 = 209x + 2260x + 101x = 2570x,\ x = 150.\ g$

125. Total mass H_2O = 18 cubes $\times \frac{30.0\ g}{cube} = 540.\ g;\ 540.\ g\ H_2O \times \frac{1\ mol\ H_2O}{18.02\ g} = 30.0\ mol\ H_2O$

Heat removed to produce ice at −5.0°C:

$$\left(\frac{4.18\ J}{g\ °C} \times 540.\ g \times 22.0°C\right) + \left(\frac{6.02 \times 10^3\ J}{mol} \times 30.0\ mol\right) + \left(\frac{2.03\ J}{g\ °C} \times 540.\ g \times 5.0°C\right)$$

$$= 4.97 \times 10^4\ J + 1.81 \times 10^5\ J + 5.5 \times 10^3\ J = 2.36 \times 10^5\ J$$

$2.36 \times 10^5\ J \times \frac{1\ g\ CF_2Cl_2}{158\ J} = 1.49 \times 10^3\ g\ CF_2Cl_2$ must be vaporized.

127. A: solid B: liquid C: vapor

D: solid + vapor E: solid + liquid + vapor

F: liquid + vapor G: liquid + vapor H: vapor

triple point: E critical point: G

Normal freezing point: Temperature at which solid-liquid line is at 1.0 atm (see following plot).

Normal boiling point: Temperature at which liquid-vapor line is at 1.0 atm (see following plot).

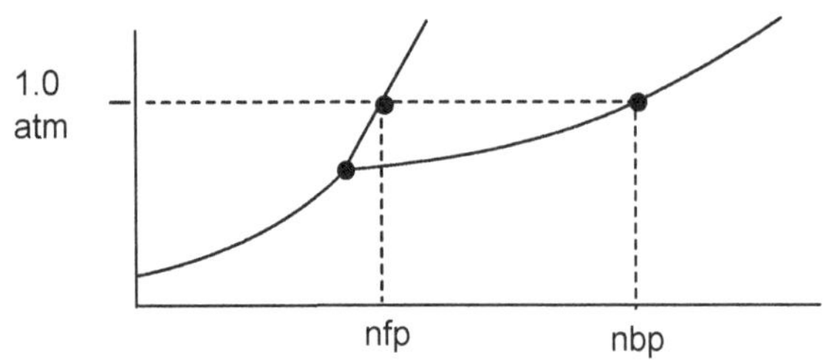

Because the solid-liquid line equilibrium has a positive slope, the solid phase is denser than the liquid phase.

129. a. two

b. Higher-pressure triple point: graphite, diamond and liquid; lower-pressure triple point at $\sim 10^7$ Pa: graphite, liquid and vapor

c. It is converted to diamond (the more dense solid form).

d. Diamond is more dense, which is why graphite can be converted to diamond by applying pressure.

131. Because the density of the liquid phase is greater than the density of the solid phase, the slope of the solid-liquid boundary line is negative (as in H_2O). With a negative slope, the melting points increase with a decrease in pressure, so the normal melting point of X should be greater than 225°C.

ChemWork Problems

133. The covalent bonds within a molecule are much stronger than the intermolecular forces between molecules. Consider water, which exhibits a very strong type of dipole force called hydrogen bonding. When water boils at 100°C, hydrogen bonding intermolecular forces are broken in the liquid resulting in $H_2O(l)$ converting to $H_2O(g)$. At 100°C, the covalent bonds within each water molecule are not broken; if they were broken, then H and O atoms would be produced. It takes a lot more energy than what is provided at 100°C to break the covalent bonds within a water molecule; the intramolecular forces within a covalent compound are much stronger than the intermolecular forces between molecules.

135. a. dipole, LD (SF_4 is a polar compound.) b. LD only (CO_2 is a nonpolar compound.)

c. H-bonding, LD (H-bonding due to the O−H bond) d. H-bonding, LD

e. dipole, LD (ICl_5 is a polar compound.) f. LD only (XeF_4 is a nonpolar compound.)

CO_2 and XeF_4 are nonpolar covalent compounds, so they only exhibit London dispersion forces. CH_3CH_2OH and HF are covalent compounds that contain either an N−H, O−H, or F−H bond, so they exhibit hydrogen bonding forces. SF_4 and ICl_5 are polar covalent compounds, so they exhibit dipole-dipole forces.

137. As the strength of the intermolecular forces increase, boiling points increase while vapor pressures at a specific temperature decrease.

a. False; the ionic forces in LiF are much stronger than molecular intermolecular forces found in H_2S.

b. True; HF is capable of H−bonding, HBr is not.

c. True; the larger Cl_2 molecule will have the stronger London dispersion forces.

d. True; HCl is a polar compound while CCl_4 is a nonpolar compound.

e. False; the ionic forces in MgO are much stronger than the molecular intermolecular forces found in CH_3CH_2OH.

139. As the physical properties indicate, the intermolecular forces are slightly stronger in D_2O than in H_2O.

141. At any temperature, the plot tells us that substance A has a higher vapor pressure than substance B, with substance C having the lowest vapor pressure. Therefore, the substance with the weakest intermolecular forces is A, and the substance with the strongest intermolecular forces is C.

NH_3 can form hydrogen-bonding interactions, whereas the others cannot. Substance C is NH_3. The other two are nonpolar compounds with only London dispersion forces. Because CH_4 is smaller than SiH_4, CH_4 will have weaker LD forces and is substance A. Therefore, substance B is SiH_4.

143. X(s, −35.0°C) → X(s, −15.0°C), ΔT = 20.0°C; let s_{solid} = specific heat capacity of X(s):

$$q_1 = s_{solid} \times m \times \Delta T = \frac{3.00\ \text{J}}{\text{g}\ {}^\circ\text{C}} \times 10.0\ \text{g} \times 20.0^\circ\text{C} = 600.\ \text{J}$$

$$\text{X(s, −15.0°C)} \rightarrow \text{X(l, −15.0°C)},\ q_2 = 10.0\ \text{g X} \times \frac{1\ \text{mol}}{100.0\ \text{g}} \times \frac{5.00\ \text{kJ}}{\text{mol}} = 0.500\ \text{kJ} = 500.\ \text{J}$$

$$\text{X(l, −15.0°C)} \rightarrow \text{X(l, 25.0°C)},\ q_3 = \frac{2.50\ \text{J}}{\text{g}\ {}^\circ\text{C}} \times 10.0\ \text{g} \times 40.0^\circ\text{C} = 1.00 \times 10^3\ \text{J}$$

$$q_{total} = q_1 + q_2 + q_3 = 600. + 500. + 1.00 \times 10^3 = 2.10 \times 10^3\ \text{J}$$

$$2.10 \times 10^3\ \text{J} \times \frac{1\ \text{min}}{450.0\ \text{J}} = 4.67\ \text{min}$$

145. $$8\ \text{corners} \times \frac{1/8\ \text{Xe}}{\text{corner}} + 1\ \text{Xe inside cell} = 2\ \text{Xe};\ \ 8\ \text{edges} \times \frac{1/4\ \text{F}}{\text{edge}} + 2\ \text{F inside cell} = 4\ \text{F}$$

The empirical formula is XeF_2.

147. B_2H_6: This compound contains only nonmetals, so it is probably a molecular solid with covalent bonding. The low boiling point confirms this.

SiO_2: This is the empirical formula for quartz, which is a network solid.

CsI: This is a metal bonded to a nonmetal, which generally form ionic solids. The electrical conductivity in aqueous solution confirms this.

W: Tungsten is a metallic solid as the conductivity data confirm.

149. In simple cubic unit cell, there are:

8 corner atoms × 1/8 atom per corner = 1 atom/unit cell

The volume of the unit cell is l^3 where l = cube edge length. For a unit cell of polonium:

$$\text{density} = \frac{\text{mass}}{\text{volume}} = 9.2\ \text{g/cm}^3 = \frac{1\ \text{Po atom} \times \frac{1\ \text{mol Po}}{6.022 \times 10^{23}\ \text{atoms}} \times \frac{209\ \text{g Po}}{\text{mol Po}}}{l^3}$$

Solving: $l = 3.35 \times 10^{-8}$ cm = cube edge length (carrying an extra sig. fig.)

In a simple cubic unit cell, the atoms touch along the cube edge l:

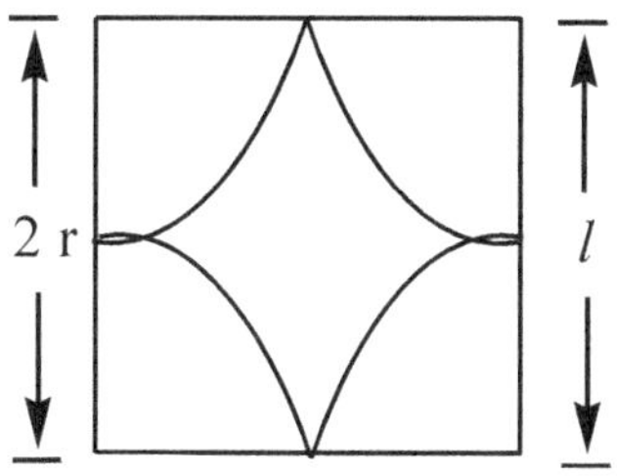

$2(\text{radius}) = 2r = l$

$2r = 3.35 \times 10^{-8}$ cm

$r = 1.7 \times 10^{-8}$ cm = 170 pm

151. $8 \text{ corner Mn ions} \times \frac{1/8 \text{ Mn}}{\text{corner}} + 1 \text{ Mn ion inside cell} = 2 \text{ Mn ions per unit cell}$

$4 \text{ face O ions} \times \frac{1/2 \text{ O}}{\text{face}} + 2 \text{ O ions inside cell} = 4 \text{ O ions per unit cell}$

The empirical formula is MnO_2. Assuming each oxygen ion has a 2− charge, then each manganese ion has a 4+ charge.

153. $24.7 \text{ g } C_6H_6 \times \frac{1 \text{ mol}}{78.11 \text{ g}} = 0.316 \text{ mol } C_6H_6$

$$P_{C_6H_6} = \frac{nRT}{V} = \frac{0.316 \text{ mol} \times \frac{0.08206 \text{ L atm}}{\text{K mol}} \times 293.2 \text{ K}}{100.0 \text{ L}} = 0.0760 \text{ atm, or } 57.8 \text{ torr}$$

155. The heat released when 500.0 g of water is cooled from 23.0 to 2.0°C is:

$$q = \frac{4.18 \text{ J}}{\text{g °C}} \times 500.0 \text{ g} \times -21.0\text{°C} = -4.39 \times 10^4 \text{ J}$$

Let n = number of ice cubes. Since each ice cube is 1 mol, the mass of ice cubes is 18.02n. The heat gained when the ice cubes at –5.0°C is converted to ice at 0.0°C is:

$$q = \frac{2.03 \text{ J}}{\text{g °C}} \times 18.02n \text{ g} \times 5.0\text{°C} = 180n \text{ J}$$

The heat gained to convert the ice cubes ice at 0.0°C to water at 0.0°C is:

$$n \times \frac{6.02 \text{ kJ}}{\text{mol}} \times \frac{1000 \text{ J}}{\text{kJ}} = 6020n$$

The heat gained when the n moles of liquid water 0.0°C is raised to 2.0°C is:

$$\frac{4.18 \text{ J}}{\text{g °C}} \times 18.02n \times 2.0\text{°C} = 150n \text{ J}$$

Now equate the magnitude of the heat loss of the water to the heat gain of the ice cubes:

$4.39 \times 10^4 = 180n + 6020n + 150n$, $n = 6.9$ ice cubes

7 ice cubes are necessary to cool 500.0 g of water from 23.0 to 2.0°C.

157. If we extend the liquid-vapor line of the water phase diagram to below the freezing point, we find that supercooled water will have a higher vapor pressure than ice at –10°C (see Figure 10.44 of the text). To achieve equilibrium, there must be a constant vapor pressure. Over time, supercooled water will be transformed through the vapor into ice in an attempt to equilibrate the vapor pressure. Eventually there will only be ice at –10°C along with $H_2O(g)$ at the vapor pressure given by the solid-vapor line in the phase diagram at –10°C.

159. Energy added to ice cubes = $5.00 \text{ min} \times \frac{60 \text{ s}}{\text{min}} \times \frac{750.\text{ J}}{\text{s}} = 2.25 \times 10^5 \text{ J} = 225 \text{ kJ}$

Energy to melt ice cubes: $403 \text{ g } H_2O \times \frac{1 \text{ mol } H_2O}{18.02 \text{ g}} \times \frac{6.02 \text{ kJ}}{\text{mol } H_2O} = 135 \text{ kJ}$

Energy remaining to heat water = 225 kJ – 135 kJ = 90. kJ

$9.0 \times 10^4 \text{ J} = s_{water} \times \text{mass} \times \Delta T = \frac{4.18 \text{ J}}{\text{g } ^\circ\text{C}} \times 403 \text{ g} \times \Delta T,\ \Delta T = 53.4^\circ\text{C}$

The final temperature of the water will increase from 0 °C to 53.4°C.

161. The critical temperature is the temperature above which the vapor cannot be liquefied no matter what pressure is applied. Since N_2 has a critical temperature below room temperature (~22°C), it cannot be liquefied at room temperature. NH_3, with a critical temperature above room temperature, can be liquefied at room temperature.

163. Molar mass of XY = $\frac{19.0 \text{ g}}{0.132 \text{ mol}} = 144 \text{ g/mol}$

X: [Kr] $5s^2 4d^{10}$; this is cadmium, Cd.

Molar mass Y = 144 – 112.4 = 32 g/mol; Y is sulfur, S.

The semiconductor is CdS. The dopant has the electron configuration of bromine, Br. Because Br has one more valence electron than S, doping with Br will produce an n-type semiconductor.

Challenge Problems

165. $\Delta H = q_p = 30.79 \text{ kJ};\ \Delta E = q_p + w,\ w = -P\Delta V$

$w = -P\Delta V = -1.00 \text{ atm}(28.90 \text{ L}) = -28.9 \text{ L atm} \times \frac{101.3 \text{ J}}{\text{L atm}} = -2930 \text{ J}$

$\Delta E = 30.79 \text{ kJ} + (-2.93 \text{ kJ}) = 27.86 \text{ kJ}$

167. A single hydrogen bond in H_2O has a strength of 21 kJ/mol. Each H_2O molecule forms two H bonds. Thus it should take 42 kJ/mol of energy to break all of the H bonds in water. Consider the phase transitions:

$$\text{Solid} \xrightarrow{6.0 \text{ kJ}} \text{liquid} \xrightarrow{40.7 \text{ kJ}} \text{vapor} \qquad \Delta H_{sub} = \Delta H_{fus} + \Delta H_{vap}$$

ΔH_{sub} = 6.0 kJ/mol + 40.7 kJ/mol = 46.7 kJ/mol; it takes a total of 46.7 kJ/mol to convert solid H_2O to vapor. This would be the amount of energy necessary to disrupt all of the intermolecular forces in ice. Thus (42 ÷ 46.7) × 100 = 90.% of the attraction in ice can be attributed to H bonding.

169. The structures of the two C_2H_6O compounds (20 valence e^-) are:

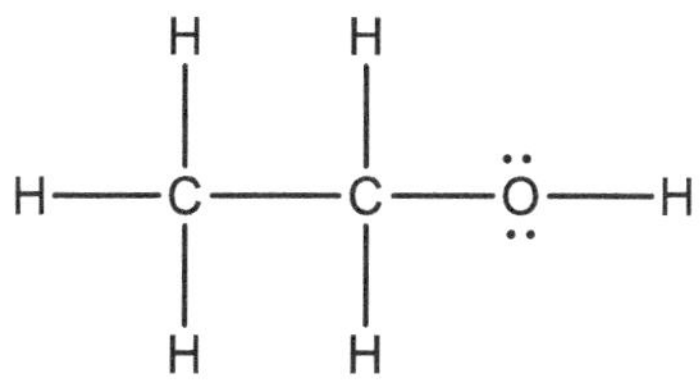
exhibits relatively strong hydrogen bonding

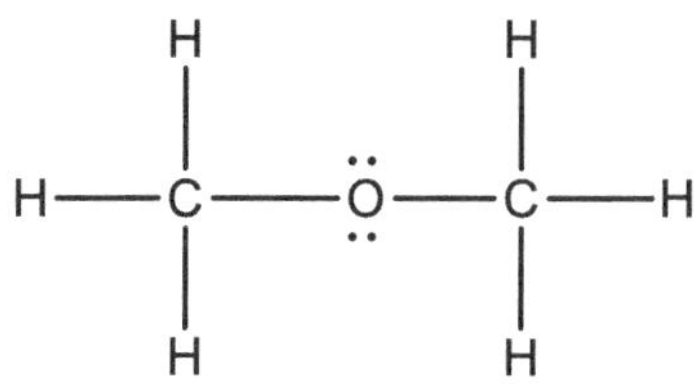
does not exhibit hydrogen bonding

The liquid will have the stronger intermolecular forces. Therefore, the first compound (ethanol) with hydrogen bonding is the liquid and the second compound (dimethyl ether) with the weaker intermolecular forces is the gas.

171. NaCl, $MgCl_2$, NaF, MgF_2, and AlF_3 all have very high melting points indicative of strong intermolecular forces. They are all ionic solids. $SiCl_4$, SiF_4, F_2, Cl_2, PF_5, and SF_6 are nonpolar covalent molecules. Only London dispersion (LD) forces are present. PCl_3 and SCl_2 are polar molecules. LD forces and dipole forces are present. In these eight molecular substances, the intermolecular forces are weak and the melting points low. $AlCl_3$ doesn't seem to fit in as well. From the melting point, there are much stronger forces present than in the nonmetal halides, but they aren't as strong as we would expect for an ionic solid. $AlCl_3$ illustrates a gradual transition from ionic to covalent bonding, from an ionic solid to discrete molecules.

173. Assuming 100.00 g:

$$28.31 \text{ g O} \times \frac{1 \text{ mol}}{16.00 \text{ g}} = 1.769 \text{ mol O}; \; 71.69 \text{ g Ti} \times \frac{1 \text{ mol}}{47.88 \text{ g}} = 1.497 \text{ mol Ti}$$

$$\frac{1.769}{1.497} = 1.182; \; \frac{1.497}{1.769} = 0.8462; \text{ the formula is } TiO_{1.182} \text{ or } Ti_{0.8462}O.$$

For $Ti_{0.8462}O$, let $x = Ti^{2+}$ per mol O^{2-} and $y = Ti^{3+}$ per mol O^{2-}. Setting up two equations and solving:

$$x + y = 0.8462 \text{ (mass balance) and } 2x + 3y = 2 \text{ (charge balance)}$$

$$2x + 3(0.8462 - x) = 2, \; x = 0.539 \text{ mol } Ti^{2+}/\text{mol } O^{2-} \text{ and } y = 0.307 \text{ mol } Ti^{3+}/\text{mol } O^{2-}$$

$$\frac{0.539}{0.8462} \times 100 = 63.7\% \text{ of the titanium ions are } Ti^{2+} \text{ and } 36.3\% \text{ are } Ti^{3+} \text{ (a 1.75 : 1 ion ratio).}$$

175. $\dfrac{\text{Density}_{\text{Mn}}}{\text{Density}_{\text{Cu}}} = \dfrac{\text{mass}_{\text{Mn}} \times \text{volume}_{\text{Cu}}}{\text{volume}_{\text{Mn}} \times \text{mass}_{\text{Cu}}} = \dfrac{\text{mass}_{\text{Mn}}}{\text{mass}_{\text{Cu}}} \times \dfrac{\text{volume}_{\text{Cu}}}{\text{volume}_{\text{Mn}}}$

The type of cubic cell formed is not important; only that Cu and Mn crystallize in the same type of cubic unit cell is important. Each cubic unit cell has a specific relationship between the cube edge length l and the radius r. In all cases $l \propto$ r. Therefore, V $\propto l^3 \propto$ r^3. For the mass ratio, we can use the molar masses of Mn and Cu since each unit cell must contain the same number of Mn and Cu atoms. Solving:

$$\frac{\text{density}_{\text{Mn}}}{\text{density}_{\text{Cu}}} = \frac{\text{mass}_{\text{Mn}}}{\text{mass}_{\text{Cu}}} \times \frac{\text{volume}_{\text{Cu}}}{\text{volume}_{\text{Mn}}} = \frac{54.94 \text{ g/mol}}{63.55 \text{ g/mol}} \times \frac{(r_{\text{Cu}})^3}{(1.056\ r_{\text{Cu}})^3}$$

$$\frac{\text{density}_{\text{Mn}}}{\text{density}_{\text{Cu}}} = 0.8645 \times \left(\frac{1}{1.056}\right)^3 = 0.7341$$

$$\text{density}_{\text{Mn}} = 0.7341 \times \text{density}_{\text{Cu}} = 0.7341 \times 8.96 \text{ g/cm}^3 = 6.58 \text{ g/cm}^3$$

177.

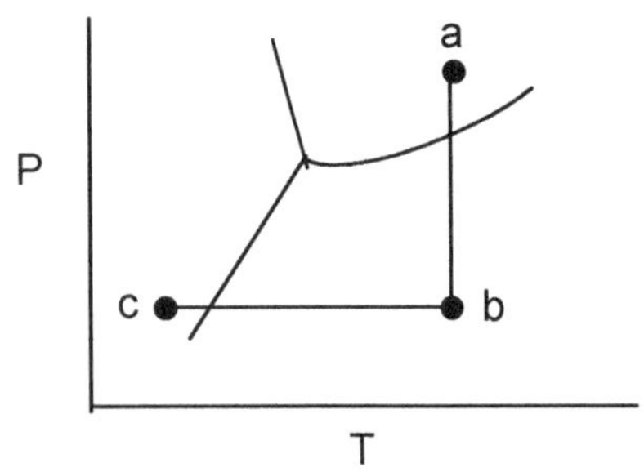

As P is lowered, we go from a to b on the phase diagram. The water boils. The boiling of water is endothermic, and the water is cooled (b → c), forming some ice. If the pump is left on, the ice will sublime until none is left. This is the basis of freeze drying.

179. For a cube: (body diagonal)^2 = (face diagonal)^2 + (cube edge length)^2

In a simple cubic structure, the atoms touch on cube edge, so the cube edge = 2r, where r = radius of sphere.

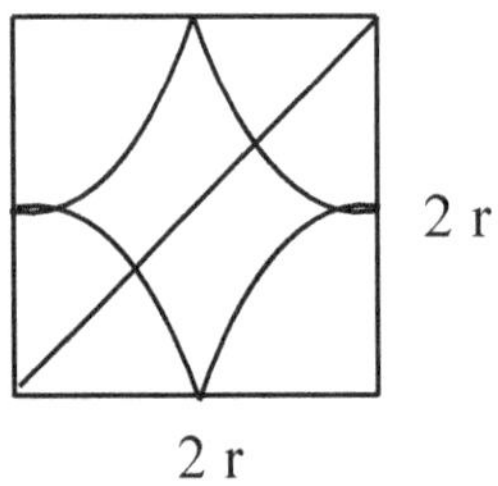

$$\text{Face diagonal} = \sqrt{(2r)^2 + (2r)^2} = \sqrt{4r^2 + 4r^2} = r\sqrt{8} = 2\sqrt{2}\ r$$

$$\text{Body diagonal} = \sqrt{(2\sqrt{2}\,r)^2 + (2r)^2} = \sqrt{12r^2} = 2\sqrt{3}\ r$$

The diameter of the hole = body diagonal – 2(radius of atoms at corners).

Diameter = $2\sqrt{3}$ r – 2r; thus the radius of the hole is: $\dfrac{2\sqrt{3}\text{ r} - 2\text{r}}{2} = (\sqrt{3} - 1)\text{r}$

The volume of the hole is $\dfrac{4}{3}\pi\left[\left(\sqrt{3} - 1\right)r\right]^3$.

CHAPTER 11

PROPERTIES OF SOLUTIONS

Solution Review

19. $$\frac{585\text{ g }C_3H_7OH \times \dfrac{1\text{ mol }C_3H_7OH}{60.09\text{ g }C_3H_7OH}}{1.00\text{ L}} = 9.74\ M$$

21. $$1.00\text{ L} \times \frac{0.040\text{ mol HCl}}{\text{L}} = 0.040\text{ mol HCl};\ 0.040\text{ mol HCl} \times \frac{1\text{ L}}{0.25\text{ mol HCl}} = 0.16\text{ L} = 160\text{ mL}$$

23. Mol $K_2S = 2.00\text{ L} \times \dfrac{1.00\text{ mol }K_2S}{\text{L}} = 2.00\text{ mol }K_2S$

$K_2S(s) \rightarrow 2\ K^+(aq) + S^{2-}(aq)$; mol $K^+ = 2(2.00) = 4.00$ mol K^+

Mol $KHSO_4 = 1.00\text{ L} \times \dfrac{1.00\text{ mol }KHSO_4}{\text{L}} = 1.00\text{ mol }KHSO_4$

$KHSO_4\ (s) \rightarrow K^+(aq) + HSO_4^-(aq)$; mol $K^+ = 1.00$ mol

$$M_{K^+} = \frac{\text{total mol }K^+}{\text{total volume}} = \frac{4.00\text{ mol} + 1.00\text{ mol}}{2.00\text{ L} + 1.00\text{ L}} = \frac{5.00\text{ mol}}{3.00\text{ L}} = 1.67\ M\ K^+$$

Questions

25. As the temperature increases, thc gas molecules will have a greater average kinetic energy. A greater fraction of the gas molecules in solution will have a kinetic energy greater than the attractive forces between the gas molecules and the solvent molecules. More gas molecules are able to escape to the vapor phase, and the solubility of the gas decreases.

27. Because the solute is volatile, both the water and solute will transfer back and forth between the two beakers. The volume in each beaker will become constant when the concentrations of solute in the beakers are equal to each other. Because the solute is less volatile than water, one would expect there to be a larger net transfer of water molecules into the right beaker than the net transfer of solute molecules into the left beaker. This results in a larger solution volume in the right beaker when equilibrium is reached, i.e., when the solute concentration is identical in each beaker.

29. No, the solution is not ideal. For an ideal solution, the strengths of intermolecular forces in solution are the same as in pure solute and pure solvent. This results in $\Delta H_{soln} = 0$ for an ideal solution. ΔH_{soln} for methanol-water is not zero. Because $\Delta H_{soln} < 0$ (heat is released), this solution shows a negative deviation from Raoult's law.

31. The main intermolecular forces are:

hexane (C_6H_{14}): London dispersion; chloroform ($CHCl_3$): dipole-dipole, London dispersion; methanol (CH_3OH): H-bonding; and H_2O: H-bonding (two places)

There is a gradual change in the nature of the intermolecular forces (weaker to stronger). Each preceding solvent is miscible in its predecessor because there is not a great change in the strengths of the intermolecular forces from one solvent to the next.

33. Only statement b is true. A substance freezes when the vapor pressure of the liquid and solid are the same. When a solute is added to water, the vapor pressure of the solution at 0°C is less than the vapor pressure of the solid, and the net result is for any ice present to convert to liquid in order to try to equalize the vapor pressures (which never can occur at 0°C). A lower temperature is needed to equalize the vapor pressure of water and ice, hence, the freezing point is depressed.

For statement a, the vapor pressure of a solution is directly related to the mole fraction of solvent (not solute) by Raoult's law. For statement c, colligative properties depend on the number of solute particles present and not on the identity of the solute. For statement d, the boiling point of water is increased because the sugar solute decreases the vapor pressure of the water; a higher temperature is required for the vapor pressure of the solution to equal the external pressure so boiling can occur.

35. Adding a solute to a solvent increases the boiling point and decreases the freezing point of the solvent. Thus the solvent is a liquid over a wider range of temperatures when a solute is dissolved.

37. Colligative properties are ones that only depend on the number of solute particles present and not on the identity of the solute. So if the boiling points of the two aqueous solutions are the same, then the two solutions have the same moles of proteins present. Other colligative properties that will be the same between the two solutions are vapor pressure, boiling point, and osmotic pressure. This assumes that molarity = molality for the solutions, which is a reasonable assumption for dilute solutions.

39. For oxidation-reduction reactions, normality is the mass of substance that can accept or furnish 1 mole of electrons; this mass is called an equivalent.

a. The reduction of Al^{3+} to Al requires 3 moles of electrons or 3 equivalents, so equivalent mass = 1/3(molar mass of Al).

b. Two equivalents are required, so equivalent mass = 1/2(molar mass of Cu).

c. Equivalent mass = molar mass of Cu because only 1 mol of electrons is required.

d. The oxidation state of oxygen goes from 0 in O_2 to –2 in H_2O. This is a gain of 2 mol electrons (2 equivalents). However, we have two moles of O in O_2, so equivalent mass = 1/4(molar mass of O_2).

41. Isotonic solutions are those which have identical osmotic pressures. Crenation and hemolysis refer to phenomena that occur when red blood cells are bathed in solutions having a mismatch in osmotic pressures inside and outside the cell. When red blood cells are in a solution having a higher osmotic pressure than that of the cells, the cells shrivel as there is a net transfer of water out of the cells. This is called crenation. Hemolysis occurs when the red blood cells are bathed in a solution having lower osmotic pressure than that inside the cell. Here, the cells rupture as there is a net transfer of water to into the red blood cells.

43. A colloid is a suspension of tiny particles is some medium. If the medium is water, these tiny, suspended particles scatter a bright light when shown through the colloidal dispersion. This scattering of light by the suspended particles is called the Tyndall effect. A solution will not scatter a bright light.

Exercises

Solution Composition

45. $0.0075 \text{ kg} \times \frac{2.5 \text{ mol NaCl}}{\text{kg}} \times \frac{58.44 \text{ g}}{\text{mol NaCl}} = 11 \text{g NaCl}$

$$\text{Mass \%} = \frac{\text{mass of NaCl}}{\text{mass of solution}} \times 100 = \frac{11 \text{ g}}{500. + 75} \times 100 = 1.9\%$$

47. Assuming 100.0 g of solution, we would have 10.0 g NaOH and 90.0 g of H_2O.

$$\text{Molality} = \frac{\text{moles solute}}{\text{kg solvent}} = \frac{10.0 \text{ g NaOH} \times \frac{1 \text{ mol NaOH}}{40.00 \text{ g}}}{0.0900 \text{ kg H}_2\text{O}} = 2.78 \text{ mol/kg}$$

$$\text{Molarity} = \frac{\text{moles solute}}{\text{L solution}} = \frac{10.0 \text{ g NaOH} \times \frac{1 \text{ mol NaOH}}{40.00 \text{ g}}}{100.0 \text{ g solution} \times \frac{1 \text{ cm}^3}{1.109 \text{ g}} \times \frac{1 \text{ L}}{1000 \text{ cm}^3}} = 2.77 \text{ mol/L}$$

49. Because the density of water is 1.00 g/mL, 100.0 mL of water has a mass of 100. g.

$$\text{Density} = \frac{\text{mass}}{\text{volume}} = \frac{10.0 \text{ g H}_3\text{PO}_4 + 100. \text{ g H}_2\text{O}}{104 \text{ mL}} = 1.06 \text{ g/mL} = 1.06 \text{ g/cm}^3$$

$$\text{Mol H}_3\text{PO}_4 = 10.0 \text{ g} \times \frac{1 \text{ mol}}{97.99 \text{ g}} = 0.102 \text{ mol H}_3\text{PO}_4$$

$$\text{Mol H}_2\text{O} = 100. \text{ g} \times \frac{1 \text{ mol}}{18.02 \text{ g}} = 5.55 \text{ mol H}_2\text{O}$$

$$\text{Mole fraction of } H_3PO_4 = \frac{0.102 \text{ mol } H_3PO_4}{(0.102 + 5.55) \text{ mol}} = 0.0180$$

$$\chi_{H_2O} = 1.000 - 0.0180 = 0.9820$$

$$\text{Molarity} = \frac{0.102 \text{ mol } H_3PO_4}{0.104 \text{ L}} = 0.981 \text{ mol/L}$$

$$\text{Molality} = \frac{0.102 \text{ mol } H_3PO_4}{0.100 \text{ kg}} = 1.02 \text{ mol/kg}$$

51. Hydrochloric acid (HCl):

$$\text{molarity} = \frac{38 \text{ g HCl}}{100. \text{ g soln}} \times \frac{1.19 \text{ g soln}}{\text{cm}^3 \text{ soln}} \times \frac{1000 \text{ cm}^3}{\text{L}} \times \frac{1 \text{ mol HCl}}{36.5 \text{ g}} = 12 \text{ mol/L}$$

$$\text{molality} = \frac{38 \text{ g HCl}}{62 \text{ g solvent}} \times \frac{1000 \text{ g}}{\text{kg}} \times \frac{1 \text{ mol HCl}}{36.5 \text{ g}} = 17 \text{ mol/kg}$$

$$38 \text{ g HCl} \times \frac{1 \text{ mol}}{36.5 \text{ g}} = 1.0 \text{ mol HCl}; \; 62 \text{ g } H_2O \times \frac{1 \text{ mol}}{18.0 \text{ g}} = 3.4 \text{ mol } H_2O$$

$$\text{mole fraction of HCl} = \chi_{HCl} = \frac{1.0}{3.4 + 1.0} = 0.23$$

Nitric acid (HNO_3):

$$\frac{70. \text{ g } HNO_3}{100. \text{ g soln}} \times \frac{1.42 \text{ g soln}}{\text{cm}^3 \text{ soln}} \times \frac{1000 \text{ cm}^3}{\text{L}} \times \frac{1 \text{ mol } HNO_3}{63.0 \text{ g}} = 16 \text{ mol/L}$$

$$\frac{70. \text{ g } HNO_3}{30. \text{ g solvent}} \times \frac{1000 \text{ g}}{\text{kg}} \times \frac{1 \text{ mol } HNO_3}{63.0 \text{ g}} = 37 \text{ mol/kg}$$

$$70. \text{ g } HNO_3 \times \frac{1 \text{ mol}}{63.0 \text{ g}} = 1.1 \text{ mol } HNO_3; \; 30. \text{ g } H_2O \times \frac{1 \text{ mol}}{18.0 \text{ g}} = 1.7 \text{ mol } H_2O$$

$$\chi_{HNO_3} = \frac{1.1}{1.7 + 1.1} = 0.39$$

Sulfuric acid (H_2SO_4):

$$\frac{95 \text{ g } H_2SO_4}{100. \text{ g soln}} \times \frac{1.84 \text{ g soln}}{\text{cm}^3 \text{ soln}} \times \frac{1000 \text{ cm}^3}{\text{L}} \times \frac{1 \text{ mol } H_2SO_4}{98.1 \text{ g } H_2SO_4} = 18 \text{ mol/L}$$

$$\frac{95 \text{ g } H_2SO_4}{5 \text{ g } H_2O} \times \frac{1000 \text{ g}}{\text{kg}} \times \frac{1 \text{ mol}}{98.1 \text{ g}} = 194 \text{ mol/kg} \approx 200 \text{ mol/kg}$$

$$95 \text{ g } H_2SO_4 \times \frac{1 \text{ mol}}{98.1 \text{ g}} = 0.97 \text{ mol } H_2SO_4; \; 5 \text{ g } H_2O \times \frac{1 \text{ mol}}{18.0 \text{ g}} = 0.3 \text{ mol } H_2O$$

$$\chi_{H_2SO_4} = \frac{0.97}{0.97 + 0.3} = 0.76$$

Acetic acid (CH_3CO_2H):

$$\frac{99\text{ g }CH_3CO_2H}{100.\text{ g soln}} \times \frac{1.05\text{ g soln}}{\text{cm}^3\text{ soln}} \times \frac{1000\text{ cm}^3}{\text{L}} \times \frac{1\text{ mol}}{60.05\text{ g}} = 17\text{ mol/L}$$

$$\frac{99\text{ g }CH_3CO_2H}{1\text{ g }H_2O} \times \frac{1000\text{ g}}{\text{kg}} \times \frac{1\text{ mol}}{60.05\text{ g}} = 1600\text{ mol/kg} \approx 2000\text{ mol/kg}$$

$$99\text{ g }CH_3CO_2H \times \frac{1\text{ mol}}{60.05\text{ g}} = 1.6\text{ mol }CH_3CO_2H;\ 1\text{ g }H_2O \times \frac{1\text{ mol}}{18.0\text{ g}} = 0.06\text{ mol }H_2O$$

$$\chi_{CH_3CO_2H} = \frac{1.6}{1.6 + 0.06} = 0.96$$

Ammonia (NH_3):

$$\frac{28\text{ g }NH_3}{100.\text{ g soln}} \times \frac{0.90\text{ g}}{\text{cm}^3} \times \frac{1000\text{ cm}^3}{\text{L}} \times \frac{1\text{ mol}}{17.0\text{ g}} = 15\text{ mol/L}$$

$$\frac{28\text{ g }NH_3}{72\text{ g }H_2O} \times \frac{1000\text{ g}}{\text{kg}} \times \frac{1\text{ mol}}{17.0\text{ g}} = 23\text{ mol/kg}$$

$$28\text{ g }NH_3 \times \frac{1\text{ mol}}{17.0\text{ g}} = 1.6\text{ mol }NH_3;\ 72\text{ g }H_2O \times \frac{1\text{ mol}}{18.0\text{ g}} - 4.0\text{ mol }H_2O$$

$$\chi_{NH_3} = \frac{1.6}{4.0 + 1.6} = 0.29$$

53. $25\text{ mL }C_5H_{12} \times \frac{0.63\text{ g}}{\text{mL}} = 16\text{ g }C_5H_{12};\ 25\text{ mL} \times \frac{0.63\text{ g}}{\text{mL}} \times \frac{1\text{ mol}}{72.15\text{ g}} = 0.22\text{ mol }C_5H_{12}$

$$45\text{ mL }C_6H_{14} \times \frac{0.66\text{ g}}{\text{mL}} = 30.\text{ g }C_6H_{14};\ 45\text{ mL} \times \frac{0.66\text{ g}}{\text{mL}} \times \frac{1\text{ mol}}{86.17\text{ g}} = 0.34\text{ mol }C_6H_{14}$$

$$\text{Mass \% pentane} = \frac{\text{mass pentane}}{\text{total mass}} \times 100 = \frac{16\text{ g}}{16\text{ g} + 30.\text{ g}} \times 100 = 35\%$$

$$\chi_{\text{pentane}} = \frac{\text{mol pentane}}{\text{total mol}} = \frac{0.22\text{ mol}}{0.22\text{ mol} + 0.34\text{ mol}} = 0.39$$

$$\text{Molality} = \frac{\text{mol pentane}}{\text{kg hexane}} = \frac{0.22\text{ mol}}{0.030\text{ kg}} = 7.3\text{ mol/kg}$$

$$\text{Molarity} = \frac{\text{mol pentane}}{\text{L solution}} = \frac{0.22\text{ mol}}{25\text{ mL} + 45\text{ mL}} \times \frac{1000\text{ mL}}{1\text{ L}} = 3.1\text{ mol/L}$$

55. If we have 100.0 mL of wine:

$$12.5\text{ mL }C_2H_5OH \times \frac{0.789\text{ g}}{\text{mL}} = 9.86\text{ g }C_2H_5OH \text{ and } 87.5\text{ mL }H_2O \times \frac{1.00\text{ g}}{\text{mL}} = 87.5\text{ g }H_2O$$

$$\text{Mass \% ethanol} = \frac{9.86\text{ g}}{87.5\text{ g} + 9.86\text{ g}} \times 100 = 10.1\%\text{ by mass}$$

$$\text{Molality} = \frac{9.86\text{ g } C_2H_5OH}{0.0875\text{ kg } H_2O} \times \frac{1\text{ mol}}{46.07\text{ g}} = 2.45\text{ mol/kg}$$

57. If we have 1.00 L of solution:

$$1.37\text{ mol citric acid} \times \frac{192.12\text{ g}}{\text{mol}} = 263\text{ g citric acid } (H_3C_6H_5O_7)$$

$$1.00 \times 10^3\text{ mL solution} \times \frac{1.10\text{ g}}{\text{mL}} = 1.10 \times 10^3\text{ g solution}$$

$$\text{Mass \% of citric acid} = \frac{263\text{ g}}{1.10 \times 10^3\text{ g}} \times 100 = 23.9\%$$

In 1.00 L of solution, we have 263 g citric acid and $(1.10 \times 10^3 - 263) = 840$ g of H_2O.

$$\text{Molality} = \frac{1.37\text{ mol citric acid}}{0.84\text{ kg } H_2O} = 1.6\text{ mol/kg}$$

$$840\text{ g } H_2O \times \frac{1\text{ mol}}{18.02\text{ g}} = 47\text{ mol } H_2O;\quad \chi_{\text{citricacid}} = \frac{1.37}{47 + 1.37} = 0.028$$

Because citric acid is a triprotic acid, the number of protons citric acid can provide is three times the molarity. Therefore, normality = 3 × molarity:

normality = 3 × 1.37 M = 4.11 N

Energetics of Solutions and Solubility

59. Using Hess's law:

$NaI(s) \rightarrow Na^+(g) + I^-(g)$	$\Delta H = -\Delta H_{LE} = -(-686\text{ kJ/mol})$
$Na^+(g) + I^-(g) \rightarrow Na^+(aq) + I^-(aq)$	$\Delta H = \Delta H_{hyd} = -694\text{ kJ/mol}$
$NaI(s) \rightarrow Na^+(aq) + I^-(aq)$	$\Delta H_{soln} = -8\text{ kJ/mol}$

ΔH_{soln} refers to the heat released or gained when a solute dissolves in a solvent. Here, an ionic compound dissolves in water.

61. Both $Al(OH)_3$ and NaOH are ionic compounds. Since the lattice energy is proportional to the charge of the ions, the lattice energy of aluminum hydroxide is greater than that of sodium hydroxide. The attraction of water molecules for Al^{3+} and OH^- cannot overcome the larger lattice energy, and $Al(OH)_3$ is insoluble. For NaOH, the favorable hydration energy is large enough to overcome the smaller lattice energy, and NaOH is soluble.

63. Water is a polar solvent and dissolves polar solutes and ionic solutes. Carbon tetrachloride (CCl_4) is a nonpolar solvent and dissolves nonpolar solutes (like dissolves like). To predict the polarity of the following molecules, draw the correct Lewis structure and then determine if the individual bond dipoles cancel or not. If the bond dipoles are arranged in such a manner that

they cancel each other out, then the molecule is nonpolar. If the bond dipoles do not cancel each other out, then the molecule is polar.

a. KrF_2, $8 + 2(7) = 22$ e^-

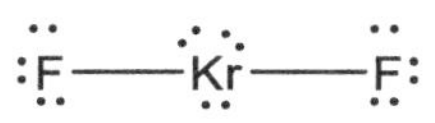

nonpolar; soluble in CCl_4

b. SF_2, $6 + 2(7) = 20$ e^-

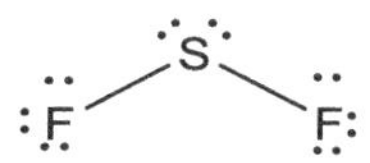

polar; soluble in H_2O

c. SO_2, $6 + 2(6) = 18$ e^-

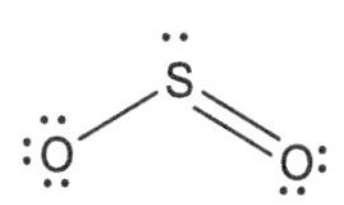

+ 1 more

polar; soluble in H_2O

d. CO_2, $4 + 2(6) = 16$ e^-

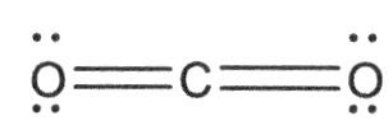

nonpolar; soluble in CCl_4

e. MgF_2 is an ionic compound so it is soluble in water.

f. CH_2O, $4 + 2(1) + 6 = 12$ e^-

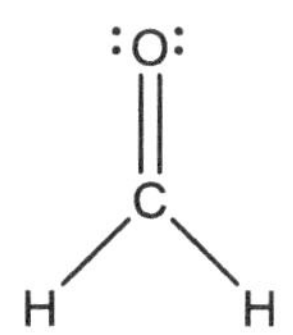

polar; soluble in H_2O

g. C_2H_4, $2(4) + 4(1) = 12$ e^-

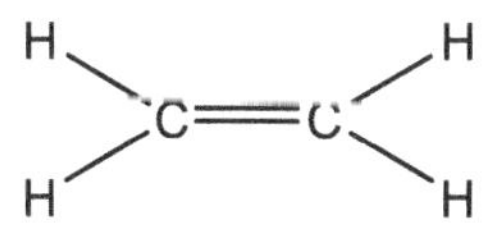

nonpolar (like all compounds made up of only carbon and hydrogen); soluble in CCl_4

65. Water exhibits H-bonding in the pure state and is classified as a polar solvent. Water will dissolve other polar solutes and ionic solutes.

a. NH_3; NH_3 is capable of H-bonding, unlike PH_3.

b. CH_3CN; CH_3CN is polar, while CH_3CH_3 is nonpolar.

c. CH_3CO_2H; CH_3CO_2H is capable of H-bonding, unlike the other compound.

67. $C = kP$, $\frac{8.21 \times 10^{-4} \text{ mol}}{\text{L}} = k \times 0.790 \text{ atm}$, $k = 1.04 \times 10^{-3}$ mol/L•atm

$C = kP$, $C = \frac{1.04 \times 10^{-4} \text{ mol}}{\text{L atm}} \times 1.10 \text{ atm} = 1.14 \times 10^{-3}$ mol/L

Vapor Pressures of Solutions

69. $P_{soln} = \chi_{C_2H_5OH} P^o_{C_2H_5OH}$; $\chi_{C_2H_5OH} = \frac{\text{moles of } C_2H_5OH}{\text{total moles in solution}}$

$$53.6 \text{ g } C_3H_8O_3 \times \frac{1 \text{ mol } C_3H_8O_3}{92.09 \text{ g}} = 0.582 \text{ mol } C_3H_8O_3$$

$$133.7 \text{ g } C_2H_5OH \times \frac{1 \text{ mol } C_2H_5OH}{46.07 \text{ g}} = 2.90 \text{ mol } C_2H_5OH$$

Total mol = 0.582 + 2.90 = 3.48 mol

$$113 \text{ torr} = \frac{2.90 \text{ mol}}{3.48 \text{ mol}} \times P^{o}_{C_2H_5OH},\ P^{o}_{C_2H_5OH} = 136 \text{ torr}$$

71. The normal boiling point of a substance is the boiling point at 1 atm pressure. So for this problem, P° = 760. torr at 34.5°C (the normal boiling point of diethyl ether).

$P = \chi P°$; 698 torr = χ(760. torr), χ = 0.918 = mole fraction of diethyl ether

73. Let x = mol octane, so mol heptane = $2x$.

$$\chi^{L}_{oct} = \frac{\text{mol octane in solution}}{\text{total mol in solution}} = \frac{x}{x + 2x} = 1/3 = 0.3333;\ \chi^{L}_{hept} = 1.0000 - 0.3333 = 0.6667$$

$$P_{oct} = \chi^{L}_{oct} P^{o}_{oct} = 0.3333(31.2 \text{ torr}) = 10.4 \text{ torr};\ P_{hept} = 0.6667(92.0 \text{ torr}) = 61.3 \text{ torr}$$

$$P_{total} = P_{oct} + P_{hept} = 10.4 + 61.3 = 71.7 \text{ torr}$$

75. a. $$25 \text{ mL } C_5H_{12} \times \frac{0.63 \text{ g}}{\text{mL}} \times \frac{1 \text{ mol}}{72.15 \text{ g}} = 0.22 \text{ mol } C_5H_{12}$$

$$45 \text{ mL } C_6H_{14} \times \frac{0.66 \text{ g}}{\text{mL}} \times \frac{1 \text{ mol}}{86.17 \text{ g}} = 0.34 \text{ mol } C_6H_{14};\ \text{total mol} = 0.22 + 0.34 = 0.56 \text{ mol}$$

$$\chi^{L}_{pen} = \frac{\text{mol pentane in solution}}{\text{total mol in solution}} = \frac{0.22 \text{ mol}}{0.56 \text{ mol}} = 0.39,\ \chi^{L}_{hex} = 1.00 - 0.39 = 0.61$$

$$P_{pen} = \chi^{L}_{pen} P^{o}_{pen} = 0.39(511 \text{ torr}) = 2.0 \times 10^2 \text{ torr};\ P_{hex} = 0.61(150. \text{ torr}) = 92 \text{ torr}$$

$$P_{total} = P_{pen} + P_{hex} = 2.0 \times 10^2 + 92 = 292 \text{ torr} = 290 \text{ torr}$$

b. From Chapter 5 on gases, the partial pressure of a gas is proportional to the number of moles of gas present (at constant volume and temperature). For the vapor phase:

$$\chi^{V}_{pen} = \frac{\text{mol pentane in vapor}}{\text{total mol vapor}} = \frac{P_{pen}}{P_{total}} = \frac{2.0 \times 10^2 \text{ torr}}{290 \text{ torr}} = 0.69$$

Note: In the *Solutions Guide*, we added V or L superscripts to the mole fraction symbol to emphasize for which value we are solving. If the L or V is omitted, then the liquid phase is assumed.

77. $P_{total} = P_{meth} + P_{prop}$, 174 torr = χ^{L}_{meth}(303 torr) + χ^{L}_{prop}(44.6 torr); $\chi^{L}_{prop} = 1.000 - \chi^{L}_{meth}$

$$174 = 303\chi^{L}_{meth} + (1.000 - \chi^{L}_{meth})44.6 \text{ torr}, \frac{129}{258} = \chi^{L}_{meth} = 0.500$$

$$\chi^{L}_{prop} = 1.000 - 0.500 = 0.500$$

79. Compared to H_2O, solution d (methanol-water) will have the highest vapor pressure since methanol is more volatile than water ($P^{o}_{H_2O}$ = 23.8 torr at 25°C). Both solution b (glucose-water) and solution c (NaCl-water) will have a lower vapor pressure than water by Raoult's law. NaCl dissolves to give Na^+ ions and Cl^- ions; glucose is a nonelectrolyte. Because there are more solute particles in solution c, the vapor pressure of solution c will be the lowest.

81. The first diagram shows positive deviation from Raoult's law. This occurs when the solute-solvent interactions are weaker than the interactions in pure solvent and pure solute. The second diagram illustrates negative deviation from Raoult's law. This occurs when the solute-solvent interactions are stronger than the interactions in pure solvent and pure solute. The third diagram illustrates an ideal solution with no deviation from Raoult's law. This occurs when the solute-solvent interactions are about equal to the pure solvent and pure solute interactions.

a. These two molecules are named acetone (CH_3COCH_3) and water. As discussed in section 11.4 on nonideal solutions, acetone-water solutions exhibit negative deviations from Raoult's law. Acetone and water have the ability to hydrogen bond with each other, which gives the solution stronger intermolecular forces as compared to the pure states of both solute and solvent. In the pure state, acetone cannot H–bond with itself. So the middle diagram illustrating negative deviations from Raoult's law is the correct choice for acetone-water solutions.

b. These two molecules are named ethanol (CH_3CH_2OH) and water. Ethanol-water solutions show positive deviations from Raoult's law. Both substances can hydrogen bond in the pure state, and they can continue this in solution. However, the solute-solvent interactions are somewhat weaker for ethanol-water solutions due to the significant nonpolar part of ethanol (CH_3–CH_2 is the nonpolar part of ethanol). This nonpolar part of ethanol weakens the intermolecular forces in solution. So the first diagram illustrating positive deviations from Raoult's law is the correct choice for ethanol-water solutions.

c. These two molecules are named heptane (C_7H_{16}) and hexane (C_6H_{14}). Heptane and hexane are very similar nonpolar substances; both are composed entirely of nonpolar C–C bonds and relatively nonpolar C–H bonds, and both have a similar size and shape. Solutions of heptane and hexane should be ideal. So the third diagram illustrating no deviation from Raoult's law is the correct choice for heptane-hexane solutions.

d. These two molecules are named heptane (C_7H_{16}) and water. The interactions between the nonpolar heptane molecules and the polar water molecules will certainly be weaker in solution as compared to the pure solvent and pure solute interactions. This results in positive deviations from Raoult's law (the first diagram).

Colligative Properties

83. Molality $= m = \dfrac{\text{mol solute}}{\text{kg solvent}} = \dfrac{27.0\text{ g } N_2H_4CO}{150.0\text{ g } H_2O} \times \dfrac{1000\text{ g}}{\text{kg}} \times \dfrac{1\text{ mol } N_2H_4CO}{60.06\text{ g } N_2H_4CO} = 3.00$ molal

$\Delta T_b = K_b m = \dfrac{0.51\ °C}{\text{molal}} \times 3.00\text{ molal} = 1.5°C$

The boiling point is raised from 100.0 to 101.5°C (assuming P = 1 atm).

85. $\Delta T_b = 77.85°C - 76.50°C = 1.35°C;\ \ m = \dfrac{\Delta T_b}{K_b} = \dfrac{1.35°C}{5.03\ °C\text{ kg/mol}} = 0.268$ mol/kg

Mol biomolecule $= 0.0150\text{ kg solvent} \times \dfrac{0.268\text{ mol hydrocarbon}}{\text{kg solvent}} = 4.02 \times 10^{-3}$ mol

From the problem, 2.00 g biomolecule was used. This mass must contain 4.02×10^{-3} mol biomolecule. The molar mass of the biomolecule is:

$$\frac{2.00\text{ g}}{4.02 \times 10^{-3}\text{ mol}} = 498\text{ g/mol}$$

87. $\Delta T_f = K_f m,\ \ \Delta T_f = 1.50°C = \dfrac{1.86\ °C}{\text{molal}} \times m,\ \ m = 0.806$ mol/kg

$0.200\text{ kg } H_2O \times \dfrac{0.806\text{ mol } C_3H_8O_3}{\text{kg } H_2O} \times \dfrac{92.09\text{ g } C_3H_8O_3}{\text{mol } C_3H_8O_3} = 14.8\text{ g } C_3H_8O_3$

89. Molality $= m = \dfrac{50.0\text{ g } C_2H_6O_2}{50.0\text{ g } H_2O} \times \dfrac{1000\text{ g}}{\text{kg}} \times \dfrac{1\text{ mol}}{62.07\text{ g}} = 16.1$ mol/kg

$\Delta T_f = K_f m = 1.86°C/\text{molal} \times 16.1\text{ molal} = 29.9°C;\ \ T_f = 0.0°C - 29.9°C = -29.9°C$

$\Delta T_b = K_b m = 0.51°C/\text{molal} \times 16.1\text{ molal} = 8.2°C;\ \ T_b = 100.0°C + 8.2°C = 108.2°C$

91. $\Delta T_f = K_f m,\ \ m = \dfrac{\Delta T_f}{K_f} = \dfrac{2.63°C}{40.\ °C\text{ kg/mol}} = \dfrac{6.6 \times 10^{-2}\text{ mol reserpine}}{\text{kg solvent}}$

The moles of reserpine present is:

$$0.0250\text{ kg solvent} \times \frac{6.6 \times 10^{-2}\text{ mol reserpine}}{\text{kg solvent}} = 1.7 \times 10^{-3}\text{ mol reserpine}$$

From the problem, 1.00 g reserpine was used, which must contain 1.7×10^{-3} mol reserpine. The molar mass of reserpine is:

$$\frac{1.00\text{ g}}{1.7 \times 10^{-3}\text{ mol}} = 590\text{ g/mol (610 g/mol if no rounding of numbers)}$$

93. a. $M = \dfrac{1.0 \text{ g protein}}{\text{L}} \times \dfrac{1 \text{ mol}}{9.0 \times 10^4 \text{ g}} = 1.1 \times 10^{-5} \text{ mol/L}; \ \pi = MRT$

$$\text{At 298 K: } \pi = \frac{1.1 \times 10^{-5} \text{ mol}}{\text{L}} \times \frac{0.08206 \text{ L atm}}{\text{K mol}} \times 298 \text{ K} \times \frac{760 \text{ torr}}{\text{atm}}, \ \pi = 0.20 \text{ torr}$$

Because d = 1.0 g/cm^3, 1.0 L solution has a mass of 1.0 kg. Because only 1.0 g of protein is present per liter of solution, 1.0 kg of H_2O is present to the correct number of significant figures, and molality equals molarity.

$$\Delta T_f = K_f m = \frac{1.86\,°\text{C}}{\text{molal}} \times 1.1 \times 10^{-5} \text{ molal} = 2.0 \times 10^{-5}\,°\text{C}$$

b. Osmotic pressure is better for determining the molar mass of large molecules. A temperature change of 10^{-5} °C is very difficult to measure. A change in height of a column of mercury by 0.2 mm (0.2 torr) is not as hard to measure precisely.

95. $M = \dfrac{\pi}{\text{RT}} = \dfrac{0.745 \text{ torr} \times \dfrac{1 \text{ atm}}{760 \text{ torr}}}{\dfrac{0.08206 \text{ L atm}}{\text{K mol}} \times 300.\text{ K}} = 3.98 \times 10^{-5} \text{ mol/L}$

$$1.00 \text{ L} \times \frac{3.98 \times 10^{-5} \text{ mol}}{\text{L}} = 3.98 \times 10^{-5} \text{ mol catalase}$$

$$\text{Molar mass} = \frac{10.00 \text{ g}}{3.98 \times 10^{-5} \text{ mol}} = 2.51 \times 10^5 \text{ g/mol}$$

97. $\pi = MRT, \ M = \dfrac{\pi}{\text{RT}} = \dfrac{15 \text{ atm}}{\dfrac{0.08206 \text{ L atm}}{\text{K mol}} \times 295 \text{ K}} = 0.62 \ M$

$$\frac{0.62 \text{ mol}}{\text{L}} \times \frac{342.30 \text{ g}}{\text{mol } C_{12}H_{22}O_{11}} = 212 \text{ g/L} \approx 210 \text{ g/L}$$

Dissolve 210 g of sucrose in some water and dilute to 1.0 L in a volumetric flask. To get 0.62 ±0.01 mol/L, we need 212 ±3 g sucrose.

Properties of Electrolyte Solutions

99. $Na_3PO_4(s) \rightarrow 3\ Na^+(aq) + PO_4^{3-}(aq)$, i = 4.0; $CaBr_2(s) \rightarrow Ca^{2+}(aq) + 2\ Br^-(aq)$, i = 3.0

$KCl(s) \rightarrow K^+(aq) + Cl^-(aq)$, i = 2.0

The effective particle concentrations of the solutions are (assuming complete dissociation):

4.0(0.010 molal) = 0.040 molal for the Na_3PO_4 solution; 3.0(0.020 molal) = 0.060 molal for the $CaBr_2$ solution; 2.0(0.020 molal) = 0.040 molal for the KCl solution; slightly greater than 0.020 molal for the HF solution because HF only partially dissociates in water (it is a weak acid).

a. The 0.010 *m* Na_3PO_4 solution and the 0.020 *m* KCl solution both have effective particle concentrations of 0.040 *m* (assuming complete dissociation), so both of these solutions should have the same boiling point as the 0.040 *m* $C_6H_{12}O_6$ solution (a nonelectrolyte).

b. $P = \chi P°$; as the solute concentration decreases, the solvent's vapor pressure increases because χ increases. Therefore, the 0.020 *m* HF solution will have the highest vapor pressure because it has the smallest effective particle concentration.

c. $\Delta T = K_f m$; the 0.020 *m* $CaBr_2$ solution has the largest effective particle concentration, so it will have the largest freezing point depression (largest ΔT).

101. a. $m = \dfrac{5.0 \text{ g NaCl}}{0.025 \text{ kg}} \times \dfrac{1 \text{ mol}}{58.44 \text{ g}} = 3.4$ molal; $NaCl(aq) \rightarrow Na^+(aq) + Cl^-(aq)$, i = 2.0

$\Delta T_f = iK_f m = 2.0 \times 1.86°C/\text{molal} \times 3.4 \text{ molal} = 13°C$; $T_f = -13°C$

$\Delta T_b = iK_b m = 2.0 \times 0.51°C/\text{molal} \times 3.4 \text{ molal} = 3.5°C$; $T_b = 103.5°C$

b. $m = \dfrac{2.0 \text{ g Al(NO}_3)_3}{0.015 \text{ kg}} \times \dfrac{1 \text{ mol}}{213.01 \text{ g}} = 0.63 \text{ mol/kg}$

$Al(NO_3)_3(aq) \rightarrow Al^{3+}(aq) + 3\ NO_3^-(aq)$, i = 4.0

$\Delta T_f = iK_f m = 4.0 \times 1.86°C/\text{molal} \times 0.63 \text{ molal} = 4.7°C$; $T_f = -4.7°C$

$\Delta T_b = iK_b m = 4.0 \times 0.51°C/\text{molal} \times 0.63 \text{ molal} = 1.3°C$; $T_b = 101.3°C$

103. $Na_3PO_4(s) \rightarrow 3\ Na^+(aq) + PO_4^{3-}(aq)$; if the solution is ideal, i = 4.0.

$\Delta T_f = iK_f m = 4.0 \times 1.86 \text{ °C kg/mol} \times 0.25 \text{ mol/kg} = 1.9°C$

If ideal, the solution would have a freezing point of −1.9°C. Since the freezing point is only −1.6°C, the solution is not ideal and there is some ion pairing in solution.

105. There are six cations and six anions in the illustration which indicates six solute formula units initially. There are a total of 10 solute particles in solution (a combined ion pair counts as one solute particle). So the value for the van't Hoff factor is:

$$i = \frac{\text{moles of particles in solution}}{\text{moles of solute dissolved}} = \frac{10}{6} = 1.67$$

107. a. $MgCl_2(s) \rightarrow Mg^{2+}(aq) + 2\ Cl^-(aq)$, i = 3.0 mol ions/mol solute

$\Delta T_f = iK_f m = 3.0 \times 1.86 \text{ °C/molal} \times 0.050 \text{ molal} = 0.28°C$

Assuming water freezes at 0.00°C, the freezing point would be −0.28°C.
freezes at 0.00°C.)

$\Delta T_b = iK_b m = 3.0 \times 0.51 \text{ °C/molal} \times 0.050 \text{ molal} = 0.077°C$; $T_b = 100.077°C$ (Assuming water boils at 100.000°C.)

b. $FeCl_3(s) \rightarrow Fe^{3+}(aq) + 3\ Cl^-(aq)$, i = 4.0 mol ions/mol solute

$\Delta T_f = iK_f m = 4.0 \times 1.86\ °C/molal \times 0.050\ molal = 0.37°C$; $T_f = -0.37°C$

$\Delta T_b = iK_b m = 4.0 \times 0.51\ °C/molal \times 0.050\ molal = 0.10°C$; $T_b = 100.10°C$

109. $\Delta T_f = iK_f m,\ i = \dfrac{\Delta T_f}{K_f m} = \dfrac{0.110°C}{1.86°C/molal \times 0.0225\ molal} = 2.63$ for 0.0225 m $CaCl_2$

$i = \dfrac{0.440}{1.86 \times 0.0910} = 2.60$ for 0.0910 m $CaCl_2$; $i = \dfrac{1.330}{1.86 \times 0.278} = 2.57$ for 0.278 m $CaCl_2$

$i_{ave} = (2.63 + 2.60 + 2.57)/3 = 2.60$

Note that i is less than the ideal value of 3.0 for $CaCl_2$. This is due to ion pairing in solution. Also note that as molality increases, i decreases. More ion pairing appears to occur as the solute concentration increases.

111. a. $T_C = 5(T_F - 32)/9 = 5(-29 - 32)/9 = -34°C$

Assuming the solubility of $CaCl_2$ is temperature independent, the molality of a saturated $CaCl_2$ solution is:

$$\frac{74.5\ g\ CaCl_2}{100.0\ g\ H_2O} \times \frac{1000\ g}{kg} \times \frac{1\ mol\ CaCl_2}{110.98\ g\ CaCl_2} = \frac{6.71\ mol\ CaCl_2}{kg\ H_2O}$$

$\Delta T_f = iK_f m = 3.00 \times 1.86\ °C\ kg/mol \times 6.71\ mol/kg = 37.4°C$

Assuming i = 3.00, a saturated solution of $CaCl_2$ can lower the freezing point of water to −37.4°C. Assuming these conditions, a saturated $CaCl_2$ solution should melt ice at −34°C (−29°F).

b. From Exercise 109, $i \approx 2.6$; $\Delta T_f = iK_f m = 2.6 \times 1.86 \times 6.71 = 32°C$; $T_f = -32°C$.

Assuming i = 2.6, a saturated $CaCl_2$ solution will not melt ice at −34°C (−29°F).

ChemWork Problems

113. Benzoic acid is capable of hydrogen-bonding, but a significant part of benzoic acid is the nonpolar benzene ring. In benzene, a hydrogen-bonded dimer forms.

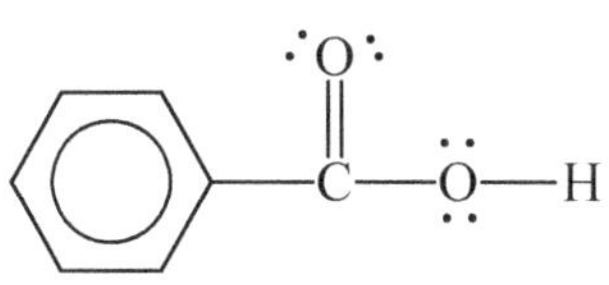

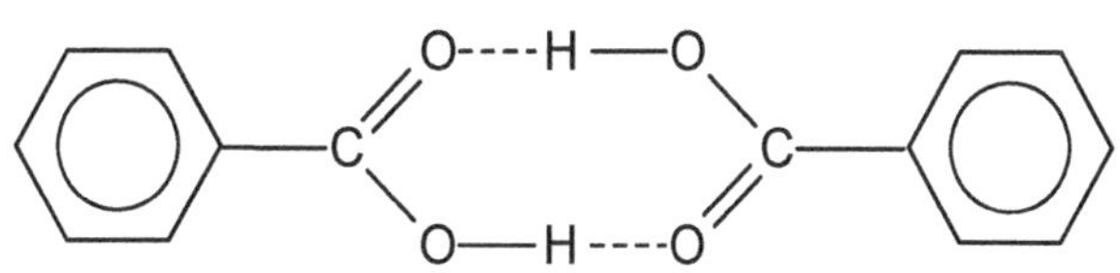

The dimer is relatively nonpolar and thus more soluble in benzene than in water. Because benzoic acid forms dimers in benzene, the effective solute particle concentration will be less than 1.0 molal. Therefore, the freezing-point depression would be less than 5.12°C ($\Delta T_f = K_f m$).

115. $$\text{Mass \% CsCl} = \frac{\text{mass CsCl}}{\text{total mass}} \times 100 = \frac{52.3\text{ g}}{52.3\text{ g} + 60.0.\text{ g}} \times 100 = 46.6\%$$

$$\text{Mol CsCl} = 52.3\text{ g} \times \frac{1\text{ mol}}{168.4\text{ g}} = 0.311\text{ mol CsCl}$$

$$\text{Mol } H_2O = 60.0\text{ g} \times \frac{1\text{ mol}}{18.02\text{ g}} = 3.33\text{ mol } H_2O$$

$$\text{Molarity} = \frac{\text{mol CsCl}}{\text{L solution}} = \frac{0.311\text{ mol CsCl}}{0.0633\text{ L}} = 4.91\text{ mol/L}$$

$$\text{Molality} = \frac{\text{mol CsCl}}{\text{kg solvent}} = \frac{0.311\text{ mol CsCl}}{0.0600\text{ kg}} = 5.18\text{ mol/kg}$$

$$\text{Mole fraction} = \chi_{CsCl} = \frac{\text{mol CsCl}}{\text{total mol}} = \frac{0.331\text{ mol}}{0.331\text{ mol} + 3.33\text{ mol}} = 0.0904$$

117.

$MX(s) \rightarrow M^+(g) + X^-(g)$	$\Delta H = -\Delta H_{\text{Lattice Energy}}$
$M^+(g) + X^-(g) \rightarrow M^+(aq) + X^-(aq)$	$\Delta H = \Delta H_{\text{hydration}}$
$MX(s) \rightarrow M^+(aq) + X^-(aq)$	$\Delta H_{\text{soln}} = -\Delta H_{\text{Lattice Energy}} + \Delta H_{\text{hydration}}$

For KF: -15 kJ/mol = $-(-804$ kJ/mol) + $\Delta H_{\text{hydration}}$, $\Delta H_{\text{hydration}} = -819$ kJ/mol

For RbF: -24 kJ/mol = $-(-768$ kJ/mol) + $\Delta H_{\text{hydration}}$, $\Delta H_{\text{hydration}} = -792$ kJ/mol

The enthalpy of hydration term refers to the interactions of the ions with water. Because KF has the more negative $\Delta H_{\text{hydration}}$ value, it forms the stronger attraction to water (answer e).

119. For ionic compounds, as the charge of the ions increases and/or the size of the ions decreases, the attraction to water (hydration) increases.

a. Fe^{3+}; smaller size, higher charge

b. Be^{2+}; smaller size

c. Cl^-; smaller size

d. SO_4^{2-}; higher charge

e. Mg^{2+}; smaller size, higher charge

f. F^-; smaller size

121. A 92-proof ethanol solution is 46% C_2H_5OH by volume. Assuming 100.0 mL of solution:

$$\text{mol ethanol} = 46\text{ mL } C_2H_5OH \times \frac{0.79\text{ g}}{\text{mL}} \times \frac{1\text{ mol } C_2H_5OH}{46.07\text{ g}} = 0.79\text{ mol } C_2H_5OH$$

$$\text{molarity} = \frac{0.79\text{ mol}}{0.1000\text{ L}} = 7.9\ M\text{ ethanol}$$

123. a. $NH_4NO_3(s) \rightarrow NH_4^+(aq) + NO_3^-(aq)\quad \Delta H_{\text{soln}} = ?$

Heat gain by dissolution process = heat loss by solution; we will keep all quantities positive to avoid sign errors. Because the temperature of the water decreased, the dissolution of NH_4NO_3 is endothermic (ΔH is positive). Mass of solution = 1.60 + 75.0 = 76.6 g.

$$\text{Heat loss by solution} = \frac{4.18\ \text{J}}{°\text{C g}} \times 76.6\ \text{g} \times (25.00°\text{C} - 23.34°\text{C}) = 532\ \text{J}$$

$$\Delta H_{soln} = \frac{532\ \text{J}}{1.60\ \text{g NH}_4\text{NO}_3} \times \frac{80.05\ \text{g NH}_4\text{NO}_3}{\text{mol NH}_4\text{NO}_3} = 2.66 \times 10^4\ \text{J/mol} = 26.6\ \text{kJ/mol}$$

b. We will use Hess's law to solve for the lattice energy. The lattice-energy equation is:

$$NH_4^+(g) + NO_3^-(g) \rightarrow NH_4NO_3(s) \quad \Delta H = \text{lattice energy}$$

$$NH_4^+(g) + NO_3^-(g) \rightarrow NH_4^+(aq) + NO_3^-(aq) \qquad \Delta H = \Delta H_{hyd} = -630.\ \text{kJ/mol}$$
$$NH_4^+(aq) + NO_3^-(aq) \rightarrow NH_4NO_3(s) \qquad \Delta H = -\Delta H_{soln} = -26.6\ \text{kJ/mol}$$

$$NH_4^+(g) + NO_3^-(g) \rightarrow NH_4NO_3(s) \qquad \Delta H = \Delta H_{hyd} - \Delta H_{soln}$$
$$\Delta H = -657\ \text{kJ/mol}$$

125. Because partial pressures are proportional to the moles of gas present, then:

$$\chi^V_{CS_2} = P_{CS_2}/P_{total}$$

$$P_{CS_2} = \chi^V_{CS_2} P_{total} = 0.855(263\ \text{torr}) = 225\ \text{torr}$$

$$P_{CS_2} = \chi^L_{CS_2} P^o_{CS_2},\ \chi^L_{CS_2} = \frac{P_{CS_2}}{P^o_{CS_2}} = \frac{225\ \text{torr}}{375\ \text{torr}} = 0.600$$

127. $50.0\ \text{g CH}_3\text{COCH}_3 \times \dfrac{1\ \text{mol}}{58.08\ \text{g}} = 0.861\ \text{mol acetone}$

$$50.0\ \text{g CH}_3\text{OH} \times \frac{1\ \text{mol}}{32.04\ \text{g}} = 1.56\ \text{mol methanol}$$

$$\chi^L_{acetone} = \frac{0.861}{0.861 + 1.56} = 0.356;\ \chi^L_{methanol} = 1.000 - \chi^L_{acetone} = 0.644$$

$$P_{total} = P_{methanol} + P_{acetone} = 0.644(143\ \text{torr}) + 0.356(271\ \text{torr}) = 92.1 + 96.5 = 188.6\ \text{torr}$$

Because partial pressures are proportional to the moles of gas present, in the vapor phase:

$$\chi^V_{acetone} = \frac{P_{acetone}}{P_{total}} = \frac{96.5\ \text{torr}}{188.6\ \text{torr}} = 0.512;\ \chi^V_{methanol} = 1.000 - 0.512 = 0.488$$

The actual vapor pressure of the solution (161 torr) is less than the calculated pressure assuming ideal behavior (188.6 torr). Therefore, the solution exhibits negative deviations from Raoult's law. This occurs when the solute-solvent interactions are stronger than in pure solute and pure solvent.

129. $\text{Molality} = m = \dfrac{\text{mol solute}}{\text{kg solvent}} = \dfrac{35.0\ \text{g solute}}{600.0\ \text{g H}_2\text{O}} \times \dfrac{1000\ \text{g}}{\text{kg}} \times \dfrac{1\ \text{mol solute}}{58.0\ \text{g solute}} = 1.01\ \text{molal}$

$$\Delta T_b = K_b m = \frac{0.51\,°C}{\text{molal}} \times 1.01 \text{ molal} = 0.52°C;\ \ 99.725\,°C + 0.52\,°C = 100.25\,°C$$

The boiling point is raised from 99.725 °C to 100.25°C.

131. $\Delta T_f = K_f m,\ m = \dfrac{\Delta T_f}{K_f} = \dfrac{0.300\,°C}{5.12\,°C\text{ kg/mol}} = \dfrac{5.86 \times 10^{-2}\text{ mol thyroxine}}{\text{kg benzene}}$

The moles of thyroxine present are:

$$0.0100\text{ kg benzene} \times \frac{5.86 \times 10^{-2}\text{ mol thyroxine}}{\text{kg benzene}} = 5.86 \times 10^{-4}\text{ mol thyroxine}$$

From the problem, 0.455 g thyroxine was used; this must contain 5.86×10^{-4} mol thyroxine. The molar mass of the thyroxine is:

$$\text{molar mass} = \frac{0.455\text{ g}}{5.86 \times 10^{-4}\text{ mol}} = 776\text{ g/mol}$$

133. Out of 100.00 g, there are:

$$31.57\text{ g C} \times \frac{1\text{ mol C}}{12.01\text{ g}} = 2.629\text{ mol C};\ \ \frac{2.629}{2.629} = 1.000$$

$$5.30\text{ g H} \times \frac{1\text{ mol H}}{1.008\text{ g}} = 5.26\text{ mol H};\ \ \frac{5.26}{2.629} = 2.00$$

$$63.13\text{ g O} \times \frac{1\text{ mol O}}{16.00\text{ g}} = 3.946\text{ mol O};\ \ \frac{3.946}{2.629} = 1.501$$

Empirical formula: $C_2H_4O_3$; use the freezing-point data to determine the molar mass.

$$m = \frac{\Delta T_f}{K_f} = \frac{5.20°C}{1.86\,°C/\text{molal}} = 2.80\text{ molal}$$

$$\text{Mol solute} = 0.0250\text{ kg} \times \frac{2.80\text{ mol solute}}{\text{kg}} = 0.0700\text{ mol solute}$$

$$\text{Molar mass} = \frac{10.56\text{ g}}{0.0700\text{ mol}} = 151\text{ g/mol}$$

The empirical formula mass of $C_2H_4O_3$ = 76.05 g/mol. Because the molar mass is about twice the empirical mass, the molecular formula is $C_4H_8O_6$, which has a molar mass of 152.10 g/mol.

Note: We use the experimental molar mass to determine the molecular formula. Knowing this, we calculate the molar mass precisely from the molecular formula using the atomic masses in the periodic table.

135. If ideal, NaCl dissociates completely, and i = 2.00. $\Delta T_f = iK_f m$; assuming water freezes at 0.00°C:

$$1.28°C = 2 \times 1.86°C\text{ kg/mol} \times m,\ \ m = 0.344\text{ mol NaCl/kg } H_2O$$

Assume an amount of solution that contains 1.00 kg of water (solvent).

$$0.344 \text{ mol NaCl} \times \frac{58.44 \text{ g}}{\text{mol}} = 20.1 \text{ g NaCl}$$

$$\text{Mass \% NaCl} = \frac{20.1 \text{ g}}{1.00 \times 10^3 \text{ g} + 20.1 \text{ g}} \times 100 = 1.97\%$$

137. $\Delta T = K_f m,\ m = \frac{\Delta T}{K_f} = \frac{2.79°\text{C}}{1.86\ °\text{C/molal}} = 1.50 \text{ molal}$

a. $\Delta T = K_b m$, $\Delta T = (0.51°\text{C/molal})(1.50 \text{ molal}) = 0.77°\text{C}$, $T_b = 100.77°\text{C}$

b. $P_{soln} = \chi_{water} P^o_{water},\ \chi_{water} = \frac{\text{mol } H_2O}{\text{mol } H_2O + \text{mol solute}}$

Assuming 1.00 kg of water, we have 1.50 mol solute, and:

$$\text{mol } H_2O = 1.00 \times 10^3 \text{ g } H_2O \times \frac{1 \text{ mol } H_2O}{18.02 \text{ g } H_2O} = 55.5 \text{ mol } H_2O$$

$$\chi_{water} = \frac{55.5 \text{ mol}}{1.50 + 55.5} = 0.974;\quad P_{soln} = (0.974)(23.76 \text{ mm Hg}) = 23.1 \text{ mm Hg}$$

c. We assumed ideal behavior in solution formation, we assumed the solute was nonvolatile, and we assumed i = 1 (no ions formed).

139. $\text{Mass of } H_2O = 160.\text{ mL} \times \frac{0.995 \text{ g}}{\text{mL}} = 159 \text{ g} = 0.159 \text{ kg}$

$$\text{Mol NaDTZ} = 0.159 \text{ kg} \times \frac{0.378 \text{ mol}}{\text{kg}} = 0.0601 \text{ mol}$$

$$\text{Molar mass of NaDTZ} = \frac{38.4 \text{ g}}{0.0601 \text{ mol}} = 639 \text{ g/mol}$$

$$P_{soln} = \chi_{H_2O} P^o_{H_2O};\quad \text{mol } H_2O = 159 \text{ g} \times \frac{1 \text{ mol}}{18.02 \text{ g}} = 8.82 \text{ mol}$$

Sodium diatrizoate is a salt because there is a metal (sodium) in the compound. From the short-hand notation for sodium diatrizoate, NaDTZ, we can assume this salt breaks up into Na^+ and DTZ^- ions. So the moles of solute particles are 2(0.0601) = 0.120 mol solute particles.

$$\chi_{H_2O} = \frac{8.82 \text{ mol}}{0.120 \text{ mol} + 8.82 \text{ mol}} = 0.987;\quad P_{soln} = 0.987 \times 34.1 \text{ torr} = 33.7 \text{ torr}$$

141. If we have 1.000 kg of water, then we have 1.00 mol of solute.

$$1000.\text{ g } H_2O \times \frac{1 \text{ mol } H_2O}{18.02 \text{ g}} = 55.49 \text{ mol } H_2O$$

$$P_{soln} = \chi_{H_2O} P^o_{H_2O} = \frac{55.49 \text{ mol}}{1.00 \text{ mol} + 55.49 \text{ mol}} \times 92.5 \text{ torr} = 90.9 \text{ torr}$$

143. $\Delta T = imK_f,\quad i = \dfrac{\Delta T}{mK_f} = \dfrac{2.79^\circ C}{\dfrac{0.250\ \text{mol}}{0.500\ \text{kg}} \times \dfrac{1.86\ ^\circ C\ \text{kg}}{\text{mol}}} = 3.00$

We have three ions in solutions, and we have twice as many anions as cations. Therefore, the formula of Q is MCl_2. Assuming 100.00 g of compound:

$$38.68\ \text{g Cl} \times \frac{1\ \text{mol Cl}}{35.45\ \text{g}} = 1.091\ \text{mol Cl}$$

$$\text{mol M} = 1.091\ \text{mol Cl} \times \frac{1\ \text{mol M}}{2\ \text{mol Cl}} = 0.5455\ \text{mol M}$$

Molar mass of M = $\dfrac{61.32\ \text{g M}}{0.5455\ \text{mol M}}$ = 112.4 g/mol; M is Cd, so Q = $CdCl_2$.

Challenge Problems

145. $m = \dfrac{\Delta T_f}{K_f} = \dfrac{5.23^\circ C}{1.86\ ^\circ C/\text{molal}} = 2.81\ \text{molal}$

$\dfrac{2.81\ \text{mol solute}}{\text{kg solvent}} = \dfrac{n}{0.0500\ \text{kg}}$, n = 0.141 mol of ions in solution

Since $NaNO_3$ and $Mg(NO_3)_2$ are strong electrolytes:

$$0.141\ \text{mol ions} = 2(x\ \text{mol of NaNO}_3) + 3[y\ \text{mol of Mg(NO}_3)_2]$$

In addition: $6.50\ \text{g} = x\ \text{mol NaNO}_3 \times \dfrac{85.00\ \text{g}}{\text{mol}} + y\ \text{mol of Mg(NO}_3)_2 \times \dfrac{148.3\ \text{g}}{\text{mol}}$

We have two equations: $2x + 3y = 0.141$ and $85.00\,x + 148.3\,y = 6.50$

Solving by simultaneous equations:

$$\begin{aligned} -85.00\,x - 127.5\,y &= -5.99 \\ 85.00\,x + 148.3\,y &= 6.50 \\ \hline 20.8\,y &= 0.51,\quad y = 0.025\ \text{mol Mg(NO}_3)_2 \end{aligned}$$

Mass of $Mg(NO_3)_2$ = $0.025\ \text{mol} \times \dfrac{148.3\ \text{g}}{\text{mol}} = 3.7\ \text{g Mg(NO}_3)_2$

Mass of $NaNO_3$ = 6.50 g − 3.7 g = 2.8 g $NaNO_3$

Mass % $NaNO_3$ = $\dfrac{2.8\ \text{g}}{6.50\ \text{g}} \times 100$ = 43% $NaNO_3$ and 57% $Mg(NO_3)_2$ by mass

147. For 30.% A by moles in the vapor, $30. = \dfrac{P_A}{P_A + P_B} \times 100$:

$$0.30 = \frac{\chi_A x}{\chi_A x + \chi_B y},\quad 0.30 = \frac{\chi_A x}{\chi_A x + (1.00 - \chi_A)y}$$

$$\chi_A x = 0.30(\chi_A x) + 0.30\,y - 0.30(\chi_A y),\ \ \chi_A x - (0.30)\chi_A x + (0.30)\chi_A y = 0.30\,y$$

$$\chi_A(x - 0.30\,x + 0.30\,y) = 0.30\,y,\ \ \chi_A = \frac{0.30\,y}{0.70\,x + 0.30\,y};\ \ \chi_B = 1.00 - \chi_A$$

Similarly, if vapor above is 50.% A: $\chi_A = \dfrac{y}{x + y};\ \ \chi_B = 1.00 - \dfrac{y}{x + y}$

If vapor above is 80.% A: $\chi_A = \dfrac{0.80\,y}{0.20\,x + 0.80\,y};\ \ \chi_B = 1.00 - \chi_A$

If the liquid solution is 30.% A by moles, $\chi_A = 0.30$.

Thus $\chi_A^V = \dfrac{P_A}{P_A + P_B} = \dfrac{0.30\,x}{0.30\,x + 0.70\,y}$ and $\chi_B^V = 1.00 - \dfrac{0.30\,x}{0.30\,x + 0.70\,y}$

If solution is 50.% A: $\chi_A^V = \dfrac{x}{x + y}$ and $\chi_B^V = 1.00 - \chi_A^V$

If solution is 80.% A: $\chi_A^V = \dfrac{0.80\,x}{0.80\,x + 0.20\,y}$ and $\chi_B^V = 1.00 - \chi_A^V$

149. $m = \dfrac{\Delta T_f}{K_f} = \dfrac{0.426^\circ C}{1.86\ ^\circ C/\text{molal}} = 0.229$ molal

Assuming a solution density = 1.00 g/mL, then 1.00 L contains 0.229 mol solute.

$NaCl \rightarrow Na^+ + Cl^-$ i = 2; so: 2(mol NaCl) + mol $C_{12}H_{22}O_{11}$ = 0.229 mol

Mass NaCl + mass $C_{12}H_{22}O_{11}$ = 20.0 g

$2n_{NaCl} + n_{C_{12}H_{22}O_{11}} = 0.229$ and $58.44(n_{NaCl}) + 342.3(n_{C_{12}H_{22}O_{11}}) = 20.0$

Solving: $n_{C_{12}H_{22}O_{11}} = 0.0425$ mol = 14.5 g and $n_{NaCl} = 0.0932$ mol = 5.45 g

Mass % $C_{12}H_{22}O_{11}$ = (14.5 g/20.0 g) × 100 = 72.5 % and 27.5% NaCl by mass

$$\chi_{C_{12}H_{22}O_{11}} = \frac{0.0425\text{ mol}}{0.0425\text{ mol} + 0.0932\text{ mol}} = 0.313$$

151. $\chi_{pen}^V = 0.15 = \dfrac{P_{pen}}{P_{total}};\ \ P_{pen} = \chi_{pen}^L P_{pen}^o;\ \ P_{total} = P_{pen} + P_{hex} = \chi_{pen}^L(511) + \chi_{hex}^L(150.)$

Because $\chi_{hex}^L = 1.000 - \chi_{pen}^L$: $P_{total} = \chi_{pen}^L(511) + (1.000 - \chi_{pen}^L)(150.) = 150. + 361\chi_{pen}^L$

$$\chi_{pen}^V = \frac{P_{pen}}{P_{total}},\ \ 0.15 = \frac{\chi_{pen}^L(511)}{150. + 361\,\chi_{pen}^L},\ \ 0.15(150. + 361\,\chi_{pen}^L) = 511\,\chi_{pen}^L$$

$$23 + 54\,\chi_{pen}^L = 511\,\chi_{pen}^L,\ \ \chi_{pen}^L = \frac{23}{457} = 0.050$$

153. $\Delta T_f = 5.51 - 2.81 = 2.70°C$; $m = \frac{\Delta T_f}{K_f} = \frac{2.70°C}{5.12\ °C/molal} = 0.527$ molal

Let x = mass of naphthalene (molar mass = 128.2 g/mol). Then $1.60 - x$ = mass of anthracene (molar mass = 178.2 g/mol).

$$\frac{x}{128.2} = \text{moles naphthalene and } \frac{1.60 - x}{178.2} = \text{moles anthracene}$$

$$\frac{0.527 \text{ mol solute}}{\text{kg solvent}} = \frac{\frac{x}{128.2} + \frac{1.60 - x}{178.2}}{0.0200 \text{ kg solvent}},\ 1.05 \times 10^{-2} = \frac{(178.2)x + 1.60(128.2) - (128.2)x}{128.2(178.2)}$$

$(50.0)x + 205 = 240.$, $(50.0)x = 240. - 205$, $(50.0)x = 35$, $x = 0.70$ g naphthalene

So the mixture is:

$$\frac{0.70 \text{ g}}{1.60 \text{ g}} \times 100 = 44\% \text{ naphthalene by mass and } 56\% \text{ anthracene by mass}$$

155. $HCO_2H \rightarrow H^+ + HCO_2^-$; only 4.2% of HCO_2H ionizes. The amount of H^+ or HCO_2^- produced is $0.042 \times 0.10\ M = 0.0042\ M$.

The amount of HCO_2H remaining in solution after ionization is $0.10\ M - 0.0042\ M = 0.10\ M$.

The total molarity of species present = $M_{HCO_2H} + M_{H^+} + M_{HCO_2^-}$

$$= 0.10 + 0.0042 + 0.0042 = 0.11\ M$$

Assuming $0.11\ M = 0.11$ molal, and assuming ample significant figures in the freezing point and boiling point of water at P = 1 atm:

$\Delta T = K_f m = 1.86°C/\text{molal} \times 0.11 \text{ molal} = 0.20°C$; freezing point = $-0.20°C$

$\Delta T = K_b m = 0.51°C/\text{molal} \times 0.11 \text{ molal} = 0.056°C$; boiling point = $100.056°C$

157. a. Assuming $MgCO_3(s)$ does not dissociate, the solute concentration in water is:

$$\frac{560\ \mu g\ MgCO_3(s)}{mL} = \frac{560 \text{ mg}}{L} = \frac{560 \times 10^{-3} \text{ g}}{L} \times \frac{1 \text{ mol } MgCO_3}{84.32 \text{ g}}$$

$$= 6.6 \times 10^{-3} \text{ mol } MgCO_3/L$$

An applied pressure of 8.0 atm will purify water up to a solute concentration of:

$$M = \frac{\pi}{RT} = \frac{8.0 \text{ atm}}{0.08206 \text{ L atm/K mol} \times 300. \text{ K}} = \frac{0.32 \text{ mol}}{L}$$

When the concentration of $MgCO_3(s)$ reaches 0.32 mol/L, the reverse osmosis unit can no longer purify the water. Let V = volume (L) of water remaining after purifying 45 L of H_2O. When V + 45 L of water has been processed, the moles of solute particles will equal:

$$6.6 \times 10^{-3} \text{ mol/L} \times (45 \text{ L} + \text{V}) = 0.32 \text{ mol/L} \times \text{V}$$

Solving: $0.30 = (0.32 - 0.0066) \times V$, $V = 0.96$ L

The minimum total volume of water that must be processed is 45 L + 0.96 L = 46 L.

Note: If $MgCO_3$ does dissociate into Mg^{2+} and CO_3^{2-} ions, then the solute concentration increases to 1.3×10^{-2} *M*, and at least 47 L of water must be processed.

b. No; a reverse osmosis system that applies 8.0 atm can only purify water with a solute concentration of less than 0.32 mol/L. Salt water has a solute concentration of 2(0.60 *M*) = 1.2 mol/L ions. The solute concentration of salt water is much too high for this reverse osmosis unit to work.

CHAPTER 12

CHEMICAL KINETICS

Questions

13. One experimental method to determine rate laws is the method of initial rates. Several experiments are carried out using different initial concentrations of reactants, and the initial rate is determined for each experiment. The results are then compared to see how the initial rate depends on the initial concentrations. This allows the orders in the rate law to be determined. The value of the rate constant is determined from the experiments once the orders are known.

    The second experimental method utilizes the fact that the integrated rate laws can be put in the form of a straight-line equation. Concentration versus time data are collected for a reactant as a reaction is run. These data are then manipulated and plotted to see which manipulation gives a straight line. From the straight-line plot we get the order of the reactant, and the slope of the line is mathematically related to k, the rate constant.

15. All these choices would affect the rate of the reaction, but only b and c affect the rate by effecting the value of the rate constant k. The value of the rate constant depends on temperature. The value of the rate constant also depends on the activation energy. A catalyst will change the value of k because the activation energy changes. Increasing the concentration (partial pressure) of either O_2 or NO does not affect the value of k, but it does increase the rate of the reaction because both concentrations appear in the rate law.

17. In a unimolecular reaction, a single reactant molecule decomposes to products. In a bimolecular reaction, two molecules collide to give products. The probability of the simultaneous collision of three molecules with enough energy and the proper orientation is very small, making termolecular steps very unlikely.

19. $$\frac{\text{Rate}_2}{\text{Rate}_1} = \frac{k[A]_2^x}{k[A]_1^x} = \left(\frac{[A]_2}{[A]_1}\right)^x$$

    The rate doubles as the concentration quadruples:

    $2 = (4)^x, \quad x = 1/2$

    The order is 1/2 (the square root of the concentration of reactant).

    For a reactant that has an order of −1 and the reactant concentration is doubled:

    $$\frac{\text{Rate}_2}{\text{Rate}_1} = (2)^{-1} = \frac{1}{2}$$

    The rate will decrease by a factor of 1/2 when the reactant concentration is doubled for a −1 order reaction. Negative orders are seen for substances that hinder or slow down a reaction.

21. Two reasons are:

(1) The collision must involve enough energy to produce the reaction; that is, the collision energy must be equal to or exceed the activation energy.

(2) The relative orientation of the reactants when they collide must allow formation of any new bonds necessary to produce products.

23. Enzymes are very efficient catalysts. As is true for all catalysts, enzymes speed up a reaction by providing an alternative pathway for reactants to convert to products. This alternative pathway has a smaller activation energy and hence, a faster rate. Also true is that catalysts are not used up in the overall chemical reaction. Once an enzyme encounters the correct reagent, the chemical reaction quickly occurs, and the enzyme is then free to catalyze another reaction. Because of the efficiency of the reaction step, only a relatively small amount of enzyme is needed to catalyze a specific reaction, no matter how complex the reaction.

25. The Arrhenius equation is $k = A\ \exp(-E_a/RT)$ or, in logarithmic form, $\ln k = -E_a/RT + \ln A$. This equation is in the form of a straight-line equation, $y = mx + b$. A graph of ln k versus 1/T yields a straight line with a slope equal to $-E_a/R$ and y-intercept of lnA. So the slope and the y-intercept of the plot can be used to determine the activation energy, E_a, and the frequency factor, *A*, for a reaction. We can then use these values to estimate k at other temperatures.

27. Increasing the concentration of A will have no effect on the rate of this reaction since the reaction is zero order in A. Increasing the concentration of B, raising the temperature, and lowering the activation energy will all increase the rate of the reaction.

29. Consider the energy plot below for a reaction with a positive ΔE. The activation energy for the reverse reaction is the energy that products must have to convert back to reactants. When a reaction has a positive ΔE, the activation energy for the reverse reaction is smaller than the activation energy for the forward reaction. Note that E_a in the plot below is the activation energy for the forward reaction. The activation energy for the reverse reaction will be smaller than the activation energy of the forward reaction by the ΔE quantity.

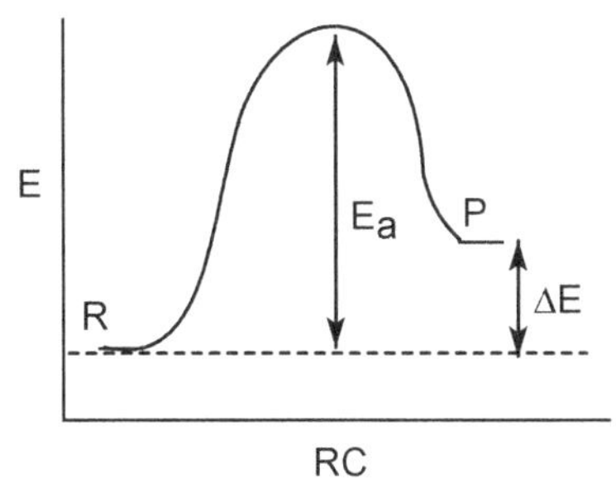

Exercises

Reaction Rates

31. The coefficients in the balanced reaction relate the rate of disappearance of reactants to the rate of production of products. From the balanced reaction, the rate of production of P_4 will be 1/4

the rate of disappearance of PH_3, and the rate of production of H_2 will be 6/4 the rate of disappearance of PH_3. By convention, all rates are given as positive values.

$$\text{Rate} = \frac{-\Delta[PH_3]}{\Delta t} = \frac{-(-0.048 \text{ mol}/2.0 \text{ L})}{\text{s}} = 2.4 \times 10^{-3} \text{ mol/L•s}$$

$$\frac{\Delta[P_4]}{\Delta t} = -\frac{1}{4}\frac{\Delta[PH_3]}{\Delta t} = 2.4 \times 10^{-3}/4 = 6.0 \times 10^{-4} \text{ mol/L•s}$$

$$\frac{\Delta[H_2]}{\Delta t} = -\frac{6}{4}\frac{\Delta[PH_3]}{\Delta t} = 6(2.4 \times 10^{-3})/4 = 3.6 \times 10^{-3} \text{ mol/L•s}$$

33. Using the coefficients in the balanced equation to relate the rates:

$$\frac{\Delta[H_2]}{\Delta t} = 3\frac{\Delta[N_2]}{\Delta t} \text{ and } \frac{\Delta[NH_3]}{\Delta t} = -2\frac{\Delta[N_2]}{\Delta t}$$

$$\text{So}: -\frac{1}{3}\frac{\Delta[H_2]}{\Delta t} = \frac{1}{2}\frac{\Delta[NH_3]}{\Delta t} \text{ or } \frac{\Delta[NH_3]}{\Delta t} = -\frac{2}{3}\frac{\Delta[H_2]}{\Delta t}$$

Ammonia is produced at a rate equal to 2/3 of the rate of consumption of hydrogen.

35. a. $\text{Average rate} = \frac{-\Delta[H_2O_2]}{\Delta t} = \frac{-(0.500\ M - 1.000\ M)}{(2.16 \times 10^4 \text{ s} - 0)} = 2.31 \times 10^{-5} \text{ mol/L•s}$

From the coefficients in the balanced equation:

$$\frac{\Delta[O_2]}{\Delta t} = -\frac{1}{2}\frac{\Delta[H_2O_2]}{\Delta t} = 1.16 \times 10^{-5} \text{ mol/L•s}$$

b. $\frac{-\Delta[H_2O_2]}{\Delta t} = \frac{-(0.250 - 0.500)\ M}{(4.32 \times 10^4 - 2.16 \times 10^4)\text{ s}} = 1.16 \times 10^{-5} \text{ mol/L•s}$

$$\frac{\Delta[O_2]}{\Delta t} = 1/2\ (1.16 \times 10^{-5}) = 5.80 \times 10^{-6} \text{ mol/L•s}$$

Notice that as time goes on in a reaction, the average rate decreases.

37. a. The units for rate are always mol/L•s.

b. Rate = k; k must have units of mol/L•s.

c. $\text{Rate} = k[A],\ \frac{\text{mol}}{\text{L s}} = k\left(\frac{\text{mol}}{\text{L}}\right)$

k must have units of s^{-1}.

d. $\text{Rate} = k[A]^2,\ \frac{\text{mol}}{\text{L s}} = k\left(\frac{\text{mol}}{\text{L}}\right)^2$

k must have units of L/mol•s.

e. L^2/mol^2•s

Rate Laws from Experimental Data: Initial Rates Method

39. a. In the first two experiments, [NO] is held constant and $[Cl_2]$ is doubled. The rate also doubled. Thus the reaction is first order with respect to Cl_2. Or mathematically, $\text{Rate} = k[NO]^x[Cl_2]^y$.

$$\frac{0.36}{0.18} = \frac{k(0.10)^x(0.20)^y}{k(0.10)^x(0.10)^y} = \frac{(0.20)^y}{(0.10)^y},\ 2.0 = 2.0^y,\ y = 1$$

We can get the dependence on NO from the second and third experiments. Here, as the NO concentration doubles (Cl_2 concentration is constant), the rate increases by a factor of four. Thus the reaction is second order with respect to NO. Or mathematically:

$$\frac{1.45}{0.36} = \frac{k(0.20)^x(0.20)}{k(0.10)^x(0.20)} = \frac{(0.20)^x}{(0.10)^x},\ 4.0 = 2.0^x,\ x = 2;\ \text{so Rate} = k[NO]^2[Cl_2].$$

Try to examine experiments where only one concentration changes at a time. The more variables that change, the harder it is to determine the orders. Also, these types of problems can usually be solved by inspection. In general, we will solve using a mathematical approach, but keep in mind that you probably can solve for the orders by simple inspection of the data.

b. The rate constant k can be determined from the experiments. From experiment 1:

$$\frac{0.18\ \text{mol}}{\text{L min}} = k\left(\frac{0.10\ \text{mol}}{\text{L}}\right)^2\left(\frac{0.10\ \text{mol}}{\text{L}}\right),\ k = 180\ \text{L}^2/\text{mol}^2\bullet\text{min}$$

From the other experiments:

$k = 180\ \text{L}^2/\text{mol}^2\bullet\text{min}$ (second exp.); $k = 180\ \text{L}^2/\text{mol}^2\bullet\text{min}$ (third exp.)

The average rate constant is $k_{mean} = 1.8 \times 10^2\ \text{L}^2/\text{mol}^2\bullet\text{min}$.

41. a. Rate = $k[NOCl]^n$; using experiments two and three:

$$\frac{2.66 \times 10^4}{6.64 \times 10^3} = \frac{k(2.0 \times 10^{16})^n}{k(1.0 \times 10^{16})^n},\ 4.01 = 2.0^n,\ n = 2;\ \text{Rate} = k[NOCl]^2$$

b. $$\frac{5.98 \times 10^4\ \text{molecules}}{\text{cm}^3\ \text{s}} = k\left(\frac{3.0 \times 10^{16}\ \text{molecules}}{\text{cm}^3}\right)^2,\ k = 6.6 \times 10^{-29}\ \text{cm}^3/\text{molecules}\bullet\text{s}$$

The other three experiments give (6.7, 6.6, and 6.6) $\times 10^{-29}$ cm^3/molecules•s, respectively. The mean value for k is 6.6×10^{-29} cm^3/molecules•s.

c. $$\frac{6.6 \times 10^{-29}\ \text{cm}^3}{\text{molecules s}} \times \frac{1\ \text{L}}{1000\ \text{cm}^3} \times \frac{6.022 \times 10^{23}\ \text{molecules}}{\text{mol}} = \frac{4.0 \times 10^{-8}\ \text{L}}{\text{mol s}}$$

43. a. Rate = $k[I^-]^x[OCl^-]^y$; $$\frac{7.91 \times 10^{-2}}{3.95 \times 10^{-2}} = \frac{k(0.12)^x(0.18)^y}{k(0.060)^x(0.18)^y} = 2.0^x,\ 2.00 = 2.0^x,\ x = 1$$

$$\frac{3.95 \times 10^{-2}}{9.88 \times 10^{-3}} = \frac{k(0.060)(0.18)^y}{k(0.030)(0.090)^y},\ 4.00 = 2.0 \times 2.0^y,\ 2.0 = 2.0^y,\ y = 1$$

Rate = $k[I^-][OCl^-]$

b. From the first experiment: $\dfrac{7.91 \times 10^{-2}\ \text{mol}}{\text{L s}} = k\left(\dfrac{0.12\ \text{mol}}{\text{L}}\right)\left(\dfrac{0.18\ \text{mol}}{\text{L}}\right)$, k = 3.7 L/mol•s

All four experiments give the same value of k to two significant figures.

c. $\text{Rate} = \dfrac{3.7\ \text{L}}{\text{mol s}} \times \dfrac{0.15\ \text{mol}}{\text{L}} \times \dfrac{0.15\ \text{mol}}{\text{L}} = 0.083\ \text{mol/L} \bullet \text{s}$

45. a. Rate = $k[Hb]^x[CO]^y$

Comparing the first two experiments, [CO] is unchanged, [Hb] doubles, and the rate doubles. Therefore, $x = 1$, and the reaction is first order in Hb. Comparing the second and third experiments, [Hb] is unchanged, [CO] triples, and the rate triples. Therefore, $y = 1$, and the reaction is first order in CO.

b. Rate = k[Hb][CO]

c. From the first experiment:

0.619 μmol/L•s = k(2.21 μmol/L)(1.00 μmol/L), k = 0.280 L/μmol•s

The second and third experiments give similar k values, so k_{mean} = 0.280 L/μmol•s

d. $\text{Rate} = k[\text{Hb}][\text{CO}] = \dfrac{0.280\ \text{L}}{\mu\text{mol s}} \times \dfrac{3.36\ \mu\text{mol}}{\text{L}} \times \dfrac{2.40\ \mu\text{mol}}{\text{L}} = 2.26\ \mu\text{mol/L} \bullet \text{s}$

47. Rate = $k[(CH_3)_3CBr]^x[OH^-]^y$; comparing the first two experiments, $[OH^-]$ is unchanged, $[(CH_3)_3CBr]$ doubles, and the rate doubles. Therefore, the reaction is first order in $(CH_3)_3CBr$. Comparing the first and third experiments, $[(CH_3)_3CBr]$ is unchanged, $[OH^-]$ triples, and the rate does not change. Therefore, the reaction is zero order in OH^-.

Rate = $k[(CH_3)_3CBr]^1$

For the first experiment:

$$\frac{1.0 \times 10^{-3}\ \text{mol}}{\text{L s}} = k\left(\frac{0.10\ \text{mol}}{\text{L}}\right),\quad k = 1.0 \times 10^{-2}\ \text{s}^{-1} = k_{mean}$$

Rate = $k[(CH_3)_3CBr]^1 = 1.0 \times 10^{-2}\ s^{-1}\ (0.30\ mol/L) = 3.0 \times 10^{-3}$ mol/L•s

Integrated Rate Laws

49. The first assumption to make is that the reaction is first order. For a first order reaction, a graph of $\ln[H_2O_2]$ versus time will yield a straight line. If this plot is not linear, then the reaction is not first order, and we make another assumption.

Time (s)	$[H_2O_2]$ (mol/L)	$\ln[H_2O_2]$
0	1.00	0.000
120.	0.91	−0.094
300.	0.78	−0.25
600.	0.59	−0.53
1200.	0.37	−0.99
1800.	0.22	−1.51
2400.	0.13	−2.04
3000.	0.082	−2.50
3600.	0.050	−3.00

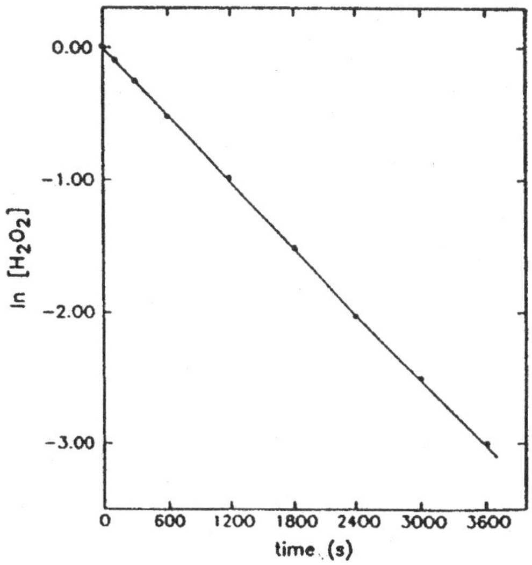

Note: We carried extra significant figures in some of the natural log values to reduce round-off error. For the plots, we will do this most of the time when the natural log function is involved.

The plot of $\ln[H_2O_2]$ versus time is linear. Thus, the reaction is first order. The rate law and integrated rate law are Rate = $k[H_2O_2]$ and $\ln[H_2O_2] = -kt + \ln[H_2O_2]_0$.

We determine the rate constant k by determining the slope of the $\ln[H_2O_2]$ versus time plot (slope = −k). Using two points on the curve gives:

$$\text{slope} = -k = \frac{\Delta y}{\Delta x} = \frac{0 - (3.00)}{0 - 3600.} = -8.3 \times 10^{-4}\ \text{s}^{-1},\quad k = 8.3 \times 10^{-4}\ \text{s}^{-1}$$

To determine $[H_2O_2]$ at 4000. s, use the integrated rate law, where $[H_2O_2]_0 = 1.00\ M$.

$$\ln[H_2O_2] = -kt + \ln[H_2O_2]_0 \quad \text{or} \quad \ln\left(\frac{[H_2O_2]}{[H_2O_2]_0}\right) = -kt$$

$$\ln\left(\frac{[H_2O_2]}{1.00}\right) = -8.3 \times 10^{-4}\ \text{s}^{-1} \times 4000.\ \text{s},\quad \ln[H_2O_2] = -3.3,\quad [H_2O_2] = e^{-3.3} = 0.037\ M$$

51. Assume the reaction is first order and see if the plot of $\ln[NO_2]$ versus time is linear. If this isn't linear, try the second-order plot of $1/[NO_2]$ versus time because second-order reactions are the next most common after first-order reactions. The data and plots follow.

Time (s)	$[NO_2]$ (M)	$\ln[NO_2]$	$1/[NO_2]$ (M^{-1})
0	0.500	−0.693	2.00
1.20×10^3	0.444	−0.812	2.25
3.00×10^3	0.381	−0.965	2.62
4.50×10^3	0.340	−1.079	2.94
9.00×10^3	0.250	−1.386	4.00
1.80×10^4	0.174	−1.749	5.75

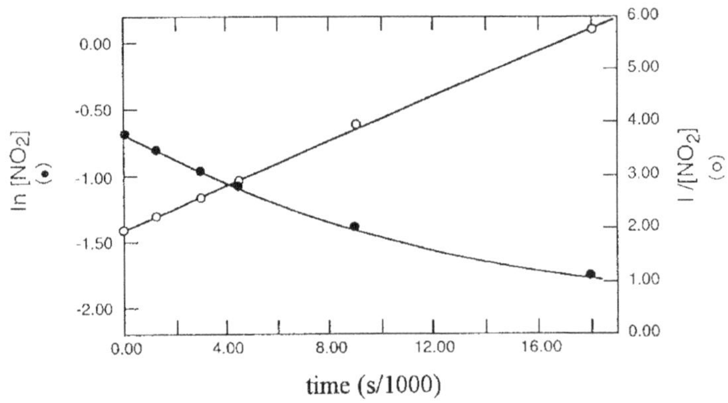

The plot of $1/[NO_2]$ versus time is linear. The reaction is second order in NO_2.

The rate law and integrated rate law are: Rate = $k[NO_2]^2$ and $\dfrac{1}{[NO_2]} = kt + \dfrac{1}{[NO_2]_0}$

The slope of the plot $1/[NO_2]$ vs. t gives the value of k. Using a couple of points on the plot:

$$\text{slope} = k = \frac{\Delta y}{\Delta x} = \frac{(5.75 - 2.00)\ M^{-1}}{(1.80 \times 10^4 - 0)\ \text{s}} = 2.08 \times 10^{-4}\ \text{L/mol•s}$$

To determine $[NO_2]$ at 2.70×10^4 s, use the integrated rate law, where $1/[NO_2]_o = 1/0.500\ M = 2.00\ M^{-1}$.

$$\frac{1}{[NO_2]} = kt + \frac{1}{[NO_2]_0},\quad \frac{1}{[NO_2]} = \frac{2.08 \times 10^{-4}\ \text{L}}{\text{mol s}} \times 2.70 \times 10^4\ \text{s} + 2.00\ M^{-1}$$

$$\frac{1}{[NO_2]} = 7.62,\quad [NO_2] = 0.131\ M$$

53. a. Because the $[C_2H_5OH]$ versus time plot was linear, the reaction is zero order in C_2H_5OH. The slope of the $[C_2H_5OH]$ versus time plot equals –k. Therefore, the rate law, the integrated rate law, and the rate constant value are: Rate = $k[C_2H_5OH]^0 = k$; $[C_2H_5OH] = -kt + [C_2H_5OH]_0$; $k = 4.00 \times 10^{-5}$ mol/L•s

b. The half-life expression for a zero-order reaction is $t_{1/2} = [A]_0/2k$.

$$t_{1/2} = \frac{[C_2H_5OH]_0}{2k} = \frac{1.25 \times 10^{-2}\ \text{mol/L}}{2 \times 4.00 \times 10^{-5}\ \text{mol/L} \bullet \text{s}} = 156\ \text{s}$$

Note: We could have used the integrated rate law to solve for $t_{1/2}$, where $[C_2H_5OH] = (1.25 \times 10^{-2}/2)$ mol/L.

c. $[C_2H_5OH] = -kt + [C_2H_5OH]_0$, 0 mol/L = $-(4.00 \times 10^{-5}$ mol/L•s)t + 1.25×10^{-2} mol/L

$$t = \frac{1.25 \times 10^{-2}\ \text{mol/L}}{4.00 \times 10^{-5}\ \text{mol/L} \bullet \text{s}} = 313\ \text{s}$$

55. The first assumption to make is that the reaction is first order. For a first-order reaction, a graph of $\ln[C_4H_6]$ versus t should yield a straight line. If this isn't linear, then try the second- order plot of $1/[C_4H_6]$ versus t. The data and the plots follow:

Time	195	604	1246	2180	6210 s
$[C_4H_6]$	1.6×10^{-2}	1.5×10^{-2}	1.3×10^{-2}	1.1×10^{-2}	0.68×10^{-2} *M*
$\ln[C_4H_6]$	–4.14	–4.20	–4.34	–4.51	–4.99
$1/[C_4H_6]$	62.5	66.7	76.9	90.9	147 M^{-1}

Note: To reduce round-off error, we carried extra significant figures in the data points.

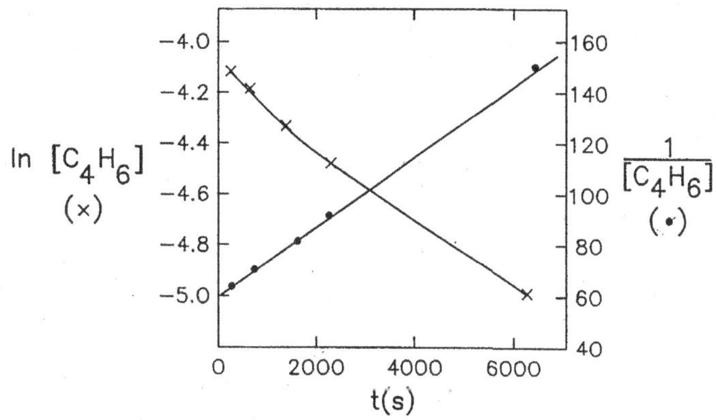

The natural log plot is not linear, so the reaction is not first order. Because the second-order plot of $1/[C_4H_6]$ versus t is linear, we can conclude that the reaction is second order in butadiene. The rate law is:

$$\text{Rate} = k[C_4H_6]^2$$

For a second-order reaction, the integrated rate law is $\dfrac{1}{[C_4H_6]} = kt + \dfrac{1}{[C_4H_6]_0}$.

The slope of the straight line equals the value of the rate constant. Using the points on the line at 1000. and 6000. s:

$$k = \text{slope} = \frac{144 \text{ L/mol} - 73 \text{ L/mol}}{6000.\text{ s} - 1000.\text{ s}} = 1.4 \times 10^{-2} \text{ L/mol}\bullet\text{s}$$

57. Because the $1/[A]$ versus time plot is linear with a positive slope, the reaction is second order with respect to A. The *y* intercept in the plot will equal $1/[A]_0$. Extending the plot, the *y* intercept will be about 10, so $1/10 = 0.1\ M = [A]_0$.

59. a. $[A] = -kt + [A]_0$; if $k = 5.0 \times 10^{-2}$ mol/L•s and $[A]_0 = 1.00 \times 10^{-3}$ *M*, then:

$$[A] = -(5.0 \times 10^{-2} \text{ mol/L}\bullet\text{s})t + 1.00 \times 10^{-3} \text{ mol/L}$$

b. $\dfrac{[A]_0}{2} = -(5.0 \times 10^{-2})t_{1/2} + [A]_0$ because at $t = t_{1/2}$, $[A] = [A]_0/2$.

$$-0.50[A]_0 = -(5.0 \times 10^{-2})t_{1/2},\ t_{1/2} = \frac{0.50(1.00 \times 10^{-3})}{5.0 \times 10^{-2}} = 1.0 \times 10^{-2}\ \text{s}$$

Note: We could have used the $t_{1/2}$ expression to solve ($t_{1/2} = \frac{[A]_0}{2k}$).

c. $[A] = -kt + [A]_0 = -(5.0 \times 10^{-2}\ \text{mol/L•s})(5.0 \times 10^{-3}\ \text{s}) + 1.00 \times 10^{-3}\ \text{mol/L}$

$[A] = 7.5 \times 10^{-4}\ \text{mol/L}$

$[A]_{reacted} = 1.00 \times 10^{-3}\ \text{mol/L} - 7.5 \times 10^{-4}\ \text{mol/L} = 2.5 \times 10^{-4}\ \text{mol/L}$

$[B]_{produced} = [A]_{reacted} = 2.5 \times 10^{-4}\ M$

61. If $[A]_0 = 100.0$, then after 65 s, 45.0% of A has reacted, or $[A] = 55.0$. For first order reactions:

$$\ln\left(\frac{[A]}{[A]_0}\right) = -kt,\ \ln\left(\frac{55.0}{100.0}\right) = -k(65\ \text{s}),\ k = 9.2 \times 10^{-3}\ \text{s}^{-1}$$

$$t_{1/2} = \frac{\ln 2}{k} = \frac{0.693}{9.2 \times 10^{-3}\ \text{s}^{-1}} = 75\ \text{s}$$

63. For a first-order reaction, the integrated rate law is $\ln([A]/[A]_0) = -kt$. Solving for k:

$$\ln\left(\frac{0.250\ \text{mol/L}}{1.00\ \text{mol/L}}\right) = -k \times 120.\ \text{s},\ k = 0.0116\ \text{s}^{-1}$$

$$\ln\left(\frac{0.350\ \text{mol/L}}{2.00\ \text{mol/L}}\right) = -0.0116\ \text{s}^{-1} \times t,\ t = 150.\ \text{s}$$

65. The integrated rate law for a second order reaction is $1/[A] = kt + 1/[A]_0$. Using the integrated rate law:

$$\frac{1}{0.020\ \text{mol/L}} = 0.40\ \text{L/mol•min} \times t + \frac{1}{0.10\ \text{mol/L}},\ t = \frac{(50. - 10.)\ \text{L/mol}}{0.40\ \text{L/mol} \bullet \text{min}} = 1.0 \times 10^2\ \text{min}$$

67. Successive half-lives double as concentration is decreased by one-half. This is consistent with second-order reactions, so assume the reaction is second order in A.

$$t_{1/2} = \frac{1}{k[A]_0},\ k = \frac{1}{t_{1/2}[A]_0} = \frac{1}{10.0\ \text{min}(0.10\ M)} = 1.0\ \text{L/mol•min}$$

a. $\frac{1}{[A]} = kt + \frac{1}{[A]_0} = \frac{1.0\ \text{L}}{\text{mol min}} \times 80.0\ \text{min} + \frac{1}{0.10\ M} = 90.\ M^{-1},\ [A] = 1.1 \times 10^{-2}\ M$

b. 30.0 min = 2 half-lives, so 25% of original A is remaining.

$[A] = 0.25(0.10\ M) = 0.025\ M$

69. The consecutive half-life values of 24 hours, then 12 hours, show a direct relationship with concentration; as the concentration decreases, the half-life decreases. Assuming the drug reaction is either zero, first, or second order, only a zero-order reaction shows this direct relationship between half-life and concentration. Therefore, assume the reaction is zero order in the drug.

$$t_{1/2} = \frac{[A]_0}{2k},\ k = \frac{[A]_0}{2t_{1/2}} = \frac{2.0 \times 10^{-3}\ \text{mol/L}}{2(24\ \text{h})} = 4.2 \times 10^{-5}\ \text{mol/L•h}$$

71. Because $[B]_0 >> [A]_0$, the concentration of B is essentially constant in this experiment. We have a pseudo-first-order reaction in A:

$$\text{Rate} = k[A][B]^2 = k'[A],\ \text{where } k' = k[B]_0^2$$

$$\ln\left(\frac{[A]}{[A]_0}\right) = -k't,\ \ln\left(\frac{2.3 \times 10^{-4}\ M}{5.0 \times 10^{-4}\ M}\right) = -k' \times 3.0\ \text{min},\ k' = 0.26\ \text{min}^{-1}$$

$$k' = k[B]_0^2,\ 0.26\ \text{min}^{-1} = k(5.0)^2,\ k = 0.010\ \text{L}^2/\text{mol}^2\text{•min}$$

73. Because $[V]_0 >> [AV]_0$, the concentration of V is essentially constant in this experiment. We have a pseudo-first-order reaction in AV:

$$\text{Rate} = k[AV][V] = k'[AV],\ \text{where } k' = k[V]_0$$

The slope of the ln[AV] versus time plot is equal to $-k'$.

$$k' = -\text{slope} = 0.32\ \text{s}^{-1};\ k = \frac{k'}{[V]_0} = \frac{0.32\ \text{s}^{-1}}{0.20\ \text{mol/L}} = 1.6\ \text{L/mol•s}$$

Reaction Mechanisms

75. For elementary reactions, the rate law can be written using the coefficients in the balanced equation to determine orders.

a. Rate = $k[CH_3NC]$

b. Rate = $k[O_3][NO]$

c. Rate = $k[O_3]$

d. Rate = $k[O_3][O]$

77. A mechanism consists of a series of elementary reactions in which the rate law for each step can be determined using the coefficients in the balanced equations. For a plausible mechanism, the rate law derived from a mechanism must agree with the rate law determined from experiment. To derive the rate law from the mechanism, the rate of the reaction is assumed to equal the rate of the slowest step in the mechanism.

Because step 1 is the rate-determining step, the rate law for this mechanism is Rate = $k[C_4H_9Br]$. To get the overall reaction, we sum all the individual steps of the mechanism. Summing all steps gives:

$$C_4H_9Br \rightarrow C_4H_9^+ + Br^-$$
$$C_4H_9^+ + H_2O \rightarrow C_4H_9OH_2^+$$
$$C_4H_9OH_2^+ + H_2O \rightarrow C_4H_9OH + H_3O^+$$

$$C_4H_9Br + 2\ H_2O \rightarrow C_4H_9OH + Br^- + H_3O^+$$

Intermediates in a mechanism are species that are neither reactants nor products but that are formed and consumed during the reaction sequence. The intermediates for this mechanism are $C_4H_9^+$ and $C_4H_9OH_2^+$.

79. The rate law is Rate = $k[NO]^2[Cl_2]$. If we assume the first step is rate-determining, we expect the rate law to be Rate = $k_1[NO][Cl_2]$. This isn't correct. If we assume the second step is rate-determining, then Rate = $k_2[NOCl_2][NO]$. To see if this agrees with experiment, we must substitute for the intermediate $NOCl_2$ concentration. Assuming a fast-equilibrium first step (rate reverse = rate forward):

$k_{-1}[NOCl_2] = k_1[NO][Cl_2]$, $[NOCl_2] = \frac{k_1}{k_{-1}}[NO][Cl_2]$; substituting into the rate equation:

$$\text{Rate} = \frac{k_2k_1}{k_{-1}}[NO]^2[Cl_2] = k[NO]^2[Cl_2] \text{ where } k = \frac{k_2k_1}{k_{-1}}$$

This is a possible mechanism with the second step the rate-determining step because the derived rate law agrees with the experimentally determined rate law.

81. In the first experiment where $[A]_0 >> [B_0]$, a plot of ln[B] vs time is linear with a negative slope. So, the reaction is first order in B. In the second experiment where $[B]_0 >> [A]_0$, a plot of 1/[A] vs time is linear with a positive slope. So, the reaction is second order in A. The rate law from experiment is Rate = $[A]^2[B]$.

Mechanism I:

Rate = $k_2[E][A]$; from the fast equilibrium first step, $k_1[A][B] = k_{-1}[E]$, $[E] = \frac{k_1}{k_{-1}}[A][B]$

Substituting: Rate = $k_2[E][A] = k_2\frac{k_1}{k_{-1}}[A][B][A] = k[A]^2[B]$; this is a possible mechanism.

Mechanism II:

Rate = $k_2[E][B]$; from the fast equilibrium first step, $k_1[A][B] = k_{-1}[E]$, $[E] = \frac{k_1}{k_{-1}}[A][B]$

Substituting: Rate = $k_2[E][B] = k_2\frac{k_1}{k_{-1}}[A][B][B] = k[A]\ [B]^2$; not a possible mechanism.

Mechanism III:

Rate = $k_1[A]^2$; not a possible mechanism

Only mechanism I gives a derived rate law that agrees with experiment. So, only mechanism I is a possible mechanism for this reaction.

Temperature Dependence of Rate Constants and the Collision Model

83. In the following plot, R = reactants, P = products, E_a = activation energy, and RC = reaction coordinate, which is the same as reaction progress. Note for this reaction that ΔE is positive because the products are at a higher energy than the reactants.

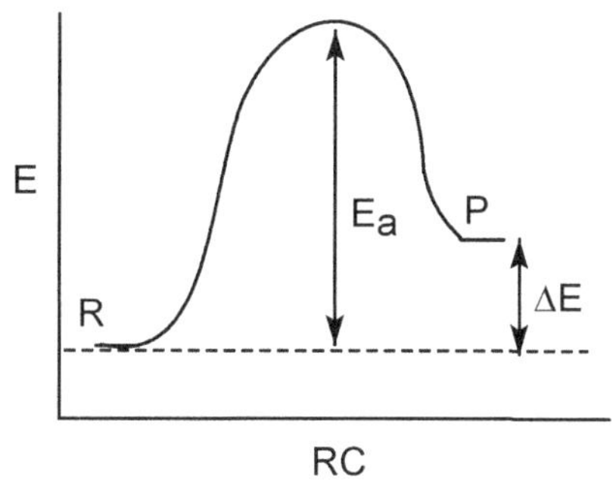

85.

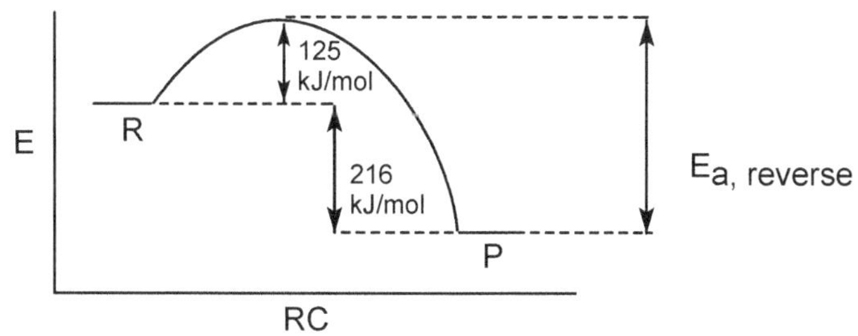

The activation energy for the reverse reaction is:

$$E_{a,\ reverse} = 216\ \text{kJ/mol} + 125\ \text{kJ/mol} = 341\ \text{kJ/mol}$$

87. The Arrhenius equation is $k = A\ \exp(-E_a/RT)$ or, in logarithmic form, $\ln k = -E_a/RT + \ln A$. Hence a graph of ln k versus 1/T should yield a straight line with a slope equal to $-E_a/R$ since the logarithmic form of the Arrhenius equation is in the form of a straight-line equation, $y = mx + b$. *Note*: We carried extra significant figures in the following ln k values in order to reduce round off error.

T (K)	1/T (K^{-1})	k (s^{-1})	ln k
338	2.96×10^{-3}	4.9×10^{-3}	−5.32
318	3.14×10^{-3}	5.0×10^{-4}	−7.60
298	3.36×10^{-3}	3.5×10^{-5}	−10.26

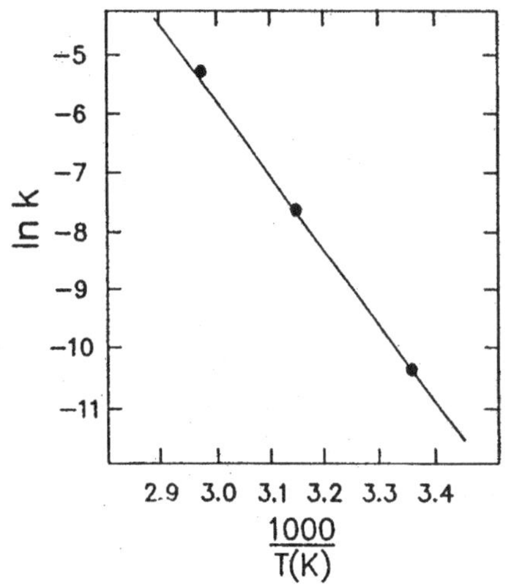

$$\text{Slope} = \frac{-10.76 - (-5.85)}{3.40 \times 10^{-3} - 3.00 \times 10^{-3}} = -1.2 \times 10^4 \text{ K} = -E_a/R$$

$$E_a = -\text{slope} \times R = 1.2 \times 10^4 \text{ K} \times \frac{8.3145 \text{ J}}{\text{K mol}}, \quad E_a = 1.0 \times 10^5 \text{ J/mol} = 1.0 \times 10^2 \text{ kJ/mol}$$

89. $k = A\exp(-E_a/RT)$ or $\ln k = \frac{-E_a}{RT} + \ln A$ (the Arrhenius equation)

For two conditions: $\ln\left(\frac{k_2}{k_1}\right) = \frac{E_a}{R}\left(\frac{1}{T_1} - \frac{1}{T_2}\right)$ (Assuming A is temperature independent.)

Let $k_1 = 3.52 \times 10^{-7}$ L/mol•s, $T_1 = 555$ K; $k_2 = ?$, $T_2 = 645$ K; $E_a = 186 \times 10^3$ J/mol

$$\ln\left(\frac{k_2}{3.52 \times 10^{-7}}\right) = \frac{1.86 \times 10^5 \text{ J/mol}}{8.3145 \text{ J/K}\bullet\text{mol}}\left(\frac{1}{555 \text{ K}} - \frac{1}{645 \text{ K}}\right) = 5.6$$

$$\frac{k_2}{3.52 \times 10^{-7}} = e^{5.6} = 270, \quad k_2 = 270(3.52 \times 10^{-7}) = 9.5 \times 10^{-5} \text{ L/mol}\bullet\text{s}$$

91. $\ln\left(\frac{k_2}{k_1}\right) = \frac{E_a}{R}\left(\frac{1}{T_1} - \frac{1}{T_2}\right)$; $\frac{k_2}{k_1} = 7.00$, $T_1 = 295$ K, $E_a = 54.0 \times 10^3$ J/mol

$$\ln(7.00) = \frac{5.4 \times 10^4 \text{ J/mol}}{8.3145 \text{ J/K}\bullet\text{mol}}\left(\frac{1}{295 \text{ K}} - \frac{1}{T_2}\right), \quad \frac{1}{295 \text{ K}} - \frac{1}{T_2} = 3.00 \times 10^{-4}$$

$$\frac{1}{T_2} = 3.09 \times 10^{-3}, \quad T_2 = 324 \text{ K} = 51°\text{C}$$

93. $H_3O^+(aq) + OH^-(aq) \rightarrow 2\ H_2O(l)$ should have the faster rate. H_3O^+ and OH^- will be electrostatically attracted to each other; Ce^{4+} and Hg_2^{2+} will repel each other. The activation energy for the Ce^{4+} and Hg_2^{2+} reaction should be a larger quantity, making it the slower reaction.

Catalysts

95.

$$\begin{aligned} C_2H_4 + H^+ &\rightarrow C_2H_5^+ \\ C_2H_5^+ + H_2O &\rightarrow C_2H_5OH_2^+ \\ C_2H_5OH_2^+ &\rightarrow C_2H_5OH + H^+ \\ \hline C_2H_4 + H_2O &\rightarrow C_2H_5OH \end{aligned}$$

C_2H_4 and H_2O are reactants, and C_2H_5OH is the product. The other species do not appear in the overall balanced equation, so they are either catalysts or intermediates. Catalysts will always appear first on the reactant side in one of the steps; intermediates always appear first on the product side of one of the steps. H^+ is a catalyst; it is a reactant in step 1 but produced in step 3. It does not appear in the overall balanced reaction. $C_2H_5^+$ and $C_2H_5OH_2^+$ are intermediates in the mechanism. These ions appear first on the product side of one of the steps but are reactants in a subsequent step.

97.

a. NO is the catalyst. NO is present in the first step of the mechanism on the reactant side, but it is not a reactant. NO is regenerated in the second step and does not appear in overall balanced equation.

b. NO_2 is an intermediate. Intermediates also never appear in the overall balanced equation. In a mechanism, intermediates always appear first on the product side, whereas catalysts always appear first on the reactant side.

c. $k = A\exp(-E_a/RT)$; $\dfrac{k_{cat}}{k_{un}} = \dfrac{A\exp[-E_a(cat)/RT]}{A\exp[-E(un)/RT]} = \exp\left[\dfrac{E_a(un) - E_a(cat)}{RT}\right]$

$$\frac{k_{cat}}{k_{un}} = \exp\left(\frac{2100\ \text{J/mol}}{8.3145\ \text{J/K} \bullet \text{mol} \times 298\ \text{K}}\right) = e^{0.85} = 2.3$$

The catalyzed reaction is approximately 2.3 times faster than the uncatalyzed reaction at 25°C.

99. The reaction at the surface of the catalyst is assumed to follow the steps:

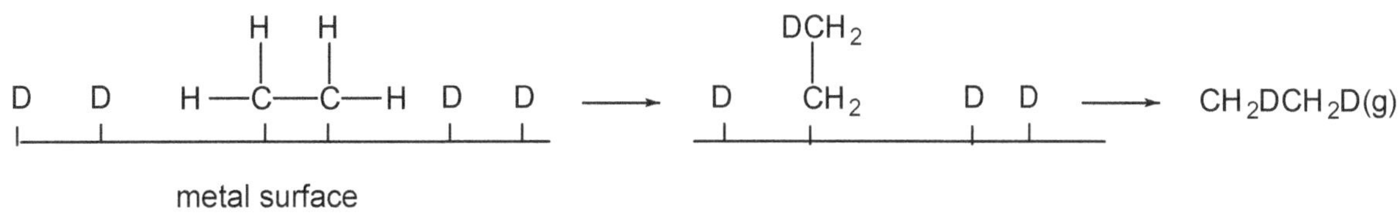

Thus $CH_2D–CH_2D$ should be the product. If the mechanism is possible, then the reaction must be:

$$C_2H_4 + D_2 \rightarrow CH_2DCH_2D$$

If we got this product, then we could conclude that this is a possible mechanism. If we got some other product, for example, CH_3CHD_2, then we would conclude that the mechanism is wrong. Even though this mechanism correctly predicts the products of the reaction, we cannot say conclusively that this is the correct mechanism; we might be able to conceive of other mechanisms that would give the same products as our proposed one.

101. The rate depends on the number of reactant molecules adsorbed on the surface of the catalyst. This quantity is proportional to the concentration of reactant. However, when all the catalyst surface sites are occupied, the rate becomes independent of the concentration of reactant.

103. Assuming the catalyzed and uncatalyzed reactions have the same form and orders, and because concentrations are assumed equal, the rates will be equal when the k values are equal.

$k = A \exp(-E_a/RT)$; $k_{cat} = k_{un}$ when $E_{a,cat}/RT_{cat} = E_{a,un}/RT_{un}$.

$$\frac{4.20 \times 10^4 \text{ J/mol}}{8.3145 \text{ J/K} \bullet \text{mol} \times 293 \text{ K}} = \frac{7.00 \times 10^4 \text{ J/mol}}{8.3145 \text{ J/K} \bullet \text{mol} \times T_{un}}, \; T_{un} = 488 \text{ K} = 215°\text{C}$$

ChemWork Problems

105. Box a has 8 NO_2 molecules. Box b has 4 NO_2 molecules, and box c has 2 NO_2 molecules. Box b represents what is present after the first half-life of the reaction, and box c represents what is present after the second half-life.

a. For first order kinetics, $t_{1/2} = 0.693/k$; the half-life for a first order reaction is concentration independent. Therefore, the time for box c, the time it takes to go through two half-lives, will be 10 + 10 = 20 minutes.

b. For second order kinetics, $t_{1/2} = 1/k[A]_0$; the half-life for a second order reaction is inversely proportional to the initial concentration. So if the first half-life is 10 minutes, the second half-life will be 20 minutes. For a second order reaction, the time for box c will be 10 + 20 = 30 minutes.

c. For zero order kinetics, $t_{1/2} = [A]_0/2k$; the half-life for a zero-order reaction is directly related to the initial concentration. So if this reaction was zero order, then the second half-life would decrease from 10 min to 5 min. The time for box c will be 10 + 5 = 15 minutes if the reaction is zero order.

107. The integrated rate law for each reaction is:

$$\ln[A] = -4.50 \times 10^{-4} \text{ s}^{-1}(t) + \ln[A]_0 \text{ and } \ln[B] = -3.70 \times 10^{-3} \text{ s}^{-1}(t) + \ln[B]_0$$

Subtracting the second equation from the first equation ($\ln[A]_0 = \ln[B]_0$):

$$\ln[A] - \ln[B] = -4.50 \times 10^{-4}(t) + 3.70 \times 10^{-3}(t), \; \ln\left(\frac{[A]}{[B]}\right) = 3.25 \times 10^{-3}(t)$$

When $[A] = 4.00\,[B]$, $\ln(4.00) = 3.25 \times 10^{-3}(t)$, $t = 427$ s.

109. From 338 K data, a plot of $\ln[N_2O_5]$ versus t is linear, and the slope $= -4.86 \times 10^{-3}$ (plot not included). This tells us the reaction is first order in N_2O_5 with $k = 4.86 \times 10^{-3}$ at 338 K. From 318 K data, the slope of $\ln[N_2O_5]$ versus t plot is equal to -4.98×10^{-4}, so $k = 4.98 \times 10^{-4}$ at 318 K. We now have two values of k at two temperatures, so we can solve for E_a.

$$\ln\left(\frac{k_2}{k_1}\right) = \frac{E_a}{R}\left(\frac{1}{T_1} - \frac{1}{T_2}\right),\quad \ln\left(\frac{4.86 \times 10^{-3}}{4.98 \times 10^{-4}}\right) = \frac{E_a}{8.3145\ \text{J/K}\bullet\text{mol}}\left(\frac{1}{318\ \text{K}} - \frac{1}{338\ \text{K}}\right)$$

$E_a = 1.0 \times 10^5$ J/mol $= 1.0 \times 10^2$ kJ/mol

111. Because the ln[A] versus time plot is linear, the reaction is first order in A. The slope of the ln[A] versus time plot equals –k. Therefore, the integrated rate law and the rate constant value are:

$\ln[A] = -kt + \ln[A]_0$; $k = 7.35 \times 10^{-3}\ s^{-1}$

From the problem, 77.1% A remains: $[A] = 0.771(1.00 \times 10^{-2}$ mol/L$) = 7.71 \times 10^{-3}$ mol/L

$$\ln\left(\frac{[A]}{[A]_0}\right) = -kt,\quad \ln\left(\frac{7.71 \times 10^{-3}\ M}{1.00 \times 10^{-2}\ M}\right) = -(7.35 \times 10^{-3}\ s^{-1})t$$

$$t = \frac{\ln(0.771)}{-7.35 \times 10^{-3}\ s^{-1}} = 35.4\ s$$

113. a.

t (s)	$[C_4H_6]$ (M)	$\ln[C_4H_6]$	$1/[C_4H_6]$ (M^{-1})
0	0.01000	–4.6052	1.000×10^2
1000.	0.00629	–5.069	1.59×10^2
2000.	0.00459	–5.384	2.18×10^2
3000.	0.00361	–5.624	2.77×10^2

The plot of $1/[C_4H_6]$ versus t is linear, thus the reaction is second order in butadiene. From the plot (not included), the integrated rate law is:

$$\frac{1}{[C_4H_6]} = (5.90 \times 10^{-2}\ \text{L/mol}\bullet\text{s})t + 100.0\ M^{-1}$$

b. When dimerization is 1.0% complete, 99.0% of C_4H_6 is left.

$[C_4H_6] = 0.990(0.01000) = 0.00990\ M$; $\dfrac{1}{0.00990} = (5.90 \times 10^{-2})t + 100.0$

$101 = (5.90 \times 10^{-2})t + 100.0$, $t = 17\ s \approx 20\ s$

c. 10.0% complete, $[C_4H_6] = 0.00900\ M$; $\dfrac{1}{0.00900} = (5.90 \times 10^{-2})t + 100.0$,

$t = 188\ s \approx 190\ s$

d. $\dfrac{1}{[C_4H_6]} = kt + \dfrac{1}{[C_4H_6]_0}$; $[C_4H_6]_0 = 0.0200\ M$; at $t = t_{1/2}$, $[C_4H_6] = 0.0100\ M$.

$\dfrac{1}{0.0100} = (5.90 \times 10^{-2})t_{1/2} + \dfrac{1}{0.0200}$, $t_{1/2} = 847\ s = 850\ s$

Or: $t_{1/2} = \frac{1}{k[A]_0} = \frac{1}{(5.90 \times 10^{-2}\ \text{L/mol} \bullet \text{s})(2.00 \times 10^{-2}\ \text{M})} = 847\ \text{s}$

e. From Exercise 12.55, $k = 1.4 \times 10^{-2}$ L/mol•s at 500. K. From this problem, $k = 5.90 \times 10^{-2}$ L/mol•s at 620. K.

$$\ln\left(\frac{k_2}{k_1}\right) = \frac{E_a}{R}\left(\frac{1}{T_1} - \frac{1}{T_2}\right),\ \ln\left(\frac{5.90 \times 10^{-2}}{1.4 \times 10^{-2}}\right) = \frac{E_a}{8.3145\ \text{J/K} \bullet \text{mol}}\left(\frac{1}{500.\ \text{K}} - \frac{1}{620.\ \text{K}}\right)$$

$12 = E_a(3.9 \times 10^{-4})$, $E_a = 3.1 \times 10^4$ J/mol = 31 kJ/mol

115. The integrated rate law for a zero-order reaction is $[A] = -kt + [A]_0$.

$$[A] = -kt + [A]_0,\ \frac{0.021\ \text{mol}}{\text{L}} = -\left(\frac{2.50 \times 10^{-2}\ \text{mol}}{\text{L s}}\right) \times t + \frac{0.0800\ \text{mol}}{\text{L}},\ t = 2.36\ \text{s}$$

117. To determine the rate of reaction, we need to calculate the value of the rate constant k. The activation energy data can be manipulated to determine k.

$$k = Ae^{-E_a/RT} = 0.850\ \text{s}^{-1} \times \exp\left(\frac{-26.2 \times 10^3\ \text{J/mol}}{8.3145\ \text{J/K} \bullet \text{mol} \times 310.2\ \text{K}}\right) = 3.29 \times 10^{-5}\ \text{s}^{-1}$$

Rate = k[acetycholine receptor-toxin complex]

$$\text{Rate} = 3.29 \times 10^{-5}\ \text{s}^{-1}\left(\frac{0.200\ \text{mol}}{\text{L}}\right) = 6.58 \times 10^{-6}\ \text{mol/L} \bullet \text{s}$$

119. For two conditions: $\ln\left(\frac{k_2}{k_1}\right) = \frac{E_a}{R}\left(\frac{1}{T_1} - \frac{1}{T_2}\right)$ (Assuming A is temperature independent.)

$$\ln\left(\frac{0.891\ \text{L/mol} \bullet \text{s}}{2.45 \times 10^{-4}\ \text{L/mol} \bullet \text{s}}\right) = \frac{E_a}{8.3145\ \text{J/K} \bullet \text{mol}}\left(\frac{1}{575\ \text{K}} - \frac{1}{781\ \text{K}}\right)$$

$8.20 = \frac{E_a}{8.3145}(4.6 \times 10^{-4})$, $E_a = 1.5 \times 10^5$ J/mol = 150 kJ/mol

Let $k_1 = 0.891$ L/mol•s, $T_1 = 781$ K; $k_2 = ?$, $T_2 = 648$ K; $E_a = 1.5 \times 10^5$ J/mol

$$\ln\left(\frac{k_2}{0.891}\right) = \frac{1.5 \times 10^5\ \text{J/mol}}{8.3145\ \text{J/K} \bullet \text{mol}}\left(\frac{1}{781\ \text{K}} - \frac{1}{648\ \text{K}}\right) = -4.7$$

$$\frac{k_2}{0.891} = e^{-4.7} = 9.1 \times 10^{-3},\ k_2 = 8.1 \times 10^{-3}\ \text{L/mol} \bullet \text{s}$$

121. a. If the interval between flashes is 16.3 s, then the rate is:

1 flash/16.3 s = $6.13 \times 10^{-2}\ \text{s}^{-1} = k$

Interval	k	T
16.3 s	6.13×10^{-2} s^{-1}	21.0°C (294.2 K)
13.0 s	7.69×10^{-2} s^{-1}	27.8°C (301.0 K)

$\ln\left(\frac{k_2}{k_1}\right) = \frac{E_a}{R}\left(\frac{1}{T_1} - \frac{1}{T_2}\right)$; solving using above data: $E_a = 2.5 \times 10^4$ J/mol = 25 kJ/mol

b. $$\ln\left(\frac{k}{6.13 \times 10^{-2}}\right) = \frac{2.5 \times 10^4 \text{ J/mol}}{8.3145 \text{ J/K} \bullet \text{mol}}\left(\frac{1}{294.2 \text{ K}} - \frac{1}{303.2 \text{ K}}\right) = 0.30$$

$k = e^{0.30} \times (6.13 \times 10^{-2}) = 8.3 \times 10^{-2}$ s^{-1}; interval = 1/k = 12 seconds.

c.

T	Interval	54-2(Intervals)
21.0°C	16.3 s	21°C
27.8°C	13.0 s	28°C
30.0°C	12 s	30.°C

This rule of thumb gives excellent agreement to two significant figures.

123. Rate = $k[A]^x[B]^y$; using data from experiment 3 and experiment 4:

$$\frac{3.46 \times 10^{-2}}{4.32 \times 10^{-3}} = \frac{k(0.24)^x(0.090)^y}{k(0.030)^x(0.090)^y} = 8.0^x,\ 8.00 = 8.0^x,\ x = 1$$

From experiments 1 and 4:

$$\frac{3.46 \times 10^{-2}}{3.46 \times 10^{-2}} = \frac{k(0.24)(0.090)^y}{k(0.12)(0.18)^y},\ 1.00 = 2.0 \times (1/2)^y,\ 1/2 = (1/2)^y,\ y = 1$$

The reaction is first order with respect to both A and B; rate = k[A][B]

From the first experiment: $\frac{3.46 \times 10^{-2} \text{ mol}}{\text{L s}} = k\left(\frac{0.12 \text{ mol}}{\text{L}}\right)\left(\frac{0.18 \text{ mol}}{\text{L}}\right)$,

k = 1.6 L/mol•s; all four experiments give the same value of k to two significant figures.

125. a. Because $[A]_0 << [B]_0$ or $[C]_0$, the B and C concentrations remain constant at 1.00 *M* for this experiment. Thus Rate = $k[A]^2[B][C] = k'[A]^2$, where $k' = k[B][C]$.

For this pseudo-second-order reaction:

$$\frac{1}{[A]} = k't + \frac{1}{[A]_0},\quad \frac{1}{3.26 \times 10^{-5}\ M} = k'(3.00 \text{ min}) + \frac{1}{1.00 \times 10^{-4}\ M}$$

k′ = 6890 L/mol•min = 115 L/mol•s

$$k' = k[B][C],\ k = \frac{k'}{[B][C]},\quad k = \frac{115 \text{ L/mol} \bullet \text{s}}{(1.00\ M)(1.00\ M)} = 115 \text{ L}^3/\text{mol}^3 \bullet \text{s}$$

b. For this pseudo-second-order reaction:

$$\text{Rate} = k'[A]^2,\ t_{1/2} = \frac{1}{k'[A]_0} = \frac{1}{115\ \text{L/mol}\bullet\text{s}(1.00 \times 10^{-4}\ \text{mol/L})} = 87.0\ \text{s}$$

c. $$\frac{1}{[A]} = k't + \frac{1}{[A]_0} = 115\ \text{L/mol}\bullet\text{s} \times 600.\ \text{s} + \frac{1}{1.00 \times 10^{-4}\ \text{mol/L}} = 7.90 \times 10^4\ \text{L/mol}$$

$[A] = 1/7.90 \times 10^4$ L/mol $= 1.27 \times 10^{-5}$ mol/L

From the stoichiometry in the balanced reaction, 1 mol of B reacts with every 3 mol of A.

Amount A reacted $= 1.00 \times 10^{-4}\ M - 1.27 \times 10^{-5}\ M = 8.7 \times 10^{-5}\ M$

$$\text{Amount B reacted} = 8.7 \times 10^{-5}\ \text{mol/L} \times \frac{1\ \text{mol B}}{3\ \text{mol A}} = 2.9 \times 10^{-5}\ M$$

$[B] = 1.00\ M - 2.9 \times 10^{-5}\ M = 1.00\ M$

As we mentioned in part a, the concentration of B (and C) remains constant because the A concentration is so small compared to the B (or C) concentration.

127. Only the plot of $\ln[Ru^{2+}]$ vs time is linear. This is the proof plot for a first order reaction. So, the reaction is first order with respect to Ru^{2+}. Answers b and c both indicate a first order reaction. The y-intercept of the $\ln[Ru^{2+}]$ vs time plot will equal the $\ln[Ru^{2+}]_0$. If the initial concentration of Ru^{2+} was 1.0 *M*, the y-intercept would be 0 [ln(1.0) = 0]. However, the y-intercept looks to be around –1.0. Because ln(0.30) = –1.2, this confirms answer b as correct (first order with an initial concentration of 0.3 *M*). Note that the concentration vs time plot confirms an initial Ru^{2+} concentration of about 0.30 *M*.

129. $$8.75\ \text{h} \times \frac{3600\ \text{s}}{\text{h}} = 3.15 \times 10^4\ \text{s};\ k = \frac{\ln 2}{t_{1/2}} = \frac{\ln 2}{3.15 \times 10^4\ \text{s}} = 2.20 \times 10^{-5}\ \text{s}^{-1}$$

The partial pressure of a gas is directly related to the concentration in mol/L. So, instead of using mol/L as the concentration units in the integrated first-order rate law, we can use partial pressures of SO_2Cl_2.

$$\ln\left(\frac{P}{P_0}\right) = -kt,\ \ln\left(\frac{P}{791\ \text{torr}}\right) = -(2.20 \times 10^{-5}\ \text{s}^{-1}) \times 12.5\ \text{h} \times \frac{3600\ \text{s}}{\text{h}}$$

$$P_{SO_2Cl_2} = 294\ \text{torr} \times \frac{1\ \text{atm}}{760\ \text{torr}} = 0.387\ \text{atm}$$

$$n = \frac{PV}{RT} = \frac{0.387\ \text{atm} \times 1.25\ \text{L}}{\frac{0.08206\ \text{L atm}}{\text{K mol}} \times 593\ \text{K}} = 9.94 \times 10^{-3}\ \text{mol}\ SO_2Cl_2$$

$$9.94 \times 10^{-3}\ \text{mol} \times \frac{6.022 \times 10^{23}\ \text{molecules}}{\text{mol}} = 5.99 \times 10^{21}\ \text{molecules}\ SO_2Cl_2$$

131. $\ln\left(\frac{k_2}{k_1}\right) = \frac{E_a}{R}\left(\frac{1}{T_1} - \frac{1}{T_2}\right)$; $\ln\left(\frac{1.7 \times 10^{-2}\ s^{-1}}{7.2 \times 10^{-4}\ s^{-1}}\right) = \frac{E_a}{8.3145\ J/K \bullet mol}\left(\frac{1}{660.\ K} - \frac{1}{720.\ K}\right)$

Solving: $E_a = 2.1 \times 10^5$ J/mol

For k at 325°C (598 K):

$$\ln\left(\frac{1.7 \times 10^{-2}\ s^{-1}}{k}\right) = \frac{2.1 \times 10^5\ J/mol}{8.3145\ J/K \bullet mol}\left(\frac{1}{598\ K} - \frac{1}{720.\ K}\right),\ k = 1.3 \times 10^{-5}\ s^{-1}$$

For three half-lives, we go from 100% → 50% → 25% → 12.5%. After three half-lives, 12.5% of the original amount of C_2H_5I remains. Partial pressures are directly related to gas concentrations in mol/L:

$$P_{C_2H_5I} = 894\ torr \times 0.125 = 112\ torr\ \text{after 3 half-lives}$$

Challenge Problems

133. Rate = $k[I^-]^x[OCl^-]^y[OH^-]^z$; comparing the first and second experiments:

$$\frac{18.7 \times 10^{-3}}{9.4 \times 10^{-3}} = \frac{k(0.0026)^x(0.012)^y(0.10)^z}{k(0.0013)^x(0.012)^y(0.10)^z},\ 2.0 = 2.0^x,\ x = 1$$

Comparing the first and third experiments:

$$\frac{9.4 \times 10^{-3}}{4.7 \times 10^{-3}} = \frac{k(0.0013)(0.012)^y(0.10)^z}{k(0.0013)(0.0060)^y(0.10)^z},\ 2.0 = 2.0^y,\ y = 1$$

Comparing the first and sixth experiments:

$$\frac{4.8 \times 10^{-3}}{9.4 \times 10^{-3}} = \frac{k(0.0013)(0.012)(0.20)^z}{k(0.0013)(0.012)(0.10)^z},\ 1/2 = 2.0^z,\ z = -1$$

Rate = $\frac{k[I^-][OCl^-]}{[OH^-]}$; the presence of OH^- decreases the rate of the reaction.

For the first experiment:

$$\frac{9.4 \times 10^{-3}\ mol}{L\ s} = k\frac{(0.0013\ mol/L)(0.012\ mol/L)}{(0.10\ mol/L)},\ k = 60.3\ s^{-1} = 60.\ s^{-1}$$

For all experiments, $k_{mean} = 60.\ s^{-1}$.

135. a. We check for first-order dependence by graphing ln[concentration] versus time for each set of data. The rate dependence on NO is determined from the first set of data because the ozone concentration is relatively large compared to the NO concentration, so $[O_3]$ is effectively constant.

Time (ms)	[NO] (molecules/cm^3)	ln[NO]
0	6.0×10^8	20.21
100.	5.0×10^8	20.03
500.	2.4×10^8	19.30
700.	1.7×10^8	18.95
1000.	9.9×10^7	18.41

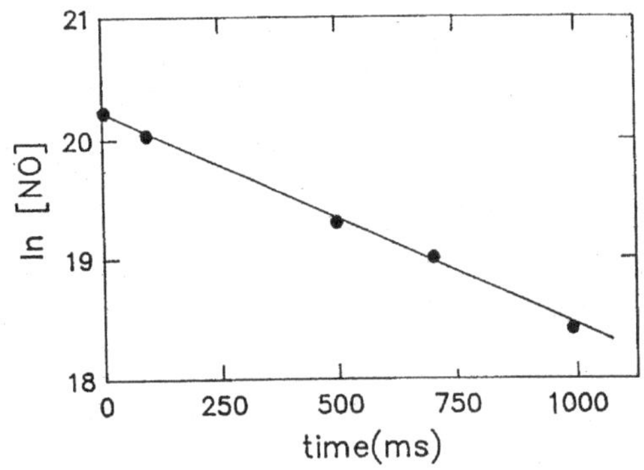

Because ln[NO] versus t is linear, the reaction is first order with respect to NO.

We follow the same procedure for ozone using the second set of data. The data and plot are:

Time (ms)	[O_3] (molecules/cm^3)	ln[O_3]
0	1.0×10^{10}	23.03
50.	8.4×10^9	22.85
100.	7.0×10^9	22.67
200.	4.9×10^9	22.31
300.	3.4×10^9	21.95

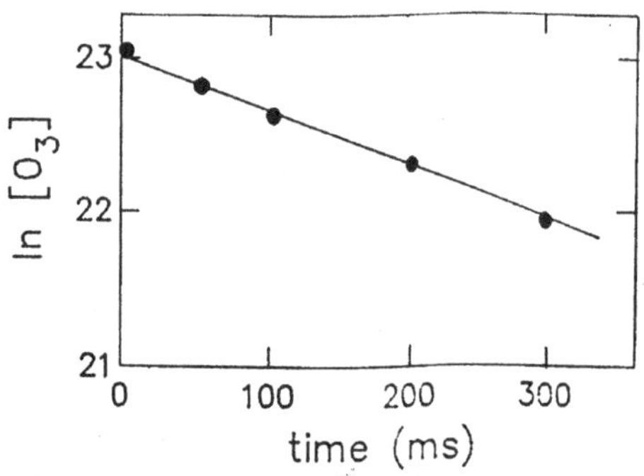

The plot of ln[O_3] versus t is linear. Hence the reaction is first order with respect to ozone.

b. Rate = k[NO][O_3] is the overall rate law.

c. For NO experiment, Rate = k′[NO] and k′ = – (slope from graph of ln[NO] versus t).

$$k' = -\text{slope} = -\frac{18.41 - 20.21}{(1000. - 0) \times 10^{-3}\ \text{s}} = 1.8\ \text{s}^{-1}$$

For ozone experiment, Rate = k″[O_3] and k″ = – (slope from ln[O_3] versus t plot).

$$k'' = -\text{slope} = -\frac{(21.95 - 23.03)}{(300. - 0) \times 10^{-3}\ \text{s}} = 3.6\ \text{s}^{-1}$$

d. From the NO experiment, Rate = $k[NO][O_3] = k'\,[NO]$ where $k' = k[O_3]$.

$k' = 1.8\ \text{s}^{-1} = k(1.0 \times 10^{14}\ \text{molecules/cm}^3)$, $k = 1.8 \times 10^{-14}\ \text{cm}^3/\text{molecules}\bullet\text{s}$

We can check this from the ozone data. Rate = $k''\,[O_3] = k[NO][O_3]$, where $k'' = k[NO]$.

$k'' = 3.6\ \text{s}^{-1} = k(2.0 \times 10^{14}\ \text{molecules/cm}^3)$, $k = 1.8 \times 10^{-14}\ \text{cm}^3/\text{molecules}\bullet\text{s}$

Both values of k agree.

137. a. Rate = $k_3[COCl][Cl_2]$; from the fast-equilibrium reactions 1 and 2:

$$\frac{[COCl]}{[Cl][CO]} = \frac{k_2}{k_{-2}},\ [COCl] = \frac{k_2}{k_{-2}}[CO][Cl]$$

$$\frac{[Cl]^2}{[Cl_2]} = \frac{k_1}{k_{-1}},\ [Cl] = \left(\frac{k_1}{k_{-1}}[Cl_2]\right)^{1/2}$$

Thus $[COCl] = \dfrac{k_2}{k_{-2}}\left(\dfrac{k_1}{k_{-1}}\right)^{1/2}[CO][Cl_2]^{1/2}$; Substituting into rate law:

$$\text{Rate} = k_3\frac{k_2}{k_{-2}}\left(\frac{k_1}{k_{-1}}\right)^{1/2}[CO][Cl_2]^{3/2} = k[CO][Cl_2]^{3/2}$$

b. Cl and COCl are intermediates.

139. $\ln\left(\dfrac{k_2}{k_1}\right) = \dfrac{E_a}{R}\left(\dfrac{1}{T_1} - \dfrac{1}{T_2}\right)$; assuming $\dfrac{\text{rate}_2}{\text{rate}_1} = \dfrac{k_2}{k_1} = 40.0$:

$$\ln(40.0) = \frac{E_a}{8.3145\ \text{J/K}\bullet\text{mol}}\left(\frac{1}{308\ \text{K}} - \frac{1}{328\ \text{K}}\right),\quad E_a = 1.55 \times 10^5\ \text{J/mol} = 155\ \text{kJ/mol}$$

(carrying an extra sig. fig.)

Note that the activation energy is close to the F_2 bond energy. Therefore, the rate-determining step probably involves breaking the F_2 bond.

$H_2(g) + F_2(g) \rightarrow 2\ HF(g)$; for every 2 moles of HF produced, only 1 mole of the reactant is used up. Therefore, to convert the data to $P_{reactant}$ versus time, $P_{reactant} = 1.00\ \text{atm} - (1/2)P_{HF}$.

$P_{reactant}$	Time
1.000 atm	0 min
0.850 atm	30.0 min
0.700 atm	65.8 min
0.550 atm	110.4 min
0.400 atm	169.1 min
0.250 atm	255.9 min

The plot of ln $P_{reactant}$ versus time (plot not included) is linear with negative slope, so the reaction is first order with respect to the limiting reagent.

For the reactant in excess, because the values of the rate constant are the same for both experiments, one can conclude that the reaction is zero order in the excess reactant.

a. For a three-step reaction with the first step limiting, the energy-level diagram could be:

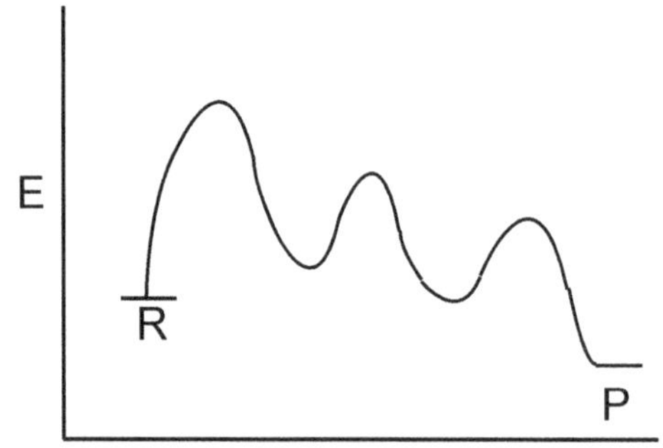

Note that the heights of the second and third humps must be lower than the first step activation energy. However, the height of the third hump could be higher than the second hump. One cannot determine this absolutely from the information in the problem.

b. We know the reaction has a slow first step, and the calculated activation energy indicates that the rate-determining step involves breaking the F_2 bond. The reaction is also first order in one of the reactants and zero order in the other reactant. All this points to F_2 being the limiting reagent. The reaction is first order in F_2, and the rate-determining step in the mechanism is $F_2 \rightarrow 2\ F$. Possible second and third steps to complete the mechanism follow.

$F_2 \rightarrow 2\ F$	slow
$F + H_2 \rightarrow HF + H$	fast
$H + F \rightarrow HF$	fast
$F_2 + H_2 \rightarrow 2\ HF$	

c. F_2 was the limiting reactant.

141. a. [B] >> [A], so [B] can be considered constant over the experiments. This gives us a pseudo-order rate-law equation.

b. Note that in each case the half-life doubles as time increases (in experiment 1, the first half-life is 40. s, the second half-life is 80. s; in experiment 2, the first half- life is 20. s, the second half-life is 40. s). This occurs only for a second-order reaction, so the reaction is second order in [A]. Between expt. 1 and expt. 2, we double [B] and the reaction rate doubles, thus it is first order in [B]. The overall rate-law equation is rate = $k[A]^2[B]$.

Using $t_{1/2} = \dfrac{1}{k[A]_0}$, we get $k = \dfrac{1}{(40.)(10.0 \times 10^{-2})} = 0.25$ L/mol•s; but this is actually k′ where Rate = $k'[A]^2$ and $k' = k[B]$.

$$k = \frac{k'}{[B]} = \frac{0.25}{5.0} = 0.050 \text{ L}^2/\text{ mol}^2\text{•s}$$

c. i. This mechanism gives the wrong stoichiometry, so it can't be correct.

ii. Rate = k[E][A]

$k_1[A][B] = k_{-1}[E]$; $[E] = \frac{k_1[A][B]}{k_{-1}}$; Rate $= \frac{k k_1}{k_{-1}}[A]^2[B]$

This mechanism gives the correct stoichiometry and gives the correct rate law. This is a possible mechanism for this reaction.

iii. Rate = $k[A]^2$

This mechanism gives the wrong derived rate law, so it can't be correct. Only mechanism ii is possible.

143. Rate = $k[A]^x[B]^y[C]^z$; during the course of experiment 1, [A] and [C] are essentially constant, and Rate = $k'[B]^y$, where $k' = k[A]_0^x[C]_0^z$.

[B] (M)	Time (s)	ln[B]	1/[B] (M^{-1})
1.0×10^{-3}	0	−6.91	1.0×10^3
2.7×10^{-4}	1.0×10^5	−8.22	3.7×10^3
1.6×10^{-4}	2.0×10^5	−8.74	6.3×10^3
1.1×10^{-4}	3.0×10^5	−9.12	9.1×10^3
8.5×10^{-5}	4.0×10^5	−9.37	12×10^3
6.9×10^{-5}	5.0×10^5	−9.58	14×10^3
5.8×10^{-5}	6.0×10^5	−9.76	17×10^3

A plot of 1/[B] versus t is linear (plot not included), so the reaction is second order in B, and the integrated rate equation is:

$1/[B] = (2.7 \times 10^{-2}$ L/mol•s)t + 1.0×10^3 L/mol; $k' = 2.7 \times 10^{-2}$ L/mol•s

For experiment 2, [B] and [C] are essentially constant, and Rate = $k''[A]^x$, where $k'' = k[B]_0^y[C]_0^z = k[B]_0^2[C]_0^z$.

[A] (M)	Time (s)	ln[A]	1/[A] (M^{-1})
1.0×10^{-2}	0	−4.61	1.0×10^2
8.9×10^{-3}	1.0	−4.95	140
5.5×10^{-3}	5.0	−5.20	180
3.8×10^{-3}	8.0	−5.57	260
2.9×10^{-3}	10.0	−5.84	340
2.0×10^{-3}	13.0	−6.21	5.0×10^2

A plot of ln[A] versus t is linear, so the reaction is first order in A, and the integrated rate law is:

$$\ln[A] = -(0.123\ s^{-1})t - 4.61;\ k'' = 0.123\ s^{-1}$$

Note: We will carry an extra significant figure in k″.

Experiment 3: [A] and [B] are constant; Rate = $k'''[C]^z$

The plot of [C] versus t is linear. Thus $z = 0$.

The overall rate law is Rate = $k[A][B]^2$.

From Experiment 1 (to determine k):

$$k' = 2.7 \times 10^{-2}\ L/mol{\bullet}s = k[A]_0^x[C]_0^z = k[A]_0 = k(2.0\ M),\ k = 1.4 \times 10^{-2}\ L^2/mol^2{\bullet}s$$

From Experiment 2: $k'' = 0.123\ s^{-1} = k[B]_0^2,\ k = \dfrac{0.123\ s^{-1}}{(3.0\ M)^2} = 1.4 \times 10^{-2}\ L^2/mol^2{\bullet}s$

Thus Rate = $k[A][B]^2$ and $k = 1.4 \times 10^{-2}\ L^2/mol^2{\bullet}s$.

CHAPTER 13

CHEMICAL EQUILIBRIUM

Questions

15. No, equilibrium is a dynamic process. Both reactions:

$$H_2O + CO \rightarrow H_2 + CO_2 \text{ and } H_2 + CO_2 \rightarrow H_2O + CO$$

are occurring at equal rates. Thus ^{14}C atoms will be distributed between CO and CO_2.

17. When the rate of the forward reaction is faster than the rate of the reverse reaction, there is a net gain of products. In essence, the reaction shifts right to get to equilibrium. As the reaction continues and there is a net change of reactants into products, the forward reaction will slow down while the reverse reaction rate will increase. Eventually the rates will be equal, and equilibrium is reached. This true for all reactions, whether $K >> 1$, $K << 1$, or $K \approx 1$. So, statement d is false. Since $\Delta n \neq 0$, $K \neq K_p$.

19. $K = [\text{products}]^y/[\text{reactants}]^x$; When a reaction contains mostly products at equilibrium, the value of K will be much larger than 1. Only answer b has a K value much larger than 1, so it is the correct answer.

21. $H_2O(g) + CO(g) \rightleftharpoons H_2(g) + CO_2(g) \quad K = \dfrac{[H_2][CO_2]}{[H_2O][CO]} = 2.0$

K is a unitless number because there are an equal number of moles of product gases as moles of reactant gases in the balanced equation. Therefore, we can use units of molecules per liter instead of moles per liter to determine K.

We need to start somewhere, so let's assume 3 molecules of CO react. If 3 molecules of CO react, then 3 molecules of H_2O must react, and 3 molecules each of H_2 and CO_2 are formed. We would have 6 – 3 = 3 molecules of CO, 8 – 3 = 5 molecules of H_2O, 0 + 3 = 3 molecules of H_2, and 0 + 3 = 3 molecules of CO_2 present. This will be an equilibrium mixture if K = 2.0:

$$K = \frac{\left(\dfrac{3 \text{ molecules } H_2}{L}\right)\left(\dfrac{3 \text{ molecules } CO_2}{L}\right)}{\left(\dfrac{5 \text{ molecules } H_2O}{L}\right)\left(\dfrac{3 \text{ molecules } CO}{L}\right)} = \frac{3}{5}$$

Because this mixture does not give a value of K = 2.0, this is not an equilibrium mixture. Let's try 4 molecules of CO reacting to reach equilibrium.

Molecules CO remaining = 6 – 4 = 2 molecules of CO

Molecules H_2O remaining = 8 – 4 = 4 molecules of H_2O

Molecules H_2 present = 0 + 4 = 4 molecules of H_2

Molecules CO_2 present = 0 + 4 = 4 molecules of CO_2

$$K = \frac{\left(\frac{4 \text{ molecules } H_2}{L}\right)\left(\frac{4 \text{ molecules } CO_2}{L}\right)}{\left(\frac{4 \text{ molecules } H_2O}{L}\right)\left(\frac{2 \text{ molecules CO}}{L}\right)} = 2.0$$

Because K = 2.0 for this reaction mixture, we are at equilibrium.

23. K and K_p are equilibrium constants, as determined by the law of mass action. For K, concentration units of mol/L are used, and for K_p, partial pressures in units of atm are used (generally). Q is called the reaction quotient. Q has the exact same form as K or K_p, but instead of equilibrium concentrations, initial concentrations are used to calculate the Q value. The use of Q is when it is compared with the K value. When Q = K (or when $Q_p = K_p$), the reaction is at equilibrium. When Q ≠ K, the reaction is not at equilibrium, and one can deduce the net change that must occur for the system to get to equilibrium.

25. $CaCO_3(s) \rightleftharpoons CaO(s) + CO_2(g)$ $K_p = P_{CO_2} = 1.04$; solids and liquids do not appear in equilibrium expressions. Once the CO_2 pressure reaches 1.04 atm, the reaction is at equilibrium. There can't be an equilibrium CO_2 pressure greater than or lesser than 1.04 atm. Answer e is correct.

27. We always try to make good assumptions that simplify the math. In some problems we can set up the problem so that the net change *x* that must occur to reach equilibrium is a small number. This comes in handy when you have expressions like 0.12 – *x* or 0.727 + 2*x*, etc. When *x* is small, we can assume that it makes little difference when subtracted from or added to some relatively big number. When this is the case, $0.12 - x \approx 0.12$ and $0.727 + 2x \approx 0.727$, etc. If the assumption holds by the 5% rule, the assumption is assumed valid. The 5% rule refers to *x* (or 2*x* or 3*x*, etc.) that is assumed small compared to some number. If *x* (or 2*x* or 3*x*, etc.) is less than 5% of the number the assumption was made against, then the assumption will be assumed valid. If the 5% rule fails to work, one can use a math procedure called the method of successive approximations to solve the quadratic or cubic equation. Of course, one could always solve the quadratic or cubic equation exactly. This is generally a last resort (and is usually not necessary unless K or $K_p \approx 1$).

29. There will be a net increase in the amount of N_2O present once equilibrium is reestablished. As $N_2O(g)$ is added, the reaction will shift right to use up some of the added N_2O. However, only some of the added N_2O will react, not all. The net effect is for N_2O to increase once equilibrium is reestablished. The amount of the NO product will also increase, but the amount of the O_2 reactant will decrease. As far as the K value is concerned, as long as the temperature didn't change, K will remain a constant value.

31. $2\ NaHCO_3(s) \rightleftharpoons Na_2CO_3(s) + CO_2(g) + H_2O(g)$ $\quad K = [CO_2][H_2O] = 0.25$

		$CO_2(g)$	$H_2O(g)$
Initial		0	0
	Let some $NaHCO_3(s)$ decompose to form x atm each of $CO_2(g)$ and $H_2O(g)$ at equilibrium.		
Change	$\rightarrow$	$+x$	$+x$
Equil.		x	x

Because solids do not appear in the equilibrium constant expression, we don't worry about the actual amounts of solids in the ICE table. Statement d is false. The mass of solid initially reacted will not affect the equilibrium position (assuming there is enough solid present to produce the equilibrium amounts of products). As soon as some of the $NaHCO_3$ reacts to give x amount of CO_2 and x amount of H_2O, the reaction is at equilibrium; it doesn't matter how much excess solid is present. From the ICE table, $[CO_2] = [H_2O] = x$ at equilibrium. For statement e, since $\Delta n \neq 0$, $K \neq K_p$.

The Equilibrium Constant

33. a. $K = \dfrac{[NO]^2}{[N_2][O_2]}$ b. $K = \dfrac{[NO_2]^2}{[N_2O_4]}$

c. $K = \dfrac{[SiCl_4][H_2]^2}{[SiH_4][Cl_2]^2}$ d. $K = \dfrac{[PCl_3]^2[Br_2]^3}{[PBr_3]^2[Cl_2]^3}$

35. $K = 1.3 \times 10^{-2} = \dfrac{[NH_3]^2}{[N_2][H_2]^3}$ for $N_2(g) + 3\ H_2(g) \rightleftharpoons 2\ NH_3(g)$.

When a reaction is reversed, then $K_{new} = 1/K_{original}$. When a reaction is multiplied through by a value of n, then $K_{new} = (K_{original})^n$.

a. $1/2\ N_2(g) + 3/2\ H_2(g) \rightleftharpoons NH_3(g)$ $\quad K' = \dfrac{[NH_3]}{[N_2]^{1/2}[H_2]^{3/2}} = K^{1/2} = (1.3 \times 10^{-2})^{1/2} = 0.11$

b. $2\ NH_3(g) \rightleftharpoons N_2(g) + 3\ H_2(g)$ $\quad K'' = \dfrac{[N_2][H_2]^3}{[NH_3]^2} = \dfrac{1}{K} = \dfrac{1}{1.3 \times 10^{-2}} = 77$

c. $NH_3(g) \rightleftharpoons 1/2\ N_2(g) + 3/2\ H_2(g)$ $\quad K''' = \dfrac{[N_2]^{1/2}[H_2]^{3/2}}{[NH_3]} = \left(\dfrac{1}{K}\right)^{1/2} = \left(\dfrac{1}{1.3 \times 10^{-2}}\right)^{1/2}$

$= 8.8$

d. $2\ N_2(g) + 6\ H_2(g) \rightleftharpoons 4\ NH_3(g)$ $\quad K'''' = \dfrac{[NH_3]^4}{[N_2]^2[H_2]^6} = (K)^2 = (1.3 \times 10^{-2})^2 = 1.7 \times 10^{-4}$

37. $2\ NO(g) + 2\ H_2(g) \rightleftharpoons N_2(g) + 2\ H_2O(g)$ $\quad K = \dfrac{[N_2][H_2O]^2}{[NO]^2[H_2]^2}$

$$K = \frac{(5.3 \times 10^{-2})(2.9 \times 10^{-3})^2}{(8.1 \times 10^{-3})^2\,(4.1 \times 10^{-5})^2} = 4.0 \times 10^6$$

39. $[NO] = \dfrac{4.5 \times 10^{-3}\ \text{mol}}{3.0\ \text{L}} = 1.5 \times 10^{-3}\,M$; $[Cl_2] = \dfrac{2.4\ \text{mol}}{3.0\ \text{L}} = 0.80\ M$

$[NOCl] = \dfrac{1.0\ \text{mol}}{3.0\ \text{L}} = 0.33\ M$; $K = \dfrac{[NO]^2[Cl_2]}{[NOCl]^2} = \dfrac{(1.5 \times 10^{-3})^2(0.80)}{(0.33)^2} = 1.7 \times 10^{-5}$

41. $K_p = \dfrac{P_{NO}^2 \times P_{O_2}}{P_{NO_2}^2} = \dfrac{(6.5 \times 10^{-5})^2(4.5 \times 10^{-5})}{(0.55)^2} = 6.3 \times 10^{-13}$

43. $K_p = K(RT)^{\Delta n}$, where Δn = sum of gaseous product coefficients − sum of gaseous reactant coefficients. For this reaction, $\Delta n = 3 - 1 = 2$.

$K = \dfrac{[CO][H_2]^2}{[CH_3OH]} = \dfrac{(0.24)(1.1)^2}{(0.15)} = 1.9$

$K_p = K(RT)^2 = 1.9(0.08206\ \text{L atm/K•mol} \times 600.\ \text{K})^2 = 4.6 \times 10^3$

45. There are many potential balanced equations for this reaction; two possible balanced equations are $2\ N_2O_5(g) \rightleftharpoons O_2(g) + 4\ NO_2(g)$ or $N_2O_5(g) \rightleftharpoons \frac{1}{2}\ O_2(g) + 2\ NO_2(g)$. So, there are many possibly correct answers to this question, depending on the balanced equation used. Only answer e corresponds to a correctly balanced equation converting N_2O_5 into products.

47. Solids and liquids do not appear in equilibrium expressions. Only gases and dissolved solutes appear in equilibrium expressions.

a. $K = \dfrac{[H_2O]}{[NH_3]^2[CO_2]}$; $K_p = \dfrac{P_{H_2O}}{P_{NH_3}^2 \times P_{CO_2}}$

b. $K = [N_2][Br_2]^3$; $K_p = P_{N_2} \times P_{Br_2}^3$

c. $K = [O_2]^3$; $K_p = P_{O_2}^3$

d. $K = \dfrac{[H_2O]}{[H_2]}$; $K_p = \dfrac{P_{H_2O}}{P_{H_2}}$

49. $K_p = K(RT)^{\Delta n}$, where Δn equals the difference in the sum of the coefficients between gaseous products and gaseous reactants (Δn = mol gaseous products − mol gaseous reactants). When $\Delta n = 0$, then $K_p = K$. In Exercise 47, only reaction d has $\Delta n = 0$, so only reaction d has $K_p = K$.

51. Solids and liquids do not appear in equilibrium expressions. Only gases and dissolved solutes appear in equilibrium expressions.

$4\ KO_2(s) + 2\ CO_2(g) \rightleftharpoons 2\ K_2CO_3(s) + 3\ O_2(g)$ $\quad K = \dfrac{[O_2]^3}{[CO_2]^2}$

$$K = \dfrac{\left(\dfrac{0.4\ \text{mol}\ O_2}{4.0\ \text{L}}\right)^3}{\left(\dfrac{4 \times 10^{-10}\ \text{mol}\ CO_2}{4.0\ \text{L}}\right)^2} = 1 \times 10^{17}$$

53. $K_p = \frac{P_{H_2}^4}{P_{H_2O}^4}$; $P_{total} = P_{H_2O} + P_{H_2}$, 36.3 torr = 15.0 torr + P_{H_2}, P_{H_2} = 21.3 torr

Because 1 atm = 760 torr: $K_p = \frac{\left(21.3 \text{ torr} \times \frac{1 \text{ atm}}{760 \text{ torr}}\right)^4}{\left(15.0 \text{ torr} \times \frac{1 \text{ atm}}{760 \text{ torr}}\right)^4} = 4.07$

Note: Solids and pure liquids are not included in K expressions.

Equilibrium Calculations

55. $H_2O(g) + Cl_2O(g) \rightarrow 2\ HOCl(g)$ $K = \frac{[HOCl]^2}{[H_2O][Cl_2O]} = 0.0900$

Use the reaction quotient Q to determine which way the reaction shifts to reach equilibrium. For the reaction quotient, initial concentrations given in a problem are used to calculate the value for Q. If Q < K, then the reaction shifts right to reach equilibrium. If Q > K, then the reaction shifts left to reach equilibrium. If Q = K, then the reaction does not shift in either direction because the reaction is already at equilibrium.

a. $Q = \frac{[HOCl]_0^2}{[H_2O]_0[Cl_2O]_0} = \frac{\left(\frac{1.0 \text{ mol}}{1.0 \text{ L}}\right)^2}{\left(\frac{0.10 \text{ mol}}{1.0 \text{ L}}\right)\left(\frac{0.10 \text{ mol}}{1.0 \text{ L}}\right)} = 1.0 \times 10^2$

Q > K, so the reaction shifts left to produce more reactants to reach equilibrium.

b. $Q = \frac{\left(\frac{0.084 \text{ mol}}{2.0 \text{ L}}\right)^2}{\left(\frac{0.98 \text{ mol}}{2.0 \text{ L}}\right)\left(\frac{0.080 \text{ mol}}{2.0 \text{ L}}\right)} = 0.090 = K$; at equilibrium

c. $Q = \frac{\left(\frac{0.25 \text{ mol}}{3.0 \text{ L}}\right)^2}{\left(\frac{0.56 \text{ mol}}{3.0 \text{ L}}\right)\left(\frac{0.0010 \text{ mol}}{3.0 \text{ L}}\right)} = 110$

Q > K, so the reaction shifts to the left to reach equilibrium.

57. $CaCO_3(s) \rightleftharpoons CaO(s) + CO_2(g)$ $K_p = P_{CO_2} = 1.04$

a. $Q = P_{CO_2}$; we only need the partial pressure of CO_2 to determine Q because solids do not appear in equilibrium expressions (or Q expressions). At this temperature, all CO_2 will be in the gas phase. Q = 2.55, so Q > K_p; the reaction will shift to the left to reach equilibrium; the mass of CaO will decrease.

b. Q = 1.04 = K_p, so the reaction is at equilibrium; mass of CaO will not change.

c. Q = 1.04 = K_p, so the reaction is at equilibrium; mass of CaO will not change.

d. Q = 0.211 < K_p; the reaction will shift to the right to reach equilibrium; mass of CaO will increase.

59. $$Q = \frac{[NH_3]_0^2}{[N_2]_0[H_2]_0^3} = \frac{(0.500\ \text{mol/L})^2}{(0.500\ \text{mol/L})(0.500\ \text{mol/L})^3} = 4.00$$

Since Q > K, the reaction shifts left to reach equilibrium.

	$3\ H_2(g)$	+	$N_2(g)$	⇌	$2\ NH_3(g)$
Initial	0.500 *M*		0.500 *M*		0.500 *M*
	Let $2x$ mol/L of NH_3 react to reach equilibrium.				
Change	$+3x$		$+x$	←	$-2x$
Equil	$0.500 + 3x$		$0.500 + x$		$0.500 - 2x$

From the ICE table, the $[NH_3]$ will be less than 0.500 *M* and the reactant concentrations will be greater than 0.500 *M*. So, NH_3 has the smallest equilibrium concentration. Also from the ICE table, H_2 will have the largest concentration at equilibrium as $0.500 + 3x > 0.500 + x$.

61. $$K = \frac{[H_2]^2[O_2]}{[H_2O]^2},\ 2.4 \times 10^{-3} = \frac{(1.9 \times 10^{-2})^2[O_2]}{(0.11)^2},\quad [O_2] = 0.080\ M$$

$$\text{Moles of } O_2 = 2.0\ \text{L} \times \frac{0.080\ \text{mol } O_2}{\text{L}} = 0.16\ \text{mol } O_2$$

63. $$SO_2(g) + NO_2(g) \rightleftharpoons SO_3(g) + NO(g) \quad K = \frac{[SO_3][NO]}{[SO_2][NO_2]}$$

To determine K, we must calculate the equilibrium concentrations. The initial concentrations are:

$$[SO_3]_0 = [NO]_0 = 0;\quad [SO_2]_0 = [NO_2]_0 = \frac{2.00\ \text{mol}}{1.00\ \text{L}} = 2.00\ M$$

Next, we determine the change required to reach equilibrium. At equilibrium, [NO] = 1.30 mol/1.00 L = 1.30 *M*. Because there was zero NO present initially, 1.30 *M* of SO_2 and 1.30 *M* NO_2 must have reacted to produce 1.30 *M* NO as well as 1.30 *M* SO_3, all required by the balanced reaction. The equilibrium concentration for each substance is the sum of the initial concentration plus the change in concentration necessary to reach equilibrium. The equilibrium concentrations are:

$$[SO_3] = [NO] = 0 + 1.30\ M = 1.30\ M;\quad [SO_2] = [NO_2] = 2.00\ M - 1.30\ M = 0.70\ M$$

We now use these equilibrium concentrations to calculate K:

$$K = \frac{[SO_3][NO]}{[SO_2][NO_2]} = \frac{(1.30)(1.30)}{(0.70)(0.70)} = 3.4$$

65. When solving equilibrium problems, a common method to summarize all the information in the problem is to set up a table. We commonly call this table an ICE table because it summarizes *i*nitial concentrations, *c*hanges that must occur to reach equilibrium, and *e*quilibrium concentrations (the sum of the initial and change columns). For the change column, we will generally use the variable x, which will be defined as the amount of reactant (or product) that must react to reach equilibrium. In this problem, the reaction must shift right to reach equilibrium because there are no products present initially. Therefore, x is defined as the amount of reactant SO_3 that reacts to reach equilibrium, and we use the coefficients in the balanced equation to relate the net change in SO_3 to the net change in SO_2 and O_2. The general ICE table for this problem is:

	$2\ SO_3(g)$	$\rightleftharpoons$	$2\ SO_2(g)$	+	$O_2(g)$	$K = \frac{[SO_2]^2[O_2]}{[SO_3]^2}$
Initial	12.0 mol/3.0 L		0		0	
	Let x mol/L of SO_3 react to reach equilibrium.					
Change	$-x$	$\rightarrow$	$+x$		$+x/2$	
Equil.	$4.0 - x$		x		$x/2$	

From the problem, we are told that the equilibrium SO_2 concentration is 3.0 mol/3.0 L = 1.0 M ($[SO_2]_e = 1.0\ M$). From the ICE table setup, $[SO_2]_e = x$, so $x = 1.0$. Solving for the other equilibrium concentrations: $[SO_3]_e = 4.0 - x = 4.0 - 1.0 = 3.0\ M$; $[O_2] = x/2 = 1.0/2 = 0.50\ M$.

$$K = \frac{[SO_2]^2[O_2]}{[SO_3]^2} = \frac{(1.0)^2(0.50)}{(3.0)^2} = 0.056$$

Alternate method: Fractions in the change column can be avoided (if you want) be defining x differently. If we were to let $2x$ mol/L of SO_3 react to reach equilibrium, then the ICE table setup is:

	$2\ SO_3(g)$	$\rightleftharpoons$	$2\ SO_2(g)$	+	$O_2(g)$	$K = \frac{[SO_2]^2[O_2]}{[SO_3]^2}$
Initial	4.0 M		0		0	
	Let $2x$ mol/L of SO_3 react to reach equilibrium.					
Change	$-2x$	$\rightarrow$	$+2x$		$+x$	
Equil.	$4.0 - 2x$		$2x$		x	

Solving: $2x = [SO_2]_e = 1.0\ M$, $x = 0.50\ M$; $[SO_3]_e = 4.0 - 2(0.50) = 3.0\ M$; $[O_2]_e = x = 0.50\ M$

These are the same equilibrium concentrations as solved for previously, thus K will be the same (as it must be). The moral of the story is to define x in a manner that is most comfortable for you. Your final answer is independent of how you define x initially.

67.

	$3\ H_2(g)$	+	$N_2(g)$	$\rightleftharpoons$	$2\ NH_3(g)$
Initial	$[H_2]_0$		$[N_2]_0$		0
	x mol/L of N_2 reacts to reach equilibrium				
Change	$-3x$		$-x$	$\rightarrow$	$+2x$
Equil	$[H_2]_0 - 3x$		$[N_2]_0 - x$		$2x$

From the problem:

$[NH_3]_e = 4.0\ M = 2x,\ x = 2.0\ M;\ [H_2]_e = 5.0\ M = [H_2]_0 - 3x;\ [N_2]_e = 8.0\ M = [N_2]_0 - x$

$5.0\ M = [H_2]_0 - 3(2.0\ M),\ [H_2]_0 = 11.0\ M;\ 8.0\ M = [N_2]_0 - 2.0\ M,\ [N_2]_0 = 10.0\ M$

69. Q = 1.00, which is less than K. The reaction shifts to the right to reach equilibrium. Summarizing the equilibrium problem in a table:

	$SO_2(g)$	+ $NO_2(g)$	$\rightleftharpoons$	$SO_3(g)$	+ $NO(g)$	K = 3.75
Initial	0.800 *M*	0.800 *M*		0.800 *M*	0.800 *M*	
	x mol/L of SO_2 reacts to reach equilibrium					
Change	$-x$	$-x$	$\rightarrow$	$+x$	$+x$	
Equil.	$0.800 - x$	$0.800 - x$		$0.800 + x$	$0.800 + x$	

Plug the equilibrium concentrations into the equilibrium constant expression:

$$K = \frac{[SO_3][NO]}{[SO_2][NO_2]},\ 3.75 = \frac{(0.800 + x)^2}{(0.800 - x)^2};\ \text{taking the square root of both sides:}$$

$$\frac{0.800 + x}{0.800 - x} = 1.94,\ 0.800 + x = 1.55 - (1.94)x,\ (2.94)x = 0.75,\ x = 0.26\ M$$

The equilibrium concentrations are:

$[SO_3] = [NO] = 0.800 + x = 0.800 + 0.26 = 1.06\ M;\ [SO_2] = [NO_2] = 0.800 - x = 0.54\ M$

71. Because only reactants are present initially, the reaction must proceed to the right to reach equilibrium. Summarizing the problem in a table:

	$Cl_2(g)$	+ $F_2(g)$	$\rightleftharpoons$	$2\ ClF(g)$	$K_p = 16 = \frac{P_{ClF}^2}{P_{Cl_2} \times P_{F_2}}$
Initial	3.00 atm	3.00 atm		0	
	Let *x* atm of Cl_2 react to reach equilibrium.				
Change	$-x$	$-x$	$\rightarrow$	$+2x$	
Equil.	$3.00 - x$	$3.00 - x$		$2x$	

$$K_p = 16 = \frac{P_{ClF}^2}{P_{Cl_2} \times P_{F_2}} = \frac{(2x)^2}{(3.00 - x)(3.00 - x)},\ \text{taking the square root of both sides:}$$

$$4.0 = \frac{2x}{3.00 - x},\ 12 - (4.0)x = 2x,\ (6.0)x = 12,\ x = 2.0\ \text{atm}$$

$P_{Cl_2} = 3.00 - 2.0 = 1.0$ atm

73. Because only reactants are present initially, the reaction must proceed to the right to reach equilibrium. Summarizing the problem in a table:

	$N_2(g)$	+	$O_2(g)$	$\rightleftharpoons$	$2\ NO(g)$	$K_p = 0.050$
Initial	0.80 atm		0.20 atm		0	
	x atm of N_2 reacts to reach equilibrium					
Change	$-x$		$-x$	$\rightarrow$	$+2x$	
Equil.	$0.80 - x$		$0.20 - x$		$2x$	

$$K_p = 0.050 = \frac{P_{NO}^2}{P_{N_2} \times P_{O_2}} = \frac{(2x)^2}{(0.80 - x)(0.20 - x)},\quad 0.050[0.16 - (1.00)x + x^2] = 4x^2$$

$$4x^2 = 8.0 \times 10^{-3} - (0.050)x + (0.050)x^2,\ (3.95)x^2 + (0.050)x - 8.0 \times 10^{-3} = 0$$

Solving using the quadratic formula (see Appendix 1 of the text):

$$x = \frac{-b \pm (b^2 - 4ac)^{1/2}}{2a} = \frac{-0.050 \pm [(0.050)]^2 - 4(3.95)(-8.0 \times 10^{-3})]^{1/2}}{2(3.95)}$$

$x = 3.9 \times 10^{-2}$ atm or $x = -5.2 \times 10^{-2}$ atm; only $x = 3.9 \times 10^{-2}$ atm makes sense (x cannot be negative), so the equilibrium NO concentration is:

$$P_{NO} = 2x = 2(3.9 \times 10^{-2}\text{ atm}) = 7.8 \times 10^{-2}\text{ atm}$$

75.

	$2\ SO_2(g)$	+	$O_2(g)$	$\rightleftharpoons$	$2\ SO_3(g)$	$K_p = 0.25$
Initial	0.50 atm		0.50 atm		0	
	$2x$ atm of SO_2 reacts to reach equilibrium					
Change	$-2x$		$-x$	$\rightarrow$	$+2x$	
Equil.	$0.50 - 2x$		$0.50 - x$		$2x$	

$$K_p = 0.25 = \frac{P_{SO_3}^2}{P_{SO_2}^2 \times P_{O_2}} = \frac{(2x)^2}{(0.50 - 2x)^2(0.50 - x)}$$

This will give a cubic equation. Graphing calculators can be used to solve this expression. If you don't have a graphing calculator, an alternative method for solving a cubic equation is to use the method of successive approximations (see Appendix 1 of the text). The first step is to guess a value for x. Because the value of K is small (K < 1), not much of the forward reaction will occur to reach equilibrium. This tells us that x is small. Let's guess that $x = 0.050$ atm. Now we take this estimated value for x and substitute it into the equation everywhere that x appears except for one. For equilibrium problems, we will substitute the estimated value for x into the denominator and then solve for the numerator value of x. We continue this process until the estimated value of x and the calculated value of x converge on the same number. This is the same answer we would get if we were to solve the cubic equation exactly. Applying the method of successive approximations and carrying extra significant figures:

$$\frac{4x^2}{[0.50 - 2(0.050)]^2[0.50 - (0.050)]} = \frac{4x^2}{(0.40)^2(0.45)} = 0.25,\ x = 0.067$$

$$\frac{4x^2}{[0.50 - 2(0.067)]^2[0.50 - (0.067)]} = \frac{4x^2}{(0.366)^2(0.433)} = 0.25,\ x = 0.060$$

$$\frac{4x^2}{(0.38)^2(0.44)} = 0.25,\ x = 0.063; \quad \frac{4x^2}{(0.374)^2(0.437)} = 0.25, \quad x = 0.062$$

The next trial gives the same value for $x = 0.062$ atm. We are done except for determining the equilibrium concentrations. They are:

$$P_{SO_2} = 0.50 - 2x = 0.50 - 2(0.062) = 0.376 = 0.38 \text{ atm}$$

$$P_{O_2} = 0.50 - x = 0.438 = 0.44 \text{ atm}; \ P_{SO_3} = 2x = 0.124 = 0.12 \text{ atm}$$

77. a. The reaction must proceed to products to reach equilibrium because only reactants are present initially. Summarizing the problem in a table:

	$2\ NOCl(g)$	$\rightleftharpoons$	$2\ NO(g)$	+	$Cl_2(g)$	$K = 1.6 \times 10^{-5}$
Initial	$\frac{2.0 \text{ mol}}{2.0 \text{ L}} = 1.0\ M$		0		0	
	$2x$ mol/L of NOCl reacts to reach equilibrium					
Change	$-2x$	$\rightarrow$	$+2x$		$+x$	
Equil.	$1.0 - 2x$		$2x$		x	

$$K = 1.6 \times 10^{-5} = \frac{[NO]^2[Cl_2]}{[NOCl]^2} = \frac{(2x)^2(x)}{(1.0 - 2x)^2}$$

If we assume that $1.0 - 2x \approx 1.0$ (from the small size of K, we know that the product concentrations will be small), then:

$$1.6 \times 10^{-5} = \frac{4x^3}{1.0^2}, \quad x = 1.6 \times 10^{-2}; \text{ now we must check the assumption.}$$

$$1.0 - 2x = 1.0 - 2(0.016) = 0.97 = 1.0 \text{ (to proper significant figures)}$$

Our error is about 3%; that is, $2x$ is 3.2% of 1.0 *M*. Generally, if the error we introduce by making simplifying assumptions is less than 5%, we go no further; the assumption is said to be valid. We call this the 5% rule. Solving for the equilibrium concentrations:

$$[NO] = 2x = 0.032\ M; \ [Cl_2] = x = 0.016\ M; \quad [NOCl] = 1.0 - 2x = 0.97\ M \approx 1.0\ M$$

Note: If we were to solve this cubic equation exactly (a longer process), we get $x = 0.016$. This is the exact same answer we determined by making a simplifying assumption. We saved time and energy. Whenever K is a very small value ($K << 1$), always make the assumption that x is small. If the assumption introduces an error of less than 5%, then the answer you calculated making the assumption will be considered the correct answer.

b.

	$2\ NOCl(g)$	$\rightleftharpoons$	$2\ NO(g)$	+	$Cl_2(g)$
Initial	$1.0\ M$		$1.0\ M$		0
	$2x$ mol/L of NOCl reacts to reach equilibrium				
Change	$-2x$	$\rightarrow$	$+2x$		$+x$
Equil.	$1.0 - 2x$		$1.0 + 2x$		x

$$1.6 \times 10^{-5} = \frac{(1.0 + 2x)^2(x)}{(1.0 - 2x)^2} = \frac{(1.0)^2(x)}{(1.0)^2} \quad \text{(assuming } 2x << 1.0\text{)}$$

$x = 1.6 \times 10^{-5}$; assumptions are great ($2x$ is $3.2 \times 10^{-3}\%$ of 1.0).

$[Cl_2] = 1.6 \times 10^{-5}$ M and [NOCl] = [NO] = 1.0 M

c.

	$2\ NOCl(g)$	$\rightleftharpoons$	$2\ NO(g)$	+	$Cl_2(g)$
Initial	2.0 M		0		1.0 M
	$2x$ mol/L of NOCl reacts to reach equilibrium				
Change	$-2x$	$\rightarrow$	$+2x$		$+x$
Equil.	$2.0 - 2x$		$2x$		$1.0 + x$

$$1.6 \times 10^{-5} = \frac{(2x)^2(1.0 + x)}{(2.0 - 2x)^2} = \frac{4x^2}{4.0} \quad \text{(assuming } x << 1.0\text{)}$$

Solving: $x = 4.0 \times 10^{-3}$; assumptions good (x is 0.4% of 1.0 and $2x$ is 0.4% of 2.0).

$[Cl_2] = 1.0 + x = 1.0$ M; $[NO] = 2(4.0 \times 10^{-3}) = 8.0 \times 10^{-3}$ M; [NOCl] = 2.0 M

79. $COCl_2(g) \rightleftharpoons CO(g) + Cl_2(g) \qquad K_p = \dfrac{P_{CO} \times P_{Cl_2}}{P_{COCl_2}} = 6.8 \times 10^{-9}$

	$COCl_2(g)$	$\rightleftharpoons$	$CO(g)$	+	$Cl_2(g)$
Initial	1.0 atm		0		0
	x atm of $COCl_2$ reacts to reach equilibrium				
Change	$-x$	$\rightarrow$	$+x$		$+x$
Equil.	$1.0 - x$		x		x

$$6.8 \times 10^{-9} = \frac{P_{CO} \times P_{Cl_2}}{P_{COCl_2}} = \frac{x^2}{1.0 - x} \approx \frac{x^2}{1.0} \quad \text{(Assuming } 1.0 - x \approx 1.0.\text{)}$$

$x = 8.2 \times 10^{-5}$ atm; assumption is good (x is $8.2 \times 10^{-3}\%$ of 1.0).

$P_{COCl_2} = 1.0 - x = 1.0 - 8.2 \times 10^{-5} = 1.0$ atm; $P_{CO} = P_{Cl_2} = x = 8.2 \times 10^{-5}$ atm

81. This is a typical equilibrium problem except that the reaction contains a solid. Whenever solids and liquids are present, we basically ignore them in the equilibrium problem.

$$NH_4OCONH_2(s) \rightleftharpoons 2\ NH_3(g) + CO_2(g) \qquad K_p = 2.9 \times 10^{-3}$$

	$NH_4OCONH_2(s)$	$\rightleftharpoons$	$2\ NH_3(g)$	+	$CO_2(g)$
Initial			0		0
	Let some NH_4OCONH_2 decomposes to produce $2x$ atm of NH_3 and x atm of CO_2.				
Change		$\rightarrow$	$+2x$		$+x$
Equil.			$2x$		x

$K_p = 2.9 \times 10^{-3} = P_{NH_3}^2 \times P_{CO_2} = (2x)^2(x) = 4x^3$

$$x = \left(\frac{2.9 \times 10^{-3}}{4}\right)^{1/3} = 9.0 \times 10^{-2} \text{ atm}; \ P_{NH_3} = 2x = 0.18 \text{ atm}; \ P_{CO_2} = x = 9.0 \times 10^{-2} \text{ atm}$$

$P_{total} = P_{NH_3} + P_{CO_2} = 0.18$ atm $+ 0.090$ atm $= 0.27$ atm

83. $C(s) + CO_2(g) \rightleftharpoons 2\ CO(g) \qquad K_p = \dfrac{P_{CO}^2}{P_{CO_2}} = 2.00$

Let x = equilibrium P_{CO} and y = equilibrium P_{CO_2}, then $P_{total} = x + y = 6.00$ and $\dfrac{x^2}{y} = 2.00.$

We have two equations and two unknowns. Solving:

$$y = \frac{x^2}{2.00};\ 6.00 = x + \frac{x^2}{2.00},\ 12.0 = 2.00x + x^2,\ x^2 + 2.00x - 12.0 = 0$$

$$x = \frac{-2.00 \pm [(2.00)^2 - 4(1)(-12.0)]^{1/2}}{2(1)},\ x = 2.61$$

$P_{CO} = x = 2.61$ atm; $P_{CO_2} = 6.00 - 2.61 = 3.39$ atm

Le Châtelier's Principle

85. a. No effect; adding more of a pure solid or pure liquid has no effect on the equilibrium position.

 b. Shifts left; HF(g) will be removed by reaction with the glass. As HF(g) is removed, the reaction will shift left to produce more HF(g).

 c. Shifts right; as $H_2O(g)$ is removed, the reaction will shift right to produce more $H_2O(g)$.

87. a. No change because there are equal numbers of reactant and product moles of gas in the balanced equation. So the equilibrium is not affected by a volume change.

 b. The mole fraction of products will increase with an increase in volume. The reaction shifts right (to products) because there are more moles of product gas (4) than moles of reactant gas (2).

 c. The mole fraction of reactants will increase with an increase in volume. In the balanced equation, there are more moles of reactant gas (2) than product gas (1); solids are ignored. When the volume increases, the reaction shifts to the side with more mol of gas.

89. a. Right b. Right c. No effect; He(g) is neither a reactant nor a product.

 d. Left; because the reaction is exothermic, heat is a product:

$$CO(g) + H_2O(g) \rightarrow H_2(g) + CO_2(g) + \text{heat}$$

 Increasing T will add heat. The equilibrium shifts to the left to use up the added heat.

 e. No effect; because the moles of gaseous reactants equals the moles of gaseous products (2 mol versus 2 mol), a change in volume will have no effect on the equilibrium.

91. a. Left b. Right c. Left

d. No effect; the reactant and product concentrations/partial pressures are unchanged.

e. No effect; because there are equal numbers of product and reactant gas molecules, a change in volume has no effect on this equilibrium position.

f. Right; a decrease in temperature will shift the equilibrium to the right because heat is a product in this reaction (as is true in all exothermic reactions).

93. Only increasing the volume (d) will cause this reaction to shift left. There are 4 mol of reactant gases vs 3 mol of product gases (solids and liquids are ignored when determining moles of gaseous reactants and products). An increase in volume will shift this reaction to the side with more moles of gas; this is the reactant side. Adding $H_2O(g)$ and increasing the temperature of this endothermic reaction will shift the reaction right. Adding Ar(g) has no effect and adding more of the solid also has no effect on the equilibrium. Solids just need to be present for a reaction to get to equilibrium; the amount of excess solid is immaterial.

95. An endothermic reaction, where heat is a reactant, will shift right to products with an increase in temperature. The amount of $NH_3(g)$ will increase as the reaction shifts right, so the smell of ammonia will increase.

ChemWork Problems

97. $K = \frac{[O_3]^2}{[O_2]^3}$, $1.8 \times 10^{-7} = \frac{[O_3]^2}{(0.062)^3}$, $[O_3] = 6.5 \times 10^{-6}\ M$

99.

$O(g) + NO(g) \rightleftharpoons NO_2(g)$	$K = 1/(6.8 \times 10^{-49}) = 1.5 \times 10^{48}$
$NO_2(g) + O_2(g) \rightleftharpoons NO(g) + O_3(g)$	$K = 1/(5.8 \times 10^{-34}) = 1.7 \times 10^{33}$
$O_2(g) + O(g) \rightleftharpoons O_3(g)$	$K = (1.5 \times 10^{48})(1.7 \times 10^{33}) = 2.6 \times 10^{81}$

101.

	$4\ NH_3(g)$	+ $3\ O_2(g)$	$\rightleftharpoons$ $2\ N_2(g)$	+ $6\ H_2O(g)$
Initial	2.0 mol/2.0 L	2.0 mol/2.0 L	0	0
	Let $4x$ mol/L of NH_3 react to reach equilibrium			
Change	$-4x$	$-3x$	$\rightarrow$ $+2x$	$+6x$
Equil.	$1.0 - 4x$	$1.0 - 3x$	$2x$	$6x$

To come up with an equilibrium concentration of $2x$ for N_2, $4x$ of NH_3 had to react. From the ICE Table, the equilibrium concentrations are:

$[NH_3] = 1.0 - 4x$, $[O_2] = 1.0 - 3x$, $[H_2O] = 6x$

103. $5.63\ g\ C_5H_6O_3 \times \frac{1\ mol\ C_5H_6O_3}{114.10\ g} = 0.0493\ mol\ C_5H_6O_3$ initially

Total moles of gas at equilibrium = $n_{total} = \dfrac{P_{total}V}{RT} = \dfrac{1.63 \text{ atm} \times 2.50 \text{ L}}{\dfrac{0.08206 \text{ L atm}}{\text{K mol}} \times 473 \text{ K}} = 0.105$ mol

	$C_5H_6O_3(g)$ $\rightleftharpoons$	$C_2H_6(g)$ +	$3\ CO(g)$
Initial	0.0493 mol	0	0
	Let x mol $C_5H_6O_3$ react to reach equilibrium.		
Change	$-x$ →	$+x$	$+3x$
Equil.	$0.0493 - x$	x	$3x$

0.105 mol total = $0.0493 - x + x + 3x = 0.0493 + 3x$, $x = 0.0186$ mol

$$K = \frac{[C_2H_6][CO]^3}{[C_5H_6O_3]} = \frac{\left[\dfrac{0.0186 \text{ mol } C_2H_6}{2.50 \text{ L}}\right]\left[\dfrac{3(0.0186) \text{ mol CO}}{2.50 \text{ L}}\right]^3}{\left[\dfrac{(0.0493 - 0.0186) \text{ mol } C_5H_6O_3}{2.50 \text{ L}}\right]} = 6.74 \times 10^{-6}$$

105. a.

	$2\ AsH_3(g)$ $\rightleftharpoons$	$2\ As(s)$ +	$3\ H_2(g)$
Initial	392.0 torr		0
Equil.	$392.0 - 2x$		$3x$

Using Dalton's law of partial pressure:

$P_{total} = 488.0 \text{ torr} = P_{AsH_3} + P_{H_2} = 392.0 - 2x + 3x$, $x = 96.0$ torr

$P_{H_2} = 3x = 3(96.0) = 288 \text{ torr} \times \dfrac{1 \text{ atm}}{760 \text{ torr}} = 0.379$ atm

b. $P_{AsH_3} = 392.0 - 2(96.0) = 200.0 \text{ torr} \times \dfrac{1 \text{ atm}}{760 \text{ torr}} = 0.2632$ atm

$$K_p = \frac{(P_{H_2})^3}{(P_{AsH_3})^2} = \frac{(0.379)^3}{(0.2632)^2} = 0.786$$

107. $NH_3(g)$ + $H_2S(g)$ $\rightleftharpoons$ $NH_4HS(s)$ $\quad K = 400. = \dfrac{1}{[NH_3][H_2S]}$

	$NH_3(g)$	$H_2S(g)$
Initial	$\dfrac{2.00 \text{ mol}}{5.00 \text{ L}}$	$\dfrac{2.00 \text{ mol}}{5.00 \text{ L}}$
	x mol/L of NH_3 reacts to reach equilibrium	
Change	$-x$	$-x$
Equil.	$0.400 - x$	$0.400 - x$

$K = 400. = \dfrac{1}{(0.400 - x)(0.400 - x)}$, $\quad 0.400 - x = \left(\dfrac{1}{400.}\right)^{1/2} = 0.0500$, $\quad x = 0.350\ M$

Moles $NH_4HS(s)$ produced = $5.00 \text{ L} \times \dfrac{0.350 \text{ mol } NH_3}{\text{L}} \times \dfrac{1 \text{ mol } NH_4HS}{\text{mol } NH_3} = 1.75$ mol

Total moles $NH_4HS(s)$ = 2.00 mol initially + 1.75 mol produced = 3.75 mol total

$$3.75 \text{ mol } NH_4HS \times \frac{51.12 \text{ g } NH_4HS}{\text{mol } NH_4HS} = 192 \text{ g } NH_4HS$$

$[H_2S]_e = 0.400\ M - x = 0.400\ M - 0.350\ M = 0.050\ M\ H_2S$

$$P_{H_2S} = \frac{n_{H_2S}RT}{V} = \frac{n_{H_2S}}{V} \times RT = \frac{0.050 \text{ mol}}{L} \times \frac{0.08206 \text{ L atm}}{\text{K mol}} \times 308.2 \text{ K} = 1.3 \text{ atm}$$

109. There is a little trick we can use to solve this problem without having to solve a quadratic equation. Because K is very large (K >> 1), the reaction will have mostly products at equilibrium. So we will let the reaction go to completion (with Fe^{3+} limiting), and then solve an equilibrium problem to determine the molarity of reactants present at equilibrium (see the following set- up).

	$Fe^{3+}(aq)$	+ $SCN^-(aq)$	$\rightleftharpoons$	$FeSCN^{2+}(aq)$	$K = 1.1 \times 10^3$
Before	0.020 *M*	0.10 *M*		0	
	Let 0.020 mol/L Fe^{3+} react completely (K is large; products dominate).				
Change	–0.020	–0.020	→	+0.020	React completely
After	0	0.08		0.020	New initial
	x mol/L $FeSCN^{2+}$ reacts to reach equilibrium				
Change	$+x$	$+x$	←	$-x$	
Equil.	x	$0.08 + x$		$0.020 - x$	

$$K = 1.1 \times 10^3 = \frac{[FeSCN^{2+}]}{[Fe^{3+}][SCN^-]} = \frac{0.020 - x}{(x)(0.08 + x)} \approx \frac{0.020}{(0.08)x}$$

$x = 2 \times 10^{-4}\ M$; x is 1% of 0.020. Assumptions are good by the 5% rule.

$x = [Fe^{3+}] = 2 \times 10^{-4}\ M$; $[SCN^-] = 0.08 + 2 \times 10^{-4} = 0.08\ M$

$[FeSCN^{2+}] = 0.020 - 2 \times 10^{-4} = 0.020\ M$

Note: At equilibrium, we do indeed have mostly products present. Our assumption to first let the reaction go to completion is good.

111.

	$SO_2Cl_2(g)$	$\rightleftharpoons$	$Cl_2(g)$	+ $SO_2(g)$	
Initial	P_0		0	0	P_0 = initial pressure of SO_2Cl_2
Change	$-x$	→	$+x$	$+x$	
Equil.	$P_0 - x$		x	x	

$P_{total} = 0.900 \text{ atm} = P_0 - x + x + x = P_0 + x$

$$\frac{x}{P_0} \times 100 = 12.5, \quad P_0 = (8.00)x$$

Solving: $0.900 = P_0 + x = (9.00)x$, $x = 0.100$ atm

$x = 0.100 \text{ atm} = P_{Cl_2} = P_{SO_2}$; $P_0 - x = 0.800 - 0.100 = 0.700 \text{ atm} = P_{SO_2Cl_2}$

$$K_p = \frac{P_{Cl_2} \times P_{SO_2}}{P_{SO_2Cl_2}} = \frac{(0.100)^2}{0.700} = 1.43 \times 10^{-2}$$

113. Add $N_2(g)$: reaction shifts right to reestablish equilibrium. $[N_2]$ will increase overall, $[H_2]$ will decrease and $[NH_3]$ will increase. Note that only some of the added N_2 will react to get back to equilibrium. Therefore, $[N_2]$ will show a net increase overall.

Remove $H_2(g)$: reaction shifts left to reestablish equilibrium. $[N_2]$ will increase, $[H_2]$ will decrease overall, and $[NH_3]$ will decrease.

Add $NH_3(g)$: reaction will shift left to reestablish equilibrium. $[N_2]$ and $[H_2]$ will increase, while $[NH_3]$ will also increase overall.

Add Ne(g) at constant volume: no change for $[N_2]$, $[H_2]$, or $[NH_3]$.

Increase temperature (add heat): because this is an exothermic reaction, heat is a product in this reaction. As temperature is increased, the reaction will shift left to reestablish equilibrium. $[N_2]$ and $[H_2]$ will increase, while $[NH_3]$ will decrease.

Decrease volume: because there are more moles of gaseous reactants as compared to gaseous products, a decrease in volume will cause this reaction to shift right to reestablish equilibrium. $[N_2]$ and $[H_2]$ will decrease, while $[NH_3]$ will increase.

Add a catalyst: no change for $[N_2]$, $[H_2]$, or $[NH_3]$.

115. $CoCl_2(s) + 6\ H_2O(g) \rightleftharpoons CoCl_2{\bullet}6H_2O(s)$; if rain is imminent, there would be a lot of water vapor in the air. Because water vapor is a reactant gas, the reaction would shift to the right and would take on the color of $CoCl_2{\bullet}6H_2O$, which is pink.

117. $H^+ + OH^- \rightarrow H_2O$; sodium hydroxide (NaOH) will react with the H^+ on the product side of the reaction. This effectively removes H^+ from the equilibrium, which will shift the reaction to the right to produce more H^+ and CrO_4^{2-}. Because more CrO_4^{2-} is produced, the solution turns yellow.

119. Assume the final reaction is $2A \rightleftharpoons B_2$ which has K = 0.01. To get to this final reaction, the previous reaction was multiplied by 2. So, multiply the final reaction by ½ to get the previous reaction. This gives $A \rightleftharpoons \frac{1}{2} B_2$. When a reaction is multiplied through by some number n, the new K value is the previous K value raised to n. So, K for this reaction is $(0.01)^{1/2} = 0.1$. Now to get the initial reaction, we need to reverse this intermediate reaction. This gives $\frac{1}{2} B_2 \rightleftharpoons A$. When a reaction is reversed, the new K value is the inverse of the old K value. So, K for the initial reaction is 1/0.1 = 10.

121.

	$N_2(g)$	+ $3\ H_2(g)$	$\rightleftharpoons$	$2\ NH_3(g)$
Initial	1.00 atm	2.00 atm		0
	x atm of N_2 reacts to reach equilibrium			
Change	$-x$	$-3x$	$\rightarrow$	$+2x$
Equil.	$1.00 - x$	$2.00 - 3x$		$2x$

$$K_p = \frac{P_{NH_3}^2}{P_{N_2} \times P_{H_2}^3}$$

From the setup: $P_{total} = 2.00 \text{ atm} = P_{N_2} + P_{H_2} + P_{NH_3}$

$$2.00 \text{ atm} = (1.00 - x) + (2.00 - 3x) + 2x = 3.00 - 2x, \quad x = 0.500$$

$P_{H_2} = 2.00 - 3x = 2.00 - 3(0.500) = 0.50 \text{ atm}$

$$K_p = \frac{(2x)^2}{(1.00 - x)(2.00 - 3x)^3} = \frac{[2(0.500)]^2}{(1.00 - 0.500)[2.00 - 3(0.500)]^3} = \frac{(1.00)^2}{(0.50)(0.50)^3} = 16$$

123.

	peptide(aq) + $H_2O(l)$	$\rightleftharpoons$	acid group(aq) +	amine group(aq) $K = 3.1 \times 10^{-5}$
Initial	$\frac{1.0 \text{ mol}}{1.0 \text{ L}} = 1.0\ M$		0	0
	x mol/L peptide reacts to reach equilibrium.			
Change	$-x$	$\rightarrow$	$+x$	$+x$
Equil.	$1.0 - x$		x	x

Note: Because water is not included in the K expression, the amount of water present initially and the amount of water that reacts are not needed to solve this problem.

$$K = 3.1 \times 10^{-5} = \frac{x(x)}{1.0 - x}, \quad 3.1 \times 10^{-5} \approx \frac{x^2}{1.0} \quad \text{(assuming } 1.0 - x \approx 1.0\text{)}$$

$x = \sqrt{3.1 \times 10^{-5}} = 5.6 \times 10^{-3}\ M$; assumption good (0.56% error).

[peptide] $= 1.0 - x = 1.0 - 5.6 \times 10^{-3} = 1.0\ M$

[acid group] = [amine group] $= x = 5.6 \times 10^{-3}\ M$

Challenge Problems

125. P_0 (for O_2) $= n_{O_2}RT/V = (6.400 \text{ g} \times 0.08206 \times 684 \text{ K})/(32.00 \text{ g/mol} \times 2.50 \text{ L}) = 4.49$ atm

	$CH_4(g)$ +	$2\ O_2(g)$	$\rightarrow$	$CO_2(g)$ +	$2\ H_2O(g)$
Change	$-x$	$-2x$	$\rightarrow$	$+x$	$+2x$

	$CH_4(g)$ +	$3/2\ O_2(g)$	$\rightarrow$	$CO(g)$ +	$2\ H_2O(g)$
Change	$-y$	$-3/2\ y$	$\rightarrow$	$+y$	$+2y$

Amount of O_2 reacted = 4.49 atm − 0.326 atm = 4.16 atm O_2

$2x + 3/2\ y = 4.16$ atm O_2 and $2x + 2y = 4.45$ atm H_2O

Solving using simultaneous equations:

$$\begin{array}{r} 2x + 2y = 4.45 \\ -2x - (3/2)y = -4.16 \\ \hline (0.50)y = 0.29 \end{array}, \quad y = 0.58 \text{ atm} = P_{CO}$$

$2x + 2(0.58) = 4.45, \quad x = 1.65 \text{ atm} = P_{CO_2}$

127. There is a little trick we can use to solve this problem to avoid solving a cubic equation. Because K for this reaction is very small (K << 1), the reaction will contain mostly reactants at equilibrium (the equilibrium position lies far to the left). We will let the products react to completion by the reverse reaction, and then we will solve the forward equilibrium problem to determine the equilibrium concentrations. Summarizing these steps in a table:

	$2\ NOCl(g)$	$\rightleftharpoons$	$2\ NO(g)$ +	$Cl_2(g)$	$K = 1.6 \times 10^{-5}$
Before	0		2.0 M	1.0 M	
	Let 1.0 mol/L Cl_2 react completely.				(K is small, reactants dominate.)
Change	+2.0	←	−2.0	−1.0	React completely
After	2.0		0	0	New initial conditions
	$2x$ mol/L of NOCl reacts to reach equilibrium				
Change	$-2x$	→	$+2x$	$+x$	
Equil.	$2.0 - 2x$		$2x$	x	

$$K = 1.6 \times 10^{-5} = \frac{(2x)^2(x)}{(2.0-2x)^2} \approx \frac{4x^3}{2.0^2} \quad \text{(assuming } 2.0 - 2x \approx 2.0\text{)}$$

$x^3 = 1.6 \times 10^{-5}$, $x = 2.5 \times 10^{-2}$ M; assumption good by the 5% rule ($2x$ is 2.5% of 2.0).

$[NOCl] = 2.0 - 0.050 = 1.95\ M = 2.0\ M$; $[NO] = 0.050\ M$; $[Cl_2] = 0.025\ M$

Note: If we do not break this problem into two parts (a stoichiometric part and an equilibrium part), then we are faced with solving a cubic equation. The setup would be:

	$2\ NOCl$	$\rightleftharpoons$	$2\ NO$	+	Cl_2
Initial	0		2.0 M		1.0 M
Change	$+2y$	←	$-2y$		$-y$
Equil.	$2y$		$2.0 - 2y$		$1.0 - y$

$$1.6 \times 10^{-5} = \frac{(2.0-2y)^2(1.0-y)}{(2y)^2}$$

If we say that y is small to simplify the problem, then:

$$1.6 \times 10^{-5} = \frac{2.0^2}{4y^2}; \text{ we get } y = 250. \text{ This is impossible!}$$

To solve this equation, we cannot make any simplifying assumptions; we have to solve the cubic equation exactly.

129.

	$N_2(g)$ +	$3\ H_2(g)$	$\rightleftharpoons$	$2\ NH_3(g)$	$K_p = 5.3 \times 10^5$
Initial	0	0		P_0	P_0 = initial pressure of NH_3
	$2x$ atm of NH_3 reacts to reach equilibrium				
Change	$+x$	$+3x$	←	$-2x$	
Equil.	x	$3x$		$P_0 - 2x$	

From problem, $P_0 - 2x = \frac{P_0}{2.00}$, so $P_0 = (4.00)x$

$$K_p = \frac{[(4.00)x - 2x]^2}{(x)(3x)^3} = \frac{[(2.00)x]^2}{(x)(3x)^3} = \frac{(4.00)x^2}{27x^4} = \frac{4.00}{27x^2} = 5.3 \times 10^5,\ x = 5.3 \times 10^{-4}\text{ atm}$$

$P_0 = (4.00)x = 4.00(5.3 \times 10^{-4}\text{ atm}) = 2.1 \times 10^{-3}$ atm

131. $N_2O_4(g) \rightleftharpoons 2\ NO_2(g)$ $\quad K_p = \frac{P^2_{NO_2}}{P_{N_2O_4}} = \frac{(1.20)^2}{0.34} = 4.2$

Doubling the volume decreases each partial pressure by a factor of 2 (P = nRT/V).

$P_{NO_2} = 0.600$ atm and $P_{N_2O_4} = 0.17$ atm are the new partial pressures.

$Q = \frac{(0.600)^2}{0.17} = 2.1$, so Q < K; equilibrium will shift to the right.

	$N_2O_4(g)$	$\rightleftharpoons$	$2\ NO_2(g)$
Initial	0.17 atm		0.600 atm
	x atm of N_2O_4 reacts to reach equilibrium		
Change	$-x$	$\rightarrow$	$+2x$
Equil.	$0.17 - x$		$0.600 + 2x$

$K_p = 4.2 = \frac{(0.600 + 2x)^2}{(0.17 - x)}$, $4x^2 + (6.6)x - 0.354 = 0$ (carrying extra significant figures)

Solving using the quadratic formula: $x = 0.052$

$P_{NO_2} = 0.600 + 2(0.052) = 0.704$ atm; $P_{N_2O_4} = 0.17 - 0.052 = 0.12$ atm

133. a. $N_2(g) + 3\ H_2(g) \rightleftharpoons 2\ NH_3(g)$; because the temperature is constant, the value of K will be the same for both container volumes. Since we now the volume in the final mixture, let's calculate K using this mixture. In this final mixture, 2 N_2 molecules, 2 H_2 molecules, and 6 NH_3 molecules are present in a 1.00 L container. Using units of molecules/L for concentrations:

$$K = \frac{[NH_3]^2}{[N_2][H_2]^3} = \frac{\left(\frac{6\ NH_3\text{ molecules}}{1.00\text{ L}}\right)^2}{\left(\frac{2\ N_2\text{ molecules}}{1.00\text{ L}}\right)\left(\frac{2\ H_2\text{ molecules}}{1.00\text{ L}}\right)^3} = 2.25\ \frac{L^2}{\text{molecules}^2}$$

For the K value in typical mol/L units for the concentrations:

$$K = 2.25\ \frac{L^2}{\text{molecules}^2} \times \left(\frac{6.022 \times 10^{23}\text{ molecules}}{\text{mol}}\right)^2 = 8.16 \times 10^{47}\ \frac{L^2}{\text{mol}^2}$$

b. Because temperature is constant, the initial mixture at the larger volume must also have $K = 2.25\ L^2/\text{molecules}^2$. In the initial mixture, there are 2 NH_3 molecules, 4 N_2 molecules, and 8 H_2 molecules in some unknown volume, V.

$$K = 2.25 = \frac{\left(\frac{2\ NH_3\ molecules}{V}\right)^2}{\left(\frac{4\ N_2\ molecules}{V}\right)\left(\frac{8\ H_2\ molecules}{V}\right)^3} = \frac{4V^2}{4(512)} = \frac{V^2}{512}$$

$V = \sqrt{2.25(512)} = 33.9$ L; the volume of the initial container would be 33.9 L.

135.

	$SO_3(g)$	$\rightleftharpoons$	$SO_2(g)$	+	$1/2\ O_2(g)$	
Initial	P_0		0		0	P_0 = initial pressure of SO_3
Change	$-x$	$\rightarrow$	$+x$		$+x/2$	
Equil.	$P_0 - x$		x		$x/2$	

Average molar mass of the mixture is:

$$\text{molar mass} = \frac{dRT}{P} = \frac{(1.60\ g/L)\ (0.08206\ L\ atm/K \bullet mol)\ (873\ K)}{1.80\ atm} = 63.7\ g/mol$$

The average molar mass is determined by:

$$\text{average molar mass} = \frac{n_{SO_3}(80.07\ g/mol) + n_{SO_2}(64.07\ g/mol) + n_{O_2}(32.00\ g/mol)}{n_{total}}$$

Because χ_A = mol fraction of component A = n_A/n_{total} = P_A/P_{total}:

$$63.7\ g/mol = = \frac{P_{SO_3}(80.07) + P_{SO_2}(64.07) + P_{O_2}(32.00)}{P_{total}}$$

$P_{total} = P_0 - x + x + x/2 = P_0 + x/2 = 1.80$ atm, $P_0 = 1.80 - x/2$

$$63.7 = \frac{(P_0 - x)(80.07) + x(64.07) + \frac{x}{2}(32.00)}{1.80}$$

$$63.7 = \frac{(1.80 - 3/2\ x)(80.07) + x(64.07) + \frac{x}{2}(32.00)}{1.80}$$

$115 = 144 - (120.1)x + (64.07)x + (16.00)x$, $(40.0)x = 29$, $x = 0.73$ atm

$P_{SO_3} = P_0 - x = 1.80 - (3/2)x = 0.71$ atm; $P_{SO_2} = 0.73$ atm; $P_{O_2} = x/2 = 0.37$ atm

$$K_p = \frac{P_{SO_2} \times P_{O_2}^{1/2}}{P_{SO_3}} = \frac{(0.73)(0.37)^{1/2}}{(0.71)} = 0.63$$

137. When exactly 100 O_2 molecules are initially present at 5000 K and 1.000 atm:

	$O_2(g)$	$\rightleftharpoons$	$2\ O(g)$
Initial	100		0
Change	−83		+166
Equil.	17		166

Mole fraction O = $\chi_O = \dfrac{166}{183} = 0.9071$ and $\chi_{O_2} = 0.0929$; $P_{O_2} = \chi_{O_2} P_{total}$ and $P_O = \chi_O P_{total}$

Because initially $P_{total} = 1.000$ atm, $P_{O_2} = 0.0929$ atm and $P_O = 0.9071$ atm.

$$K_p = \frac{P_O^2}{P_{O_2}} = \frac{(0.9071)^2}{0.0929} = 8.86 \text{ atm}$$

At 95.0% O_2 dissociated, let x = initial partial pressure of O_2 and y = amount (atm) of O_2 that dissociates to reach equilibrium.

	O_2	$\rightleftharpoons$	2 O
Initial	x		0
Change	$-y$	$\rightarrow$	$+2y$
Equil.	$x - y$		$2y$

$\dfrac{(2y)^2}{x - y} = 8.86$; $\dfrac{y}{x} \times 100 = 95.0$; we have two equations and two unknowns.

Solving: $x = 0.123$ atm and $y = 0.117$ atm; $P_{total} = (x - y) + 2y = 0.240$ atm

139. a. Because the density (mass/volume) decreased while the mass remained constant (mass is conserved in a chemical reaction), the volume must have increased as reactants were converted to products. The volume increased because the number of moles of gas increased ($V \propto n$ at constant T and P).

$$\frac{\text{Density (initial)}}{\text{Density (equil.)}} = \frac{4.495 \text{ g/L}}{4.086 \text{ g/L}} = 1.100 = \frac{V_{equil.}}{V_{initial}} = \frac{n_{equil.}}{n_{initial}}$$

Assuming an initial volume of 1.000 L:

$$4.495 \text{ g NOBr} \times \frac{1 \text{ mol NOBr}}{109.91 \text{ g}} = 0.04090 \text{ mol NOBr initially}$$

	2 NOBr(g)	$\rightleftharpoons$	2 NO(g)	+ Br_2(g)
Initial	0.04090 mol		0	0
Change	$-2x$	$\rightarrow$	$+2x$	$+x$
Equil.	$0.04090 - 2x$		$2x$	x

$$\frac{n_{equil.}}{n_{initial}} = \frac{0.04090 - 2x + 2x + x}{0.04090} = 1.100; \quad \text{solving: } x = 0.00409 \text{ mol}$$

If the initial volume is 1.000 L, then the equilibrium volume will be 1.110(1.000 L) = 1.110 L. Solving for the equilibrium concentrations:

$$[\text{NOBr}] = \frac{0.03272 \text{ mol}}{1.100 \text{ L}} = 0.02975\ M; \quad [\text{NO}] = \frac{0.00818 \text{ mol}}{1.100 \text{ L}} = 0.00744\ M$$

$$[\text{Br}_2] = \frac{0.00409 \text{ mol}}{1.100 \text{ L}} = 0.00372\ M$$

$$K = \frac{(0.00744)^2(0.00372)}{(0.02975)^2} = 2.33 \times 10^{-4}$$

b. The argon gas will increase the volume of the container. This is because the container is a constant-pressure system, and if the number of moles increases at constant T and P, the volume must increase. An increase in volume will dilute the concentrations of all gaseous reactants and gaseous products. Because there are more moles of product gases versus reactant gases (3 mol versus 2 mol), the dilution will decrease the numerator of K more than the denominator will decrease. This causes Q < K and the reaction shifts right to get back to equilibrium. Because temperature was unchanged, the value of K will not change. K is a constant as long as temperature is constant.

141. Initial moles VCl_4 = 6.6834 g VCl_4 × 1 mol VCl_4/192.74 g VCl_4 = 3.4676×10^{-2} mol VCl_4

Total molality of solute particles = $im = \frac{\Delta T}{K_f} = \frac{5.97\ ^{\circ}C}{29.8\ ^{\circ}C\ kg/mol} = 0.200$ mol/kg

Because we have 0.1000 kg CCl_4, the total moles of solute particles present is:

0.200 mol/kg(0.1000 kg) = 0.0200 mol

	$2\ VCl_4$	$\rightleftharpoons$	V_2Cl_8	$K = \frac{[V_2Cl_8]}{[VCl_4]^2}$
Initial	3.4676×10^{-2} mol		0	
	$2x$ mol VCl_4 reacts to reach equilibrium			
Equil.	$3.4676 \times 10^{-2} - 2x$		x	

Total moles solute particles = 0.0200 mol = mol VCl_4 + mol V_2Cl_8 = $3.4676 \times 10^{-2} - 2x + x$

$0.0200 = 3.4676 \times 10^{-2} - x,\ x = 0.0147$ mol

At equilibrium, we have 0.0147 mol V_2Cl_8 and 0.0200 – 0.0147 = 0.0053 mol VCl_4. To determine the equilibrium constant, we need the total volume of solution in order to calculate equilibrium concentrations. The total mass of solution is 100.0 g + 6.6834 g = 106.7 g.

Total volume = 106.7 g × 1 cm^3/1.696 g = 62.91 cm^3 = 0.06291 L

The equilibrium concentrations are:

$$[V_2Cl_8] = \frac{0.0147\ mol}{0.06291\ L} = 0.234\ mol/L;\ [VCl_4] = \frac{0.0053\ mol}{0.06291\ L} = 0.084\ mol/L$$

$$K = \frac{[V_2Cl_8]}{[VCl_4]^2} = \frac{0.234}{(0.084)^2} = 33$$

CHAPTER 14

ACIDS AND BASES

Questions

21. Acids are proton (H^+) donors, and bases are proton acceptors.

HCO_3^- as an acid: $HCO_3^-(aq) + H_2O(l) \rightleftharpoons CO_3^{2-}(aq) + H_3O^+(aq)$

HCO_3^- as a base: $HCO_3^-(aq) + H_2O(l) \rightleftharpoons H_2CO_3(aq) + OH^-(aq)$

$H_2PO_4^-$ as an acid: $H_2PO_4^- + H_2O(l) \rightleftharpoons HPO_4^{2-}(aq) + H_3O^+(aq)$

$H_2PO_4^-$ as a base: $H_2PO_4^- + H_2O(l) \rightleftharpoons H_3PO_4(aq) + OH^-(aq)$

23. Basic solutions (at 25°C) have an $[OH^-] > 1.0 \times 10^{-7}$ *M*, which gives a pOH < 7.0. Because $[H^+][OH^-] = 1.0 \times 10^{-14}$ and pH + pOH = 14.00 for any aqueous solution at 25°C, a basic solution must also have $[H^+] < 1.0 \times 10^{-7}$ *M* and pH > 7.00. From these relationships, the solutions in parts b, c, and d are basic solutions. The solution in part a will have a pH < 7.0 (pH = 14.00 – 11.21 = 2.79) and is therefore not basic (solution is acidic).

25. The K_a reaction always refers to an acid reacting with water to form H_3O^+ and the conjugate base of the acid. We often leave out water in the reaction since it isn't included in the K expression. Only the reaction in answer d is a reaction associated with the K_a equilibrium constant. Answer c is a reaction associated with the K_b equilibrium constant (a base reacting with water to produce OH^- and the conjugate acid of the base).

27. 10.78 (4 S.F.); 6.78 (3 S.F.); 0.78 (2 S.F.); a pH value is a logarithm. The numbers to the left of the decimal point identify the power of 10 to which $[H^+]$ is expressed in scientific notation, for example, 10^{-11}, 10^{-7}, 10^{-1}. The number of decimal places in a pH value identifies the number of significant figures in $[H^+]$. In all three pH values, the $[H^+]$ should be expressed only to two significant figures because these pH values have only two decimal places.

29.

NH_3	+	NH_3	$\rightleftharpoons$	NH_2^-	+	NH_4^+
Acid		Base		Conjugate Base		Conjugate Acid

One of the NH_3 molecules acts as a base and accepts a proton to form NH_4^+. The other NH_3 molecule acts as an acid and donates a proton to form NH_2^-. NH_4^+ is the conjugate acid of the NH_3 base. In the reverse reaction, NH_4^+ donates a proton. NH_2^- is the conjugate base of the NH_3 acid. In the reverse reaction, NH_2^- accepts a proton. Conjugate acid-base pairs only differ by a H^+ in the formula.

31. a. These are solutions of strong acids like HCl, HBr, HI, HNO_3, H_2SO_4, and $HClO_4$. So 0.10 *M* solutions of any of the acids would be examples of a strong electrolyte solution that is very acidic.

b. These are solutions containing salts of the conjugate acids of the bases in Table 14.3. These conjugate acids are all weak acids, and they are cations with a 1+ charge. NH_4Cl, $CH_3NH_3NO_3$, and $C_2H_5NH_3Br$ are three examples of this type of slightly acidic salts. Note that the anions used to form these salts are conjugate bases of strong acids; this is so because they have no acidic or basic properties in water (except for HSO_4^-, which has weak acid properties).

c. These are solutions of strong bases like LiOH, NaOH, KOH, RbOH, CsOH, $Ca(OH)_2$, $Sr(OH)_2$, and $Ba(OH)_2$. All these strong bases are strong electrolytes.

d. These are solutions containing salts of the conjugate bases of the neutrally charged weak acids in Table 14.2. These conjugate bases are all weak bases, and they are anions with a 1− charge. Three examples of this type of slightly basic salts are $NaClO_2$, $KC_2H_3O_2$, and CaF_2. The cations used to form these salts are Li^+, Na^+, K^+, Rb^+, Cs^+, Ca^{2+}, Sr^{2+}, and Ba^{2+} because these cations have no acidic or basic properties in water. Notice that these are the cations of the strong bases you should memorize.

e. There are two ways to make a neutral salt solutions. The easiest way is to combine a conjugate base of a strong acid (except for HSO_4^-) with one of the cations from a strong base. These ions have no acidic/basic properties in water, so salts of these ions are neutral. Three examples are NaCl, KNO_3, and SrI_2. Another type of strong electrolyte that can produce neutral solutions are salts that contain an ion with weak acid properties combined with an ion of opposite charge having weak base properties. If the K_a for the weak acid ion is equal to the K_b for the weak base ion, then the salt will produce a neutral solution. The most common example of this type of salt is ammonium acetate ($NH_4C_2H_3O_2$). For this salt, K_a for NH_4^+ = K_b for $C_2H_3O_2^- = 5.6 \times 10^{-10}$. This salt at any concentration produces a neutral solution.

33. a. $H_2O(l) + H_2O(l) \rightleftharpoons H_3O^+(aq) + OH^-(aq)$ or

$H_2O(l) \rightleftharpoons H^+(aq) + OH^-(aq) \quad K = K_w = [H^+][OH^-]$

b. $HF(aq) + H_2O(l) \rightleftharpoons F^-(aq) + H_3O^+(aq)$ or

$$HF(aq) \rightleftharpoons H^+(aq) + F^-(aq) \quad K = K_a = \frac{[H^+][F^-]}{[HF]}$$

c. $$C_5H_5N(aq) + H_2O(l) \rightleftharpoons C_5H_5NH^+(aq) + OH^-(aq) \quad K = K_b = \frac{[C_5H_5NH^+][OH^-]}{[C_5H_5N]}$$

35. a. H_2SO_3 b. $HClO_3$ c. H_3PO_3

NaOH and KOH are soluble ionic compounds composed of Na^+ and K^+ cations and OH^- anions. All soluble ionic compounds dissolve to form the ions from which they are formed. In oxyacids, the compounds are all covalent compounds in which electrons are shared to form bonds (unlike ionic compounds). When these compounds are dissolved in water, the covalent bond between oxygen and hydrogen breaks to form H^+ ions.

37. a. This expression holds true for solutions of monoprotic strong acids having a concentration greater than 1.0×10^{-6} *M*. 0.10 *M* HCl, 7.8 *M* HNO_3, and 3.6×10^{-4} *M* $HClO_4$ are examples where this expression holds true.

b. This expression holds true for solutions of weak acids where the two normal assumptions hold. The two assumptions are that water does not contribute enough H^+ to solution to make a difference, and that the acid is less than 5% dissociated in water (from the assumption that *x* is small compared to some number). This expression will generally hold true for solutions of weak acids having a K_a value less than 1×10^{-4}, if there is a significant amount of weak acid present. Three example solutions are 1.5 *M* $HC_2H_3O_2$, 0.10 *M* HOCl, and 0.72 *M* HCN.

c. This expression holds true for strong bases that donate 2 OH^- ions per formula unit. As long as the concentration of the base is above 5×10^{-7} *M*, this expression will hold true. Three examples are 5.0×10^{-3} *M* $Ca(OH)_2$, 2.1×10^{-4} *M* $Sr(OH)_2$, and 9.1×10^{-5} *M* $Ba(OH)_2$.

d. This expression holds true for solutions of weak bases where the two normal assumptions hold. The assumptions are that the OH^- contribution from water is negligible and that and that the base is less than 5% ionized in water (for the 5% rule to hold). For the 5% rule to hold, you generally need bases with $K_b < 1 \times 10^{-4}$, and concentrations of weak base greater than 0.10 *M*. Three examples are 0.10 *M* NH_3, 0.54 *M* $C_6H_5NH_2$, and 1.1 *M* C_5H_5N.

39. Answer c is best. For conjugate acid-base pairs, $K_a \times K_b = K_w = 1.0 \times 10^{-14}$ (at 25°C). A weak acid has a K_a value between 1 and 1.0×10^{-14}. Applying $K_a \times K_b = K_w = 1.0 \times 10^{-14}$, the K_b value for the conjugate base must be between 1.0×10^{-14} and 1. Bases with K_b values between 1 and 1.0×10^{-14} are classified as weak bases.

41. One reason HF is a weak acid is that the H–F bond is unusually strong and is difficult to break. This contributes significantly to the reluctance of the HF molecules to dissociate in water.

Exercises

Nature of Acids and Bases

43. a. $HClO_4(aq) + H_2O(l) \rightarrow H_3O^+(aq) + ClO_4^-(aq)$. Only the forward reaction is indicated because $HClO_4$ is a strong acid and is basically 100% dissociated in water. For acids, the dissociation reaction is commonly written without water as a reactant. The common abbreviation for this reaction is $HClO_4(aq) \rightarrow H^+(aq) + ClO_4^-(aq)$. This reaction is also called the K_a reaction because the equilibrium constant for this reaction is designated as K_a.

b. Propanoic acid is a weak acid, so it is only partially dissociated in water. The dissociation reaction is:

$$CH_3CH_2CO_2H(aq) + H_2O(l) \rightleftharpoons H_3O^+(aq) + CH_3CH_2CO_2^-(aq) \quad \text{or}$$

$$CH_3CH_2CO_2H(aq) \rightleftharpoons H^+(aq) + CH_3CH_2CO_2^-(aq).$$

c. NH_4^+ is a weak acid. Like propanoic acid, the dissociation reaction is:

$$NH_4^+(aq) + H_2O(l) \rightleftharpoons H_3O^+(aq) + NH_3(aq) \quad \text{or} \quad NH_4^+(aq) \rightleftharpoons H^+(aq) + NH_3(aq)$$

45. An acid is a proton (H^+) donor, and a base is a proton acceptor. A conjugate acid-base pair differs by only a proton (H^+).

	Acid	Base	Conjugate Base of Acid	Conjugate Acid of Base
a.	H_2CO_3	H_2O	HCO_3^-	H_3O^+
b.	$C_5H_5NH^+$	H_2O	C_5H_5N	H_3O^+
c.	$C_5H_5NH^+$	HCO_3^-	C_5H_5N	H_2CO_3

47. Strong acids have a $K_a >> 1$, and weak acids have $K_a < 1$. Table 14.2 in the text lists some K_a values for weak acids. K_a values for strong acids are hard to determine, so they are not listed in the text. However, there are only a few common strong acids so, if you memorize the strong acids, then all other acids will be weak acids. The strong acids to memorize are HCl, HBr, HI, HNO_3, $HClO_4$, and H_2SO_4.

a. $HClO_4$ is a strong acid.

b. HOCl is a weak acid ($K_a = 3.5 \times 10^{-8}$).

c. H_2SO_4 is a strong acid.

d. H_2SO_3 is a weak diprotic acid because the K_{a1} and K_{a2} values are much less than 1.

49. The K_a value is directly related to acid strength. As K_a increases, acid strength increases. For water, use K_w when comparing the acid strength of water to other species. The K_a values are:

$HClO_4$: strong acid ($K_a >> 1$); $HClO_2$: $K_a = 1.2 \times 10^{-2}$

NH_4^+: $K_a = 5.6 \times 10^{-10}$; H_2O: $K_a = K_w = 1.0 \times 10^{-14}$

From the K_a values, the ordering is $HClO_4 > HClO_2 > NH_4^+ > H_2O$.

51. a. HCl is a strong acid, and water is a very weak acid with $K_a = K_w = 1.0 \times 10^{-14}$. HCl is a much stronger acid than H_2O.

b. H_2O, $K_a = K_w = 1.0 \times 10^{-14}$; HNO_2, $K_a = 4.0 \times 10^{-4}$; HNO_2 is a stronger acid than H_2O because K_a for $HNO_2 > K_w$ for H_2O.

c. HOC_6H_5 , $K_a = 1.6 \times 10^{-10}$; HCN, $K_a = 6.2 \times 10^{-10}$; HCN is a slightly stronger acid than HOC_6H_5 because K_a for HCN > K_a for HOC_6H_5.

Autoionization of Water and the pH Scale

53. At 25°C, the relationship $[H^+][OH^-] = K_w = 1.0 \times 10^{-14}$ always holds for aqueous solutions. When $[H^+]$ is greater than 1.0×10^{-7} *M* (pH < 7.0), the solution is acidic; when $[H^+]$ is less than 1.0×10^{-7} *M* (pH > 7.0), the solution is basic; when $[H^+] = 1.0 \times 10^{-7}$ *M* (pH = 7.0), the solution is neutral. In terms of $[OH^-]$, an acidic solution has $[OH^-] < 1.0 \times 10^{-7}$ *M* (pOH > 7.0), a basic solution has $[OH^-] > 1.0 \times 10^{-7}$ *M* (pOH < 7.0), and a neutral solution has $[OH^-] = 1.0 \times 10^{-7}$ *M* (pOH = 7.0). At 25°C, pH + pOH = 14.00.

a. $[OH^-] = \dfrac{K_w}{[H^+]} = \dfrac{1.0 \times 10^{-14}}{1.0 \times 10^{-7}} = 1.0 \times 10^{-7}\ M$; the solution is neutral.

$pH = -\log[H^+] = -\log(1.0 \times 10^{-7}) = 7.00$; $pOH = 14.00 - 7.00 = 7.00$

b. $[OH^-] = \dfrac{1.0 \times 10^{-14}}{8.3 \times 10^{-16}} = 12\ M$; the solution is basic.

$pH = -\log(8.3 \times 10^{-16}) = 15.08$; $pOH = 14.00 - 15.08 = -1.08$

c. $[OH^-] = \dfrac{1.0 \times 10^{-14}}{12} = 8.3 \times 10^{-16}\ M$; the solution is acidic.

$pH = -\log(12) = -1.08$; $pOH = 14.00 - (-1.08) = 15.08$

d. $[OH^-] = \dfrac{1.0 \times 10^{-14}}{5.4 \times 10^{-5}} = 1.9 \times 10^{-10}\ M$; the solution is acidic.

$pH = -\log(5.4 \times 10^{-5}) = 4.27$; $pOH = 14.00 - 4.27 = 9.73$

Note that pH is greater than 14.00 when $[OH^-]$ is greater than 1.0 *M* (an extremely basic solution). Also note the the pH is negative when $[H^+]$ is greater than 1.0 *M* (an extremely acidic solution).

55. a. Because the value of the equilibrium constant increases as the temperature increases, the reaction is endothermic. In endothermic reactions, heat is a reactant, so an increase in temperature (heat) shifts the reaction to produce more products and increases K in the process.

b. $H_2O(l) \rightleftharpoons H^+(aq) + OH^-(aq)$ $K_w = 5.47 \times 10^{-14} = [H^+][OH^-]$ at 50.°C

In pure water $[H^+] = [OH^-]$, so $5.47 \times 10^{-14} = [H^+]^2$, $[H^+] = 2.34 \times 10^{-7}\ M = [OH^-]$

57. a. $[H^+] = 10^{-pH}$, $[H^+] = 10^{-7.40} = 4.0 \times 10^{-8}\ M$

$pOH = 14.00 - pH = 14.00 - 7.40 = 6.60$; $[OH^-] = 10^{-pOH} = 10^{-6.60} = 2.5 \times 10^{-7}\ M$

or $[OH^-] = \dfrac{K_w}{[H^+]} = \dfrac{1.0 \times 10^{-14}}{4.0 \times 10^{-8}} = 2.5 \times 10^{-7}\ M$; this solution is basic since pH > 7.00.

b. $[H^+] = 10^{-15.3} = 5 \times 10^{-16}$ *M*; pOH = 14.00 – 15.3 = –1.3; $[OH^-] = 10^{-(-1.3)} = 20$ *M*; basic

c. $[H^+] = 10^{-(-1.0)} = 10$ *M*; pOH = 14.0 – (–1.0) = 15.0; $[OH^-] = 10^{-15.0} =$ 1×10^{-15} *M*; acidic

d. $[H^+] = 10^{-3.20} = 6.3 \times 10^{-4}$ *M*; pOH = 14.00 – 3.20 = 10.80; $[OH^-] = 10^{-10.80} =$ 1.6×10^{-11} *M*; acidic

e. $[OH^-] = 10^{-5.0} = 1 \times 10^{-5}$ *M*; pH = 14.0 – pOH = 14.0 – 5.0 = 9.0; $[H^+] = 10^{-9.0}$ = 1×10^{-9} *M*; basic

f. $[OH^-] = 10^{-9.60} = 2.5 \times 10^{-10}$ *M*; pH = 14.00 – 9.60 = 4.40; $[H^+] = 10^{-4.40} = 4.0 \times 10^{-5}$ *M*; acidic

59. pOH = 14.0 – pH = 14.0 – 2.1 = 11.9; $[H^+] = 10^{-pH} = 10^{-2.1} = 8 \times 10^{-3}$ *M* (1 sig. fig.)

$$[OH^-] = \frac{K_w}{[H^+]} = \frac{1.0 \times 10^{-14}}{8 \times 10^{-3}} = 1 \times 10^{-12}\,M \text{ or } [OH^-] = 10^{-pOH} = 10^{-11.9} = 1 \times 10^{-12}\,M$$

The sample of gastric juice is acidic because the pH is less than 7.00 at 25°C.

Solutions of Acids

61. HI is a strong acid (100% dissociated in water) and the rest are weak acids. Statement e is false. From the acid strength listing, HNO_2 is a stronger weak acid than HCN, so the K_a value for HNO_2 is greater than the K_a value for HCN. A 0.10 *M* solution of HNO_2 will be more acidic than that of HCN, which means the HNO_2 solution will have the lower pH. For statement c, the stronger the acid the weaker the conjugate base. Because HI is a strong acid while HNO_2 is a weak acid, NO_2^- will be a stronger base than I^-. For statement d, the acid strength ordering indicates that HOCl is a stronger acid than HCN, so solutions of HOCl are more acidic meaning they have a higher $[H^+]$.

63. All the acids in this problem are strong acids that are always assumed to completely dissociate in water. The general dissociation reaction for a strong acid is $HA(aq) \rightarrow H^+(aq) + A^-(aq)$, where A^- is the conjugate base of the strong acid HA. For 0.250 *M* solutions of these strong acids, 0.250 *M* H^+ and 0.250 *M* A^- are present when the acids completely dissociate. The amount of H^+ donated from water will be insignificant in this problem since H_2O is a very weak acid.

a. Major species present after dissociation = H^+, ClO_4^-, and H_2O

pH = $-\log[H^+] = -\log(0.250) = 0.602$

b. Major species = H^+, NO_3^-, and H_2O; pH = 0.602

65. Strong acids are assumed to completely dissociate in water, for example, $HCl(aq) + H_2O(l) \rightarrow H_3O^+(aq) + Cl^-(aq)$ or $HCl(aq) \rightarrow H^+(aq) + Cl^-(aq)$.

a. A 0.10 *M* HCl solution gives 0.10 *M* H^+ and 0.10 *M* Cl^- because HCl completely dissociates. The amount of H^+ from H_2O will be insignificant.

$pH = -\log[H^+] = -\log(0.10) = 1.00$

b. 5.0 *M* H^+ is produced when 5.0 *M* $HClO_4$ completely dissociates. The amount of H^+ from H_2O will be insignificant. $pH = -\log(5.0) = -0.70$ (Negative pH values just indicate very concentrated acid solutions.)

c. 1.0×10^{-11} *M* H^+ is produced when 1.0×10^{-11} *M* HI completely dissociates. If you take the negative log of 1.0×10^{-11}, this gives pH = 11.00. This is impossible! We dissolved an acid in water and got a basic pH. What we must consider in this problem is that water by itself donates 1.0×10^{-7} *M* H^+. We can normally ignore the small amount of H^+ from H_2O except when we have a very dilute solution of an acid (as in the case here). Therefore, the pH is that of neutral water (pH = 7.00) because the amount of HI present is insignificant.

67. $[H^+] = 10^{-pH} = 10^{-4.25} = 5.6 \times 10^{-5}$ *M*. Because HBr is a strong acid, a 5.6×10^{-5} *M* HBr solution is necessary to produce a pH = 4.25 solution.

69. HCl is a strong acid. $[H^+] = 10^{-1.50} = 3.16 \times 10^{-2}$ *M* (carrying one extra sig. fig.)

$$M_1V_1 = M_2V_2, \quad V_1 = \frac{M_2V_2}{M_1} = \frac{3.16 \times 10^{-2}\ \text{mol/L} \times 1.6\ \text{L}}{12\ \text{mol/L}} = 4.2 \times 10^{-3}\ \text{L}$$

Add 4.2 mL of 12 *M* HCl to water with mixing; add enough water to make 1600 mL of solution. The resulting solution will have $[H^+] = 3.2 \times 10^{-2}$ *M* and pH = 1.50.

71. HA is a weak acid, so write out the K_a reaction and set-up the general ICE table.

	HA	$\rightleftharpoons$	H^+	+	A^-
Initial	1.0 mol/1.0 L		~0		0
	Let x mol/L HA dissociate to reach equilibrium				
Change	$-x$	$\rightarrow$	$+x$		$+x$
Equil.	$1.0 - x$		x		x

Since HA is a weak acid, the amount of the weak acid that dissociates, x, will typically be less than 5%. Assuming this, $[HA] > [H^+]$ and $[A^-]$. Since this is an acidic solution, $[H^+] > [OH^-]$. And from the ICE table, $[H^+] = [A^-] = x$. Only answer d is a correct statement.

73. a. HNO_2 ($K_a = 4.0 \times 10^{-4}$) and H_2O ($K_a = K_w = 1.0 \times 10^{-14}$) are the major species. HNO_2 is a much stronger acid than H_2O, so it is the major source of H^+. However, HNO_2 is a weak acid ($K_a < 1$), so it only partially dissociates in water. We must solve an equilibrium problem to determine $[H^+]$. In the Solutions Guide, we will summarize the *i*nitial, *c*hange, and *e*quilibrium concentrations into one table called the ICE table. Solving the weak acid problem:

	HNO_2	$\rightleftharpoons$	H^+	+	NO_2^-
Initial	0.250 M		~0		0
	x mol/L HNO_2 dissociates to reach equilibrium				
Change	$-x$	$\rightarrow$	$+x$		$+x$
Equil.	$0.250 - x$		x		x

$$K_a = \frac{[H^+][NO_2^-]}{[HNO_2]} = 4.0 \times 10^{-4} = \frac{x^2}{0.250 - x}$$; if we assume $x << 0.250$, then:

$$4.0 \times 10^{-4} \approx \frac{x^2}{0.250},\quad x = \sqrt{4.0 \times 10^{-4}(0.250)} = 0.010\ M$$

We must check the assumption: $\frac{x}{0.250} \times 100 = \frac{0.010}{0.250} \times 100 = 4.0\%$

All the assumptions are good. The H^+ contribution from water ($1 \times 10^{-7}\ M$) is negligible, and x is small compared to 0.250 (percent error = 4.0%). If the percent error is less than 5% for an assumption, we will consider it a valid assumption (called the 5% rule). Finishing the problem:

$x = 0.010\ M = [H^+]$; $pH = -\log(0.010) = 2.00$

b. CH_3CO_2H or $HC_2H_3O_2$ ($K_a = 1.8 \times 10^{-5}$) and H_2O ($K_a = K_w = 1.0 \times 10^{-14}$) are the major species. CH_3CO_2H is the major source of H^+. Solving the weak acid problem:

	CH_3CO_2H	$\rightleftharpoons$	H^+	+	$CH_3CO_2^-$
Initial	0.250 M		~0		0
	x mol/L CH_3CO_2H dissociates to reach equilibrium				
Change	$-x$	$\rightarrow$	$+x$		$+x$
Equil.	$0.250 - x$		x		x

$$K_a = \frac{[H^+][CH_3CO_2^-]}{[CH_3CO_2H]},\quad 1.8 \times 10^{-5} = \frac{x^2}{0.250 - x} \approx \frac{x^2}{0.250}$$ (assuming $x << 0.250$)

$x = 2.1 \times 10^{-3}\ M$; checking assumption: $\frac{2.1 \times 10^{-3}}{0.250} \times 100 = 0.84\%$. Assumptions good.

$[H^+] = x = 2.1 \times 10^{-3}\ M$; $pH = -\log(2.1 \times 10^{-3}) = 2.68$

75. This is a weak acid in water. Solving the weak acid problem:

	HF	$\rightleftharpoons$	H^+	+	F^-	$K_a = 7.2 \times 10^{-4}$
Initial	0.020 M		~0		0	
	x mol/L HF dissociates to reach equilibrium					
Change	$-x$	$\rightarrow$	$+x$		$+x$	
Equil.	$0.020 - x$		x		x	

$$K_a = 7.2 \times 10^{-4} = \frac{[H^+][F^-]}{[HF]} = \frac{x^2}{0.020 - x} \approx \frac{x^2}{0.020}$$ (assuming $x << 0.020$)

$x = [H^+] = 3.8 \times 10^{-3}\ M$; check assumptions:

$$\frac{x}{0.020} \times 100 = \frac{3.8 \times 10^{-3}}{0.020} \times 100 = 19\%$$

The assumption $x << 0.020$ is not good (x is more than 5% of 0.020). We must solve $x^2/(0.020 - x) = 7.2 \times 10^{-4}$ exactly by using either the quadratic formula or the method of successive approximations (see Appendix 1 of the text). Using successive approximations, we let 0.016 M be a new approximation for [HF]. That is, in the denominator try $x = 0.0038$ (the value of x we calculated making the normal assumption) so that $0.020 - 0.0038 = 0.016$; then solve for a new value of x in the numerator.

$$\frac{x^2}{0.020 - x} \approx \frac{x^2}{0.016} = 7.2 \times 10^{-4},\ x = 3.4 \times 10^{-3}$$

We use this new value of x to further refine our estimate of [HF], that is, $0.020 - x = 0.020 - 0.0034 = 0.0166$ (carrying an extra sig. fig.).

$$\frac{x^2}{0.020 - x} \approx \frac{x^2}{0.0166} = 7.2 \times 10^{-4},\ x = 3.5 \times 10^{-3}$$

We repeat until we get a self-consistent answer. This would be the same answer we would get solving exactly using the quadratic equation. In this case it is, $x = 3.5 \times 10^{-3}$. Thus:

$[H^+] = [F^-] = x = 3.5 \times 10^{-3}\ M$; $[OH^-] = K_w/[H^+] = 2.9 \times 10^{-12}\ M$

$[HF] = 0.020 - x = 0.020 - 0.0035 = 0.017\ M$; pH = 2.46

Note: When the 5% assumption fails, use whichever method you are most comfortable with to solve exactly. The method of successive approximations is probably fastest when the percent error is less than ~25% (unless you have a graphing calculator).

77. $HC_3H_5O_2$ ($K_a = 1.3 \times 10^{-5}$) and H_2O ($K_a = K_w = 1.0 \times 10^{-14}$) are the major species present. $HC_3H_5O_2$ will be the dominant producer of H^+ because $HC_3H_5O_2$ is a stronger acid than H_2O. Solving the weak acid problem:

	$HC_3H_5O_2$	$\rightleftharpoons$	H^+	+	$C_3H_5O_2^-$
Initial	0.100 M		~0		0
	x mol/L $HC_3H_5O_2$ dissociates to reach equilibrium				
Change	$-x$	→	$+x$		$+x$
Equil.	$0.100 - x$		x		x

$$K_a = 1.3 \times 10^{-5} = \frac{[H^+][C_3H_5O_2^-]}{[HC_3H_5O_2]} = \frac{x^2}{0.100 - x} \approx \frac{x^2}{0.100}$$

$x = [H^+] = 1.1 \times 10^{-3} M$; pH $= -\log(1.1 \times 10^{-3}) = 2.96$

Assumption follows the 5% rule (x is 1.1% of 0.100).

$[H^+] = [C_3H_5O_2^-] = 1.1 \times 10^{-3} M$; $[OH^-] = K_w/[H^+] = 9.1 \times 10^{-12} M$

$[HC_3H_5O_2] = 0.100 - 1.1 \times 10^{-3} = 0.099\ M$

$$\text{Percent dissociation} = \frac{[H^+]}{[HC_3H_5O_2]_0} \times 100 = \frac{1.1 \times 10^{-3}}{0.100} \times 100 = 1.1\%$$

79. $$[HC_9H_7O_4] = \frac{2 \text{ tablets} \times \dfrac{0.325 \text{ g } HC_9H_7O_4}{\text{tablet}} \times \dfrac{1 \text{ mol } HC_9H_7O_4}{180.15 \text{ g}}}{0.237 \text{ L}} = 0.0152\ M$$

	$HC_9H_7O_4$	$\rightleftharpoons$	H^+	+	$C_9H_7O_4^-$
Initial	$0.0152\ M$		~0		0
	x mol/L $HC_9H_7O_4$ dissociates to reach equilibrium				
Change	$-x$	$\rightarrow$	$+x$		$+x$
Equil.	$0.0152 - x$		x		x

$$K_a = 3.3 \times 10^{-4} = \frac{[H^+][C_9H_7O_4^-]}{[HC_9H_7O_4]} = \frac{x^2}{0.0152 - x} \approx \frac{x^2}{0.0152},\ x = 2.2 \times 10^{-3}\ M$$

Assumption that $0.0152 - x \approx 0.0152$ fails the 5% rule: $\dfrac{2.2 \times 10^{-3}}{0.0152} \times 100 = 14\%$

Using successive approximations or the quadratic equation gives an exact answer of $x = 2.1 \times 10^{-3}\ M$.

$[H^+] = x = 2.1 \times 10^{-3} M$; $pH = -\log(2.1 \times 10^{-3}) = 2.68$

81.

	HA	$\rightleftharpoons$	H^+	+	A^-
Initial	$0.10\ M$		~0		0
	Let x mol/L HA dissociate to reach equilibrium				
Change	$-x$	$\rightarrow$	$+x$		$+x$
Equil.	$0.10 - x$		x		x

From the ICE table, $[H^+] = [A^-]$. Since pH = 3.0, $[H^+] = [A^-] = 10^{-3.0} = 1 \times 10^{-3}\ M$.

83. HF and HOC_6H_5 are both weak acids with K_a values of 7.2×10^{-4} and 1.6×10^{-10}, respectively. Since the K_a value for HF is much greater than the K_a value for HOC_6H_5, HF will be the dominant producer of H^+ (we can ignore the amount of H^+ produced from HOC_6H_5 because it will be insignificant).

	HF	$\rightleftharpoons$	H^+	+	F^-
Initial	$1.0\ M$		~0		0
	x mol/L HF dissociates to reach equilibrium				
Change	$-x$	$\rightarrow$	$+x$		$+x$
Equil.	$1.0 - x$		x		x

$$K_a = 7.2 \times 10^{-4} = \frac{[H^+][F^-]}{[HF]} = \frac{x^2}{1.0 - x} \approx \frac{x^2}{1.0}$$

$x = [H^+] = 2.7 \times 10^{-2}\ M$; pH = $-\log(2.7 \times 10^{-2}) = 1.57$; assumptions good.

Solving for $[OC_6H_5^-]$ using $HOC_6H_5 \rightleftharpoons H^+ + OC_6H_5^-$ equilibrium:

$$K_a = 1.6 \times 10^{-10} = \frac{[H^+][OC_6H_5^-]}{[HOC_6H_5]} = \frac{(2.7 \times 10^{-2})[OC_6H_5^-]}{1.0},\ [OC_6H_5^-] = 5.9 \times 10^{-9}\ M$$

Note that this answer indicates that only $5.9 \times 10^{-9}\ M$ HOC_6H_5 dissociates, which confirms that HF is truly the only significant producer of H^+ in this solution.

85. In all parts of this problem, acetic acid ($HC_2H_3O_2$) is the best weak acid present. We must solve a weak acid problem.

a.

	$HC_2H_3O_2$	$\rightleftharpoons$	H^+	+	$C_2H_3O_2^-$
Initial	0.50 M		~0		0
	x mol/L $HC_2H_3O_2$ dissociates to reach equilibrium				
Change	$-x$	$\rightarrow$	$+x$		$+x$
Equil.	$0.50 - x$		x		x

$$K_a = 1.8 \times 10^{-5} = \frac{[H^+][C_2H_3O_2^-]}{[HC_2H_3O_2]} = \frac{x^2}{0.50 - x} \approx \frac{x^2}{0.50}$$

$x = [H^+] = [C_2H_3O_2^-] = 3.0 \times 10^{-3}\ M$; assumptions good.

$$\text{Percent dissociation} = \frac{[H^+]}{[HC_2H_3O_2]_0} \times 100 = \frac{3.0 \times 10^{-3}}{0.50} \times 100 = 0.60\%$$

b. The setup for solutions b and c are similar to solution a except that the final equation is different because the new concentration of $HC_2H_3O_2$ is different.

$$K_a = 1.8 \times 10^{-5} = \frac{x^2}{0.050 - x} \approx \frac{x^2}{0.050}$$

$x = [H^+] = [C_2H_3O_2^-] = 9.5 \times 10^{-4}\ M$; assumptions good.

$$\text{Percent dissociation} = \frac{9.5 \times 10^{-4}}{0.050} \times 100 = 1.9\%$$

c. $$K_a = 1.8 \times 10^{-5} = \frac{x^2}{0.0050 - x} \approx \frac{x^2}{0.0050}$$

$x = [H^+] = [C_2H_3O_2^-] = 3.0 \times 10^{-4}\ M$; check assumptions.

Assumption that x is negligible is borderline (6.0% error). We should solve exactly. Using the method of successive approximations (see Appendix 1 of the text):

$$1.8 \times 10^{-5} = \frac{x^2}{0.0050 - (3.0 \times 10^{-4})} = \frac{x^2}{0.0047},\ x = 2.9 \times 10^{-4}$$

Next trial also gives $x = 2.9 \times 10^{-4}$.

$$\text{Percent dissociation} = \frac{2.9 \times 10^{-4}}{5.0 \times 10^{-3}} \times 100 = 5.8\%$$

d. As we dilute a solution, all concentrations are decreased. Dilution will shift the equilibrium to the side with the greater number of particles. For example, suppose we double the volume of an equilibrium mixture of a weak acid by adding water; then:

$$Q = \frac{\left(\frac{[H^+]_{eq}}{2}\right)\left(\frac{[X^-]_{eq}}{2}\right)}{\left(\frac{[HX]_{eq}}{2}\right)} = \frac{1}{2}K_a$$

$Q < K_a$, so the equilibrium shifts to the right or toward a greater percent dissociation.

e. $[H^+]$ depends on the initial concentration of weak acid and on how much weak acid dissociates. For solutions a-c, the initial concentration of acid decreases more rapidly than the percent dissociation increases. Thus $[H^+]$ decreases.

87. Let HA symbolize the weak acid. Set up the problem like a typical weak acid equilibrium problem.

	HA	$\rightleftharpoons$	H^+	+	A^-
Initial	0.15 *M*		~0		0
	x mol/L HA dissociates to reach equilibrium				
Change	$-x$	$\rightarrow$	$+x$		$+x$
Equil.	$0.15 - x$		x		x

If the acid is 3.0% dissociated, then $x = [H^+]$ is 3.0% of 0.15: $x = 0.030 \times (0.15\ M) = 4.5 \times 10^{-3}\ M$. Now that we know the value of x, we can solve for K_a.

$$K_a = \frac{[H^+][A^-]}{[HA]} = \frac{x^2}{0.15 - x} = \frac{(4.5 \times 10^{-3})^2}{0.15 - (4.5 \times 10^{-3})} = 1.4 \times 10^{-4}$$

89. $HClO_4$ is a strong acid with $[H^+] = 0.040\ M$. This equals the $[H^+]$ in the trichloroacetic acid solution. Set up the problem using the K_a equilibrium reaction for CCl_3CO_2H.

	CCl_3CO_2H	$\rightleftharpoons$	H^+	+	$CCl_3CO_2^-$
Initial	0.050 *M*		~0		0
Equil.	$0.050 - x$		x		x

$$K_a = \frac{[H^+][CCl_3CO_2^-]}{[CCl_3CO_2H]} = \frac{x^2}{0.050 - x};\quad \text{from the problem, } x = [H^+] = 4.0 \times 10^{-2}\ M$$

$$K_a = \frac{(4.0 \times 10^{-2})^2}{0.050 - (4.0 \times 10^{-2})} = 0.16$$

91. HCl is a strong acid, so $[H^+] = 0.010\ M$. Set up the ICE table using the K_a reaction for HF.

	HF	$\rightleftharpoons$	H^+	+	F^-	
Initial	C		~0		0	where C = $[HF]_0$
	x mol/L HF dissociates to reach equilibrium					
Change	$-x$	$\rightarrow$	$+x$		$+x$	
Equil.	$C - x$		x		x	

$$K_a = 1.8 \times 10^{-4} = \frac{[H^+][F^-]}{[HF]} = \frac{x^2}{C - x},\ \text{where } x = [H^+] = 0.010\ M$$

$$7.2 \times 10^{-4} = \frac{[H^+]^2}{C - [H^+]} = \frac{(0.010)^2}{C - 0.010},\ C - 0.010 = \frac{1.0 \times 10^{-4}}{7.2 \times 10^{-4}},\ C = 0.15\ M$$

A 0.15 *M* HF solution will have $[H^+] = 0.010\ M$.

93. $[HA]_0 = \dfrac{1.0\ \text{mol}}{2.0\ \text{L}} = 0.50$ mol/L; solve using the K_a equilibrium reaction.

	HA	$\rightleftharpoons$	H^+	+	A^-
Initial	0.50 *M*		~0		0
Equil.	$0.50 - x$		x		x

$$K_a = \frac{[H^+][A^-]}{[HA]} = \frac{x^2}{0.50 - x};\ \text{in this problem, } [HA] = 0.45\ M \text{ so:}$$

$$[HA] = 0.45\ M = 0.50\ M - x,\ x = 0.05\ M;\quad K_a = \frac{(0.05)^2}{0.45} = 6 \times 10^{-3}$$

Solutions of Bases

95. All K_b reactions refer to the base reacting with water to produce the conjugate acid of the base and OH^-.

a. $NH_3(aq) + H_2O(l) \rightleftharpoons NH_4^+(aq) + OH^-(aq)$ $\qquad K_b = \dfrac{[NH_4^+][OH^-]}{[NH_3]}$

b. $C_5H_5N(aq) + H_2O(l) \rightleftharpoons C_5H_5NH^+(aq) + OH^-(aq)$ $\qquad K_b = \dfrac{[C_5H_5NH^+][OH^-]}{[C_5H_5N]}$

97. NO_3^-: Because HNO_3 is a strong acid, NO_3^- is a terrible base ($K_b << K_w$). All conjugate bases of strong acids have no base strength.

H_2O: $K_b = K_w = 1.0 \times 10^{-14}$; NH_3: $K_b = 1.8 \times 10^{-5}$; C_5H_5N: $K_b = 1.7 \times 10^{-9}$

Base strength = $NH_3 > C_5H_5N > H_2O > NO_3^-$ (As K_b increases, base strength increases.) Excluding water, the acid list contains the conjugate acids of the bases in the initial list. In general, the stronger the base, the weaker is the conjugate acid. *Note*: Even though NH_4^+ and $C_5H_5NH^+$ are conjugate acids of weak bases, they are still weak acids with K_a values between K_w and 1. Prove this to yourself by calculating the K_a values for NH_4^+ and $C_5H_5NH^+$ ($K_a = K_w/K_b$).

Acid strength = $HNO_3 > C_5H_5NH^+ > NH_4^+ > H_2O$

99. $NaOH(aq) \rightarrow Na^+(aq) + OH^-(aq)$; NaOH is a strong base that completely dissociates into Na^+ and OH^-. The initial concentration of NaOH will equal the concentration of OH^- donated by NaOH.

a. $[OH^-] = 0.10\ M$; $pOH = -\log[OH^-] = -\log(0.10) = 1.00$

$pH = 14.00 - pOH = 14.00 - 1.00 = 13.00$

Note that H_2O is also present, but the amount of OH^- produced by H_2O will be insignificant as compared to the 0.10 M OH^- produced from the NaOH.

b. The $[OH^-]$ concentration donated by the NaOH is $1.0 \times 10^{-10}\ M$. Water by itself donates $1.0 \times 10^{-7}\ M$. In this exercise, water is the major OH^- contributor, and $[OH^-] = 1.0 \times 10^{-7}\ M$.

$pOH = -\log(1.0 \times 10^{-7}) = 7.00$; $pH = 14.00 - 7.00 = 7.00$

c. $[OH^-] = 2.0\ M$; $pOH = -\log(2.0) = -0.30$; $pH = 14.00 - (-0.30) = 14.30$

101. a. Major species: K^+, OH^-, H_2O (KOH is a strong base.)

$[OH^-] = 0.015\ M$, $pOH = -\log(0.015) = 1.82$; $pH = 14.00 - pOH = 12.18$

b. Major species: Ba^{2+}, OH^-, H_2O; $Ba(OH)_2(aq) \rightarrow Ba^{2+}(aq) + 2\ OH^-(aq)$; because each mole of the strong base $Ba(OH)_2$ dissolves in water to produce two mol OH^-, $[OH^-] = 2(0.015\ M) = 0.030\ M$.

$pOH = -\log(0.030) = 1.52$; $pH = 14.00 - 1.52 = 12.48$

103. $pOH = 14.00 - 11.56 = 2.44$; $[OH^-] = [KOH] = 10^{-2.44} = 3.6 \times 10^{-3}\ M$

$$0.8000\ L \times \frac{3.6 \times 10^{-3}\ \text{mol KOH}}{L} \times \frac{56.11\ \text{g KOH}}{\text{mol KOH}} = 0.16\ \text{g KOH}$$

105. For a weak base in water, write out the K_b reaction then set-up the ICE table.

	$B(aq)$ + $H_2O(l)$	$\rightleftharpoons$	$BH^+(aq)$	+	$OH^-(aq)$
Initial	1.0 mol/1.0 L		0		~0
	Let x mol/L B react with H_2O to reach equilibrium				
Change	$-x$	$\rightarrow$	$+x$		$+x$
Equil.	$1.0 - x$		x		x

$$K_b = \frac{[BH^+][OH^-]}{[B]} = \frac{x^2}{1.0 - x}$$

In a typical weak base problem, it is assumed that $1.0 - x \approx 1.0$. Assuming this assumption is good, then x will be less than 5% of 1.0 (which is typical for most weak bases). B will have the largest concentration, the concentrations of BH^+ and OH^- will be equal and will have the next largest equilibrium concentration, and H^+ will have the smallest equilibrium concentration since this is a basic solution. Only statement c is true ($[BH^+] = [OH^-] = x$).

107. NH_3 is a weak base with $K_b = 1.8 \times 10^{-5}$. The major species present will be NH_3 and H_2O ($K_b = K_w = 1.0 \times 10^{-14}$). Because NH_3 has a much larger K_b value than H_2O, NH_3 is the stronger base present and will be the major producer of OH^-. To determine the amount of OH^- produced from NH_3, we must perform an equilibrium calculation using the K_b reaction for NH_3.

	$NH_3(aq)$ + $H_2O(l)$	$\rightleftharpoons$	$NH_4^+(aq)$	+	$OH^-(aq)$
Initial	0.150 M		0		~0
	x mol/L NH_3 reacts with H_2O to reach equilibrium				
Change	$-x$	$\longrightarrow$	$+x$		$+x$
Equil.	$0.150 - x$		x		x

$$K_b = 1.8 \times 10^{-5} = \frac{[NH_4^+][OH^-]}{[NH_3]} = \frac{x^2}{0.150 - x} \approx \frac{x^2}{0.150} \quad \text{(assuming } x \ll 0.150\text{)}$$

$x = [OH^-] = 1.6 \times 10^{-3}\ M$; check assumptions: x is 1.1% of 0.150, so the assumption $0.150 - x \approx 0.150$ is valid by the 5% rule. Also, the contribution of OH^- from water will be insignificant (which will usually be the case). Finishing the problem:

$$pOH = -\log[OH^-] = -\log(1.6 \times 10^{-3}\ M) = 2.80;\ pH = 14.00 - pOH = 14.00 - 2.80 = 11.20.$$

109. These are solutions of weak bases in water.

a. $C_6H_5NH_2 + H_2O \rightleftharpoons C_6H_5NH_3^+ + OH^- \qquad K_b = 3.8 \times 10^{-10}$

	$C_6H_5NH_2$		$C_6H_5NH_3^+$	OH^-
Initial	0.40 M		0	~0
	x mol/L of $C_6H_5NH_2$ reacts with H_2O to reach equilibrium			
Change	$-x$	$\longrightarrow$	$+x$	$+x$
Equil.	$0.40 - x$		x	x

$$3.8 \times 10^{-10} = \frac{x^2}{0.40 - x} \approx \frac{x^2}{0.40},\quad x = [OH^-] = 1.2 \times 10^{-5}\ M;\ \text{assumptions good.}$$

$[H^+] = K_w/[OH^-] = 8.3 \times 10^{-10}\ M$; pH = 9.08

b. $CH_3NH_2 + H_2O \rightleftharpoons CH_3NH_3^+ + OH^- \qquad K_b = 4.38 \times 10^{-4}$

	CH_3NH_2	$CH_3NH_3^+$	OH^-
Initial	0.40 M	0	~0
Equil.	$0.40 - x$	x	x

$$K_b = 4.38 \times 10^{-4} = \frac{x^2}{0.40 - x} \approx \frac{x^2}{0.40},\quad x = 1.3 \times 10^{-2}\ M;\ \text{assumptions good.}$$

$[OH^-] = 1.3 \times 10^{-2}\ M$; $[H^+] = K_w/[OH^-] = 7.7 \times 10^{-13}\ M$; pH = 12.11

111. This is a solution of a weak base in water. We must solve the weak base equilibrium problem.

	$C_2H_5NH_2$ + H_2O	$\rightleftharpoons$	$C_2H_5NH_3^+$	+ OH^-	$K_b = 5.6 \times 10^{-4}$
Initial	0.20 M		0	~0	
	x mol/L $C_2H_5NH_2$ reacts with H_2O to reach equilibrium				
Change	$-x$	$\rightarrow$	$+x$	$+x$	
Equil.	$0.20 - x$		x	x	

$$K_b = 5.6 \times 10^{-4} = \frac{[C_2H_5NH_3^+][OH^-]}{[C_2H_5NH_2]} = \frac{x^2}{0.20 - x} \approx \frac{x^2}{0.20} \quad \text{(assuming } x \ll 0.20\text{)}$$

$$x = 1.1 \times 10^{-2}; \text{ checking assumption: } \frac{1.1 \times 10^{-2}}{0.20} \times 100 = 5.5\%$$

The assumption fails the 5% rule. We must solve exactly using either the quadratic equation or the method of successive approximations (see Appendix 1 of the text). Using successive approximations and carrying extra significant figures:

$$\frac{x^2}{0.20 - 0.011} = \frac{x^2}{0.189} = 5.6 \times 10^{-4},\ x = 1.0 \times 10^{-2}\ M \text{ (consistent answer)}$$

$$x = [OH^-] = 1.0 \times 10^{-2}\ M;\ [H^+] = \frac{K_w}{[OH^-]} = \frac{1.0 \times 10^{-14}}{1.0 \times 10^{-2}} = 1.0 \times 10^{-12}\ M;\ pH = 12.00$$

113. To solve for percent ionization, we first solve the weak base equilibrium problem.

a.

	NH_3 + H_2O	$\rightleftharpoons$	NH_4^+	+ OH^-	$K_b = 1.8 \times 10^{-5}$
Initial	0.10 M		0	~0	
Equil.	$0.10 - x$		x	x	

$$K_b = 1.8 \times 10^{-5} = \frac{x^2}{0.10 - x} \approx \frac{x^2}{0.10},\ x = [OH^-] = 1.3 \times 10^{-3}\ M; \text{ assumptions good.}$$

$$\text{Percent ionization} = \frac{x}{[NH_3]_0} \times 100 = \frac{1.3 \times 10^{-3}\ M}{0.10\ M} \times 100 = 1.3\%$$

b.

	NH_3 + H_2O	$\rightleftharpoons$	NH_4^+	+ OH^-
Initial	0.010 M		0	~0
Equil.	$0.010 - x$		x	x

$$1.8 \times 10^{-5} = \frac{x^2}{0.010 - x} \approx \frac{x^2}{0.010},\ x = [OH^-] = 4.2 \times 10^{-4}\ M; \text{ assumptions good.}$$

$$\text{Percent ionization} = \frac{4.2 \times 10^{-4}}{0.010} \times 100 = 4.2\%$$

Note: For the same base, the percent ionization increases as the initial concentration of base decreases.

c.

	CH_3NH_2 + H_2O	$\rightleftharpoons$	$CH_3NH_3^+$	+ OH^-	$K_b = 4.38 \times 10^{-4}$
Initial	0.10 M		0	~0	
Equil.	$0.10 - x$		x	x	

$4.38 \times 10^{-4} = \frac{x^2}{0.10 - x} \approx \frac{x^2}{0.10}$, $x = 6.6 \times 10^{-3}$; assumption fails the 5% rule (x is 6.6% of 0.10). Using successive approximations and carrying extra significant figures:

$$\frac{x^2}{0.10 - 0.0066} = \frac{x^2}{0.093} = 4.38 \times 10^{-4},\ x = 6.4 \times 10^{-3} \quad \text{(consistent answer)}$$

$$\text{Percent ionization} = \frac{6.4 \times 10^{-3}}{0.10} \times 100 = 6.4\%$$

115. Using the K_b reaction to solve where PT = p-toluidine ($CH_3C_6H_4NH_2$):

	PT	+	H_2O	$\rightleftharpoons$	PTH^+	+	OH^-
Initial	0.016 M				0		~0
	x mol/L of PT reacts with H_2O to reach equilibrium						
Change	$-x$			$\rightarrow$	$+x$		$+x$
Equil.	$0.016 - x$				x		x

$$K_b = \frac{[PTH^+][OH^-]}{[PT]} = \frac{x^2}{0.016 - x}$$

Because pH = 8.60: pOH = 14.00 − 8.60 = 5.40 and $[OH^-] = x = 10^{-5.40} = 4.0 \times 10^{-6}\ M$

$$K_b = \frac{(4.0 \times 10^{-6})^2}{0.016 - (4.0 \times 10^{-6})} = 1.0 \times 10^{-9}$$

Polyprotic Acids

117. $H_2SO_3(aq) \rightleftharpoons HSO_3^-(aq) + H^+(aq)$ $\qquad K_{a_1} = \frac{[HSO_3^-][H^+]}{[H_2SO_3]}$

$HSO_3^-(aq) \rightleftharpoons SO_3^{2-}(aq) + H^+(aq)$ $\qquad K_{a_2} = \frac{[SO_3^{2-}][H^+]}{[HSO_3^-]}$

119. For H_2CO_3, The relevant reactions are:

$H_2CO_3 \rightleftharpoons H^+ + HCO_3^-$ $\quad K_{a_1} = 4.3 \times 10^{-7}$; $HCO_3^- \rightleftharpoons H^+ + CO_3^{2-}$ $\quad K_{a_2} = 5.6 \times 10^{-11}$

H_2CO_3 is a weak acid, and since $K_{a_1} >> K_{a_2}$, only the first dissociation step makes an important contribution to $[H^+]$.

	H_2CO_3	$\rightleftharpoons$	H^+	+	HCO_3^-
Initial	0.10 M		~0		0
Equil.	$0.10 - x$		x		x

$$K_{a_1} = 4.3 \times 10^{-7} = \frac{[H^+][HCO_3^-]}{[H_2CO_3]} = \frac{x^2}{0.10 - x}$$

When solving this, $x = [H^+]$ will be much less than 0.10 *M* since H_2CO_3 is a weak acid with $K_a \sim 10^{-7}$.

For H_2SO_4, the first dissociation occurs to completion. So, a 0.10 *M* H_2SO_4 solution gives 0.10 *M* HSO_4^- and 0.10 *M* H^+. The hydrogen sulfate ion (HSO_4^-) is a pretty good weak acid with $K_{a_2} = 1.2 \times 10^{-2}$. We also need to consider this equilibrium for additional H^+ production:

	HSO_4^-	$\rightleftharpoons$	H^+	+	SO_4^{2-}
Initial	0.10 *M*		0.10 *M*		0
	x mol/L HSO_4^- dissociates to reach equilibrium				
Change	$-x$	$\rightarrow$	$+x$		$+x$
Equil.	$0.10 - x$		$0.10 + x$		x

Since H_2SO_4 is both a strong acid and a pretty good weak acid problem in one, *x* will not be insignificant compared to 0.10, so the $[H^+]$ will be greater than 0.10 *M*.

From the set-ups above, the only true statement is c. Both the K_a values for H_2SO_4 are greater than the K_a values for H_2CO_3. Hence H_2SO_4 is a better acid than H_2CO_3, and HSO_4^- is a better acid than HCO_3^-. This means that more of HSO_4^- will dissociate than HCO_3^-, so more of the conjugate base of HSO_4^- will be present at equilibrium. Therefore at equilibrium, $[SO_4^{2-}] > [CO_3^{2-}]$. For statements a and d, H_2SO_4 is a much better acid than H_2CO_3, so H_2SO_4 solution will have the higher $[H^+]$ and lower pH. For statement b, the ICE table tell us that the concentration of H^+ will be greater than 0.10 *M*. And for statement e, since H_2SO_4 is a strong acid while H_2CO_3 is a weak acid, the equilibrium concentration of H_2CO_3 will be larger than the equilibrium concentration of H_2SO_4.

121. For $H_2C_6H_6O_6$. $K_{a_1} = 7.9 \times 10^{-5}$ and $K_{a_2} = 1.6 \times 10^{-12}$. Because $K_{a_1} >> K_{a_2}$, the amount of H^+ produced by the K_{a_2} reaction will be negligible.

$$[H_2C_6H_6O_6]_0 = \frac{0.500 \text{ g} \times \dfrac{1 \text{ mol } H_2C_6H_6O_6}{176.12 \text{ g}}}{0.2000 \text{ L}} = 0.0142\ M$$

	$H_2C_6H_6O_6(aq)$	$\rightleftharpoons$	$HC_6H_6O_6^-(aq)$	+	$H^+(aq)$	$K_{a_1} = 7.9 \times 10^{-5}$
Initial	0.0142 *M*		0		~0	
Equil.	$0.0142 - x$		x		x	

$$K_{a_1} = 7.9 \times 10^{-5} = \frac{x^2}{0.0142 - x} \approx \frac{x^2}{0.0142},\ x = 1.1 \times 10^{-3};\ \text{assumption fails the 5\% rule.}$$

Solving by the method of successive approximations:

$$7.9 \times 10^{-5} = \frac{x^2}{0.0142 - 1.1 \times 10^{-3}},\ x = 1.0 \times 10^{-3}\ M \text{ (consistent answer)}$$

Because H^+ produced by the K_{a_2} reaction will be negligible, $[H^+] = 1.0 \times 10^{-3}$ and pH = 3.00.

123. Because K_{a_2} for H_2S is so small, we can ignore the H^+ contribution from the K_{a_2} reaction.

	H_2S	$\rightleftharpoons$	H^+	HS^-	$K_{a_1} = 1.0 \times 10^{-7}$
Initial	$0.10\ M$		~0	0	
Equil.	$0.10 - x$		x	x	

$$K_{a_1} = 1.0 \times 10^{-7} = \frac{x^2}{0.10 - x} \approx \frac{x^2}{0.10},\quad x = [H^+] = 1.0 \times 10^{-4};\ \text{assumptions good.}$$

$$pH = -\log(1.0 \times 10^{-4}) = 4.00$$

Use the K_{a_2} reaction to determine $[S^{2-}]$.

	HS^-	$\rightleftharpoons$	H^+	+	S^{2-}
Initial	$1.0 \times 10^{-4}\ M$		$1.0 \times 10^{-4}\ M$		0
Equil.	$1.0 \times 10^{-4} - x$		$1.0 \times 10^{-4} + x$		x

$$K_{a_2} = 1.0 \times 10^{-19} = \frac{(1.0 \times 10^{-4} + x)x}{(1.0 \times 10^{-4} - x)} \approx \frac{(1.0 \times 10^{-4})x}{1.0 \times 10^{-4}}$$

$x = [S^{2-}] = 1.0 \times 10^{-19}\ M$; assumptions good.

125. The dominant H^+ producer is the strong acid H_2SO_4. A 2.0 M H_2SO_4 solution produces 2.0 M HSO_4^- and 2.0 M H^+. However, HSO_4^- is a weak acid that could also add H^+ to the solution.

	HSO_4^-	$\rightleftharpoons$	H^+	+	SO_4^{2-}
Initial	$2.0\ M$		$2.0\ M$		0
	x mol/L HSO_4^- dissociates to reach equilibrium				
Change	$-x$	$\rightarrow$	$+x$		$+x$
Equil.	$2.0 - x$		$2.0 + x$		x

$$K_{a_2} = 1.2 \times 10^{-2} = \frac{[H^+][SO_4^{2-}]}{[HSO_4^-]} = \frac{(2.0 + x)x}{2.0 - x} \approx \frac{2.0(x)}{2.0},\quad x = 1.2 \times 10^{-2}\ M$$

Because x is 0.60% of 2.0, the assumption is valid by the 5% rule. The amount of additional H^+ from HSO_4^- is $1.2 \times 10^{-2}\ M$. The total amount of H^+ present is:

$$[H^+] = 2.0 + (1.2 \times 10^{-2}) = 2.0\ M;\ pH = -\log(2.0) = -0.30$$

Note: In this problem, H^+ from HSO_4^- could have been ignored. However, this is not usually the case in more dilute solutions of H_2SO_4.

Acid-Base Properties of Salts

127. $KHSO_4 \rightarrow K^+ + HSO_4^-$; K^+ has no acidic or basic properties as is true for all alkali metal ions. HSO_4^- is a weak acid with $K_a = 0.012$. This solution will be acidic since a weak acid is present.

$HONH_3Cl \rightarrow HONH_3^+ + Cl^-$; $HONH_3^+$ is the conjugate acid of the weak base $HONH_2$ (K_b = 1.1×10^{-8}). The conjugate acids of weak bases are weak acids. So, $HONH_3^+$ is an acidic ion. Cl^- is the conjugate base of the strong acid HCl. Cl^- has no basic properties. This is an acidic salt due to the $HONH_3^+$ ion.

$C_5H_5NHNO_3 \rightarrow C_5H_5H^+ + NO_3^-$; $C_5H_5NH^+$ is the conjugate acid of the weak base C_5H_5N (K_b = 1.7×10^{-9}). The conjugate acids of weak bases are weak acids. So, $C_5H_5NH^+$ is an acidic ion. NO_3^- is the conjugate base of the strong acid HNO_3. NO_3^- has no basic properties. This is an acidic salt due to the $C_5H_5NH^+$ ion.

H_2NNH_2 is a weak base with $K_b = 3.0 \times 10^{-6}$. The solution will be basic.

$NaHCO_3 \rightarrow Na^+ + HCO_3^-$; Na^+ has no acidic or basic properties, while HCO_3^- is a weak acid with $K_a = 5.6 \times 10^{-11}$. This solution will be acidic since a weak acid is present.

From the discussion above, all solutions except for H_2NNH_2 will be acidic. Appendix 5 of the text has pertinent K_a and K_b values.

129. One difficult aspect of acid-base chemistry is recognizing what types of species are present in solution, that is, whether a species is a strong acid, strong base, weak acid, weak base, or a neutral species. Below are some ideas and generalizations to keep in mind that will help in recognizing types of species present.

a. Memorize the following strong acids: HCl, HBr, HI, HNO_3, $HClO_4$, and H_2SO_4

b. Memorize the following strong bases: LiOH, NaOH, KOH, RbOH, CsOH, $Ca(OH)_2$, $Sr(OH)_2$, and $Ba(OH)_2$

c. Weak acids have a K_a value of less than 1 but greater than K_w. Some weak acids are listed in Table 14.2 of the text. Weak bases have a K_b value of less than 1 but greater than K_w. Some weak bases are listed in Table 14.3 of the text.

d. Conjugate bases of weak acids are weak bases; that is, all have a K_b value of less than 1 but greater than K_w. Some examples of these are the conjugate bases of the weak acids listed in Table 14.2 of the text.

e. Conjugate acids of weak bases are weak acids; that is, all have a K_a value of less than 1 but greater than K_w. Some examples of these are the conjugate acids of the weak bases listed in Table 14.3 of the text.

f. Alkali metal ions (Li^+, Na^+, K^+, Rb^+, Cs^+) and heavier alkaline earth metal ions (Ca^{2+}, Sr^{2+}, Ba^{2+}) have no acidic or basic properties in water.

g. All conjugate bases of strong acids (Cl^-, Br^-, I^-, NO_3^-, ClO_4^-, HSO_4^-) have no basic properties in water ($K_b << K_w$), and only HSO_4^- has any acidic properties in water.

Let's apply these ideas to this problem to see what type of species are present. The letters in parenthesis is(are) the generalization(s) above that identifies the species.

KOH: Strong base (b)

KNO_3: Neutral; K^+ and NO_3^- have no acidic/basic properties (f and g).

KCN: CN^- is a weak base, $K_b = K_w/K_{a,\,HCN} = 1.0 \times 10^{-14}/6.2 \times 10^{-10} = 1.6 \times 10^{-5}$ (c and d). Ignore K^+ (f).

NH_4Cl: NH_4^+ is a weak acid, $K_a = 5.6 \times 10^{-10}$ (c and e). Ignore Cl^- (g).

HCl: Strong acid (a)

The most acidic solution will be the strong acid solution, with the weak acid solution less acidic. The most basic solution will be the strong base solution, with the weak base solution less basic. The KNO_3 solution will be neutral at pH = 7.00.

Most acidic → most basic: HCl > NH_4Cl > KNO_3 > KCN > KOH

131. Answer e is true. NH_3 is a weak base with $K_b = 1.8 \times 10^{-5}$. $NaC_2H_3O_2$ is a basic salt due to the $C_2H_3O_2^-$ ion. This ion is the conjugate base of the weak acid $HC_2H_3O_2$. The K_b value for $C_2H_3O_2^-$ is $K_w/K_a = 1.0 \times 10^{-14}/1.8 \times 10^{-5} = 5.6 \times 10^{-10}$. Because the K_b value for NH_3 is larger than the K_b value for the $C_2H_3O_2^-$ ion, the NH_3 solution is more basic with the higher pH.

In answer a, $NaC_2H_3O_2$ is a basic salt while NH_4Cl is an acidic salt due to the NH_4^+ ion. This ion is a weak acid since it is the conjugate acid a weak base. So, the NH_4Cl solution has a lower pH. In answer b, NH_3 is a stronger weak base than $C_2H_3O_2^-$ (see above). In part c and d, K_a for $HC_2H_3O_2$ is 1.8×10^{-5} and K_a for $NH_4^+ = K_w/K_b = 1.0 \times 10^{-14}/1.8 \times 10^{-5} = 5.6 \times 10^{-10}$. $HC_2H_3O_2$ has the larger K_a value, making it a better weak acid than NH_4^+. The $HC_2H_3O_2$ solution will have the lower pH.

133. a. KCl is a soluble ionic compound that dissolves in water to produce $K^+(aq)$ and $Cl^-(aq)$. K^+ (like the other alkali metal cations) has no acidic or basic properties. Cl^- is the conjugate base of the strong acid HCl. Cl^- has no basic (or acidic) properties. Therefore, a solution of KCl will be neutral because neither of the ions has any acidic or basic properties. The 1.0 *M* KCl solution has $[H^+] = [OH^-] = 1.0 \times 10^{-7}$ *M* and pH = pOH = 7.00.

b. $KC_2H_3O_2$ is also a soluble ionic compound that dissolves in water to produce $K^+(aq)$ and $C_2H_3O_2^-(aq)$. The difference between the KCl solution and the $KC_2H_3O_2$ solution is that $C_2H_3O_2^-$ does have basic properties in water, unlike Cl^-. $C_2H_3O_2^-$ is the conjugate base of the weak acid $HC_2H_3O_2$, and as is true for all conjugate bases of weak acids, $C_2H_3O_2^-$ is a weak base in water. We must solve an equilibrium problem in order to determine the amount of OH^- this weak base produces in water.

$$C_2H_3O_2^- + H_2O \rightleftharpoons HC_2H_3O_2 + OH^- \qquad K_b = \frac{K_w}{K_{a,\,C_2H_3O_2}} = \frac{1.0 \times 10^{-14}}{1.8 \times 10^{-5}}$$

	$C_2H_3O_2^-$		$HC_2H_3O_2$	OH^-	
Initial	1.0 *M*		0	~0	$K_b = 5.6 \times 10^{-10}$
	x mol/L of $C_2H_3O_2^-$ reacts with H_2O to reach equilibrium				
Change	$-x$	→	$+x$	$+x$	
Equil.	$1.0 - x$		x	x	

$$K_b = 5.6 \times 10^{-10} = \frac{[HC_2H_3O_2][OH^-]}{[C_2H_3O_2^-]},\quad 5.6 \times 10^{-10} = \frac{x^2}{1.0 - x} \approx \frac{x^2}{1.0}$$

$x = [OH^-] = 2.4 \times 10^{-5}\ M$; assumptions good

pOH = 4.62; pH = 14.00 – 4.62 = 9.38; $[H^+] = 10^{-9.38} = 4.2 \times 10^{-10}\ M$

135. a. $CH_3NH_3Cl \rightarrow CH_3NH_3^+ + Cl^-$: $CH_3NH_3^+$ is a weak acid. Cl^- is the conjugate base of a strong acid. Cl^- has no basic (or acidic) properties.

$$CH_3NH_3^+ \rightleftharpoons CH_3NH_2 + H^+ \quad K_a = \frac{[CH_3NH_2][H^+]}{[CH_3NH_3^+]} = \frac{K_w}{K_b} = \frac{1.00 \times 10^{-14}}{4.38 \times 10^{-4}} = 2.28 \times 10^{-11}$$

	$CH_3NH_3^+$	$\rightleftharpoons$	CH_3NH_2	+	H^+
Initial	0.10 *M*		0		~0
	x mol/L $CH_3NH_3^+$ dissociates to reach equilibrium				
Change	$-x$	$\rightarrow$	$+x$		$+x$
Equil.	$0.10 - x$		x		x

$$K_a = 2.28 \times 10^{-11} = \frac{x^2}{0.10 - x} \approx \frac{x^2}{0.10} \quad \text{(assuming } x << 0.10\text{)}$$

$x = [H^+] = 1.5 \times 10^{-6}\ M$; pH = 5.82; assumptions good.

b. $NaCN \rightarrow Na^+ + CN^-$: CN^- is a weak base. Na^+ has no acidic (or basic) properties.

$$CN^- + H_2O \rightleftharpoons HCN + OH^- \quad K_b = \frac{K_w}{K_a} = \frac{1.0 \times 10^{-14}}{6.2 \times 10^{-10}}$$

	CN^- + H_2O	$\rightleftharpoons$	HCN	+	OH^-	
Initial	0.050 *M*		0		~0	$K_b = 1.6 \times 10^{-5}$
	x mol/L CN^- reacts with H_2O to reach equilibrium					
Change	$-x$	$\rightarrow$	$+x$		$+x$	
Equil.	$0.050 - x$		x		x	

$$K_b = 1.6 \times 10^{-5} = \frac{[HCN][OH^-]}{[CN^-]} = \frac{x^2}{0.050 - x} \approx \frac{x^2}{0.050}$$

$x = [OH^-] = 8.9 \times 10^{-4}\ M$; pOH = 3.05; pH = 10.95; assumptions good.

137. $NaN_3 \rightarrow Na^+ + N_3^-$; azide ($N_3^-$) is a weak base because it is the conjugate base of a weak acid. All conjugate bases of weak acids are weak bases ($K_w < K_b < 1$). Ignore Na^+.

$$N_3^- + H_2O \rightleftharpoons HN_3 + OH^- \quad K_b = \frac{K_w}{K_a} = \frac{1.0 \times 10^{-14}}{1.9 \times 10^{-5}} = 5.3 \times 10^{-10}$$

	N_3^- + H_2O	$\rightleftharpoons$	HN_3	+	OH^-
Initial	0.010 *M*		0		~0
	x mol/L of N_3^- reacts with H_2O to reach equilibrium				
Change	$-x$	$\rightarrow$	$+x$		$+x$
Equil.	$0.010 - x$		x		x

$K_b = \dfrac{[HN_3][OH^-]}{[N_3^-]}$, $5.3 \times 10^{-10} = \dfrac{x^2}{0.010 - x} \approx \dfrac{x^2}{0.010}$ (assuming $x << 0.010$)

$x = [OH^-] = 2.3 \times 10^{-6}$ *M*; $[H^+] = \dfrac{1.0 \times 10^{-14}}{2.3 \times 10^{-6}} = 4.3 \times 10^{-9}$ *M;* assumptions good.

$[HN_3] = [OH^-] = 2.3 \times 10^{-6}$ *M*; $[Na^+] = 0.010$ *M*; $[N_3^-] = 0.010 - 2.3 \times 10^{-6} = 0.010$ *M*

139. All these salts contain Na^+, which has no acidic/basic properties, and a conjugate base of a weak acid (except for NaCl, where Cl^- is a neutral species). All conjugate bases of weak acids are weak bases since K_b values for these species are between K_w and 1. To identify the species, we will use the data given to determine the K_b value for the weak conjugate base. From the K_b value and data in Table 14.2 of the text, we can identify the conjugate base present by calculating the K_a value for the weak acid. We will use A^- as an abbreviation for the weak conjugate base.

	A^- + H_2O	$\rightleftharpoons$	HA	+	OH^-
Initial	0.100 mol/1.00 L		0		~0
	x mol/L A^- reacts with H_2O to reach equilibrium				
Change	$-x$	→	$+x$		$+x$
Equil.	$0.100 - x$		x		x

$K_b = \dfrac{[HA][OH^-]}{[A^-]} = \dfrac{x^2}{0.100 - x}$; from the problem, pH = 8.07:

pOH = 14.00 – 8.07 = 5.93; $[OH^-] = x = 10^{-5.93} = 1.2 \times 10^{-6}$ *M*

$K_b = \dfrac{(1.2 \times 10^{-6})^2}{0.100 - (1.2 \times 10^{-6})} = 1.4 \times 10^{-11} = K_b$ value for the conjugate base of a weak acid.

The K_a value for the weak acid equals K_w/K_b: $K_a = \dfrac{1.0 \times 10^{-14}}{1.4 \times 10^{-11}} = 7.1 \times 10^{-4}$

From Table 14.2 of the text, this K_a value is closest to HF. Therefore, the unknown salt is NaF.

141. B^- is a weak base. Use the weak base data to determine K_b for B^-.

	B^- + H_2O	$\rightleftharpoons$	HB	+	OH^-
Initial	0.050 *M*		0		~0
Equil.	$0.050 - x$		x		x

From pH = 9.00: pOH = 5.00, $[OH^-] = 10^{-5.00} = 1.0 \times 10^{-5}$ *M* = *x*.

$K_b = \dfrac{[HB][OH^-]}{[B^-]} = \dfrac{x^2}{0.050 - x} = \dfrac{(1.0 \times 10^{-5})^2}{0.050 - (1.0 \times 10^{-5})} = 2.0 \times 10^{-9}$

Because B^- is a weak base, HB will be a weak acid. Solve the weak acid problem.

	HB	⇌	H^+	+	B^-
Initial	0.010 M		~0		0
Equil.	$0.010 - x$		x		x

$$K_a = \frac{K_w}{K_b} = \frac{1.0 \times 10^{-14}}{2.0 \times 10^{-9}},\quad 5.0 \times 10^{-6} = \frac{x^2}{0.010 - x} \approx \frac{x^2}{0.010}$$

$x = [H^+] = 2.2 \times 10^{-4}\ M$; pH = 3.66; assumptions good.

143. Major species present: $Al(H_2O)_6^{3+}$ ($K_a = 1.4 \times 10^{-5}$), NO_3^- (neutral), and H_2O ($K_w = 1.0 \times 10^{-14}$); $Al(H_2O)_6^{3+}$ is a stronger acid than water, so it will be the dominant H^+ producer.

	$Al(H_2O)_6^{3+}$	⇌	$Al(H_2O)_5(OH)^{2+}$	+	H^+
Initial	0.050 M		0		~0
	x mol/L $Al(H_2O)_6^{3+}$ dissociates to reach equilibrium				
Change	$-x$	→	$+x$		$+x$
Equil.	$0.050 - x$		x		x

$$K_a = 1.4 \times 10^{-5} = \frac{[Al(H_2O)_5(OH)^{2+}][H^+]}{[Al(H_2O)_6^{3+}]} = \frac{x^2}{0.050 - x} \approx \frac{x^2}{0.050}$$

$x = 8.4 \times 10^{-4}\ M = [H^+]$; pH = $-\log(8.4 \times 10^{-4}) = 3.08$; assumptions good.

145. Reference Table 14.6 of the text and the solution to Exercise 129 for some generalizations on acid-base properties of salts.

a. $NaNO_3 \rightarrow Na^+ + NO_3^-$ neutral; neither species has any acidic/basic properties.

b. $NaNO_2 \rightarrow Na^+ + NO_2^-$ basic; NO_2^- is a weak base, and Na^+ has no effect on pH.

$$NO_2^- + H_2O \rightleftharpoons HNO_2 + OH^- \quad K_b = \frac{K_w}{K_{a,\,HNO_2}} = \frac{1.0 \times 10^{-14}}{4.0 \times 10^{-4}} = 2.5 \times 10^{-11}$$

c. $C_5H_5NHClO_4 \rightarrow C_5H_5NH^+ + ClO_4^-$ acidic; $C_5H_5NH^+$ is a weak acid, and ClO_4^- has no effect on pH.

$$C_5H_5NH^+ \rightleftharpoons H^+ + C_5H_5N \quad K_a = \frac{K_w}{K_{b,\,C_5H_5N}} = \frac{1.0 \times 10^{-14}}{1.7 \times 10^{-9}} = 5.9 \times 10^{-6}$$

d. $NH_4NO_2 \rightarrow NH_4^+ + NO_2^-$ acidic; NH_4^+ is a weak acid ($K_a = 5.6 \times 10^{-10}$), and NO_2^- is a weak base ($K_b = 2.5 \times 10^{-11}$). Because $K_{a,\,NH_4^+} > K_{b,\,NO_2^-}$, the solution is acidic.

$NH_4^+ \rightleftharpoons H^+ + NH_3$ $K_a = 5.6 \times 10^{-10}$; $NO_2^- + H_2O \rightleftharpoons HNO_2 + OH^-$ $K_b = 2.5 \times 10^{-11}$

e. $KOCl \rightarrow K^+ + OCl^-$ basic; OCl^- is a weak base, and K^+ has no effect on pH.

$$OCl^- + H_2O \rightleftharpoons HOCl + OH^- \quad K_b = \frac{K_w}{K_{a,\,HOCl}} = \frac{1.0 \times 10^{-14}}{3.5 \times 10^{-8}} = 2.9 \times 10^{-7}$$

f. $NH_4OCl \rightarrow NH_4^+ + OCl^-$ basic; NH_4^+ is a weak acid, and OCl^- is a weak base. Because $K_{b, OCl^-} > K_{a, NH_4^+}$, the solution is basic.

$NH_4^+ \rightleftharpoons NH_3 + H^+$ $K_a = 5.6 \times 10^{-10}$; $OCl^- + H_2O \rightleftharpoons HOCl + OH^-$ $K_b = 2.9 \times 10^{-7}$

Relationships Between Structure and Strengths of Acids and Bases

147. a. $HIO_3 < HBrO_3$; as the electronegativity of the central atom increases, acid strength increases.

b. $HNO_2 < HNO_3$; as the number of oxygen atoms attached to the central nitrogen atom increases, acid strength increases.

c. $HOI < HOCl$; same reasoning as in a.

d. $H_3PO_3 < H_3PO_4$; same reasoning as in b.

149. a. $H_2O < H_2S < H_2Se$; as the strength of the H–X bond decreases, acid strength increases.

b. $CH_3CO_2H < FCH_2CO_2H < F_2CHCO_2H < F_3CCO_2H$; as the electronegativity of neighboring atoms increases, acid strength increases.

c. $NH_4^+ < HONH_3^+$; same reason as in b.

d. $NH_4^+ < PH_4^+$; same reason as in a.

151. In general, metal oxides form basic solutions when dissolved in water, and nonmetal oxides form acidic solutions in water.

a. Basic; $CaO(s) + H_2O(l) \rightarrow Ca(OH)_2(aq)$; $Ca(OH)_2$ is a strong base.

b. Acidic; $SO_2(g) + H_2O(l) \rightarrow H_2SO_3(aq)$; H_2SO_3 is a weak diprotic acid.

c. Acidic; $Cl_2O(g) + H_2O(l) \rightarrow 2\ HOCl(aq)$; HOCl is a weak acid.

Lewis Acids and Bases

153. A Lewis base is an electron pair donor, and a Lewis acid is an electron pair acceptor.

a. $B(OH)_3$, acid; H_2O, base b. Ag^+, acid; NH_3, base c. BF_3, acid; F^-, base

155. $Al(OH)_3(s) + 3\ H^+(aq) \rightarrow Al^{3+}(aq) + 3\ H_2O(l)$ (Brønsted-Lowry base, H^+ acceptor)

$Al(OH)_3(s) + OH^-(aq) \rightarrow Al(OH)_4^-(aq)$ (Lewis acid, electron pair acceptor)

157. Fe^{3+} should be the stronger Lewis acid. Fe^{3+} is smaller and has a greater positive charge. Because of this, Fe^{3+} will be more strongly attracted to lone pairs of electrons as compared to Fe^{2+}.

ChemWork Problems

159. $H_2O(l) \rightleftharpoons H^+(aq) + OH^-(aq) \quad K_w = [H^+][OH^-]$

In a neutral solution, $[H^+] = [OH^-]$, so answer c is true. The value of K_w depends on temperature. At 25°C, $K_w = 1.0 \times 10^{-14}$, which gives a pH = pOH = 7.00. However, at different temperatures, the value of K_w is different, giving different pH values for neutral solutions. For answer b, the pH of the solution when moles of base = moles of acid depends on the identity of the acid and base; it can produce a neutral solution, but it certainly doesn't have to be neutral.

161. At pH = 2.000, $[H^+] = 10^{-2.000} = 1.00 \times 10^{-2}\ M$

At pH = 4.000, $[H^+] = 10^{-4.000} = 1.00 \times 10^{-4}\ M$

$$\text{Moles } H^+ \text{ present} = 0.0100\text{ L} \times \frac{0.0100\text{ mol } H^+}{\text{L}} = 1.00 \times 10^{-4}\text{ mol } H^+$$

Let V = total volume of solution at pH = 4.000:

$$1.00 \times 10^{-4}\text{ mol/L} = \frac{1.00 \times 10^{-4}\text{ mol } H^+}{V}, \quad V = 1.00\text{ L}$$

Volume of water added = 1.00 L – 0.0100 L = 0.99 L = 990 mL

163. $K_w = [H^+][OH^-]$; pH + pOH = 14.00; $HA(aq) \rightleftharpoons H^+(aq) + A^-(aq) \quad K_a = \dfrac{[H^+][A^-]}{[HA]}$

As acid strength increase, more of the acid dissociates to reach equilibrium. The results in more H^+ being produce, a larger value for K_a, and the solution becoming more acidic.

a. $[H^+]$ will increase.

b. pH will decrease as solution becomes more acidic.

c. $[OH^-]$ will decrease.

d. pOH will increase.

e. K_a will increase.

165. HA is a stronger weak acid than HB since the pH of the 1.0 *M* acid solution is lower. This means that K_a for HA is greater than K_a for HB. In general, the weaker the acid, the stronger the conjugate base. So, B^- will be a better base than A^-. Answers a and d are true. For answer b, if HA was a strong acid, then the pH would be 0.00 [= –log(1.0)]. And for answer c, the most acidic solution has the highest $[H^+]$; this would be the solution with HA.

167. The light bulb is bright because a strong electrolyte is present; that is, a solute is present that dissolves to produce a lot of ions in solution. The pH meter value of 4.6 indicates that a weak acid is present. (If a strong acid were present, the pH would be close to zero.) Of the possible substances, only HCl (strong acid), NaOH (strong base), and NH_4Cl are strong electrolytes. Of these three substances, only NH_4Cl contains a weak acid (the HCl solution would have a pH close to zero, and the NaOH solution would have a pH close to 14.0). NH_4Cl dissociates into NH_4^+ and Cl^- ions when dissolved in water. Cl^- is the conjugate base of a strong acid, so it has no basic (or acidic properties) in water. NH_4^+, however, is the conjugate acid of the weak base NH_3, so NH_4^+ is a weak acid and would produce a solution with a pH = 4.6 when the concentration is ~1.0 *M*. NH_4Cl is the solute.

169. When a patient's breathing becomes too slow or weak, not enough $CO_2(g)$ is exhaled causing a buildup of dissolved CO_2 in the blood. From the equilibrium in Exercise 168, as the CO_2 concentration in the blood builds up, the equilibrium is shifted to the right which results in more H^+ in the blood and a lowering of the blood pH.

171. a. In the lungs there is a lot of O_2, and the equilibrium favors $Hb(O_2)_4$. In the cells there is a lower concentration of O_2, and the equilibrium favors HbH_4^{4+}.

b. CO_2 is a weak acid, $CO_2 + H_2O \rightleftharpoons HCO_3^- + H^+$. Removing CO_2 essentially decreases H^+, which causes the hemoglobin reaction to shift right. $Hb(O_2)_4$ is then favored, and O_2 is not released by hemoglobin in the cells. Breathing into a paper bag increases CO_2 in the blood, thus increasing $[H^+]$, which shifts the hemoglobin reaction left.

c. CO_2 builds up in the blood, and it becomes too acidic, driving the hemoglobin equilibrium to the left. Hemoglobin can't bind O_2 as strongly in the lungs. Bicarbonate ion acts as a base in water and neutralizes the excess acidity.

173.

	HBz	$\rightleftharpoons$	H^+	+	Bz^-
Initial	C		~0		0
	x mol/L HBz dissociates to reach equilibrium				
Change	$-x$	$\rightarrow$	$+x$		$+x$
Equil.	$C - x$		x		x

$HBz = C_6H_5CO_2H$

$C = [HBz]_0$ = concentration of HBz that dissolves to give saturated solution.

$$K_a = \frac{[H^+][Bz^-]}{[HBz]} = 6.4 \times 10^{-5} = \frac{x^2}{C - x}, \text{ where } x = [H^+]$$

$$6.4 \times 10^{-5} = \frac{[H^+]^2}{C - [H^+]}; \text{ pH} = 2.80; \ [H^+] = 10^{-2.80} = 1.6 \times 10^{-3}\ M$$

$$C - (1.6 \times 10^{-3}) = \frac{(1.6 \times 10^{-3})^2}{6.4 \times 10^{-5}} = 4.0 \times 10^{-2}$$

$$C = (4.0 \times 10^{-2}) + (1.6 \times 10^{-3}) = 4.2 \times 10^{-2}\ M$$

The molar solubility of $C_6H_5CO_2H$ is 4.2×10^{-2} mol/L.

175. For this problem we will abbreviate $CH_2{=}CHCO_2H$ as Hacr and $CH_2{=}CHCO_2^-$ as acr^-.

a. Solving the weak acid problem:

	Hacr	$\rightleftharpoons$	H^+	+	acr^-	$K_a = 5.6 \times 10^{-5}$
Initial	0.10 *M*		~0		0	
Equil.	$0.10 - x$		x		x	

$$\frac{x^2}{0.10 - x} = 5.6 \times 10^{-5} \approx \frac{x^2}{0.10}, \ x = [H^+] = 2.4 \times 10^{-3}\ M; \text{ pH} = 2.62; \text{ assumptions good.}$$

b. Percent dissociation $= \dfrac{[H^+]}{[Hacr]_0} \times 100 = \dfrac{2.4 \times 10^{-3}}{0.10} \times 100 = 2.4\%$

c. acr^- is a weak base and the major source of OH^- in this solution.

$$acr^- + H_2O \rightleftharpoons Hacr + OH^- \qquad K_b = \frac{K_w}{K_a} = \frac{1.0 \times 10^{-14}}{5.6 \times 10^{-5}}$$

	acr^-	Hacr	OH^-	
Initial	0.050 *M*	0	~0	$K_b = 1.8 \times 10^{-10}$
Equil.	$0.050 - x$	x	x	

$$K_b = \frac{[Hacr][OH^-]}{[acr^-]}, \quad 1.8 \times 10^{-10} = \frac{x^2}{0.050 - x} \approx \frac{x^2}{0.050}$$

$x = [OH^-] = 3.0 \times 10^{-6}$ *M*; pOH = 5.52; pH = 8.48; assumptions good.

177. a. HA is a weak acid. Most of the acid is present as HA molecules; only one set of H^+ and A^- ions is present. In a strong acid, all the acid would be dissociated into H^+ and A^- ions.

b. This picture is the result of 1 out of 10 HA molecules dissociating.

$$\text{Percent dissociation} = \frac{1}{10} \times 100 = 10\% \text{ (an exact number)}$$

$$HA \rightleftharpoons H^+ + A^- \qquad K_a = \frac{[H^+][A^-]}{[HA]}$$

	HA	H^+	A^-
Initial	0.20 *M*	~0	0
	x mol/L HA dissociates to reach equilibrium		
Change	$-x$ →	$+x$	$+x$
Equil.	$0.20 - x$	x	x

$[H^+] = [A^-] = x = 0.10 \times 0.20\ M = 0.020\ M$; $[HA] = 0.20 - 0.020 = 0.18\ M$

$$K_a = \frac{(0.020)^2}{0.18} = 2.2 \times 10^{-3}$$

179. $$HONH_2 + H_2O \rightleftharpoons HONH_3^+ + OH^- \qquad K_b = 1.1 \times 10^{-8}$$

	$HONH_2$	$HONH_3^+$	OH^-	
Initial	I	0	~0	$I = [HONH_2]_0$
Equil.	$I - x$	x	x	

$$K_b = 1.1 \times 10^{-8} = \frac{x^2}{I - x}$$

From problem, pH = 10.00, so pOH = 4.00 and $x = [OH^-] = 1.0 \times 10^{-4}\ M$.

$$1.1 \times 10^{-8} = \frac{(1.0 \times 10^{-4})^2}{I - (1.0 \times 10^{-4})}, \quad I = 0.91\ M$$

$$\text{Mass } HONH_2 = 0.2500\ L \times \frac{0.91\ mol\ HONH_2}{L} \times \frac{33.03\ g\ HONH_2}{mol\ HONH_2} = 7.5\ g\ HONH_2$$

181. a.

	$Fe(H_2O)_6^{3+}$ + H_2O	$\rightleftharpoons$	$Fe(H_2O)_5(OH)^{2+}$	+	H_3O^+
Initial	0.10 *M*		0		~0
Equil.	$0.10 - x$		x		x

$$K_a = \frac{[Fe(H_2O)_5(OH)^{2+}][H_3O^+]}{[Fe(H_2O)_6^{3+}]},\ 6.0 \times 10^{-3} = \frac{x^2}{0.10 - x} \approx \frac{x^2}{0.10}$$

$x = 2.4 \times 10^{-2}$; assumption is poor (x is 24% of 0.10). Using successive approximations:

$$\frac{x^2}{0.10 - 0.024} = 6.0 \times 10^{-3},\ x = 0.021$$

$$\frac{x^2}{0.10 - 0.021} = 6.0 \times 10^{-3},\ x = 0.022;\ \frac{x^2}{0.10 - 0.022} = 6.0 \times 10^{-3},\ x = 0.022$$

$x = [H^+] = 0.022\ M$; pH = 1.66

b. Because of the lower charge, $Fe^{2+}(aq)$ will not be as strong an acid as $Fe^{3+}(aq)$. A solution of iron(II) nitrate will be less acidic (have a higher pH) than a solution with the same concentration of iron(III) nitrate.

183. a. $CaBr_2$ is a soluble ionic compound that dissolves in water to produce $Ca^{2+}(aq)$ and $Br^-(aq)$. Ca^{2+} has no acidic or basic properties. Br^- is the conjugate base of the strong acid HBr. Br^- has no basic (or acidic) properties. Therefore, a solution of $CaBr_2$ will be neutral because neither of the ions has any acidic or basic properties. The 1.2 *M* $CaBr_2$ solution has $[H^+] = [OH^-] = 1.0 \times 10^{-7}$ *M* and pH = pOH = 7.00.

b. $C_6H_5NH_3NO_3 \rightarrow C_6H_5NH_3^+ + NO_3^-$; $C_6H_5NH_3^+$ is the conjugate acid of the weak base $C_6H_5NH_2$ ($K_b = 3.8 \times 10^{-10}$). As is true for all conjugate acids of weak bases, $C_6H_5NH_3^+$ is a weak acid. NO_3^- has no basic (or acidic) properties. Ignore NO_3^-. Solving the weak acid problem:

	$C_6H_5NH_3^+$	$\rightleftharpoons$	$C_6H_5NH_2$	+	H^+
Initial	0.84 *M*		0		~0
	x mol/L $C_6H_5NH_3^+$ dissociates to reach equilibrium				
Change	$-x$	$\rightarrow$	$+x$		$+x$
Equil.	$0.84 - x$		x		x

$K_a = K_w/3.8 \times 10^{-10} = 2.6 \times 10^{-5}$

$$K_a = 2.6 \times 10^{-5} = \frac{[C_6H_5NH_2][H^+]}{[C_6H_5NH_3^+]} = \frac{x^2}{0.84 - x} \approx \frac{x^2}{0.84}\ \text{(assuming } x \ll 0.84\text{)}$$

$x = [H^+] = 4.7 \times 10^{-3}\ M$; pH = $-\log[H^+]$ = 2.33; assumptions good.

c. $KC_7H_5O_2$ is a soluble ionic compound that dissolves in water to produce $K^+(aq)$ and $C_7H_5O_2^-(aq)$. K^+ has no acidic (or basic) properties. $C_7H_5O_2^-$ is the conjugate base of the weak acid $HC_7H_5O_2$, and as is true for all conjugate bases of weak acids, $C_7H_5O_2^-$ is a weak base in water. We must solve an equilibrium problem in order to determine the amount of OH^- this weak base produces in water.

	$C_7H_5O_2^-$ + H_2O	$\rightleftharpoons$ $HC_7H_5O_2$	+ OH^-	$K_b = \frac{K_w}{K_{a,\,C_7H_5O_2}} = \frac{1.0 \times 10^{-14}}{6.4 \times 10^{-5}}$
Initial	0.57 M	0	~0	$K_b = 1.6 \times 10^{-10}$
	x mol/L of $C_7H_5O_2^-$ reacts with H_2O to reach equilibrium			
Change	$-x$ →	$+x$	$+x$	
Equil.	$0.57 - x$	x	x	

$$K_b = 1.6 \times 10^{-10} = \frac{[HC_7H_5O_2][OH^-]}{[C_7H_5O_2^-]},\ 1.6 \times 10^{-10} = \frac{x^2}{0.57 - x} \approx \frac{x^2}{0.57}$$

$x = [OH^-] = 9.5 \times 10^{-6}\ M$; assumptions good

pOH = $-\log[OH^-] = 5.02$; pH = 14.00 − 5.02 = 8.98

185. The solution is acidic from $HSO_4^- \rightleftharpoons H^+ + SO_4^{2-}$. Solving the weak acid problem:

	HSO_4^-	$\rightleftharpoons$ H^+	+ SO_4^{2-}	$K_a = 1.2 \times 10^{-2}$
Initial	0.10 M	~0	0	
Equil.	$0.10 - x$	x	x	

$$1.2 \times 10^{-2} = \frac{[H^+][SO_4^{2-}]}{[HSO_4^-]} = \frac{x^2}{0.10 - x} \approx \frac{x^2}{0.10},\ x = 0.035$$

Assumption is not good (x is 35% of 0.10). Using successive approximations:

$$\frac{x^2}{0.10 - x} = \frac{x^2}{0.10 - 0.035} = 1.2 \times 10^{-2},\ x = 0.028$$

$$\frac{x^2}{0.10 - 0.028} = 1.2 \times 10^{-2},\ x = 0.029;\quad \frac{x^2}{0.10 - 0.029} = 1.2 \times 10^{-2},\ x = 0.029$$

$x = [H^+] = 0.029\ M$; pH = 1.54

187. a. The initial concentrations are halved since equal volumes of the two solutions are mixed. Use the K_a reaction for $HC_2H_3O_2$ to determine the amount of H^+ produced from this weak acid.

	$HC_2H_3O_2$	$\rightleftharpoons$ H^+	+ $C_2H_3O_2^-$
Initial	0.100 M	$5.00 \times 10^{-4}\ M$	0
Equil.	$0.100 - x$	$5.00 \times 10^{-4} + x$	x

$$K_a = 1.8 \times 10^{-5} = \frac{x(5.00 \times 10^{-4} + x)}{0.100 - x} \approx \frac{x(5.00 \times 10^{-4})}{0.100}$$

$x = 3.6 \times 10^{-3}$; assumption is horrible. Using the quadratic formula:

$$x^2 + (5.18 \times 10^{-4})x - 1.8 \times 10^{-6} = 0$$

$x = 1.1 \times 10^{-3}\ M$; $[H^+] = 5.00 \times 10^{-4} + x = 1.6 \times 10^{-3}\ M$; pH = 2.80

b. $x = [C_2H_3O_2^-] = 1.1 \times 10^{-3}\ M$

189. The typical ICE table set-up to solve a weak acid problem is:

	HA	$\rightleftharpoons$	H^+	+	A^-	HA = weak acid
Initial	1.0 *M*		~0		0	
	x mol/L HA dissociates to reach equilibrium					
Change	$-x$	$\rightarrow$	$+x$		$+x$	
Equil.	$1.0 - x$		x		x	

$K_a = \dfrac{[H^+][A^-]}{[HA]} = \dfrac{x^2}{1.0 - x} \approx \dfrac{x^2}{1.0}$ when the 5% rule applies.

We want x to be less than or equal to 5.0% of 1.0 to follow the 5% rule:

$x = 0.050 \times 1.0\ M = 0.050\ M$; this gives: $K_a = \dfrac{x^2}{1.0} = \dfrac{(0.050)^2}{1.0} = 2.5 \times 10^{-3}$

When $K_a \leq 2.5 \times 10^{-3}$, a 1.0 *M* weak acid solution will follow the 5% rule.

191. For H_3PO_4, $K_{a_1} = 7.5 \times 10^{-3}$, $K_{a_2} = 6.2 \times 10^{-8}$, and $K_{a_3} = 4.8 \times 10^{-13}$. Because K_{a_1} is much larger than K_{a_2} and K_{a_3}, the dominant H^+ producer is H_3PO_4, and the H^+ contributed from $H_2PO_4^-$ and HPO_4^{2-} can be ignored. Solving the weak acid problem in the typical manner.

	H_3PO_4	$\rightleftharpoons$	$H_2PO_4^-$	+	H^+	$K_{a_1} = 7.5 \times 10^{-3}$
Initial	0.007 *M*		0		~0	
Equil.	$0.007 - x$		x		x	

$$K_{a_1} = 7.5 \times 10^{-3} = \frac{[H_2PO_4^-][H^+]}{[H_3PO_4]} = \frac{x^2}{0.007 - x} \approx \frac{x^2}{0.007}$$

$x = 7.2 \times 10^{-3}$; assumption is horrible because x is more than 100% of 0.007. We will use the quadratic equation to solve exactly and carry extra significant figures through the calculation.

$$7.5 \times 10^{-3} = \frac{x^2}{0.007 - x},\ x^2 = 5.25 \times 10^{-5} - (7.5 \times 10^{-3})x,\ x^2 + (7.5 \times 10^{-3})x - 5.25 \times 10^{-5} = 0$$

$$x = [H^+] = \frac{-7.5 \times 10^{-3} \pm [(7.5 \times 10^{-3})^2 - 4(1)(-5.25 \times 10^{-5})]^{1/2}}{2(1)} = 4.4 \times 10^{-3} = 4 \times 10^{-3}\ M$$

$pH = -\log(4 \times 10^{-3}) = 2.4$

Challenge Problems

193. Because this is a very dilute solution of NaOH, we must worry about the amount of OH^- donated from the autoionization of water.

$NaOH \rightarrow Na^+ + OH^-$

$H_2O \rightleftharpoons H^+ + OH^-\quad K_w = [H^+][OH^-] = 1.0 \times 10^{-14}$

This solution, like all solutions, must be charge balanced; that is, [positive charge] = [negative charge]. For this problem, the charge balance equation is:

$$[Na^+] + [H^+] = [OH^-], \text{ where } [Na^+] = 1.0 \times 10^{-7}\ M \text{ and } [H^+] = \frac{K_w}{[OH^-]}$$

Substituting into the charge balance equation:

$$1.0 \times 10^{-7} + \frac{1.0 \times 10^{-14}}{[OH^-]} = [OH^-],\quad [OH^-]^2 - (1.0 \times 10^{-7})[OH^-] - 1.0 \times 10^{-14} = 0$$

Using the quadratic formula to solve:

$$[OH^-] = \frac{-(-1.0 \times 10^{-7}) \pm [(-1.0 \times 10^{-7})^2 - 4(1)(-1.0 \times 10^{-14})]^{1/2}}{2(1)}$$

$$[OH^-] = 1.6 \times 10^{-7}\ M;\ pOH = -\log(1.6 \times 10^{-7}) = 6.80;\ pH = 7.20$$

195.

	HA	⇌	H^+	+	A^-	$K_a = 1.00 \times 10^{-6}$
Initial	C		~0		0	$C = [HA]_0$, for pH = 4.000,
Equil.	$C - 1.00 \times 10^{-4}$		1.00×10^{-4}		1.00×10^{-4}	$x = [H^+] = 1.00 \times 10^{-4}\ M$

$$K_a = \frac{(1.00 \times 10^{-4})^2}{(C - 1.00 \times 10^{-4})} = 1.00 \times 10^{-6};\ \text{solving: } C = 0.0101\ M$$

The solution initially contains 50.0×10^{-3} L × 0.0101 mol/L = 5.05×10^{-4} mol HA. We then dilute to a total volume V in liters. The resulting pH = 5.000, so $[H^+] = 1.00 \times 10^{-5}$. In the typical weak acid problem, $x = [H^+]$, so:

	HA	⇌	H^+	+	A^-
Initial	5.05×10^{-4} mol/V		~0		0
Equil.	$(5.05 \times 10^{-4}/V) - (1.00 \times 10^{-5})$		1.00×10^{-5}		1.00×10^{-5}

$$K_a = \frac{(1.00 \times 10^{-5})^2}{(5.05 \times 10^{-4}/V) - (1.00 \times 10^{-5})} = 1.00 \times 10^{-6}$$

$$1.00 \times 10^{-4} = (5.05 \times 10^{-4}/V) - 1.00 \times 10^{-5}$$

V = 4.59 L; 50.0 mL are present initially, so we need to add 4540 mL of water.

197. Major species present are H_2O, $C_5H_5NH^+$ [$K_a = K_w/K_{b,\,C_5H_5N} = (1.0 \times 10^{-14})/(1.7 \times 10^{-9}) = 5.9 \times 10^{-6}$], and F^- [$K_b = K_w/K_{a,\,HF} = (1.0 \times 10^{-14})/(7.2 \times 10^{-4}) = 1.4 \times 10^{-11}$]. The reaction to consider is the best acid present ($C_5H_5NH^+$) reacting with the best base present (F^-). Let's solve by first setting up an ICE table.

	$C_5H_5NH^+(aq)$	+	$F^-(aq)$	⇌	$C_5H_5N(aq)$	+	$HF(aq)$
Initial	0.200 M		0.200 M		0		0
Change	$-x$		$-x$	→	$+x$		$+x$
Equil.	$0.200 - x$		$0.200 - x$		x		x

$$K = K_{a,\,C_5H_5NH^+} \times \frac{1}{K_{a,\,HF}} = 5.9 \times 10^{-6} \times \frac{1}{7.2 \times 10^{-4}} = 8.2 \times 10^{-3}$$

$$K = \frac{[C_5H_5N][HF]}{[C_5H_5NH^+][F^-]},\quad 8.2 \times 10^{-3} = \frac{x^2}{(0.200 - x)^2}$$; taking the square root of both sides:

$$0.091 = \frac{x}{0.200 - x},\quad x = 0.018 - (0.091)x,\ x = 0.016\ M$$

From the setup to the problem, $x = [C_5H_5N] = [HF] = 0.016\ M$, and $0.200 - x = 0.200 - 0.016 = 0.184\ M = [C_5H_5NH^+] = [F^-]$. To solve for the $[H^+]$, we can use either the K_a equilibrium for $C_5H_5NH^+$ or the K_a equilibrium for HF. Using $C_5H_5NH^+$ data:

$$K_{a,\,C_5H_5NH^+} = 5.9 \times 10^{-6} = \frac{[C_5H_5N][H^+]}{[C_5H_5NH^+]} = \frac{(0.016)[H^+]}{0.184},\quad [H^+] = 6.8 \times 10^{-5}\ M$$

$$pH = -\log(6.8 \times 10^{-5}) = 4.17$$

As one would expect, because the K_a for the weak acid is larger than the K_b for the weak base, a solution of this salt should be acidic.

199. Because NH_3 is so concentrated, we need to calculate the OH^- contribution from the weak base NH_3.

	NH_3 + H_2O	$\rightleftharpoons$	NH_4^+	+	OH^-	$K_b = 1.8 \times 10^{-5}$
Initial	$15.0\ M$		0		$0.0100\ M$	(Assume no volume change.)
Equil.	$15.0 - x$		x		$0.0100 + x$	

$$K_b = 1.8 \times 10^{-5} = \frac{x(0.0100 + x)}{15.0 - x} \approx \frac{x(0.0100)}{15.0},\quad x = 0.027;$$ assumption is horrible

(x is 270% of 0.0100).

Using the quadratic formula:

$$(1.8 \times 10^{-5})(15.0 - x) = (0.0100)x + x^2,\quad x^2 + (0.0100)x - 2.7 \times 10^{-4} = 0$$

$$x = 1.2 \times 10^{-2}\ M,\quad [OH^-] = (1.2 \times 10^{-2}) + 0.0100 = 0.022\ M$$

201. $$1.000\ L \times \frac{1.00 \times 10^{-4}\ \text{mol HA}}{L} = 1.00 \times 10^{-4}\ \text{mol HA}$$

25.0% dissociation gives:

$$\text{moles } H^+ = 0.250 \times (1.00 \times 10^{-4}) = 2.50 \times 10^{-5}\ \text{mol}$$

$$\text{moles } A^- = 0.250 \times (1.00 \times 10^{-4}) = 2.50 \times 10^{-5}\ \text{mol}$$

$$\text{moles HA} = 0.750 \times (1.00 \times 10^{-4}) = 7.50 \times 10^{-5}\ \text{mol}$$

$$1.00 \times 10^{-4} = K_a = \frac{[H^+][A^-]}{[HA]} = \frac{\left(\frac{2.50 \times 10^{-5}}{V}\right)\left(\frac{2.50 \times 10^{-5}}{V}\right)}{\left(\frac{7.50 \times 10^{-5}}{V}\right)}$$

$$1.00 \times 10^{-4} = \frac{(2.50 \times 10^{-5})^2}{(7.50 \times 10^{-5})(V)}, \quad V = \frac{(2.50 \times 10^{-5})^2}{(1.00 \times 10^{-4})(7.50 \times 10^{-5})} = 0.0833 \text{ L} = 83.3 \text{ mL}$$

The volume goes from 1000. mL to 83.3 mL, so 917 mL of water evaporated.

203. PO_4^{3-} is the conjugate base of HPO_4^{2-}. The K_a value for HPO_4^{2-} is $K_{a_3} = 4.8 \times 10^{-13}$.

$$PO_4^{3-}(aq) + H_2O(l) \rightleftharpoons HPO_4^{2-}(aq) + OH^-(aq) \quad K_b = \frac{K_w}{K_{a_3}} = \frac{1.0 \times 10^{-14}}{4.8 \times 10^{-13}} = 0.021$$

HPO_4^{2-} is the conjugate base of $H_2PO_4^-$ ($K_{a_2} = 6.2 \times 10^{-8}$).

$$HPO_4^{2-} + H_2O \rightleftharpoons H_2PO_4^- + OH^- \quad K_b = \frac{K_w}{K_{a_1}} = \frac{1.0 \times 10^{-14}}{6.2 \times 10^{-8}} = 1.6 \times 10^{-7}$$

$H_2PO_4^-$ is the conjugate base of H_3PO_4 ($K_{a_1} = 7.5 \times 10^{-3}$).

$$H_2PO_4^- + H_2O \rightleftharpoons H_3PO_4 + OH^- \quad K_b = \frac{K_w}{K_{a_1}} = \frac{1.0 \times 10^{-14}}{7.5 \times 10^{-3}} = 1.3 \times 10^{-12}$$

From the K_b values, PO_4^{3-} is the strongest base. This is expected because PO_4^{3-} is the conjugate base of the weakest acid (HPO_4^{2-}).

205. a. $NH_4(HCO_3) \rightarrow NH_4^+ + HCO_3^-$

$$K_{a, NH_4^+} = \frac{1.0 \times 10^{-14}}{1.8 \times 10^{-5}} = 5.6 \times 10^{-10}; \quad K_{b, HCO_3^-} = \frac{K_w}{K_{a_1}} = \frac{1.0 \times 10^{-14}}{4.3 \times 10^{-7}} = 2.3 \times 10^{-8}$$

The solution is basic because HCO_3^- is a stronger base than NH_4^+ is as an acid. The acidic properties of HCO_3^- were ignored because K_{a_2} is very small (4.8×10^{-11}).

b. $NaH_2PO_4 \rightarrow Na^+ + H_2PO_4^-$; ignore Na^+.

$$K_{a_2, H_2PO_4^-} = 6.2 \times 10^{-8}; \quad K_{b, H_2PO_4^-} = \frac{K_w}{K_{a_1}} = \frac{1.0 \times 10^{-14}}{7.5 \times 10^{-3}} = 1.3 \times 10^{-12}$$

Solution is acidic because $K_a > K_b$.

c. $Na_2HPO_4 \rightarrow 2\ Na^+ + HPO_4^{2-}$; ignore Na^+.

$$K_{a_3, HPO_4^{2-}} = 4.8 \times 10^{-13}; \quad K_{b, HPO_4^{2-}} = \frac{K_w}{K_{a_2}} = \frac{1.0 \times 10^{-14}}{6.2 \times 10^{-8}} = 1.6 \times 10^{-7}$$

Solution is basic because $K_b > K_a$.

d. $NH_4(H_2PO_4) \rightarrow NH_4^+ + H_2PO_4^-$

NH_4^+ is a weak acid, and $H_2PO_4^-$ is also acidic (see part b). Solution with both ions present will be acidic.

e. $NH_4(HCO_2) \rightarrow NH_4^+ + HCO_2^-$; from Appendix 5, $K_{a,\,HCO_2H} = 1.8 \times 10^{-4}$.

$$K_{a,\,NH_4^+} = 5.6 \times 10^{-10};\ K_{b,\,HCO_2^-} = \frac{K_w}{K_a} = \frac{1.0 \times 10^{-14}}{1.8 \times 10^{-4}} = 5.6 \times 10^{-11}$$

Solution is acidic because NH_4^+ is a stronger acid than HCO_2^- is as a base.

207. $$\text{Molality} = m = \frac{0.100\text{ g} \times \frac{1\text{ mol}}{100.0\text{ g}}}{0.5000\text{ kg}} = 2.00 \times 10^{-3}\text{ mol/kg} \approx 2.00 \times 10^{-3}\text{ mol/L}$$

$\Delta T_f = iK_f m$, $0.0056°C = i(1.86°C/\text{molal})(2.00 \times 10^{-3}\text{ molal})$, $i = 1.5$

If i = 1.0, the percent dissociation of the acid = 0%, and if i = 2.0, the percent dissociation of the acid = 100%. Because i = 1.5, the weak acid is 50.% dissociated.

$$HA \rightleftharpoons H^+ + A^- \qquad K_a = \frac{[H^+][A^-]}{[HA]}$$

Because the weak acid is 50.% dissociated:

$$[H^+] = [A^-] = [HA]_0 \times 0.50 = 2.00 \times 10^{-3}\,M \times 0.50 = 1.0 \times 10^{-3}\,M$$

$$[HA] = [HA]_0 - \text{amount HA reacted} = 2.00 \times 10^{-3}\,M - 1.0 \times 10^{-3}\,M = 1.0 \times 10^{-3}\,M$$

$$K_a = \frac{[H^+][A^-]}{[HA]} = \frac{(1.0 \times 10^{-3})(1.0 \times 10^{-3})}{1.0 \times 10^{-3}} = 1.0 \times 10^{-3}$$

209. $$\text{Molar mass} = \frac{dRT}{P} = \frac{5.11\text{ g/L} \times \frac{0.08206\text{ L atm}}{\text{K mol}} \times 298\text{ K}}{1.00\text{ atm}} = 125\text{ g/mol}$$

$$[HA]_0 = \frac{1.50\text{ g} \times \frac{1\text{ mol}}{125\text{ g}}}{0.100\text{ L}} = 0.120\,M;\ \text{pH} = 1.80,\ [H^+] = 10^{-1.80} = 1.6 \times 10^{-2}\,M$$

	HA	⇌	H^+	+	A^-	
Initial	$0.120\,M$		~0		0	
Equil.	$0.120 - x$		x		x	where $x = [H^+] = 1.6 \times 10^{-2}\,M$

$$K_a = \frac{[H^+][A^-]}{[HA]} = \frac{(1.6 \times 10^{-2})^2}{0.120 - 0.016} = 2.5 \times 10^{-3}$$

CHAPTER 15

ACID-BASE EQUILIBRIA

Questions

13. When an acid dissociates, ions are produced. The common ion effect is observed when one of the product ions in a particular equilibrium is added from an outside source. For a weak acid dissociating to its conjugate base and H^+, the common ion would be the conjugate base; this would be added by dissolving a soluble salt of the conjugate base into the acid solution. The presence of the conjugate base from an outside source shifts the equilibrium to the left so less acid dissociates.

15. The more weak acid and conjugate base present, the more H^+ and/or OH^- that can be absorbed by the buffer without significant pH change. When the concentrations of weak acid and conjugate base are equal (so that pH = pK_a), the buffer system is equally efficient at absorbing either H^+ or OH^-. If the buffer is overloaded with weak acid or with conjugate base, then the buffer is not equally efficient at absorbing either H^+ or OH^-.

17. If a strong acid and/or strong base is/are present, react them first, react the best base with the best acid present, and assume the reaction goes to completion. After the strong acid and/or strong base has reacted to completion, take stock of what remains in solution to determine what to do next. Strong acids and strong bases are great at what they do, either donating a proton or accepting a proton. So always react the strong acid and/or strong base first.

19. Only statement e is false. At the halfway point to equivalence for a weak base-strong acid titration, a buffer solution consisting of equal amounts of the weak base and its conjugate acid are present. At the halfway point, pH = pK_a for the conjugate acid. If the K_a value for the conjugate acid is greater than 1×10^{-7}, then the pH will be acidic at the halfway point to equivalence. Weak bases with K_b values less than 1×10^{-7} have conjugate acids with K_a values greater than 1×10^{-7}; these weak bases have an acidic pH at the halfway point ($K_w = K_a \times K_b$ for conjugate acid-base pairs).

21. a. Let's call the acid HA, which is a weak acid. When HA is present in the beakers, it exists in the undissociated form, making it a weak acid. A strong acid would exist as separate H^+ and A^- ions.

    b. Beaker a contains 4 HA molecules and 2 A^- ions, beaker b contains 6 A^- ions, beaker c contains 6 HA molecules, beaker d contains 6 A^- and 6 OH^- ions, and beaker e contains 3 HA molecules and 3 A^- ions. $HA + OH^- \rightarrow A^- + H_2O$; this is the neutralization reaction that occurs when OH^- is added. We start off the titration with a beaker full of weak acid (beaker c). When some OH^- is added, we convert some weak acid HA into its conjugate base A^- (beaker a). At the halfway point to equivalence, we have converted exactly one-half of the initial amount of acid present into its conjugate base (beaker e). We finally reach the equivalence point when we have added just enough OH^- to convert all of the acid present initially into its conjugate base (beaker b). Past the equivalence point, we have added an excess of OH^-, so we have excess OH^- present as well as the conjugate base of the acid produced from the neutralization reaction (beaker d). The order of the beakers from start to finish is:

beaker c $\rightarrow$ beaker a $\rightarrow$ beaker e $\rightarrow$ beaker b $\rightarrow$ beaker d

c. pH = pK_a when a buffer solution is present that has equal concentrations of the weak acid and conjugate base. This is beaker e.

d. The equivalence point is when just enough OH^- has been added to exactly react with all the acid present initially. This is beaker b.

e. Past the equivalence point, the pH is dictated by the concentration of excess OH^- added from the strong base. We can ignore the amount of hydroxide added by the weak conjugate base that is also present. This is beaker d.

23. The three key points to emphasize in your sketch are the initial pH, the pH at the halfway point to equivalence, and the pH at the equivalence point. For all the weak bases titrated, pH = pK_a at the halfway point to equivalence (50.0 mL HCl added) because [weak base] = [conjugate acid] at this point. Here, the weak base with $K_b = 1 \times 10^{-5}$ has a conjugate acid with $K_a = 1 \times 10^{-9}$, so pH = 9.0 at the halfway point. The weak base with $K_b = 1 \times 10^{-10}$ has a pH = 4.0 at the halfway point to equivalence. For the initial pH, the strong base has the highest pH (most basic), whereas the weakest base has the lowest pH (least basic). At the equivalence point (100.0 mL HCl added), the strong base titration has pH = 7.0. The weak bases titrated have acidic pH's because the conjugate acids of the weak bases titrated are the major species present. The weakest base has the strongest conjugate acid so its pH will be lowest (most acidic) at the equivalence point.

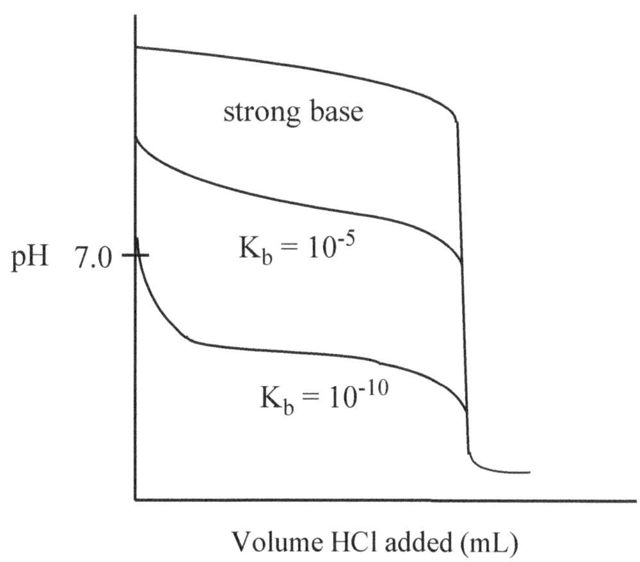

25. H_3AsO_4 is a triprotic acid, so it has three equivalence points. The initial titration reaction is $H_3AsO_4 + OH^- \rightarrow H_2AsO_4^- + H_2O$. Because the concentration of the H_3AsO_4 and NaOH solutions are equal, it will take 100.0 mL of NaOH to reach the first equivalence point. The second equivalence point where $H_2AsO_4^-$ is titrated will occur at 200.0 mL of NaOH added, and the third equivalence point where $HAsO_4^{2-}$ is titrated will occur at 300.0 mL of NaOH added.

At 50.0 mL of NaOH added, this is the first halfway point to equivalence where H_3AsO_4 and $H_2AsO_4^-$ are the major species present (along with H_2O). At this point, we have a buffer solution where $[H_3AsO_4] = [H_2AsO_4^-]$ and $pH = pK_{a1} = -\log(5.5 \times 10^{-3}) = 2.26$. At 150.0 mL of NaOH added, this is the second halfway point to equivalence where $H_2AsO_4^-$ and $HAsO_4^{2-}$ are the major species present (along with H_2O). At this point, we have a buffer solution where $[H_2AsO_4^-] = [HAsO_4^{2-}]$ and $pH = pK_{a2} = -\log(1.7 \times 10^{-7}) = 6.77$. At the third halfway point to equivalence, $[HAsO_4^{2-}] = [AsO_4^{3-}]$ and $pH = pK_{a3}$. This third halfway point will be at 250.0 mL of NaOH added.

Exercises

Buffers

27. Only the third (lower) beaker represents a buffer solution. A weak acid and its conjugate base must both be present in large quantities to have a buffer solution. This is only the case in the third beaker. The first beaker represents a beaker full of strong acid which is 100% dissociated. The second beaker represents a weak acid solution. In a weak acid solution, only a small fraction of the acid is dissociated. In this representation, 1/10 of the weak acid has dissociated. The only B^- present in this beaker is from the dissociation of the weak acid. A buffer solution has B^- added from another source.

29. When strong acid or strong base is added to a bicarbonate-carbonate mixture, the strong acid(base) is neutralized. The reaction goes to completion, resulting in the strong acid(base) being replaced with a weak acid(base), resulting in a new buffer solution. The reactions are:

$$H^+(aq) + CO_3^{2-}(aq) \rightarrow HCO_3^-(aq);\quad OH^- + HCO_3^-(aq) \rightarrow CO_3^{2-}(aq) + H_2O(l)$$

31. a. This is a weak acid problem. Let $HC_3H_5O_2$ = HOPr and $C_3H_5O_2^-$ = OPr^-.

	HOPr(aq)	⇌	H^+(aq)	+	OPr^-(aq)	$K_a = 1.3 \times 10^{-5}$
Initial	0.100 *M*		~0		0	
	x mol/L HOPr dissociates to reach equilibrium					
Change	$-x$	→	$+x$		$+x$	
Equil.	$0.100 - x$		x		x	

$$K_a = 1.3 \times 10^{-5} = \frac{[H^+][OPr^-]}{[HOPr]} = \frac{x^2}{0.100 - x} \approx \frac{x^2}{0.100}$$

$x = [H^+] = 1.1 \times 10^{-3}\ M$; pH = 2.96; assumptions good by the 5% rule.

b. This is a weak base problem.

	$OPr^-(aq)$ + $H_2O(l)$	⇌	$HOPr(aq)$	+ $OH^-(aq)$	$K_b = \frac{K_w}{K_a} = 7.7 \times 10^{-10}$
Initial	0.100 M		0	~0	
	x mol/L OPr^- reacts with H_2O to reach equilibrium				
Change	$-x$	→	$+x$	$+x$	
Equil.	$0.100 - x$		x	x	

$$K_b = 7.7 \times 10^{-10} = \frac{[HOPr][OH^-]}{[OPr^-]} = \frac{x^2}{0.100 - x} \approx \frac{x^2}{0.100}$$

$x = [OH^-] = 8.8 \times 10^{-6}\ M$; pOH = 5.06; pH = 8.94; assumptions good.

c. Pure H_2O, $[H^+] = [OH^-] = 1.0 \times 10^{-7}\ M$; pH = 7.00

d. This solution contains a weak acid and its conjugate base. This is a buffer solution. We will solve for the pH through the weak acid equilibrium reaction.

	$HOPr(aq)$	⇌	$H^+(aq)$	+ $OPr^-(aq)$	$K_a = 1.3 \times 10^{-5}$
Initial	0.100 M		~0	0.100 M	
	x mol/L HOPr dissociates to reach equilibrium				
Change	$-x$	→	$+x$	$+x$	
Equil.	$0.100 - x$		x	$0.100 + x$	

$$1.3 \times 10^{-5} = \frac{(0.100 + x)(x)}{0.100 - x} \approx \frac{(0.100)(x)}{0.100} = x = [H^+]$$

$[H^+] = 1.3 \times 10^{-5}\ M$; pH = 4.89; assumptions good.

Alternately, we can use the Henderson-Hasselbalch equation to calculate the pH of buffer solutions.

$$pH = pK_a + \log\frac{[base]}{[acid]} = pK_a + \log\left(\frac{0.100}{0.100}\right) = pK_a = -\log(1.3 \times 10^{-5}) = 4.89$$

The Henderson-Hasselbalch equation will be valid when an assumption of the type $0.1 + x \approx 0.1$ that we just made in this problem is valid. From a practical standpoint, this will almost always be true for useful buffer solutions. If the assumption is not valid, the solution will have such a low buffering capacity it will not be of any use to control the pH. *Note*: The Henderson-Hasselbalch equation can <u>only</u> be used to solve for the pH of buffer solutions.

33. 0.100 M $HC_3H_5O_2$: % dissociation $= \frac{[H^+]}{[HC_3H_5O_2]_0} \times 100 = \frac{1.1 \times 10^{-3}\ M}{0.100\ M} \times 100 = 1.1\%$

0.100 M $HC_3H_5O_2$ + 0.100 M $NaC_3H_5O_2$: % dissociation $= \frac{1.3 \times 10^{-5}}{0.100} \times 100 = 1.3 \times 10^{-2}\ \%$

The percent dissociation of the acid decreases from 1.1% to 1.3×10^{-2} % (a factor of 85) when $C_3H_5O_2^-$ is present. This is known as the common ion effect. The presence of the conjugate base of the weak acid inhibits the acid dissociation reaction.

35. a. We have a weak acid ($HOPr = HC_3H_5O_2$) and a strong acid (HCl) present. The amount of H^+ donated by the weak acid will be negligible. To prove it, consider the weak acid equilibrium reaction:

	HOPr	⇌	H^+	+	OPr^-	$K_a = 1.3 \times 10^{-5}$
Initial	0.100 *M*		0.020 *M*		0	
	x mol/L HOPr dissociates to reach equilibrium					
Change	$-x$	→	$+x$		$+x$	
Equil.	$0.100 - x$		$0.020 + x$		x	

$[H^+] = 0.020 + x \approx 0.020\ M$; pH = 1.70; assumption good ($x = 6.5 \times 10^{-5}$ is << 0.020).

Note: The H^+ contribution from the weak acid HOPr was negligible. The pH of the solution can be determined by only considering the amount of strong acid present.

b. Added H^+ reacts completely with the best base present, OPr^-.

	OPr^-	+	H^+	→	HOPr	
Before	0.100 *M*		0.020 *M*		0	
Change	−0.020		−0.020	→	+0.020	Reacts completely
After	0.080		0		0.020 *M*	

After reaction, a weak acid, HOPr , and its conjugate base, OPr^-, are present. This is a buffer solution. Using the Henderson-Hasselbalch equation where $pK_a = -\log(1.3 \times 10^{-5}) = 4.89$:

$$pH = pK_a + \log\frac{[\text{base}]}{[\text{acid}]} = 4.89 + \log\frac{(0.080)}{(0.020)} = 5.49;\ \text{assumptions good.}$$

c. This is a strong acid problem. $[H^+] = 0.020\ M$; pH = 1.70

d. Added H^+ reacts completely with the best base present, OPr^-.

	OPr^-	+	H^+	→	HOPr	
Before	0.100 *M*		0.020 *M*		0.100 *M*	
Change	−0.020		−0.020	→	+0.020	Reacts completely
After	0.080		0		0.120	

A buffer solution results (weak acid + conjugate base). Using the Henderson-Hasselbalch equation:

$$pH = pK_a + \log\frac{[\text{base}]}{[\text{acid}]} = 4.89 + \log\frac{(0.080)}{(0.120)} = 4.71;\ \text{assumptions good.}$$

37. a. OH^- will react completely with the best acid present, HOPr.

	HOPr	+	OH^-	→	OPr^-	+	H_2O	
Before	0.100 *M*		0.020 *M*		0			
Change	−0.020		−0.020	→	+0.020			Reacts completely
After	0.080		0		0.020			

A buffer solution results after the reaction. Using the Henderson-Hasselbalch equation:

$$pH = pK_a + \log\frac{[base]}{[acid]} = 4.89 + \log\frac{(0.020)}{(0.080)} = 4.29;\ \text{assumptions good.}$$

b. We have a weak base and a strong base present at the same time. The amount of OH^- added by the weak base will be negligible. To prove it, let's consider the weak base equilibrium:

	OPr^- + H_2O	⇌	$HOPr$	+	OH^-	$K_b = 7.7 \times 10^{-10}$
Initial	0.100 M		0		0.020 M	
	x mol/L OPr^- reacts with H_2O to reach equilibrium					
Change	$-x$	→	$+x$		$+x$	
Equil.	$0.100 - x$		x		$0.020 + x$	

$[OH^-] = 0.020 + x \approx 0.020\ M$; pOH = 1.70; pH = 12.30; assumption good.

Note: The OH^- contribution from the weak base OPr^- was negligible ($x = 3.9 \times 10^{-9}\ M$ as compared to 0.020 M OH^- from the strong base). The pH can be determined by only con-sidering the amount of strong base present.

c. This is a strong base in water. $[OH^-] = 0.020\ M$; pOH = 1.70; pH = 12.30

d. OH^- will react completely with HOPr, the best acid present.

	$HOPr$	+	OH^-	→	OPr^-	+	H_2O	
Before	0.100 M		0.020 M		0.100 M			
Change	−0.020		−0.020	→	+0.020			Reacts completely
After	0.080		0		0.120			

Using the Henderson-Hasselbalch equation to solve for the pH of the resulting buffer solution:

$$pH = pK_a + \log\frac{[base]}{[acid]} = 4.89 + \log\frac{(0.120)}{(0.080)} = 5.07;\ \text{assumptions good.}$$

39. Consider all the results to Exercises 31, 35, and 37:

Solution	Initial pH	After Added H^+	After Added OH^-
a	2.96	1.70	4.29
b	8.94	5.49	12.30
c	7.00	1.70	12.30
d	4.89	4.71	5.07

The solution in Exercise 31d is a buffer; it contains both a weak acid ($HC_3H_5O_2$) and a weak base ($C_3H_5O_2^-$). Solution d shows the greatest resistance to changes in pH when either a strong acid or a strong base is added, which is the primary property of buffers.

41. Statement c is false. Added OH^- will convert HCN into CN^-. So, the concentration of HCN will decrease and the concentration of CN^- will increase. For statement a, the buffer solution of HCN and NaCN will have a higher pH than just a solution of the acid HCN. For statement b, pH = pK_a ($[H^+] = K_a$) when the weak acid concentration equals the conjugate base concentration. For statement d, adding more of the base component of the buffer will increase the pH.

43. Major species: HNO_2, NO_2^- and Na^+. Na^+ has no acidic or basic properties. One appropriate equilibrium reaction you can use is the K_a reaction of HNO_2, which contains both HNO_2 and NO_2^-. However, you could also use the K_b reaction for NO_2^- and come up with the same answer. Solving the equilibrium problem (called a buffer problem):

	HNO_2	$\rightleftharpoons$	NO_2^-	+	H^+
Initial	1.00 *M*		1.00 *M*		~0
	x mol/L HNO_2 dissociates to reach equilibrium				
Change	$-x$	$\rightarrow$	$+x$		$+x$
Equil.	$1.00 - x$		$1.00 + x$		x

$$K_a = 4.0 \times 10^{-4} = \frac{[NO_2^-][H^+]}{[HNO_2]} = \frac{(1.00 + x)(x)}{1.00 - x} \approx \frac{(1.00)(x)}{1.00} \quad \text{(assuming } x << 1.00)$$

$x = 4.0 \times 10^{-4}\ M = [H^+]$; assumptions good ($x$ is 4.0×10^{-2} % of 1.00).

$$pH = -\log(4.0 \times 10^{-4}) = 3.40$$

Note: We would get the same answer using the Henderson-Hasselbalch equation. Use whichever method you prefer.

45. Major species after NaOH added: HNO_2, NO_2^-, Na^+, and OH^-. The OH^- from the strong base will react with the best acid present (HNO_2). Any reaction involving a strong base is assumed to go to completion. Because all species present are in the same volume of solution, we can use molarity units to do the stoichiometry part of the problem (instead of moles). The stoichiometry problem is:

	OH^-	+	HNO_2	$\rightarrow$	NO_2^- + H_2O	
Before	0.10 mol/1.00 L		1.00 *M*		1.00 *M*	
Change	−0.10 *M*		−0.10 *M*	$\rightarrow$	+0.10 *M*	Reacts completely
After	0		0.90		1.10	

After all the OH^- reacts, we are left with a solution containing a weak acid (HNO_2) and its conjugate base (NO_2^-). This is what we call a buffer problem. We will solve this buffer problem using the K_a equilibrium reaction. One could also use the K_b equilibrium reaction or use the Henderson-Hasselbalch equation to solve for the pH.

	HNO_2	$\rightleftharpoons$	NO_2^-	+	H^+
Initial	0.90 *M*		1.10 *M*		~0
	x mol/L HNO_2 dissociates to reach equilibrium				
Change	$-x$	$\rightarrow$	$+x$		$+x$
Equil.	$0.90 - x$		$1.10 + x$		x

$$K_a = 4.0 \times 10^{-4} = \frac{(1.10 + x)(x)}{0.90 - x} \approx \frac{(1.10)(x)}{0.90}, \quad x = [H^+] = 3.3 \times 10^{-4}\ M;\ pH = 3.48;$$

assumptions good.

Note: The added NaOH to this buffer solution changes the pH only from 3.40 to 3.48. If the NaOH were added to 1.0 L of pure water, the pH would change from 7.00 to 13.00.

Major species after HCl added: HNO_2, NO_2^-, H^+, Na^+, Cl^-; the added H^+ from the strong acid will react completely with the best base present (NO_2^-).

	H^+	+	NO_2^-	$\rightarrow$	HNO_2	
Before	$\frac{0.20\ \text{mol}}{1.00\ \text{L}}$		1.00 *M*		1.00 *M*	
Change	–0.20 *M*		–0.20 *M*	$\rightarrow$	+0.20 *M*	Reacts completely
After	0		0.80		1.20	

After all the H^+ has reacted, we have a buffer solution (a solution containing a weak acid and its conjugate base). Solving the buffer problem:

	HNO_2	$\rightleftharpoons$	NO_2^-	+	H^+
Initial	1.20 *M*		0.80 *M*		0
Equil.	$1.20 - x$		$0.80 + x$		$+x$

$$K_a = 4.0 \times 10^{-4} = \frac{(0.80 + x)(x)}{1.20 - x} \approx \frac{(0.80)(x)}{1.20}, \quad x = [H^+] = 6.0 \times 10^{-4}\ M;\ pH = 3.22;$$

assumptions good.

Note: The added HCl to this buffer solution changes the pH only from 3.40 to 3.22. If the HCl were added to 1.0 L of pure water, the pH would change from 7.00 to 0.70.

47. K_a for $C_5H_5NH^+ = 1.0 \times 10^{-14}/1.7 \times 10^{-9} = 5.9 \times 10^{-6}$; $pK_a = -\log(5.9 \times 10^{-6}) = 5.23$

A solution with equal concentrations of weak base and conjugate acid ($[C_5H_5N] = [C_5H_5NH^+]$) will have $[H^+] = K_a = 5.9 \times 10^{-6}$ *M* and pH = 5.23. Statement d is false. If a solution has a larger weak base concentration than conjugate acid concentration ($[C_5H_5N] > [C_5H_5NH^+]$), then the buffer will have $[H^+] < K_a = 5.9 \times 10^{-6}$ *M* and pH > 5.23. Statement b is false. If the concentration of conjugate acid is greater than the weak base concentration ($[C_5H_5N] < [C_5H_5NH^+]$), then the buffer will have $[H^+] > K_a = 5.9 \times 10^{-6}$ *M* and pH < 5.23. Statement c is true. Statement a is false as the solution with the larger concentrations will have the greater capacity.

49. a. $HC_2H_3O_2 \rightleftharpoons H^+ + C_2H_3O_2^-$ $\quad K_a = 1.8 \times 10^{-5}$

	$HC_2H_3O_2$	$\rightleftharpoons$	H^+	+	$C_2H_3O_2^-$
Initial	0.10 *M*		~0		0.25 *M*
	x mol/L $HC_2H_3O_2$ dissociates to reach equilibrium				
Change	$-x$	$\rightarrow$	$+x$		$+x$
Equil.	$0.10 - x$		x		$0.25 + x$

$$1.8 \times 10^{-5} = \frac{x(0.25 + x)}{(0.10 - x)} \approx \frac{x(0.25)}{0.10} \quad \text{(assuming } 0.25 + x \approx 0.25 \text{ and } 0.10 - x \approx 0.10)$$

$x = [H^+] = 7.2 \times 10^{-6}\ M$; pH = 5.14; assumptions good by the 5% rule.

Alternatively, we can use the Henderson-Hasselbalch equation:

$$pH = pK_a + \log\frac{[\text{base}]}{[\text{acid}]}, \text{ where } pK_a = -\log(1.8 \times 10^{-5}) = 4.74$$

$$pH = 4.74 + \log\frac{(0.25)}{(0.10)} = 4.74 + 0.40 = 5.14$$

The Henderson-Hasselbalch equation will be valid when assumptions of the type, $0.10 - x \approx 0.10$, that we just made are valid. From a practical standpoint, this will almost always be true for useful buffer solutions. *Note*: The Henderson-Hasselbalch equation can only be used to solve for the pH of buffer solutions.

b. $pH = 4.74 + \log\frac{(0.10)}{(0.25)} = 4.74 + (-0.40) = 4.34$

c. $pH = 4.74 + \log\frac{(0.20)}{(0.080)} = 4.74 + 0.40 = 5.14$

d. $pH = 4.74 + \log\frac{(0.080)}{(0.20)} = 4.74 + (-0.40) = 4.34$

51. $$[HC_7H_5O_2] = \frac{21.5 \text{ g } HC_7H_5O_2 \times \frac{1 \text{ mol } HC_7H_5O_2}{122.12 \text{ g}}}{0.2000 \text{ L}} = 0.880\ M$$

$$[C_7H_5O_2^-] = \frac{37.7 \text{ g } NaC_7H_5O_2 \times \frac{1 \text{ mol } NaC_7H_5O_2}{144.10 \text{ g}} \times \frac{1 \text{ mol } C_7H_5O_2^-}{\text{mol } NaC_7H_5O_2}}{0.2000 \text{ L}} = 1.31\ M$$

We have a buffer solution since we have both a weak acid and its conjugate base present at the same time. One can use the K_a reaction or the K_b reaction to solve. We will use the K_a reaction for the acid component of the buffer.

	$HC_7H_5O_2$	$\rightleftharpoons$	H^+	+	$C_7H_5O_2^-$
Initial	0.880 M		~0		1.31 M
	x mol/L of $HC_7H_5O_2$ dissociates to reach equilibrium				
Change	$-x$	$\rightarrow$	$+x$		$+x$
Equil.	$0.880 - x$		x		$1.31 + x$

$$K_a = 6.4 \times 10^{-5} = \frac{x(1.31 + x)}{0.880 - x} \approx \frac{x(1.31)}{0.880}, \quad x = [H^+] = 4.3 \times 10^{-5}\ M$$

$pH = -\log(4.3 \times 10^{-5}) = 4.37$; assumptions good.

Alternatively, we can use the Henderson-Hasselbalch equation to calculate the pH of buffer solutions.

$$pH = pK_a + \log\frac{[base]}{[acid]} = pK_a + \log\frac{[C_7H_5O_2^-]}{[HC_7H_5O_2]}$$

$$pH = -\log(6.4 \times 10^{-5}) + \log\left(\frac{1.31}{0.880}\right) = 4.19 + 0.173 = 4.36$$

Within round-off error, this is the same answer we calculated solving the equilibrium problem using the K_a reaction.

The Henderson-Hasselbalch equation will be valid when an assumption of the type $1.31 + x \approx 1.31$ that we just made in this problem is valid. From a practical standpoint, this will almost always be true for useful buffer solutions. If the assumption is not valid, the solution will have such a low buffering capacity that it will be of no use to control the pH. *Note*: The Henderson-Hasselbalch equation can only be used to solve for the pH of buffer solutions.

53. $[H^+]$ added $= \dfrac{0.010 \text{ mol}}{0.2500 \text{ L}} = 0.040\ M$; the added H^+ reacts completely with NH_3 to form NH_4^+.

a.

	NH_3	+	H^+	$\rightarrow$	NH_4^+	
Before	0.050 *M*		0.040 *M*		0.15 *M*	
Change	−0.040		−0.040	$\rightarrow$	+0.040	Reacts completely
After	0.010		0		0.19	

A buffer solution still exists after H^+ reacts completely. Using the Henderson-Hasselbalch equation:

$$pH = pK_a + \log\frac{[NH_3]}{[NH_4^+]} = -\log(5.6 \times 10^{-10}) + \log\left(\frac{0.010}{0.19}\right) = 9.25 + (-1.28) = 7.97$$

b.

	NH_3	+	H^+	$\rightarrow$	NH_4^+	
Before	0.50 *M*		0.040 *M*		1.50 *M*	
Change	−0.040		−0.040	$\rightarrow$	+0.040	Reacts completely
After	0.46		0		1.54	

A buffer solution still exists. $pH = pK_a + \log\dfrac{[NH_3]}{[NH_4^+]}$, $9.25 + \log\left(\dfrac{0.46}{1.54}\right) = 8.73$

The two buffers differ in their capacity and not their initial pH (both buffers had an initial pH = 8.77). Solution b has the greatest capacity since it has the largest concentrations of weak acid and conjugate base. Buffers with greater capacities will be able to absorb more added H^+ or OH^-.

55. $HSO_3^- \rightleftharpoons H^+ + SO_3^{2-}$ $K_a = 1.0 \times 10^{-7}$, $pK_a = -\log(1.0 \times 10^{-7}) = 7.00$

$$pH = pK_a + \log\frac{[base]}{[acid]},\ 7.25 = 7.00 + \log\frac{[SO_3^{2-}]}{[HSO_3^-]}$$

For pH = 7.25, we need the log term to be positive. This will only happen when $[SO_3^{2-}]$ > $[HSO_3^-]$. Another way to think about this is to compare the pK_a and pH values. Here, we want a pH on the "basic" side of the pK_a value, i.e., we want the pH to be greater than the pK_a value. To get a pH on the "basic" side of the pK_a, we need more base than acid in the buffer. Hence, we need a larger concentration of SO_3^{2-} than HSO_3^- to achieve a pH = 7.25.

$$7.25 = 7.00 + \log\frac{1.0\,M}{[HSO_3^-]},\quad 10^{0.25} = \frac{1.0\,M}{[HSO_3^-]},\quad [HSO_3^-] = 0.56\,M$$

57. pK_b for $H_2NNH_2 = -\log(3.0 \times 10^{-6}) = 5.52$; pK_a for $H_2NNH_3^+ = 14.00 - 5.52 = 8.48$

$$pH = pK_a + \log\frac{[H_2NNH_2]}{[H_2NNH_3^+]},\quad 8.00 = 8.48 + \log\frac{1.87\,M}{[H_2NNH_3^+]}$$

$$-0.48 = \log\frac{1.87\,M}{[H_2NNH_3^+]},\quad [C_6H_5NH_3^+] = 5.6\,M$$

59. $$pH = pK_a + \log\frac{[C_2H_3O_2^-]}{[HC_2H_3O_2]};\quad pK_a = -\log(1.8 \times 10^{-5}) = 4.74$$

Because the buffer components, $C_2H_3O_2^-$ and $HC_2H_3O_2$, are both in the same volume of solution, the concentration ratio of $[C_2H_3O_2^-]$: $[HC_2H_3O_2]$ will equal the mole ratio of mol $C_2H_3O_2^-$ to mol $HC_2H_3O_2$.

$$5.00 = 4.74 + \log\frac{\text{mol } C_2H_3O_2^-}{\text{mol } HC_2H_3O_2};\quad \text{mol } HC_2H_3O_2 = 0.5000\text{ L} \times \frac{0.200\text{ mol}}{\text{L}} = 0.100\text{ mol}$$

$$0.26 = \log\frac{\text{mol } C_2H_3O_2^-}{0.100\text{ mol}},\quad \frac{\text{mol } C_2H_3O_2^-}{0.100} = 10^{0.26} = 1.8,\quad \text{mol } C_2H_3O_2^- = 0.18\text{ mol}$$

$$\text{Mass } NaC_2H_3O_2 = 0.18\text{ mol } NaC_2H_3O_2 \times \frac{82.03\text{ g}}{\text{mol}} = 15\text{ g } NaC_2H_3O_2$$

61. $$C_5H_5NH^+ \rightleftharpoons H^+ + C_5H_5N \quad K_a = \frac{K_w}{K_b} = \frac{1.0 \times 10^{-14}}{1.7 \times 10^{-9}} = 5.9 \times 10^{-6}$$

$pK_a = -\log(5.9 \times 10^{-6}) = 5.23$

We will use the Henderson-Hasselbalch equation to calculate the concentration ratio necessary for each buffer.

$$pH = pK_a + \log\frac{[\text{base}]}{[\text{acid}]},\quad pH = 5.23 + \log\frac{[C_5H_5N]}{[C_5H_5NH^+]}$$

a. $$4.50 = 5.23 + \log\frac{[C_5H_5N]}{[C_5H_5NH^+]},\quad \frac{[C_5H_5N]}{[C_5H_5NH^+]} = 10^{-0.73} = 0.19$$

b. $$5.00 = 5.23 + \log\frac{[C_5H_5N]}{[C_5H_5NH^+]},\quad \frac{[C_5H_5N]}{[C_5H_5NH^+]} = 10^{-0.23} = 0.59$$

c. $$5.23 = 5.23 + \log\frac{[C_5H_5N]}{[C_5H_5NH^+]},\quad \frac{[C_5H_5N]}{[C_5H_5NH^+]} = 10^{0.0} = 1.0$$

d. $5.50 = 5.23 + \log\frac{[C_5H_5N]}{[C_5H_5NH^+]}$, $\frac{[C_5H_5N]}{[C_5H_5NH^+]} = 10^{0.27} = 1.9$

63. $pH = pK_a + \log\frac{[base]}{[acid]}$; in a buffer, when [acid] = [base], pH = pK_a. When [acid] > [base], pH < pK_a, and when [base] > [acid], pH > pK_a. For the H_2CO_3/ HCO_3^- buffer, we want pH = pK_a, so $[H_2CO_3] = [HCO_3^-]$. For the $H_2PO_4^-$/ HPO_4^{2-} buffer, we want pH < pK_a, so $[H_2PO_4^-] > [HPO_4^{2-}]$. Answer b is correct.

65. $pH = pK_a + \log\frac{[HCO_3^-]}{[H_2CO_3]}$, $7.40 = -\log(4.3 \times 10^{-7}) + \log\frac{[HCO_3^-]}{0.0012}$

$\log\frac{[HCO_3^-]}{0.0012} = 7.40 - 6.37 = 1.03$, $\frac{[HCO_3^-]}{0.0012} = 10^{1.03}$, $[HCO_3^-] = 1.3 \times 10^{-2}\ M$

67. A best buffer has large and equal quantities of weak acid and conjugate base. Because [acid] = [base] for a best buffer, $pH = pK_a + \log\frac{[base]}{[acid]} = pK_a + 0 = pK_a$ ($pH \approx pK_a$ for a best buffer).

The best acid choice for a pH = 7.00 buffer would be the weak acid with a pK_a close to 7.0 or $K_a \approx 1 \times 10^{-7}$. HOCl is the best choice in Table 14.2 ($K_a = 3.5 \times 10^{-8}$; $pK_a = 7.46$). To make this buffer, we need to calculate the [base] : [acid] ratio.

$$7.00 = 7.46 + \log\frac{[base]}{[acid]},\quad \frac{[OCl^-]}{[HOCl]} = 10^{-0.46} = 0.35$$

Any OCl^-/HOCl buffer in a concentration ratio of 0.35 : 1 will have a pH = 7.00. One possibility is [NaOCl] = 0.35 *M* and [HOCl] = 1.0 *M*.

69. K_a for $H_2NNH_3^+ = K_w/K_{b,\,H_2NNH_2} = 1.0 \times 10^{-14}/3.0 \times 10^{-6} = 3.3 \times 10^{-9}$

$$pH = pK_a + \log\frac{[H_2NNH_2]}{[H_2NNH_3^+]} = -\log(3.3 \times 10^{-9}) + \log\left(\frac{0.40}{0.80}\right) = 8.48 + (-0.30) = 8.18$$

pH = pK_a for a buffer when [acid] = [base]. Here, the acid ($H_2NNH_3^+$) concentration needs to decrease, while the base (H_2NNH_2) concentration needs to increase in order for $[H_2NNH_3^+] = [H_2NNH_2]$. Both of these changes are accomplished by adding a strong base (like NaOH) to the original buffer. The added OH^- from the strong base converts the acid component of the buffer into the conjugate base. Here, the reaction is $H_2NNH_3^+ + OH^- \rightarrow H_2NNH_2 + H_2O$. Because a strong base is reacting, the reaction is assumed to go to completion. The following set-up determines the number of moles of OH^-(*x*) that must be added so that mol $H_2NNH_3^+$ = mol H_2NNH_2. When mol acid = mol base in a buffer, then [acid] = [base] and pH = pK_a.

	$H_2NNH_3^+$	+ OH^-	→	H_2NNH_2	+ H_2O
Before	1.0 L × 0.80 mol/L	*x*		1.0 L × 0.40 mol/L	
Change	$-x$	$-x$	→	$+x$	Reacts completely
After	$0.80 - x$	0		$0.40 + x$	

We want mol $H_2NNH_3^+$ = mol H_2NNH_2. Therefore:

$$0.80 - x = 0.40 + x, \quad 2x = 0.40, \quad x = 0.20 \text{ mol } OH^-$$

When 0.20 mol OH^- is added to the initial buffer, mol $H_2NNH_3^+$ is decreased to 0.60 mol, while mol H_2NNH_2 is increased to 0.60 mol. Therefore, 0.20 mol of NaOH must be added to the initial buffer solution to produce a solution where pH = pK_a.

71. The reaction $OH^- + CH_3NH_3^+ \rightarrow CH_3NH_2 + H_2O$ goes to completion for solutions a, c, and d (no reaction occurs between the species in solution b because both species are bases). After the OH^- reacts completely, there must be both $CH_3NH_3^+$ and CH_3NH_2 in solution for it to be a buffer. The important components of each solution (after the OH^- reacts completely) is(are):

a. 0.05 *M* CH_3NH_2 (no $CH_3NH_3^+$ remains, no buffer)

b. 0.05 *M* OH^- and 0.1 *M* CH_3NH_2 (two bases present, no buffer)

c. 0.05 *M* OH^- and 0.05 *M* CH_3NH_2 (too much OH^- added, no $CH_3NH_3^+$ remains, no buffer)

d. 0.05 *M* CH_3NH_2 and 0.05 *M* $CH_3NH_3^+$ (a buffer solution results)

Only the combination in mixture d results in a buffer. Note that the concentrations are halved from the initial values. This is so because equal volumes of two solutions were added together, which halves the concentrations.

73. Added OH^- converts $HC_2H_3O_2$ into $C_2H_3O_2^-$: $HC_2H_3O_2 + OH^- \rightarrow C_2H_3O_2^- + H_2O$

From this reaction, the moles of $C_2H_3O_2^-$ produced equal the moles of OH^- added. Also, the total concentration of acetic acid plus acetate ion must equal 2.0 *M* (assuming no volume change on addition of NaOH). Summarizing for each solution:

$[C_2H_3O_2^-] + [HC_2H_3O] = 2.0\ M$ and $[C_2H_3O_2^-]$ produced = $[OH^-]$ added

a. $pH = pK_a + \log\frac{[C_2H_3O_2^-]}{[HC_2H_3O_2]}$; for pH = pK_a, $\log\frac{[C_2H_3O_2^-]}{[HC_2H_3O_2]} = 0$

Therefore, $\frac{[C_2H_3O_2^-]}{[HC_2H_3O_2]} = 1.0$ and $[C_2H_3O_2^-] = [HC_2H_3O_2]$.

Because $[C_2H_3O_2^-] + [HC_2H_3O_2] = 2.0\ M$:

$$[C_2H_3O_2^-] = [HC_2H_3O_2] = 1.0\ M = [OH^-] \text{ added}$$

To produce a 1.0 *M* $C_2H_3O_2^-$ solution, we need to add 1.0 mol of NaOH to 1.0 L of the 2.0 *M* $HC_2H_3O_2$ solution. The resulting solution will have pH = pK_a = 4.74.

b. $4.00 = 4.74 + \log\frac{[C_2H_3O_2^-]}{[HC_2H_3O_2]}$, $\frac{[C_2H_3O_2^-]}{[HC_2H_3O_2]} = 10^{-0.74} = 0.18$

$[C_2H_3O_2^-] = 0.18[HC_2H_3O_2]$ or $[HC_2H_3O_2] = 5.6[C_2H_3O_2^-]$

Because $[C_2H_3O_2^-] + [HC_2H_3O_2] = 2.0\ M$:

$$[C_2H_3O_2^-] + 5.6[C_2H_3O_2^-] = 2.0\ M,\ [C_2H_3O_2^-] = \frac{2.0}{6.6} = 0.30\ M = [OH^-] \text{ added}$$

We need to add 0.30 mol of NaOH to 1.0 L of 2.0 M $HC_2H_3O_2$ solution to produce 0.30 M $C_2H_3O_2^-$. The resulting solution will have pH = 4.00.

c. $5.00 = 4.74 + \log\frac{[C_2H_3O_2^-]}{[HC_2H_3O_2]},\ \frac{[C_2H_3O_2^-]}{[HC_2H_3O_2]} = 10^{0.26} = 1.8$

$1.8[HC_2H_3O_2] = [C_2H_3O_2^-]$ or $[HC_2H_3O_2] = 0.56[C_2H_3O_2^-]$

$1.56[C_2H_3O_2^-] = 2.0\ M,\ [C_2H_3O_2^-] = 1.3\ M = [OH^-]$ added

We need to add 1.3 mol of NaOH to 1.0 L of 2.0 M $HC_2H_3O_2$ to produce a solution with pH = 5.00.

Acid-Base Titrations

75.

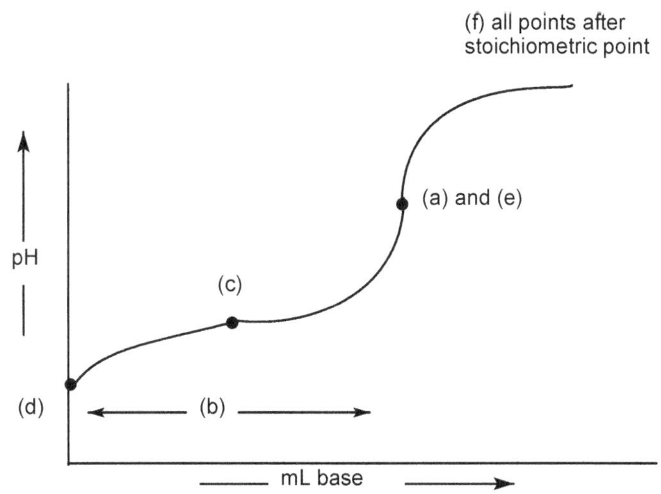

$HA + OH^- \rightarrow A^- + H_2O$; added OH^- from the strong base converts the weak acid HA into its conjugate base A^-. Initially, before any OH^- is added (point d), HA is the dominant species present. After OH^- is added, both HA and A^- are present, and a buffer solution results (region b). At the equivalence point (points a and e), exactly enough OH^- has been added to convert all the weak acid HA into its conjugate base A^-. Past the equivalence point (region f), excess OH^- is present. For the answer to part b, we included almost the entire buffer region. The maximum buffer region (or the region which is the best buffer solution) is around the halfway point to equivalence (point c). At this point, enough OH^- has been added to convert exactly one-half of the weak acid present initially into its conjugate base, so $[HA] = [A^-]$ and pH = pK_a. A "best" buffer has about equal concentrations of weak acid and conjugate base present.

77. This is a strong acid ($HClO_4$) titrated by a strong base (KOH). Added OH^- from the strong base will react completely with the H^+ present from the strong acid to produce H_2O.

a. Only strong acid present. $[H^+] = 0.200\ M$; pH = 0.699

b. $\text{mmol OH}^- \text{ added} = 10.0 \text{ mL} \times \dfrac{0.100 \text{ mmol OH}^-}{\text{mL}} = 1.00 \text{ mmol OH}^-$

$\text{mmol H}^+ \text{ present} = 40.0 \text{ mL} \times \dfrac{0.200 \text{ mmol H}^+}{\text{mL}} = 8.00 \text{ mmol H}^+$

Note: The units millimoles are usually easier numbers to work with. The units for molarity are moles per liter but are also equal to millimoles per milliliter.

	H^+	+	OH^-	→ H_2O	
Before	8.00 mmol		1.00 mmol		
Change	−1.00 mmol		−1.00 mmol		Reacts completely
After	7.00 mmol		0		

The excess H^+ determines the pH. $[H^+]_{excess} = \dfrac{7.00 \text{ mmol H}^+}{40.0 \text{ mL} + 10.0 \text{ mL}} = 0.140\ M$

pH = −log(0.140) = 0.854

c. mmol OH^- added = 40.0 mL × 0.100 M = 4.00 mmol OH^-

	H^+	+	OH^-	→ H_2O
Before	8.00 mmol		4.00 mmol	
After	4.00 mmol		0	

$[H^+]_{excess} = \dfrac{4.00 \text{ mmol}}{(40.0 + 40.0) \text{ mL}} = 0.0500\ M$; pH = 1.301

d. mmol OH^- added = 80.0 mL × 0.100 M = 8.00 mmol OH^-; this is the equivalence point because we have added just enough OH^- to react with all the acid present. For a strong acid-strong base titration, pH = 7.00 at the equivalence point because only neutral species are present (K^+, ClO_4^-, H_2O).

e. mmol OH^- added = 100.0 mL × 0.100 M = 10.0 mmol OH^-

	H^+	+	OH^-	→ H_2O
Before	8.00 mmol		10.0 mmol	
After	0		2.0 mmol	

Past the equivalence point, the pH is determined by the excess OH^- present.

$[OH^-]_{excess} = \dfrac{2.0 \text{ mmol}}{(40.0 + 100.0) \text{ mL}} = 0.014\ M$; pOH = 1.85; pH = 12.15

79. This is a strong base, $Ba(OH)_2$, titrated by a strong acid, HCl. The added strong acid will neutralize the OH^- from the strong base. As is always the case when a strong acid and/or strong base reacts, the reaction is assumed to go to completion.

a. Only a strong base is present, but it breaks up into 2 moles of OH^- ions for every mole of $Ba(OH)_2$. $[OH^-] = 2 \times 0.100\ M = 0.200\ M$; pOH = 0.699; pH = 13.301

b. $\text{mmol OH}^- \text{ present} = 80.0 \text{ mL} \times \dfrac{0.100 \text{ mmol Ba(OH)}_2}{\text{mL}} \times \dfrac{2 \text{ mmol OH}^-}{\text{mmol Ba(OH)}_2}$

$= 16.0 \text{ mmol OH}^-$

$\text{mmol H}^+ \text{ added} = 20.0 \text{ mL} \times \dfrac{0.400 \text{ mmol H}^+}{\text{mL}} = 8.00 \text{ mmol H}^+$

	OH^- +	H^+ → H_2O	
Before	16.0 mmol	8.00 mmol	
Change	–8.00 mmol	–8.00 mmol	Reacts completely
After	8.0 mmol	0	

$[OH^-]_{excess} = \dfrac{8.0 \text{ mmol OH}^-}{80.0 \text{ mL} + 20.0 \text{ mL}} = 0.080\ M$; pOH = 1.10; pH = 12.90

c. $\text{mmol H}^+ \text{ added} = 30.0 \text{ mL} \times 0.400\ M = 12.0 \text{ mmol H}^+$

	OH^- +	H^+ → H_2O
Before	16.0 mmol	12.0 mmol
After	4.0 mmol	0

$[OH^-]_{excess} = \dfrac{4.0 \text{ mmol OH}^-}{(80.0 + 30.0) \text{ mL}} = 0.036\ M$; pOH = 1.44; pH = 12.56

d. $\text{mmol H}^+ \text{ added} = 40.0 \text{ mL} \times 0.400\ M = 16.0 \text{ mmol H}^+$; this is the equivalence point. Because the H^+ will exactly neutralize the OH^- from the strong base, all we have in solution is Ba^{2+}, Cl^-, and H_2O. All are neutral species, so pH = 7.00.

e. $\text{mmol H}^+ \text{ added} = 80.0 \text{ mL} \times 0.400\ M = 32.0 \text{ mmol H}^+$

	OH^- +	H^+ → H_2O
Before	16.0 mmol	32.0 mmol
After	0	16.0 mmol

$[H^+]_{excess} = \dfrac{16.0 \text{ mmol H}^+}{(80.0 + 80.0) \text{ mL}} = 0.100\ M$; pH = 1.000

81. This is a weak acid ($HC_2H_3O_2$) titrated by a strong base (KOH).

a. Only weak acid is present. Solving the weak acid problem:

	$HC_2H_3O_2$	⇌	H^+	+ $C_2H_3O_2^-$
Initial	0.200 M		~0	0
	x mol/L $HC_2H_3O_2$ dissociates to reach equilibrium			
Change	$-x$	→	$+x$	$+x$
Equil.	$0.200 - x$		x	x

$$K_a = 1.8 \times 10^{-5} = \frac{x^2}{0.200 - x} \approx \frac{x^2}{0.200},\ x = [H^+] = 1.9 \times 10^{-3}\ M$$

pH = 2.72; assumptions good.

b. The added OH^- will react completely with the best acid present, $HC_2H_3O_2$.

$$\text{mmol } HC_2H_3O_2 \text{ present} = 100.0 \text{ mL} \times \frac{0.200 \text{ mmol } HC_2H_3O_2}{\text{mL}} = 20.0 \text{ mmol } HC_2H_3O_2$$

$$\text{mmol } OH^- \text{ added} = 50.0 \text{ mL} \times \frac{0.100 \text{ mmol } OH^-}{\text{mL}} = 5.00 \text{ mmol } OH^-$$

	$HC_2H_3O_2$	+	OH^-	→	$C_2H_3O_2^-$ + H_2O	
Before	20.0 mmol		5.00 mmol		0	
Change	−5.00 mmol		−5.00 mmol	→	+5.00 mmol	Reacts Completely
After	15.0 mmol		0		5.00 mmol	

After reaction of all the strong base, we have a buffer solution containing a weak acid ($HC_2H_3O_2$) and its conjugate base ($C_2H_3O_2^-$). We will use the Henderson-Hasselbalch equation to solve for the pH.

$$pH = pK_a + \log\frac{[C_2H_3O_2^-]}{[HC_2H_3O_2]} = -\log(1.8 \times 10^{-5}) + \log\left(\frac{5.00 \text{ mmol}/V_T}{15.0 \text{ mmol}/V_T}\right),$$ where V_T = total volume

$$pH = 4.74 + \log\left(\frac{5.00}{15.0}\right) = 4.74 + (-0.477) = 4.26$$

Note that the total volume cancels in the Henderson-Hasselbalch equation. For the [base]/[acid] term, the mole ratio equals the concentration ratio because the components of the buffer are always in the same volume of solution.

c. mmol OH^- added = 100.0 mL × (0.100 mmol OH^-/mL) = 10.0 mmol OH^-; the same amount (20.0 mmol) of $HC_2H_3O_2$ is present as before (it doesn't change). As before, let the OH^- react to completion, then see what is remaining in solution after this reaction.

	$HC_2H_3O_2$	+	OH^-	→	$C_2H_3O_2^-$ + H_2O
Before	20.0 mmol		10.0 mmol		0
After	10.0 mmol		0		10.0 mmol

A buffer solution results after reaction. Because $[C_2H_3O_2^-] = [HC_2H_3O_2]$ = 10.0 mmol/total volume, pH = pK_a. This is always true at the halfway point to equivalence for a weak acid-strong base titration, pH = pK_a.

$$pH = -\log(1.8 \times 10^{-5}) = 4.74$$

d. mmol OH^- added = 150.0 mL × 0.100 M = 15.0 mmol OH^-. Added OH^- reacts completely with the weak acid.

	$HC_2H_3O_2$	+	OH^-	→	$C_2H_3O_2^-$	+	H_2O
Before	20.0 mmol		15.0 mmol		0		
After	5.0 mmol		0		15.0 mmol		

We have a buffer solution after all the OH^- reacts to completion. Using the Henderson-Hasselbalch equation:

$$pH = 4.74 + \log\frac{[C_2H_3O_2^-]}{[HC_2H_3O_2]} = 4.74 + \log\left(\frac{15.0 \text{ mmol}}{5.0 \text{ mmol}}\right)$$

$$pH = 4.74 + 0.48 = 5.22$$

e. mmol OH^- added = 200.00 mL × 0.100 M = 20.0 mmol OH^-; as before, let the added OH^- react to completion with the weak acid; then see what is in solution after this reaction.

	$HC_2H_3O_2$	+	OH^-	→	$C_2H_3O_2^-$	+	H_2O
Before	20.0 mmol		20.0 mmol		0		
After	0		0		20.0 mmol		

This is the equivalence point. Enough OH^- has been added to exactly neutralize all the weak acid present initially. All that remains that affects the pH at the equivalence point is the conjugate base of the weak acid ($C_2H_3O_2^-$). This is a weak base equilibrium problem.

$$C_2H_3O_2^- + H_2O \rightleftharpoons HC_2H_3O_2 + OH^- \qquad K_b = \frac{K_w}{K_b} = \frac{1.0 \times 10^{-14}}{1.8 \times 10^{-5}}$$

	$C_2H_3O_2^-$		$HC_2H_3O_2$	OH^-
Initial	20.0 mmol/300.0 mL		0	0
	x mol/L $C_2H_3O_2^-$ reacts with H_2O to reach equilibrium			
Change	$-x$	→	$+x$	$+x$
Equil.	$0.0667 - x$		x	x

$K_b = 5.6 \times 10^{-10}$

$$K_b = 5.6 \times 10^{-10} = \frac{x^2}{0.0667 - x} \approx \frac{x^2}{0.0667}, \quad x = [OH^-] = 6.1 \times 10^{-6}\ M$$

pOH = 5.21; pH = 8.79; assumptions good.

f. mmol OH^- added = 250.0 mL × 0.100 M = 25.0 mmol OH^-

	$HC_2H_3O_2$	+	OH^-	→	$C_2H_3O_2^-$	+	H_2O
Before	20.0 mmol		25.0 mmol		0		
After	0		5.0 mmol		20.0 mmol		

After the titration reaction, we have a solution containing excess OH^- and a weak base $C_2H_3O_2^-$. When a strong base and a weak base are both present, assume that the amount of OH^- added from the weak base will be minimal; that is, the pH past the equivalence point is determined by the amount of excess strong base.

$$[OH^-]_{excess} = \frac{5.0 \text{ mmol}}{100.0 \text{ mL} + 250.0 \text{ mL}} = 0.014\ M; \ pOH = 1.85; \ pH = 12.15$$

83. This is a weak base (H_2NNH_2) titrated by a strong acid (HNO_3). To calculate the pH at the various points, let the strong acid react completely with the weak base present; then see what is in solution.

a. Only a weak base is present. Solve the weak base equilibrium problem.

	H_2NNH_2 + H_2O	$\rightleftharpoons$ $H_2NNH_3^+$	+ OH^-
Initial	0.100 M	0	~0
Equil.	$0.100 - x$	x	x

$$K_b = 3.0 \times 10^{-6} = \frac{x^2}{0.100 - x} \approx \frac{x^2}{0.100}, \quad x = [OH^-] = 5.5 \times 10^{-4}\ M$$

pOH = 3.26; pH = 10.74; assumptions good.

b. $\text{mmol } H_2NNH_2 \text{ present} = 100.0 \text{ mL} \times \dfrac{0.100 \text{ mmol } H_2NNH_2}{\text{mL}} = 10.0 \text{ mmol } H_2NNH_2$

$\text{mmol } H^+ \text{ added} = 20.0 \text{ mL} \times \dfrac{0.200 \text{ mmol } H^+}{\text{mL}} = 4.00 \text{ mmol } H^+$

	H_2NNH_2	+ H^+	$\rightarrow$ $H_2NNH_3^+$	
Before	10.0 mmol	4.00 mmol	0	
Change	–4.00 mmol	–4.00 mmol	→ +4.00 mmol	Reacts completely
After	6.0 mmol	0	4.00 mmol	

A buffer solution results after the titration reaction. Solving using the Henderson-Hasselbalch equation:

$$pH = pK_a + \log\frac{[\text{base}]}{[\text{acid}]}; \quad K_a = \frac{K_w}{K_b} = \frac{1.0 \times 10^{-14}}{3.0 \times 10^{-6}} = 3.3 \times 10^{-9}$$

$$pH = -\log(3.3 \times 10^{-9}) + \log\left(\frac{6.0 \text{ mmol}/V_T}{4.00 \text{ mmol}/V_T}\right), \text{ where } V_T = \text{total volume (cancels out)}$$

$$pH = 8.48 + \log(1.5) = 8.48 + 0.18 = 8.66$$

c. $\text{mmol } H^+ \text{ added} = 25.0 \text{ mL} \times 0.200\ M = 5.00 \text{ mmol } H^+$

	H_2NNH_2	+ H^+	$\rightarrow$ $H_2NNH_3^+$
Before	10.0 mmol	5.00 mmol	0
After	5.0 mmol	0	5.00 mmol

This is the halfway point to equivalence, where $[H_2NNH_3^+] = [H_2NNH_2]$. At this point, $pH = pK_a$ (which is characteristic of the halfway point for any weak base-strong acid titration).

$$pH = -\log(3.3 \times 10^{-9}) = 8.48$$

d. mmol H^+ added = 40.0 mL × 0.200 M = 8.00 mmol H^+

	H_2NNH_2	+	H^+	→	$H_2NNH_3^+$
Before	10.0 mmol		8.00 mmol		0
After	2.0 mmol		0		8.00 mmol

A buffer solution results.

$$pH = pK_a + \log\frac{[base]}{[acid]} = 8.48 + \log\left(\frac{2.0\ mmol/V_T}{8.00\ mmol/V_T}\right) = 8.48 + (-0.60) = 7.88$$

e. mmol H^+ added = 50.0 mL × 0.200 M = 10.0 mmol H^+

	H_2NNH_2	+	H^+	→	$H_2NNH_3^+$
Before	10.0 mmol		10.0 mmol		0
After	0		0		10.0 mmol

As is always the case in a weak base-strong acid titration, the pH at the equivalence point is acidic because only a weak acid ($H_2NNH_3^+$) is present. Solving the weak acid equilibrium problem:

	$H_2NNH_3^+$	⇌	H^+	+	H_2NNH_2
Initial	10.0 mmol/150.0 mL		0		0
Equil.	$0.0667 - x$		x		x

$$K_a = 3.3 \times 10^{-9} = \frac{x^2}{0.0667 - x} \approx \frac{x^2}{0.0667}, \quad x = [H^+] = 1.5 \times 10^{-5}\ M$$

pH = 4.82; assumptions good.

f. mmol H^+ added = 100.0 mL × 0.200 M = 20.0 mmol H^+

	H_2NNH_2	+	H^+	→	$H_2NNH_3^+$
Before	10.0 mmol		20.0 mmol		0
After	0		10.0 mmol		10.0 mmol

Two acids are present past the equivalence point, but the excess H^+ will determine the pH of the solution since $H_2NNH_3^+$ is a weak acid.

$$[H^+]_{excess} = \frac{10.0\ mmol}{100.0\ mL + 100.0\ mL} = 0.0500\ M;\ pH = 1.301$$

85. We will do sample calculations for the various parts of the titration. All results are summarized in Table 15.1 at the end of Exercise 87.

At the beginning of the titration, only the weak acid $HC_3H_5O_3$ is present. Let HLac = $HC_3H_5O_3$ and $Lac^- = C_3H_5O_3^-$.

	HLac	$\rightleftharpoons$	H^+	+	Lac^-	$K_a = 10^{-3.86} = 1.4 \times 10^{-4}$
Initial	0.100 *M*		~0		0	
	x mol/L HLac dissociates to reach equilibrium					
Change	$-x$	$\rightarrow$	$+x$		$+x$	
Equil.	$0.100 - x$		x		x	

$$1.4 \times 10^{-4} = \frac{x^2}{0.100 - x} \approx \frac{x^2}{0.100}, \quad x = [H^+] = 3.7 \times 10^{-3}\ M;\ \ pH = 2.43;\ \ \text{assumptions good.}$$

Up to the stoichiometric point, we calculate the pH using the Henderson-Hasselbalch equation. This is the buffer region. For example, at 4.0 mL of NaOH added:

$$\text{initial mmol HLac present} = 25.0\text{ mL} \times \frac{0.100\text{ mmol}}{\text{mL}} = 2.50\text{ mmol HLac}$$

$$\text{mmol OH}^-\text{ added} = 4.0\text{ mL} \times \frac{0.100\text{ mmol}}{\text{mL}} = 0.40\text{ mmol OH}^-$$

Note: The units millimoles are usually easier numbers to work with. The units for molarity are moles per liter but are also equal to millimoles per milliliter.

The 0.40 mmol of added OH^- converts 0.40 mmol HLac to 0.40 mmol Lac^- according to the equation:

$HLac + OH^- \rightarrow Lac^- + H_2O$ Reacts completely since a strong base is added.

mmol HLac remaining = 2.50 – 0.40 = 2.10 mmol; mmol Lac^- produced = 0.40 mmol

We have a buffer solution. Using the Henderson-Hasselbalch equation where $pK_a = 3.86$:

$$pH = pK_a + \log\frac{[Lac^-]}{[HLac]} = 3.86 + \log\frac{(0.40)}{(2.10)}$$

(Total volume cancels, so we can use the ratio of moles or millimoles.)

$$pH = 3.86 - 0.72 = 3.14$$

Other points in the buffer region are calculated in a similar fashion. Perform a stoichiometry problem first, followed by a buffer problem. The buffer region includes all points up to and including 24.9 mL OH^- added.

At the stoichiometric point (25.0 mL OH^- added), we have added enough OH^- to convert all the HLac (2.50 mmol) into its conjugate base (Lac^-). All that is present is a weak base. To determine the pH, we perform a weak base calculation.

$$[Lac^-]_0 = \frac{2.50\text{ mmol}}{25.0\text{ mL} + 25.0\text{ mL}} = 0.0500\ M$$

	$Lac^- + H_2O$	$\rightleftharpoons$	HLac	+	OH^-	$K_b = \frac{1.0 \times 10^{-14}}{1.4 \times 10^{-4}} = 7.1 \times 10^{-11}$
Initial	0.0500 *M*		0		0	
	x mol/L Lac^- reacts with H_2O to reach equilibrium					
Change	$-x$	$\rightarrow$	$+x$		$+x$	
Equil.	$0.0500 - x$		x		x	

$$K_b = \frac{x^2}{0.0500 - x} \approx \frac{x^2}{0.0500} = 7.1 \times 10^{-11}$$

$x = [OH^-] = 1.9 \times 10^{-6}$ *M*; pOH = 5.72; pH = 8.28; assumptions good.

Past the stoichiometric point, we have added more than 2.50 mmol of NaOH. The pH will be determined by the excess OH^- ion present. An example of this calculation follows.

At 25.1 mL: OH^- added $= 25.1\text{ mL} \times \frac{0.100\text{ mmol}}{\text{mL}} = 2.51\text{ mmol } OH^-$

2.50 mmol OH^- neutralizes all the weak acid present. The remainder is excess OH^-.

Excess $OH^- = 2.51 - 2.50 = 0.01$ mmol OH^-

$$[OH^-]_{excess} = \frac{0.01\text{ mmol}}{(25.0 + 25.1)\text{ mL}} = 2 \times 10^{-4}\ M;\ \text{pOH} = 3.7;\ \text{pH} = 10.3$$

All results are listed in Table 15.1 at the end of the solution to Exercise 87.

87. At beginning of the titration, only the weak base NH_3 is present. As always, solve for the pH using the K_b reaction for NH_3.

	$NH_3 + H_2O$	$\rightleftharpoons$	NH_4^+	+	OH^-	$K_b = 1.8 \times 10^{-5}$
Initial	0.100 *M*		0		~0	
Equil.	$0.100 - x$		x		x	

$$K_b = \frac{x^2}{0.100 - x} \approx \frac{x^2}{0.100} = 1.8 \times 10^{-5}$$

$x = [OH^-] = 1.3 \times 10^{-3}$ *M*; pOH = 2.89; pH = 11.11; assumptions good.

In the buffer region (4.0 – 24.9 mL), we can use the Henderson-Hasselbalch equation:

$$K_a = \frac{1.0 \times 10^{-14}}{1.8 \times 10^{-5}} = 5.6 \times 10^{-10};\ pK_a = 9.25;\ \text{pH} = 9.25 + \log\frac{[NH_3]}{[NH_4^+]}$$

We must determine the amounts of NH_3 and NH_4^+ present after the added H^+ reacts completely with the NH_3. For example, after 8.0 mL HCl added:

$$\text{initial mmol } NH_3 \text{ present} = 25.0\text{ mL} \times \frac{0.100\text{ mmol}}{\text{mL}} = 2.50\text{ mmol } NH_3$$

$$\text{mmol } H^+ \text{ added} = 8.0 \text{ mL} \times \frac{0.100 \text{ mmol}}{\text{mL}} = 0.80 \text{ mmol } H^+$$

Added H^+ reacts with NH_3 to completion: $NH_3 + H^+ \rightarrow NH_4^+$

mmol NH_3 remaining = 2.50 – 0.80 = 1.70 mmol; mmol NH_4^+ produced = 0.80 mmol

$pH = 9.25 + \log\frac{1.70}{0.80} = 9.58$ (Mole ratios can be used since the total volume cancels.)

Other points in the buffer region are calculated in similar fashion. Results are summarized in Table 15.1 that follows.

At the stoichiometric point (25.0 mL H^+ added), just enough HCl has been added to convert all the weak base (NH_3) into its conjugate acid (NH_4^+). Perform a weak acid calculation.

$[NH_4^+]_0 = 2.50 \text{ mmol}/50.0 \text{ mL} = 0.0500$ M

	NH_4^+	$\rightleftharpoons$	H^+	+	NH_3	$K_a = 5.6 \times 10^{-10}$
Initial	0.0500 M		0		0	
Equil.	0.0500 - x		x		x	

$$5.6 \times 10^{-10} = \frac{x^2}{0.0500 - x} \approx \frac{x^2}{0.0500}, \quad x = [H^+] = 5.3 \times 10^{-6} M; \; pH = 5.28$$

Assumptions good.

Beyond the stoichiometric point, the pH is determined by the excess H^+. For example, at 28.0 mL of H^+ added:

$$H^+ \text{ added} = 28.0 \text{ mL} \times \frac{0.100 \text{ mmol}}{\text{mL}} = 2.80 \text{ mmol } H^+$$

Excess H^+ = 2.80 mmol – 2.50 mmol = 0.30 mmol excess H^+

$$[H^+]_{excess} = \frac{0.30 \text{ mmol}}{(25.0 + 28.0) \text{ mL}} = 5.7 \times 10^{-3} M; \; pH = 2.24$$

All results are summarized in the following Table 15.1.

Table 15.1 Summary of pH Results for Exercises 85 and 87 (Plot below.)

Titrant mL	Exercise 85	Exercise 87
0.0	2.43	11.11
4.0	3.14	9.97
8.0	3.53	9.58
12.5	3.86	9.25
20.0	4.46	8.65
24.0	5.24	7.87
24.5	5.6	7.6
24.9	6.3	6.9
25.0	8.28	5.28
25.1	10.3	3.7
26.0	11.30	2.71
28.0	11.75	2.24
30.0	11.96	2.04

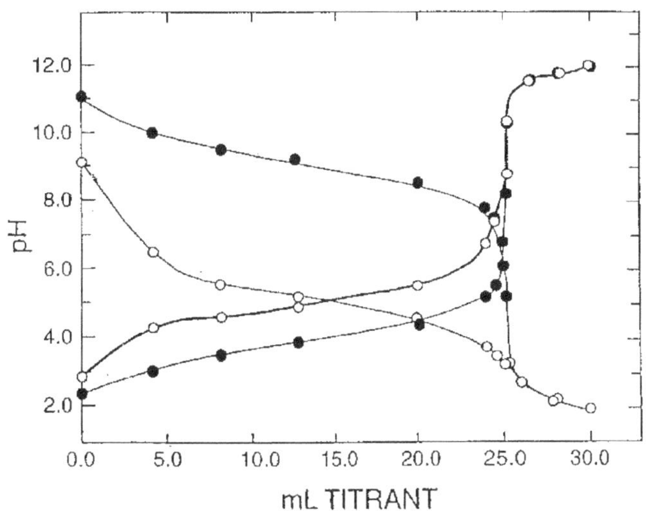

89. For the titration of HCl by RbOH, the pH is very acidic at the halfway point and pH = 7.00 at the equivalence point for this strong-acid strong base titration. The other titrations are weak acid-strong base titrations. For these titrations, pH = pK_a at the halfway point to equivalence since a buffer solution having equal concentrations of weak acid and conjugate base are present. The acid with the smallest K_a value will have the highest pH at the halfway point. This is HCN.

At the equivalence point of a weak acid-strong base titration, the major species present with acid-base properties is the conjugate base of the weak acid titrated. The titration with the highest pH (most basic pH) will have the conjugate base with the largest K_b value. For conjugate acid-base pairs, the weakest acid has the strongest conjugate base. Of the weak acids titrated, HCN is the weakest acid. So HCN will have the strongest conjugate base and the highest pH at the equivalence point.

91. $75.0 \text{ mL} \times \dfrac{0.10 \text{ mmol}}{\text{mL}} = 7.5 \text{ mmol HA}$; $30.0 \text{ mL} \times \dfrac{0.10 \text{ mmol}}{\text{mL}} = 3.0 \text{ mmol } OH^-$ added

The added strong base reacts to completion with the weak acid to form the conjugate base of the weak acid and H_2O.

	HA	+	OH^-	→	A^-	+	H_2O
Before	7.5 mmol		3.0 mmol		0		
After	4.5 mmol		0		3.0 mmol		

A buffer results after the OH^- reacts to completion. Using the Henderson-Hasselbalch equation:

$$pH = pK_a + \log\frac{[A^-]}{[HA]},\ 5.50 = pK_a + \log\left(\frac{3.0 \text{ mmol}/105.0 \text{ mL}}{4.5 \text{ mmol}/105.0 \text{ mL}}\right)$$

$pK_a = 5.50 - \log(3.0/4.5) = 5.50 - (-0.18) = 5.68$; $K_a = 10^{-5.68} = 2.1 \times 10^{-6}$

Indicators

93. $HIn \rightleftharpoons In^- + H^+ \quad K_a = \dfrac{[In^-][H^+]}{[HIn]} = 1.0 \times 10^{-9}$

a. In a very acid solution, the HIn form dominates, so the solution will be yellow.

b. The color change occurs when the concentration of the more dominant form is approximately ten times as great as the less dominant form of the indicator.

$$\frac{[HIn]}{[In^-]} = \frac{10}{1};\ K_a = 1.0 \times 10^{-9} = \left(\frac{1}{10}\right)[H^+],\ [H^+] = 1 \times 10^{-8}\ M;\ pH = 8.0 \text{ at color change}$$

c. This is way past the equivalence point (100.0 mL OH^- added), so the solution is very basic, and the In^- form of the indicator dominates. The solution will be blue.

95. At the equivalence point, P^{2-} is the major species. P^{2-} is a weak base in water because it is the conjugate base of a weak acid.

	P^{2-}	+	H_2O	⇌	HP^-	+	OH^-
Initial	$\dfrac{0.5 \text{ g}}{0.1 \text{ L}} \times \dfrac{1 \text{ mol}}{204.2 \text{ g}} = 0.024\ M$				0		~0 (carry extra sig. fig.)
Equil.	$0.024 - x$				x		x

$$K_b = \frac{[HP^-][OH^-]}{P^{2-}} = \frac{K_w}{K_a} = \frac{1.0 \times 10^{-14}}{10^{-5.51}},\ 3.2 \times 10^{-9} = \frac{x^2}{0.024 - x} \approx \frac{x^2}{0.024}$$

$x = [OH^-] = 8.8 \times 10^{-6}\ M$; pOH = 5.1; pH = 8.9; assumptions good.

Phenolphthalein would be the best indicator for this titration because it changes color at around $pH \approx 9$ (from acid color to base color).

97. When choosing an indicator, we want the color change of the indicator to occur approximately at the pH of the equivalence point. Because the pH generally changes very rapidly at the equivalence point, we don't have to be exact. This is especially true for strong acid-strong base titrations. The following are some indicators where the color change occurs at about the pH of the equivalence point:

Exercise	pH at Eq. Pt.	Indicator
77	7.00	bromthymol blue or phenol red
81	8.79	o-cresolphthalein or phenolphthalein

99.

Exercise	pH at Eq. Pt.	Indicator
85	8.28	o-cresolphthalein or phenolphthalein
87	5.28	bromcresol green

101. pH > 5 for bromcresol green to be blue. pH < 8 for thymol blue to be yellow. The pH is between 5 and 8.

103. a. yellow b. green (Both yellow and blue forms are present.) c. yellow d. blue

Polyprotic Acids

105. The first titration plot (from 0 – 100.0 mL) corresponds to the titration of H_2A by OH^-. The reaction is $H_2A + OH^- \rightarrow HA^- + H_2O$. After all the H_2A has been reacted, the second titration (from 100.0 – 200.0 mL) corresponds to the titration of HA^- by OH^-. The reaction is $HA^- + OH^- \rightarrow A^{2-} + H_2O$.

a. At 100.0 mL of NaOH, just enough OH^- has been added to react completely with all the H_2A present (mol OH^- added = mol H_2A present initially). From the balanced equation, the mol of HA^- produced will equal the mol of H_2A present initially. Because mol of HA^- present at 100.0 mL OH^- added equals the mol of H_2A present initially, exactly 100.0 mL more of NaOH must be added to react with all the HA^-. The volume of NaOH added to reach the second equivalence point equals 100.0 mL + 100.0 mL = 200.0 mL.

b. $H_2A + OH^- \rightarrow HA^- + H_2O$ is the reaction occurring from 0 – 100.0 mL NaOH added.

$HA^- + OH^- \rightarrow A^{2-} + H_2O$ is the reaction occurring from 100.0 – 200.0 mL NaOH added.

i. No reaction has taken place, so H_2A and H_2O are the major species.

ii. Adding OH^- converts H_2A into HA^-. The major species between 0 and 100.0 mL NaOH added are H_2A, HA^-, H_2O, and Na^+.

iii. At 100.0 mL NaOH added, mol of OH^- = mol H_2A, so all of the H_2A present initially has been converted into HA^-. The major species are HA^-, H_2O, and Na^+.

iv. Between 100.0 and 200.0 mL NaOH added, the OH^- converts HA^- into A^{2-}. The major species are HA^-, A^{2-}, H_2O, and Na^+.

v. At the second equivalence point (200.0 mL), just enough OH^- has been added to convert all the HA^- into A^{2-}. The major species are A^{2-}, H_2O, and Na^+.

vi. Past 200.0 mL NaOH added, excess OH^- is present. The major species are OH^-, A^{2-}, H_2O, and Na^+.

c. 50.0 mL of NaOH added corresponds to the first halfway point to equivalence. Exactly one-half of the H_2A present initially has been converted into its conjugate base HA^-, so $[H_2A] = [HA^-]$ in this buffer solution.

$$H_2A \rightleftharpoons HA^- + H^+ \qquad K_{a_1} = \frac{[HA^-][H^+]}{[H_2A]}$$

When $[HA^-] = [H_2A]$, then $K_{a_1} = [H^+]$ or $pK_{a_1} = pH$.

Here, pH = 4.0, so $pK_{a_1} = 4.0$ and $K_{a_1} = 10^{-4.0} = 1 \times 10^{-4}$.

150.0 mL of NaOH added correspond to the second halfway point to equivalence, where $[HA^-] = [A^{2-}]$ in this buffer solution.

$$HA^- \rightleftharpoons A^{2-} + H^+ \qquad K_{a_2} = \frac{[A^{2-}][H^+]}{[HA^-]}$$

When $[A^{2-}] = [HA^-]$, then $K_{a_2} = [H^+]$ or $pK_{a_2} = pH$.

Here, pH = 8.0, so $pK_{a_2} = 8.0$ and $K_{a_2} = 10^{-8.0} = 1 \times 10^{-8}$.

107. For a triprotic titration, there are three titrations to consider. In the first titration, added OH^- converts H_3A into H_2A^-. Because the molarities of H_3A and KOH are equal, the first equivalence point will be at 50.0 mL of KOH added. So from 0 – 50.0 mL of KOH added, the first proton is removed from the triprotic acid. From 50.0 – 100.0 mL of KOH added, OH^- converts H_2A^- into HA^{2-}, and from 100.0 – 150.0 mL of KOH added, OH^- converts HA^{2-} into A^{3-}.

a. 125.0 mL of KOH added is the third halfway point to equivalence where $[HA^{2-}] = [A^{3-}]$. At this point we have a buffer solution where pH = pK_{a_3}. pH = $-\log(1.0 \times 10^{-11})$ = 11.00.

b. $pK_{a_1} = -\log(5.0 \times 10^{-4}) = 3.30$; pH = pK_{a_1} = 3.30 at the first halfway point to equivalence. This occurs at 25.0 mL of KOH added. Here, we have a buffer solution where $[H_3A] = [H_2A^-]$, so pH = pK_{a_1}.

c. 75.0 mL of KOH added is the second halfway point to equivalence where $[H_2A^-] = [HA^{2-}]$. At this buffer point, pH = pK_{a_2}. pH = $-\log(1.0 \times 10^{-8})$ = 8.00; the pH will be basic at 75.0 mL of KOH added.

ChemWork Problems

109. $NH_3 + H_2O \rightleftharpoons NH_4^+ + OH^-$ $K_b = \dfrac{[NH_4^+][OH^-]}{[NH_3]}$; taking the −log of the K_b expression:

$$-\log K_b = -\log[OH^-] - \log\frac{[NH_4^+]}{[NH_3]}, \quad -\log[OH^-] = -\log K_b + \log\frac{[NH_4^+]}{[NH_3]}$$

$$pOH = pK_b + \log\frac{[NH_4^+]}{[NH_3]} \text{ or } pOH = pK_b + \log\frac{[acid]}{[base]}$$

111. a. The optimum pH for a buffer is when pH = pK_a. At this pH a buffer will have equal neutralization capacity for both added acid and base. As shown next, because the pK_a for $TRISH^+$ is 8.1, the optimal buffer pH is about 8.1.

$K_b = 1.19 \times 10^{-6}$; $K_a = K_w/K_b = 8.40 \times 10^{-9}$; $pK_a = -\log(8.40 \times 10^{-9}) = 8.076$

b. $pH = pK_a + \log\dfrac{[TRIS]}{[TRISH^+]}$, $7.00 = 8.076 + \log\dfrac{[TRIS]}{[TRISH^+]}$

$$\frac{[TRIS]}{[TRISH^+]} = 10^{-1.08} = 0.083 \quad \text{(at pH = 7.00)}$$

$$9.00 = 8.076 + \log\frac{[TRIS]}{[TRISH^+]}, \quad \frac{[TRIS]}{[TRISH^+]} = 10^{0.92} = 8.3 \quad \text{(at pH = 9.00)}$$

c. $\dfrac{50.0 \text{ g TRIS}}{2.0 \text{ L}} \times \dfrac{1 \text{ mol}}{121.14 \text{ g}} = 0.206\ M = 0.21\ M = [TRIS]$

$$\frac{65.0 \text{ g TRISHCl}}{2.0 \text{ L}} \times \frac{1 \text{ mol}}{157.60 \text{ g}} = 0.206\ M = 0.21\ M = [TRISHCl] = [TRISH^+]$$

$$pH = pK_a + \log\frac{[TRIS]}{[TRISH^+]} = 8.076 + \log\frac{(0.21)}{(0.21)} = 8.08$$

The amount of H^+ added from HCl is: $(0.50 \times 10^{-3} \text{ L}) \times 12 \text{ mol/L} = 6.0 \times 10^{-3}$ mol H^+

The H^+ from HCl will convert TRIS into $TRISH^+$. The reaction is:

	TRIS	+	H^+	→	$TRISH^+$	
Before	0.21 M		$\dfrac{6.0 \times 10^{-3}}{0.2005} = 0.030\ M$		0.21 M	
Change	−0.030		−0.030	→	+0.030	Reacts completely
After	0.18		0		0.24	

Now use the Henderson-Hasselbalch equation to solve this buffer problem.

$$pH = 8.076 + \log\left(\frac{0.18}{0.24}\right) = 7.95$$

113. A best buffer is when pH ≈ pK_a; these solutions have about equal concentrations of weak acid and conjugate base. Therefore, choose combinations that yield a buffer where pH ≈ pK_a; that is, look for acids whose pK_a is closest to the pH.

a. Potassium fluoride + HCl will yield a buffer consisting of HF (pK_a = 3.14) and F^-.

b. Benzoic acid + NaOH will yield a buffer consisting of benzoic acid (pK_a = 4.19) and benzoate anion.

c. Sodium acetate + acetic acid (pK_a = 4.74) is the best choice for pH = 5.0 buffer since acetic acid has a pK_a value closest to 5.0.

d. HOCl and NaOH: This is the best choice to produce a conjugate acid-base pair with pH = 7.0. This mixture would yield a buffer consisting of HOCl (pK_a = 7.46) and OCl^-. Actually, the best choice for a pH = 7.0 buffer is an equimolar mixture of ammonium chloride and sodium acetate. NH_4^+ is a weak acid ($K_a = 5.6 \times 10^{-10}$), and $C_2H_3O_2^-$ is a weak base ($K_b = 5.6 \times 10^{-10}$). A mixture of the two will give a buffer at pH = 7.0 because the weak acid and weak base are the same strengths (K_a for NH_4^+ = K_b for $C_2H_3O_2^-$). $NH_4C_2H_3O_2$ is commercially available, and its solutions are used for pH = 7.0 buffers.

e. Ammonium chloride + NaOH will yield a buffer consisting of NH_4^+ (pK_a = 9.26) and NH_3.

115. a. $\text{pH} = \text{p}K_a + \log\frac{[\text{base}]}{[\text{acid}]},\ 7.15 = -\log(6.2 \times 10^{-8}) + \log\frac{[\text{HPO}_4^{\,2-}]}{[\text{H}_2\text{PO}_4^{\,-}]}$

$$7.15 = 7.21 + \log\frac{[\text{HPO}_4^{\,2-}]}{[\text{H}_2\text{PO}_4^{\,-}]},\ \frac{[\text{HPO}_4^{\,2-}]}{[\text{H}_2\text{PO}_4^{\,-}]} = 10^{-0.06} = 0.9,\ \frac{[\text{H}_2\text{PO}_4^{\,-}]}{[\text{HPO}_4^{\,2-}]} = \frac{1}{0.9} = 1.1 \approx 1$$

b. A best buffer has approximately equal concentrations of weak acid and conjugate base, so pH ≈ pK_a for a best buffer. The pK_a value for a $H_3PO_4/H_2PO_4^-$ buffer is $-\log(7.5 \times 10^{-3})$ = 2.12. A pH of 7.15 is too high for a $H_3PO_4/H_2PO_4^-$ buffer to be effective. At this high of pH, there would be so little H_3PO_4 present that we could hardly consider it a buffer; this solution would not be effective in resisting pH changes, especially when a strong base is added.

117. $\text{pH} = \text{p}K_a + \log\frac{[\text{C}_7\text{H}_4\text{O}_2\text{F}^-]}{[\text{C}_7\text{H}_5\text{O}_2\text{F}]} = 2.90 + \log\left[\frac{(55.0\text{ mL} \times 0.472\ M)/130.0\text{ mL}}{(75.0\text{ mL} \times 0.275\ M)/130.0\text{ mL}}\right]$

$$\text{pH} = 2.90 + \log\left(\frac{26.0}{20.6}\right) = 2.90 + 0.101 = 3.00$$

119. a. 1.00 L × 0.100 mol/L = 0.100 mol HCl added to reach stoichiometric point.

The 10.00-g sample must have contained 0.100 mol of NaA. $\frac{10.00\text{ g}}{0.100\text{ mol}} = 100.\text{ g/mol}$

b. 500.0 mL of HCl added represents the halfway point to equivalence. Thus pH = pK_a = 5.00 and $K_a = 1.0 \times 10^{-5}$. At the equivalence point, enough H^+ has been added to convert all the A^- present initially into HA. The concentration of HA at the equivalence point is:

$$[HA]_0 = \frac{0.100 \text{ mol}}{1.10 \text{ L}} = 0.0909\ M$$

	HA	$\rightleftharpoons$	H^+	+	A^-	$K_a = 1.0 \times 10^{-5}$
Initial	$0.0909\ M$		0		0	
Equil.	$0.0909 - x$		x		x	

$$K_a = 1.0 \times 10^{-5} = \frac{x^2}{0.0909 - x} \approx \frac{x^2}{0.0909}$$

$x = 9.5 \times 10^{-4}\ M = [H^+]$; pH = 3.02; assumptions good.

121. a. $HC_2H_3O_2 + OH^- \rightleftharpoons C_2H_3O_2^- + H_2O$

$$K_{eq} = \frac{[C_2H_3O_2^-]}{[HC_2H_3O_2][OH^-]} \times \frac{[H^+]}{[H^+]} = \frac{K_{a,\,HC_2H_3O_2}}{K_w} = \frac{1.8 \times 10^{-5}}{1.0 \times 10^{-14}} = 1.8 \times 10^9$$

Note: This reaction is the reverse of the K_b reaction for $C_2H_3O_2^-$.

b. $C_2H_3O_2^- + H^+ \rightleftharpoons HC_2H_3O_2$ $\quad K_{eq} = \frac{[HC_2H_3O_2]}{[H^+][C_2H_3O_2^-]} = \frac{1}{K_{a,\,HC_2H_3O_2}} = 5.6 \times 10^4$

c. $HCl + NaOH \rightarrow NaCl + H_2O$

Net ionic equation is $H^+ + OH^- \rightleftharpoons H_2O$; $K_{eq} = \frac{1}{K_w} = 1.0 \times 10^{14}$

123. a. This is a weak acid-strong base titration. At the halfway point to equivalence, [weak acid] = [conjugate base], so pH = pK_a (always for a weak acid-strong base titration).

$pH = -\log(6.4 \times 10^{-5}) = 4.19$

mmol $HC_7H_5O_2$ present = 100.0 mL × 0.10 M = 10. mmol $HC_7H_5O_2$. For the equivalence point, 10. mmol of OH^- must be added. The volume of OH^- added to reach the equivalence point is:

$$10.\text{ mmol } OH^- \times \frac{1 \text{ mL}}{0.10 \text{ mmol } OH^-} = 1.0 \times 10^2 \text{ mL } OH^-$$

At the equivalence point, 10. mmol of $HC_7H_5O_2$ is neutralized by 10. mmol of OH^- to produce 10. mmol of $C_7H_5O_2^-$. This is a weak base. The total volume of the solution is 100.0 mL + 1.0×10^2 mL = 2.0×10^2 mL. Solving the weak base equilibrium problem:

	$C_7H_5O_2^-$ + H_2O	$\rightleftharpoons$	$HC_7H_5O_2$	+	OH^-	$K_b = \frac{1.0 \times 10^{-14}}{6.4 \times 10^{-5}} = 1.6 \times 10^{-10}$
Initial	10. mmol/2.0×10^2 mL		0		0	
Equil.	$0.050 - x$		x		x	

$$K_b = 1.6 \times 10^{-10} = \frac{x^2}{0.050 - x} \approx \frac{x^2}{0.050},\ x = [OH^-] = 2.8 \times 10^{-6}\ M$$

pOH = 5.55; pH = 8.45; assumptions good.

b. At the halfway point to equivalence for a weak base-strong acid titration, pH = pK_a because [weak base] = [conjugate acid].

$$K_a = \frac{K_w}{K_b} = \frac{1.0 \times 10^{-14}}{5.6 \times 10^{-4}} = 1.8 \times 10^{-11};\ pH = pK_a = -\log(1.8 \times 10^{-11}) = 10.74$$

For the equivalence point (mmol acid added = mmol base present):

mmol $C_2H_5NH_2$ present = 100.0 mL × 0.10 *M* = 10. mmol $C_2H_5NH_2$

$$\text{mL } H^+ \text{ added} = 10.\ \text{mmol } H^+ \times \frac{1\ \text{mL}}{0.20\ \text{mmol } H^+} = 50.\ \text{mL } H^+$$

The strong acid added completely converts the weak base into its conjugate acid. Therefore, at the equivalence point, $[C_2H_5NH_3^+]_0$ = 10. mmol/(100.0 + 50.) mL = 0.067 *M*. Solving the weak acid equilibrium problem:

	$C_2H_5NH_3^+$	$\rightleftharpoons$	H^+	+	$C_2H_5NH_2$
Initial	0.067 *M*		0		0
Equil.	0.067 − *x*		*x*		*x*

$$K_a = 1.8 \times 10^{-11} = \frac{x^2}{0.067 - x} \approx \frac{x^2}{0.067},\ x = [H^+] = 1.1 \times 10^{-6}\ M$$

pH = 5.96; assumptions good.

c. In a strong acid-strong base titration, the halfway point has no special significance other than that exactly one-half of the original amount of acid present has been neutralized.

mmol H^+ present = 100.0 mL × 0.50 *M* = 50. mmol H^+

$$\text{mL } OH^- \text{ added} = 25\ \text{mmol } OH^- \times \frac{1\ \text{mL}}{0.25\ \text{mmol}} = 1.0 \times 10^2\ \text{mL } OH^-$$

	H^+	+	OH^-	→	H_2O
Before	50. mmol		25 mmol		
After	25 mmol		0		

$$[H^+]_{excess} = \frac{25\ \text{mmol}}{(100.0 + 1.0 \times 10^2)\ \text{mL}} = 0.13\ M;\ pH = 0.89$$

At the equivalence point of a strong acid-strong base titration, only neutral species are present (Na^+, Cl^-, and H_2O), so the pH = 7.00.

125. In the final solution: $[H^+] = 10^{-2.15} = 7.1 \times 10^{-3}$ *M*

Beginning mmol HCl = 500.0 mL × 0.200 mmol/mL = 100. mmol HCl

Amount of HCl that reacts with NaOH = 1.50×10^{-2} mmol/mL × V

$$\frac{7.1 \times 10^{-3}\ \text{mmol}}{\text{mL}} = \frac{\text{final mmol H}^+}{\text{total volume}} = \frac{100. - (0.0150)V}{500.0 + V}$$

$3.6 + (7.1 \times 10^{-3})V = 100. - (1.50 \times 10^{-2})V$, $(2.21 \times 10^{-2})V = 100. - 3.6$

$V = 4.36 \times 10^3$ mL = 4.36 L = 4.4 L NaOH

127. $HC_2H_3O_2 \rightleftharpoons H^+ + C_2H_3O_2^-$; let C_0 = initial concentration of $HC_2H_3O_2$

From normal weak acid setup: $K_a = 1.8 \times 10^{-5} = \dfrac{[H^+][C_2H_3O_2^-]}{[HC_2H_3O_2]} = \dfrac{[H^+]^2}{C_0 - [H^+]}$

$[H^+] = 10^{-2.68} = 2.1 \times 10^{-3}$ *M*; $1.8 \times 10^{-5} = \dfrac{(2.1 \times 10^{-3})^2}{C_0 - (2.1 \times 10^{-3})}$, $C_0 = 0.25$ *M*

25.0 mL × 0.25 mmol/mL = 6.3 mmol $HC_2H_3O_2$

Need 6.3 mmol KOH = $V_{KOH} \times 0.0975$ mmol/mL, V_{KOH} = 65 mL

129. $HA + OH^- \rightarrow A^- + H_2O$, where HA = acetylsalicylic acid (assuming it is a monoprotic acid).

$$\text{mmol HA present} = 27.36\ \text{mL OH}^- \times \frac{0.5106\ \text{mmol OH}^-}{\text{mL OH}^-} \times \frac{1\ \text{mmol HA}}{\text{mmol OH}^-} = 13.97\ \text{mmol HA}$$

$$\text{Molar mass of HA} = \frac{2.51\ \text{g HA}}{13.97 \times 10^{-3}\ \text{mol HA}} = 180.\ \text{g/mol}$$

To determine the K_a value, use the pH data. After complete neutralization of acetylsalicylic acid by OH^-, we have 13.97 mmol of A^- produced from the neutralization reaction. A^- will react completely with the added H^+ and re-form acetylsalicylic acid HA.

$$\text{mmol H}^+\ \text{added} = 13.68\ \text{mL} \times \frac{0.5106\ \text{mmol H}^+}{\text{mL}} = 6.985\ \text{mmol H}^+$$

	A^-	+	H^+	$\rightarrow$	HA	
Before	13.97 mmol		6.985 mmol		0	
Change	−6.985		−6.985	$\rightarrow$	+6.985	Reacts completely
After	6.985 mmol		0		6.985 mmol	

We have back titrated this solution to the halfway point to equivalence, where pH = pK_a (assuming HA is a weak acid). This is true because after H^+ reacts completely, equal millimoles of HA and A^- are present, which only occurs at the halfway point to equivalence. Assuming acetylsalicylic acid is a weak monoprotic acid, then pH = pK_a = 3.48. $K_a = 10^{-3.48} = 3.3 \times 10^{-4}$.

131. 50.0 mL × 0.100 *M* = 5.00 mmol NaOH initially

At pH = 10.50, pOH = 3.50, $[OH^-] = 10^{-3.50} = 3.2 \times 10^{-4}$ *M*

mmol OH^- remaining = 3.2×10^{-4} mmol/mL × 73.75 mL = 2.4×10^{-2} mmol

mmol OH^- that reacted = 5.00 – 0.024 = 4.98 mmol

Because the weak acid is monoprotic, 23.75 mL of the weak acid solution contains 4.98 mmol HA.

$$[HA]_0 = \frac{4.98 \text{ mmol}}{23.75 \text{ mL}} = 0.210\ M$$

133. At equivalence point: 16.00 mL × 0.125 mmol/mL = 2.00 mmol OH^- added; there must be 2.00 mmol HX present initially.

$HX + OH^- \rightarrow X^- + H_2O$ (neutralization rection)

2.00 mL NaOH added = 2.00 mL × 0.125 mmol/mL = 0.250 mmol OH^-; 0.250 mmol of OH^- added will convert 0.250 mmol HX into 0.250 mmol X^-. Remaining HX = 2.00 – 0.250 = 1.75 mmol HX; this is a buffer solution where $[H^+] = 10^{-6.912} = 1.22 \times 10^{-7}$ *M.* Because total volume cancels:

$$K_a = \frac{[H^+][X^-]}{[HX]} = \frac{1.22 \times 10^{-7}(0.250/V_T)}{1.75/V_T} = \frac{1.22 \times 10^{-7}(0.250)}{1.75} = 1.74 \times 10^{-8}$$

Note: We could solve for K_a using the Henderson-Hasselbalch equation.

135. Since we have added two solutions together, the concentrations of each reagent has changed. What hasn't changed is the moles or mmoles of each reagent. Let's determine the mmol of each reagent present by multiplying the volume in mL by the molarity in units of mmol/mL.

100.0 mL × 0.100 *M* = 10.0 mmol NaF; 100.0 mL × 0.025 *M* = 2.5 mmol HCl

Let the added H^+ from HCl react completely with the best base present, F^-. $H^+ + F^- \rightarrow HF$; 2.5 mmol H^+ converts 2.5 mmol F^- into 2.5 mmol HF. After the reaction, a buffer solution results containing 2.5 mmol HF and (10.0 – 2.5 =) 7.5 mmol F^- in 200.0 mL of solution.

$$pH = pK_a + \log\frac{[F^-]}{[HF]} = 3.14 + \log\left(\frac{7.5 \text{ mmol}/200.0 \text{ mL}}{2.5 \text{ mmol}/200.0 \text{ mL}}\right) = 3.62; \text{ assumptions good.}$$

Challenge Problems

137. mmol $HC_3H_5O_2$ present initially = 45.0 mL × $\frac{0.750 \text{ mmol}}{\text{mL}}$ = 33.8 mmol $HC_3H_5O_2$

mmol $C_3H_5O_2^-$ present initially = 55.0 mL × $\frac{0.700 \text{ mmol}}{\text{mL}}$ = 38.5 mmol $C_3H_5O_2^-$

The initial pH of the buffer is:

$$pH = pK_a + \log\frac{[C_3H_5O_2^-]}{[HC_3H_5O_2]} = -\log(1.3 \times 10^{-5}) + \log\frac{\frac{38.5 \text{ mmol}}{100.0 \text{ mL}}}{\frac{33.8 \text{ mmol}}{100.0 \text{ mL}}} = 4.89 + \log\frac{38.5}{33.8} = 4.95$$

Note: Because the buffer components are in the same volume of solution, we can use the mole (or millimole) ratio in the Henderson-Hasselbalch equation to solve for pH instead of using the concentration ratio of $[C_3H_5O_2^-] : [HC_3H_5O_2]$.

When NaOH is added, the pH will increase, and the added OH^- will convert $HC_3H_5O_2$ into $C_3H_5O_2^-$. The pH after addition of OH^- increases by 2.5%, so the resulting pH is:

$$4.95 + 0.025(4.95) = 5.07$$

At this pH, a buffer solution still exists, and the millimole ratio between $C_3H_5O_2^-$ and $HC_3H_5O_2$ is:

$$pH = pK_a + \log\frac{\text{mmol } C_3H_5O_2^-}{\text{mmol } HC_3H_5O_2}, \quad 5.07 = 4.89 + \log\frac{\text{mmol } C_3H_5O_2^-}{\text{mmol } HC_3H_5O_2}$$

$$\frac{\text{mmol } C_3H_5O_2^-}{\text{mmol } HC_3H_5O_2} = 10^{0.18} = 1.5$$

Let x = mmol OH^- added to increase pH to 5.07. Because OH^- will essentially react to completion with $HC_3H_5O_2$, the setup to the problem using millimoles is:

	$HC_3H_5O_2$	+	OH^-	$\rightarrow$	$C_3H_5O_2^-$	
Before	33.8 mmol		x mmol		38.5 mmol	
Change	$-x$		$-x$	$\rightarrow$	$+x$	Reacts completely
After	$33.8 - x$		0		$38.5 + x$	

$$\frac{\text{mmol } C_3H_5O_2^-}{\text{mmol } HC_3H_5O_2} = 1.5 = \frac{38.5 + x}{33.8 - x}, \quad 1.5(33.8 - x) = 38.5 + x, \quad x = 4.9 \text{ mmol } OH^- \text{ added}$$

The volume of NaOH necessary to raise the pH by 2.5% is:

$$4.9 \text{ mmol NaOH} \times \frac{1 \text{ mL}}{0.10 \text{ mmol NaOH}} = 49 \text{ mL}$$

49 mL of 0.10 M NaOH must be added to increase the pH by 2.5%.

139. For HOCl, $K_a = 3.5 \times 10^{-8}$ and $pK_a = -\log(3.5 \times 10^{-8}) = 7.46$. This will be a buffer solution because the pH is close to the pK_a value.

$$pH = pK_a + \log\frac{[OCl^-]}{[HOCl]}, \quad 8.00 = 7.46 + \log\frac{[OCl^-]}{[HOCl]}, \quad \frac{[OCl^-]}{[HOCl]} = 10^{0.54} = 3.5$$

1.00 L × 0.0500 M = 0.0500 mol HOCl initially. Added OH^- converts HOCl into OCl^-. The total moles of OCl^- and HOCl must equal 0.0500 mol. Solving where n = moles:

$n_{OCl^-} + n_{HOCl} = 0.0500$ and $n_{OCl^-} = (3.5)n_{HOCl}$

$(4.5)n_{HOCl} = 0.0500$, $n_{HOCl} = 0.011$ mol; $n_{OCl^-} = 0.039$ mol

Need to add 0.039 mol NaOH to produce 0.039 mol OCl^-.

0.039 mol = V × 0.0100 *M*, V = 3.9 L NaOH; *Note:* Normal buffer assumptions hold.

141. $pH = pK_a + \log\frac{[(CH_3)_2AsO_2^-]}{[(CH_3)_2AsO_2H]}$, $6.60 = 6.19 + \log\frac{[(CH_3)_2AsO_2^-]}{[(CH_3)_2AsO_2H]}$

$\frac{[(CH_3)_2AsO_2^-]}{[(CH_3)_2AsO_2H]} = 10^{0.41} = 2.6$, $[(CH_3)_2AsO_2^-] = 2.6\,[(CH_3)_2AsO_2H]$

$[(CH_3)_2AsO_2^-] + [(CH_3)_2AsO_2H] = 0.25$; $3.6\,[(CH_3)_2AsO_2H] = 0.25$

$[(CH_3)_2AsO_2H] = 0.069$ *M* and $[(CH_3)_2AsO_2^-] = 0.18$ *M*

$$0.500\text{ L} \times \frac{0.069\text{ mol }(CH_3)_2AsO_2H}{\text{L}} \times \frac{138.0\text{ g }(CH_3)_2AsO_2H}{\text{mol}} = 4.8\text{ g cacodylic acid}$$

$$0.500\text{ L} \times \frac{0.18\text{ mol }(CH_3)_2AsO_2Na}{\text{L}} \times \frac{160.0\text{ g }(CH_3)_2AsO_2Na}{\text{mol}} = 14\text{ g sodium cacodylate}$$

143. a. Na^+ is present in all solutions. The added H^+ from HCl reacts completely with CO_3^{2-} to convert it into HCO_3^- (points A-C). After all of the CO_3^{2-} is reacted (after point C, the first equivalence point), H^+ then reacts completely with the next best base present, HCO_3^- (points C-E). Point E represents the second equivalence point. The major species present at the various points after H^+ reacts completely follow.

A. CO_3^{2-}, H_2O, Na^+

B. CO_3^{2-}, HCO_3^-, H_2O, Cl^-, Na^+

C. HCO_3^-, H_2O, Cl^-, Na^+

D. HCO_3^-, CO_2 (H_2CO_3), H_2O, Cl^-, Na^+

E. CO_2 (H_2CO_3), H_2O, Cl^-, Na^+

F. H^+ (excess), CO_2 (H_2CO_3), H_2O, Cl^-. Na^+

b. $H_2CO_3 \rightleftharpoons HCO_3^- + H^+$ $\quad K_{a_1} = 4.3 \times 10^{-7}$

$HCO_3^- \rightleftharpoons CO_3^{2-} + H^+$ $\quad K_{a_2} = 5.6 \times 10^{-11}$

The first titration reaction occurring between points A and C is:

$$H^+ + CO_3^{2-} \rightarrow HCO_3^-$$

At point B, enough H^+ has been added to convert one-half of the CO_3^{2-} into its conjugate acid. At this halfway point to equivalence, $[CO_3^{2-}] = [HCO_3^-]$. For this buffer solution,

$pH = pK_{a_2} = -\log(5.6 \times 10^{-11}) = 10.25$

The second titration reaction occurring between points C and E is:

$$H^+ + HCO_3^- \rightarrow H_2CO_3$$

Point D is the second halfway point to equivalence, where $[HCO_3^-] = [H_2CO_3]$. Here, $pH = pK_{a_1} = -\log(4.3 \times 10^{-7}) = 6.37$.

145. Mmol $(C_2H_5)_3N$ = 100.0 mL × 0.50 M = 50. mmol

Mmol H^+ = 50.0 mL × 0.30 M = 15 mmol; mmol OH^- = 50.0 mL × 0.30 M = 15 mmol

Mmol $(C_2H_5)_3NH^+$ = 100.0 mL × 0.50 M = 50. mmol

Note that K^+ and ClO_4^- have no acidic and basic properties.

a. Major species: $(C_2H_5)_3N$, H^+, and ClO_4^-; let the added H^+ from the strong acid react to completion with the best base present. This is $(C_2H_5)_3N$.

	$(C_2H_5)_3N$	+	H^+	→	$(C_2H_5)_3NH^+$
Before	50. mmol		15 mmol		0
After	35 mmol		0		15 mmol

A buffer solution is present after all the H^+ reacts. Using the Henderson-Hasselbalch equation to solve for the pH (let V_T = total volume of solution):

$$pH = pK_a + \log\frac{[\text{base}]}{[\text{acid}]} \quad \text{where } pK_a = -\log(1.0 \times 10^{-14}/4.0 \times 10^{-4}) = 10.60$$

$$pH = 10.60 + \log\left(\frac{35 \text{ mmol}/V_T}{15 \text{ mmol}/V_T}\right) = 10.60 + 0.37 = 10.97$$

b. Major species: $(C_2H_5)_3N$, K^+, OH^-, $(C_2H_5)_3NH^+$, and ClO_4^-; let the added OH^- from the strong base react to completion with the best acid present. This is $(C_2H_5)_3NH^+$.

	$(C_2H_5)_3NH^+$	+	OH^-	→	$(C_2H_5)_3N$	+	H_2O
Before	50. mmol		15 mmol		50. mmol		
After	35 mmol		0		65 mmol		

A buffer solution is present after all the OH^- reacts.

$$pH = pK_a + \log\frac{[\text{base}]}{[\text{acid}]} = 10.60 + \log\left(\frac{65 \text{ mmol}/V_T}{35 \text{ mmol}/V_T}\right) = 10.60 + (0.27) = 10.87$$

c. Major species: $(C_2H_5)_3N$, H^+, ClO_4^-, K^+, OH^-, and $(C_2H_5)_3NH^+$; React the best acid, H^+, with the best base, OH^-. Since a strong acid and a strong base is reacting, the reaction is assumed to go to completion.

	OH^-	+	H^+	→	H_2O
Before	15. mmol		15 mmol		
After	0		0		

Major species after reaction: $(C_2H_5)_3N$, ClO_4^-, K^+, and $(C_2H_5)_3NH^+$. We have 50. mmol of the weak base $(C_2H_5)_3N$, 50. mmol of its conjugate acid $(C_2H_5)_3NH^+$, and some ions (ClO_4^- and K^+) that have no acidic or basic properties. We have a buffer solution with equal amounts of the weak base and conjugate acid, so pH = pK_a of the conjugate acid = 10.60.

147. An indicator changes color at pH ≈ pK_a ±1. The result from each indicator tells us something about the pH. The conclusions are summarized below:

Results from	pH
bromphenol blue	≥~5.0
bromcresol purple	≤~5.0
bromcresol green *	pH ≈ pK_a ≈ 4.8
alizarin	≤~5.5

*For bromcresol green, the resultant color is green. This is a combination of the extremes (yellow and blue). This occurs when pH ≈ pK_a of the indicator.

From the indicator results, the pH of the solution is about 5.0. We solve for K_a by setting up the typical weak acid problem.

	HX	⇌	H^+	+	X^-
Initial	1.0 *M*		~0		0
Equil.	$1.0 - x$		x		x

$$K_a = \frac{[H^+][X^-]}{[HX]} = \frac{x^2}{1.0 - x}; \text{ because pH} \approx 5.0,\ [H^+] = x \approx 1 \times 10^{-5}\,M.$$

$$K_a \approx \frac{(1 \times 10^{-5})^2}{1.0 - 1 \times 10^{-5}} \approx 1 \times 10^{-10}$$

CHAPTER 16

SOLUBILITY AND COMPLEX ION EQUILIBRIA

Questions

13. K_{sp} values can only be directly compared to determine relative solubilities when the salts produce the same number of ions (have the same stoichiometry). Here, Ag_2S and CuS do not produce the same number of ions when they dissolve, so each has a different mathematical relationship between the K_{sp} value and the molar solubility. To determine which salt has the larger molar solubility, you must do the actual calculations and compare the two molar solubility values.

15. i. This is the result when you have a salt that breaks up into two ions. Examples of these salts include AgCl, $SrSO_4$, $BaCrO_4$, and $ZnCO_3$.

    ii. This is the result when you have a salt that breaks up into three ions, either two cations and one anion or one cation and two anions. Some examples are SrF_2, Hg_2I_2, and Ag_2SO_4.

    iii. This is the result when you have a salt that breaks up into four ions, either three cations and one anion (Ag_3PO_4) or one cation and three anions (ignoring the hydroxides, there are no examples of this type of salt in Table 16.1).

    iv. This is the result when you have a salt that breaks up into five ions, either three cations and two anions [$Sr_3(PO_4)_2$] or two cations and three anions (no examples of this type of salt are in Table 16.1).

17. For the K_{sp} reaction of a salt dissolving into its respective ions, a common ion would be one of the ions in the salt added from an outside source. When a common ion (a product in the K_{sp} reaction) is present, the K_{sp} equilibrium shifts to the left, resulting in less of the salt dissolving into its ions (solubility decreases).

19. Some people would automatically think that an increase in temperature would increase the solubility of a salt. This is not always the case as some salts show a decrease in solubility as temperature increases. The two major methods used to increase solubility of a salt both involve removing one of the ions in the salt by reaction. If the salt has an ion with basic properties, adding H^+ will increase the solubility of the salt because the added H^+ will react with the basic ion, thus removing it from solution. More salt dissolves to make up for the lost ion. Some examples of salts with basic ions are AgF, $CaCO_3$, and $Al(OH)_3$. The other way to remove an ion is to form a complex ion. For example, the Ag^+ ion in silver salts forms the complex ion $Ag(NH_3)_2^+$ as ammonia is added. Silver salts increase their solubility as NH_3 is added because the Ag^+ ion is removed through complex ion formation.

21. Parts ii and ii have a common ion present before $Ag_2CrO_4(s)$ dissolve. In part ii, the common ion is Ag^+, and in part iii, the common is CrO_4^{2-}. The presence of a common decreases the solubility of $Ag_2CrO_4(s)$. Hence, the molar solubility of Ag_2CrO_4 in water (i) will have the largest value.

23. For conjugate acid-base pairs, the weaker the acid, the stronger is the conjugate base. Because HX is a stronger acid (has a larger K_a value) than HY, Y^- will be a stronger base than X^-. In acidic solution, Y^- will have a greater affinity for the H^+ ions. Therefore, AgY(s) will be more soluble in acidic solution because more Y^- will be removed through reaction with H^+, which will cause more AgY(s) to dissolve.

25. The $C_2H_3O_2^-$ anion is the conjugate base of the weak acid $HC_2H_3O_2$. In acidic solution, $C_2H_3O_2^-$ will react with H^+ to form $HC_2H_3O_2$. Because of this reaction, $AgC_2H_3O_2(s)$ will be more soluble with addition of HNO_3. Ag^+ forms the complex ion $Ag(NH_3)_2^+$. Because of the formation of the complex ion, $AgC_2H_3O_2(s)$ will be more soluble with addition of NH_3.

27. In 2.0 *M* NH_3, the soluble complex ion $Cu(NH_3)_4^{2+}$ forms, which increases the solubility of $CuCO_3(s)$. The reaction is $CuCO_3(s) + 4\,NH_3 \rightleftharpoons Cu(NH_3)_4^{2+} + CO^{2-}$. In 2.0 *M* NH_4NO_3, NH_3 is only formed by the dissociation of the weak acid NH_4^+. There is not enough NH_3 produced by this reaction to dissolve $CuCO_3(s)$ by the formation of the complex ion.

Exercises

Solubility Equilibria

29. a. $AgC_2H_3O_2(s) \rightleftharpoons Ag^+(aq) + C_2H_3O_2^-(aq) \quad K_{sp} = [Ag^+][C_2H_3O_2^-]$

b. $Al(OH)_3(s) \rightleftharpoons Al^{3+}(aq) + 3\,OH^-(aq) \quad K_{sp} = [Al^{3+}][OH^-]^3$

c. $Ca_3(PO_4)_2(s) \rightleftharpoons 3\,Ca^{2+}(aq) + 2\,PO_4^{3-}(aq) \quad K_{sp} = [Ca^{2+}]^3[PO_4^{3-}]^2$

31. In our setup, s = solubility of the ionic solid in mol/L. This is defined as the maximum amount of a salt that can dissolve. Because solids do not appear in the K_{sp} expression, we do not need to worry about their initial and equilibrium amounts.

a.

	$CaC_2O_4(s)$	$\rightleftharpoons$	$Ca^{2+}(aq)$	+	$C_2O_4^{2-}(aq)$
Initial			0		0
	s mol/L of $CaC_2O_4(s)$ dissolves to reach equilibrium				
Change	$-s$	$\rightarrow$	$+s$		$+s$
Equil.			s		s

From the problem, $s = 4.8 \times 10^{-5}$ mol/L.

$K_{sp} = [Ca^{2+}][C_2O_4^{2-}] = (s)(s) = s^2, \quad K_{sp} = (4.8 \times 10^{-5})^2 = 2.3 \times 10^{-9}$

b.

	$Cu(IO_3)_2(s)$	$\rightleftharpoons$	$Cu^{2+}(aq)$	+	$2\ IO_3^-(aq)$
Initial			0		0
	s mol/L of $Cu(IO_3)_2(s)$ dissolves to reach equilibrium				
Change	$-s$	$\rightarrow$	$+s$		$+2s$
Equil.			s		$2s$

$K_{sp} = [Cu^{2+}][IO_3^-]^2 = (s)(2s)^2 = 4s^3$, $K_{sp} = 4(3.3 \times 10^{-3})^3 = 1.4 \times 10^{-7}$

33. $\text{Solubility} = s = \dfrac{0.14\text{ g Ni(OH)}_2}{\text{L}} \times \dfrac{1\text{ mol Ni(OH)}_2}{92.71\text{ g Ni(OH)}_2} = 1.5 \times 10^{-3}\text{ mol/L}$

	$Ni(OH)_2(s)$	$\rightleftharpoons$	$Ni^{2+}(aq)$	+	$2\ OH^-(aq)$
Initial			0		$1.0 \times 10^{-7}\ M$ (from water)
	s mol/L of $Ni(OH)_2$ dissolves to reach equilibrium				
Change	$-s$	$\rightarrow$	$+s$		$+2s$
Equil.			s		$1.0 \times 10^{-7} + 2s$

From the calculated molar solubility, $1.0 \times 10^{-7} + 2s \approx 2s$.

$K_{sp} = [Ni^{2+}][OH^-]^2 = s(2s)^2 = 4s^3$, $K_{sp} = 4(1.5 \times 10^{-3})^3 = 1.4 \times 10^{-8}$

35.

	$PbBr_2(s)$	$\rightleftharpoons$	$Pb^{2+}(aq)$	+	$2\ Br^-(aq)$
Initial			0		0
	s mol/L of $PbBr_2(s)$ dissolves to reach equilibrium				
Change	$-s$	$\rightarrow$	$+s$		$+2s$
Equil.			s		$2s$

From the problem, $s = [Pb^{2+}] = 2.14 \times 10^{-2}\ M$. So:

$K_{sp} = [Pb^{2+}][Br^-]^2 = s(2s)^2 = 4s^3$, $K_{sp} = 4(2.14 \times 10^{-2})^3 = 3.92 \times 10^{-5}$

37.

	$Ce(IO_3)_3(s)$	$\rightleftharpoons$	$Ce^{3+}(aq)$	+	$3\ IO_3^-(aq)$
Initial			0		0
	s mol/L of $Ce(IO_3)_3(s)$ dissolves to reach equilibrium				
Change	$-s$	$\rightarrow$	$+s$		$+3s$
Equil.			s		$3s$

From the problem, $3s = [IO_3^-] = 5.7 \times 10^{-3} M$, $s = 1.9 \times 10^{-3}\ M$. Solving for K_{sp}:

$K_{sp} = [Ce^{3+}][IO_3^-]^3 = s(3s)^3 = 27s^4$, $K_{sp} = 27(1.9 \times 10^{-3})^4 = 3.5 \times 10^{-10}$

39. In our setup, s = solubility in mol/L. Because solids do not appear in the K_{sp} expression, we do not need to worry about their initial or equilibrium amounts.

a.

	$Ag_3PO_4(s)$	$\rightleftharpoons$	$3\ Ag^+(aq)$	+	$PO_4^{3-}(aq)$
Initial			0		0
	s mol/L of $Ag_3PO_4(s)$ dissolves to reach equilibrium				
Change	$-s$	$\rightarrow$	$+3s$		$+s$
Equil.			$3s$		s

$K_{sp} = 1.8 \times 10^{-18} = [Ag^+]^3[PO_4^{3-}] = (3s)^3(s) = 27s^4$

$27s^4 = 1.8 \times 10^{-18}$, $s = (6.7 \times 10^{-20})^{1/4} = 1.6 \times 10^{-5}$ mol/L = molar solubility

b.

	$CaCO_3(s)$	$\rightleftharpoons$	$Ca^{2+}(aq)$	+	$CO_3^{2-}(aq)$
Initial	s = solubility (mol/L)		0		0
Equil.			s		s

$K_{sp} = 8.7 \times 10^{-9} = [Ca^{2+}][CO_3^{2-}] = s^2$, $s = 9.3 \times 10^{-5}$ mol/L

c.

	$Hg_2Cl_2(s)$	$\rightleftharpoons$	$Hg_2^{2+}(aq)$	+	$2\ Cl^-(aq)$
Initial	s = solubility (mol/L)		0		0
Equil.			s		$2s$

$K_{sp} = 1.1 \times 10^{-18} = [Hg_2^{2+}][Cl^-]^2 = (s)(2s)^2 = 4s^3$, $s = 6.5 \times 10^{-7}$ mol/L

41. KBT dissolves to form the potassium ion (K^+) and the bitartrate ion (abbreviated as BT^-).

	$KBT(s)$	$\rightleftharpoons$	$K^+(aq)$	+	$BT^-(aq)$	$K_{sp} = 3.8 \times 10^{-4}$
Initial	s = solubility (mol/L)		0		0	
Equil.			s		s	

$3.8 \times 10^{-4} = [K^+][BT^-] = s(s) = s^2$, $s = 1.9 \times 10^{-2}$ mol/L

$$0.2500\ \text{L} \times \frac{1.9 \times 10^{-2}\ \text{mol KBT}}{\text{L}} \times \frac{188.2\ \text{g KBT}}{\text{mol}} = 0.89\ \text{g KBT}$$

43.

	$Cd(OH)_2(s)$	$\rightleftharpoons$	$Cd^{2+}(aq)$	+	$2\ OH^-(aq)$	$K_{sp} = 5.9 \times 10^{-15}$
Initial	s = solubility (mol/L)		0		$1.0 \times 10^{-7}\ M$	
Equil.			s		$1.0 \times 10^{-7} + 2s$	

$K_{sp} = [Cd^{2+}][OH^-]^2 = s(1.0 \times 10^{-7} + 2s)^2$; assume that $1.0 \times 10^{-7} + 2s \approx 2s$, then:

$K_{sp} = 5.9 \times 10^{-15} = s(2s)^2 = 4s^3$, $s = 1.1 \times 10^{-5}$ mol/L

Assumption is good (1.0×10^{-7} is 0.4% of $2s$). Molar solubility = 1.1×10^{-5} mol/L

45. Let s = solubility of $Al(OH)_3$ in mol/L. *Note*: Because solids do not appear in the K_{sp} expression, we do not need to worry about their initial or equilibrium amounts.

	$Al(OH)_3(s)$	$\rightleftharpoons$	$Al^{3+}(aq)$	+	$3\ OH^-(aq)$
Initial			0		$1.0 \times 10^{-7}\ M$ (from water)
	s mol/L of $Al(OH)_3(s)$ dissolves to reach equilibrium = molar solubility				
Change	$-s$	$\rightarrow$	$+s$		$+3s$
Equil.			s		$1.0 \times 10^{-7} + 3s$

$K_{sp} = 2 \times 10^{-32} = [Al^{3+}][OH^-]^3 = (s)(1.0 \times 10^{-7} + 3s)^3 \approx s(1.0 \times 10^{-7})^3$

$s = \dfrac{2 \times 10^{-32}}{1.0 \times 10^{-21}} = 2 \times 10^{-11}$ mol/L; assumption good ($1.0 \times 10^{-7} + 3s \approx 1.0 \times 10^{-7}$).

47.

	$M_3Y_4(s)$	$\rightleftharpoons$	$3\ M^{4+}(aq)$	+	$4\ Y^{3-}(aq)$
Initial			0		0
	s mol/L of $M_3Y_4(s)$ dissolves to reach equilibrium = molar solubility				
Change	$-s$	$\rightarrow$	$+3s$		$+4s$
Equil.			$3s$		$4s$

$K_{sp} = [M^{4+}]^3[Y^{3-}]^4 = (3s)^3(4s)^4 = 6912s^7$

49. a. Because both solids dissolve to produce three ions in solution, we can compare values of K_{sp} to determine relative solubility. Because the K_{sp} for CaF_2 is the smallest, $CaF_2(s)$ has the smallest molar solubility.

b. We must calculate molar solubilities because each salt yields a different number of ions when it dissolves.

	$Ca_3(PO_4)_2(s)$	$\rightleftharpoons$	$3\ Ca^{2+}(aq)$	+	$2\ PO_4^{3-}(aq)$	$K_{sp} = 1.3 \times 10^{-32}$
Initial	s = solubility (mol/L)		0		0	
Equil.			$3s$		$2s$	

$K_{sp} = [Ca^{2+}]^3[PO_4^{3-}]^2 = (3s)^3(2s)^2 = 108s^5$, $s = (1.3 \times 10^{-32}/108)^{1/5} = 1.6 \times 10^{-7}$ mol/L

	$FePO_4(s)$	$\rightleftharpoons$	$Fe^{3+}(aq)$	+	$PO_4^{3-}(aq)$	$K_{sp} = 1.0 \times 10^{-22}$
Initial	s = solubility (mol/L)		0		0	
Equil.			s		s	

$K_{sp} = [Fe^{3+}][PO_4^{3-}] = s^2$, $s = \sqrt{1.0 \times 10^{-22}} = 1.0 \times 10^{-11}$ mol/L

$FePO_4$ has the smallest molar solubility.

51.

	$Sr_3(PO_4)_2(s)$	$\rightleftharpoons$	$3\ Sr^{2+}(aq)$	+	$2\ PO_4^{3-}(aq)$	$K_{sp} = 1 \times 10^{-31}$
Initial	s = solubility (mol/L)		0		0	
Equil.			$3s$		$2s$	

$K_{sp} = [Sr^{2+}]^3[PO_4^{3-}]^2 = (3s)^3(2s)^2 = 108s^5$, $s = (1 \times 10^{-31}/108)^{1/5} = 2 \times 10^{-7}$ mol/L

	MnS(s)	⇌	Mn^{2+}(aq)	+	S^{2-}(aq)
Equil.	s = solubility (mol/L)		s		s

$K_{sp} = 2 \times 10^{-13} = [Mn^{2+}][S^{2-}] = s^2$, $s = 4 \times 10^{-7}$ mol/L; MnS has a greater solubility.

	FeS(s)	⇌	Fe^{2+}(aq)	+	S^{2-}(aq)
Equil.	s = solubility (mol/L)		s		s

$K_{sp} = 4 \times 10^{-19} = [Fe^{2+}][S^{2-}] = s^2$, $s = 6 \times 10^{-10}$ mol/L

FeS does not have a greater molar solubility than $Sr_3(PO_4)_2$. NiS and CoS have smaller K_{sp} values than FeS, so they also do not have a greater molar solubility. Only MnS has a greater molar solubility than $Sr_3(PO_4)_2$.

53. a.

	$Fe(OH)_3$(s)	⇌	Fe^{3+}(aq)	+	3 OH^-(aq)
Initial			0		1×10^{-7} *M* (from water)
	s mol/L of $Fe(OH)_3$(s) dissolves to reach equilibrium = molar solubility				
Change	$-s$	→	$+s$		$+3s$
Equil.			s		$1 \times 10^{-7} + 3s$

$K_{sp} = 4 \times 10^{-38} = [Fe^{3+}][OH^-]^3 = (s)(1 \times 10^{-7} + 3s)^3 \approx s(1 \times 10^{-7})^3$

$s = 4 \times 10^{-17}$ mol/L; assumption good ($3s << 1 \times 10^{-7}$)

b.

$Fe(OH)_3(s) \rightleftharpoons Fe^{3+}(aq) + 3\ OH^-(aq)$ pH = 5.0, $[OH^-] = 1 \times 10^{-9}$ *M*

	$Fe(OH)_3$(s)	⇌	Fe^{3+}(aq)	3 OH^-(aq)	
Initial			0	1×10^{-9} *M*	(buffered)
	s mol/L dissolves to reach equilibrium				
Change	$-s$	→	$+s$		(assume no pH change in buffer)
Equil.			s	1×10^{-9}	

$K_{sp} = 4 \times 10^{-38} = [Fe^{3+}][OH^-]^3 = (s)(1 \times 10^{-9})^3$, $s = 4 \times 10^{-11}$ mol/L = molar solubility

c.

$Fe(OH)_3(s) \rightleftharpoons Fe^{3+}(aq) + 3\ OH^-(aq)$ pH = 11.0, $[OH^-] = 1 \times 10^{-3}$ *M*

	$Fe(OH)_3$(s)	⇌	Fe^{3+}(aq)	3 OH^-(aq)	
Initial			0	0.001 *M*	(buffered)
	s mol/L dissolves to reach equilibrium				
Change	$-s$	→	$+s$		(assume no pH change)
Equil.			s	0.001	

$K_{sp} = 4 \times 10^{-38} = [Fe^{3+}][OH^-]^3 = (s)(0.001)^3$, $s = 4 \times 10^{-29}$ mol/L = molar solubility

Note: As $[OH^-]$ increases, solubility decreases. This is the common ion effect.

55. a.

	$Ag_2SO_4(s)$	$\rightleftharpoons$	$2\ Ag^+(aq)$	+	$SO_4^{2-}(aq)$
Initial	s = solubility (mol/L)		0		0
Equil.			$2s$		s

$K_{sp} = 1.2 \times 10^{-5} = [Ag^+]^2[SO_4^{2-}] = (2s)^2s = 4s^3,\ s = 1.4 \times 10^{-2}$ mol/L

b.

	$Ag_2SO_4(s)$	$\rightleftharpoons$	$2\ Ag^+(aq)$	+	$SO_4^{2-}(aq)$
Initial	s = solubility (mol/L)		0.10 M		0
Equil.			$0.10 + 2s$		s

$K_{sp} = 1.2 \times 10^{-5} = (0.10 + 2s)^2(s) \approx (0.10)^2(s),\ s = 1.2 \times 10^{-3}$ mol/L; assumption good.

c.

	$Ag_2SO_4(s)$	$\rightleftharpoons$	$2\ Ag^+(aq)$	+	$SO_4^{2-}(aq)$
Initial	s = solubility (mol/L)		0		0.20 M
Equil			$2s$		$0.20 + s$

$1.2 \times 10^{-5} = (2s)^2(0.20 + s) \approx 4s^2(0.20),\ s = 3.9 \times 10^{-3}$ mol/L; assumption good.

Note: Comparing the solubilities of parts b and c to that of part a illustrates that the solubility of a salt decreases when a common ion is present.

57.

	$Ag_3PO_4(s)$	$\rightleftharpoons$	$3\ Ag^+(aq)$	+	$PO_4^{3-}(aq)$
Initial			0		0.10 M
	s mol/L of Ag_3PO_4 (s) dissolves to reach equilibrium				
Change	$-s$	$\rightarrow$	$+3s$		$+s$
Equil.			$3s$		$0.10 + s$

$K_{sp} = 1.8 \times 10^{-18} = [Ag^+]^3[PO_4^{3-}] = (3s)^3(0.10 + s)$

Assuming $0.10 + s \approx 0.10$: $1.8 \times 10^{-18} = (3s)^3(0.10) = 27s^3(0.10)$

Assumption good; s = molar solubility = 8.7×10^{-7} mol/L

59.

	$Ce(IO_3)_3(s)$	$\rightleftharpoons$	$Ce^{3+}(aq)$	+	$3\ IO_3^-(aq)$
Initial	s = solubility (mol/L)		0		0.20 M
Equil.			s		$0.20 + 3s$

$K_{sp} = [Ce^{3+}][IO_3^-]^3 = s(0.20 + 3s)^3$

From the problem, $s = 4.4 \times 10^{-8}$ mol/L; solving for K_{sp}:

$$K_{sp} = (4.4 \times 10^{-8})[0.20 + 3(4.4 \times 10^{-8})]^3 = 3.5 \times 10^{-10}$$

61. If the anion in the salt can act as a base in water, then the solubility of the salt will increase as the solution becomes more acidic. Added H^+ will react with the base, forming the conjugate acid. As the basic anion is removed, more of the salt will dissolve to replenish the basic anion. The salts with basic anions are Ag_3PO_4, $CaCO_3$, $CdCO_3$ and $Sr_3(PO_4)_2$. Hg_2Cl_2 and PbI_2 do not

have any pH dependence because Cl^- and I^- are terrible bases (the conjugate bases of a strong acids).

$$Ag_3PO_4(s) + H^+(aq) \rightarrow 3\ Ag^+(aq) + HPO_4^{2-}(aq) \xrightarrow{\text{excess } H^+} 3\ Ag^+(aq) + H_3PO_4(aq)$$

$$CaCO_3(s) + H^+ \rightarrow Ca^{2+} + HCO_3^- \xrightarrow{\text{excess } H^+} Ca^{2+} + H_2CO_3\ \ [H_2O(l) + CO_2(g)]$$

$$CdCO_3(s) + H^+ \rightarrow Cd^{2+} + HCO_3^- \xrightarrow{\text{excess } H^+} Cd^{2+} + H_2CO_3\ \ [H_2O(l) + CO_2(g)]$$

$$Sr_3(PO_4)_2(s) + 2\ H^+ \rightarrow 3\ Sr^{2+} + 2\ HPO_4^{2-} \xrightarrow{\text{excess } H^+} 3\ Sr^{2+} + 2\ H_3PO_4$$

Precipitation Conditions

63.

	$ZnS(s)$	$\rightleftharpoons$	$Zn^{2+}(aq)$	$+\ S^{2-}(aq)$	$K_{sp} = [Zn^{2+}][S^{2-}]$
Initial	s = solubility (mol/L)		$0.050\ M$	0	
Equil.			$0.050 + s$	s	

$K_{sp} = 2.5 \times 10^{-22} = (0.050 + s)(s) \approx (0.050)s,\ \ s = 5.0 \times 10^{-21}$ mol/L; assumption good.

$$\text{Mass ZnS that dissolves} = 0.3000\ \text{L} \times \frac{5.0 \times 10^{-21}\ \text{mol ZnS}}{\text{L}} \times \frac{97.45\ \text{g ZnS}}{\text{mol}} = 1.5 \times 10^{-19}\ \text{g}$$

65. $Sr_3(PO_4)_2(s) \rightleftharpoons 3\ Sr^{2+}(aq) + 2\ PO_4^{3-}(aq) \qquad K_{sp} = 1.0 \times 10^{-31}$

$Q = [Sr^{2+}]_0^3[PO_4^{3-}]_0^2 = (1.0 \times 10^{-6})^3(5.0 \times 10^{-7})^2 = 2.5 \times 10^{-31}$

Because $Q > K_{sp}$, $Sr_3(PO_4)_2(s)$ will form as a precipitate.

67. $PbF_2(s) \rightleftharpoons Pb^{2+}(aq) + 2\ F^-(aq) \qquad K_{sp} = 4 \times 10^{-8}$

$$[Pb^{2+}]_0 = \frac{\text{mmol } Pb^{2+}(aq)}{\text{total mL solution}} = \frac{100.0\ \text{mL} \times \dfrac{1.0 \times 10^{-2}\ \text{mmol } Pb^{2+}}{\text{mL}}}{100.0\ \text{mL} + 100.0\ \text{mL}} = 5.0 \times 10^{-3}\ M$$

$$[F^-]_0 = \frac{\text{mmol } F^-}{\text{total mL solution}} = \frac{100.0\ \text{mL} \times \dfrac{1.0 \times 10^{-3}\ \text{mmol } F^-}{\text{mL}}}{200.0\ \text{mL}} = 5.0 \times 10^{-4}\ M$$

$Q = [Pb^{2+}]_0[F^-]_0^2 = (5.0 \times 10^{-3})(5.0 \times 10^{-4})^2 = 1.3 \times 10^{-9}$

Because $Q < K_{sp}$, $PbF_2(s)$ will not form as a precipitate.

69. $[Na_2SO_4] = [SO_4^{2-}] = \dfrac{0.0050\ \text{mol}}{0.5000\ \text{L}} = 0.010\ M$

Precipitation occurs when Q > K. For each salt, let's calculate Q, then compare this value to K_{sp} for each salt.

For $BaSO_4$ ($K_{sp} = 1.5 \times 10^{-10}$):

$Q = [Ba^{2+}]_0[SO_4^{2-}]_0 = (1.5 \times 10^{-3})(0.010) = 1.5 \times 10^{-5}$; $Q > K_{sp}$, so precipitate forms.

For $CaSO_4$ ($K_{sp} = 6.1 \times 10^{-5}$):

$Q = [Ca^{2+}]_0[SO_4^{2-}]_0 = (1.5 \times 10^{-3})(0.010) = 1.5 \times 10^{-5}$; $Q < K_{sp}$, so no precipitate forms.

Only $BaSO_4(s)$ will form.

71. The concentrations of ions are large, so Q will be greater than K_{sp}, and $BaC_2O_4(s)$ will form. To solve this problem, we will assume that the precipitation reaction goes to completion; then we will solve an equilibrium problem to get the actual ion concentrations. This makes the math reasonable.

$$100.\ \text{mL} \times \frac{0.200\ \text{mmol}\ K_2C_2O_4}{\text{mL}} = 20.0\ \text{mmol}\ K_2C_2O_4$$

$$150.\ \text{mL} \times \frac{0.250\ \text{mmol}\ BaBr_2}{\text{mL}} = 37.5\ \text{mmol}\ BaBr_2$$

	$Ba^{2+}(aq)$	+	$C_2O_4^{2-}(aq)$	$\rightarrow$	$BaC_2O_4(s)$	$K = 1/K_{sp} >> 1$
Before	37.5 mmol		20.0 mmol		0	
Change	–20.0		–20.0	$\rightarrow$	+20.0	Reacts completely (K is large).
After	17.5		0		20.0	

New initial concentrations (after complete precipitation) are:

$$[Ba^{2+}] = \frac{17.5\ \text{mmol}}{250.\ \text{mL}} = 7.00 \times 10^{-2}\ M;\ [C_2O_4^{2-}] = 0\ M$$

$$[K^+] = \frac{2(20.0\ \text{mmol})}{250.\ \text{mL}} = 0.160\ M;\ [Br^-] = \frac{2(37.5\ \text{mmol})}{250.\ \text{mL}} = 0.300\ M$$

For K^+ and Br^-, these are also the final concentrations. We can't have 0 M $C_2O_4^{2-}$. For Ba^{2+} and $C_2O_4^{2-}$, we need to perform an equilibrium calculation.

	$BaC_2O_4(s)$	$\rightleftharpoons$	$Ba^{2+}(aq)$	+	$C_2O_4^{2-}(aq)$	$K_{sp} = 2.3 \times 10^{-8}$
Initial			0.0700 M		0	
	s mol/L of $BaC_2O_4(s)$ dissolves to reach equilibrium					
Equil.			$0.0700 + s$		s	

$K_{sp} = 2.3 \times 10^{-8} = [Ba^{2+}][C_2O_4^{2-}] = (0.0700 + s)(s) \approx (0.0700)s$

$s = [C_2O_4^{2-}] = 3.3 \times 10^{-7}$ mol/L; $[Ba^{2+}] = 0.0700\ M$; assumption good ($s << 0.0700$).

73. $[Ag^+]_0 = \dfrac{0.0500\text{ L} \times \dfrac{2.00\text{ mol}}{\text{L}}}{0.1000\text{ L}} = 1.00\ M$; $[CO_3^{2-}]_0 = \dfrac{0.0500\text{ L} \times \dfrac{3.00\text{ mol}}{\text{L}}}{0.1000\text{ L}} = 1.50\ M$

Note: Because we added equal volumes of each solution together, the total volume of the solution doubled, which halved the concentrations.

We have very large concentrations of the Ag^+ and CO_3^{2-} ions. Because the value of K_{sp} for $Ag_2CO_3(s)$ is very small ($K_{sp} << 1$), the concentrations of the ions cannot both be large values. To solve this complicated problem, let the ions react to completion to form $Ag_2CO_3(s)$, then solve a back equilibrium problem to determine the equilibrium concentrations of the ions. In the stoichiometry part of the problem, Ag^+ is the limiting reagent.

	$Ag_2CO_3(s)$	$\rightleftharpoons$	$2\ Ag^+$	+	CO_3^{2-}	$K_{sp} = 8.1 \times 10^{-12}$
Before			$1.00\ M$		$1.50\ M$	
	Let 1.00 mol/L Ag^+ react with CO_3^{2-} to completion because $K_{sp} << 1$.					
Change		$\leftarrow$	-1.00		-0.500	Reacts completely
After			0		1.00	New initial
	s mol/L $Ag_2CO_3(s)$ dissolves to reach equilibrium					
Change	$-s$	$\rightarrow$	$+2s$		$+s$	
Equil.			$2s$		$1.00 + s$	

$K_{sp} = 8.1 \times 10^{-12} = [Ag^+]^2[CO_3^{2-}] = (2s)^2(1.00 + s) \approx 4s^2(1.00)$

Solving: $s = 1.4 \times 10^{-6}$ mol/L; assumption good (s is $1.4 \times 10^{-4}\%$ of 1.00).

$[Ag^+] = 2(1.4 \times 10^{-6}) = 2.8 \times 10^{-6}\ M$; $[CO_3^{2-}] = 1.00 + 1.4 \times 10^{-6} = 1.00\ M$

75. $Ag_3PO_4(s) \rightleftharpoons 3\ Ag^+(aq) + PO_4^{3-}(aq)$; when Q is greater than K_{sp}, precipitation will occur. We will calculate the $[Ag^+]_0$ necessary for $Q = K_{sp}$. Any $[Ag^+]_0$ greater than this calculated number will cause precipitation of $Ag_3PO_4(s)$. In this problem, $[PO_4^{3-}]_0 = [Na_3PO_4]_0 = 1.0 \times 10^{-5}\ M$.

$K_{sp} = 1.8 \times 10^{-18}$; $Q = 1.8 \times 10^{-18} = [Ag^+]_0^3[PO_4^{3-}]_0 = [Ag^+]_0^3(1.0 \times 10^{-5}\ M)$

$[Ag^+]_0 = \left(\dfrac{1.8 \times 10^{-18}}{1.0 \times 10^{-5}}\right)^{1/3}$, $[Ag^+]_0 = 5.6 \times 10^{-5}\ M$

When $[Ag^+]_0 = [AgNO_3]_0$ is greater than $5.6 \times 10^{-5}\ M$, precipitation of $Ag_3PO_4(s)$ will occur.

77. For each salt, determine the $[OH^-]_0$ necessary for $Q = K_{sp}$. Precipitation occurs when $Q > K_{sp}$. This will occur when the $[OH^-]_0$ is greater than the number calculated for $Q = K_{sp}$. The salt that requires the smallest amount of OH^- will precipitate first as OH^- is added gradually.

For AgOH ($K_{sp} = 2 \times 10^{-8}$):

$Q = K_{sp} = 2 \times 10^{-8} = [Ag^+]_0[OH^-]_0 = (0.010)[OH^-]_0$, $[OH^-]_0 = 2 \times 10^{-6}\ M$

For $Sc(OH)_3$ ($K_{sp} = 1 \times 10^{-14}$):

$Q = K_{sp} = 1 \times 10^{-14} = [Sc^{3+}]_0[OH^-]_0^3 = (0.010)[OH^-]_0^3$, $[OH^-]_0 = 1 \times 10^{-4}\ M$

AgOH(s) will precipitate when the OH^- concentration is a little larger than 2×10^{-6} *M*, while $Sc(OH)_3(s)$ will form when the OH^- concentration is a little larger than 1×10^{-4} *M*. Because precipitation of AgOH(s) requires the smaller amount of OH^-, AgOH will precipitate first and will form when the $[OH^-]_0 > 2 \times 10^{-6}$ *M*.

79. For each lead salt, we will calculate the $[Pb^{2+}]_0$ necessary for $Q = K_{sp}$. Any $[Pb^{2+}]_0$ greater than this value will cause precipitation of the salt ($Q > K_{sp}$).

$PbF_2(s) \rightleftharpoons Pb^{2+}(aq) + 2\ F^-(aq) \quad K_{sp} = 4 \times 10^{-8};\ Q = 4 \times 10^{-8} = [Pb^{2+}]_0[F^-]_0^2$

$$[Pb^{2+}]_0 = \frac{4 \times 10^{-8}}{(1 \times 10^{-4})^2} = 4\ M$$

$PbS(s) \rightleftharpoons Pb^{2+}(aq) + S^{2-}(aq) \quad K_{sp} = 7 \times 10^{-29};\ Q = 7 \times 10^{-29} = [Pb^{2+}]_0[S^{2-}]_0$

$$[Pb^{2+}]_0 = \frac{7 \times 10^{-29}}{1 \times 10^{-4}} = 7 \times 10^{-25}\ M$$

$Pb_3(PO_4)_2(s) \rightleftharpoons 3\ Pb^{2+}(aq) + 2\ PO_4^{3-}(aq) \quad K_{sp} = 1 \times 10^{-54}$

$Q = 1 \times 10^{-54} = [Pb^{2+}]_0^3[PO_4^{\ 3-}]_0^2$

$$[Pb^{2+}]_0 = \left[\frac{1 \times 10^{-54}}{(1 \times 10^{-4})^2}\right]^{1/3} = 5 \times 10^{-16}\ M$$

From the calculated $[Pb^{2+}]_0$, the least soluble salt is PbS(s), and it will form first. $Pb_3(PO_4)_2(s)$ will form second, and $PbF_2(s)$ will form last because it requires the largest $[Pb^{2+}]_0$ in order for precipitation to occur.

Complex Ion Equilibria

81. a.

$Ni^{2+} + CN^- \rightleftharpoons NiCN^+$	K_1
$NiCN^+ + CN^- \rightleftharpoons Ni(CN)_2$	K_2
$Ni(CN)_2 + CN^- \rightleftharpoons Ni(CN)_3^-$	K_3
$Ni(CN)_3^- + CN^- \rightleftharpoons Ni(CN)_4^{2-}$	K_4
$Ni^{2+}(aq) + 4\ CN^-(aq) \rightleftharpoons Ni(CN)_4^{2-}(aq)$	$K_f = K_1K_2K_3K_4$

Note: The various K constants are included for your information. Each CN^- adds with a corresponding K value associated with that reaction. The overall formation constant K_f for the overall reaction is equal to the product of all the stepwise K values.

b.

$V^{3+} + C_2O_4^{2-} \rightleftharpoons VC_2O_4^+$	K_1
$VC_2O_4^+ + C_2O_4^{2-} \rightleftharpoons V(C_2O_4)_2^-$	K_2
$V(C_2O_4)_2^- + C_2O_4^{2-} \rightleftharpoons V(C_2O_4)_3^{3-}$	K_3
$V^{3+}(aq) + 3\ C_2O_4^{2-}(aq) \rightleftharpoons V(C_2O_4)_3^{3-}(aq)$	$K_f = K_1K_2K_3$

83. $Fe^{3+}(aq) + 6\ CN^-(aq) \rightleftharpoons Fe(CN)_6^{3-}$ $\quad K = \dfrac{[Fe(CN)_6^{3-}]}{[Fe^{3+}][CN^-]^6}$

$$K = \frac{1.5 \times 10^{-3}}{(8.5 \times 10^{-40})(0.11)^6} = 1.0 \times 10^{42}$$

85. $Hg^{2+}(aq) + 2\ I^-(aq) \rightleftharpoons HgI_2(s)$; $\quad HgI_2(s) + 2\ I^-(aq) \rightleftharpoons HgI_4^{2-}(aq)$

orange ppt (HgI₂) — soluble complex ion (HgI₄²⁻)

87. The formation constant for HgI_4^{2-} is an extremely large number. Because of this, we will let the Hg^{2+} and I^- ions present initially react to completion and then solve an equilibrium problem to determine the Hg^{2+} concentration.

	$Hg^{2+}(aq)$	+ $4\ I^-(aq)$	$\rightleftharpoons$	$HgI_4^{2-}(aq)$	$K = 1.0 \times 10^{30}$
Before	0.010 *M*	0.78 *M*		0	
Change	−0.010	−0.040	→	+0.010	Reacts completely (K is large).
After	0	0.74		0.010	New initial
	x mol/L HgI_4^{2-} dissociates to reach equilibrium				
Change	$+x$	$+4x$	←	$-x$	
Equil.	x	$0.74 + 4x$		$0.010 - x$	

$$K = 1.0 \times 10^{30} = \frac{[HgI_4^{2-}]}{[Hg^{2+}][I^-]^4} = \frac{(0.010 - x)}{(x)(0.74 + 4x)^4};\ \text{making normal assumptions:}$$

$$1.0 \times 10^{30} = \frac{(0.010)}{(x)(0.74)^4},\quad x = [Hg^{2+}] = 3.3 \times 10^{-32}\ M;\quad \text{assumptions good.}$$

Note: 3.3×10^{-32} mol/L corresponds to one Hg^{2+} ion per 5×10^7 L. It is very reasonable to approach this problem in two steps. The reaction does essentially go to completion.

89. $[X^-]_0 = 5.00\ M$ and $[Cu^+]_0 = 1.0 \times 10^{-3}\ M$ because equal volumes of each reagent are mixed.

Because the K values are much greater than 1, assume the reaction goes completely to CuX_3^{2-}, and then solve an equilibrium problem.

	$Cu^+(aq)$	+ $3\ X^-(aq)$	$\rightleftharpoons$ $CuX_3^{2-}(aq)$	$K = K_1 \times K_2 \times K_3$
Before	$1.0 \times 10^{-3}\ M$	5.00 *M*	0	$K = 1.0 \times 10^9$
After	0	$5.00 - 3(10^{-3}) \approx 5.00$	1.0×10^{-3}	Reacts completely
Equil.	x	$5.00 + 3x$	$1.0 \times 10^{-3} - x$	

$$K = \frac{(1.0 \times 10^{-3} - x)}{(x)(5.00 + 3x)^3} = 1.0 \times 10^9 \approx \frac{1.0 \times 10^{-3}}{(x)(5.00)^3},\quad x = [Cu^+] = 8.0 \times 10^{-15}\ M$$

Assumptions good.

$[CuX_3^{2-}] = 1.0 \times 10^{-3} - 8.0 \times 10^{-15} = 1.0 \times 10^{-3}\ M$

$$K_3 = \frac{[CuX_3^{2-}]}{[CuX_2^-][X^-]} = 1.0 \times 10^3 = \frac{(1.0 \times 10^{-3})}{[CuX_2^-](5.00)}, \quad [CuX_2^-] = 2.0 \times 10^{-7}\ M$$

Summarizing:

$[CuX_3^{2-}] = 1.0 \times 10^{-3}\ M$ (answer a)

$[CuX_2^-] = 2.0 \times 10^{-7}\ M$ (answer b)

$[Cu^{2+}] = 8.0 \times 10^{-15}\ M$ (answer c)

91. a.

	$AgI(s)$	$\rightleftharpoons$	$Ag^+(aq)$	+	$I^-(aq)$	$K_{sp} = [Ag^+][I^-] = 1.5 \times 10^{-16}$
Initial	s = solubility (mol/L)		0		0	
Equil.			s		s	

$K_{sp} = 1.5 \times 10^{-16} = s^2$, $s = 1.2 \times 10^{-8}$ mol/L

b.

$AgI(s) \rightleftharpoons Ag^+ + I^-$ $\qquad K_{sp} = 1.5 \times 10^{-16}$

$Ag^+ + 2\ NH_3 \rightleftharpoons Ag(NH_3)_2^+$ $\qquad K_f = 1.7 \times 10^7$

$AgI(s) + 2\ NH_3(aq) \rightleftharpoons Ag(NH_3)_2^+(aq) + I^-(aq)$ $\qquad K = K_{sp} \times K_f = 2.6 \times 10^{-9}$

	$AgI(s)$ + $2\ NH_3$	$\rightleftharpoons$	$Ag(NH_3)_2^+$	+ I^-
Initial	3.0 M		0	0
	s mol/L of AgI(s) dissolves to reach equilibrium = molar solubility			
Equil.	$3.0 - 2s$		s	s

$$K = \frac{[Ag(NH_3)_2^+][I^-]}{[NH_3]^2} = \frac{s^2}{(3.0 - 2s)^2} = 2.6 \times 10^{-9} \approx \frac{s^2}{(3.0)^2},\quad s = 1.5 \times 10^{-4}\ \text{mol/L}$$

Assumption good.

c. The presence of NH_3 increases the solubility of AgI. Added NH_3 removes Ag^+ from solution by forming the complex ion, $Ag(NH_3)_2^+$. As Ag^+ is removed, more AgI(s) will dissolve to replenish the Ag^+ concentration.

93.

$AgCl(s) \rightleftharpoons Ag^+ + Cl^-$ $\qquad K_{sp} = 1.6 \times 10^{-10}$

$Ag^+ + 2\ NH_3 \rightleftharpoons Ag(NH_3)_2^+$ $\qquad K_f = 1.7 \times 10^7$

$AgCl(s) + 2\ NH_3(aq) \rightleftharpoons Ag(NH_3)_2^+(aq) + Cl^-(aq)$ $\qquad K = K_{sp} \times K_f = 2.7 \times 10^{-3}$

	$AgCl(s)$ + $2\ NH_3$	$\rightleftharpoons$	$Ag(NH_3)_2^+$	+ Cl^-
Initial	1.0 M		0	0
	s mol/L of AgCl(s) dissolves to reach equilibrium = molar solubility			
Equil.	$1.0 - 2s$		s	s

$$K = 2.7 \times 10^{-3} = \frac{[Ag(NH_3)_2^+][Cl^-]}{[NH_3]^2} = \frac{s^2}{(1.0 - 2s)^2};\ \text{taking the square root:}$$

$$\frac{s}{1.0 - 2s} = (2.7 \times 10^{-3})^{1/2} = 5.2 \times 10^{-2}, \ s = 4.7 \times 10^{-2} \text{ mol/L}$$

In pure water, the solubility of AgCl(s) is $(1.6 \times 10^{-10})^{1/2} = 1.3 \times 10^{-5}$ mol/L. Notice how the presence of NH_3 increases the solubility of AgCl(s) by over a factor of 3500.

95. Test tube 1: Added Cl^- reacts with Ag^+ to form a silver chloride precipitate. The net ionic equation is $Ag^+(aq) + Cl^-(aq) \rightarrow AgCl(s)$. Test tube 2: Added NH_3 reacts with Ag^+ ions to form a soluble complex ion, $Ag(NH_3)_2^+$. As this complex ion forms, Ag^+ is removed from the solution, which causes the AgCl(s) to dissolve. When enough NH_3 is added, all the silver chloride precipitate will dissolve. The equation is $AgCl(s) + 2\ NH_3(aq) \rightarrow Ag(NH_3)_2^+(aq) + Cl^-(aq)$. Test tube 3: Added H^+ reacts with the weak base, NH_3, to form NH_4^+. As NH_3 is removed from the $Ag(NH_3)_2^+$ complex ion, Ag^+ ions are released to solution and can then react with Cl^- to re-form AgCl(s). The equations are $Ag(NH_3)_2^+(aq) + 2\ H^+(aq) \rightarrow Ag^+(aq) + 2\ NH_4^+(aq)$ and $Ag^+(aq) + Cl^-(aq) \rightarrow AgCl(s)$.

ChemWork Problems

97.

	$AgCl(s)$	$\rightleftharpoons$	$Ag^+(aq)$	+	$Cl^-(aq)$
Initial	s = solubility (mol/L)		0		0
Equil.			s		s

$K_{sp} = [Ag^+][Cl^-] = s(s) = s^2$; the expression given in the problem is incorrect.

	$PbCl_2(s)$	$\rightleftharpoons$	$Pb^{2+}(aq)$	+	$2\ Cl^-(aq)$
Initial	s = solubility (mol/L)		0		0
Equil.			s		2s

$K_{sp} = [Pb^{2+}][Cl^-]^2 = s(2s)^2 = 4s^3$; the expression given in the problem is incorrect.

	$Ag_3PO_4(s)$	$\rightleftharpoons$	$3\ Ag^+(aq)$	+	$PO_4^{3-}(aq)$
Initial	s = solubility (mol/L)		0		0
Equil.			$3s$		s

$K_{sp} = [Ag^+]^3[PO_4^{3-}] = (3s)^3(s) = 27s^4$; the expression given in the problem is incorrect.

	$Pb_3(PO_4)_2(s)$	$\rightleftharpoons$	$3\ Pb^{2+}(aq)$	+	$2\ PO_4^{3-}(aq)$
Initial	s = solubility (mol/L)		0		0
Equil.			$3s$		$2s$

$K_{sp} = [Pb^{2+}]^3[PO_4^{3-}]^2 = (3s)^3(2s)^2 = 108s^5$; the expression given in the problem is incorrect.

All the expressions are incorrect.

99. Mol Ag^+ added $= 0.200 \text{ L} \times \dfrac{0.24 \text{ mol AgNO}_3}{\text{L}} \times \dfrac{1 \text{ mol Ag}^+}{\text{mol AgNO}_3} = 0.048 \text{ mol Ag}^+$

The added Ag^+ will react with the halogen ions to form a precipitate. Because the K_{sp} values are small, we can assume these precipitation reactions go to completion. The order of precipitation will be AgI(s) first (the least soluble compound since K_{sp} is the smallest), followed by AgBr(s), with AgCl(s) forming last [AgCl(s) is the most soluble compound listed since it has the largest K_{sp}].

Let the Ag^+ react with I^- to completion.

	$Ag^+(aq)$ +	$I^-(aq)$ →	AgI(s)	$K = 1/K_{sp} >> 1$
Before	0.048 mol	0.018 mol	0	
Change	−0.018	−0.018	+0.018	I^- is limiting.
After	0.030 mol	0	0.018 mol	

Let the Ag^+ remaining react next with Br^- to completion.

	$Ag^+(aq)$ +	$Br^-(aq)$ →	AgBr(s)	$K = 1/K_{sp} >> 1$
Before	0.030 mol	0.018 mol	0	
Change	−0.018	−0.018	+0.018	Br^- is limiting.
After	0.012 mol	0	0.018 mol	

Finally, let the remaining Ag^+ react with Cl^- to completion.

	$Ag^+(aq)$ +	$Cl^-(aq)$ →	AgCl(s)	$K = 1/K_{sp} >> 1$
Before	0.012 mol	0.018 mol	0	
Change	−0.012	−0.012	+0.012	Ag^+ is limiting.
After	0	0.006 mol	0.012 mol	

Some of the AgCl will redissolve to produce some Ag^+ ions; we can't have $[Ag^+] = 0\ M$. Calculating how much AgCl(s) redissolves:

	AgCl(s)	⇌	$Ag^+(aq)$ +	$Cl^-(aq)$	$K_{sp} = 1.6 \times 10^{-10}$
Initial			0	0.006 mol/0.200 L = 0.03 M	
	s mol/L of AgCl dissolves to reach equilibrium				
Change	$-s$	→	$+s$	$+s$	
Equil.			s	$0.03 + s$	

$$K_{sp} = 1.6 \times 10^{-10} = [Ag^+][Cl^-] = s(0.03 + s) \approx (0.03)s$$

$s = 5 \times 10^{-9}$ mol/L; the assumption that $0.03 + s \approx 0.03$ is good.

Mol AgCl present = 0.012 mol − 5×10^{-9} mol = 0.012 mol

$$\text{Mass AgCl present} = 0.012 \text{ mol AgCl} \times \frac{143.4 \text{ g}}{\text{mol AgCl}} = 1.7 \text{ g AgCl}$$

$[Ag^+] = s = 5 \times 10^{-9}$ mol/L

101.

	$M_3(PO_4)_2(s)$	$\rightleftharpoons$	$3\ M^{2+}(aq)$	+	$2\ PO_4^{3-}(aq)$
Initial	s = solubility (mol/L)		0		0
Equil.			$3s$		$2s$

$K_{sp} = [M^{2+}]^3[PO_4^{3-}]^2 = (3s)^3(2s)^2 = 108s^5$; $K_{sp} = 108(6.2 \times 10^{-12})^5 = 9.9 \times 10^{-55}$.

From appendix A5.4, $Pb_3(PO_4)_2$ has a K_{sp} value closest to 9.9×10^{-55}, so M is Pb^{2+}. Now solve the common ion part of the problem.

	$Pb_3(PO_4)_2(s)$	$\rightleftharpoons$	$3\ Pb^{2+}(aq)$	+	$2\ PO_4^{3-}(aq)$
Initial	s = solubility (mol/L)		0		0.10 M
Equil.			$3s$		$0.10 + 2s$

$K_{sp} = 9.9 \times 10^{-55} = [Pb^{2+}]^3[PO_4^{3-}]^2 = (3s)^3(0.10 + 2s)^2 \approx 27s^3(0.10)^2$

$s = 1.5 \times 10^{-18}$ mol/L; the assumption is good.

103.

	$Ca_5(PO_4)_3OH(s)$	$\rightleftharpoons$	$5\ Ca^{2+}$	+	$3\ PO_4^{3-}$	+	OH^-	
Initial	s = solubility (mol/L)		0		0		1.0×10^{-7}	from water
Equil.			$5s$		$3s$		$s + 1.0 \times 10^{-7} \approx s$	

$K_{sp} = 6.8 \times 10^{-37} = [Ca^{2+}]^5[PO_4^{3-}]^3[OH^-] = (5s)^5(3s)^3(s)$

$6.8 \times 10^{-37} = (3125)(27)s^9$, $s = 2.7 \times 10^{-5}$ mol/L; assumption is good.

The solubility of hydroxyapatite will increase as the solution gets more acidic because both phosphate and hydroxide can react with H^+.

	$Ca_5(PO_4)_3F(s)$	$\rightleftharpoons$	$5\ Ca^{2+}(aq)$	+	$3\ PO_4^{3-}(aq)$	+	$F^-(aq)$
Initial	s = solubility (mol/L)		0		0		0
Equil.			$5s$		$3s$		s

$K_{sp} = 1 \times 10^{-60} = (5s)^5(3s)^3(s) = (3125)(27)s^9$, $s = 6 \times 10^{-8}$ mol/L

The hydroxyapatite in tooth enamel is converted to the less soluble fluorapatite by fluoride-treated water. The less soluble fluorapatite is more difficult to remove, making teeth less susceptible to decay.

105. $Al(OH)_3(s) \rightleftharpoons Al^{3+}(aq) + 3\ OH^-(aq)$ $\qquad K_{sp} = 2 \times 10^{-32}$

$Q = 2 \times 10^{-32} = [Al^{3+}]_0[OH^-]_0^3 = (0.2)[OH^-]_0^3$, $[OH^-]_0 = 4.6 \times 10^{-11}$ (carrying extra sig. fig.)

pOH = $-\log(4.6 \times 10^{-11}) = 10.3$; when the pOH of the solution equals 10.3, $K_{sp} = Q$. For precipitation, we want $Q > K_{sp}$. This will occur when $[OH^-]_0 > 4.6 \times 10^{-11}$ or when pOH < 10.3. Because pH + pOH = 14.00, precipitation of $Al(OH)_3(s)$ will begin when pH > 3.7 because this corresponds to a solution with pOH < 10.3.

107. $$s = \text{solubility} = \frac{0.24 \text{ g PbI}_2 \times \frac{1 \text{ mol PbI}_2}{461.0 \text{ g}}}{0.2000 \text{ L}} = 2.6 \times 10^{-3}\ M$$

	$PbI_2(s)$	$\rightleftharpoons$	$Pb^{2+}(aq)$	+	$2\ I^-(aq)$	$K_{sp} = [Pb^{2+}][I^-]^2$
Initial	s = solubility (mol/L)		0		0	
Equil.			s		$2s$	

$K_{sp} = s(2s)^2 = 4s^3$, $K_{sp} = 4(2.6 \times 10^{-3})^3 = 7.0 \times 10^{-8}$

109. a.

	$AgOH(s)$	$\rightleftharpoons$	$Ag^+(aq)$ +	$OH^-(aq)$	$K_{sp} = [Ag^+][OH^-] = 2.0 \times 10^{-8}$
Initial	s = solubility (mol/L)		0	~0	(Ignoring OH^- from water.)
Equil.			s	s	

$K_{sp} = 2.0 \times 10^{-8} = s(s) = s^2$, $s = 1.4 \times 10^{-4}$ mol/L

Assumption good, the amount of OH^- from the autoionization of H_2O can be ignored.

$[OH^-] = s = 1.4 \times 10^{-4}$ mol/L; pOH = 3.85, pH = 10.15

b. $Cd(OH)_2(s) \rightleftharpoons Cd^{2+}(aq) + 2\ OH^-(aq)$ $K_{sp} = [Cd^{2+}][OH^-]^2 = 5.9 \times 10^{-15}$

Equil. s $2s$

$K_{sp} = 5.9 \times 10^{-15} = 4s^3$, $s = 1.1 \times 10^{-5}$ mol/L; asssumption good.

$[OH^-] = 2(1.1 \times 10^{-5}) = 2.2 \times 10^{-5}\ M$; pOH = 4.66, pH = 9.34

c. $Pb(OH)_2(s) \rightleftharpoons Pb^{2+}(aq) + 2\ OH^-(aq)$ $K_{sp} = [Pb^{2+}][OH^-]^2 = 1.2 \times 10^{-15}$

Equil. s $2s$

$K_{sp} = 1.2 \times 10^{-15} = 4s^3$, $s = 6.7 \times 10^{-6}$ mol/L; assumption good.

$[OH^-] = 2(6.7 \times 10^{-6}) = 1.3 \times 10^{-5}$ mol/L; pOH = 4.89, pH = 9.11

111. $Cu(OH)_2(s) \rightleftharpoons Cu^{2+}(aq) + 2\ OH^-(aq)$; adding either Cu^{2+} or OH^- will shift this equilibrium to the left causing more precipitate to form. So, answers a and b are not possible.

$Cu^{2+}(aq) + 4\ NH_3(aq) \rightleftharpoons Cu(NH_3)_4^{2+}(aq)$; when NH_3 is added, Cu^{2+} is removed from solution as it reacts with NH_3 to form the soluble complex ion $Cu(NH_3)_4^{2+}$. As Cu^{2+} is removed, this forces the $Cu(OH)_2(s)$ equilibrium to shift right to replenish the lost Cu^{2+}. Adding NH_3 will cause the precipitate to dissolve (answer c is correct).

$H^+(aq) + OH^-(aq) \rightarrow H_2O(l)$; when H^+ is added from the strong acid HNO_3, it reacts with OH^- to form H_2O. As OH^- is removed, this forces the $Cu(OH)_2(s)$ equilibrium to shift right to replenish the lost OH^-. Adding HNO_3 will also cause the precipitate to dissolve (answer d is correct).

113.

$$Mn^{2+} + C_2O_4^{2-} \rightleftharpoons MnC_2O_4 \qquad K_1 = 7.9 \times 10^3$$
$$MnC_2O_4 + C_2O_4^{2-} \rightleftharpoons Mn(C_2O_4)_2^{2-} \qquad K_2 = 7.9 \times 10^1$$

$$Mn^{2+}(aq) + 2\ C_2O_4^{2-}(aq) \rightleftharpoons Mn(C_2O_4)_2^{2-}(aq) \qquad K_f = K_1K_2 = 6.2 \times 10^5$$

115. a.

	$Pb(OH)_2(s)$	$\rightleftharpoons$	Pb^{2+}	+	$2\ OH^-$
Initial	s = solubility (mol/L)		0		1.0×10^{-7} M from water
Equil.			s		$1.0 \times 10^{-7} + 2s$

$K_{sp} = 1.2 \times 10^{-15} = [Pb^{2+}][OH^-]^2 = s(1.0 \times 10^{-7} + 2s)^2 \approx s(2s^2) = 4s^3$

$s = [Pb^{2+}] = 6.7 \times 10^{-6}\ M$; assumption is good by the 5% rule.

b.

	$Pb(OH)_2(s)$	$\rightleftharpoons$	Pb^{2+}	+	$2\ OH^-$	
Initial			0		0.10 M	pH = 13.00, $[OH^-] = 0.10\ M$
	s mol/L $Pb(OH)_2(s)$ dissolves to reach equilibrium					
Equil.			s		0.10	(Buffered solution)

$1.2 \times 10^{-15} = (s)(0.10)^2$, $s = [Pb^{2+}] = 1.2 \times 10^{-13}\ M$

c. We need to calculate the Pb^{2+} concentration in equilibrium with $EDTA^{4-}$. Since K is large for the formation of $PbEDTA^{2-}$, let the reaction go to completion and then solve an equilibrium problem to get the Pb^{2+} concentration.

	Pb^{2+}	+	$EDTA^{4-}$	$\rightleftharpoons$	$PbEDTA^{2-}$	$K = 1.1 \times 10^{18}$
Before	0.010 M		0.050 M		0	
	0.010 mol/L Pb^{2+} reacts completely (large K)					
Change	–0.010		–0.010	$\rightarrow$	+0.010	Reacts completely
After	0		0.040		0.010	New initial
	x mol/L $PbEDTA^{2-}$ dissociates to reach equilibrium					
Equil.	x		$0.040 + x$		$0.010 - x$	

$$1.1 \times 10^{18} = \frac{(0.010 - x)}{(x)(0.040 + x)} \approx \frac{0.010}{x(0.040)},\ x = [Pb^{2+}] = 2.3 \times 10^{-19}\ M;\ \text{assumptions good.}$$

Now calculate the solubility quotient for $Pb(OH)_2$ to see if precipitation occurs. The concentration of OH^- is 0.10 M since we have a solution buffered at pH = 13.00.

$Q = [Pb^{2+}]_0[OH^-]_0^2 = (2.3 \times 10^{-19})(0.10)^2 = 2.3 \times 10^{-21} < K_{sp}\ (1.2 \times 10^{-15})$

$Pb(OH)_2(s)$ will not form since Q is less than K_{sp}.

117. a.

$$Cu(OH)_2 \rightleftharpoons Cu^{2+} + 2\ OH^- \qquad K_{sp} = 1.6 \times 10^{-19}$$
$$Cu^{2+} + 4\ NH_3 \rightleftharpoons Cu(NH_3)_4^{2+} \qquad K_f = 1.0 \times 10^{13}$$

$$Cu(OH)_2(s) + 4\ NH_3(aq) \rightleftharpoons Cu(NH_3)_4^{2+}(aq) + 2\ OH^-(aq) \qquad K = K_{sp}K_f = 1.6 \times 10^{-6}$$

b.

	$Cu(OH)_2(s)$ +	$4\ NH_3$	$\rightleftharpoons$	$Cu(NH_3)_4^{2+}$ +	$2\ OH^-$	$K = 1.6 \times 10^{-6}$
Initial		5.0 M		0	0.0095 M	
	s mol/L $Cu(OH)_2$ dissolves to reach equilibrium					
Equil.		$5.0 - 4s$		s	$0.0095 + 2s$	

$$K = 1.6 \times 10^{-6} = \frac{[Cu(NH_3)_4^{2+}][OH^-]^2}{[NH_3]^4} = \frac{s(0.0095 + 2s)^2}{(5.0 - 4s)^4}$$

If s is small: $1.6 \times 10^{-6} = \dfrac{s(0.0095)^2}{(5.0)^4}$, $s = 11.$ mol/L

Assumptions are horrible. We will solve the problem by successive approximations.

$s_{calc} = \dfrac{1.6 \times 10^{-6}(5.0 - 4s_{guess})^4}{(0.0095 + 2s_{guess})^2}$; the results from six trials are:

s_{guess}: 0.10, 0.050, 0.060, 0.055, 0.056

s_{calc}: 1.6×10^{-2}, 0.071, 0.049, 0.058, 0.056

Thus the solubility of $Cu(OH)_2$ is 0.056 mol/L in 5.0 M NH_3.

119. From the problem, $[Hg^{2+}]_0 = 0.088$ M and $[I^-]_0 = 5.0$ M. Because the K values are large, assume all reactions go to completion, and then solve an equilibrium problem.

$$Hg^{2+}(aq) + 4\ I^-(aq) \rightleftharpoons HgI_4^{2-}(aq) \qquad K = K_1K_2K_3K_4 = 1.0 \times 10^{30}$$

	$Hg^{2+}(aq)$	$4\ I^-(aq)$	$HgI_4^{2-}(aq)$	
Before	0.088M	5.0 M	0	
Change	−0.088	−4(0.088)	+0.088	
After	0	4.6 M	0.088 M	Assume reacts completely.
Equil.	x	$4.6 + 4x$	$0.088 - x$	

$$K = 1.0 \times 10^{30} = \frac{[HgI_4^{2-}]}{[Hg^{2+}][I^-]^4} = \frac{0.088 - x}{x(4.6 + 4x)^4} \approx \frac{0.088}{x(4.6)^4}$$

$x = [Hg^{2+}] = 2.0 \times 10^{-34}$ M; assumptions good. $[I^-] = 4.6$ M; $[HgI_4^{2-}] = 0.088$ M

a. $[HgI_4^{2-}] = 0.088$ M; b. $[I^-] = 4.6$ M; c. $[Hg^{2+}] = 2.0 \times 10^{-34}$ M

121. Solubility $= s = \dfrac{15\text{ g } Ba(NO_3)_2}{L} \times \dfrac{1\text{ mol } Ba(NO_3)_2}{261.3\text{ g } Ba(NO_3)_2} = 0.057$ mol/L

$$Ba(NO_3)_2(s) \rightleftharpoons Ba^{2+}(aq) + 2\ NO_3^-(aq) \qquad K_{sp} = [Ba^{2+}][NO_3^-]^2$$

		Ba^{2+}	NO_3^-
Initial	s = solubility (mol/L)	0	0
Equil.		s	$2s$

$K_{sp} = s(2s)^2 = 4s^3$, $K_{sp} = 4(0.057)^3 = 7.4 \times 10^{-4}$

Challenge Problems

123. $$M_3X_2(s) \rightleftharpoons 3\ M^{2+}(aq) + 2\ X^{3-}(aq) \qquad K_{sp} = [M^{2+}]^3[X^{3-}]^2$$

		M^{2+}	X^{3-}
Initial	s = solubility (mol/L)	0	0
Equil.		$3s$	$2s$

$K_{sp} = (3s)^3(2s)^2 = 108s^5$; total ion concentration = $3s + 2s = 5s$

$$\pi = iMRT,\ iM = \text{total ion concentration} = \frac{\pi}{RT} = \frac{2.64 \times 10^{-2}\ \text{atm}}{0.08206\ \text{L atm/K}\bullet\text{mol} \times 298\ \text{K}}$$

$$= 1.08 \times 10^{-3}\ \text{mol/L}$$

$5s = 1.08 \times 10^{-3}$ mol/L, $s = 2.16 \times 10^{-4}$ mol/L; $K_{sp} = 108s^5 = 108(2.16 \times 10^{-4})^5$

$K_{sp} = 5.08 \times 10^{-17}$

125. a.

$CuBr(s) \rightleftharpoons Cu^+ + Br^-$	$K_{sp} = 1.0 \times 10^{-5}$
$Cu+ + 3\ CN^- \rightleftharpoons Cu(CN)_3^{2-}$	$K_f = 1.0 \times 10^{11}$
$CuBr(s) + 3\ CN^- \rightleftharpoons Cu(CN)_3^{2-} + Br^-$	$K = 1.0 \times 10^6$

Because K is large, assume that enough CuBr(s) dissolves to completely use up the 1.0 *M* CN^-; then solve the back equilibrium problem to determine the equilibrium concentrations.

	CuBr(s)	+ 3 CN^-	$\rightleftharpoons$	$Cu(CN)_3^{2-}$	+ Br^-
Before	x	1.0 *M*		0	0
	x mol/L of CuBr(s) dissolves to react completely with 1.0 *M* CN^-				
Change	$-x$	$-3x$	$\rightarrow$	$+x$	$+x$
After	0	$1.0 - 3x$		x	x

For reaction to go to completion, $1.0 - 3x = 0$ and $x = 0.33$ mol/L. Now solve the back-equilibrium problem.

	CuBr(s) + 3 CN^-	$\rightleftharpoons$	$Cu(CN)_3^{2-}$	+ Br^-
Initial	0		0.33 *M*	0.33 *M*
	Let *y* mol/L of $Cu(CN)_3^{2-}$ react to reach equilibrium.			
Change	$+3y$	$\leftarrow$	$-y$	$-y$
Equil.	3y		$0.33 - y$	$0.33 - y$

$$K = 1.0 \times 10^6 = \frac{(0.33 - y)^2}{(3y)^3} \approx \frac{(0.33)^2}{27y^3},\ y = 1.6 \times 10^{-3}\ M;\ \text{assumptions good.}$$

Of the initial 1.0 *M* CN^-, only $3(1.6 \times 10^{-3}) = 4.8 \times 10^{-3}$ *M* is present at equilibrium. Indeed, enough CuBr(s) did dissolve to essentially remove the initial 1.0 *M* CN^-. This amount, 0.33 mol/L, is the solubility of CuBr(s) in 1.0 *M* NaCN.

b. $[Br^-] = 0.33 - y = 0.33 - 1.6 \times 10^{-3} = 0.33\ M$

c. $[CN^-] = 3y = 3(1.6 \times 10^{-3}) = 4.8 \times 10^{-3}\ M$

127. a.

	AgBr(s)	$\rightleftharpoons$	$Ag^+(aq)$ +	$Br^-(aq)$	$K_{sp} = [Ag^+][Br^-] = 5.0 \times 10^{-13}$
Initial	s = solubility (mol/L)		0	0	
Equil.			s	s	

$K_{sp} = 5.0 \times 10^{-13} = s^2$, $s = 7.1 \times 10^{-7}$ mol/L

b.

$$\begin{array}{lr} AgBr(s) \rightleftharpoons Ag^+ + Br^- & K_{sp} = 5.0 \times 10^{-13} \\ Ag^+ + 2\,NH_3 \rightleftharpoons Ag(NH_3)_2^+ & K_f = 1.7 \times 10^7 \\ \hline AgBr(s) + 2\,NH_3(aq) \rightleftharpoons Ag(NH_3)_2^+(aq) + Br^-(aq) & K = K_{sp} \times K_f = 8.5 \times 10^{-6} \end{array}$$

	AgBr(s) +	$2\,NH_3$	$\rightleftharpoons$ $Ag(NH_3)_2^+$	+ Br^-
Initial		3.0 *M*	0	0
	s mol/L of AgBr(s) dissolves to reach equilibrium = molar solubility			
Equil.		$3.0 - 2s$	s	s

$$K = \frac{[Ag(NH_3)_2^+][Br^-]}{[NH_3]^2} = \frac{s^2}{(3.0-2s)^2} = 8.5 \times 10^{-6} \approx \frac{s^2}{(3.0)^2},\ s = 8.7 \times 10^{-3}\ \text{mol/L}$$

Assumption good.

c. The presence of NH_3 increases the solubility of AgBr. Added NH_3 removes Ag^+ from solution by forming the complex ion, $Ag(NH_3)_2^+$. As Ag^+ is removed, more AgBr(s) will dissolve to replenish the Ag^+ concentration.

d. $$\text{Mass AgBr} = 0.2500\ \text{L} \times \frac{8.7 \times 10^{-3}\ \text{mol AgBr}}{\text{L}} \times \frac{187.8\ \text{g AgBr}}{\text{mol AgBr}} = 0.41\ \text{g AgBr}$$

e. Added HNO_3 will have no effect on the AgBr(s) solubility in pure water. Neither H^+ nor NO_3^- react with Ag^+ or Br^- ions. Br^- is the conjugate base of the strong acid HBr, so it is a terrible base. However, added HNO_3 will reduce the solubility of AgBr(s) in the ammonia solution. NH_3 is a weak base ($K_b = 1.8 \times 10^{-5}$). Added H^+ will react with NH_3 to form NH_4^+. As NH_3 is removed, a smaller amount of the $Ag(NH_3)_2^+$ complex ion will form, resulting in a smaller amount of AgBr(s) that will dissolve.

129.

$$\begin{array}{lr} AgCN(s) \rightleftharpoons Ag^+(aq) + CN^-(aq) & K_{sp} = 2.2 \times 10^{-12} \\ H^+(aq) + CN^-(aq) \rightleftharpoons HCN(aq) & K = 1/K_{a,\,HCN} = 1.6 \times 10^9 \\ \hline AgCN(s) + H^+(aq) \rightleftharpoons Ag^+(aq) + HCN(aq) & K = 2.2 \times 10^{-12}(1.6 \times 10^9) = 3.5 \times 10^{-3} \end{array}$$

	AgCN(s) +	$H^+(aq)$	$\rightleftharpoons$ $Ag^+(aq)$	+ HCN(aq)
Initial		1.0 *M*	0	0
	s mol/L AgCN(s) dissolves to reach equilibrium			
Equil.		$1.0 - s$	s	s

$$3.5 \times 10^{-3} = \frac{[Ag^+][HCN]}{[H^+]} = \frac{s(s)}{1.0 - s} \approx \frac{s^2}{1.0},\ s = 5.9 \times 10^{-2}$$

Assumption fails the 5% rule (*s* is 5.9% of 1.0 *M*). Using the method of successive approximations:

$$3.5 \times 10^{-3} = \frac{s^2}{1.0 - 0.059},\ s = 5.7 \times 10^{-2}$$

$$3.5 \times 10^{-3} = \frac{s^2}{1.0 - 0.057},\ s = 5.7 \times 10^{-2} \text{ (consistent answer)}$$

The molar solubility of AgCN(s) in 1.0 M H^+ is 5.7×10^{-2} mol/L.

131. $H_2S(aq) \rightleftharpoons H^+(aq) + HS^-(aq)$ $\quad K_{a_1} = 1.0 \times 10^{-7}$

$HS^-(aq) \rightleftharpoons H^+(aq) + S^{2-}(aq)$ $\quad K_{a_2} = 1 \times 10^{-19}$

$$H_2S(aq) \rightleftharpoons 2\,H^+(aq) + S^{2-}(aq) \qquad K = K_{a_1} \times K_{a_2} = 1 \times 10^{-26} = \frac{[H^+]^2[S^{2-}]}{[H_2S]}$$

Because K is very small, only a tiny fraction of the H_2S will react. At equilibrium, $[H_2S] = 0.10\ M$ and $[H^+] = 1 \times 10^{-3}$.

$$[S^{2-}] = \frac{K[H_2S]}{[H^+]^2} = \frac{(1 \times 10^{-26})(0.10)}{(1 \times 10^{-3})^2} = 1 \times 10^{-21}\ M$$

$$NiS(s) \rightleftharpoons Ni^{2+}(aq) + S^{2-}(aq) \qquad K_{sp} = [Ni^{2+}][S^{2-}] = 3 \times 10^{-21}$$

Precipitation of NiS will occur when $Q > K_{sp}$. We will calculate $[Ni^{2+}]$ for $Q = K_{sp}$.

$$Q = K_{sp} = [Ni^{2+}][S^{2-}] = 3.0 \times 10^{-21},\ [Ni^{2+}] = \frac{3.0 \times 10^{-21}}{1 \times 10^{-21}} = 3\ M = \text{maximum concentration}$$

133. $Mg^{2+} + P_3O_{10}^{5-} \rightleftharpoons MgP_3O_{10}^{3-}$ $\quad K = 4.0 \times 10^8$

$$[Mg^{2+}]_0 = \frac{50. \times 10^{-3}\ g}{L} \times \frac{1\ mol}{24.31\ g} = 2.1 \times 10^{-3}\ M$$

$$[P_3O_{10}^{5-}]_0 = \frac{40.\ g\ Na_5P_3O_{10}}{L} \times \frac{1\ mol}{367.86\ g} = 0.11\ M$$

Assume the reaction goes to completion because K is large. Then solve the back-equilibrium problem to determine the small amount of Mg^{2+} present.

	Mg^{2+}	+ $P_3O_{10}^{5-}$	$\rightleftharpoons$	$MgP_3O_{10}^{3-}$	
Before	$2.1 \times 10^{-3}\ M$	$0.11\ M$		0	
Change	-2.1×10^{-3}	-2.1×10^{-3}	$\rightarrow$	$+2.1 \times 10^{-3}$	React completely
After	0	0.11		2.1×10^{-3}	New initial condition
	x mol/L $MgP_3O_{10}^{3-}$ dissociates to reach equilibrium				
Change	$+x$	$+x$	$\leftarrow$	$-x$	
Equil.	x	$0.11 + x$		$2.1 \times 10^{-3} - x$	

$$K = 4.0 \times 10^8 = \frac{[MgP_3O_{10}^{3-}]}{[Mg^{2+}][P_3O_{10}^{5-}]} = \frac{2.1 \times 10^{-3} - x}{x(0.11 + x)} \quad (\text{assume } x << 2.1 \times 10^{-3})$$

$$4.0 \times 10^8 \approx \frac{2.1 \times 10^{-3}}{x(0.11)},\ x = [Mg^{2+}] = 4.8 \times 10^{-11}\ M;\ \text{assumptions good.}$$

135. a.

	$SrF_2(s)$	$\rightleftharpoons$	$Sr^{2+}(aq)$	+	$2\ F^-(aq)$
Initial			0		0
	s mol/L SrF_2 dissolves to reach equilibrium				
Equil.			*s*		2*s*

$[Sr^{2+}][F^-]^2 = K_{sp} = 7.9 \times 10^{-10} = 4s^3$, $s = 5.8 \times 10^{-4}$ mol/L in pure water

b. Greater, because some of the F^- would react with water:

$$F^- + H_2O \rightleftharpoons HF + OH^- \quad K_b = \frac{K_w}{K_{a,\,HF}} = 1.4 \times 10^{-11}$$

This lowers the concentration of F^-, forcing more SrF_2 to dissolve.

c. $SrF_2(s) \rightleftharpoons Sr^{2+} + 2\ F^-$ $\quad K_{sp} = 7.9 \times 10^{-10} = [Sr^{2+}][F^-]^2$

Let s = solubility = $[Sr^{2+}]$; then $2s$ = total F^- concentration.

Since F^- is a weak base, some of the F^- is converted into HF. Therefore:

total F^- concentration = $2s = [F^-] + [HF]$

$$HF \rightleftharpoons H^+ + F^- \quad K_a = 7.2 \times 10^{-4} = \frac{[H^+][F^-]}{[HF]} = \frac{1.0 \times 10^{-2}[F^-]}{[HF]} \quad \text{(since pH = 2.00 buffer)}$$

$$7.2 \times 10^{-2} = \frac{[F^-]}{[HF]}, \quad [HF] = 14[F^-]; \text{ solving:}$$

$[Sr^{2+}] = s$; $2s = [F^-] + [HF] = [F^-] + 14[F^-]$, $2s = 15[F^-]$, $[F^-] = 2s/15$

$$K_{sp} = 7.9 \times 10^{-10} = [Sr^{2+}][F^-]^2 = (s)\left(\frac{2s}{15}\right)^2, \quad s = 3.5 \times 10^{-3} \text{ mol/L in pH = 2.00 solution}$$

137. Major species: H^+, HSO_4^-, Ba^{2+}, NO_3^-, and H_2O; Ba^{2+} will react with the SO_4^{2-} produced from the K_a reaction for HSO_4^-.

$$HSO_4^- \rightleftharpoons H^+ + SO_4^{2-} \qquad K_{a_2} = 1.2 \times 10^{-2}$$
$$Ba^{2+} + SO_4^{2-} \rightleftharpoons BaSO_4(s) \qquad K = 1/K_{sp} = 1/(1.5 \times 10^{-9}) = 6.7 \times 10^8$$

$$Ba^{2+} + HSO_4^- \rightleftharpoons H^+ + BaSO_4(s) \qquad K_{overall} = (1.2 \times 10^{-2}) \times (6.7 \times 10^8) = 8.0 \times 10^6$$

Because $K_{overall}$ is so large, the reaction essentially goes to completion. Because H_2SO_4 is a strong acid, $[HSO_4^-]_0 = [H^+]_0 = 0.10\ M$.

	Ba^{2+}	+ HSO_4^-	$\rightleftharpoons$	H^+	+ $BaSO_4(s)$	
Before	0.30 *M*	0.10 *M*		0.10 *M*		
Change	–0.10	–0.10	→	+0.10		
After	0.20	0		0.20 *M*		New initial
Change	+*x*	+*x*		–*x*		
Equil.	0.20 + *x*	*x*		0.20 – *x*		

$$K = 8.0 \times 10^6 = \frac{0.20 - x}{(0.20 + x)x} \approx \frac{0.20}{0.20(x)}, \quad x = 1.3 \times 10^{-7}\ M; \text{ assumptions good.}$$

$[H^+] = 0.20 - 1.3 \times 10^{-7} = 0.20\ M$; pH $= -\log(0.20) = 0.70$

$[Ba^{2+}] = 0.20 + 1.3 \times 10^{-7} = 0.20\ M$

From the initial reaction essentially going to completion, 1.0 L(0.10 mol HSO_4^-/L) = 0.10 mol HSO_4^- reacted; this will produce 0.10 mol $BaSO_4$(s). Only 1.3×10^{-7} mol $BaSO_4$ dissolves to reach equilibrium, so 0.10 mol $BaSO_4$(s) is produced.

0.10 mol $BaSO_4$ × 233.4 g/mol = 23 g $BaSO_4$ forms

CHAPTER 17

SPONTANEITY, ENTROPY, AND FREE ENERGY

Questions

17. Living organisms need an external source of energy to carry out these processes. Green plants use the energy from sunlight to produce glucose from carbon dioxide and water by photosynthesis. In the human body, the energy released from the metabolism of glucose helps drive the synthesis of proteins. For all processes combined, ΔS_{univ} must be greater than zero (the second law).

19. As a process occurs, ΔS_{univ} will increase; ΔS_{univ} cannot decrease. Time, like ΔS_{univ}, only goes in one direction.

21. This reaction is kinetically slow but thermodynamically favorable ($\Delta G < 0$). Thermodynamics only tells us if a reaction can occur. To answer the question will it occur, one also needs to consider the kinetics (speed of reaction). The ultraviolet light provides the activation energy for this slow reaction to occur.

23. $\Delta S_{surr} = -\Delta H/T$; heat flow ($\Delta H$) into or out of the system dictates ΔS_{surr}. If heat flows into the surroundings, the random motions of the surroundings increase, and the entropy of the surroundings increases. The opposite is true when heat flows from the surroundings into the system (an endothermic reaction). Although the driving force described here really results from the change in entropy of the surroundings, it is often described in terms of energy. Nature tends to seek the lowest possible energy.

25 a. Positional probability increases; there is a greater volume accessible to the randomly moving gas molecules, which increases disorder.

b. The positional probability doesn't change. There is no change in volume and thus no change in the numbers of positions of the molecules.

c. Positional probability decreases because the volume decreases (P and V are inversely related).

27. This equation refers to breaking the bond in the H_2 molecule. To break a bond, energy must be added; this is an endothermic process. The moles of gas increase as the bond is broken to produce the individual H atoms. Because of the increase in moles of gas, ΔS is positive. When ΔH and ΔS are both positive, the process is spontaneous above some temperature where the favorable ΔS term dominates. This reaction is spontaneous at high temperatures.

29. Note that these substances are not in the solid or gaseous state but are in the aqueous state; water molecules are also present. There is an apparent increase in ordering when these ions are placed in water as compared to the separated state. The hydrating water molecules must be in a highly ordered arrangement when surrounding these anions.

31. One can determine ΔS° and ΔH° for the reaction using the standard entropies and standard enthalpies of formation in Appendix 4; then use the equation ΔG° = ΔH° – TΔS°. One can also use the standard free energies of formation in Appendix 4. And finally, one can use Hess's law to calculate ΔG°. Here, reactions having known ΔG° values are manipulated to determine ΔG° for a different reaction.

For temperatures other than 25°C, ΔG° is estimated using the ΔG° = ΔH° – TΔS° equation. The assumptions made are that the ΔH° and ΔS° values determined from Appendix 4 data are temperature independent. We use the same ΔH° and ΔS° values as determined when T = 25°C; then we plug in the new temperature in Kelvin into the equation to estimate ΔG° at the new temperature.

33. The light source for the first reaction is necessary for kinetic reasons. The first reaction is just too slow to occur unless a light source is available. The kinetics of a reaction are independent of the thermodynamics of a reaction. Even though the first reaction is more favorable thermodynamically (assuming standard conditions), it is unfavorable for kinetic reasons. The second reaction has a negative ΔG° value and is a fast reaction, so the second reaction which occurs very quickly is favored both kinetically and thermodynamically. When considering if a reaction will occur, thermodynamics and kinetics must both be considered.

35. Only statement c is false. The forward reaction is spontaneous when $\Delta G < 0$. This occurs when the free energy of the products is lower than the free energy of the reactants.

37. ΔG° = –RTln(K); ΔG° = ΔH° – TΔS°; because the reaction is spontaneous at standard conditions, ΔG° (and ΔG) must be negative. If $\Delta G° < 0$, then K must be greater than 1 as dictated by the ΔG° = –RTln(K) equation. Statement c is false. Because Δn is negative for this reaction, ΔS° will be negative. We know $\Delta G° < 0$, so ΔH° must be negative as dictated mathematically be the ΔG° = ΔH° – TΔS° equation.

Exercises

Spontaneity, Entropy, and the Second Law of Thermodynamics: Free Energy

39. a, b, and c; from our own experiences, salt water, colored water, and rust form without any outside intervention. It takes an outside energy source to clean a bedroom, so this process is not spontaneous.

41. Possible arrangements for one molecule:

Both are equally probable.

Possible arrangements for two molecules:

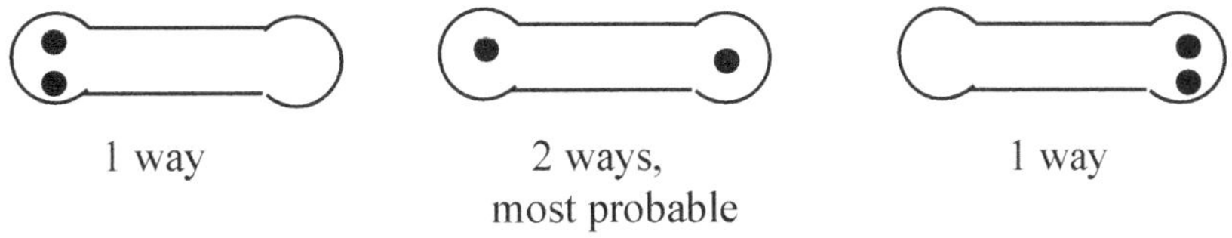

Possible arrangement for three molecules:

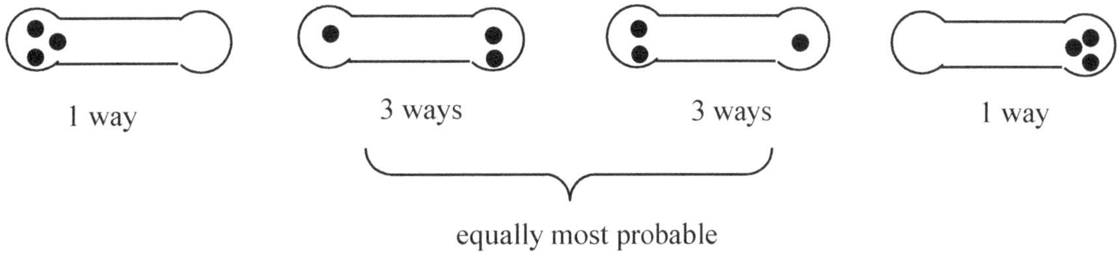

43. We draw all the possible arrangements of the two particles in the three levels.

2 kJ	___	___	x	___	x	xx
1 kJ	___	x	___	xx	x	___
0 kJ	xx	x	x	___	___	___
Total E =	0 kJ	1 kJ	2 kJ	2 kJ	3 kJ	4 kJ

The most likely total energy is 2 kJ.

45. a. H_2 at 100°C and 0.5 atm; higher temperature and lower pressure means greater volume and hence larger positional probability.

b. N_2 at STP; N_2 at STP has the greater volume because P is smaller and T is larger.

c. $H_2O(l)$ has a larger positional probability than $H_2O(s)$.

47. a. Boiling a liquid requires heat. Hence this is an endothermic process. All endothermic processes decrease the entropy of the surroundings (ΔS_{surr} is negative).

b. This is an exothermic process. Heat is released when gas molecules slow down enough to form the solid. In exothermic processes, the entropy of the surroundings increases (ΔS_{surr} is positive).

49. $\Delta G = \Delta H - T\Delta S$; when ΔG is negative, then the process will be spontaneous.

a. $\Delta G = \Delta H - T\Delta S = 25 \times 10^3 \text{ J} - (300.\text{ K})(5.0 \text{ J/K}) = 24{,}000 \text{ J}$; not spontaneous

b. $\Delta G = 25{,}000 \text{ J} - (300.\text{ K})(100.\text{ J/K}) = -5000 \text{ J}$; spontaneous

c. Without calculating ΔG, we know this reaction will be spontaneous at all temperatures. ΔH is negative and ΔS is positive ($-T\Delta S < 0$). ΔG will always be less than zero with these sign combinations for ΔH and ΔS.

d. $\Delta G = -1.0 \times 10^4$ J – (200. K)(–40. J/K) = –2000 J; spontaneous

51. At the boiling point, ΔG = 0, so ΔH = TΔS.

$$\Delta S = \frac{\Delta H}{T} = \frac{27.5 \text{ kJ/mol}}{(273 + 35) \text{ K}} = 8.93 \times 10^{-2} \text{ kJ/K•mol} = 89.3 \text{ J/K•mol}$$

53. $C_2H_5OH(l) \rightarrow C_2H_5OH(g)$; at the boiling point, ΔG = 0 and $\Delta S_{univ} = 0$. For the vaporization process, ΔS is a positive value, whereas ΔH is a negative value. To calculate ΔS_{sys}, we will determine ΔS_{surr} from ΔH and the temperature; then $\Delta S_{sys} = -\Delta S_{surr}$ for a system at equilibrium.

$$\Delta S_{surr} = \frac{-\Delta H}{T} = \frac{38.7 \times 10^3 \text{ J/mol}}{351 \text{ K}} = -110. \text{ J/K•mol}$$

$$\Delta S_{sys} = -\Delta S_{surr} = -(-110.) = 110. \text{ J/K•mol}$$

Chemical Reactions: Entropy Changes and Free Energy

55. a. Decrease in positional probability; ΔS° will be negative. There is only one way to arrange 12 gas molecules all in one bulb, but there are many more ways to arrange the gas molecules equally distributed in each flask.

b. Decrease in positional probability; ΔS° is negative for the liquid to solid phase change.

c. Decrease in positional probability; ΔS° is negative because the moles of gas decreased when going from reactants to products (3 moles → 0 moles). Changes in the moles of gas present as reactants are converted to products dictates predicting positional probability. The gaseous phase always has the larger positional probability associated with it.

d. Increase in positional probability; ΔS° is positive for the liquid to gas phase change. The gas phase always has the larger positional probability.

57. a. $C_{graphite}(s)$; diamond has a more ordered structure (has a smaller positional probability) than graphite.

b. $C_2H_5OH(g)$; the gaseous state is more disordered (has a larger positional probability) than the liquid state.

c. $CO_2(g)$; the gaseous state is more disordered (has a larger positional probability) than the solid state.

59. a. $2\ H_2S(g) + SO_2(g) \rightarrow 3\ S_{rhombic}(s) + 2\ H_2O(g)$; because there are more molecules of reactant gases than product molecules of gas ($\Delta n = 2 - 3 < 0$), ΔS° will be negative.

$$\Delta S° = \sum n_p S°_{products} - \sum n_r S°_{reactants}$$

$\Delta S° = [3 \text{ mol } S_{rhombic}(s)\ (32 \text{ J/K•mol}) + 2 \text{ mol } H_2O(g)\ (189 \text{ J/K•mol})]$

$- [2 \text{ mol } H_2S(g)\ (206 \text{ J/K•mol}) + 1 \text{ mol } SO_2(g)\ (248 \text{ J/K•mol})]$

$\Delta S° = 474 \text{ J/K} - 660. \text{ J/K} = -186 \text{ J/K}$

b. $2\ SO_3(g) \rightarrow 2\ SO_2(g) + O_2(g)$; because Δn of gases is positive (Δn = 3 – 2), ΔS° will be positive.

$\Delta S = 2 \text{ mol}(248 \text{ J/K•mol}) + 1 \text{ mol}(205 \text{ J/K•mol}) - [2 \text{ mol}(257 \text{ J/K•mol})] = 187 \text{ J/K}$

c. $Fe_2O_3(s) + 3\ H_2(g) \rightarrow 2\ Fe(s) + 3\ H_2O(g)$; because Δn of gases = 0 (Δn = 3 – 3), we can't easily predict if ΔS° will be positive or negative.

$\Delta S = 2 \text{ mol}(27 \text{ J/K•mol}) + 3 \text{ mol}(189 \text{ J/K•mol}) -$

$[1 \text{ mol}(90. \text{ J/K•mol}) + 3 \text{ mol } (141 \text{ J/K•mol})] = 138 \text{ J/K}$

61. $C_2H_2(g) + 4\ F_2(g) \rightarrow 2\ CF_4(g) + H_2(g)$; $\Delta S° = 2S°_{CF_4} + S°_{H_2} - [S°_{C_2H_2} + 4S°_{F_2}]$

$-358 \text{ J/K} = (2 \text{ mol})S°_{CF_4} + 131 \text{ J/K} - [201 \text{ J/K} + 4(203 \text{ J/K})]$, $S°_{CF_4} = 262 \text{ J/K•mol}$

63. a. $S_{rhombic} \rightarrow S_{monoclinic}$; this phase transition is spontaneous (ΔG < 0) at temperatures above 95°C. ΔG = ΔH – TΔS; for ΔG to be negative only above a certain temperature, then ΔH is positive and ΔS is positive (see Table 17.5 of text).

b. Because ΔS is positive, $S_{rhombic}$ is the more ordered crystalline structure (has the smaller positional probability).

65. a.

	$CH_4(g)$	+ $2\ O_2(g)$	→ $CO_2(g)$	+ $2\ H_2O(g)$	
ΔH_f°	–75 kJ/mol	0	–393.5	–242	
ΔG_f°	–51 kJ/mol	0	–394	–229	Data from Appendix 4
S°	186 J/K•mol	205	214	189	

$\Delta H° = \sum n_p \Delta H°_{f,\ products} - \sum n_r \Delta H°_{f,\ reactants}$; $\Delta S° = \sum n_p S°_{products} - \sum n_r S°_{reactants}$

$\Delta H° = 2 \text{ mol}(-242 \text{ kJ/mol}) + 1 \text{ mol}(-393.5 \text{ kJ/mol}) - [1 \text{ mol}(-75 \text{ kJ/mol})] = -803 \text{ kJ}$

$\Delta S° = 2 \text{ mol}(189 \text{ J/K•mol}) + 1 \text{ mol}(214 \text{ J/K•mol})$

$- [1 \text{ mol}(186 \text{ J/K•mol}) + 2 \text{ mol}(205 \text{ J/K•mol})] = -4 \text{ J/K}$

There are two ways to get ΔG°. We can use ΔG° = ΔH° – TΔS° (be careful of units):

$$\Delta G° = \Delta H° - T\Delta S° = -803 \times 10^3 \text{ J} - (298 \text{ K})(-4 \text{ J/K}) = -8.018 \times 10^5 \text{ J} = -802 \text{ kJ}$$

or we can use ΔG_f° values, where $\Delta G° = \sum n_p \Delta G°_{f,\ products} - \sum n_r \Delta G°_{f,\ reactants}$:

$\Delta G° = 2\text{ mol}(-229\text{ kJ/mol}) + 1\text{ mol}(-394\text{ kJ/mol}) - [1\text{ mol}(-51\text{ kJ/mol})]$

$\Delta G° = -801$ kJ (Answers are the same within round off error.)

b. $6\ CO_2(g) + 6\ H_2O(l) \rightarrow C_6H_{12}O_6(s) + 6\ O_2(g)$

	$6\ CO_2(g)$	$6\ H_2O(l)$	$C_6H_{12}O_6(s)$	$6\ O_2(g)$
ΔH_f°	−393.5 kJ/mol	−286	−1275	0
S°	214 J/K•mol	70.	212	20

$\Delta H° = -1275 - [6{-286}) + 6(-393.5)] = 2802$ kJ

$\Delta S° = 6(205) + 212 - [6(214) + 6(70.)] = -262$ J/K

$\Delta G° = 2802\text{ kJ} - (298\text{ K})(-0.262\text{ kJ/K}) = 2880.$ kJ

c. $P_4O_{10}(s) + 6\ H_2O(l) \rightarrow 4\ H_3PO_4(s)$

	$P_4O_{10}(s)$	$6\ H_2O(l)$	$4\ H_3PO_4(s)$
ΔH_f° (kJ/mol)	−2984	−286	−1279
S° (J/K•mol)	229	70.	110.

$\Delta H° = 4\text{ mol}(-1279\text{ kJ/mol}) - [1\text{ mol}(-2984\text{ kJ/mol}) + 6\text{ mol}(-286\text{ kJ/mol})] = -416$ kJ

$\Delta S° = 4(110.) - [229 + 6(70.)] = -209$ J/K

$\Delta G° = \Delta H° - T\Delta S° = -416\text{ kJ} - (298\text{ K})(-0.209\text{ kJ/K}) = -354$ kJ

d. $HCl(g) + NH_3(g) \rightarrow NH_4Cl(s)$

	$HCl(g)$	$NH_3(g)$	$NH_4Cl(s)$
ΔH_f° (kJ/mol)	−92	−46	−314
S° (J/K•mol)	187	193	96

$\Delta H° = -314 - [-92 - 46] = -176$ kJ; $\Delta S° = 96 - [187 + 193] = -284$ J/K

$\Delta G° = \Delta H° - T\Delta S° = -176\text{ kJ} - (298\text{ K})(-0.284\text{ kJ/K}) = -91$ kJ

67. a. $\Delta H° = 2(-46\text{ kJ}) = -92$ kJ; $\Delta S° = 2(193\text{ J/K}) - [3(131\text{ J/K}) + 192\text{ J/K}] = -199$ J/K

$\Delta G° = \Delta H° - T\Delta S° = -92\text{ kJ} - 298\text{ K}(-0.199\text{ kJ/K}) = -33$ kJ

b. ΔG° is negative, so this reaction is spontaneous at standard conditions.

c. $\Delta G° = 0$ when $T = \dfrac{\Delta H°}{\Delta S°} = \dfrac{-92\text{ kJ}}{-0.199\text{ kJ/K}} = 460$ K

At T < 460 K and standard pressures (1 atm), the favorable ΔH° term dominates, and the reaction is spontaneous (ΔG° < 0).

69. $C_2H_2(g) + H_2(g) \rightarrow C_2H_4(g)$; $\Delta G° = 0$ when $T = \dfrac{\Delta H°}{\Delta S°} = \dfrac{-174 \text{ kJ}}{-0.112 \text{ kJ/K}} = 1550 \text{ K} = 1280°C$

At T < 1550 K (1280°C) and standard pressures (1 atm), the favorable ΔH° term dominates, and the reaction is spontaneous (ΔG° < 0).

71. $H_2O(l) \rightarrow H_2O(g)$; ΔG° = 0 at the boiling point of water at 1 atm and 100. °C.

$$\Delta H° = T\Delta S°,\ \Delta S° = \frac{\Delta H°}{T} = \frac{40.6 \times 10^3 \text{ J/mol}}{373 \text{ K}} = 109 \text{ J/K•mol}$$

At 90.°C: ΔG° = ΔH° – TΔS° = 40.6 kJ/mol – (363 K)(0.109 kJ/K•mol) = 1.0 kJ/mol

As expected, ΔG° > 0 at temperatures below the boiling point of water at 1 atm (process is nonspontaneous).

At 110.°C: ΔG° = ΔH° – TΔS° = 40.6 kJ/mol – (383 K)(0.109 J/K•mol) = –1.1 kJ/mol

When ΔG° < 0, the boiling of water is spontaneous at 1 atm, and T > 100. °C (as expected).

73.

$CH_4(g) \rightarrow 2\ H_2(g) + C(s)$	ΔG° = – (–51 kJ)
$2\ H_2(g) + O_2(g) \rightarrow 2\ H_2O(l)$	ΔG° = –474 kJ)
$C(s) + O_2(g) \rightarrow CO_2(g)$	ΔG° = –394 kJ
$CH_4(g) + 2\ O_2(g) \rightarrow 2\ H_2O(l) + CO_2(g)$	ΔG° = –817 kJ

75. $\Delta G° = \Sigma n_p \Delta G^o_{f,\text{ products}} - \Sigma n_r \Delta G^o_{f,\text{ reactants}}$, $-374 \text{ kJ} = -1105 \text{ kJ} - \Delta G^o_{f,\text{ SF}_4}$

$\Delta G^o_{f,\text{ SF}_4} = -731$ kJ/mol

77. a. ΔG° = 2 mol(0) + 3 mol(–229 kJ/mol) – [1 mol(–740. kJ/mol) + 3 mol(0)] = 53 kJ

b. Because ΔG° is positive, this reaction is not spontaneous at standard conditions and 298 K.

c. $\Delta G° = \Delta H° - T\Delta S°,\ \Delta S° = \dfrac{\Delta H° - \Delta G°}{T} = \dfrac{100.\text{ kJ} - 53 \text{ kJ}}{298 \text{ K}} = 0.16 \text{ kJ/K}$

We need to solve for the temperature when ΔG° = 0:

$$\Delta G° = 0 = \Delta H° - T\Delta S°,\ \Delta H° = T\Delta S°,\ T = \frac{\Delta H°}{\Delta S°} = \frac{100.\text{ kJ}}{0.16 \text{ kJ/K}} = 630 \text{ K}$$

This reaction will be spontaneous (ΔG < 0) at T > 630 K, where the favorable entropy term will dominate.

Free Energy: Pressure Dependence and Equilibrium

79. ΔG = ΔG° + RT ln Q; for this reaction: $\Delta G = \Delta G° + RT \ln \dfrac{P^2_{HCl} \times P_{Br_2}}{P^2_{HBr} \times P_{Cl_2}}$

$$\Delta G = -81.0 \text{ kJ} + \frac{8.3145 \text{ J/K} \bullet \text{mol}}{1000 \text{ J/kJ}}(298 \text{ K}) \ln \frac{(0.500)^2 (0.200)}{(1.00)^2 (5.00)}$$

$\Delta G = -81.0 \text{ kJ} + (-11.4 \text{ kJ} = -92.4 \text{ kJ}$

81. $\Delta G = \Delta G° + RT \ln Q = \Delta G° + RT \ln \dfrac{P_{N_2O_4}}{P^2_{NO_2}}$

$\Delta G° = 1 \text{ mol}(98 \text{ kJ/mol}) - 2 \text{ mol}(52 \text{ kJ/mol}) = -6 \text{ kJ}$

a. These are standard conditions, so $\Delta G = \Delta G°$ because $Q = 1$ and $\ln Q = 0$. Because $\Delta G°$ is negative, the forward reaction is spontaneous. The reaction shifts right to reach equilibrium.

b. $\Delta G = -6 \times 10^3 \text{ J} + 8.3145 \text{ J/K•mol } (298 \text{ K}) \ln \dfrac{0.50}{(0.21)^2}$

$\Delta G = -6 \times 10^3 \text{ J} + 6.0 \times 10^3 \text{ J} = 0$

Because $\Delta G = 0$, this reaction is at equilibrium (no shift).

c. $\Delta G = -6 \times 10^3 \text{ J} + 8.3145 \text{ J/K•mol } (298 \text{ K}) \ln \dfrac{1.6}{(0.29)^2}$

$\Delta G = -6 \times 10^3 \text{ J} + 7.3 \times 10^3 \text{ J} = 1.3 \times 10^3 \text{ J} = 1 \times 10^3 \text{ J}$

Because ΔG is positive, the reverse reaction is spontaneous, and the reaction shifts to the left to reach equilibrium.

83. $NO(g) + O_3(g) \rightleftharpoons NO_2(g) + O_2(g)$; $\Delta G° = \Sigma n_p \Delta G^o_{f,\text{products}} - \Sigma n_r \Delta G^o_{f,\text{reactants}}$

$\Delta G° = 1 \text{ mol}(52 \text{ kJ/mol}) - [1 \text{ mol}(87 \text{ kJ/mol}) + 1 \text{ mol}(163 \text{ kJ/mol})] = -198 \text{ kJ}$

$$\Delta G° = -RT \ln K,\ K = \exp\frac{-\Delta G^o}{RT} = \exp\left[\frac{-(-1.98 \times 10^5 \text{ J})}{8.3145 \text{ J/K} \bullet \text{mol}(298 \text{ K})}\right] = e^{79.912} = 5.07 \times 10^{34}$$

Note: When determining exponents, we will round off after the calculation is complete. This helps eliminate excessive round off error.

85. At 25.0°C: $\Delta G° = \Delta H° - T\Delta S° = -58.03 \times 10^3 \text{ J/mol} - (298.2 \text{ K})(-176.6 \text{ J/K•mol})$
$= -5.37 \times 10^3 \text{ J/mol}$

$$\Delta G° = -RT \ln K,\ \ln K = \frac{-\Delta G^o}{RT} = \exp\left[\frac{-(-5.37 \times 10^3 \text{ J/mol})}{(8.3145 \text{ J/K} \bullet \text{mol})(298.2 \text{ K})}\right] = 2.166$$

$K = e^{2.166} = 8.72$

At 100.0°C: $\Delta G° = -58.03 \times 10^3 \text{ J/mol} - (373.2 \text{ K})(-176.6 \text{ J/K•mol}) = 7.88 \times 10^3 \text{ J/mol}$

$$\ln K = \frac{-(7.88 \times 10^3 \text{ J/mol})}{(8.3145 \text{ J/K} \bullet \text{mol})(373.2 \text{ K})} = -2.540, \ K = e^{-2.540} = 0.0789$$

Note: When determining exponents, we will round off after the calculation is complete. This helps eliminate excessive round off error.

87. When reactions are added together, the equilibrium constants are multiplied together to determine the K value for the final reaction.

$H_2(g) + O_2(g) \rightleftharpoons H_2O_2(g)$	$K = 2.3 \times 10^6$
$H_2O(g) \rightleftharpoons H_2(g) + 1/2O_2(g)$	$K = (1.8 \times 10^{37})^{-1/2}$
$H_2O(g) + 1/2O_2(g) \rightleftharpoons H_2O_2(g)$	$K = 2.3 \times 10^6(1.8 \times 10^{37})^{-1/2} = 5.4 \times 10^{-13}$

$$\Delta G° = -RT \ln K = \frac{-8.3145 \text{ J}}{\text{K} \bullet \text{mol}}(600.\text{ K}) \ln(5.4 \times 10^{-13}) = 1.4 \times 10^5 \text{ J/mol} = 140 \text{ kJ/mol}$$

89. a. $\Delta G° = -RT \ln K, \ K = \exp\left[\frac{-(-30{,}500 \text{ J})}{8.3145 \text{ J/K} \bullet \text{mol} \times 298 \text{ K}}\right] = 2.22 \times 10^5$

b. $C_6H_{12}O_6(s) + 6\ O_2(g) \rightarrow 6\ CO_2(g) + 6\ H_2O(l)$

$\Delta G° = 6 \text{ mol}(-394 \text{ kJ/mol}) + 6 \text{ mol}(-237 \text{ kJ/mol}) - 1 \text{ mol}(-911 \text{ kJ/mol}) = -2875 \text{ kJ}$

$\frac{2875 \text{ kJ}}{\text{mol glucose}} \times \frac{1 \text{ mol ATP}}{30.5 \text{ kJ}} = 94.3$ mol ATP; 94.3 molecules ATP/molecule glucose

This is an overstatement. The assumption that all the free energy goes into this reaction is false. Only 38 moles of ATP are produced by metabolism of 1 mole of glucose.

91. $K = \frac{P_{NF_3}^2}{P_{N_2} \times P_{F_2}^3} = \frac{(0.48)^2}{0.021(0.063)^3} = 4.4 \times 10^4$

$\Delta G°_{800} = -RT \ln K = -8.3145 \text{ J/K•mol}(800.\text{ K}) \ln(4.4 \times 10^4) = -7.1 \times 10^4 \text{ J/mol} = -71 \text{ kJ/mol}$

93. The equilibrium concentration of C is greater than the initial concentration, so the reaction shifts right to reach equilibrium. The general ICE table for this problem is:

	A(aq)	+	2 B(aq)	$\rightleftharpoons$	3 C(aq)
Initial	1.00 *M*		1.00 *M*		1.00 *M*
	Let *x* mol/L of A react to reach equilibrium				
Change	$-x$		$-2x$	$\rightarrow$	$+3x$
Equil	$1.00 - x$		$1.00 - 2x$		$1.00 + 3x$

In the problem, we are told that $[C]_e = 1.96\ M$. From the set-up, $[C]_e = 1.00 + 3x = 1.96\ M$. Solving: $x = 0.32$ mol/L. The other concentrations are $[A] = 1.00 - 0.32 = 0.68$ mol/L and $[B]_e = 1.00 - 2(0.32) = 0.36$ mol/L. Calculating K:

$$K = \frac{[C]^3}{[A][B]^2} = \frac{(1.96)^3}{0.68(0.36)^2} = 85$$

$\Delta G° = -RT \ln K = -8.3145$ J/K•mol(298 K) ln(85) = -1.1×10^4 J/mol = -11 kJ/mol

95. The equation $\ln K = \frac{-\Delta H°}{R}\left(\frac{1}{T}\right) + \frac{\Delta S°}{R}$ is in the form of a straight line equation ($y = mx + b$). A graph of ln K versus 1/T will yield a straight line with slope = $m = -\Delta H°/R$ and a y intercept = $b = \Delta S°/R$.

From the plot:

$$\text{slope} = \frac{\Delta y}{\Delta x} = \frac{0 - 40.}{3.0 \times 10^{-3}\ K^{-1} - 0} = -1.3 \times 10^4\ K$$

-1.3×10^4 K = $-\Delta H°/R$, $\Delta H° = 1.3 \times 10^4$ K × 8.3145 J/K•mol = 1.1×10^5 J/mol

y intercept = 40. = $\Delta S°/R$, $\Delta S°$ = 40. × 8.3145 J/K•mol = 330 J/K•mol

As seen here, when $\Delta H°$ is positive, the slope of the ln K versus 1/T plot is negative. When $\Delta H°$ is negative as in an exothermic process, then the slope of the ln K versus 1/T plot will be positive (slope = $-\Delta H°/R$).

97. $\Delta G° = -RT \ln K$; $\Delta G° = \Delta H° - T\Delta S°$; at 25°C, K < 1 so $\Delta G°$ must be positive. At 227 °C, K > 1 so $\Delta G°$ must be negative. As temperature increased, the sign of $\Delta G°$ changed from a positive value to a negative value. This only occurs when the signs of $\Delta H°$ and $\Delta S°$ are both positive for a reaction. So, at 25°C, $\Delta G°$, $\Delta H°$, and $\Delta S°$ are all positive.

99. $NO(g) + 1/2\ O_2(g) \rightarrow NO_2(g)$

$\Delta G° = -RT \ln K = -8.3145$ J/K•mol(298 K) ln(1.50×10^6) = -3.52×10^4 J/mol

$$\Delta G° = \Delta H° - T\Delta S,\ \Delta S° = \frac{\Delta H° - \Delta G°}{T} = \frac{-57.0\ kJ - (-35.2\ kJ)}{298\ K} = -0.0732\ kJ/K = -73.2\ J/K$$

$2\ NO(g) + O_2(g) \rightarrow 2\ NO_2(g)$; the coefficients of this reaction are double the original reaction. So, $\Delta S° = 2(-73.2) = -146$ J/K.

ChemWork Problems

101. From Appendix 4, S° = 198 J/K•mol for CO(g) and S° = 27 J/K•mol for Fe(s).

Let S_l° = S° for $Fe(CO)_5(l)$ and S_g° = S° for $Fe(CO)_5(g)$.

$\Delta S°$ = –677 J/K = 1 mol(S_l°) – [1 mol (27 J/K•mol) + 5 mol(198 J/ K•mol]

S_l° = 340. J/K•mol

$\Delta S° = 107\ J/K = 1\ mol\ (S^{o}_{g}) - 1\ mol\ (340.\ J/K{\bullet}mol)$

$S^{o}_{g} = S°$ for $Fe(CO)_5(g) = 447\ J/K{\bullet}mol$

103. $H_2O(s) \rightarrow H_2O(l)$; this is an endothermic process because energy must be added to melt an ice cube. So, ΔH is positive. There is an increase in positional probability as a solid is converted into a liquid, so ΔS is positive. At –10°C, melting an ice cube is not spontaneous, so ΔG is positive.

105. A negative value for $\Delta S°$ indicates a process that has a decrease in positional probability. Processes b, c, and d all are expected to have negative values for $\Delta S°$. In these processes, the number of moles of gaseous molecules decreases as reactants are converted to products. Whenever this occurs, positional probability decreases. Processes a and e have positive $\Delta S°$ values. In process a, the number of gaseous molecules increase as reactants are converted to products, so positional probability increases. In process e, the liquid state has a larger positional probability as compared to the solid state.

107. It appears that the sum of the two processes has no net change. This is not so. By the second law of thermodynamics, ΔS_{univ} must have increased even though it looks as if we have gone through a cyclic process.

109. $C_2H_4(g) + H_2O(g) \rightarrow CH_3CH_2OH(l)$

$\Delta H° = -278 - (52 - 242) = -88\ kJ$; $\Delta S° = 161 - (219 + 189) = -247\ J/K$

When $\Delta G° = 0$, $\Delta H° = T\Delta S°$, so $T = \dfrac{\Delta H°}{\Delta S°} = \dfrac{-88 \times 10^3\ J}{-247\ J/K} = 360\ K.$

Since the signs of $\Delta H°$ and $\Delta S°$ are both negative, this reaction at standard concentrations will be spontaneous at temperatures below 360 K (where the favorable $\Delta H°$ term will dominate).

$C_2H_6(g) + H_2O(g) \rightarrow CH_3CH_2OH(l) + H_2(g)$

$\Delta H° = -278 - (-84.7 - 242) = 49\ kJ$; $\Delta S° = 131 + 161 - (229.5 + 189) = -127\ J/K$

Both $\Delta H°$ and $\Delta S°$ have unfavorable signs. $\Delta G°$ can never be negative when $\Delta H°$ is positive and $\Delta S°$ is negative. So this reaction can never be spontaneous at standard conditions.

Thus the reaction $C_2H_4(g) + H_2O(g) \rightarrow C_2H_5OH(l)$ would be preferred at standard conditions.

111. a.

$\Delta H° = 185.0 - 286 - (-53 + 135.1) = -183\ kJ$; $\Delta S° = 274 + 70. - (242 + 202) = -100.\ J/K$

$\Delta G° = \Delta H° - T\Delta S° = -183\ kJ - 298\ K(-0.100\ kJ/K) = -153\ kJ$

b. $HC{\equiv}CH(g) + HCN(g) \rightarrow CH_2{=}CHCN(g)$

$\Delta H° = 185.0 - (135.1 + 227) = -177$ kJ; $\Delta S° = 274 - (202 + 201) = -129$ J/K

T = 70.°C = 343 K; $\Delta G^\circ_{343} = \Delta H° - T\Delta S° = -177 \text{ kJ} - 343 \text{ K}(-0.129 \text{ kJ/K})$

$\Delta G^\circ_{343} = -177 \text{ kJ} + 44 \text{ kJ} = -133 \text{ kJ}$

c. $4\ CH_2{=}CHCH_3(g) + 6\ NO(g) \rightarrow 4\ CH_2{=}CHCN(g) + 6\ H_2O(g) + N_2(g)$

$\Delta H° = 6(-242) + 4(185.0) - [4(20.9) + 6(90.)] = -1336$ kJ

$\Delta S° = 192 + 6(189) + 4(274) - [6(211) + 4(266.9)] = 88$ J/K

T = 700.°C = 973 K; $\Delta G^\circ_{973} = \Delta H° - T\Delta S° = -1336 \text{ kJ} - 973 \text{ K}(0.088 \text{ kJ/K})$

$\Delta G^\circ_{973} = -1336 \text{ kJ} - 86 \text{ kJ} = -1422 \text{ kJ}$

113. solid I $\rightarrow$ solid II; equilibrium occurs when $\Delta G = 0$.

$$\Delta G = \Delta H - T\Delta S,\ \ \Delta H = T\Delta S,\ \ T = \Delta H/\Delta S = \frac{-743.1 \text{ J/mol}}{-17.0 \text{ J/K} \bullet \text{mol}} = 43.7 \text{ K} = -229.5°\text{C}$$

115. $Ba(NO_3)_2(s) \rightleftharpoons Ba^{2+}(aq) + 2\ NO_3^-(aq)$ $K = K_{sp}$; $\Delta G° = -561 + 2(-109) - (-797) = 18$ kJ

$$\Delta G° = -RT \ln K_{sp},\ \ \ln K_{sp} = \frac{-\Delta G^\circ}{RT} = \frac{-18{,}000 \text{ J}}{8.3145 \text{ J/K} \bullet \text{mol}\,(298 \text{ K})} = -7.26$$

$K_{sp} = e^{-7.26} = 7.0 \times 10^{-4}$

117. $\Delta G° = -RT \ln K$; $\Delta G = \Delta H - T\Delta S$; With K < 1, $\Delta G°$ must be positive. The reaction is spontaneous, so ΔG must be negative. This indicates that the reaction shifts right to reach equilibrium. With a negative ΔG but a positive ΔH (the reaction is endothermic), ΔS must be positive (favorable) to give a negative ΔG value. Statement c is false.

119.

$HgbO_2 \rightarrow Hgb + O_2$	$\Delta G° = -(-70 \text{ kJ})$
$Hgb + CO \rightarrow HgbCO$	$\Delta G° = -80 \text{ kJ}$
$HgbO_2 + CO \rightarrow HgbCO + O_2$	$\Delta G° = -10 \text{ kJ}$

$$\Delta G° = -RT \ln K,\ \ K = \exp\left(\frac{-\Delta G^\circ}{RT}\right) = \exp\left[\frac{-(-10 \times 10^3 \text{ J})}{(8.3145 \text{ J/K} \bullet \text{mol})(298 \text{ K})}\right] = 60$$

121. $H_2O(l) \rightleftharpoons H^+(aq) + OH^-(aq)$; $\Delta G = \Delta G° + RT \ln [H^+][OH^-]$

$\Delta G = 79.9 \times 10^3 \text{ J/mol} + (8.3145 \text{ J/K}\bullet\text{mol})(298 \text{ K}) \ln [0.71(0.15)]$

$\Delta G = 7.99 \times 10^4 \text{ J/mol} - 5.5 \times 10^3 \text{ J/mol} = 7.44 \times 10^4 \text{ J/mol} = 74.4 \text{ kJ/mol}$

123. From Section 17.9 of the text: $\ln K = \dfrac{-\Delta H^\circ}{RT} + \dfrac{\Delta S^\circ}{R}$

For two sets of K and T:

$$\ln K_1 = \frac{-\Delta H^\circ}{R}\left(\frac{1}{T_1}\right) + \frac{\Delta S}{R};\quad \ln K_2 = \frac{-\Delta H^\circ}{R}\left(\frac{1}{T_2}\right) + \frac{\Delta S^\circ}{R}$$

Subtracting the first expression from the second:

$$\ln K_2 - \ln K_1 = \frac{\Delta H^\circ}{R}\left(\frac{1}{T_1} - \frac{1}{T_2}\right) \text{ or } \ln\frac{K_2}{K_1} = \frac{\Delta H^\circ}{R}\left(\frac{1}{T_1} - \frac{1}{T_2}\right)$$

Let $K_2/K_1 = 6.67$ (from the problem), so that $T_2 = 350.0$ K and $T_1 = 300.0$ K.

$$\ln 6.67 = \frac{\Delta H^\circ}{8.3145 \text{ J/K} \bullet \text{mol}}\left(\frac{1}{300.0 \text{ K}} - \frac{1}{350.0 \text{ K}}\right)$$

125. a. False; there is a decrease in the number of moles of gaseous molecules, so ΔS° is negative (unfavorable). Because ΔG° is negative, the reaction must be exothermic (favorable).

b. True

c. False; heat is a product in this exothermic reaction. As heat is added (temperature increases), this reaction will shift left by Le Châtelier's principle. This results in a decrease in the value of K. At equilibrium at the higher temperature, reactant concentrations will increase and the product concentration will decrease. Therefore, the ratio of $[PCl_5]/[PCl_3]$ will decrease with an increase in temperature.

d. False; when the signs of ΔH° and ΔS° are both negative, the reaction will only be spontaneous at temperatures below some value (at low temperatures) where the favorable ΔH° term dominates.

e. True; from $\Delta G^\circ = -RT \ln K$, when $\Delta G^\circ < 0$, K must be greater than 1.

127. ΔS is more favorable (less negative) for reaction 2 than for reaction 1, resulting in $K_2 > K_1$. In reaction 1, seven particles in solution are forming one particle in solution. In reaction 2, four particles are forming one, which results in a smaller decrease in positional probability than for reaction 1.

129. $\Delta G^\circ = -RT \ln K$; when $K = 1.00$, $\Delta G^\circ = 0$ since $\ln 1.00 = 0$. $\Delta G^\circ = 0 = \Delta H^\circ - T\Delta S^\circ$

$\Delta H^\circ = 3(-242 \text{ kJ}) - [-826 \text{ kJ}] = 100.\text{ kJ}$; $\Delta S^\circ = [2(27 \text{ J/K}) + 3(189 \text{ J/K})] - [90.\text{ J/K} + 3(131 \text{ J/K})] = 138 \text{ J/K}$

$$\Delta H^\circ = T\Delta S^\circ,\ T = \frac{\Delta H^\circ}{\Delta S^\circ} = \frac{100.\text{ kJ}}{0.138 \text{ kJ/K}} = 725 \text{ K}$$

Challenge Problems

131 a. Vessel 1: At 0°C, this system is at equilibrium, so $\Delta S_{univ} = 0$ and $\Delta S = \Delta S_{surr}$. Because the vessel is perfectly insulated, q = 0, so $\Delta S_{surr} = 0 = \Delta S_{sys}$.

b. Vessel 2: The presence of salt in water lowers the freezing point of water to a tem-perature below 0°C. In vessel 2, the conversion of ice into water will be spontaneous at 0°C, so $\Delta S_{univ} > 0$. Because the vessel is perfectly insulated, $\Delta S_{surr} = 0$. Therefore, ΔS_{sys} must be positive ($\Delta S > 0$) in order for ΔS_{univ} to be positive.

133. $3\ O_2(g) \rightleftharpoons 2\ O_3(g)$; $\Delta H° = 2(143\ kJ) = 286\ kJ$; $\Delta G° = 2(163\ kJ) = 326\ kJ$

$$\ln K = \frac{-\Delta G^\circ}{RT} = \frac{-326 \times 10^3\ J}{(8.3145\ J/K \bullet mol)(298\ K)} = -131.573,\ K = e^{-131.573} = 7.22 \times 10^{-58}$$

We need the value of K at 230. K. From Section 17.9 of the text:

$$\ln K = \frac{-\Delta G^\circ}{RT} + \frac{\Delta S^\circ}{R}$$

For two sets of K and T:

$$\ln K_1 = \frac{-\Delta H^\circ}{R}\left(\frac{1}{T_1}\right) + \frac{\Delta S}{R};\ \ln K_2 = \frac{-\Delta H^\circ}{R}\left(\frac{1}{T_2}\right) + \frac{\Delta S^\circ}{R}$$

Subtracting the first expression from the second:

$$\ln K_2 - \ln K_1 = \frac{\Delta H^\circ}{R}\left(\frac{1}{T_1} - \frac{1}{T_2}\right) \text{ or } \ln\frac{K_2}{K_1} = \frac{\Delta H^\circ}{R}\left(\frac{1}{T_1} - \frac{1}{T_2}\right)$$

Let $K_2 = 7.22 \times 10^{-58}$, $T_2 = 298$ K; $K_1 = K_{230}$, $T_1 = 230.$ K; $\Delta H° = 286 \times 10^3$ J

$$\ln\frac{7.22 \times 10^{-58}}{K_{230}} = \frac{286 \times 10^3}{8.3145}\left(\frac{1}{230.} - \frac{1}{298}\right) = 34.13$$

$$\frac{7.22 \times 10^{-58}}{K_{230}} = e^{34.13} = 6.6 \times 10^{14},\ K_{230} = 1.1 \times 10^{-72}$$

$$K_{230} = 1.1 \times 10^{-72} = \frac{P^2_{O_3}}{P^3_{O_3}} = \frac{P^2_{O_3}}{(1.0 \times 10^{-3}\ atm)^3},\ P_{O_3} = 3.3 \times 10^{-41}\ atm$$

The volume occupied by one molecule of ozone is:

$$V = \frac{nRT}{P} = \frac{(1/6.022 \times 10^{23}\ mol)(0.08206\ L\ atm/K \bullet mol)(230.\ K)}{(3.3 \times 10^{-41}\ atm)} = 9.5 \times 10^{17}\ L$$

Equilibrium is probably not maintained under these conditions. When only two ozone molecules are in a volume of 9.5×10^{17} L, the reaction is not at equilibrium. Under these conditions, Q > K, and the reaction shifts left. But with only 2 ozone molecules in this huge volume, it is extremely unlikely that they will collide with each other. At these conditions, the concentration of ozone is not large enough to maintain equilibrium.

135. a. From the plot, the activation energy of the reverse reaction is $E_a + (-\Delta G°) = E_a - \Delta G°$ ($\Delta G°$ is a negative number as drawn in the diagram).

$$k_f = A\exp\left(\frac{-E_a}{RT}\right) \text{ and } k_r = A\exp\left[\frac{-(E_a - \Delta G°)}{RT}\right],\ \frac{k_f}{k_r} = \frac{A\exp\left(\frac{-E_a}{RT}\right)}{A\exp\left[\frac{-(E_a - \Delta G°)}{RT}\right]}$$

If the A factors are equal: $\dfrac{k_f}{k_r} = \exp\left[\dfrac{-E_a}{RT} + \dfrac{(E_a - \Delta G°)}{RT}\right] = \exp\left(\dfrac{-\Delta G°}{RT}\right)$

From $\Delta G° = -RT\ln K$, $K = \exp\left(\dfrac{-\Delta G°}{RT}\right)$; because K and $\dfrac{k_f}{k_r}$ are both equal to the same expression, $K = k_f/k_r$.

b. A catalyst will lower the activation energy for both the forward and reverse reactions (but not change $\Delta G°$). Therefore, a catalyst must increase the rate of both the forward and reverse reactions.

137. a. $\Delta G° - G_B^o - G_A^o = 11{,}718 - 8996 = 2722$ J

$$K = \exp\left(\frac{-\Delta G°}{RT}\right) = \exp\left[\frac{-2722\text{ J}}{(8.3145\text{ J/K}\bullet\text{mol})(298\text{ K})}\right] = 0.333$$

b. When $Q = 1.00 > K$, the reaction shifts left. Let x = atm of B(g), which reacts to reach equilibrium.

	A(g)	⇌	B(g)
Initial	1.00 atm		1.00 atm
Equil.	$1.00 + x$		$1.00 - x$

$$K = \frac{P_B}{P_A} = \frac{1.00 - x}{1.00 + x} = 0.333,\ 1.00 - x = 0.333 + (0.333)x,\ x = 0.50\text{ atm}$$

$P_B = 1.00 - 0.50 = 0.50$ atm; $P_A = 1.00 + 0.50 = 1.50$ atm

c. $\Delta G = \Delta G° + RT\ln Q = \Delta G° + RT\ln(P_B/P_A)$

$\Delta G = 2722\text{ J} + (8.3145)(298)\ln(0.50/1.50) = 2722\text{ J} - 2722\text{ J} = 0$ (carrying extra sig. figs.)

139. $K = P_{CO_2}$; to ensure Ag_2CO_3 from decomposing, P_{CO_2} should be greater than K.

From Exercise 95, $\ln K = \dfrac{\Delta H°}{RT} + \dfrac{\Delta S°}{R}$. For two conditions of K and T, the equation is:

$$\ln\frac{K_2}{K_1} = \frac{\Delta H^\circ}{R}\left(\frac{1}{T_1} + \frac{1}{T_2}\right)$$

Let $T_1 = 25°C = 298$ K, $K_1 = 6.23 \times 10^{-3}$ torr; $T_2 = 110.°C = 383$ K, $K_2 = ?$

$$\ln\frac{K_2}{6.23 \times 10^{-3}\ \text{torr}} = \frac{79.14 \times 10^3\ \text{J/mol}}{8.3145\ \text{J/K} \bullet \text{mol}}\left(\frac{1}{298\ \text{K}} - \frac{1}{383\ \text{K}}\right)$$

$$\ln\frac{K_2}{6.23 \times 10^{-3}} = 7.1,\quad \frac{K_2}{6.23 \times 10^{-3}} = e^{7.1} = 1.2 \times 10^3,\quad K_2 = 7.5\ \text{torr}$$

To prevent decomposition of Ag_2CO_3, the partial pressure of CO_2 should be greater than 7.5 torr.

141. $NaCl(s) \rightleftharpoons Na^+(aq) + Cl^-(aq) \qquad K = K_{sp} = [Na^+][Cl^-]$

$\Delta G° = [(-262\ \text{kJ}) + (-131\ \text{kJ})] - (-384\ \text{kJ}) = -9\ \text{kJ} = -9000\ \text{J}$

$$\Delta G° = -RT \ln K_{sp},\quad K_{sp} = \exp\left[\frac{-(-9000\ \text{J})}{8.3145\ \text{J/K} \bullet \text{mol} \times 298\ \text{K}}\right] = 38 = 40$$

	$NaCl(s)$	$\rightleftharpoons$	$Na^+(aq)$	+	$Cl^-(aq)$	$K_{sp} = 40$
Initial	s = solubility (mol/L)		0		0	
Equil.			s		s	

$K_{sp} = 40 = s(s)$, $s = (40)^{1/2} = 6.3 = 6\ M = [Cl^-]$

143.

	$HX(aq)$	$\rightleftharpoons$	$H^+(aq)$	+	$X^-(aq)$	$K_a = \frac{[H^+][X^-]}{[HX]}$
Initial	0.10 M		~0		0	
Equil.	$0.10 - x$		x		x	

From problem, $x = [H^+] = 10^{-5.83} = 1.5 \times 10^{-6}$; $K_a = \frac{(1.5 \times 10^{-6})^2}{0.10 - 1.5 \times 10^{-6}} = 2.3 \times 10^{-11}$

$\Delta G° = -RT \ln K = -8.3145\ \text{J/K•mol}(298\ \text{K}) \ln(2.3 \times 10^{-11}) = 6.1 \times 10^4\ \text{J/mol} = 61\ \text{kJ/mol}$

145. $\ln K = \frac{-\Delta G°}{RT} = \frac{-(2.0 \times 10^4\ \text{J})}{8.3145\ \text{J/K} \bullet \text{mol}(308\ \text{K})} = -7.81,\ K = e^{-7.81} = 4.1 \times 10^{-4}$

	$2\ NOCl(g)$	$\rightleftharpoons$	$2\ NO(g)$ +	$Cl_2(g)$
Initial	2.0 atm		0	0
	$2x$ atm of NOCl reacts to reach equilibrium			
Change	$-2x$	→	$+2x$	$+x$
Equil.	$2.0 - 2x$		$2x$	x

$$K = 4.1 \times 10^{-4} = \frac{(2x)^2(x)}{(2.0 - 2x)^2} \approx \frac{4x^3}{4.0},\ x = 7.4 \times 10^{-2}\ \text{atm}$$

Assumption fails the 5% rule ($2x$ is 7.4% of 2.0). Using successive approximations:

$$4.1 \times 10^{-4} = \frac{(2x)^2(x)}{[2.0 - 2(7.4 \times 10^{-2})]^2},\ x = 7.1 \times 10^{-2}\ \text{atm}\ \text{(This answers repeats itself.)}$$

$P_{NO} = 2x = 2(7.1 \times 10^{-2}\ \text{atm}) = 0.14$ atm

147. $\Delta G° = \Delta H° - T\Delta S° = -28.0 \times 10^3\ \text{J} - 298\ \text{K}(-175\ \text{J/K}) = 24{,}200\ \text{J}$

$$\Delta G° = -RT \ln K,\ \ln K = \frac{-\Delta G°}{RT} = \frac{-24{,}000\ \text{J}}{8.3145\ \text{J/K} \bullet \text{mol} \times 298\ \text{K}} = -9.767$$

$K = e^{-9.767} = 5.73 \times 10^{-5}$

	B	+	H_2O	$\rightleftharpoons$	BH^+	+	OH^-	$K = K_b = 5.73 \times 10^{-5}$
Initial	0.125 M				0		~0	
Equil.	$0.125 - x$				x		x	

$$K_b = 5.73 \times 10^{-5} = \frac{[BH^+][OH^-]}{[B]} = \frac{x^2}{0.125 - x} \approx \frac{x^2}{0.125},\ x = [OH^-] = 2.68 \times 10^{-3}\ M$$

pH = $-\log(2.68 \times 10^{-3}) = 2.572$; pOH = $14.000 - 2.572 = 11.428$; assumptions good

CHAPTER 18

ELECTROCHEMISTRY

Review of Oxidation-Reduction Reactions

21. Oxidation: increase in oxidation number; loss of electrons

    Reduction: decrease in oxidation number; gain of electrons

23. The species oxidized shows an increase in oxidation numbers and is called the reducing agent. The species reduced shows a decrease in oxidation numbers and is called the oxidizing agent. The pertinent oxidation numbers are listed by the substance oxidized and the substance reduced.

	Redox?	Ox. Agent	Red. Agent	Substance Oxidized	Substance Reduced
a.	Yes	H_2O	CH_4	CH_4 (C, $-4 \rightarrow +2$)	H_2O (H, $+1 \rightarrow 0$)
b.	Yes	$AgNO_3$	Cu	Cu ($0 \rightarrow +2$)	$AgNO_3$ (Ag, $+1 \rightarrow 0$)
c.	Yes	HCl	Zn	Zn ($0 \rightarrow +2$)	HCl (H, $+1 \rightarrow 0$)

d. No; there is no change in any of the oxidation numbers.

25. Review Chapter 4.10 of the text for rules on balancing oxidation-reduction reactions.

a. $Cr \rightarrow Cr^{3+} + 3\ e^-$

$$NO_3^- \rightarrow NO$$
$$4\ H^+ + NO_3^- \rightarrow NO + 2\ H_2O$$
$$3\ e^- + 4\ H^+ + NO_3^- \rightarrow NO + 2\ H_2O$$

$$Cr \rightarrow Cr^{3+} + 3\ e^-$$
$$3\ e^- + 4\ H^+ + NO_3^- \rightarrow NO + 2\ H_2O$$
$$4\ H^+(aq) + NO_3^-(aq) + Cr(s) \rightarrow Cr^{3+}(aq) + NO(g) + 2\ H_2O(l)$$

b. $(Ce^{4+} + e^- \rightarrow Ce^{3+}) \times 6$

$$CH_3OH \rightarrow CO_2$$
$$H_2O + CH_3OH \rightarrow CO_2 + 6\ H^+$$
$$H_2O + CH_3OH \rightarrow CO_2 + 6\ H^+ + 6\ e^-$$

$$6\ Ce^{4+} + 6\ e^- \rightarrow 6\ Ce^{3+}$$
$$H_2O + CH_3OH \rightarrow CO_2 + 6\ H^+ + 6\ e^-$$
$$H_2O(l) + CH_3OH(aq) + 6\ Ce^{4+}(aq) \rightarrow 6\ Ce^{3+}(aq) + CO_2(g) + 6\ H^+(aq)$$

c. $SO_3^{2-} \rightarrow SO_4^{2-}$

$(H_2O + SO_3^{2-} \rightarrow SO_4^{2-} + 2\,H^+ + 2\,e^-) \times 5$

$MnO_4^- \rightarrow Mn^{2+}$

$(5\,e^- + 8\,H^+ + MnO_4^- \rightarrow Mn^{2+} + 4\,H_2O) \times 2$

$$5\,H_2O + 5\,SO_3^{2-} \rightarrow 5\,SO_4^{2-} + 10\,H^+ + 10\,e^-$$
$$10\,e^- + 16\,H^+ + 2\,MnO_4^- \rightarrow 2\,Mn^{2+} + 8\,H_2O$$
$$6\,H^+(aq) + 2\,MnO_4^-(aq) + 5\,SO_3^{2-}(aq) \rightarrow 5\,SO_4^{2-}(aq) + 2\,Mn^{2+}(aq) + 3\,H_2O(l)$$

Questions

27. Electrochemistry is the study of the interchange of chemical and electrical energy. A redox (oxidation-reduction) reaction is a reaction in which one or more electrons are transferred. In a galvanic cell, a spontaneous redox reaction occurs that produces an electric current. In an electrolytic cell, electricity is used to force a nonspontaneous redox reaction to occur.

29. Magnesium is an alkaline earth metal; Mg will oxidize to Mg^{2+}. The oxidation state of hydrogen in HCl is +1. To be reduced, the oxidation state of H must decrease. The obvious choice for the hydrogen product is $H_2(g)$, where hydrogen has a zero oxidation state. The balanced reaction is $Mg(s) + 2HCl(aq) \rightarrow MgCl_2(aq) + H_2(g)$. Mg goes from the 0 to the +2 oxidation state by losing two electrons. Each H atom goes from the +1 to the 0 oxidation state by gaining one electron. Since there are two H atoms in the balanced equation, then a total of two electrons are gained by the H atoms. Hence two electrons are transferred in the balanced reaction. When the electrons are transferred directly from Mg to H^+, no work is obtained. To harness this reaction to do useful work, we must control the flow of electrons through a wire. This is accomplished by making a galvanic cell that separates the reduction reaction from the oxidation reaction to control the flow of electrons through a wire to produce a voltage.

31. Statements c and d are false. Reducing agents cause reduction to occur by something in the formula being oxidized. Oxidizing agents cause oxidation to occur by something in the formula being reduced. A good oxidizing agent has a large, positive reduction potential. Oxidation occurs at the anode.

33. An extensive property is one that depends directly on how many times the reaction occurs. The free energy change for a reaction depends on whether 1 mole of product is produced, or 2 moles of product is produced or 1 million moles of product is produced. This is not the case for cell potentials, which do not depend on how many times a reaction occurs. The equation that relates ΔG to E is $\Delta G = -nFE$. It is the n term that converts the intensive property E into the extensive property ΔG. n is the number of moles of electrons transferred in the balanced reaction that ΔG is associated with.

35. A potential hazard when jump starting a car is the possibility for the electrolysis of $H_2O(l)$ to occur. When $H_2O(l)$ is electrolyzed, the products are the explosive gas mixture of $H_2(g)$ and $O_2(g)$. A spark produced during jump-starting a car could ignite any $H_2(g)$ and $O_2(g)$ produced. Grounding the jumper cable far from the battery minimizes the risk of a spark nearby the battery, where $H_2(g)$ and $O_2(g)$ could be accumulating.

37. You need to know the identity of the metal so that you know which molar mass to use. You need to know the oxidation state of the metal ion in the salt so that the moles of electrons transferred can be determined. And finally, you need to know the amount of current and the time the current was passed through the electrolytic cell. If you know these four quantities, then the mass of metal plated out can be calculated.

39. All the statements are true. Corrosion is the spontaneous oxidation of some metal (iron) coupled with the reduction of oxygen. For statement a, cars rust more easily in high-moisture areas (the humid areas) because water is a reactant in the reduction half-reaction as well as providing a medium for ion migration (a salt bridge of sorts). For statement d, paint and protective oxides are great at limiting corrosion only if the paint or protective oxide completely covers the surface. As soon as a metal surface is exposed, corrosion will occur. For statement e, salting roads adds ions to the corrosion process, which increases the conductivity of the aqueous solution and, in turn, accelerates corrosion.

41. $Na^+ + e^- \rightarrow Na$ $E° = -2.71$ V; $2\ H_2O + 2\ e^- \rightarrow H_2 + 2\ OH^-$ $E° = -0.83$ V; in molten NaCl, no water is present, just Na^+ and Cl^- ions. The reaction that occurs at the cathode is the reduction of Na^+ as the applied external potential is increased. In aqueous solution, H_2O is now present, and from the potentials, water is more easily reduced than Na^+. In aqueous NaCl, the reduction of water occurs at the cathode producing $H_2(g)$. Answer c is correct.

Exercises

Galvanic Cells, Cell Potentials, Standard Reduction Potentials, and Free Energy

43. The reducing agent causes reduction to occur; it does this by containing the species which is oxidized. Oxidation occurs at the anode, so the reducing agent will be in the anode compartment. The oxidizing agent causes oxidation to occur; it does this by containing the species which is reduced. Reduction occurs at the cathode, so the oxidizing agent will be in the cathode compartment. Electron flow is always from the anode compartment to the cathode compartment.

45. A typical galvanic cell diagram is:

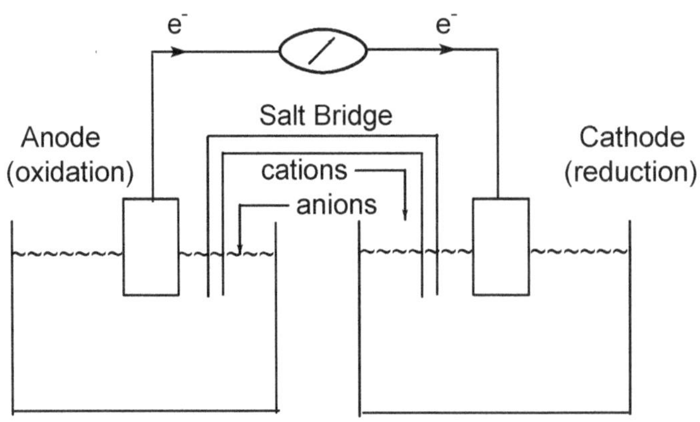

The diagram for all cells will look like this. The contents of each half-cell compartment will be identified for each reaction, with all solute concentrations at 1.0 *M* and all gases at 1.0 atm. For Exercises 45 and 46, the flow of ions through the salt bridge was not asked for in the questions. If asked, however, cations always flow into the cathode compartment, and anions always flow into the anode compartment. This is required to keep each compartment electrically neutral.

a. Table 18.1 of the text lists balanced reduction half-reactions for many substances. For this overall reaction, we need the Cl_2 to Cl^- reduction half-reaction and the Cr^{3+} to $Cr_2O_7^{2-}$ oxidation half-reaction. Manipulating these two half-reactions gives the overall balanced equation.

$$(Cl_2 + 2\ e^- \rightarrow 2\ Cl^-) \times 3$$
$$7\ H_2O + 2\ Cr^{3+} \rightarrow Cr_2O_7^{2-} + 14\ H^+ + 6\ e^-$$
$$\overline{7\ H_2O(l) + 2\ Cr^{3+}(aq) + 3\ Cl_2(g) \rightarrow Cr_2O_7^{2-}\ (aq) + 6\ Cl^-(aq) + 14\ H^+(aq)}$$

The contents of each compartment are:

Cathode: Pt electrode; Cl_2 bubbled into solution, Cl^- in solution

Anode: Pt electrode; Cr^{3+}, H^+, and $Cr_2O_7^{2-}$ in solution

We need a nonreactive metal to use as the electrode in each case, since all the reactants and products are in solution. Pt is a common choice. Another possibility is graphite.

b.
$$Cu^{2+} + 2\ e^- \rightarrow Cu$$
$$Mg \rightarrow Mg^{2+} + 2e^-$$
$$\overline{Cu^{2+}(aq) + Mg(s) \rightarrow Cu(s) + Mg^{2+}(aq)}$$

Cathode: Cu electrode; Cu^{2+} in solution; anode: Mg electrode; Mg^{2+} in solution

47. To determine E° for the overall cell reaction, we must add the standard reduction potential to the standard oxidation potential ($E^\circ_{cell} = E^\circ_{red} + E^\circ_{ox}$). Reference Table 18.1 for values of standard reduction potentials. Remember that $E^\circ_{ox} = -E^\circ_{red}$ and that standard potentials are <u>not</u> multiplied by the integer used to obtain the overall balanced equation.

45a. $E^\circ_{cell} = E^\circ_{Cl_2 \rightarrow Cl^-} + E^\circ_{Cr^{3+} \rightarrow Cr_2O_7^{2-}} = 1.36\ V + (-1.33\ V) = 0.03\ V$

45b. $E^\circ_{cell} = E^\circ_{Cu^{2+} \rightarrow Cu} + E^\circ_{Mg \rightarrow Mg^{2+}} = 0.34\ V + 2.37\ V = 2.71\ V$

49.
$$Br_2 + 2\ e^- \rightarrow 2\ Br^- \qquad E^\circ = 1.09\ V$$
$$Pb \rightarrow Pb^{2+} + 2\ e^- \qquad -E^\circ = 0.13\ V$$
$$\overline{Br_2(aq) + Pb(s) \rightarrow Pb^{2+}(aq) + 2\ Br^-(aq) \qquad E^\circ_{cell} = 1.22\ V}$$

Statements c, d, and e are true. Pb^{2+} is a product in the spontaneous cell reaction, so its concentration will increase as the reaction proceeds. The bromine half-cell is the cathode, and cations from the salt bridge flow to the cathode. Electrons travel from the lead half-cell to the bromine half-cell (from the anode to the cathode).

51. Reference the answer to Exercise 45 for a typical galvanic cell design. The contents of each half-cell compartment are identified below with all solute concentrations at 1.0 *M* and all gases at 1.0 atm. For each pair of half-reactions, the half-reaction with the largest (most positive) standard reduction potential will be the cathode reaction, and the half-reaction with the smallest (most negative) reduction potential will be reversed to become the anode reaction. Only this

combination gives a spontaneous overall reaction, i.e., a reaction with a positive overall standard cell potential. Note that in a galvanic cell as illustrated in Exercise 37 the cations in the salt bridge migrate to the cathode, and the anions migrate to the anode.

a.

$$\begin{array}{lr} Cl_2 + 2\,e^- \rightarrow 2\,Cl^- & E° = 1.36\text{ V} \\ 2\,Br^- \rightarrow Br_2 + 2\,e^- & -E° = -1.09\text{ V} \\ \hline Cl_2(g) + 2\,Br^-(aq) \rightarrow Br_2(aq) + 2\,Cl^-(aq) & E^{o}_{cell} = 0.27\text{ V} \end{array}$$

The contents of each compartment are:

Cathode: Pt electrode; $Cl_2(g)$ bubbled in, Cl^- in solution

Anode: Pt electrode; Br_2 and Br^- in solution

b.

$$\begin{array}{lr} (2\,e^- + 2\,H^+ + IO_4^- \rightarrow IO_3^- + H_2O) \times 5 & E° = 1.60\text{ V} \\ (4\,H_2O + Mn^{2+} \rightarrow MnO_4^- + 8\,H^+ + 5\,e^-) \times 2 & -E° = -1.51\text{ V} \\ \hline 10\,H^+ + 5\,IO_4^- + 8\,H_2O + 2\,Mn^{2+} \rightarrow 5\,IO_3^- + 5\,H_2O + 2\,MnO_4^- + 16\,H^+ & E^{o}_{cell} = 0.09\text{ V} \end{array}$$

This simplifies to:

$$3\,H_2O(l) + 5\,IO_4^-(aq) + 2\,Mn^{2+}(aq) \rightarrow 5\,IO_3^-(aq) + 2\,MnO_4^-(aq) + 6\,H^+(aq) \qquad E^{o}_{cell} = 0.09\text{ V}$$

Cathode: Pt electrode; IO_4^-, IO_3^-, and H_2SO_4 (as a source of H^+) in solution

Anode: Pt electrode; Mn^{2+}, MnO_4^-, and H_2SO_4 in solution

53. In standard line notation, the anode is listed first, and the cathode is listed last. A double line separates the two compartments. By convention, the electrodes are on the ends with all solutes and gases toward the middle. A single line is used to indicate a phase change. We also included all concentrations.

45a. $Pt \mid Cr^{3+}$ (1.0 *M*), $Cr_2O_7^{2-}$ (1.0 *M*), H^+ (1.0 *M*) $\|$ Cl_2 (1.0 atm) $\mid Cl^-$ (1.0 *M*) $\mid Pt$

45b. $Mg \mid Mg^{2+}$ (1.0 *M*) $\|$ Cu^{2+} (1.0 *M*) $\mid Cu$

51a. $Pt \mid Br^-$ (1.0 *M*), Br_2 (1.0 *M*) $\|$ Cl_2 (1.0 atm) $\mid Cl^-$ (1.0 *M*) $\mid Pt$

51b. $Pt \mid Mn^{2+}$ (1.0 *M*), MnO_4^- (1.0 *M*), H^+ (1.0 *M*) $\|$ IO_4^- (1.0 *M*), H^+ (1.0 *M*), IO_3^- (1.0 *M*) $\mid Pt$

55. Locate the pertinent half-reactions in Table 18.1, and then figure which combination will give a positive standard cell potential. In all cases, the anode compartment contains the species with the smallest standard reduction potential. For part a, the copper compartment is the anode, and in part b, the cadmium compartment is the anode.

a.

$$\begin{array}{lr} Au^{3+} + 3\,e^- \rightarrow Au & E° = 1.50\text{ V} \\ (Cu^+ \rightarrow Cu^{2+} + e^-) \times 3 & -E° = -0.16\text{ V} \\ \hline Au^{3+}(aq) + 3\,Cu^+(aq) \rightarrow Au(s) + 3\,Cu^{2+}(aq) & E^{o}_{cell} = 1.34\text{ V} \end{array}$$

b. $(VO_2^+ + 2\,H^+ + e^- \rightarrow VO^{2+} + H_2O) \times 2$ $E° = 1.00$ V

$Cd \rightarrow Cd^{2+} + 2e^-$ $-E° = 0.40$ V

$2\,VO_2^+(aq) + 4\,H^+(aq) + Cd(s) \rightarrow 2\,VO^{2+}(aq) + 2\,H_2O(l) + Cd^{2+}(aq)$ $E^o_{cell} = 1.40$ V

57. a. $(5\,e^- + 8\,H^+ + MnO_4^- \rightarrow Mn^{2+} + 4\,H_2O) \times 2$ $E° = 1.51$ V

$(2\,I^- \rightarrow I_2 + 2\,e^-) \times 5$ $-E° = -0.54$ V

$16\,H^+(aq) + 2\,MnO_4^-(aq) + 10\,I^-(aq) \rightarrow 5\,I_2(aq) + 2\,Mn^{2+}(aq) + 8\,H_2O(l)$ $E^o_{cell} = 0.97$ V

This reaction is spontaneous at standard conditions because $E^o_{cell} > 0$.

b. $(5\,e^- + 8\,H^+ + MnO_4^- \rightarrow Mn^{2+} + 4\,H_2O) \times 2$ $E° = 1.51$ V

$(2\,F^- \rightarrow F_2 + 2\,e^-) \times 5$ $-E° = -2.87$ V

$16\,H^+(aq) + 2\,MnO_4^-(aq) + 10\,F^-(aq) \rightarrow 5\,F_2(aq) + 2\,Mn^{2+}(aq) + 8\,H_2O(l)$ $E^o_{cell} = -1.36$ V

This reaction is not spontaneous at standard conditions because $E^o_{cell} < 0$.

59. $Cl_2 + 2\,e^- \rightarrow 2\,Cl^-$ $E° = 1.36$ V

$(ClO_2^- \rightarrow ClO_2 + e^-) \times 2$ $-E° = -0.954$ V

$2\,ClO_2^-(aq) + Cl_2(g) \rightarrow 2\,ClO_2(aq) + 2\,Cl^-(aq)$ $E^o_{cell} = 0.41\text{ V} = 0.41\text{ J/C}$

$\Delta G° = -nFE^o_{cell} = -(2\text{ mol e}^-)(96{,}485\text{ C/mol e}^-)(0.41\text{ J/C}) = -7.9 \times 10^4\text{ J} = -79\text{ kJ}$

61. Because the cells are at standard conditions, $w_{max} = \Delta G = \Delta G° = -nFE^o_{cell}$. See Exercise 55 for the balanced overall equations and for E^o_{cell}.

55a. $w_{max} = -(3\text{ mol e}^-)(96{,}485\text{ C/mol e}^-)(1.34\text{ J/C}) = -3.88 \times 10^5\text{ J} = -388\text{ kJ}$

55b. $w_{max} = -(2\text{ mol e}^-)(96{,}485\text{ C/mol e}^-)(1.40\text{ J/C}) = -2.70 \times 10^5\text{ J} = -270.\text{ kJ}$

63. $2\,H_2O + 2\,e^- \rightarrow H_2 + 2\,OH^-$

$\Delta G° = \Sigma n_p \Delta G^o_{f,\text{ products}} - \Sigma n_r \Delta G^o_{f,\text{ reactants}} = 2(-157) - [2(-237)] = 160.\text{ kJ}$

$$\Delta G° = -nFE°,\ E° = \frac{-\Delta G°}{nF} = \frac{-1.60 \times 10^5\text{ J}}{(2\text{ mol e}^-)(96{,}485\text{ C/mol e}^-)} = -0.829\text{ J/C} = -0.829\text{ V}$$

The two values agree to two significant figures (–0.83 V in Table 18.1).

65. $\Delta G° = [6\text{ mol}(-394\text{ kJ/mol}) + 6\text{ mol}(-237\text{ kJ/mol}] - [1\text{ mol}(-911\text{ kJ/mol} + 6\text{ mol}(0)]$
$= -2875\text{ kJ}$

Carbon is oxidized in this combustion reaction. In $C_6H_{12}O_6$, H has a +1 oxidation state, and oxygen has a –2 oxidation, so $6(x) + 12(+1) + 6(-2) = 0$, x = oxidation state of C in $C_6H_{12}O_6$ = 0. In CO_2, O has an oxidation state of –2, so $y + 2(-2) = 0$, y = oxidation state of C in CO_2 = +4. Carbon goes from the 0 oxidation state in $C_6H_{12}O_6$ to the +4 oxidation state in CO_2, so each carbon atom loses 4 electrons. Because the balanced reaction has 6 mol of carbon, 6(4) = 24 mol electrons are transferred in the balanced equation.

$$\Delta G° = -nFE°, \; E° = \frac{-\Delta G°}{nF} = \frac{-(-2875 \times 10^3 \text{ J})}{(24 \text{ mol e}^-)(96{,}485 \text{ C/mol e}^-)} = 1.24 \text{ J/C} = 1.24 \text{ V}$$

67. Good oxidizing agents are easily reduced. Oxidizing agents are on the left side of the reduction half-reactions listed in Table 18.1. We look for the largest, most positive standard reduction potentials to correspond to the best oxidizing agents. The ordering from worst to best oxidizing agents is:

	Mg^{2+} <	Fe^{2+} <	Fe^{3+} <	$Cr_2O_7^{2-}$ <	Cl_2 <	MnO_4^-
E°(V)	−2.37	−0.44	0.77	1.33	1.36	1.68

69. a. $2\,H^+ + 2\,e^- \rightarrow H_2$ $E° = 0.00$ V; $Cu \rightarrow Cu^{2+} + 2\,e^-$ $-E° = -0.34$ V

$E°_{cell} = -0.34$ V; no, H^+ cannot oxidize Cu to Cu^{2+} at standard conditions ($E°_{cell} < 0$).

b. $Fe^{3+} + e^- \rightarrow Fe^{2+}$ $E° = 0.77$ V; $2\,I^- \rightarrow I_2 + 2\,e^-$ $-E° = -0.54$ V

$E°_{cell} = 0.77 - 0.54 = 0.23$ V; yes, Fe^{3+} can oxidize I^- to I_2.

c. $H_2 \rightarrow 2\,H^+ + 2\,e^-$ $-E° = 0.00$ V; $Ag^+ + e^- \rightarrow Ag$ $E° = 0.80$ V

$E°_{cell} = 0.80$ V; yes, H_2 can reduce Ag^+ to Ag at standard conditions ($E°_{cell} > 0$).

71. The reduction of Cu^{2+} has $E° = 0.34$ V and the reduction of Al^{3+} has $E° = -1.66$ V. We need a substance that has an oxidation potential that when added to 0.34 V gives a positive value, yet when added to —-1.66 V, gives a negative value. From the options available, the oxidation of Fe to Fe^{2+} has $-E = 0.44$ V. When the oxidation of Fe^{2+} is coupled with the reduction of Cu^{2+}, $E°_{cell}$ will be positive, so the reaction is spontaneous at standard conditions. When the oxidation of Fe^{2+} is coupled with the reduction of Al^{3+}, $E°_{cell}$ will be negative, and this reaction will not be spontaneous at standard conditions. Answer c is correct.

73.

$Cl_2 + 2\,e^- \rightarrow 2\,Cl^-$	$E° = 1.36$ V	$Ag^+ + e^- \rightarrow Ag$	$E° = 0.80$ V
$Pb^{2+} + 2\,e^- \rightarrow Pb$	$E° = -0.13$ V	$Zn^{2+} + 2\,e^- \rightarrow Zn$	$E° = -0.76$ V
$Na^+ + e^- \rightarrow Na$	$E° = -2.71$ V		

a. Oxidizing agents (species reduced) are on the left side of the preceding reduction half-reactions. Of the species available, Ag^+ would be the best oxidizing agent since it has the largest E° value. Note that Cl_2 is a better oxidizing agent than Ag^+, but it is not one of the choices listed.

b. Reducing agents (species oxidized) are on the right side of the reduction half-reactions. Of the species available, Zn would be the best reducing agent since it has the largest −E° value.

c. $SO_4^{2-} + 4\,H^+ + 2\,e^- \rightarrow H_2SO_3 + H_2O$ $E° = 0.20$ V; SO_4^{2-} can oxidize Pb and Zn at standard conditions. When SO_4^{2-} is coupled with these reagents, $E°_{cell}$ is positive.

d. $Al \rightarrow Al^{3+} + 3\,e^-$ $-E° = 1.66$ V; Al can reduce Ag^+ and Zn^{2+} at standard conditions because $E°_{cell} > 0$.

75. a. $2\ Br^- \rightarrow Br_2 + 2\ e^-$ $-E° = -1.09$ V; $2\ Cl^- \rightarrow Cl_2 + 2\ e^-$ $-E° = -1.36$ V; $E° > 1.09$ V to oxidize Br^-; $E° < 1.36$ V to not oxidize Cl^-; $Cr_2O_7^{2-}$, O_2, MnO_2, and IO_3^- are all possible since when all of these oxidizing agents are coupled with Br^-, $E^\circ_{cell} > 0$, and when coupled with Cl^-, $E^\circ_{cell} < 0$ (assuming standard conditions).

b. $Mn \rightarrow Mn^{2+} + 2\ e^-$ $-E° = 1.18$; $Ni \rightarrow Ni^{2+} + 2\ e^-$ $-E° = 0.23$ V; any oxidizing agent with $-0.23\ V > E° > -1.18$ V will work. $PbSO_4$, Cd^{2+}, Fe^{2+}, Cr^{3+}, Zn^{2+}, and H_2O will be able to oxidize Mn but not Ni (assuming standard conditions).

The Nernst Equation

77.

$$\begin{array}{lr} H_2O_2 + 2\ H^+ + 2\ e^- \rightarrow 2\ H_2O & E° = 1.78\ V \\ (Ag \rightarrow Ag^+ + e^-) \times 2 & -E° = -0.80\ V \\ \hline H_2O_2(aq) + 2\ H^+(aq) + 2\ Ag(s) \rightarrow 2\ H_2O(l) + 2\ Ag^+(aq) & E^\circ_{cell} = 0.98\ V \end{array}$$

a. A galvanic cell is based on spontaneous redox reactions. At standard conditions, this reaction produces a voltage of 0.98 V. Any change in concentration that increases the tendency of the forward reaction to occur will increase the cell potential. Conversely, any change in concentration that decreases the tendency of the forward reaction to occur (increases the tendency of the reverse reaction to occur) will decrease the cell potential. Using Le Chatelier's principle, increasing the reactant concentrations of H_2O_2 and H^+ from 1.0 to 2.0 *M* will drive the forward reaction further to right (will further increase the tendency of the forward reaction to occur). Therefore, E_{cell} will be greater than E°_{cell}.

b. Here, we decreased the reactant concentration of H^+ and increased the product concentration of Ag^+ from the standard conditions. This decreases the tendency of the forward reaction to occur, which will decrease E_{cell} as compared to E°_{cell} ($E_{cell} < E^\circ_{cell}$).

79. For concentration cells, the driving force for the reaction is the difference in ion concentrations between the anode and cathode. In order to equalize the ion concentrations, the anode always has the smaller ion concentration. The general setup for this concentration cell is:

$$\begin{array}{llr} \text{Cathode:} & Ag^+(x\ M) + e^- \rightarrow Ag & E° = 0.80\ V \\ \text{Anode:} & Ag \rightarrow Ag^+\ (y\ M) + e^- & -E° = -0.80\ V \\ \hline & Ag^+(\text{cathode}, x\ M) \rightarrow Ag^+\ (\text{anode}, y\ M) & E^\circ_{cell} = 0.00\ V \end{array}$$

$$E_{cell} = E^\circ_{cell} - \frac{0.0591}{n} \log Q = \frac{-0.0591}{1} \log \frac{[Ag^+]_{anode}}{[Ag^+]_{cathode}}$$

For each concentration cell, we will calculate the cell potential using the preceding equation. Remember that the anode always has the smaller ion concentration.

a. Both compartments are at standard conditions ($[Ag^+] = 1.0\ M$), so $E_{cell} = E^\circ_{cell} = 0$ V. No voltage is produced since no reaction occurs. Concentration cells only produce a voltage when the ion concentrations are <u>not</u> equal.

b. Cathode = 2.0 M Ag^+; anode = 1.0 M Ag^+; electron flow is always from the anode to the cathode, so electrons flow to the right in the diagram.

$$E_{cell} = \frac{-0.0591}{n}\log\frac{[Ag^+]_{anode}}{[Ag^+]_{cathode}} = \frac{-0.0591}{1}\log\frac{1.0}{2.0} = 0.018\text{ V}$$

c. Cathode = 1.0 M Ag^+; anode = 0.10 M Ag^+; electrons flow to the left in the diagram.

$$E_{cell} = \frac{-0.0591}{n}\log\frac{[Ag^+]_{anode}}{[Ag^+]_{cathode}} = \frac{-0.0591}{1}\log\frac{0.10}{1.0} = 0.059\text{ V}$$

d. Cathode = 1.0 M Ag^+; anode = 4.0×10^{-5} M Ag^+; electrons flow to the left in the diagram.

$$E_{cell} = \frac{-0.0591}{n}\log\frac{4.0\times 10^{-5}}{1.0} = 0.26\text{ V}$$

e. The ion concentrations are the same; thus $\log([Ag^+]_{anode}/[Ag^+]_{cathode}) = \log(1.0) = 0$ and E_{cell} = 0. No electron flow occurs.

81. a. As is the case for all concentration cells, $E^\circ_{cell} = 0$, and the smaller ion concentration is always in the anode compartment. Here, n = 3 as three electrons are needed to convert Al^{3+} to Al and vice versa. The general Nernst equation for this Al concentration cell is:

$$E_{cell} = E^\circ_{cell} - \frac{0.0591}{n}\log Q = \frac{-0.0591}{3}\log\frac{[Al^{3+}]_{anode}}{[Al^{3+}]_{cathode}}$$

$$E_{cell} = \frac{-0.0591}{3}\log\frac{1.0\times 10^{-4}}{1.5} = 0.082\text{ V}$$

b. The overall cell reaction is Al^{3+} (1.5 M) $\rightarrow$ Al^{3+}(1.0×10^{-4} M). In general, the larger the difference in ion concentrations between the anode and cathode, the greater the cell potential. If the cathode Al^{3+} concentration is decreased from 1.5 M, the cell potential will potential will decrease. You can also think of this in terms of Le Châtelier's principle. Any change that increases the tendency of the forward reaction will increase the cell potential. Decreasing the Al^{3+} concentration at the cathode shifts the reaction left, thus decreasing the cell potential.

c. Increasing the Al^{3+} concentration at the anode decreases the tendency of the forward reaction to occur, thus the cell potential will decrease.

d. The amount of a solid does not affect the equilibrium of a reaction. So, using a more massive Al electrode will have no effect.

83. n = 2 for this reaction (lead goes from Pb $\rightarrow$ Pb^{2+} in $PbSO_4$).

$$E = E^\circ - \frac{0.0591}{2}\log\frac{1}{[H^+]^2[HSO_4^-]^2} = 2.04\text{ V} - \frac{0.0591}{2}\log\frac{1}{(4.5)^2(4.5)^2}$$

$$E = 2.04\text{ V} + 0.077\text{ V} = 2.12\text{ V}$$

85.

$$\begin{array}{lr} (Ag^+ + e^- \rightarrow Ag) \times 2 & E° = 0.80\ V \\ Zn \rightarrow Zn^{2+} + 2\ e^- & -E° = 0.76\ V \\ \hline 2\ Ag^+(aq) + Zn(s) \rightarrow Zn^{2+}(aq) + 2\ Ag(s) & E^o_{cell} = 1.56\ V \end{array}$$

Because Zn^{2+} is a product in the reaction, the Zn^{2+} concentration increases from 1.00 to 1.20 *M*. This means that the reactant concentration of Ag^+ must decrease from 1.00 to 0.60 *M* (from the 1 : 2 mole ratio between Zn^{2+} and Ag^+ in the balanced reaction).

$$E_{cell} = E^o_{cell} - \frac{0.0591}{n}\log Q = 1.56\ V - \frac{0.0591}{2}\log\frac{[Zn^{2+}]}{[Ag^+]^2}$$

$$E_{cell} = 1.56\ V - \frac{0.0591}{2}\log\frac{1.20}{(0.60)^2} = 1.56\ V - 0.020\ V = 1.54\ V$$

87. See Exercises 45, 47, and 51 for balanced reactions and standard cell potentials. Balanced reactions are necessary to determine n, the moles of electrons transferred.

45a. $7\ H_2O + 2\ Cr^{3+} + 3\ Cl_2 \rightarrow Cr_2O_7^{2-} + 6\ Cl^- + 14\ H^+ \quad E^o_{cell} = 0.03\ V = 0.03\ J/C$

$$\Delta G° = -nFE^o_{cell} = -(6\ mol\ e^-)(96{,}485\ C/mol\ e^-)(0.03\ J/C) = -1.7 \times 10^4\ J = -20\ kJ$$

$$E_{cell} = E^o_{cell} - \frac{0.0591}{n}\log Q;\ \text{at equilibrium, } E_{cell} = 0 \text{ and } Q = K\text{, so:}$$

$$E^o_{cell} = \frac{0.0591}{n}\log K,\ \log K = \frac{nE°}{0.0591} = \frac{6(0.03)}{0.0591} = 3.05,\ K = 10^{3.05} = 1 \times 10^3$$

Note: When determining exponents, we will round off to the correct number of significant figures after the calculation is complete in order to help eliminate excessive round-off error.

45b. $\Delta G° = -(2\ mol\ e^-)(96{,}485\ C/mol\ e^-)(2.71\ J/C) = -5.23 \times 10^5\ J = -523\ kJ$

$$\log K = \frac{2(2.71)}{0.0591} = 91.709,\ K = 5.12 \times 10^{91}$$

51a. $\Delta G° = -(2\ mol\ e^-)(96{,}485\ C/mol\ ^-)(0.27\ J/C) = -5.21 \times 10^4\ J = -52\ kJ$

$$\log K = \frac{2(0.27)}{0.0591} = 9.14,\ K = 1.4 \times 10^9$$

51b. $\Delta G° = -(10\ mol\ e^-)(96{,}485\ C/mol\ e^-)(0.09\ J/C) = -8.7 \times 10^4\ J = -90\ kJ$

$$\log K = \frac{10(0.09)}{0.0591} = 15.23,\ K = 2 \times 10^{15}$$

89. a.

$$\begin{array}{lr} Fe^{2+} + 2\ e^- \rightarrow Fe & E° = -0.44\ V \\ Zn \rightarrow Zn^{2+} + 2\ e^- & -E° = 0.76\ V \\ \hline Fe^{2+}(aq) + Zn(s) \rightarrow Zn^{2+}(aq) + Fe(s) & E^o_{cell} = 0.32\ V = 0.32\ J/C \end{array}$$

b. $\Delta G° = -nFE^o_{cell} = -(2\ mol\ e^-)(96{,}485\ C/mol\ e^-)(0.32\ J/C) = -6.2 \times 10^4\ J = -62\ kJ$

$$E^{\circ}_{cell} = \frac{0.0591}{n}\log K,\ \log K = \frac{nE^{\circ}}{0.0591} = \frac{2(0.32)}{0.0591} = 10.83,\ K = 10^{10.83} = 6.8 \times 10^{10}$$

c. $$E_{cell} = E^{\circ}_{cell} - \frac{0.0591}{n}\log Q = 0.32\text{ V} - \frac{0.0591}{n}\log\frac{[Zn^{2+}]}{[Fe^{2+}]}$$

$$E_{cell} = 0.32 - \frac{0.0591}{2}\log\frac{0.10}{1.0 \times 10^{-5}} = 0.32 - 0.12 = 0.20\text{ V}$$

91. $Cu^{2+}(aq) + H_2(g) \rightarrow 2\ H^+(aq) + Cu(s)$ $E^{\circ}_{cell} = 0.34\text{ V} - 0.00\text{ V} = 0.34\text{ V}$; n = 2 mol electrons

$P_{H_2} = 1.0$ atm and $[H^+] = 1.0\ M$: $E_{cell} = E^{\circ}_{cell} - \frac{0.0591}{n}\log\frac{1}{[Cu^{2+}]}$

a. $$E_{cell} = 0.34\text{ V} - \frac{0.0591}{2}\log\frac{1}{2.5 \times 10^{-4}} = 0.34\text{ V} - 0.11\text{V} = 0.23\text{ V}$$

b. $$0.195\text{ V} = 0.34\text{ V} - \frac{0.0591}{2}\log\frac{1}{[Cu^{2+}]},\ \log\frac{1}{[Cu^{2+}]} = 4.91,\ [Cu^{2+}] = 10^{-4.91}$$

$$= 1.2 \times 10^{-5}\ M$$

Note: When determining exponents, we will carry extra significant figures.

93. $Cu^{2+}(aq) + H_2(g) \rightarrow 2\ H^+(aq) + Cu(s)$ $E^{\circ}_{cell} = 0.34\text{ V} - 0.00\text{ V} = 0.34\text{ V}$; n = 2

$P_{H_2} = 1.0$ atm and $[H^+] = 1.0\ M$: $E_{cell} = E^{\circ}_{cell} - \frac{0.0591}{2}\log\frac{1}{[Cu^{2+}]}$

Use the K_{sp} expression to calculate the Cu^{2+} concentration in the cell.

$Cu(OH)_2(s) \rightleftharpoons Cu^{2+}(aq) + 2\ OH^-(aq)$ $K_{sp} = 1.6 \times 10^{-19} = [Cu^{2+}][OH^-]^2$

From problem, $[OH^-] = 0.10\ M$, so: $[Cu^{2+}] = \frac{1.6 \times 10^{-19}}{(0.10)^2} = 1.6 \times 10^{-17}\ M$

$$E_{cell} = E^{\circ}_{cell} - \frac{0.0591}{2}\log\frac{1}{[Cu^{2+}]} = 0.34\text{ V} - \frac{0.0591}{2}\log\frac{1}{1.6 \times 10^{-17}}$$

$E_{cell} = 0.34 - 0.50 = -0.16$ V

Because $E_{cell} < 0$, the forward reaction is not spontaneous, but the reverse reaction is spontaneous. The Cu electrode becomes the anode and $E_{cell} = 0.16$ V for the reverse reaction. The cell reaction is $2\ H^+(aq) + Cu(s) \rightarrow Cu^{2+}(aq) + H_2(g)$.

95. Cathode: $M^{2+} + 2e^- \rightarrow M(s)$ $E^{\circ} = -0.31$ V

Anode: $M(s) \rightarrow M^{2+} + 2e^-$ $-E^{\circ} = 0.31$ V

M^{2+} (cathode) $\rightarrow M^{2+}$ (anode) $E^{\circ}_{cell} = 0.00$ V

$$E_{cell} = 0.44\text{ V} = 0.00\text{ V} - \frac{0.0591}{2}\log\frac{[M^{2+}]_{anode}}{[M^{2+}]_{cathode}},\ 0.44 = -\frac{0.0591}{2}\log\frac{[M^{2+}]_{anode}}{1.0}$$

$$\log [M^{2+}]_{anode} = -\frac{2(0.44)}{0.0591} = -14.89, \ [M^{2+}]_{anode} = 1.3 \times 10^{-15} \ M$$

Because we started with equal numbers of moles of SO_4^{2-} and M^{2+}, $[M^{2+}] = [SO_4^{2-}]$ at equilibrium.

$$K_{sp} = [M^{2+}][SO_4^{2-}] = (1.3 \times 10^{-15})^2 = 1.7 \times 10^{-30}$$

97. a. Possible reaction: $I_2(s) + 2\ Cl^-(aq) \rightarrow 2\ I^-(aq) + Cl_2(g)$ $E^\circ_{cell} = 0.54\ V - 1.36\ V = -0.82\ V$

This reaction is not spontaneous at standard conditions because $E^\circ_{cell} < 0$; no reaction occurs.

b. Possible reaction: $Cl_2(g) + 2\ I^-(aq) \rightarrow I_2(s) + 2\ Cl^-(aq)$ $E^\circ_{cell} = 0.82$ V; this reaction is spontaneous at standard conditions because $E^\circ_{cell} > 0$. The reaction will occur.

$$Cl_2(g) + 2\ I^-(aq) \rightarrow I_2(s) + 2\ Cl^-(aq) \quad E^\circ_{cell} = 0.82\ V = 0.82\ J/C$$

$$\Delta G^\circ = -nFE^\circ_{cell} = -(2\ mol\ e^-)(96{,}485\ C/mol\ e^-)(0.82\ J/C) = -1.6 \times 10^5\ J = -160\ kJ$$

$$E^\circ = \frac{0.0591}{n} \log K, \ \log K = \frac{nE^\circ}{0.0591} = \frac{2(0.82)}{0.0591} = 27.75, \ K = 10^{27.75} = 5.6 \times 10^{27}$$

c. Possible reaction: $2\ Ag(s) + Cu^{2+}(aq) \rightarrow Cu(s) + 2\ Ag^+(aq)$ $E^\circ_{cell} = -0.46$ V; no reaction occurs.

d. Fe^{2+} can be oxidized or reduced. The other species present are H^+, SO_4^{2-}, H_2O, and O_2 from air. Only O_2 in the presence of H^+ has a large enough standard reduction potential to oxidize Fe^{2+} to Fe^{3+} (resulting in $E^\circ_{cell} > 0$). All other combinations, including the possible reduction of Fe^{2+}, give negative cell potentials. The spontaneous reaction is:

$$4\ Fe^{2+}(aq) + 4\ H^+(aq) + O_2(g) \rightarrow 4\ Fe^{3+}(aq) + 2\ H_2O(l) \quad E^\circ_{cell} = 1.23 - 0.77 = 0.46\ V$$

$$\Delta G^\circ = -nFE^\circ_{cell} = -(4\ mol\ e^-)(96{,}485\ C/mol\ e^-)(0.46\ J/C)(1\ kJ/1000\ J) = -180\ kJ$$

$$\log K = \frac{4(0.46)}{0.0591} = 31.13, \ K = 1.3 \times 10^{31}$$

99.

$$(Cr^{2+} \rightarrow Cr^{3+} + e^-) \times 2$$
$$Co^{2+} + 2\ e^- \rightarrow Co$$
$$\overline{2\ Cr^{2+}(aq) + Co^{2+}(aq) \rightarrow 2\ Cr^{3+}(aq) + Co(s)}$$

$$E^\circ_{cell} = \frac{0.0591}{n} \log K = \frac{0.0591}{2} \log(2.79 \times 10^7) = 0.220\ V$$

$$E = E^\circ - \frac{0.0591}{n} \log \frac{[Cr^{3+}]^2}{[Cr^{2+}]^2[Co^{2+}]} = 0.220\ V - \frac{0.0591}{2} \log \frac{(2.0)^2}{(0.30)^2(0.20)} = 0.151\ V$$

$$\Delta G = -nFE = -(2\ mol\ e^-)(96{,}485\ C/mol\ e^-)(0.151\ J/C) = -2.91 \times 10^4\ J = -29.1\ kJ$$

101. The K_{sp} reaction is $Cd(s) \rightleftharpoons Cd^{2+}(aq) + S^{2-}(aq)$ $K = K_{sp}$. Manipulate the given equations so that when added together we get the K_{sp} reaction. Then we can use the value of E^o_{cell} for the reaction to determine K_{sp} (by using the equation $\log K = nE°/0.0591$).

$$\begin{array}{lll} CdS + 2\,e^- \rightarrow Cd + S^{2-} & E° = -1.21 \text{ V} & \\ Cd \rightarrow Cd^{2+} + 2\,e^- & -E° = 0.402 \text{ V} & \\ \hline CdS(s) \rightarrow Cd^{2+}(aq) + S^{2-}(aq) & E^o_{cell} = -0.81 \text{ V} & K_{sp} = ? \end{array}$$

$$\log K_{sp} = \frac{nE^o}{0.0591} = \frac{2(-0.81)}{0.0591} = -27.41,\ K_{sp} = 10^{-27.41} = 3.9 \times 10^{-28}$$

103.
$$\begin{array}{ll} e^- + AgI \rightarrow Ag + I^- & E^o_{AgI} = ? \\ Ag \rightarrow Ag^+ + e^- & -E° = -0.80 \text{ V} \\ \hline AgI(s) \rightarrow Ag^+(aq) + I^-(aq) & E^o_{cell} = E^o_{AgI} - 0.80,\ \ K = K_{sp} = 1.5 \times 10^{-16} \end{array}$$

For this overall reaction:

$$E^o_{cell} = \frac{0.0591}{n} \log K_{sp} = \frac{0.0591}{1} \log(1.5 \times 10^{-16}) = -0.94 \text{ V}$$

$$E^o_{cell} = -0.94 \text{ V} = E^o_{AgI} - 0.80 \text{ V},\ E^o_{AgI} = -0.94 + 0.80 = -0.14 \text{ V}$$

Electrolysis

105. a. $Al^{3+} + 3\,e^- \rightarrow Al$; 3 mol e^- are needed to produce 1 mol Al from Al^{3+}.

$$1.0 \times 10^3 \text{ g Al} \times \frac{1 \text{ mol Al}}{26.98 \text{ g Al}} \times \frac{3 \text{ mol } e^-}{\text{mol Al}} \times \frac{96{,}485 \text{ C}}{\text{mol } e^-} \times \frac{1 \text{ s}}{100.0 \text{ C}} = 1.07 \times 10^5 \text{ s}$$

$$= 30. \text{ hours}$$

b. $$1.0 \text{ g Ni} \times \frac{1 \text{ mol Ni}}{58.69 \text{ g Ni}} \times \frac{2 \text{ mol } e^-}{\text{mol Ni}} \times \frac{96{,}485 \text{ C}}{\text{mol } e^-} \times \frac{1 \text{ s}}{100.0 \text{ C}} = 33 \text{ s}$$

c. $$5.0 \text{ mol Ag} \times \frac{1 \text{ mol } e^-}{\text{mol Ag}} \times \frac{96{,}485 \text{ C}}{\text{mol } e^-} \times \frac{1 \text{ s}}{100.0 \text{ C}} = 4.8 \times 10^3 \text{ s} = 1.3 \text{ hours}$$

107. $$15 \text{ A} = \frac{15 \text{ C}}{\text{s}} \times \frac{60 \text{ s}}{\text{min}} \times \frac{60 \text{ min}}{\text{h}} = 5.4 \times 10^4 \text{ C of charge passed in 1 hour}$$

a. $$5.4 \times 10^4 \text{ C} \times \frac{1 \text{ mol } e^-}{96{,}485 \text{ C}} \times \frac{1 \text{ mol Co}}{2 \text{ mol } e^-} \times \frac{58.93 \text{ g Co}}{\text{mol Co}} = 16 \text{ g Co}$$

b. $$5.4 \times 10^4 \text{ C} \times \frac{1 \text{ mol } e^-}{96{,}485 \text{ C}} \times \frac{1 \text{ mol Hf}}{4 \text{ mol } e^-} \times \frac{178.5 \text{ g Hf}}{\text{mol Hf}} = 25 \text{ g Hf}$$

c. $2\,I^- \rightarrow I_2 + 2\,e^-$; $$5.4 \times 10^4 \text{ C} \times \frac{1 \text{ mol } e^-}{96{,}485 \text{ C}} \times \frac{1 \text{ mol } I_2}{2 \text{ mol } e^-} \times \frac{253.8 \text{ g } I_2}{\text{mol } I_2} = 71 \text{ g } I_2$$

d. $CrO_3(l) \rightarrow Cr^{6+} + 3\ O^{2-}$; 6 mol e^- are needed to produce 1 mol Cr from molten CrO_3.

$$5.4 \times 10^4 \text{ C} \times \frac{1 \text{ mol e}^-}{96{,}485 \text{ C}} \times \frac{1 \text{ mol Cr}}{6 \text{ mol e}^-} \times \frac{52.00 \text{ g Cr}}{\text{mol Cr}} = 4.9 \text{ g Cr}$$

109. Al is in the +3 oxidation in Al_2O_3, so 3 mol e^- are needed to convert Al^{3+} into Al(s).

$$2.00 \text{ h} \times \frac{60 \text{ min}}{\text{h}} \times \frac{60 \text{ s}}{\text{min}} \times \frac{1.00 \times 10^6 \text{ C}}{\text{s}} \times \frac{1 \text{ mol e}^-}{96{,}485 \text{ C}} \times \frac{1 \text{ mol Al}}{3 \text{ mol e}^-} \times \frac{26.98 \text{ g Al}}{\text{mol Al}} = 6.71 \times 10^5 \text{ g}$$

111. From the MCl_2 formula, M forms +2 ions. Two mol of e^- are transferred to form the metal M.

$$\text{Moles of M} = 748 \text{ s} \times \frac{5.00 \text{ C}}{\text{s}} \times \frac{1 \text{ mol e}^-}{96{,}485 \text{ C}} \times \frac{1 \text{ mol M}}{2 \text{ mol e}^-} \times \frac{1 \text{ mol e}^-}{96{,}485 \text{ C}} = 1.94 \times 10^{-2} \text{ mol M}$$

$$\text{Molar mass of M} = \frac{0.471 \text{ g M}}{1.94 \times 10^{-2} \text{ mol M}} = 24.3 \text{ g/mol; M is magnesium, Mg.}$$

113. $$\text{Moles of e}^- = 50.0 \text{ min} \times \frac{60 \text{ s}}{\text{min}} \times \frac{2.50 \text{ C}}{\text{s}} \times \frac{1 \text{ mol e}^-}{96{,}485 \text{ C}} = 7.77 \times 10^{-2} \text{ mol e}^-$$

$$\text{Moles of Ru} = 2.618 \text{ g Ru} \times \frac{1 \text{ mol Ru}}{101.1 \text{ g Ru}} = 2.590 \times 10^{-2} \text{ mol Ru}$$

$$\frac{\text{Moles of e}^-}{\text{Moles of Ru}} = \frac{7.77 \times 10^{-2} \text{ mol e}^-}{2.590 \times 10^{-2} \text{ mol Ru}} = 3.00\text{; the charge on the ruthenium ions is +3.}$$

$(Ru^{3+} + 3\ e^- \rightarrow Ru)$

115. F_2 is produced at the anode: $2\ F^- \rightarrow F_2 + 2\ e^-$

$$2.00 \text{ h} \times \frac{60 \text{ min}}{\text{h}} \times \frac{60 \text{ s}}{\text{min}} \times \frac{10.0 \text{ C}}{\text{s}} \times \frac{1 \text{ mol e}^-}{96{,}485 \text{ C}} = 0.746 \text{ mol e}^-$$

$$0.746 \text{ mol e}^- \times \frac{1 \text{ mol F}_2}{2 \text{ mol e}^-} = 0.373 \text{ mol F}_2\text{; } PV = nRT, \ V = \frac{nRT}{P}$$

$$V = \frac{(0.373 \text{ mol})(0.08206 \text{ L atm/K} \bullet \text{mol})(298 \text{ K})}{1.00 \text{ atm}} = 9.12 \text{ L F}_2$$

K is produced at the cathode: $K^+ + e^- \rightarrow K$

$$0.746 \text{ mol e}^- \times \frac{1 \text{ mol K}}{\text{mol e}^-} \times \frac{39.10 \text{ g K}}{\text{mol K}} = 29.2 \text{ g K}$$

117. $Al^{3+} + 3\ e^- \rightarrow Al$; 3 mol e^- are needed to produce Al from Al^{3+}

$$2000 \text{ lb Al} \times \frac{453.6 \text{ g}}{\text{lb}} \times \frac{1 \text{ mol Al}}{26.98 \text{ g}} \times \frac{3 \text{ mol e}^-}{\text{mol Al}} \times \frac{96{,}485 \text{ C}}{\text{mol e}^-} = 1 \times 10^{10} \text{ C of electricity needed}$$

$$\frac{1 \times 10^{10} \text{ C}}{24 \text{ h}} \times \frac{1 \text{ h}}{60 \text{ min}} \times \frac{1 \text{ min}}{60 \text{ s}} = 1 \times 10^5 \text{ C/s} = 1 \times 10^5 \text{ A}$$

119. $2.30 \text{ min} \times \frac{60 \text{ s}}{\text{min}} = 138 \text{ s}$; $138 \text{ s} \times \frac{2.00 \text{ C}}{\text{s}} \times \frac{1 \text{ mol e}^-}{96{,}485 \text{ C}} \times \frac{1 \text{ mol Ag}}{\text{mol e}^-} = 2.86 \times 10^{-3} \text{ mol Ag}$

$[Ag^+] = 2.86 \times 10^{-3}$ mol Ag^+/0.250 L = 1.14×10^{-2} *M*

121. $Au^{3+} + 3e^- \rightarrow Au$ E° = 1.50 V $Ni^{2+} + 2e^- \rightarrow Ni$ E° = –0.23 V

$Ag^+ + e^- \rightarrow Ag$ E° = 0.80 V $Cd^{2+} + 2e^- \rightarrow Cd$ E° = –0.40 V

$2 H_2O + 2e^- \rightarrow H_2 + 2 OH^-$ E° = –0.83 V (Water can also be reduced.)

Au(s) will plate out first since it has the most positive reduction potential, followed by Ag(s), which is followed by Ni(s), and finally Cd(s) will plate out last since it has the most negative reduction potential of the metals listed. Water will not interfere with the plating process.

123. The species present in solution are Ca^{2+}, I^-, and H_2O. Ca^{2+} and H_2O can be reduced, while I^- and H_2O can be oxidized. For the cathode reaction, H_2O has a more positive reduction potential than Ca^{2+} (–0.83 V vs. –2.37 V), so the cathode reaction will be the reduction of water. For the anode reaction, I^- has a more positive oxidation potential than H_2O (–0.54 V vs. –1.23 V), so the anode reaction will be the oxidation of I^-. The reactions that will occur are:

Cathode: $2 H_2O + 2e^- \rightarrow H_2 + 2 OH^-$; anode: $2 I^- \rightarrow I_2 + 2e^-$

Answer c is correct as I_2 will be produced at the anode and H_2 and OH^- will be produced at the cathode.

125. Species present: Na^+, SO_4^{2-}, and H_2O. H_2O and Na^+ can be reduced and H_2O and SO_4^{2-} can be oxidized. From the potentials, H_2O is the most easily reduced and the most easily oxidized species present. This is the case because water, of the species present, has the most positive reduction potential as well as the most positive oxidation potential. The reactions are:

Cathode: $2 H_2O + 2e^- \rightarrow H_2(g) + 2 OH^-$; anode: $2 H_2O \rightarrow O_2(g) + 4 H^+ + 4e^-$

127. Reduction occurs at the cathode, and oxidation occurs at the anode. First, determine all the species present; then look up pertinent reduction and/or oxidation potentials in Table 18.1 for all these species. The cathode reaction will be the reaction with the most positive reduction potential, and the anode reaction will be the reaction with the most positive oxidation potential.

a. Species present: Ni^{2+} and Br^-; Ni^{2+} can be reduced to Ni, and Br^- can be oxidized to Br_2 (from Table 18.1). The reactions are:

Cathode: $Ni^{2+} + 2e^- \rightarrow Ni$ E° = –0.23 V

Anode: $2 Br^- \rightarrow Br_2 + 2e^-$ –E° = –1.09 V

b. Species present: Al^{3+} and F^-; Al^{3+} can be reduced, and F^- can be oxidized. The reactions are:

Cathode: $Al^{3+} + 3e^- \rightarrow Al$ E° = –1.66 V

Anode: $2 F^- \rightarrow F_2 + 2e^-$ –E° = –2.87 V

c. Species present: Mn^{2+} and I^-; Mn^{2+} can be reduced, and I^- can be oxidized. The reactions are:

Cathode: $Mn^{2+} + 2\ e^- \rightarrow Mn$ $\quad E° = -1.18$ V

Anode: $2\ I^- \rightarrow I_2 + 2\ e^-$ $\quad -E° = -0.54$ V

129. These are all in aqueous solutions, so we must also consider the reduction and oxidation of H_2O in addition to the potential redox reactions of the ions present. For the cathode reaction, the species with the most positive reduction potential will be reduced, and for the anode reaction, the species with the most positive oxidation potential will be oxidized.

a. Species present: Ni^{2+}, Br^-, and H_2O. Possible cathode reactions are:

$Ni^{2+} + 2e^- \rightarrow Ni$ $\quad E° = -0.23$ V

$2\ H_2O + 2\ e^- \rightarrow H_2 + 2\ OH^-$ $\quad E° = -0.83$ V

Because it is easier to reduce Ni^{2+} than H_2O (assuming standard conditions), Ni^{2+} will be reduced by the preceding cathode reaction.

Possible anode reactions are:

$2\ Br^- \rightarrow Br_2 + 2\ e^-$ $\quad -E° = -1.09$ V

$2\ H_2O \rightarrow O_2 + 4\ H^+ + 4\ e^-$ $\quad -E° = -1.23$ V

Because Br^- is easier to oxidize than H_2O (assuming standard conditions), Br^- will be oxidized by the preceding anode reaction.

b. Species present: Al^{3+}, F^-, and H_2O; Al^{3+} and H_2O can be reduced. The reduction potentials are $E° = -1.66$ V for Al^{3+} and $E° = -0.83$ V for H_2O (assuming standard conditions). H_2O will be reduced at the cathode ($2\ H_2O + 2\ e^- \rightarrow H_2 + 2\ OH^-$).

F^- and H_2O can be oxidized. The oxidation potentials are $-E° = -2.87$ V for F^- and $-E° = -1.23$ V for H_2O (assuming standard conditions). From the potentials, we would predict H_2O to be oxidized at the anode ($2\ H_2O \rightarrow O_2 + 4\ H^+ + 4\ e^-$).

c. Species present: Mn^{2+}, I^-, and H_2O; Mn^{2+} and H_2O can be reduced. The possible cathode reactions are:

$Mn^{2+} + 2\ e^- \rightarrow Mn$ $\quad E° = -1.18$ V

$2\ H_2O + 2\ e^- \rightarrow H_2 + 2\ OH^-$ $\quad E° = -0.83$ V

Reduction of H_2O will occur at the cathode since $E^{\circ}_{H_2O}$ is most positive.

I^- and H_2O can be oxidized. The possible anode reactions are:

$2\ I^- \rightarrow I_2 + 2\ e^-$ $\quad -E° = -0.54$ V

$2\ H_2O \rightarrow O_2 + 4\ H^+ + 4\ e^-$ $\quad -E° = -1.23$ V

Oxidation of I^- will occur at the anode since $-E^{\circ}_{I^-}$ is most positive.

ChemWork Problems

131. The half-reaction for the SCE is:

$$Hg_2Cl_2 + 2\ e^- \rightarrow 2\ Hg + 2\ Cl^- \qquad E_{SCE} = 0.242\ V$$

For a spontaneous reaction to occur, E_{cell} must be positive. Using the standard reduction potentials in Table 18.1 and the given the SCE potential, deduce which combination will produce a positive overall cell potential.

a. $Cu^{2+} + 2\ e^- \rightarrow Cu \qquad E° = 0.34\ V$

$E_{cell} = 0.34 - 0.242 = 0.10$ V; SCE is the anode.

b. $Fe^{3+} + e^- \rightarrow Fe^{2+} \qquad E° = 0.77\ V$

$E_{cell} = 0.77 - 0.242 = 0.53$ V; SCE is the anode.

c. $AgCl + e^- \rightarrow Ag + Cl^- \quad E° = 0.22\ V$

$E_{cell} = 0.242 - 0.22 = 0.02$ V; SCE is the cathode.

d. $Al^{3+} + 3\ e^- \rightarrow Al \qquad E° = -1.66\ V$

$E_{cell} = 0.242 + 1.66 = 1.90$ V; SCE is the cathode.

e. $Ni^{2+} + 2\ e^- \rightarrow Ni \qquad E° = -0.23\ V$

$E_{cell} = 0.242 + 0.23 = 0.47$ V; SCE is the cathode.

133. $Ag^+(aq) + Cu(s) \rightarrow Cu^{2+}(aq) + 2\ Ag(s) \quad E^o_{cell} = 0.80 - 0.34\ V = 0.46\ V$; a galvanic cell produces a voltage as the forward reaction occurs. Any stress that increases the tendency of the forward reaction to occur will increase the cell potential, whereas a stress that decreases the tendency of the forward reaction to occur will decrease the cell potential.

a. Added Cu^{2+} (a product ion) will decrease the tendency of the forward reaction to occur, which will decrease the cell potential.

b. Added NH_3 removes Cu^{2+} in the form of $Cu(NH_3)_4^{2+}$. Because a product ion is removed, this will increase the tendency of the forward reaction to occur, which will increase the cell potential.

c. Added Cl^- removes Ag^+ in the form of AgCl(s). Because a reactant ion is removed, this will decrease the tendency of the forward reaction to occur, which will decrease the cell potential.

d. $Q_1 = \frac{[Cu^{2+}]_0}{[Ag^+]_0^2}$; as the volume of solution is doubled, each concentration is halved.

$$Q_2 = \frac{1/2\ [Cu^{2+}]_0}{(1/2\ [Ag^+]_0)^2} = \frac{2[Cu^{2+}]_0}{[Ag^+]_0^2} = 2Q_1$$

The reaction quotient is doubled because the concentrations are halved. Because reactions are spontaneous when $Q < K$, and because Q increases when the solution volume doubles, the reaction is closer to equilibrium, which will decrease the cell potential.

e. Because Ag(s) is not a reactant in this spontaneous reaction, and because solids do not appear in the reaction quotient expressions, replacing the silver electrode with a platinum electrode will have no effect on the cell potential.

135. a.

$(Fe^{2+} + 2\ e^- \rightarrow Fe) \times 3$	$E° = -0.44$ V
$(La \rightarrow La^{3+} + 3\ e^-) \times 2$	$-E° = 2.37$ V
$3\ Fe^{2+}(aq) + 2\ La(s) \rightarrow 3\ Fe(s) + 2\ La^{3+}(aq)$	$E^\circ_{cell} = 1.93$ V

The standard cell potential would be 1.93 V.

b. The oxidizing agent is Fe^{2+}.

c. The anode would be composed of a La electrode and 1.0 *M* La^{3+}.

d. The electrons flow from the La/La^{3+} compartment to the Fe^{2+}/Fe compartment (from the anode to the cathode.)

e. Six electrons are transferred per unit of cell reaction.

f. $E = E° - \frac{0.0591}{n}\log Q,\ E_{cell} = 1.93\ V - \frac{0.0591}{6}\log\frac{[La^{3+}]^2}{[Fe^{2+}]^3}$

$$E_{cell} = 1.93 - \frac{0.0591}{6}\log\frac{(3.00\times10^{-3})^2}{(2.00\times10^{-4})^3},\ E_{cell} = 1.93 - 0.0596 = 1.87\ V$$

137. a.

$(Au^{3+} + 3\ e^- \rightarrow Au) \times 2$	$E° = 1.50$ V
$(Mg \rightarrow Mg^{2+} + 2\ e^-) \times 3$	$-E° = 2.37$ V
$2\ Au^{3+}(aq) + 3\ Mg(s) \rightarrow 2\ Au(s) + 3\ Mg^{2+}(aq)$	$E^\circ_{cell} = 3.87$ V

b. $E_{cell} = 3.87\ V - \frac{0.0591}{6}\log\frac{[Mg^{2+}]^3}{[Au^{3+}]^2},\ 4.01 = 3.87 - \frac{0.0591}{6}\log\frac{(1.0\times10^{-5})^3}{[Au^{3+}]^2}$

$$\frac{(1.0\times10^{-5})^3}{[Au^{3+}]^2} = 10^{-14.21},\ [Au^{3+}] = 0.40\ M$$

139. a. $\Delta G° = \sum n_p \Delta G^\circ_{f,\ products} - \sum n_r \Delta G^\circ_{f,\ reactants} = 2(-480.) + 3(86) - [3(-40.)] = -582$ kJ

From oxidation numbers, n = 6. $\Delta G° = -nFE°,\ E° = \frac{-\Delta G°}{nF} = \frac{-(-582{,}000\ J)}{6(96{,}485)\ C} = 1.01\ V$

$$\log K = \frac{nE°}{0.0591} = \frac{6(1.01)}{0.0591} = 102.538,\ K = 10^{102.538} = 3.45\times10^{102}$$

b.

$$
\begin{array}{ll}
2\ e^- + Ag_2S \rightarrow 2\ Ag + S^{2-}) \times 3 & E^\circ_{Ag_2S} = ? \\
(Al \rightarrow Al^{3+} + 3\ e^-) \times 2 & -E^\circ = 1.66\ V \\
\hline
3\ Ag_2S(s) + 2\ Al(s) \rightarrow 6\ Ag(s) + 3\ S^{2-}(aq) + 2\ Al^{3+}(aq) & E^\circ_{cell} = 1.01\ V = E^\circ_{Ag_2S} + 1.66V
\end{array}
$$

$E^\circ_{Ag_2S} = 1.01\ V - 1.66\ V = -0.65\ V$

141. $E^\circ_{cell} = 0.400\ V - 0.240\ V = 0.160\ V;\ E = E^\circ - \frac{0.0591}{n}\log Q$

$0.180 = 0.160 - \frac{0.0591}{n}\log(9.32 \times 10^{-3})\ \ 0.020 = \frac{0.120}{n},\ n = 6$

Six moles of electrons are transferred in the overall balanced reaction. We now must figure out how to get 6 mol e^- into the overall balanced equation. The two possibilities are to have ion charges of +1 and +6 or +2 and +3; only these two combinations yield a 6 when common multiples are determined when adding the reduction half-reaction to the oxidation half-reaction. Because N forms +2 charged ions, M must form for +3 charged ions. The overall cell reaction can now be determined.

$$
\begin{array}{ll}
(M^{3+} + 3\ e^- \rightarrow M) \times 2 & E^\circ = 0.400\ V \\
(N \rightarrow N^{2+} + 2\ e^-) \times 3 & -E^\circ = -0.240\ V \\
\hline
2\ M^{3+}(aq) + 3\ N(s) \rightarrow 3\ N^{2+}(aq) + 2\ M(s) & E^\circ_{cell} = 0.160\ V
\end{array}
$$

$Q = 9.32 \times 10^{-3} = \frac{[N^{2+}]_0^3}{[M^{3+}]_0^2} = \frac{(0.10)^3}{[M^{3+}]^2},\ [M^{3+}] = 0.33\ M$

$w_{max} = \Delta G = -nFE = -6(96{,}485)(0.180) = -1.04 \times 10^5\ J = -104\ kJ$

The maximum amount of work this cell could produce is 104 kJ.

143. a. In a concentration cell, the anode always has the smaller ion concentration. So the cell with AgCl(s) at the bottom has the smaller $[Ag^+]$ and is the anode, while the compartment with $[Ag^+] = 1.0\ M$ is the cathode. Electron flow is from the anode to the cathode.

b. $AgCl(s) \rightleftharpoons Ag^+(aq) + Cl^-(aq)$ $K_{sp} = [Ag^+][Cl^-]$; use the concentration cell potential of 0.58 V to calculate the silver ion concentration in the anode compartment. For a concentration cell, E° = 0, and for Ag^+/Ag half-reactions, n = 1.

$$E = 0.58\ V = 0 - \frac{0.0591}{1}\log\frac{[Ag^+]_{anode}}{[Ag^+]_{cathode}}$$

$$\frac{-0.58}{0.0591} = \log\frac{[Ag^+]_{anode}}{1.0},\ [Ag^+]_{anode} = 1.5 \times 10^{-10}\ M$$

At the anode, we have $[Ag^+] = 1.5 \times 10^{-10}\ M$ and $[Cl^-] = 1.0\ M$:

$$K_{sp} = [Ag^+][Cl^-] = 1.5 \times 10^{-10}(1.0) = 1.5 \times 10^{-10}$$

145. Aluminum forms a durable oxide coating over its surface. Once the HCl dissolves this oxide coating, Al is exposed to H^+ and is easily oxidized to Al^{3+}, i.e., the Al foil disappears after the oxide coating is dissolved.

147. H_2O_2 as an oxidizing agent: $H_2O_2 + 2\ H^+ + 2\ e^- \rightarrow 2\ H_2O \quad E° = 1.78\ V$

H_2O_2 as a reducing agent: $H_2O_2 \rightarrow O_2 + 2\ H^+ + 2\ e^- \quad -E° = -0.68\ V$

From the more positive reduction potential, H_2O_2 is a better oxidizing agent than it is a reducing agent at standard conditions.

149. For C_2H_5OH, H has a +1 oxidation state, and O has a −2 oxidation state. This dictates a −2 oxidation state for C. For CO_2, O has a −2 oxidation state, so carbon has a +4 oxidation state. Six moles of electrons are transferred per mole of carbon oxidized (C goes from −2 → +4). Two moles of carbon are in the balanced reaction, so n = 12.

$w_{max} = -1320\ kJ = \Delta G = -nFE, \quad -1320 \times 10^3\ J = -nFE = -(12\ mol\ e^-)(96{,}485\ C/mol\ e^-)E$

$E = 1.14\ J/C = 1.14\ V$

151.

$$\begin{array}{ll} O_2 + 2\ H_2O + 4\ e^- \rightarrow 4\ OH^- & E° = 0.40\ V \\ (H_2 + 2\ OH^- \rightarrow 2\ H_2O + 2\ e^-) \times 2 & -E° = 0.83\ V \\ \hline 2\ H_2(g) + O_2(g) \rightarrow 2\ H_2O(l) & E°_{cell} = 1.23\ V = 1.23\ J/C \end{array}$$

Because standard conditions are assumed, $w_{max} = \Delta G°$ for 2 mol H_2O produced.

$\Delta G° = -nFE°_{cell} = -(4\ mol\ e^-)(96{,}485\ C/mol\ e^-)(1.23\ J/C) = -475{,}000\ J = -475\ kJ$

For 1.00×10^3 g H_2O produced, w_{max} is:

$$1.00 \times 10^3\ g\ H_2O \times \frac{1\ mol\ H_2O}{18.02\ g\ H_2O} \times \frac{-475\ kJ}{2\ mol\ H_2O} = -13{,}200\ kJ = w_{max}$$

The work done can be no larger than the free energy change. The best that could happen is that all of the free energy released would go into doing work, but this does not occur in any real process because there is always waste energy in a real process. Fuel cells are more efficient in converting chemical energy into electrical energy; they are also less massive. The major disadvantage is that they are expensive. In addition, $H_2(g)$ and $O_2(g)$ are an explosive mixture if ignited; much more so than fossil fuels.

153.

$$\begin{array}{l} (CO + O^{2-} \rightarrow CO_2 + 2\ e^-) \times 2 \\ O_2 + 4\ e^- \rightarrow 2\ O^{2-} \\ \hline CO + O_2 \rightarrow 2\ CO_2 \end{array}$$

$$\Delta G = -nFE, \quad E = \frac{-\Delta G°}{nF} = \frac{-(-380 \times 10^3\ J)}{(4\ mol\ e^-)(96{,}485\ C/mol\ e^-)} = 0.98\ V$$

155.

$$\frac{150. \times 10^3\ g\ C_6H_8N_2}{h} \times \frac{1\ h}{60\ min} \times \frac{1\ min}{60\ s} \times \frac{1\ mol\ C_6H_8N_2}{108.14\ g\ C_6H_8N_2} \times \frac{2\ mol\ e^-}{mol\ C_6H_8N_2} \times \frac{96{,}485\ C}{mol\ e^-}$$

$= 7.44 \times 10^4$ C/s, or a current of 7.44×10^4 A

157. $15\text{ kWh} = \frac{15000\text{ J h}}{\text{s}} \times \frac{60\text{ s}}{\text{min}} \times \frac{60\text{ min}}{\text{h}} = 5.4 \times 10^7\text{ J or } 5.4 \times 10^4\text{ kJ}$ (Hall-Heroult process)

To melt 1.0 kg Al requires: $1.0 \times 10^3\text{ g Al} \times \frac{1\text{ mol Al}}{26.98\text{ g}} \times \frac{10.7\text{ kJ}}{\text{mol Al}} = 4.0 \times 10^2\text{ kJ}$

It is feasible to recycle Al by melting the metal because, in theory, it takes less than 1% of the energy required to produce the same amount of Al by the Hall-Heroult process.

159. Pd is in the +2 oxidation state in $PdCl_2$. $Pd^{2+} + 2e^- \rightarrow Pd$

$$0.1064\text{ g Pd} \times \frac{1\text{ mol Pd}}{106.4\text{ g}} \times \frac{2\text{ mol e}^-}{\text{mol Pd}} \times \frac{96{,}485\text{ C}}{\text{mol e}^-} = 193.0\text{ C of electricity needed}$$

$$\text{Current} = \frac{193.0\text{ C}}{48.6\text{ s}} = 3.97\text{ A}$$

161. Chromium(III) nitrate $[Cr(NO_3)_3]$ has chromium in the +3 oxidation state.

$$1.15\text{ g Cr} \times \frac{1\text{ mol Cr}}{52.00\text{ g}} \times \frac{3\text{ mol e}^-}{\text{mol Cr}} \times \frac{96{,}485\text{ C}}{\text{mol e}^-} = 6.40 \times 10^3\text{ C of charge}$$

For the Os cell, 6.40×10^3 C of charge also was passed.

$$3.15\text{ g Os} \times \frac{1\text{ mol Os}}{190.2\text{ g}} = 0.0166\text{ mol Os};\ 6.40 \times 10^3\text{ C} \times \frac{1\text{ mol e}^-}{96{,}485\text{ C}} = 0.0663\text{ mol e}^-$$

$$\frac{\text{Mol e}^-}{\text{Mol Os}} = \frac{0.0663}{0.0166} = 3.99 \approx 4$$

This salt is composed of Os^{4+} and NO_3^- ions. The compound is $Os(NO_3)_4$, osmium(IV) nitrate.

For the third cell, identify X by determining its molar mass. Two moles of electrons are transferred when X^{2+} is reduced to X.

$$\text{Molar mass} = \frac{2.11\text{ g X}}{6.40 \times 10^3\text{ C} \times \frac{1\text{ mol e}^-}{96{,}485\text{ C}} \times \frac{1\text{ mol X}}{2\text{ mol e}^-}} = 63.6\text{ g/mol}$$

This is copper (Cu), which has an electron configuration of $[Ar]4s^1 3d^{10}$.

Challenge Problems

163. a. $3\text{ e}^- + 4\text{ H}^+ + NO_3^- \rightarrow NO + 2\text{ H}_2O \qquad E° = 0.96\text{ V}$

Nitric acid can oxidize Co to Co^{2+} ($E^\circ_{cell} > 0$), but is not strong enough to oxidize Co to Co^{3+} ($E^\circ_{cell} < 0$). Co^{2+} is the primary product assuming standard conditions.

b. Concentrated nitric acid is about 16 mol/L. $[H^+] = [NO_3^-] = 16\ M$; assume $P_{NO} = 1$ atm.

$$E = 0.96\text{ V} - \frac{0.0591}{3}\log\frac{P_{NO}}{[H^+]^4[NO_3^-]} = 0.96 - \frac{0.0591}{3}\log\frac{1}{(16)^5}$$

E = 0.96 – (–0.12) = 1.08 V ; no, concentrated nitric acid still will only be able to oxidize Co to Co^{2+}.

165. $\Delta G° = -nFE° = \Delta H° - T\Delta S°,\ E° = \frac{T\Delta S°}{nF} - \frac{\Delta H°}{nF}$

If we graph E° versus T we should get a straight line ($y = mx + b$). The slope of the line is equal to ΔS°/nF, and the y intercept is equal to –ΔH°/nF. From the preceding equation, E° will have a small temperature dependence when ΔS° is close to zero.

167.

$$(Ag^+ + e^- \rightarrow Ag) \times 2 \qquad E° = 0.80\text{ V}$$
$$Pb \rightarrow Pb^{2+} + 2e^- \qquad -E° = -(-0.13)$$

$$2Ag^+(aq) + Pb(s) \rightarrow 2Ag(s) + Pb^{2+}(aq) \qquad E°_{cell} = 0.93\text{ V}$$

$$E = E° - \frac{0.0591}{n}\log\frac{[Pb^{2+}]}{[Ag^+]^2},\ 0.83\text{ V} = 0.93\text{ V} - \frac{0.0591}{n}\log\frac{(1.8)}{[Ag^+]^2}$$

$$\log\frac{(1.8)}{[Ag^+]^2} = \frac{0.10(2)}{0.0591} = 3.4,\ \frac{(1.8)}{[Ag^+]^2} = 10^{3.4},\ [Ag^+] = 0.027\ M$$

	$Ag_2SO_4(s)$	⇌	$2Ag^+(aq)$	+	$SO_4^{2-}(aq)$	$K_{sp} = [Ag^+]^2[SO_4^{2-}]$
Initial	s = solubility (mol/L)		0		0	
Equil.			$2s$		s	

From problem: $2s = 0.027\ M,\ s = 0.027/2;\ K_{sp} = (2s)^2(s) = (0.027)^2(0.027/2) = 9.8 \times 10^{-6}$

169.

$$2H^+ + 2e^- \rightarrow H_2 \qquad E° = 0.000\text{ V}$$
$$Fe \rightarrow Fe^{2+} + 2e^- \qquad -E° = -(-0.440\text{V})$$

$$2H^+(aq) + Fe(s) \rightarrow H_2(g) + Fe^{3+}(aq) \qquad E°_{cell} = 0.440\text{ V}$$

$$E_{cell} = E°_{cell} - \frac{0.0591}{n}\log Q,\text{ where n = 2 and } Q = \frac{P_{H_2} \times [Fe^{3+}]}{[H^+]^2}$$

To determine K_a for the weak acid, first use the electrochemical data to determine the H^+ concentration in the half-cell containing the weak acid.

$$0.333\text{ V} = 0.440\text{ V} - \frac{0.0591}{2}\log\frac{1.00\text{ atm}(1.00 \times 10^{-3}\ M)}{[H^+]^2}$$

$$\frac{0.107(2)}{0.0591} = \log\frac{1.0 \times 10^{-3}}{[H^+]^2},\ \frac{1.0 \times 10^{-3}}{[H^+]^2} = 10^{3.621} = 4.18 \times 10^3,\ [H^+] = 4.89 \times 10^{-4}\ M$$

Now we can solve for the K_a value of the weak acid HA through the normal setup for a weak acid problem.

	$HA(aq)$	$\rightleftharpoons$	$H^+(aq)$	+	$A^-(aq)$	$K_a = \frac{[H^+][A^-]}{[HA]}$
Initial	1.00 *M*		~0		0	
Equil.	$1.00 - x$		x		x	

$$K_a = \frac{x^2}{1.00 - x}, \text{ where } x = [H^+] = 4.89 \times 10^{-4}\ M,\ K_a = \frac{(4.89 \times 10^{-4})^2}{1.00 - 4.89 \times 10^{-4}} = 2.39 \times 10^{-7}$$

171. a. $E_{cell} = E_{ref} + 0.05916\ pH$, $0.480\ V = 0.250\ V + 0.05916\ pH$

$$pH = \frac{0.480 - 0.250}{0.05916} = 3.888; \text{ uncertainty} = \pm 1\ mV = \pm\ 0.001\ V$$

$$pH_{max} = \frac{0.481 - 0.250}{0.05916} = 3.905; \quad pH_{min} = \frac{0.479 - 0.250}{0.05916} = 3.871$$

Thus, if the uncertainty in potential is ±0.001 V, then the uncertainty in pH is ±0.017, or about ±0.02 pH units. For this measurement, $[H^+] = 10^{-3.888} = 1.29 \times 10^{-4}\ M$. For an error of +1 mV, $[H^+] = 10^{-3.905} = 1.24 \times 10^{-4}\ M$. For an error of −1 mV, $[H^+] = 10^{-3.871} = 1.35 \times 10^{-4}\ M$. So the uncertainty in $[H^+]$ is $\pm 0.06 \times 10^{-4}\ M = \pm 6 \times 10^{-6}\ M$.

b. From part a, we will be within ±0.02 pH units if we measure the potential to the nearest ±0.001 V (±1 mV).

173. a.

$$(Ag^+ + e^- \rightarrow Ag) \times 2 \qquad E° = 0.80\ V$$
$$Cu \rightarrow Cu^{2+} + 2\ e^- \qquad -E° = -0.34\ V$$

$$2\ Ag^+(aq) + Cu(s) \rightarrow 2\ Ag(s) + Cu^{2+}(aq) \qquad E^\circ_{cell} = 0.46\ V$$

$$E_{cell} = E^\circ_{cell} - \frac{0.0591}{n} \log Q, \text{ where } n = 2 \text{ and } Q = \frac{[Cu^{2+}]}{[Ag^+]^2}.$$

To calculate E_{cell}, we need to use the K_{sp} data to determine $[Ag^+]$.

	$AgCl(s)$	$\rightleftharpoons$	$Ag^+(aq)$	+	$Cl^-(aq)$	$K_{sp} = 1.6 \times 10^{-10} = [Ag^+][Cl^-]$
Initial	s = solubility (mol/L)		0		0	
Equil.			s		s	

$K_{sp} = 1.6 \times 10^{-10} = s^2$, $s = [Ag^+] = 1.3 \times 10^{-5}$ mol/L

$$E_{cell} = 0.46\ V - \frac{0.0591}{2} \log \frac{2.0}{(1.3 \times 10^{-5})^2} = 0.46\ V - 0.30 = 0.16\ V$$

b. $Cu^{2+}(aq) + 4\ NH_3(aq) \rightleftharpoons Cu(NH_4)_4^{2+}(aq) \quad K = 1.0 \times 10^{13} = \frac{[Cu(NH_3)_4^{2+}]}{[Cu^{2+}][NH_3]^4}$

Because K is very large for the formation of $Cu(NH_3)_4^{2+}$, the forward reaction is dominant. At equilibrium, essentially all the 2.0 *M* Cu^{2+} will react to form 2.0 *M* $Cu(NH_3)_4^{2+}$. This reaction requires 8.0 *M* NH_3 to react with all the Cu^{2+} in the balanced equation. Therefore, the moles of NH_3 added to 1.0-L solution will be larger than 8.0 mol since some NH_3 must be present at equilibrium. To calculate how much NH_3 is present at equilibrium, we need to use the electrochemical data to determine the Cu^{2+} concentration.

$$E_{cell} = E^{o}_{cell} - \frac{0.0591}{n}\log Q,\ 0.52\ V = 0.46\ V - \frac{0.0591}{2}\log\frac{[Cu^{2+}]}{(1.3\times10^{-5})^2}$$

$$\log\frac{[Cu^{2+}]}{(1.3\times10^{-5})^2} = \frac{-0.06(2)}{0.0591} = -2.03,\ \frac{[Cu^{2+}]}{(1.3\times10^{-5})^2} = 10^{-2.03} = 9.3\times10^{-3}$$

$[Cu^{2+}] = 1.6 \times 10^{-12} = 2 \times 10^{-12}$ *M*

(We carried extra significant figures in the calculation.)

Note: Our assumption that the 2.0 *M* Cu^{2+} essentially reacts to completion is excellent because only 2×10^{-12} *M* Cu^{2+} remains after this reaction. Now we can solve for the equilibrium $[NH_3]$.

$$K = 1.0\times10^{13} = \frac{[Cu(NH_3)_4^{2+}]}{[Cu^{2+}][NH_3]^4} = \frac{(2.0)}{(2\times10^{-12})[NH_3]^4},\quad [NH_3] = 0.6\ M$$

Because 1.0 L of solution is present, 0.6 mol NH_3 remains at equilibrium. The total moles of NH_3 added is 0.6 mol plus the 8.0 mol NH_3 necessary to form 2.0 *M* $Cu(NH_3)_4^{2+}$. Therefore, 8.0 + 0.6 = 8.6 mol NH_3 was added.

175. $2\ Ag^+(aq) + Ni(s) \rightarrow Ni^{2+}(aq) + Ag(s)$; the cell is dead at equilibrium (E = 0).

$E^{o}_{cell} - 0.80\ V + 0.23\ V - 1.03\ V$

$$0 = 1.03\ V - \frac{0.0591}{2}\log K,\ K = 7.18\times10^{34}$$

K is very large. Let the forward reaction go to completion.

	$2\ Ag^+(aq)$ + $Ni(s)$ →	$Ni^{2+}(aq)$ + $2\ Ag(s)$	$K = [Ni^{2+}]/[Ag^+]^2 = 7.18 \times 10^{34}$
Before	1.0 *M*	1.0 *M*	
Change	−1.0 →	+0.50	
After	0	1.5 *M*	

Now solve the back-equilibrium problem.

	$2\ Ag^+(aq) + Ni(s)$ ⇌	$Ni^{2+}(aq)$ + $2\ Ag(s)$
Initial	0	1.5 *M*
Change	$+2x$ ←	$-x$
Equil.	$2x$	$1.5 - x$

$$K = 7.18\times10^{34} = \frac{1.5-x}{(2x)^2} \approx \frac{1.5}{(2x)^2};\ \text{solving, } x = 2.3\times10^{-18}\ M.\ \text{ Assumptions good.}$$

$[Ag^+] = 2x = 4.6 \times 10^{-18}$ *M*; $[Ni^{2+}] = 1.5 - 2.3 \times 10^{-18} = 1.5$ *M*

177.

$(Ag^+ + e^- \rightarrow Ag) \times 2$	$E° = 0.80\ V$
$Cd \rightarrow Cd^{2+} + 2\ e^-$	$-E° = 0.40\ V$
$2\ Ag^+(aq) + Cd(s) \rightarrow Cd^{2+}(aq) + 2\ Ag(s)$	$E^{o}_{cell} = 1.20\ V$

Overall complex ion reaction:

$$Ag^+(aq) + 2\,NH_3(aq) \rightarrow Ag(NH_3)_2^+(aq) \quad K = K_1K_2 = 2.1 \times 10^3(8.2 \times 10^3) = 1.7 \times 10^7$$

Because K is large, we will let the reaction go to completion and then solve the back-equilibrium problem.

	$Ag^+(aq)$	+ $2\,NH_3(aq)$	$\rightleftharpoons$	$Ag(NH_3)_2^+(aq)$	$K = 1.7 \times 10^7$
Before	1.00 *M*	15.0 *M*		0	
After	0	13.0		1.00	New initial
Change	x	$+2x$	$\leftarrow$	$-x$	
Equil.	x	$13.0 + 2x$		$1.00 - x$	

$$K = \frac{[Ag(NH_3)_2^+]}{[Ag^+][NH_3]^2};\quad 1.7 \times 10^7 = \frac{1.00 - x}{x(13.0 + 2x)^2} \approx \frac{1.00}{x(13.0)^2}$$

Solving: $x = 3.5 \times 10^{-10}\ M = [Ag^+]$; assumptions good.

$$E = E° - \frac{0.0591}{2}\log\frac{[Cd^{2+}]}{[Ag^+]^2} = 1.20\ V - \frac{0.0591}{2}\log\left[\frac{1.0}{(3.5 \times 10^{-10})^2}\right]$$

$$E = 1.20 - 0.56 = 0.64\ V$$

CHAPTER 19

THE NUCLEUS: A CHEMIST'S VIEW

Questions

3. Characteristic frequencies of energies emitted in a nuclear reaction suggest that discrete energy levels exist in the nucleus. The extra stability of certain numbers of nucleons and the predominance of nuclei with even numbers of nucleons suggest that the nuclear structure might be described by using quantum numbers.

5. Radiotracers generally have short half-lives. The radioactivity from the radiotracers is monitored to study a specific area of the body. However, we don't want long-lived radioactivity in order to minimize potential damage to healthy tissue by the radioactivity.

7. Beta-particle production has the net effect of turning a neutron into a proton. Radioactive nuclei having too many neutrons typically undergo β-particle decay. Positron production has the net effect of turning a proton into a neutron. Nuclei having too many protons typically undergo positron decay.

9. The deuterium nuclei are positively charges particles. For the two nuclei to bind together and release energy, they must get very close together. Because the deuterium nuclei are both positively charged, they repel each other electrostatically. The increase in energy as two deuterium nuclei are brought together is due to the electrostatic repulsion of the two positively charge nuclei for each other. To overcome the electrostatic barrier, the two nuclei must be travelling at extremely fast velocities when they collide. Temperatures of 4×10^7 K are required to give the nuclei enough energy upon collision to bind together.

11. Carbon-14 has 8 neutrons. Some other nuclides with 8 protons are ^{13}B, ^{15}N, ^{16}O, and ^{17}F.

13. The transuranium elements are the elements having more protons than uranium. They are synthesized by bombarding heavier nuclei with neutrons and positive ions in a particle accelerator.

15. $\Delta E = \Delta mc^2$; the key difference is the mass change when going from reactants to products. In chemical reactions, the mass change is indiscernible. In nuclear processes, the mass change is discernible. It is the conversion of this discernible mass change into energy that results in the huge energies associated with nuclear processes.

17. The temperatures of fusion reactions are so high that all physical containers would be destroyed. At these high temperatures, most of the electrons are stripped from the atoms. A plasma of gaseous ions is formed that can be controlled by magnetic fields.

19. Somatic damage is immediate damage to the organism itself, resulting in sickness or death. Genetic damage is damage to the genetic material which can be passed on to future generations. The organism will not feel immediate consequences from genetic damage, but its offspring may be damaged.

Exercises

Radioactive Decay and Nuclear Transformations

21. All nuclear reactions must be charge balanced and mass balanced. To charge balance, balance the sum of the atomic numbers on each side of the reaction, and to mass balance, balance the sum of the mass numbers on each side of the reaction.

a. ${}^{3}_{1}\text{H} \rightarrow {}^{3}_{2}\text{He} + {}^{0}_{-1}\text{e}$

b. ${}^{8}_{3}\text{Li} \rightarrow {}^{8}_{4}\text{Be} + {}^{0}_{-1}\text{e}$

${}^{8}_{4}\text{Be} \rightarrow 2\ {}^{4}_{2}\text{He}$

${}^{8}_{3}\text{Li} \rightarrow 2\ {}^{4}_{2}\text{He} + {}^{0}_{-1}\text{e}$

c. ${}^{7}_{4}\text{Be} + {}^{0}_{-1}\text{e} \rightarrow {}^{7}_{3}\text{Li}$

d. ${}^{8}_{5}\text{B} \rightarrow {}^{8}_{4}\text{Be} + {}^{0}_{+1}\text{e}$

23. All nuclear reactions must be charge balanced and mass balanced. To charge balance, balance the sum of the atomic numbers on each side of the reaction, and to mass balance, balance the sum of the mass numbers on each side of the reaction.

a. ${}^{238}_{92}\text{U} \rightarrow {}^{4}_{2}\text{He} + {}^{234}_{90}\text{Th}$; this is alpha-particle production.

b. ${}^{234}_{90}\text{Th} \rightarrow {}^{234}_{91}\text{Pa} + {}^{0}_{-1}\text{e}$; this is β-particle production.

25. a. ${}^{68}_{31}\text{Ga} + {}^{0}_{-1}\text{e} \rightarrow {}^{68}_{30}\text{Zn}$

b. ${}^{62}_{29}\text{Cu} \rightarrow {}^{0}_{+1}\text{e} + {}^{62}_{28}\text{Ni}$

c. ${}^{212}_{87}\text{Fr} \rightarrow {}^{4}_{2}\text{He} + {}^{208}_{85}\text{At}$

d. ${}^{129}_{51}\text{Sb} \rightarrow {}^{0}_{-1}\text{e} + {}^{129}_{52}\text{Te}$

27. ${}^{75}_{33}\text{As} + {}^{1}_{0}\text{n} \rightarrow {}^{76}_{33}\text{As} \rightarrow {}^{76}_{34}\text{Se} + {}^{0}_{-1}\text{e}$; ${}^{76}_{34}\text{Se}$ is the final product.

29. ${}^{235}_{92}\text{U} \rightarrow {}^{207}_{82}\text{Pb} + ?\ {}^{4}_{2}\text{He} + ?\ {}^{0}_{-1}\text{e}$

From the two possible decay processes, only alpha-particle decay changes the mass number. So the mass number change of 28 from 235 to 207 must be done in the decay series by seven alpha particles. The atomic number change of 10 from 92 to 82 is due to both alpha-particle production and beta-particle production. However, because we know that seven alpha-particles are in the complete decay process, we must have four beta-particle decays in order to balance the atomic number. The complete decay series is summarized as:

$${}^{235}_{92}\text{U} \rightarrow {}^{207}_{82}\text{Pb} + 7\ {}^{4}_{2}\text{He} + 4\ {}^{0}_{-1}\text{e}$$

31. a. $^{241}_{95}\text{Am} \rightarrow {}^{4}_{2}\text{He} + {}^{237}_{93}\text{Np}$

b. $^{241}_{95}\text{Am} \rightarrow 8\,{}^{4}_{2}\text{He} + 4\,{}^{0}_{-1}\text{e} + {}^{209}_{83}\text{Bi}$; the final product is $^{209}_{83}\text{Bi}$.

c. $^{241}_{95}\text{Am} \rightarrow {}^{237}_{93}\text{Np} + \alpha \rightarrow {}^{233}_{91}\text{Pa} + \alpha \rightarrow {}^{233}_{92}\text{U} + \beta \rightarrow {}^{229}_{90}\text{Th} + \alpha \rightarrow {}^{225}_{88}\text{Ra} + \alpha$

$\downarrow$

$^{213}_{84}\text{Po} + \beta \leftarrow {}^{213}_{83}\text{Bi} + \alpha \leftarrow {}^{217}_{85}\text{At} + \alpha \leftarrow {}^{221}_{87}\text{Fr} + \alpha \leftarrow {}^{225}_{89}\text{Ac} + \beta$

$\downarrow$

$^{209}_{82}\text{Pb} + \alpha \rightarrow {}^{209}_{83}\text{Bi} + \beta$

The intermediate radionuclides are:

$^{237}_{93}\text{Np}$, $^{233}_{91}\text{Pa}$, $^{233}_{92}\text{U}$, $^{229}_{90}\text{Th}$, $^{225}_{88}\text{Ra}$, $^{225}_{89}\text{Ac}$, $^{221}_{87}\text{Fr}$, $^{217}_{85}\text{At}$, $^{213}_{83}\text{Bi}$, $^{213}_{84}\text{Po}$, and $^{209}_{82}\text{Pb}$

33. Reference Table 19.2 of the text for potential radioactive decay processes. ^{8}B and ^{9}B contain too many protons or too few neutrons. Electron capture or positron production are both possible decay mechanisms that increase the neutron to proton ratio. Alpha particle production also increases the neutron to proton ratio, but it is not likely for these light nuclei. ^{12}B and ^{13}B contain too many neutrons or too few protons. Beta-particle production lowers the neutron to proton ratio, so we expect ^{12}B and ^{13}B to be β-emitters.

35. a. $^{249}_{98}\text{Cf} + {}^{18}_{8}\text{O} \rightarrow {}^{263}_{106}\text{Sg} + 4\,{}^{1}_{0}\text{n}$ b. $^{259}_{104}\text{Rf}$; $^{263}_{106}\text{Sg} \rightarrow {}^{4}_{2}\text{He} + {}^{259}_{104}\text{Rf}$

Kinetics of Radioactive Decay

37. For $t_{1/2}$ = 12,000 yr:

$$k = \frac{\ln 2}{t_{1/2}} = \frac{0.693}{12{,}000\text{ yr}} \times \frac{1\text{ yr}}{365\text{ d}} \times \frac{1\text{ d}}{24\text{ h}} \times \frac{1\text{ h}}{3600\text{ s}} = 1.8 \times 10^{-12}\text{ s}^{-1}$$

$$\text{Rate} = kN = 1.8 \times 10^{-12}\text{ s}^{-1} \times 6.02 \times 10^{23}\text{ nuclei} = 1.1 \times 10^{12}\text{ decays/s}$$

For $t_{1/2}$ = 12 h:

$$k = \frac{\ln 2}{t_{1/2}} = \frac{0.693}{12\text{ h}} \times \frac{1\text{ h}}{3600\text{ s}} = 1.6 \times 10^{-5}\text{ s}^{-1}$$

$$\text{Rate} = kN = 1.6 \times 10^{-5}\text{ s}^{-1} \times 6.02 \times 10^{23}\text{ nuclei} = 9.6 \times 10^{18}\text{ decays/s}$$

For $t_{1/2}$ = 12 s:

$$\text{Rate} = kN = \frac{0.693}{12\text{ s}} \times 6.02 \times 10^{23}\text{ nuclei} = 3.5 \times 10^{22}\text{ decays/s}$$

39. All radioactive decay follows first-order kinetics where $t_{1/2} = (\ln 2)/k$.

$$t_{1/2} = \frac{\ln 2}{k} = \frac{0.6931}{4.91 \times 10^{-11}\ \text{yr}^{-1}} = 1.41 \times 10^{10}\ \text{yr}$$

41. 12.5% of ^{40}K will remain after 3 half-lives:

$$100\% \xrightarrow[t_{1/2}]{} 50\% \xrightarrow[t_{1/2}]{} 25\% \xrightarrow[t_{1/2}]{} 12.5\%$$

So, each half-life is $3.9 \times 10^9 / 3 = 1.3 \times 10^9$ yr.

43. $1.0\ \text{day} \times \frac{24\ \text{h}}{\text{day}} \times \frac{1\ \text{half-life}}{6\ \text{h}} = 4$ half-lives;

$$100\% \xrightarrow[t_{1/2}]{} 50\% \xrightarrow[t_{1/2}]{} 25\% \xrightarrow[t_{1/2}]{} 12.5\% \xrightarrow[t_{1/2}]{} 6.25\%$$

After 4 half-lives, 6.25% of the sample remains: 0.0625(150.0 mg) = 9.4 mg ^{99}Tc remains

45. Units for N and N_0 are usually number of nuclei but can also be grams if the units are identical for both N and N_0. In this problem, m_0 = the initial mass of $^{47}Ca^{2+}$ to be ordered.

$$k = \frac{\ln 2}{t_{1/2}};\ \ln\left(\frac{N}{N_0}\right) = -kt = \frac{-(0.693)t}{t_{1/2}},\ \ln\left(\frac{5.0\ \mu\text{g Ca}^{2+}}{m_0}\right) = \frac{-0.693(2.0\ \text{d})}{4.5\ \text{d}} = -0.31$$

$\frac{5.0}{m_0} = e^{-0.31} = 0.73$, $m_0 = 6.8\ \mu g$ of $^{47}Ca^{2+}$ needed initially

$6.8\ \mu g\ ^{47}Ca^{2+} \times \frac{107.0\ \mu g\ ^{47}CaCO_3}{47.0\ \mu g\ ^{47}Ca^{2+}} = 15\ \mu g\ ^{47}CaCO_3$ should be ordered at the minimum.

47. $t = 78.0$ yr; $k = \frac{\ln 2}{t_{1/2}}$; $\ln\left(\frac{N}{N_0}\right) = -kt = \frac{-(0.6931)\ 78.0\ \text{yr}}{28.9\ \text{yr}} = -1.87$, $\left(\frac{N}{N_0}\right) = e^{-1.87} = 0.154$

15.4% of the ^{90}Sr remains as of July 16, 2023.

49. $175\ \text{mg}\ Na_3{}^{32}PO_4 \times \frac{32.0\ \text{mg}\ ^{32}P}{165.0\ \text{mg}\ Na_3{}^{32}PO_4} = 33.9\ \text{mg}\ ^{32}P$; $k = \frac{\ln 2}{t_{1/2}}$

$\ln\left(\frac{N}{N_0}\right) = -kt = \frac{-(0.6931)t}{t_{1/2}},\ \ln\left(\frac{m}{33.9\ \text{mg}}\right) = \frac{-0.6931(35.0\ \text{d})}{14.3\ \text{d}}$; carrying extra sig. figs.:

$\ln(m) = -1.696 + 3.523 = 1.827$, $m = e^{1.827} = 6.22$ mg ^{32}P remains

51. $$\ln\left(\frac{N}{N_0}\right) = -kt = \frac{-(\ln 2)t}{t_{1/2}},\ \ln\left(\frac{1.0\ \text{g}}{m_0}\right) = \frac{-0.693\left(3.0\ \text{d} \times \frac{24\ \text{h}}{\text{d}} \times \frac{60\ \text{min}}{\text{h}}\right)}{1.0 \times 10^3\ \text{min}}$$

$\ln\left(\frac{1.0\ \text{g}}{m_0}\right) = -3.0$, $\frac{1.0}{m_0} = e^{-3.0}$, $m_0 = 20.$ g ^{82}Br needed

$$20.\ g\ ^{82}Br \times \frac{1\ mol\ ^{82}Br}{82.0\ g} \times \frac{1\ mol\ Na^{82}Br}{mol\ ^{82}Br} \times \frac{105.0\ g\ Na^{82}Br}{mol\ Na^{82}Br} = 26\ g\ Na^{82}Br$$

53. $$N = 180\ lb \times \frac{453.6\ g}{lb} \times \frac{18\ g\ C}{100\ g\ body} \times \frac{1.6 \times 10^{-10}\ g\ ^{14}C}{100\ g\ C} \times \frac{1\ mol\ ^{14}C}{14\ g\ ^{14}C}$$

$$\times \frac{6.022 \times 10^{23}\ nuclei\ ^{14}C}{mol\ ^{14}C} = 1.0 \times 10^{15}\ \text{carbon-14 nuclei}$$

$$Rate = kN;\ k = \frac{\ln 2}{t_{1/2}} = \frac{0.693}{5730\ yr} \times \frac{1\ yr}{365\ d} \times \frac{1\ d}{24\ h} \times \frac{1\ h}{3600\ s} = 3.8 \times 10^{-12}\ s^{-1}$$

$$Rate = kN;\ k = 3.8 \times 10^{-12}\ s^{-1}(1.0 \times 10^{15}\ ^{14}C\ nuclei) = 3800\ decays/s$$

A typical 180 lb person produces 3800 beta particles each second.

55. $$k = \frac{\ln 2}{t_{1/2}};\ \ln\left(\frac{N}{N_0}\right) = -kt = \frac{-(0.693)t}{t_{1/2}},\ \ln\left(\frac{N}{13.6}\right) = \frac{-0.693(15{,}000\ yr)}{5730\ yr} = -1.8$$

$$\frac{N}{13.6} = e^{-1.8} = 0.17,\ N = 13.6 \times 0.17 = 2.3\ \text{counts per minute per g of C}$$

If we had 10. mg C, we would see:

$$10.\ mg \times \frac{1\ g}{1000\ mg} \times \frac{2.3\ counts}{min\ g} = \frac{0.023\ counts}{min}$$

It would take roughly 40 min to see a single disintegration. This is too long to wait, and the background radiation would probably be much greater than the ^{14}C activity. Thus ^{14}C dating is not practical for very small samples.

57. Assuming 1.000 g ^{238}U present in a sample, then 0.688 g ^{206}Pb is present. Because 1 mol ^{206}Pb is produced per mol ^{238}U decayed:

$$^{238}U\ decayed = 0.688\ g\ Pb \times \frac{1\ mol\ Pb}{206\ g\ Pb} \times \frac{1\ mol\ U}{mol\ Pb} \times \frac{238\ g\ U}{mol\ U} = 0.795\ g\ ^{238}U$$

Original mass ^{238}U present = 1.000 g + 0.795 g = 1.795 g ^{238}U

$$\ln\left(\frac{N}{N_0}\right) = -kt = \frac{-(\ln 2)t}{t_{1/2}},\ \ln\left(\frac{1.000\ g}{1.795\ g}\right) = \frac{-0.693(t)}{4.5 \times 10^9\ yr},\quad t = 3.8 \times 10^9\ yr$$

Energy Changes in Nuclear Reactions

59. $$\Delta E = \Delta mc^2,\ \Delta m = \frac{\Delta E}{c^2} = \frac{3.9 \times 10^{23}\ kg\ m^2/s^2}{(3.00 \times 10^8\ m/s)^2} = 4.3 \times 10^6\ kg$$

The sun loses 4.3×10^6 kg of mass each second. *Note*: 1 J = 1 kg m^2/s^2

61. For ^{127}I (53e, 53p, and 74n), mass defect = Δm = [mass of ^{127}I nucleus] – [mass of 53 protons + mass of 73 neutrons]. The atomic masses given include the mass of the electrons. Let m_e = mass of electron.

$\Delta m = (126.9004 \text{ u} - 53m_e) - [53(1.00782 - m_e) + 74(1.00866)]$; mass of electrons ($m_e$) cancel.

$\Delta m = 126.9004 - [53(1.00782) + 74(1.00866)] = -1.1549 \text{ u}$ = mass defect for ^{127}I

For 1 mol of ^{127}I, the mass defect is –1.1549 g.

63. We need to determine the mass defect Δm between the mass of the nucleus and the mass of the individual parts that make up the nucleus. Once Δm is known, we can then calculate ΔE (the binding energy) using $E = mc^2$. *Note*: $1 \text{ J} = 1 \text{ kg m}^2/\text{s}^2$.

For $^{232}_{94}Pu$ (94 e, 94 p, 138 n):

mass of ^{232}Pu nucleus = 3.85285×10^{-22} g – mass of 94 electrons

mass of ^{232}Pu nucleus = $3.85285 \times 10^{-22} \text{ g} - 94(9.10939 \times 10^{-28}) \text{ g} = 3.85199 \times 10^{-22}$ g

$\Delta m = 3.85199 \times 10^{-22}$ g – (mass of 94 protons + mass of 138 neutrons)

$$\Delta m = 3.85199 \times 10^{-22} \text{ g} - [94(1.67262 \times 10^{-24}) + 138(1.67493 \times 10^{-24})] \text{ g}$$
$$= -3.168 \times 10^{-24} \text{ g}$$

For 1 mol of nuclei: $\Delta m = -3.168 \times 10^{-24}$ g/nuclei $\times$ 6.0221×10^{23} nuclei/mol
$$= -1.908 \text{ g/mol}$$

$$\Delta E = \Delta mc^2 = (-1.908 \times 10^{-3} \text{ kg/mol})(2.9979 \times 10^8 \text{ m/s})^2 = -1.715 \times 10^{14} \text{ J/mol}$$

For $^{231}_{91}Pa$ (91 e, 91 p, 140 n):

mass of ^{231}Pa nucleus = $3.83616 \times 10^{-22} \text{ g} - 91(9.10939 \times 10^{-28}) \text{ g} = 3.83533 \times 10^{-22}$ g

$$\Delta m = 3.83533 \times 10^{-22} \text{ g} - [91(1.67262 \times 10^{-24}) + 140(1.67493 \times 10^{-24})] \text{ g}$$
$$= -3.166 \times 10^{-24} \text{ g}$$

$$\Delta E = \Delta mc^2 = \frac{-3.166 \times 10^{-27} \text{ kg}}{\text{nuclei}} \times \frac{6.0221 \times 10^{23} \text{ nuclei}}{\text{mol}} \times \left(\frac{2.9979 \times 10^8 \text{ m}}{\text{s}}\right)^2$$
$$= -1.714 \times 10^{14} \text{ J/mol}$$

65. For ^{12}C (6e, 6p, and 6n), mass defect = Δm = [mass of ^{12}C nucleus] – [mass of 6 protons + mass of 6 neutrons]. The atomic masses given include the mass of the electrons. Let m_e = mass of electron.

$\Delta m = (12.00000 \text{ u} - 6m_e) - [6(1.00782 - m_e) + 6(1.00866)]$; mass of electrons cancel.

$\Delta m = 12.00000 - [6(1.00782) + 6(1.00866)] = -0.09888 \text{ u}$

$$\Delta E = \Delta mc^2 = -0.09888 \text{ u} \times \frac{1.6605 \times 10^{-27} \text{ kg}}{\text{u}} \times (2.9979 \times 10^8 \text{ m/s})^2 = -1.476 \times 10^{-11} \text{ J}$$

$$\frac{\text{Binding energy}}{\text{Nucleon}} = \frac{1.476 \times 10^{-11} \text{ J}}{12 \text{ nucleons}} = 1.230 \times 10^{-12} \text{ J/nucleon}$$

For ^{235}U (92e, 92p, and 143n):

$$\Delta m = (235.0439 - 92m_e) - [92(1.00782 - m_e) + 143(1.00866)] = -1.9139 \text{ u}$$

$$\Delta E = \Delta mc^2 = -1.9139 \times \frac{1.66054 \times 10^{-27} \text{ kg}}{\text{u}} \times (2.99792 \times 10^8 \text{ m/s})^2 = -2.8563 \times 10^{-10} \text{ J}$$

$$\frac{\text{Binding energy}}{\text{Nucleon}} = \frac{2.8563 \times 10^{-10} \text{ J}}{235 \text{ nucleons}} = 1.2154 \times 10^{-12} \text{ J/nucleon}$$

Because ^{56}Fe is the most stable known nucleus, the binding energy per nucleon for ^{56}Fe (1.408×10^{-12} J/nucleon) will be larger than that of ^{12}C or ^{235}U (see Figure 19.9 of the text).

67. $\Delta E = \Delta mc^2$, $\Delta m = \dfrac{\Delta E}{c^2} = \dfrac{-3.086 \times 10^{12} \text{ J}}{(2.9979 \times 10^8 \text{ m/s})^2} = -3.434 \times 10^{-5} \text{ kg}$

The mass defect for 1 mol of ^6Li is -3.434×10^{-5} kg = −0.03434 g. The mass defect for one ^6Li nuclei is −0.03434 u.

Let m_{Li} = mass of ^6Li nucleus; an ^6Li nucleus has 3p and 3n.

Mass defect = −0.03434 u = $m_{Li} - (3m_p + 3m_n) = m_{Li} - [3(1.00728 \text{ u}) + 3(1.00866 \text{ u})]$

m_{Li} = 6.01348 u

Mass of ^6Li atom = 6.01348 u + $3m_e$ = 6.01348 + $3(5.49 \times 10^{-4}$ u) = 6.01513 u
(includes mass of 3 e^-)

69. ${}^1_1H + {}^1_1H \rightarrow {}^2_1H + {}^{0}_{+1}e$; $\Delta m = (2.01410 \text{ u} - m_e + m_e) - 2(1.00782 \text{ u} - m_e)$

$\Delta m = 2.01410 - 2(1.00782) + 2(0.000549) = -4.4 \times 10^{-4}$ u for two protons reacting

When 2 mol of protons undergoes fusion, $\Delta m = -4.4 \times 10^{-4}$ g.

$\Delta E = \Delta mc^2 = -4.4 \times 10^{-7} \text{ kg} \times (3.00 \times 10^8 \text{ m/s})^2 = -4.0 \times 10^{10}$ J

$$\frac{-4.0 \times 10^{10} \text{ J}}{2 \text{ mol protons}} \times \frac{1 \text{ mol}}{1.01 \text{ g}} = -2.0 \times 10^{10} \text{ J/g of hydrogen nuclei}$$

Detection, Uses, and Health Effects of Radiation

71. The Geiger-Müller tube has a certain response time. After the gas in the tube ionizes to produce a "count," sometime must elapse for the gas to return to an electrically neutral state. The response of the tube levels off because at high activities, radioactive particles are entering the tube faster than the tube can respond to them.

73. Water is produced in this reaction by removing an OH group from one substance and an H from the other substance. There are two ways to do this:

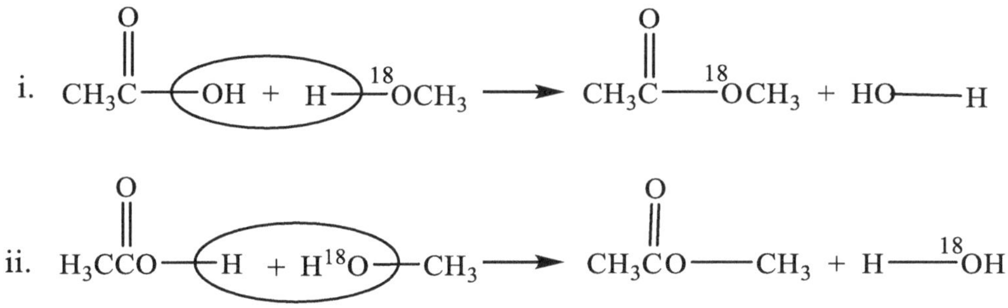

Because the water produced is not radioactive, methyl acetate forms by the first reaction where all the oxygen-18 ends up in methyl acetate.

75. $^{235}_{92}U + ^{1}_{0}n \rightarrow ^{144}_{58}Ce + ^{90}_{38}Sr + ?\ ^{1}_{0}n + ?\ ^{0}_{-1}e$; to balance the atomic number, we need 4 beta-particles, and to balance the mass number, we need 2 neutrons.

77. Release of Sr is probably more harmful. Xe is chemically unreactive. Strontium is in the same family as calcium and could be absorbed and concentrated in the body in a fashion similar to Ca. This puts the radioactive Sr in the bones; red blood cells are produced in bone marrow. Xe would not be readily incorporated into the body.

The chemical properties determine where a radioactive material may be concentrated in the body or how easily it may be excreted. The length of time of exposure and what is exposed to radiation significantly affects the health hazard. (See Exercise 64 for a specific example.)

ChemWork Problems

79. The most abundant isotope is generally the most stable isotope. The periodic table predicts that the most stable isotopes for exercises a-d are ^{39}K, ^{56}Fe, ^{23}Na, and ^{204}Tl. (Reference Table 19.2 of the text for potential decay processes.)

a. Unstable; ^{45}K has too many neutrons and will undergo beta-particle production.

b. Stable

c. Unstable; ^{20}Na has too few neutrons and will most likely undergo electron capture or positron production. Alpha-particle production makes too severe of a change to be a likely decay process for the relatively light ^{20}Na nuclei. Alpha-particle production usually occurs for heavy nuclei.

d. Unstable; ^{194}Tl has too few neutrons and will undergo electron capture, positron production, and/or alpha-particle production.

81. The equations for the nuclear reactions are:

$^{239}_{94}Pu \rightarrow ^{235}_{92}U + ^{4}_{2}He$; $^{214}_{82}Pb \rightarrow ^{214}_{83}Bi + ^{0}_{-1}e$; $^{60}_{27}Co \rightarrow ^{60}_{28}Ni + ^{0}_{-1}e$

$^{99}_{43}Tc \rightarrow ^{99}_{44}Ru + ^{0}_{-1}e$; $^{239}_{93}Np \rightarrow ^{239}_{94}Pu + ^{0}_{-1}e$

83. $^{249}_{97}\text{Bk} + ^{22}_{10}\text{Ne} \rightarrow ^{267}_{107}\text{Bh} + ?$; this equation is charge balanced, but it is not mass balanced. The products are off by 4 mass units. The only possibility to account for the 4 mass units is to have 4 neutrons produced. The balanced equation is:

$$^{249}_{97}\text{Bk} + ^{22}_{10}\text{Ne} \rightarrow ^{267}_{107}\text{Bh} + 4\,^{1}_{0}\text{n}$$

$$\ln\left(\frac{N}{N_0}\right) = -kt = \frac{-(0.6931)t}{t_{1/2}},\quad \ln\left(\frac{11}{199}\right) = \frac{-(0.6931)t}{15.0\text{ s}},\quad t = 62.7\text{ s}$$

Bh: $[\text{Rn}]7s^2 5f^{14} 6d^5$ is the expected electron configuration.

85. $t_{1/2} = 5730$ yr; $k = (\ln 2)/t_{1/2}$; $\ln(N/N_0) = -kt$; $\ln\frac{11.2}{15.3} = \frac{-(\ln 2)t}{5730\text{ yr}}$, $t = 2580$ yr

The age of the icon is 2580 yr.

87. The third-life will be the time required for the number of nuclides to reach one-third of the original value ($N_0/3$).

$$\ln\left(\frac{N}{N_0}\right) = -kt = \frac{-(0.6931)t}{t_{1/2}},\quad \ln\left(\frac{1}{3}\right) = \frac{-(0.6931)t}{31.4\text{ yr}},\quad t = 49.8\text{ yr}$$

The third-life of this nuclide is 49.8 years.

89. $$\ln\left(\frac{N}{N_0}\right) = -kt = \frac{-(\ln 2)t}{12.3\text{ yr}},\quad \ln\left(\frac{0.17 \times N_0}{N_0}\right) = -(5.64 \times 10^{-2})t,\quad t = 31.4\text{ yr}$$

It takes 31.4 years for the tritium to decay to 17% of the original amount. Hence the watch stopped fluorescing enough to be read in 1975 (1944 + 31.4).

91. Iron-56 has the largest mass defect per nucleon (see Fig 19.9). The mass defect per nucleon decreases from ^{56}Fe as nuclei get lighter and get heavier. Of the nuclei listed, ^{34}S is the nuclei with a mass closest to ^{56}Fe, so it will have the largest mass defect per nucleon. Answer e is correct.

93. $^{58}_{26}\text{Fe} + 2\,^{1}_{0}\text{n} \rightarrow ^{60}_{27}\text{Co} + ?$; in order to balance the equation, the missing particle has no mass and a charge of 1−; this is an electron.

An atom of $^{60}_{27}\text{Co}$ has 27 e, 27 p, and 33 n. The mass defect of the ^{60}Co nucleus is:

$$\Delta m = (59.9338 - 27m_e) - [27(1.00782 - m_e) + 33(1.00866)] = -0.5631\text{ u}$$

$$\Delta E = \Delta mc^2 = -0.5631\text{ u} \times \frac{1.6605 \times 10^{-27}\text{ kg}}{\text{u}} \times (2.9979 \times 10^8\text{ m/s})^2 = -8.403 \times 10^{-11}\text{ J}$$

$$\frac{\text{Binding energy}}{\text{Nucleon}} = \frac{8.403 \times 10^{-11}\text{ J}}{60\text{ nucleons}} = 1.401 \times 10^{-12}\text{ J/nucleon}$$

The emitted particle was an electron, which has a mass of 9.109×10^{-31} kg. The deBroglie wavelength is:

$$\lambda = \frac{h}{mv} = \frac{6.626 \times 10^{-34}\ \text{J s}}{9.109 \times 10^{-31}\ \text{kg} \times (0.90 \times 2.998 \times 10^{8}\ \text{m/s})} = 2.7 \times 10^{-12}\ \text{m}$$

95. $$20{,}000\ \text{ton TNT} \times \frac{4 \times 10^{9}\ \text{J}}{\text{ton TNT}} \times \frac{1\ \text{mol}\ ^{235}\text{U}}{2 \times 10^{13}\ \text{J}} \times \frac{235\ \text{g}\ ^{235}\text{U}}{\text{mol}\ ^{235}\text{U}} = 940\ \text{g}\ ^{235}\text{U} \approx 900\ \text{g}\ ^{235}\text{U}$$

This assumes that all of the ^{235}U undergoes fission.

97. Mass of nucleus = atomic mass – mass of electron = 2.01410 u – 0.000549 u = 2.01355 u

$$u_{rms} = \left(\frac{3\,RT}{M}\right)^{1/2} = \left(\frac{3(8.3145\ \text{J/K} \bullet \text{mol})(4 \times 10^{7}\ \text{K})}{2.01355\ \text{g}(1\ \text{kg}/1000\ \text{g})}\right)^{1/2} = 7 \times 10^{5}\ \text{m/s}$$

$$KE_{avg} = \frac{1}{2}mu^2 = \frac{1}{2}\left(2.01355\ \text{u} \times \frac{1.66 \times 10^{-27}\ \text{kg}}{\text{u}}\right)(7 \times 10^{5}\ \text{m/s})^2 = 8 \times 10^{-16}\ \text{J/nuclei}$$

We could have used $KE_{ave} = (3/2)RT$ to determine the same average kinetic energy.

99. All evolved oxygen in O_2 comes from water and not from carbon dioxide.

Challenge Problems

101. $$k = \frac{\ln 2}{t_{1/2}};\quad \ln\left(\frac{N}{N_0}\right) = -kt = \frac{-(0.693)t}{t_{1/2}}$$

For ^{238}U: $$\ln\left(\frac{N}{N_0}\right) = \frac{-(0.693)(4.5 \times 10^{9}\ \text{yr})}{4.5 \times 10^{9}\ \text{yr}} = -0.693,\quad \frac{N}{N_0} = e^{-0.693} = 0.50$$

For ^{235}U: $$\ln\left(\frac{N}{N_0}\right) = \frac{-(0.693)(4.5 \times 10^{9}\ \text{yr})}{7.1 \times 10^{8}\ \text{yr}} = -4.39,\quad \frac{N}{N_0} = e^{-4.39} = 0.012$$

If we have a current sample of 10,000 uranium nuclei, 9928 nuclei of ^{238}U and 72 nuclei of ^{235}U are present. Now let's calculate the initial number of nuclei that must have been present 4.5×10^9 years ago to produce these 10,000 uranium nuclei.

For ^{238}U: $$\frac{N}{N_0} = 0.50,\ N_0 = \frac{N}{0.50} = \frac{9928\ \text{nuclei}}{0.50} = 2.0 \times 10^{4}\ ^{238}\text{U nuclei}$$

For ^{235}U: $$N_0 = \frac{N}{0.012} = \frac{72\ \text{nuclei}}{0.012} = 6.0 \times 10^{3}\ ^{235}\text{U nuclei}$$

So 4.5 billion years ago, the 10,000-nuclei sample of uranium was composed of 2.0×10^4 ^{238}U nuclei and 6.0×10^3 ^{235}U nuclei. The percent composition 4.5 billion years ago would have been:

$$\frac{2.0 \times 10^{4}\ ^{238}\text{U nuclei}}{(6.0 \times 10^{3} + 2.0 \times 10^{4})\ \text{total nuclei}} \times 100 = 77\%\ ^{238}\text{U and } 23\%\ ^{235}\text{U}$$

103. Assuming that the radionuclide is long-lived enough that no significant decay occurs during the time of the experiment, the total counts of radioactivity injected are:

$$0.10\text{ mL} \times \frac{5.0 \times 10^3\text{ cpm}}{\text{mL}} = 5.0 \times 10^2\text{ cpm}$$

Assuming that the total activity is uniformly distributed only in the rat's blood, the blood volume is:

$$V \times \frac{48\text{ cpm}}{\text{mL}} = 5.0 \times 10^2\text{ cpm},\ \ V = 10.4\text{ mL} = 10.\text{ mL}$$

105. a. ^{12}C; it takes part in the first step of the reaction but is regenerated in the last step. ^{12}C is not consumed, so it is not a reactant.

b. ^{13}N, ^{13}C, ^{14}N, ^{15}O, and ^{15}N are the intermediates.

c. $4\,^1_1H \rightarrow\ ^4_2He + 2\,^0_{+1}e$; $\Delta m = 4.00260\text{ u} - 2\text{ m}_e + 2\text{ m}_e - [4(1.00782\text{ u} - \text{m}_e)]$

$\Delta m = 4.00260 - 4(1.00782) + 4(0.000549) = -0.02648$ u for four protons reacting

For 4 mol of protons, $\Delta m = -0.02648$ g, and ΔE for the reaction is:

$$\Delta E = \Delta mc^2 = -2.648 \times 10^{-5}\text{ kg} \times (2.9979 \times 10^8\text{ m/s})^2 = -2.380 \times 10^{12}\text{ J}$$

For 1 mol of protons reacting: $\dfrac{-2.380 \times 10^{12}\text{ J}}{4\text{ mol }^1H} = -5.950 \times 10^{11}\text{ J/mol }^1H$

107. Moles of $I^- = \dfrac{33\text{ counts}}{\text{min}} \times \dfrac{1\text{ mol I} \bullet \text{min}}{5.0 \times 10^{11}\text{ counts}} = 6.6 \times 10^{-11}\text{ mol I}^-$

$$[I^-] = \frac{6.6 \times 10^{-11}\text{ mol I}^-}{0.150\text{ L}} = 4.4 \times 10^{-10}\text{ mol/L}$$

	$Hg_2I_2(s)$	$\rightleftharpoons$	$Hg_2^{2+}(aq)$	+	$2\ I^-(aq)$	$K_{sp} = [Hg_2^{2+}][I^-]^2$
Initial	s = solubility (mol/L)		0		0	
Equil.			s		$2s$	

From the problem, $2s = 4.4 \times 10^{-10}$ mol/L, $s = 2.2 \times 10^{-10}$ mol/L.

$K_{sp} = (s)(2s)^2 = (2.2 \times 10^{-10})(4.4 \times 10^{-10})^2 = 4.3 \times 10^{-29}$

CHAPTER 20

THE REPRESENTATIVE ELEMENTS

Questions

3. The gravity of the earth is not strong enough to keep the light H_2 molecules in the atmosphere.

5. Calcium is found in the structural material that make up bones and teeth. Magnesium plays a vital role in metabolism and in muscle function.

7. For Groups 1A-3A, the small sizes of H (as compared to Li), Be (as compared to Mg), and B (as compared to Al) seem to be the reason why these elements have nonmetallic properties, while others in the Groups 1A-3A are strictly metallic. The small sizes of H, Be, and B also cause these species to polarize the electron cloud in nonmetals, thus forcing a sharing of electrons when bonding occurs. For Groups 4A-6A, a major difference between the first and second members of a group is the ability to form π bonds. The smaller elements form stable π bonds, while the larger elements are not capable of good overlap between parallel p orbitals and, in turn, do not form strong π bonds. For Group 7A, the small size of F as compared to Cl is used to explain the low electron affinity of F and the weakness of the F–F bond.

9. Solids have stronger intermolecular forces than liquids. To maximize the hydrogen bonding in the solid phase, ice is forced into an open structure. This open structure is why $H_2O(s)$ is less dense than $H_2O(l)$.

11. Group 1A and 2A metals are all easily oxidized. They must be produced in the absence of materials (H_2O, O_2) that are capable of oxidizing them.

13. The reaction $N_2(g) + 3H_2(g) \rightleftharpoons 2NH_3(g)$ is exothermic. Thus, the value of the equilibrium constant K decreases as the temperature increases. Lower temperatures are favored for maximum yield of ammonia. However, at lower temperatures the rate is slow; without a catalyst the rate is too slow for the process to be feasible. The discovery of a catalyst increased the rate of reaction at a lower temperature favored by thermodynamics.

15. Boranes are covalent compounds made up of boron and hydrogen. The B–H bonds in boranes are relatively weak and are very reactive. When boranes are reacted with oxygen, the product bonds are much stronger than the reactant bonds. When this is the case, very exothermic reactions result which is a common trait for good rocket fuels.

17. Atomic size decreases from left to right across the periodic table. This means the alkali metals have the largest radius. The electronegativity trend is to increase from left to right across the periodic table. The noble gases are the group furthest to the right in the periodic table. However, electronegativity is a bonding term. Since noble gases are generally unreactive, they don't have electronegativity values. The group having the highest electronegativity values is the halogens.

Exercises

Group 1A Elements

19. a. $\Delta H° = -110.5 - [-75 + (-242)] = 207$ kJ; $\Delta S° = 198 + 3(131) - [186 + 189] = 216$ J/K

b. $\Delta G° = \Delta H° - T\Delta S°$; $\Delta G° = 0$ when $T = \dfrac{\Delta H^{o}}{\Delta S^{o}} = \dfrac{207 \times 10^{3}\ J}{216\ J/K} = 958\ K$

At T > 958 K and standard pressures, the favorable ΔS° term dominates, and the reaction is spontaneous (ΔG° < 0).

21. The atomic radius of H is much smaller than the atomic radius of Li. The small H atom has a much greater attraction for electrons than Li. This attraction for electrons results in H acting as a nonmetal when it forms compounds. Li acts as a typical metal when forming compounds; Li loses an electron to form 1+ charged cations in its ionic compounds.

23. $4\ Li(s) + O_2(g) \rightarrow 2\ Li_2O(s)$

$2\ Li(s) + S(s) \rightarrow Li_2S(s)$; $2\ Li(s) + Cl_2(g) \rightarrow 2\ LiCl(s)$

$12\ Li(s) + P_4(s) \rightarrow 4\ Li_3P(s)$; $2\ Li(s) + H_2(g) \rightarrow 2\ LiH(s)$

$2\ Li(s) + 2\ H_2O(l) \rightarrow 2\ LiOH(aq) + H_2(g)$; $2\ Li(s) + 2\ HCl(aq) \rightarrow 2\ LiCl(aq) + H_2(g)$

25. When lithium reacts with excess oxygen, Li_2O forms, which is composed of Li^+ and O^{2-} ions. This is called an oxide salt. When sodium reacts with oxygen, Na_2O_2 forms, which is composed of Na^+ and O_2^{2-}ions. This is called a peroxide salt. When potassium (or rubidium or cesium) reacts with oxygen, KO_2 forms, which is composed of K^+ and O_2^- ions. For your information, this is called a superoxide salt. So the three types of alkali metal oxides that can form differ in the oxygen anion part of the formula (O^{2-} versus O_2^{2-} versus O_2^-). Each of these anions has unique bonding arrangements and oxidation states.

27. The small size of the Li^+ cation results in a much greater attraction to water. The attraction to water is not so great for the other alkali metal ions. Thus, lithium salts tend to absorb water.

Group 2A Elements

29. $CaCO_3(s) + H_2SO_4(aq) \rightarrow CaSO_4(aq) + H_2O(l) + CO_2(g)$

31. $Mg^{2+} + 2\ e^- \rightarrow Mg$

$$1.00 \times 10^6\ g\ Mg \times \frac{1\ mol\ Mg}{24.31\ g} \times \frac{2\ mol\ e^-}{mol\ Mg} \times \frac{96,465\ C}{mol\ e^-} \times \frac{1\ s}{5.00 \times 10^4\ C} \times \frac{1\ h}{3600\ s} = 44.1\ hours$$

33. Beryllium has a small size and a large electronegativity as compared to the other alkaline earth metals. The electronegativity of Be is so high that it does not readily give up electrons to nonmetals, as is the case for the other alkaline earth metals. Instead, Be has significant covalent character in its bonds; it prefers to share valence electrons rather than give them up to form ionic bonds.

35.

	$CaCO_3(s)$	$\rightleftharpoons$	$Ca^{2+}(aq)$	+	$CO_3^{2-}(aq)$
Initial	s = solubility (mol/L)		0		0
Equil.			s		s

$K_{sp} = 8.7 \times 10^{-9} = [Ca^{2+}][CO_3^{2-}] = s^2$, $s = 9.3 \times 10^{-5}$ mol/L

Group 3A Elements

37. Nh: $[Rn]7s^25f^{14}6d^{10}7p^1$; Nh falls below Tl in the periodic table. Like Tl, we would expect Nh to form +1 and +3 oxidation states in its compounds.

39. $B_2H_6(g) + 3\ O_2(g) \rightarrow 2\ B(OH)_3(s)$

41. $2\ Ga(s) + 3\ F_2(g) \rightarrow 2\ GaF_3(s)$; $4\ Ga(s) + 3\ O_2(g) \rightarrow 2\ Ga_2O_3(s)$

$2Ga(s) + 3\ S(s) \rightarrow Ga_2S_3(s)$; $2\ Ga(s) + 6\ HCl(aq) \rightarrow 2\ GaCl_3(aq) + 3\ H_2(g)$

43. An amphoteric substance is one that can behave as either an acid or as a base. Al_2O_3 dissolves in both acidic and basic solutions. The reactions are:

$$Al_2O_3(s) + 6\ H^+(aq) \rightarrow 2Al^{3+}(aq) + 3\ H_2O(l)$$

$$Al_2O_3(s) + 2\ OH^-(aq) + 3\ H_2O(l) \rightarrow 2\ Al(OH)_4^-(aq)$$

45. Compounds called boranes have three-centered bonds. Three-centered bonds occur when a single H atom forms bridging bonds between two boron atoms. The bonds have two electrons bonding all three atoms together. The bond is electron-deficient and makes boranes very reactive.

Group 4A Elements

47. Compounds containing Si–Si single and multiple bonds are rare, unlike compounds of carbon. The bond strengths of the Si–Si and C–C single bonds are similar. The difference in bonding properties must be for other reasons. One reason is that silicon does not form strong π bonds, unlike carbon. Another reason is that silicon forms particularly strong sigma bonds to oxygen, resulting in compounds with Si–O bonds instead of Si–Si bonds.

49. 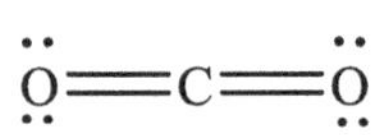

The darker green orbitals about carbon are sp hybrid orbitals. The lighter green orbitals about each oxygen are sp^2 hybrid orbitals, and the gold orbitals about all of the atoms are unhybridized p atomic orbitals. In each double bond in CO_2, one sigma and one π bond exists. The two carbon-oxygen sigma bonds are formed from overlap of sp hybrid orbitals from carbon with a sp^2 hybrid orbital from each oxygen. The two carbon-oxygen π bonds are formed from side-to-side overlap of the unhybridized p atomic orbitals from carbon with an unhybridized p atomic orbital from each oxygen. These two π bonds are oriented perpendicular to each other as illustrated in the figure.

51. a. $SiO_2(s) + 2\ C(s) \rightarrow Si(s) + 2\ CO(g)$

b. $SiCl_4(l) + 2\ Mg(s) \rightarrow Si(s) + 2\ MgCl_2(s)$

c. $Na_2SiF_6(s) + 4\ Na(s) \rightarrow Si(s) + 6\ NaF(s)$

53. Pb_3O_4: we assign −2 for the oxidation state of O. The sum of the oxidation states of Pb must be +8. We get this if two of the lead atoms are Pb(II) and one is Pb(IV). Therefore, the mole ratio of lead(II) to lead(IV) is 2 : 1.

55. The electronegativity of Pb^{4+} is larger than the electronegativity of Pb^{2+}. With the higher electronegativity, Pb^{4+} forms covalent compounds with nonmetals, unlike the ionic compounds formed by Pb^{2+}. $PbCl_4$: 4 + 4(7) = 32 valence electrons:

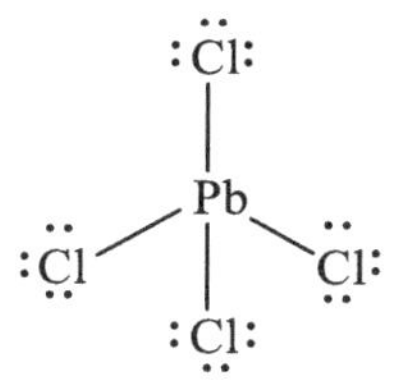

The central Pb atom exhibits a tetrahedral arrangement of the bonded atoms giving expected bond angles of 109.5°.

Group 5A Elements

57. NO_4^{3-}

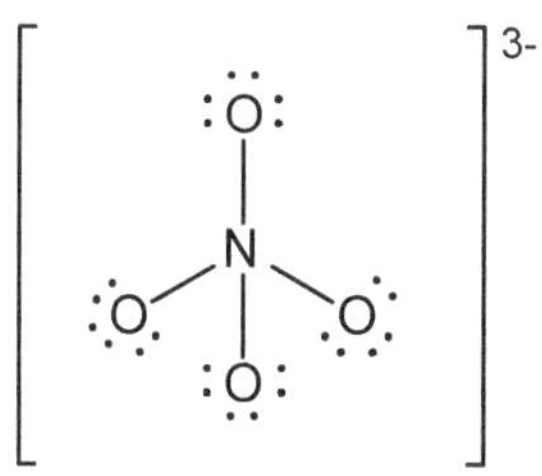

Both NO_4^{3-} and PO_4^{3-} have 32 valence electrons, so both have similar Lewis structures. From the Lewis structure for NO_4^{3-}, the central N atom has a tetrahedral arrangement of electron pairs. N is small. There is probably not enough room for all 4 oxygen atoms around N. P is larger; thus PO_4^{3-} is stable.

PO_3^-

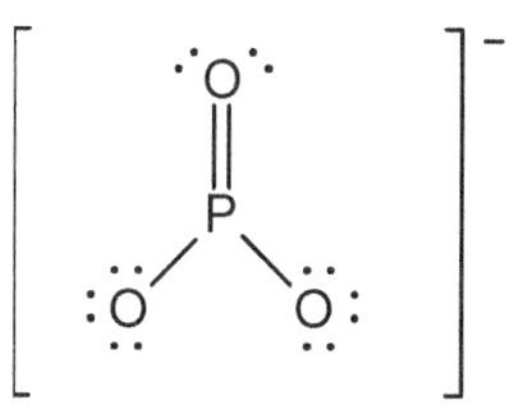

PO_3^- and NO_3^- each have 24 valence electrons, so both have similar Lewis structures. From the Lewis structure for PO_3^-, PO_3^- has a trigonal planar arrangement of electron pairs about the central P atom (two single bonds and one double bond). P=O bonds are not particularly stable, while N=O bonds are stable. Thus NO_3^- is stable.

59. Production of bismuth:

$$2\ Bi_2S_3(s) + 9\ O_2(g) \rightarrow 2\ Bi_2O_3(s) + 6\ SO_2(g);\ \ 2\ Bi_2O_3(s) + 3\ C(s) \rightarrow 4\ Bi(s) + 3\ CO_2(g)$$

Production of antimony:

$$2\ Sb_2S_3(s) + 9\ O_2(g) \rightarrow 2\ Sb_2O_3(s) + 6\ SO_2(g);\quad 2\ Sb_2O_3(s) + 3\ C(s) \rightarrow 4\ Sb(s) + 3\ CO_2(g)$$

61. NH_3, $5 + 3(1) = 8\ e^-$

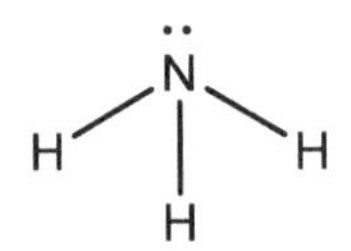

Trigonal pyramid; sp^3

$AsCl_5$, $5 + 5(7) = 40\ e^-$

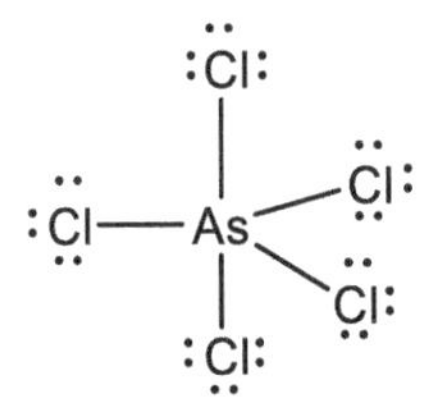

Trigonal bipyramid; dsp^3

PF_6^-, $5 + 6(7) + 1 = 48\ e^-$

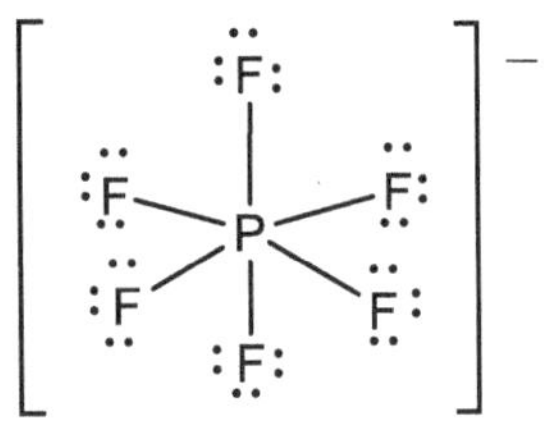

Octahedral; d^2sp^3

Nitrogen does not have low-energy d orbitals it can use to expand its octet. Both NF_5 and NCl_6^- would require nitrogen to have more than 8 valence electrons around it; this never happens.

63. Nitrogen can form strong pi bonds with itself. This is not the case for the other larger atoms in Group 5A which do not form strong pi bonds because of poor side-to-side overlap of the larger valence p atomic orbitals. The P_4, As_4, and Sb_4 molecules are composed entirely of single bonds.

65. $1/2\ N_2(g) + 1/2\ O_2(g) \rightarrow NO(g)$ $\Delta G° = \Delta G^\circ_{f,\ NO} = 87$ kJ/mol; by definition, ΔG°_f for a compound equals the free energy change that would accompany the formation of 1 mol of that compound from its elements in their standard states. NO (and some other oxides of nitrogen) have weaker bonds as compared to the triple bond of N_2 and the double bond of O_2. Because of this, NO (and some other oxides of nitrogen) have higher (positive) standard free energies of formation as compared to the relatively stable N_2 and O_2 molecules.

67. The pollution provides sources of nitrogen and phosphorus nutrients so that the algae can grow. The algae consume oxygen, which decrease the dissolved oxygen levels below that required for other aquatic life to survive, and fish die.

69. The acidic hydrogens in the oxyacids of phosphorus all are bonded to oxygen. The hydrogens bonded directly to phosphorus are not acidic. H_3PO_4 has three oxygen-bonded hydrogens, and it is a triprotic acid. H_3PO_3 has only two of the hydrogens bonded to oxygen, and it is a diprotic acid. The third oxyacid of phosphorus, H_3PO_2, has only one of the hydrogens bonded to an oxygen; it is a monoprotic acid.

Group 6A Elements

71. O=O–O $\rightarrow$ O=O + O

Break O–O bond: $\Delta H = \frac{146\text{ kJ}}{\text{mol}} \times \frac{1\text{ mol}}{6.022 \times 10^{23}} = 2.42 \times 10^{-22}\text{ kJ} = 2.42 \times 10^{-19}\text{ J}$

A photon of light must contain at least 2.42×10^{-19} J to break one O–O bond.

$$E_{photon} = \frac{hc}{\lambda},\ \lambda = \frac{(6.626 \times 10^{-34}\text{ J s})(2.998 \times 10^{8}\text{ m/s})}{2.42 \times 10^{-19}\text{ J}} = 8.21 \times 10^{-7}\text{ m} = 821\text{ nm}$$

73. $H_2SeO_4(aq) + 3\ SO_2(g) \rightarrow Se(s) + 3\ SO_3(g) + H_2O(l)$

75. In the upper atmosphere, O_3 acts as a filter for ultraviolet (UV) radiation:

$$O_3(g) \xrightarrow{h\nu} O_2(g) + O(g)$$

O_3 is also a powerful oxidizing agent. It irritates the lungs and eyes, and at high concentration, it is toxic. The smell of a "spring thunderstorm" is O_3 formed during lightning discharges. Toxic materials don't necessarily smell bad. For example, HCN smells like almonds.

77. $\Delta G = \Delta H - T\Delta S$; $S_{rhombic}(s) \rightarrow S_{monoclinic}(s)$; from the problem, rhombic sulfur converts to monoclinic sulfur only at high temperatures. For this to be true, the signs on ΔH and ΔS must both be positive. At high enough temperatures, the favorable ΔS term starts to dominate over the unfavorable ΔH term, and the process becomes spontaneous. Because the sign of ΔS is positive for the process $S_{rhombic}(s) \rightarrow S_{monoclinic}(s)$, $S_{monoclinic}$ has the more disordered structure (has the larger positional probability).

79. O_2: $(\sigma_{2s})^2(\sigma_{2s}{*})^2(\sigma_{2p})^2(\pi_{2p})^4(\pi_{2p}{*})^2$; the MO electron configuration of O_2 has two unpaired electrons in the degenerate π antibonding (π^*_{2p}) orbitals. A substance with unpaired electrons is paramagnetic (see Figure 9.39).

Group 7A Elements

81. O_2F_2 has 2(6) + 2(7) = 26 valence e^-; from the following Lewis structure, each oxygen atom has a tetrahedral arrangement of electron pairs. Therefore, bond angles are ≈109.5° and each O is sp^3 hybridized.

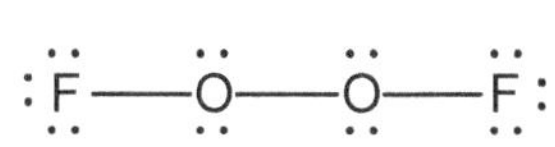

Formal charge	0	0	0	0
Oxidation state	–1	+1	+1	–1

Oxidation states are more useful. We are forced to assign +1 as the oxidation state for oxygen. Oxygen is very electronegative, and +1 is not a stable oxidation state for O.

83. SF_2, $6 + 2(7) = 20\ e^-$ SF_4, $6 + 4(7) = 34\ e^-$ SF_6, $6 + 6(7) = 48\ e^-$

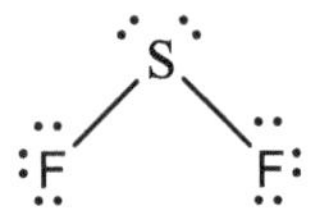

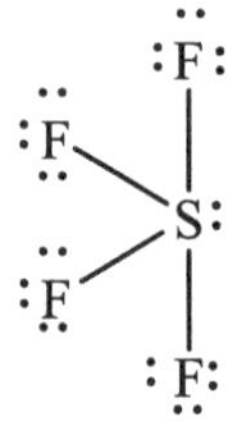

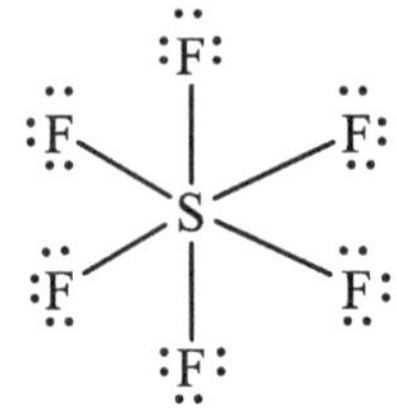

V-shaped; $<109.5°$ See-saw; ≈ 90, ≈ 120 Octahedral; 90°

OF_4 would have the same Lewis structure as SF_4. To form OF_4, the central oxygen atom must expand its octet. O is too small and doesn't have low-energy d orbitals available to expand its octet. Therefore, OF_4 would not be a stable compound.

85. Sn^{4+} has a relatively large electronegativity compared to Sn^{2+}. The compounds of Sn^{4+} have significant covalent character unlike the ionic compounds of Sn^{2+}. The ionic forces that hold ionic compounds together in the condensed phases are always much stronger than the covalent intermolecular forces that hold covalent compounds together. Hence the covalent $SnCl_4$ compound has a much smaller boiling point as compared to the ionic $SnCl_2$ compound.

87. The oxyacid strength increases as the number of oxygens in the formula increase. Therefore, the order of the oxyacids from weakest to strongest acid is $HOCl < HClO_2 < HClO_3 < HClO_4$.

Group 8A Elements

89. Helium is formed from the α-decay of radioactive elements. The α particles pick up electrons from the environment to form helium atoms.

91. Xe has one more valence electron than I. Thus, the isoelectric species will have I plus one extra electron substituted for Xe, giving a species with a net minus one charge.

a. IO_4^- b. IO_3^- c. IF_2^- d. IF_4^- e. IF_6^-

93. Helium is unreactive and doesn't combine with any other elements. It is a very light gas and would easily escape the earth's gravitational pull as the planet was formed.

95. One would expect RnF_2, RnF_4, and maybe RnF_6 to form in a fashion similar to XeF_2, XeF_4, and XeF_6.

ChemWork Problems

97. a. $\Delta H° = -110.5 - (-242) = 132$ kJ; $\Delta S° = 198 + 131 - [6 + 189] = 134$ J/K

b. $\Delta G° = \Delta H° - T\Delta S°$; $\Delta G° = 0$ when $T = \dfrac{\Delta H°}{\Delta S°} = \dfrac{132 \times 10^3\ J}{134\ J/K} = 985\ K$

At T > 985 K and standard pressures, the favorable ΔS° term dominates, and the reaction is spontaneous ($\Delta G° < 0$).

99.

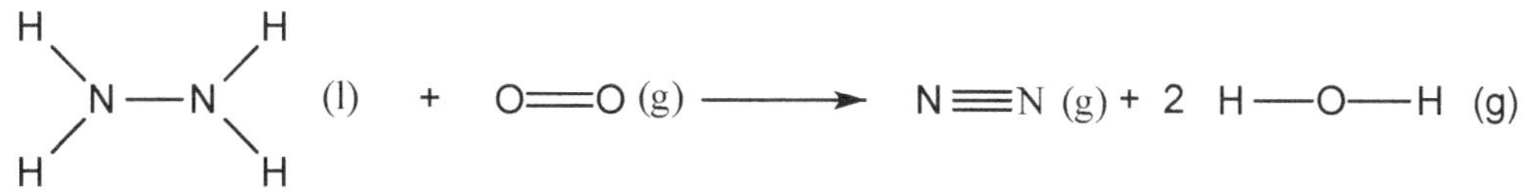

Bonds broken:	Bonds formed:
1 N–N (160. kJ/mol)	1 N≡N (941 kJ/mol)
4 N–H (391 kJ/mol)	2 × 2 O–H (467 kJ/mol)
1 O=O (495 kJ/mol)	

$\Delta H = 160. + 4(391) + 495 - [941 + 4(467)] = 2219 \text{ kJ} - 2809 \text{ kJ} = -590. \text{ kJ}$

101. Ga(I): $[Ar]4s^2 3d^{10}$, no unpaired e^-; Ga(III): $[Ar]3d^{10}$, no unpaired e^-

Ga(II): $[Ar]4s^1 3d^{10}$, 1 unpaired e^-; note that the s electrons are lost before the d electrons.

If the compound contained Ga(II), it would be paramagnetic, and if the compound contained Ga(I) and Ga(III), it would be diamagnetic. This can be determined easily by measuring the mass of a sample in the presence and in the absence of a magnetic field. Paramagnetic compounds will have an apparent increase in mass in a magnetic field.

103. $NaH(s) + H_2O(l) \rightarrow Na^+(aq) + OH^-(aq) + H_2(g)$; NaH is an ionic compound composed of Na^+ and H^- ions. Oxidation-reduction: The oxidation state of hydrogen is −1 in NaH, +1 in H_2O, and zero in H_2. Hydrogen is oxidized when it goes from NaH to H_2 (from $-1 \rightarrow 0$) and hydrogen is reduced when it goes from H_2O to H_2 (from $+1 \rightarrow 0$). In this reaction, an electron is transferred from the hydride ion to a hydrogen in water when forming H_2. Acid-base: A proton is transferred from an acid, H_2O, to a base, H^-, forming the conjugate base of water, OH^-, and the conjugate acid of H^-, H_2.

105. pH = 9.42, pOH = 14.00 − 9.42 = 4.58, $[OH^-] = 10^{-4.58} = 2.6 \times 10^{-5}\ M$

	$Mg(OH)_2(s)$	⇌	$Mg^{2+}(aq)$	+	$2\ OH^-(aq)$	
Initial	s = solubility (mol/L)		0		$2.6 \times 10^{-5}\ M$	(from buffer)
Equil.			s		$2.6 \times 10^{-5}\ M$	(pH constant)

$K_{sp} = [Mg^{2+}][OH^-]^2$, $8.9 \times 10^{-12} = s(2.6 \times 10^{-5})^2$, s = solubility = 0.013 mol/L

107. a. $AgCl(s) \xrightarrow{h\nu} Ag(s) + Cl$; the reactive chlorine atom is trapped in the crystal. When light is removed, Cl reacts with silver atoms to re-form AgCl; i.e., the reverse reaction occurs. In pure AgCl, the Cl atoms escape, making the reverse reaction impossible.

b. Over time, chlorine is lost, and the dark silver metal is permanent.

109. $Tl^{3+} + 2\,e^- \rightarrow Tl^+$ $\quad E° = 1.25$ V

$3\,I^- \rightarrow I_3^- + 2\,e^-$ $\quad -E° = -0.55$ V

$Tl^{3+} + 3\,I^- \rightarrow Tl^+ + I_3^-$ $\quad E^o_{cell} = 0.70$ V $\quad$ (Spontaneous because $E^o_{cell} > 0$.)

In solution, Tl^{3+} can oxidize I^- to I_3^-. Thus we expect TlI_3 to be thallium(I) triiodide.

111. a. SF_6, 48 valence electrons

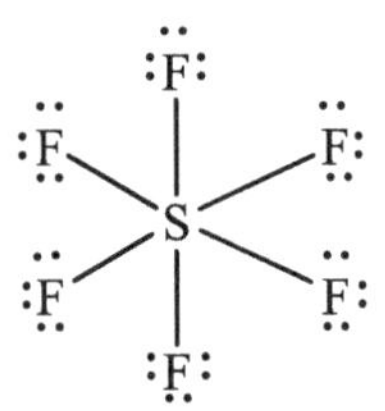

S is d^2sp^3 hybridized.

b. ClF_3, 28 valence electrons

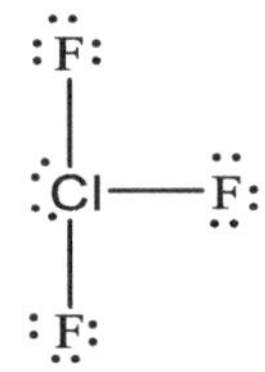

Cl is dsp^3 hybridized.

c. $GeCl_4$, 32 valence electrons

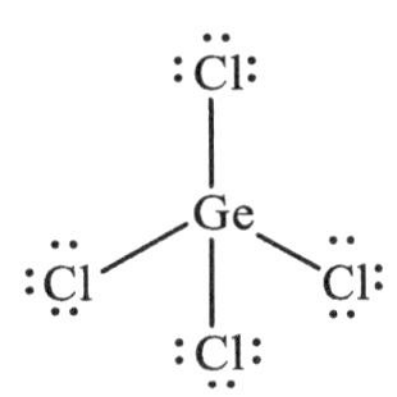

Ge is sp^3 hybridized.

d. XeF_4, 36 valence electrons

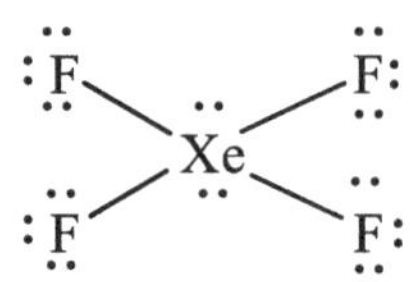

Xe is d^2sp^3 hybridized.

113. Strontium and calcium are both alkaline earth metals, so both have similar chemical properties. Because milk is a good source of calcium, strontium could replace some calcium in milk without much difficulty.

115. +6 oxidation state: SO_4^{2-}, SO_3, SF_6

+4 oxidation state: SO_3^{2-}, SO_2, SF_4

+2 oxidation state: SCl_2

0 oxidation state: S_8 and all other elemental forms of sulfur

–2 oxidation state: H_2S, Na_2S

117. a. $-307 \text{ kJ} = (-1136 + x) - [(-254 \text{ kJ}) + 3(-96 \text{ kJ})]$, $x = \Delta H^o_{f,\,NI_3} = 287$ kJ/mol

b. IF_2^+, $7 + 2(7) - 1 = 20\ e^-$

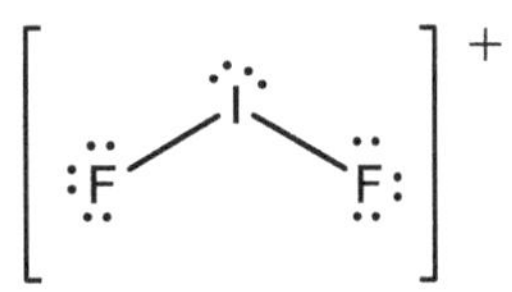

V-shaped; sp^3

BF_4^-, $3 + 4(7) + 1 = 32\ e^-$

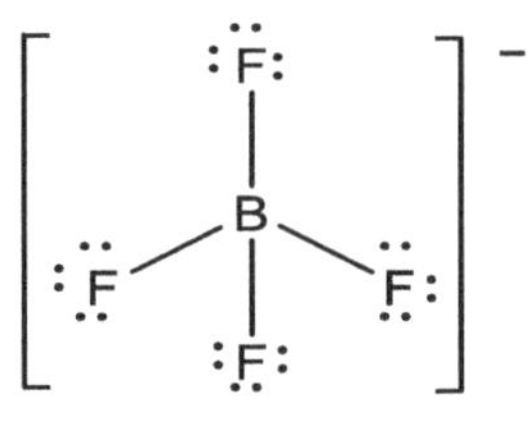

Tetrahedral; sp^3

Challenge Problems

119. The reaction is $X(s) + 2H_2O(l) \rightarrow H_2(g) + X(OH)_2(aq)$.

$$\text{Mol X} = \text{mol } H_2 = \frac{PV}{RT} = \frac{1.00 \text{ atm} \times 6.10 \text{ L}}{\frac{0.08206 \text{ L atm}}{\text{K mol}} \times 298 \text{ K}} = 0.249 \text{ mol}$$

$$\text{Molar mass X} = \frac{10.00 \text{ g X}}{0.249 \text{ mol X}} = 40.2 \text{ g/mol; X is Ca.}$$

$Ca(s) + 2\ H_2O(l) \rightarrow H_2(g) + Ca(OH)_2(aq)$; $Ca(OH)_2$ is a strong base.

$$[OH^-] = \frac{10.00 \text{ g Ca} \times \frac{1 \text{ mol Ca}}{40.08 \text{ g}} \times \frac{1 \text{ mol Ca(OH)}_2}{\text{mol Ca}} \times \frac{2 \text{ mol OH}^-}{\text{mol Ca(OH)}_2}}{10.0 \text{ L}} = 0.0499\ M$$

pOH = –log(0.0499) = 1.302, pH = 14.000 – 1.302 = 12.698

121. $PbX_4 \rightarrow PbX_2 + X_2$; from the equation, mol PbX_4 = mol PbX_2. Let x = molar mass of the halogen. Setting up an equation where mol PbX_4 = mol PbX_2:

$$\frac{25.00 \text{ g}}{207.2 + 4x} = \frac{16.12 \text{ g}}{207.2 + 2x}; \text{ solving, } x = 127.1; \text{ the halogen is iodine, I.}$$

123. For the reaction:

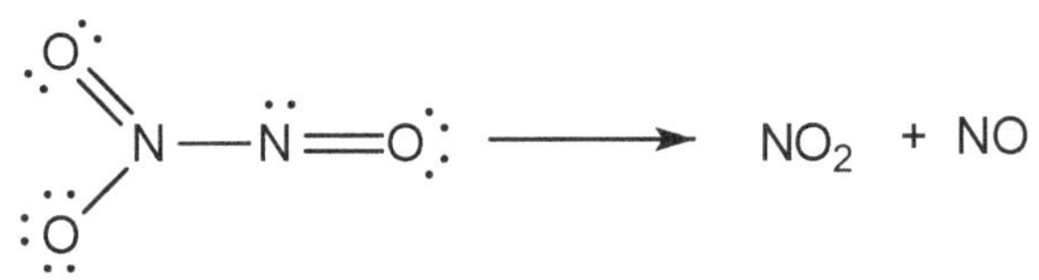

the activation energy must in some way involve breaking a nitrogen-nitrogen single bond.

For the reaction:

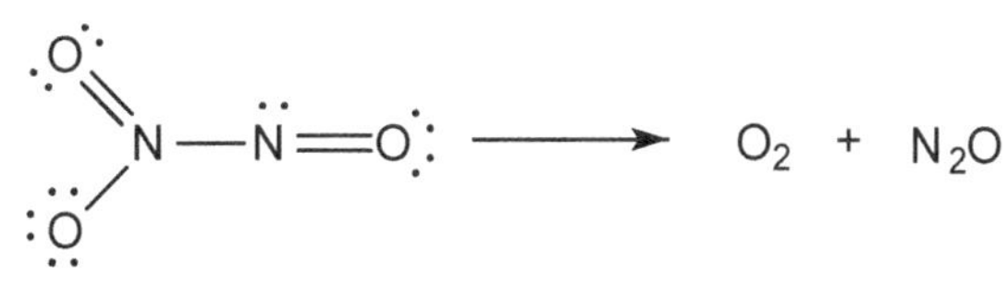

at some point nitrogen-oxygen bonds must be broken. N–N single bonds (160. kJ/mol) are weaker than N–O single bonds (201 kJ/mol). In addition, resonance structures indicate that there is more double-bond character in the N–O bonds than in the N–N bond. Thus, NO_2 and NO are preferred by kinetics because of the lower activation energy.

125. $NH_3 + NH_3 \rightleftharpoons NH_4^+ + NH_2^-$ $\quad K = [NH_4^+][NH_2^-] = 1.8 \times 10^{-12}$

NH_3 is the solvent, so it is not included in the K expression. In a neutral solution of ammonia:

$[NH_4^+] = [NH_2^-]$; $1.8 \times 10^{-12} = [NH_4^+]^2$, $[NH_4^+] = 1.3 \times 10^{-6}\ M = [NH_2^-]$

We could abbreviate this autoionization as: $NH_3 \rightleftharpoons H^+ + NH_2^-$, where $[H^+] = [NH_4^+]$.

This abbreviation is synonymous with the abbreviation used for the autoionization of water ($H_2O \rightleftharpoons H^+ + OH^-$). So $pH = pNH_4^+ = -\log(1.3 \times 10^{-6}) = 5.89$.

127. Let n_{SO_2} = initial moles SO_2 present. The reaction is summarized in the following table (O_2 is in excess).

	$2\ SO_2$	+ $O_2(g)$	→ $2\ SO_3(g)$
Initial	n_{SO_2}	2.00 mol	0
Change	$-n_{SO_2}$	$-n_{SO_2}/2$	$+n_{SO_2}$
Final	0	$2.00 - n_{SO_2}/2$	n_{SO_2}

Density = d = mass/volume; let d_i = initial density of gas mixture and d_f = final density of gas mixture after reaction. Because mass is conserved in a chemical reaction, $mass_i = mass_f$.

$$\frac{d_f}{d_i} = \frac{mass_f/V_f}{mass_i/V_i} = \frac{V_i}{V_f}$$

At constant P and T, $V \propto n$, so $\frac{d_f}{d_i} = \frac{V_i}{V_f} = \frac{n_i}{n_f}$; setting up an equation:

$$\frac{d_f}{d_i} = \frac{0.8471\ g/L}{0.8000\ g/L} = 1.059,\quad 1.059 = \frac{n_i}{n_f} = \frac{n_{SO_2} + 2.00}{(2.00 - n_{SO_2}/2) + n_{SO_2}} = \frac{n_{SO_2} + 2.00}{2.00 + n_{SO_2}/2}$$

Solving: $n_{SO_2} = 0.25$ mol; so, 0.25 moles of SO_3 is formed.

$$0.25\ mol\ SO_3 \times \frac{80.07\ g\ SO_3}{mol} = 20.\ g\ SO_3$$

129. Table 20.2 lists the mass percents of various elements in the human body. If we consider the mass percents through sulfur, that will cover 99.5% of the body mass, which is fine for a reasonable estimate. 150 lb × 454 g/lb = 68,000 g. We will carry an extra significant figure in some of the following calculations.

Moles of O = 0.650 × 68,000 g × 1 mol O/16.00 g O = 2760 mol

Moles of C = 0.180 × 68,000 g × 1 mol C/12.01 g C = 1020 mol

Moles of H = 0.100 × 68,000 g × 1 mol H/1.008 g H = 6750 mol

Moles of N = 0.030 × 68,000 g × 1 mol N/14.01 g N = 150 mol

Moles of Ca = 0.014 × 68,000 g × 1 mol Ca/40.08 g Ca = 24 mol

Moles of P = 0.010 × 68,000 g × 1 mol P/30.97 g P = 22 mol

Moles of Mg = 0.0050 × 68,000 g × 1 mol Mg/24.31 g Mg = 14 mol

Moles of K = 0.0034 × 68,000 g × 1 mol K/39.10 g K = 5.9 mol

Moles of S = 0.0026 × 68,000 g × 1 mol S/32.07 g S = 5.5 mol

Total moles of elements in 150-lb body = 10,750 mol atoms

$$10{,}750 \text{ mol atoms} \times \frac{6.022 \times 10^{23} \text{ atoms}}{\text{mol atoms}} = 6.474 \times 10^{27} \text{ atoms} \approx 6.5 \times 10^{27} \text{ atoms}$$

131. a. Because the hydroxide ion has a 1− charge, Te has a +6 oxidation state.

b. Assuming Te is limiting:

$$(0.545 \text{ cm})^3 \times \frac{6.240 \text{ g}}{\text{cm}^3} \times \frac{1 \text{ mol Te}}{127.6} \times \frac{1 \text{ mol TeF}_6}{\text{mol Te}} = 7.92 \times 10^{-3} \text{ mol TeF}_6$$

Assuming F_2 is limiting:

$$\text{Mol F}_2 = n = \frac{PV}{RT} = \frac{1.06 \text{ atm} \times 2.34 \text{ L}}{0.08206 \text{ L atm/K} \bullet \text{mol} \times 298 \text{ K}} = 0.101 \text{ mol F}_2$$

$$0.101 \text{ mol F}_2 \times \frac{1 \text{ mol TeF}_6}{3 \text{ mol F}_2} = 3.37 \times 10^{-2} \text{ mol TeF}_6$$

Because Te produces the smaller amount of product, Te is limiting and 7.92×10^{-3} mol TeF_6 can be produced. From the first equation given in the question, the moles of TeF_6 reacted equals the moles of $Te(OH)_6$ produced. So 7.92×10^{-3} mol $Te(OH)_6$ can be produced.

$$[\text{Te(OH)}_6] = \frac{7.92 \times 10^{-3} \text{ mol Te(OH)}_6}{0.115 \text{ L}} = 6.89 \times 10^{-2} M$$

Because $K_{a_1} > K_{a_2}$, the amount of protons produced by the K_{a_2} reaction will be insignificant.

	$Te(OH)_6$	⇌	$Te(OH)_5O^-$	+	H^+	$K_{a_1} = 10^{-7.68} = 2.1 \times 10^{-8}$
Initial	0.0689 M		0		~0	
Equil.	$0.0689 - x$		x		x	

$$K_{a_1} = 2.1 \times 10^{-8} = \frac{x^2}{0.0689 - x} \approx \frac{x^2}{0.0689}, \quad x = [H^+] = 3.8 \times 10^{-5} M$$

pH = $-\log(3.8 \times 10^{-5}) = 4.42$; assumptions good.

CHAPTER 21

TRANSITION METALS AND COORDINATION CHEMISTRY

Questions

7. The lanthanide elements are located just before the 5d transition metals. The lanthanide contraction is the steady decrease in the atomic radii of the lanthanide elements when going from left to right across the periodic table. As a result of the lanthanide contraction, the sizes of the 4d and 5d elements are very similar. This leads to a greater similarity in the chemistry of the 4d and 5d elements in each vertical group.

9.

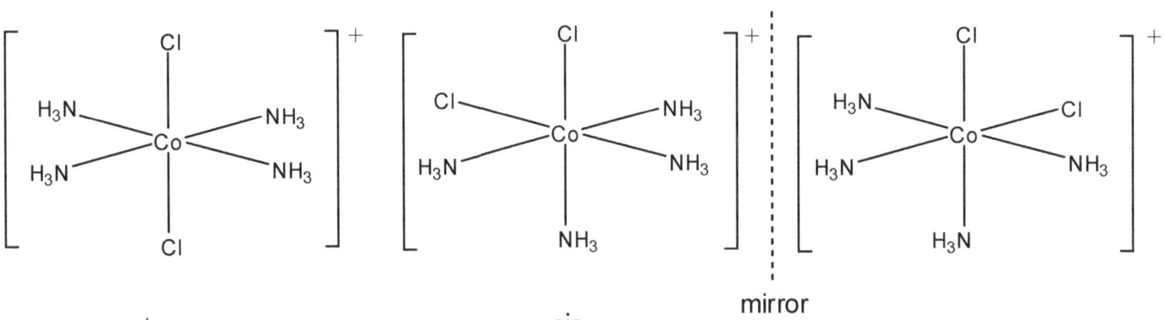

trans (mirror image is superimposable) | cis | The mirror image of the cis isomer is also superimposable.

No; both the trans and the cis forms of $Co(NH_3)_4Cl_2^+$ have mirror images that are superimposable. For the cis form, the mirror image only needs a 90° rotation to produce the original structure. Hence neither the trans nor cis form is optically active.

11. $Fe_2O_3(s) + 6\ H_2C_2O_4(aq) \rightarrow 2\ Fe(C_2O_4)_3^{3-}(aq) + 3\ H_2O(l) + 6\ H^+(aq)$; the oxalate anion forms a soluble complex ion with iron in rust (Fe_2O_3), which allows rust stains to be removed.

13. a. $CoCl_4^{2-}$; Co^{2+}: $4s^03d^7$; all tetrahedral complexes are a weak field (high spin).

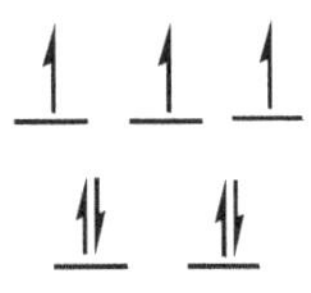

small Δ

$CoCl_4^{2-}$ is an example of a weak-field case having three unpaired electrons.

b. $Co(CN)_6^{3-}$: Co^{3+} : $4s^03d^6$; because CN^- is a strong-field ligand, $Co(CN)_6^{3-}$ will be a strong-field case (low-spin case).

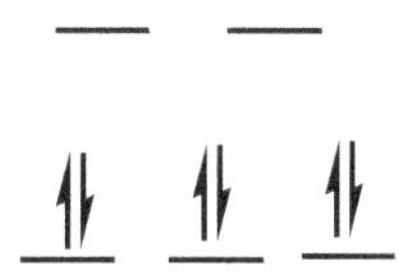

large Δ

CN^- is a strong-field ligand, so $Co(CN)_6^{3-}$ will be a low-spin case having zero unpaired electrons.

15. From Table 21.15, the red octahedral $Co(H_2O)_6^{2+}$ complex ion absorbs blue-green light ($\lambda \approx$ 490 nm), whereas the blue tetrahedral $CoCl_4^{2-}$ complex ion absorbs orange light ($\lambda \approx 600$ nm). Because tetrahedral complexes have a d-orbital splitting much less than octahedral complexes, one would expect the tetrahedral complex to have a smaller energy difference between split d orbitals. This translates into longer-wavelength light absorbed ($E = hc/\lambda$) for tetrahedral complex ions compared to octahedral complex ions. Information from Table 21.15 confirms this.

17. Linkage isomers differ in the way that the ligand bonds to the metal. SCN^- can bond through the sulfur or through the nitrogen atom. NO_2^- can bond through the nitrogen or through the oxygen atom. OCN^- can bond through the oxygen or through the nitrogen atom. N_3^-, $NH_2CH_2CH_2NH_2$, and I^- are not capable of linkage isomerism.

19. Sc^{3+} has no electrons in d orbitals. Ti^{3+} and V^{3+} have d electrons present. The color of transition metal complexes results from electron transfer between split d orbitals. If no d electrons are present, no electron transfer can occur, and the compounds are not colored.

21. The compound is composed of CO and Ni. From the name, the formula would be $Ni(CO)_4$. Since the CO ligands are neutral in charge, nickel has an oxidation state of zero in $Ni(CO)_4$.

23. At high altitudes, the oxygen content of air is lower, so less oxyhemoglobin is formed, which diminishes the transport of oxygen in the blood. A serious illness called high-altitude sickness can result from the decrease of O_2 in the blood. High-altitude acclimatization is the phenomenon that occurs with time in the human body in response to the lower amounts of oxyhemoglobin in the blood. This response is to produce more hemoglobin and hence, increase the oxyhemoglobin in the blood. High-altitude acclimatization takes several weeks to take hold for people moving from lower altitudes to higher altitudes.

25. Roasting is the process of converting sulfide minerals to oxides by heating in air at temperatures below their melting points. Smelting is the process of reducing metal cations to their free metals.

Exercises

Transition Metals and Coordination Compounds

27. a. Ni: $[Ar]4s^23d^8$

    b. Cd: $[Kr]5s^24d^{10}$

    c. Zr: $[Kr]5s^24d^2$

    d. Os: $[Xe]6s^24f^{14}5d^6$

29. Transition metal ions lose the s electrons before the d electrons.

a. Ti: $[Ar]4s^23d^2$	b. Re: $[Xe]6s^24f^{14}5d^5$	c. Ir: $[Xe]6s^24f^{14}5d^7$
Ti^{2+}: $[Ar]3d^2$	Re^{2+}: $[Xe]4f^{14}5d^5$	Ir^{2+}: $[Xe]4f^{14}5d^7$
Ti^{4+}: [Ar] or $[Ne]3s^23p^6$	Re^{3+}: $[Xe]4f^{14}5d^4$	Ir^{3+}: $[Xe]4f^{14}5d^6$

31. a. With K^+ and CN^- ions present, iron has a 3+ charge. Fe^{3+}: $[Ar]3d^5$

b. With a Cl^- ion and neutral NH_3 molecules present, silver has a 1+ charge. Ag^+: $[Kr]4d^{10}$

c. With Br^- ions and neutral H_2O molecules present, nickel has a 2+ charge. Ni^{2+}: $[Ar]3d^8$

d. With NO_2^- ions, an I^- ion, and neutral H_2O molecules present, chromium has a 3+ charge. Cr^{3+}: $[Ar]3d^3$

33. a. molybdenum(IV) sulfide; molybdenum(VI) oxide

b. MoS_2, +4; MoO_3, +6; $(NH_4)_2Mo_2O_7$, +6; $(NH_4)_6Mo_7O_{24}{\bullet}4H_2O$, +6

35. NH_3 is a weak base which produces OH^- ions in solution. The white precipitate is $Cu(OH)_2(s)$.

$$Cu^{2+}(aq) + 2\ OH^-(aq) \rightarrow Cu(OH)_2(s)$$

With excess NH_3 present, Cu^{2+} forms a soluble complex ion, $Cu(NH_3)_4^{2+}$.

$$Cu(OH)_2(s) + 4\ NH_3(aq) \rightarrow Cu(NH_3)_4^{2+}(aq) + 2\ OH^-(aq)$$

37. Only $[Cr(NH_3)_6]Cl_3$ will form a precipitate since only this compound will have Cl^- ions in solution. The Cl^- ions in the other compounds are ligands and are bound to the central Cr^{3+} ion. The Cl^- ions in $[Cr(NH_3)_6]Cl_3$ are counter ions needed to produce a neutral compound, while the NH_3 molecules are the ligands bound to Cr^{3+}.

39. To determine the oxidation state of the metal, you must know the charges of the various common ligands (see Table 21.12 of the text).

a. hexacyanomanganate(II) ion

b. cis-tetraamminedichlorocobalt(III) ion

c. pentaamminechlorocobalt(II) ion

41. a. hexaamminecobalt(II) chloride

b. hexaaquacobalt(III) iodide

c. potassium tetrachloroplatinate(II)

d. potassium hexachloroplatinate(II)

e. pentaamminechlorocobalt(III) chloride

f. triamminetrinitrocobalt(III)

43. a. $K_2[CoCl_4]$

b. $[Pt(H_2O)(CO)_3]Br_2$

c. $Na_3[Fe(CN)_2(C_2O_4)_2]$

d. $[Cr(NH_3)_3Cl(H_2NCH_2CH_2NH_2)]I_2$

45. a.

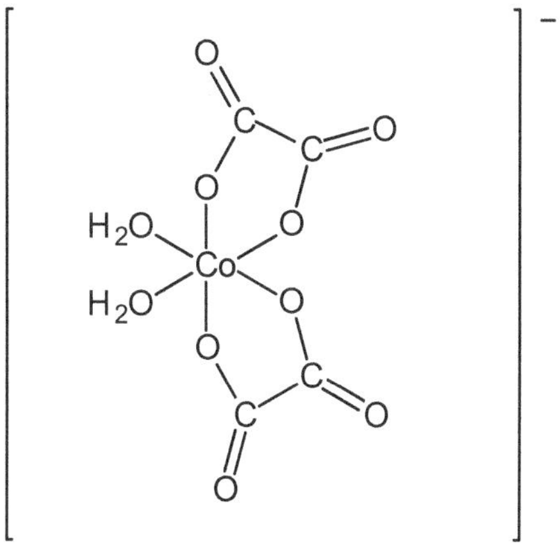

cis

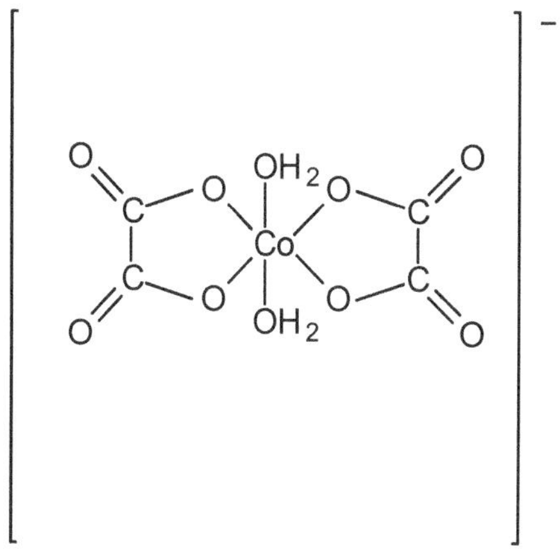

trans

Note: $C_2O_4^{2-}$ is a bidentate ligand. Bidentate ligands bond to the metal at two positions that are 90° apart from each other in octahedral complexes. Bidentate ligands do not bond to the metal at positions 180° apart.

b.

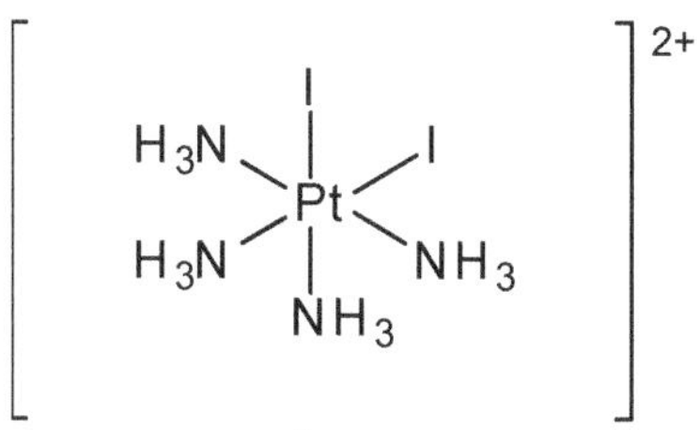

cis

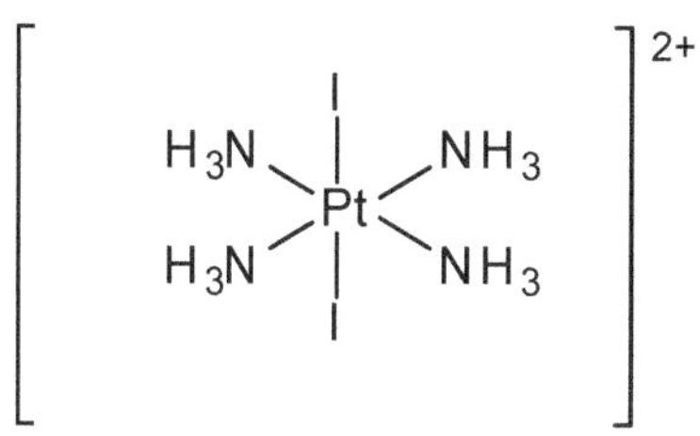

trans

c.

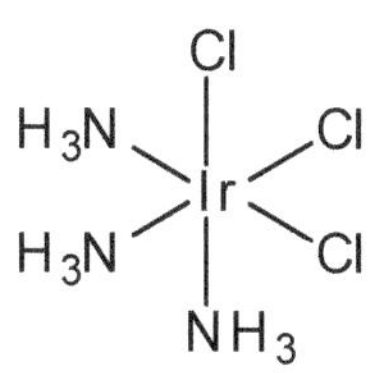

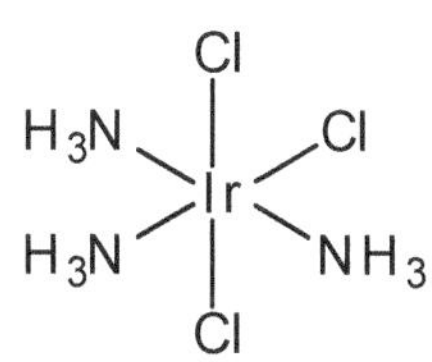

d.

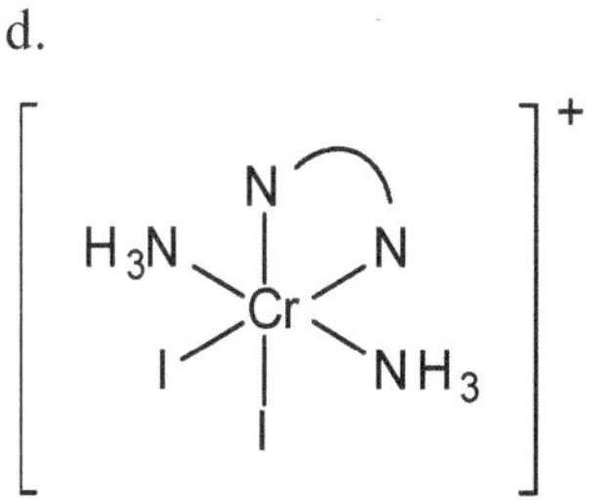

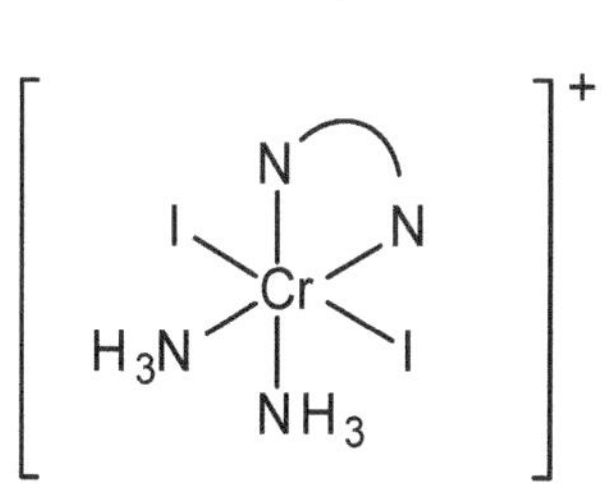

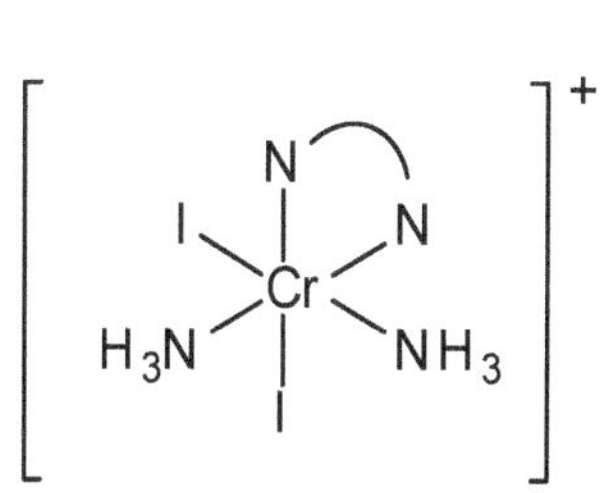

Note: en = N ⌒ N are abbreviations for the bidentate ligand ethylenediamine ($H_2NCH_2CH_2NH_2$).

47. monodentate

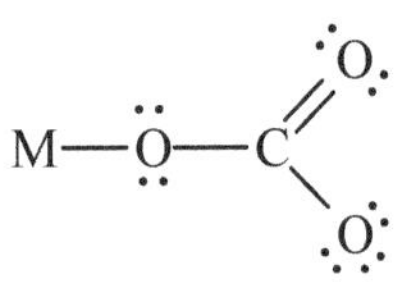

bidentate

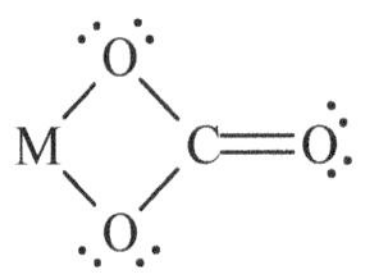

bridging

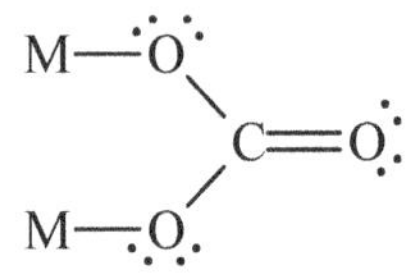

49. a. 2; forms bonds through the lone pairs on the two oxygen atoms.

b. 3; forms bonds through the lone pairs on the three nitrogen atoms.

c. 4; forms bonds through the two nitrogen atoms and the two oxygen atoms.

d. 4; forms bonds through the four nitrogen atoms.

51.

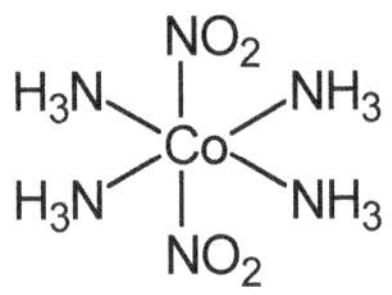

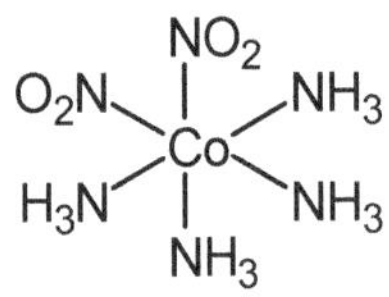

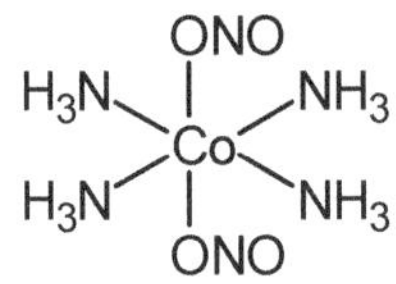

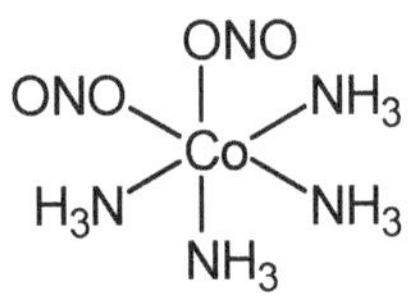

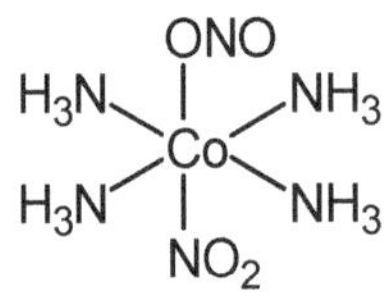

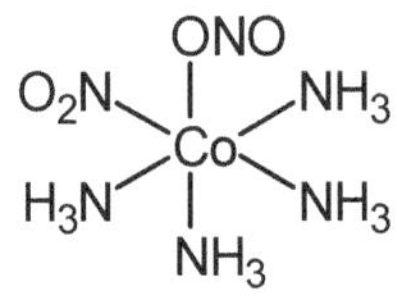

53. The formula for the complex ion is $Co(H_2O)_4F_2^+$ (one of the fluorides is a counter ion to the +1 complex ion). The compound can have cis and trans isomers, depending on the relative positions of the fluoride ligands in the complex ion. If the fluorides are 180° apart, this is the trans isomer. If the fluorides are 90° apart, this is the cis isomer. Both the cis and trans complex ions have superimposable mirror images, so they do not have optical isomers. Lastly, linkage isomerism is not possible with H_2O and F^- ligands.

55. Like the molecules discussed in Figures 21.16 and 21.17 of the text, $Cr(acac)_3$ and cis-$Cr(acac)_2(H_2O)_2$ are optically active. The mirror images of these two complexes are nonsuperimposable. There is a plane of symmetry in trans-$Cr(acac)_2(H_2O)_2$, so it is not optically active. A molecule with a plane of symmetry is never optically active because the mirror images are always superimposable. A plane of symmetry is a plane through a molecule where one side reflects the other side of the molecule.

Bonding, Color, and Magnetism in Coordination Compounds

57. NH_3 and H_2O are neutral charged ligands, while chloride and bromide are 1− charged ligands. The metal ions in the three compounds are Cr^{3+}: $[Ar]d^3$, Co^{3+}: $[Ar]d^6$, and Fe^{3+}: $[Ar]d^5$.

a. With five electrons each in a different orbital, this diagram is for the weak-field $[Fe(H_2O)_6]^{3+}$ complex ion.

b. With three electrons, this diagram is for the $[Cr(NH_3)_5Cl]^{2+}$ complex ion.

c. With six electrons all paired up, this diagram is for the strong-field $[Co(NH_3)_4Br_2]^+$ complex ion.

59. a. Fe^{2+}: $[Ar]3d^6$

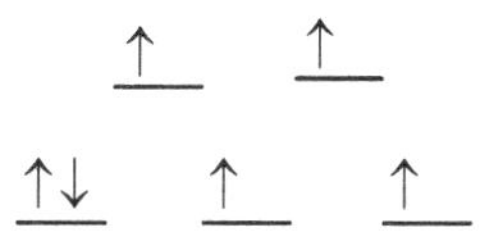

High spin, small Δ

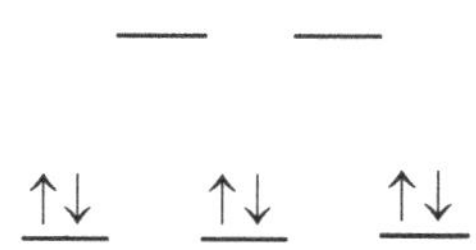

Low spin, large Δ

b. Fe^{3+}: $[Ar]3d^5$

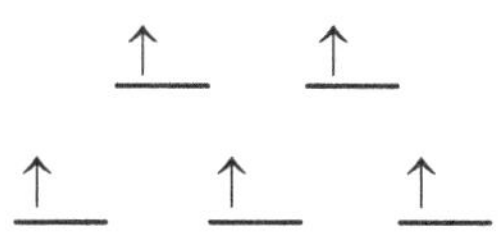

High spin, small Δ

c. Ni^{2+}: $[Ar]3d^8$

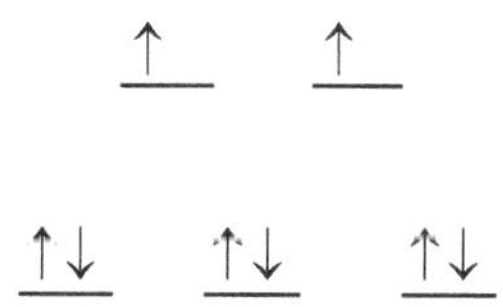

61. CN^- is a strong field ligand. The metal ions in the complexes are Ni^{2+}: $[Ar]3d^8$, Ti^{3+}: $[Ar]3d^1$, Fe^{3+}: $[Ar]3d^5$, Co^{3+}: $[Ar]3d^6$, and Cr^{3+}: $[Ar]3d^3$. A metal ion must have an even number of d electrons to have a possibility of being diamagnetic. So, we only need to consider $Co(CN)_6^{3-}$ which has a d^6 metal ion and $Ni(CN)_6^{4-}$ which has a d^8 metal ion.

Co^{3+}: $[Ar]3d^6$, no unpaired e^-

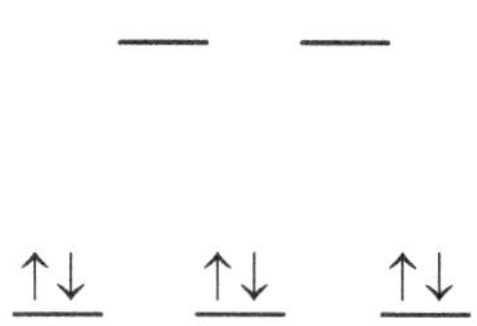

Low spin, large Δ

Ni^{2+}: $[Ar]3d^8$, 2 unpaired e^-

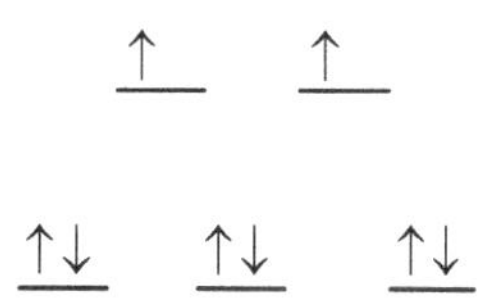

Only $Co(CN)_6^{3-}$ will be diamagnetic (answer d).

63. Because fluorine has a −1 charge as a ligand, chromium has a +2 oxidation state in CrF_6^{4-}. The electron configuration of Cr^{2+} is $[Ar]3d^4$. For four unpaired electrons, this must be a weak-field (high-spin) case where the splitting of the d-orbitals is small and the number of unpaired electrons is maximized. The crystal field diagram for this ion is:

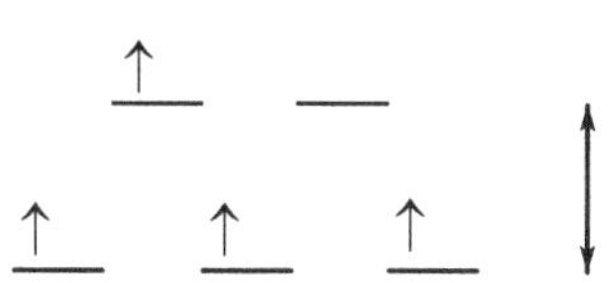

small Δ

65. To determine the crystal field diagrams, you need to determine the oxidation state of the transition metal, which can only be determined if you know the charges of the ligands (see Table 21.12). The electron configurations and the crystal field diagrams follow.

a. Ru^{2+}: $[Kr]4d^6$, no unpaired e^-

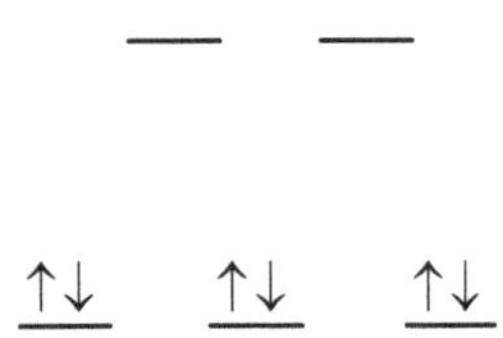

Low spin, large Δ

b. Ni^{2+}: $[Ar]3d^8$, 2 unpaired e^-

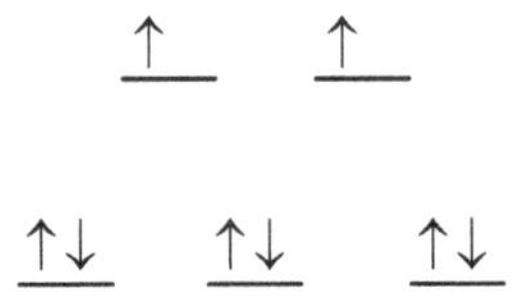

c. V^{3+}: $[Ar]3d^2$, 2 unpaired e^-

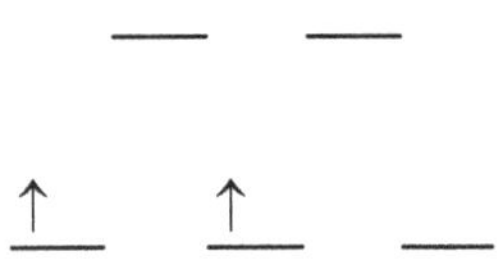

Note: Ni^{2+} must have 2 unpaired electrons, whether high-spin or low-spin, and V^{3+} must have 2 unpaired electrons, whether high-spin or low-spin.

67. Co: $[Ar]4s^2 3d^7$; cobalt is a d^6 is $Co(NH_3)_6^{3+}$ and a d^7 in $Co(NH_3)_6^{2+}$.

Co^{3+}: $[Ar]3d^6$
Strong field

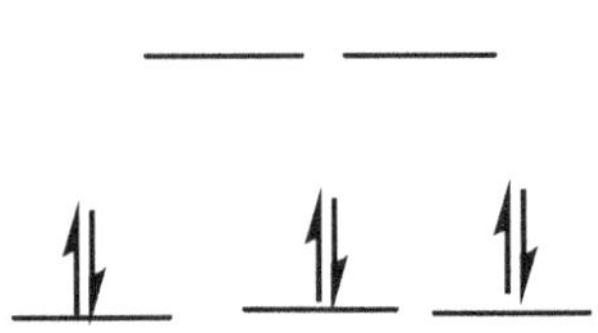

Co^{2+}: $[Ar]3d^7$
Weak field

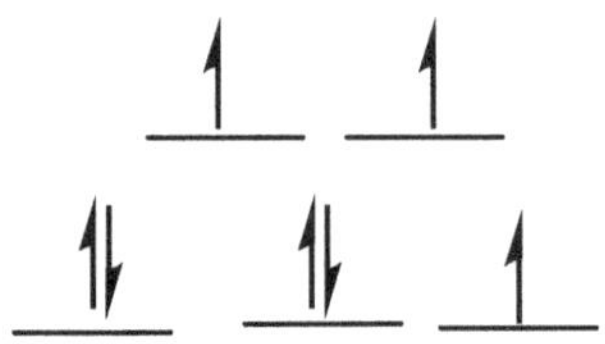

From the diagrams, $Co(NH_3)_6^{3+}$ has 0 unpaired electrons while $Co(NH_3)_6^{2+}$ has 3 unpaired electrons. The complex ion $Co(NH_3)_6^{2+}$ has 3 more unpaired electrons than $Co(NH_3)_6^{3+}$ (asnswer c).

69. All have octahedral Co^{3+} ions, so the difference in d orbital splitting and the wavelength of light absorbed only depends on the ligands. From the spectrochemical series, the order of the ligands from strongest to weakest field is $CN^- > en > H_2O > I^-$. The strongest-field ligand produces the greatest d-orbital splitting (largest Δ) and will absorb light having the smallest wavelength. The weakest-field ligand produces the smallest Δ and absorbs light having the longest wavelength. The order is:

$Co(CN)_6^{3-} < Co(en)_3^{3+} < Co(H_2O)_6^{3+} < CoI_6^{3-}$
shortest λ absorbed longest λ absorbed

71. From Table 21.15 of the text, the violet complex ion absorbs yellow-green light ($\lambda \approx 570$ nm), the yellow complex ion absorbs blue light ($\lambda \approx 450$ nm), and the green complex ion absorbs red light ($\lambda \approx 650$ nm). The spectrochemical series shows that NH_3 is a stronger-field ligand than H_2O, which is a stronger-field ligand than Cl^-. Therefore, $Cr(NH_3)_6^{3+}$ will have the largest d-orbital splitting and will absorb the lowest-wavelength electromagnetic radiation ($\lambda \approx 450$ nm) since energy and wavelength are inversely related ($\lambda = hc/E$). Thus the yellow solution contains the $Cr(NH_3)_6^{3+}$ complex ion. Similarly, we would expect the $Cr(H_2O)_4Cl_2^+$ complex ion to have the smallest d-orbital splitting since it contains the weakest-field ligands. The green solution with the longest wavelength of absorbed light contains the $Cr(H_2O)_4Cl_2^+$ complex ion. This leaves the violet solution, which contains the $Cr(H_2O)_6^{3+}$ complex ion. This makes sense because we would expect $Cr(H_2O)_6^{3+}$ to absorb light of a wavelength between that of $Cr(NH_3)_6^{3+}$ and $Cr(H_2O)_4Cl_2^+$.

73. $CoBr_6^{4-}$ has an octahedral structure, and $CoBr_4^{2-}$ has a tetrahedral structure (as do most Co^{2+} complexes with four ligands). Coordination complexes absorb electromagnetic radiation (EMR) of energy equal to the energy difference between the split d-orbitals. Because the tetrahedral d-orbital splitting is less than one-half the octahedral d-orbital splitting, tetrahedral complexes will absorb lower energy EMR, which corresponds to longer-wavelength EMR ($E = hc/\lambda$). Therefore, $CoBr_6^{2-}$ will absorb EMR having a wavelength shorter than 3.4×10^{-6} m.

75. Assuming the four ligands lie along the x and y-axis, the square planar crystal field diagram is:

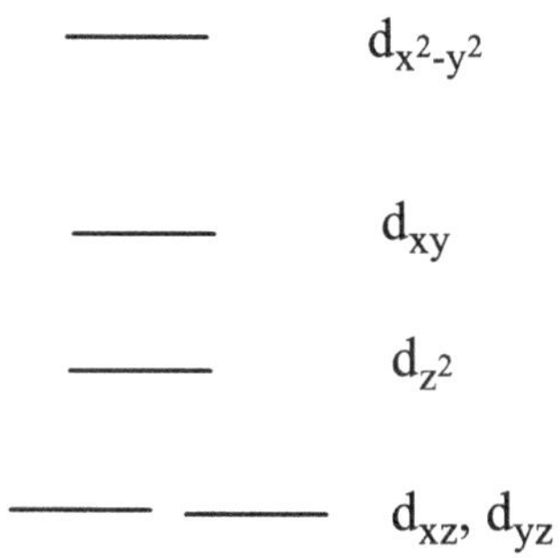

Answer b is true. The four ligands point directly at the four lobes of $d_{x^2-y^2}$ orbital, making it the highest energy orbital. Since the $d_{x^2-y^2}$ orbital is by far the largest energy orbital, d^8 complex ions are diamagnetic.

77. Because the ligands are Cl^-, iron is in the +3 oxidation state. Fe^{3+}: $[Ar]3d^5$

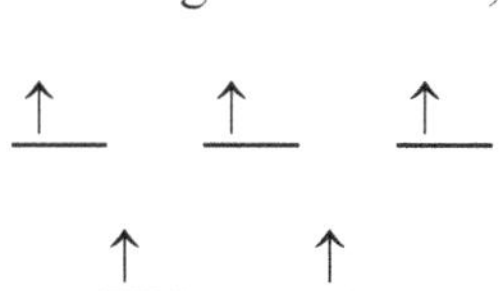

Because all tetrahedral complexes are high spin, there are 5 unpaired electrons in $FeCl_4^-$.

Metallurgy

79. a. To avoid fractions, let's first calculate ΔH for the reaction:

$$6\ FeO(s) + 6\ CO(g) \rightarrow 6\ Fe(s) + 6\ CO_2(g)$$

$6\ FeO + 2\ CO_2 \rightarrow 2\ Fe_3O_4 + 2\ CO$	$\Delta H° = -2(18\ kJ)$
$2\ Fe_3O_4 + CO_2 \rightarrow 3\ Fe_2O_3 + CO$	$\Delta H° = -(-39\ kJ)$
$3\ Fe_2O_3 + 9\ CO \rightarrow 6\ Fe + 9\ CO_2$	$\Delta H° = 3(-23\ kJ)$
$6\ FeO(s) + 6\ CO(g) \rightarrow 6\ Fe(s) + 6\ CO_2(g)$	$\Delta H° = -66\ kJ$

So for: $FeO(s) + CO(g) \rightarrow Fe(s) + CO_2(g)$ $\qquad \Delta H° = \frac{-66\ kJ}{6} = -11\ kJ$

b. $\Delta H° = 2(-110.5\ kJ) - (-393.5\ kJ + 0) = 172.5\ kJ$

$\Delta S° = 2(198\ J/K) - (214\ J/K + 6\ J/K) = 176\ J/K$

$$\Delta G° = \Delta H° - T\Delta S°,\ \Delta G° = 0 \text{ when } T = \frac{\Delta H°}{\Delta S°} = \frac{172.5\ kJ}{0.176\ kJ/K} = 980.\ K$$

Due to the favorable ΔS° term, this reaction is spontaneous at T > 980. K. From Figure 21.36 of the text, this reaction takes place in the blast furnace at temperatures greater than 980. K, as required by thermodynamics.

81. Fe_2O_3: iron has a +3 oxidation state; Fe_3O_4: iron has a +8/3 oxidation state. The three iron ions in Fe_3O_4 must have a total charge of +8. The only combination that works is to have two Fe^{3+} ions and one Fe^{2+} ion per formula unit. This makes sense from the other formula for magnetite, $FeO{\bullet}Fe_2O_3$. FeO has an Fe^{2+} ion, and Fe_2O_3 has two Fe^{3+} ions.

83. Review Section 4.11 for balancing reactions in basic solution by the half-reaction method.

$$(2\ CN^- + Ag \rightarrow Ag(CN)^{2-} + e^-) \times 4$$
$$4\ e^- + O_2 + 4\ H^+ \rightarrow 2\ H_2O$$
$$8\ CN^- + 4\ Ag + O_2 + 4\ H^+ \rightarrow 4\ Ag(CN)_2^- + 2\ H_2O$$

Adding 4 OH^- to both sides and canceling out 2 H_2O on both sides of the equation gives the balanced equation:

$$8\ CN^-(aq) + 4\ Ag(s) + O_2(g) + 2\ H_2O(l) \rightarrow 4\ Ag(CN)_2^-(aq) + 4\ OH^-(aq)$$

ChemWork Problems

85. $Hg^{2+}(aq) + 2\ I^-(aq) \rightarrow HgI_2(s)$, orange precipitate

$HgI_2(s) + 2\ I^-(aq) \rightarrow HgI_4^{2-}(aq)$, soluble complex ion

Hg^{2+} is a d^{10} ion. Color is the result of electron transfer between split d orbitals. This cannot occur for the filled d orbitals in Hg^{2+}. Therefore, we would not expect Hg^{2+} complex ions to form colored solutions.

87. $0.112 \text{ g } Eu_2O_3 \times \dfrac{304.0 \text{ g Eu}}{352.0 \text{ g } Eu_2O_3} = 0.0967 \text{ g Eu}$; mass % Eu = $\dfrac{0.0967 \text{ g}}{0.286 \text{ g}} \times 100 = 33.8\%$ Eu

Mass % O = 100.00 − (33.8 + 40.1 + 4.71) = 21.4% O

Assuming 100.00 g of compound:

$$33.8 \text{ g Eu} \times \frac{1 \text{ mol}}{152.0 \text{ g}} = 0.222 \text{ mol Eu};\ \ 40.1 \text{ g C} \times \frac{1 \text{ mol}}{12.01 \text{ g}} = 3.34 \text{ mol C}$$

$$4.71 \text{ g H} \times \frac{1 \text{ mol}}{1.008 \text{ g}} = 4.67 \text{ mol H};\ \ 21.4 \text{ g O} \times \frac{1 \text{ mol}}{16.00 \text{ g}} = 1.34 \text{ mol O}$$

$$\frac{3.34}{0.222} = 15.0,\ \ \frac{4.67}{0.222} = 21.0,\ \ \frac{1.34}{0.222} = 6.04$$

The molecular formula is $EuC_{15}H_{21}O_6$. Because each acac$^-$ ligand has a formula of $C_5H_7O_2^-$, an abbreviated molecular formula is $Eu(acac)_3$.

89. The five possible geometrical isomers for $MA_2B_2C_2$ are:

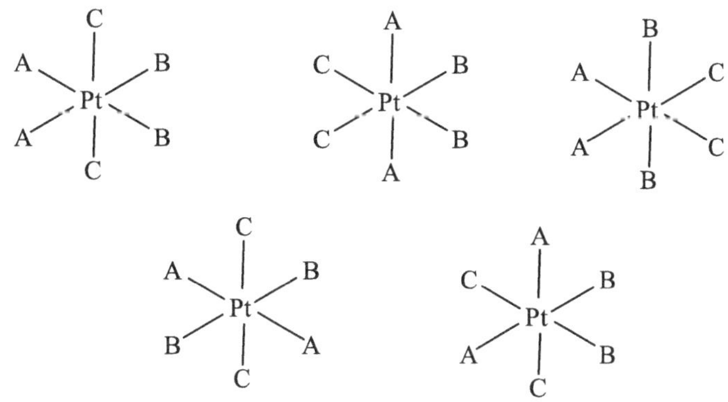

The first three isomers have 90° bond angles between two of the three different ligands and a 180° bond angle between the third ligand. The other two isomers have 90° bond angles between the various ligands or 180° bond angles between the various ligands.

91.

$(Au(CN)_2^- + e^- \rightarrow Au + 2\ CN^-) \times 2$	E° = −0.60 V
$Zn + 4\ CN^- \rightarrow Zn(CN)_4^{2-} + 2\ e^-$	−E° = 1.26 V
$2\ Au(CN)_2^-(aq) + Zn(s) \rightarrow 2\ Au(s) + Zn(CN)_4^{2-}(aq)$	$E^\circ_{cell} = 0.66$ V

$$\Delta G^\circ = -nFE^\circ_{cell} = -(2 \text{ mol } e^-)(96{,}485 \text{ C/mol } e^-)(0.66 \text{ J/C}) = -1.3 \times 10^5 \text{ J} = -130 \text{ kJ}$$

$$E^\circ = \frac{0.0591}{n}\log K,\ \ \log K = \frac{nE^\circ}{0.0591} = \frac{2(0.66)}{0.0591} = 22.34,\ \ K = 10^{22.34} = 2.2 \times 10^{22}$$

Note: We carried extra significant figures to determine K.

93. There are four geometrical isomers (labeled i-iv). Isomers iii and iv are optically active, and the nonsuperimposable mirror images are shown.

i.

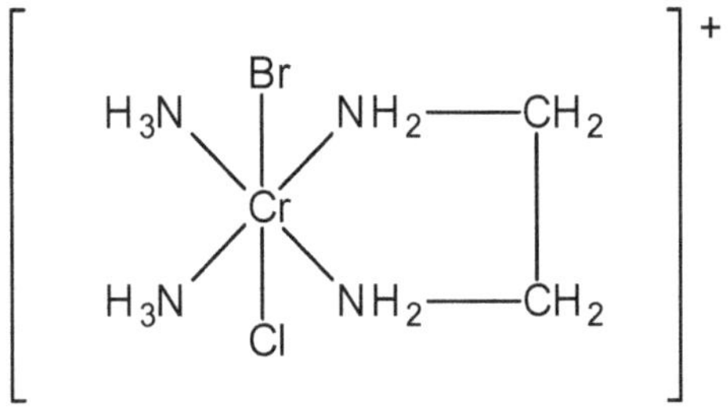

ii.

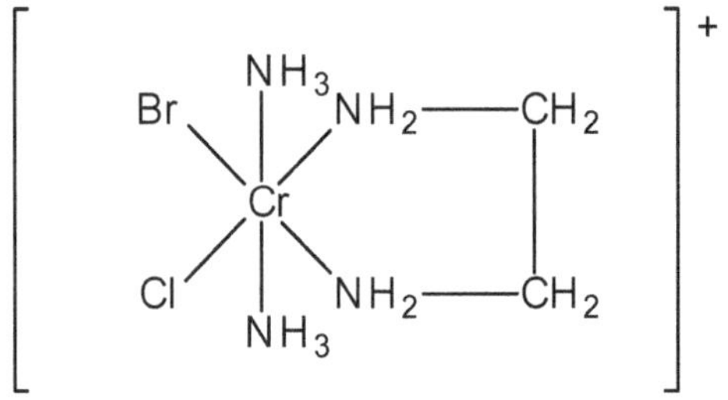

iii.

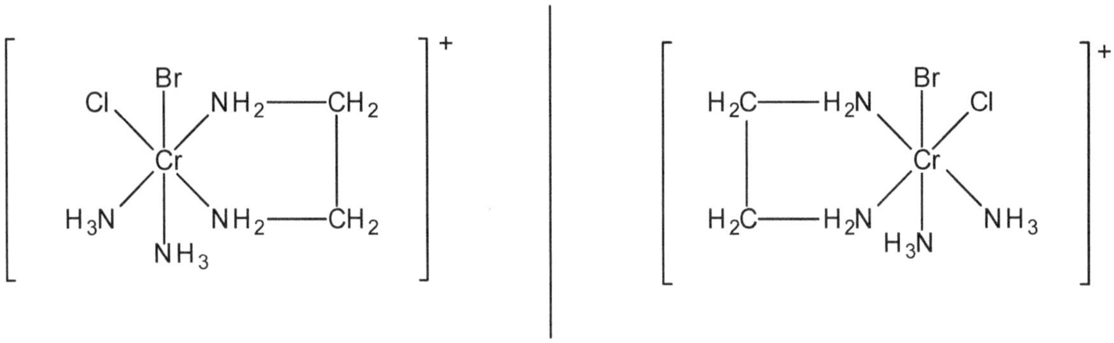

optically active　　mirror　　mirror image of iii (nonsuperimposable)

iv.

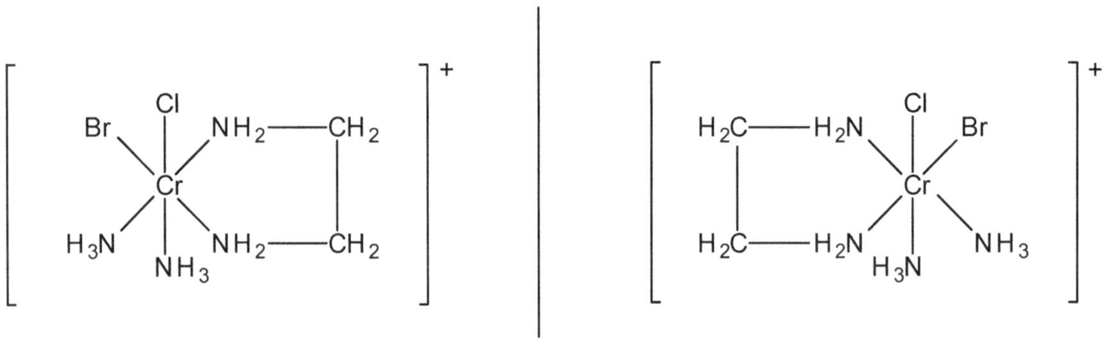

optically active　　mirror　　mirror image of iv (nonsuperimposable)

95. Ti^{2+}: $[Ar]3d^2$; Ti^{4+}: $[Ar]3d^0$; Cu^+: $[Ar]3d^{10}$; Cu^{2+}: $[Ar]3d^9$; Pd^{2+}: $[Ar]4d^8$; V^{2+}: $[Ar]3d^3$; Cr^{2+}: $[Ar]3d^4$; Cr^{3+}: $[Ar]3d^3$; consider the crystal field diagram for V^{2+} and Pd^{2+}:

V^{2+}: $[Ar]3d^3$

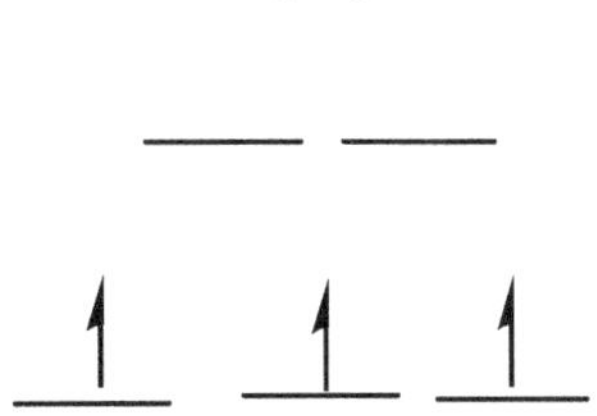

Pd^{2+}: $[Ar]3d^8$

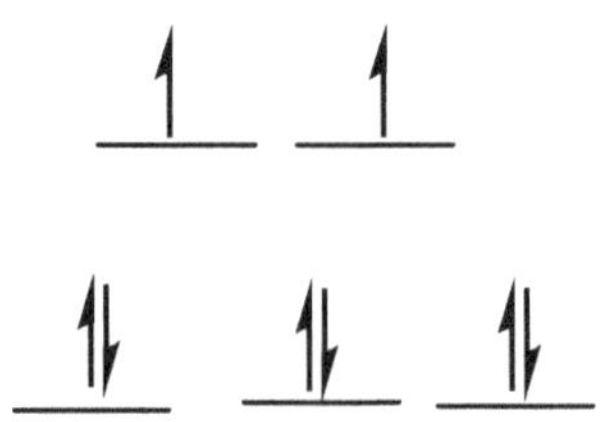

Any octahedral complex ion having 3 or fewer d electrons will have the same numbered of unpaired electrons in a weak or a strong field. The same is true for complex ions have 8 or more d electrons. Except for Cr^{2+}, which is a d^4, all the ions have the same number of unpaired electrons whether in a strong or a weak field.

97. a. Zn^{2+} has a $3d^{10}$ electron configuration. This diagram is for a d^{10} octahedral complex ion, not a d^{10} tetrahedral complex as is given. So, this diagram is incorrect.

b. This diagram is correct. In this complex ion, Mn has a +3 oxidation state and an $[Ar]3d^4$ electron configuration. This strong field diagram for a d^4 ion is shown.

c. This diagram is correct. In this complex ion, Ni^{2+} is present which has a $3d^8$ electron configuration. The diagram shown is correct for a diamagnetic square planar complex.

99. a. $[Co(C_5H_5N)_6]Cl_3$ b. $[Cr(NH_3)_5I]I_2$ c. $[Ni(NH_2CH_2CH_2NH_2)_3]Br_2$

d. $K_2[Ni(CN)_4]$ e. $[Pt(NH_3)_4Cl_2][PtCl_4]$

101. Because each compound contains an octahedral complex ion, the formulas for the compounds are $[Co(NH_3)_6]I_3$, $[Pt(NH_3)_4I_2]I_2$, $Na_2[PtI_6]$, and $[Cr(NH_3)_4I_2]I$. Note that in some cases the I^- ions are ligands bound to the transition metal ion as required for a coordination number of 6, while in other cases the I^- ions are counter ions required to balance the charge of the complex ion. The $AgNO_3$ solution will only precipitate the I^- counterions and will not precipitate the I^- ligands. Therefore, 3 moles of AgI will precipitate per mole of $[Co(NH_3)_6]I_3$, 2 moles of AgI will precipitate per mole of $[Pt(NH_3)_4I_2]I_2$, 0 moles of AgI will precipitate per mole of $Na_2[PtI_6]$, and 1 mole of AgI will precipitate per mole of $[Cr(NH_3)_4I_2]I$.

103. Octahedral complex ions with d^5 or d^6 metal ions have 4 more unpaired electrons in the high-spin (weak filed) as compared to the low-spin (weak field) case:

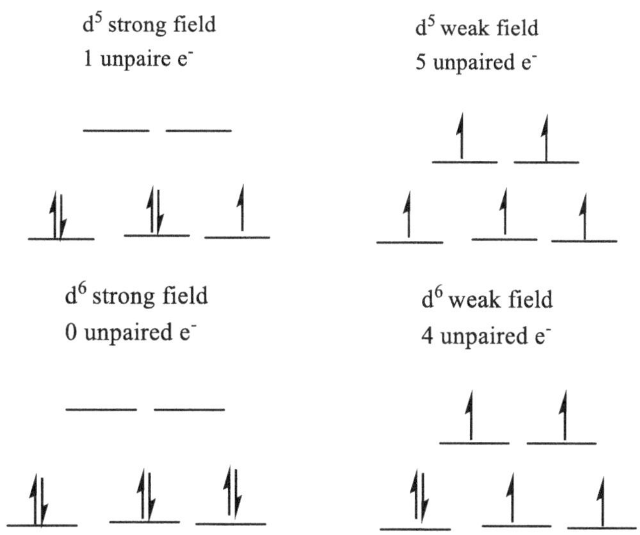

Mn^{2+} and Fe^{3+} both have $[Ar]3d^5$ configurations, and Fe^{2+} and Co^{3+} both have $[Ar]3d^6$ configurations. These four ions have 4 more unpaired electrons in a high spin vs a low spin case. The other ions are not d^5 or d^6 metal ions.

105.

$$\begin{array}{lr} HbO_2 \rightarrow Hb + O_2 & \Delta G° = -(-70 \text{ kJ}) \\ Hb + CO \rightarrow HbCO & \Delta G° = -80 \text{ kJ} \\ \hline HbO_2(aq) + CO(g) \rightarrow HbCO(aq) + O_2(g) & \Delta G° = -10 \text{ kJ} \end{array}$$

$$\Delta G° = -RT \ln K, \quad K = \exp\left(\frac{-\Delta G°}{RT}\right) = \exp\left[\frac{-(-10 \times 10^3 \text{ J})}{(8.3145 \text{ J/K} \bullet \text{mol})(298 \text{ kJ})}\right] = 60$$

107. a. Because O is in the −2 oxidation state, iron must be in the +6 oxidation state. Fe^{6+}: $[Ar]3d^2$.

b. Using the half-reaction method of balancing redox reactions, the balanced equation is:

$$10\ H^+(aq) + 2\ FeO_4^{2-}(aq) + 2\ NH_3(aq) \rightarrow 2\ Fe^{3+}(aq) + N_2(g) + 8\ H_2O(l)$$

$$0.0250 \text{ L} \times \frac{0.243 \text{ mol } FeO_4^{2-}}{\text{L}} \times \frac{1 \text{ mol } N_2}{2 \text{ mol } FeO_4^{2-}} = 3.04 \times 10^{-3} \text{ mol } N_2$$

(if FeO_4^{2-} limiting)

$$0.0550 \text{ L} \times \frac{1.45 \text{ mol } NH_3}{\text{L}} \times \frac{1 \text{ mol } N_2}{2 \text{ mol } NH_3} = 3.99 \times 10^{-2} \text{ mol } N_2 \text{ (if } NH_3 \text{ limiting)}$$

Because the FeO_4^{2-} reagent produces the smaller quantity of N_2, FeO_4^{2-} is limiting and 3.04×10^{-3} mol N_2 can be produced.

$$V_{N_2} = \frac{nRT}{P} = \frac{3.04 \times 10^{-3} \text{ mol } N_2 \times \frac{0.08206 \text{ L atm}}{\text{K mol}} \times 298 \text{ K}}{1.50 \text{ atm}}$$

$$V_{N_2} = 0.0496 \text{ L} = 49.6 \text{ mL } N_2$$

Challenge Problems

109. $Ni^{2+} = d^8$; if ligands A and B produced very similar crystal fields, the cis-$[NiA_2B_4]^{2+}$ complex ion would give the following octahedral crystal field diagram for a d^8 ion:

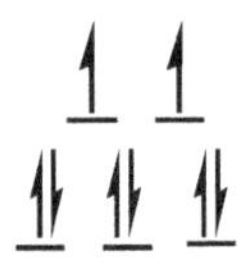

This is paramagnetic.

Because it is given that the complex ion is diamagnetic, the A and B ligands must produce different crystal fields, giving a unique d-orbital splitting diagram that would result in a diamagnetic species.

111. a. Consider the following electrochemical cell:

$$\begin{array}{lr} Co^{3+} + e^- \rightarrow Co^{2+} & E° = 1.82 \text{ V} \\ Co(en)_3^{2+} \rightarrow Co(en)_3^{3+} + e^- & -E° = ? \\ \hline Co^{3+} + Co(en)_3^{2+} \rightarrow Co^{2+} + Co(en)_3^{3+} & E°_{cell} = 1.82 - E° \end{array}$$

The equilibrium constant for this overall reaction is:

$Co^{3+} + 3\ en \rightarrow Co(en)_3^{3+}$	$K_1 = 2.0 \times 10^{47}$
$Co(en)_3^{2+} \rightarrow Co^{2+} + 3\ en$	$K_2 = 1/1.5 \times 10^{12}$
$Co^{3+} + Co(en)_3^{2+} \rightarrow Co(en)_3^{3+} + Co^{2+}$	$K = K_1K_2 = \dfrac{2.0 \times 10^{47}}{1.5 \times 10^{12}} = 1.3 \times 10^{35}$

From the Nernst equation for the overall reaction:

$$E^\circ_{cell} = \frac{0.0591}{n} \log K = \frac{0.0591}{1} \log(1.3 \times 10^{35}),\ E^\circ_{cell} = 2.08\ V$$

$$E^\circ_{cell} = 1.82 - E^\circ = 2.08\ V,\ E^\circ = 1.82\ V - 2.08\ V = -0.26\ V$$

b. The stronger oxidizing agent will be the more easily reduced species and will have the more positive standard reduction potential. From the reduction potentials, Co^{3+} (E° = 1.82 V) is a much stronger oxidizing agent than $Co(en)_3^{3+}$ (E° = –0.26 V).

c. In aqueous solution, Co^{3+} forms the hydrated transition metal complex $Co(H_2O)_6^{3+}$. In both complexes, $Co(H_2O)_6^{3+}$ and $Co(en)_3^{3+}$, cobalt exists as Co^{3+}, which has 6 d electrons. Assuming a strong-field case for each complex ion, the d-orbital splitting diagram for each is:

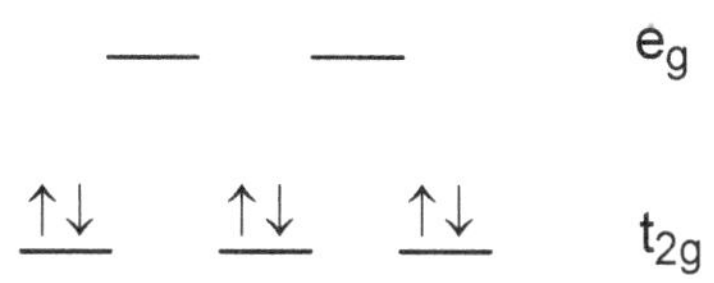

When each complex gains an electron, the electron enters a higher energy e_g orbital. Since en is a stronger-field ligand than H_2O, the d-orbital splitting is larger for $Co(en)_3^{3+}$, and it takes more energy to add an electron to $Co(en)_3^{3+}$ than to $Co(H_2O)_6^{3+}$. Therefore, it is more favorable for $Co(H_2O)_6^{3+}$ to gain an electron than for $Co(en)_3^{3+}$ to gain an electron.

113. No; in all three cases, six bonds are formed between Ni^{2+} and nitrogen, so ΔH values should be similar. ΔS° for formation of the complex ion is most negative for 6 NH_3 molecules reacting with a metal ion (7 independent species become 1). For penten reacting with a metal ion, 2 independent species become 1, so ΔS° is least negative of all three of the reactions. Thus the chelate effect occurs because the more bonds a chelating agent can form to the metal, the less unfavorable ΔS° becomes for the formation of the complex ion, and the larger the formation constant.

115.

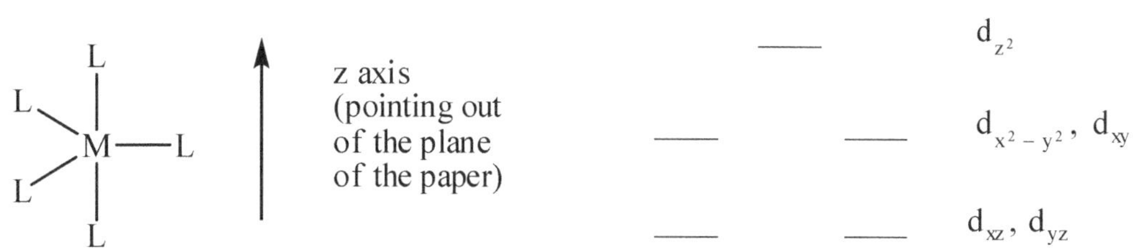

The d_{z^2} orbital will be destabilized much more than in the trigonal planar case (see Exercise 114). The d_{z^2} orbital has electron density on the z axis directed at the two axial ligands. The $d_{x^2-y^2}$ and d_{xy} orbitals are in the plane of the three trigonal planar ligands and should be destabilized a lesser amount than the d_{z^2} orbital; only a portion of the electron density in the $d_{x^2-y^2}$ and d_{xy} orbitals is directed at the ligands. The d_{xz} and d_{yz} orbitals will be destabilized the least since the electron density is directed between the ligands.

117. Ni^{2+}: $[Ar]3d^8$; the coordinate system for trans-$[Ni(NH_3)_2(CN)_4]^{2-}$ is shown below. Because CN^- produces a much stronger crystal field, it will dominate the d-orbital splitting. From the coordinate system, the CN^- ligands are in a square planar arrangement. Therefore, the diagram will most likely resemble the square planar diagram given in Figure 21.28. Note that the relative position of d_{z^2} orbital is hard to predict. With the NH_3 ligands on the z axis, we will assume the d_{z^2} orbital is destabilized more than the d_{xy} orbital. However, this is only an assumption. It could be that the d_{xy} orbital is destabilized more.

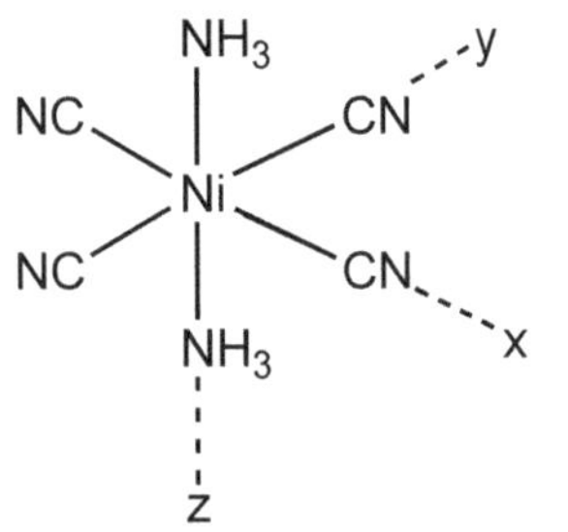

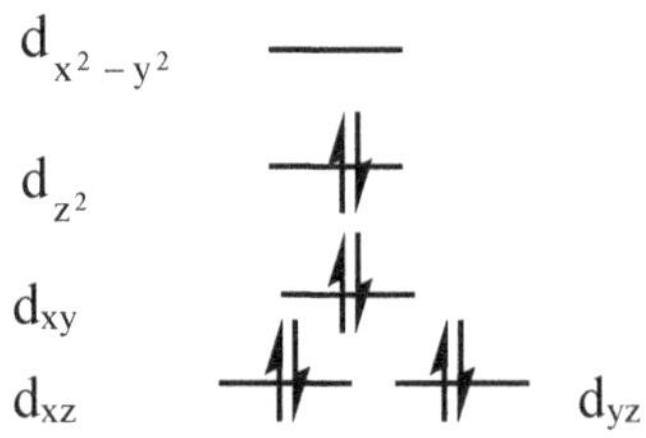

119.

$$AgBr(s) \rightleftharpoons Ag^+ + Br^- \qquad K_{sp} = 5.0 \times 10^{-13}$$
$$Ag^+ + 2\,NH_3 \rightleftharpoons Ag(NH_3)_2^+ \qquad K_f = 1.7 \times 10^7$$
$$AgBr(s) + 2\,NH_3(aq) \rightleftharpoons Ag(NH_3)_2^+(aq) + Br^-(aq) \qquad K = K_{sp} \times K_f = 8.5 \times 10^{-6}$$

	AgBr(s)	+	2 NH_3	$\rightleftharpoons$	$Ag(NH_3)_2^+$	+	Br^-
Initial			3.0 *M*		0		0

If 0.10 mol of AgBr dissolves in 1.0 L, then the concentration changes would be:

Change	−0.10	−0.20	→	+0.10	+0.10
After		2.8 *M*		0.10 *M*	0.10 *M*

To check to see if all the solid dissolves, let's see what the reaction quotient indicates using these concentrations.

$$Q = \frac{[Ag(NH_3)_2^+]_o[Br^-]_o}{[NH_3]_o^2} = \frac{(0.10)(0.10)}{(2.8)^2} = 1.3 \times 10^{-3}$$

Since Q > K (8.5×10^{-6}), the reaction will shift left to establish equilibrium, causing AgBr(s) to form. No, 0.10 mol of AgBr will not dissolve in 1.0 L of 3.0 *M* NH_3.

121. i. $0.0203 \text{ g CrO}_3 \times \frac{52.00 \text{ g Cr}}{100.0 \text{ g CrO}_3} = 0.0106 \text{ g Cr}$; $\% \text{ Cr} = \frac{0.0106 \text{ g}}{0.105 \text{ g}} \times 100 = 10.1\% \text{ Cr}$

ii. $32.93 \text{ mL HCl} \times \frac{0.100 \text{ mmol HCl}}{\text{mL}} \times \frac{1 \text{ mmol NH}_3}{\text{mmol HCl}} \times \frac{17.03 \text{ mg NH}_3}{\text{mmol}} = 56.1 \text{ mg NH}_3$

$\% \text{ NH}_3 = \frac{56.1 \text{ mg}}{341 \text{ mg}} \times 100 = 16.5\% \text{ NH}_3$

iii. 73.53% + 16.5% + 10.1% = 100.1%; the compound must be composed of only Cr, NH_3, and I.
Out of 100.00 g of compound:

$10.1 \text{ g Cr} \times \frac{1 \text{ mol}}{52.00 \text{ g}} = 0.194 \text{ mol}$; $\frac{0.194}{0.194} = 1.00$

$16.5 \text{ g NH}_3 \times \frac{1 \text{ mol}}{17.03 \text{ g}} = 0.969 \text{ mol}$; $\frac{0.969}{0.194} = 4.99$

$73.53 \text{ g I} \times \frac{1 \text{ mol}}{126.9 \text{ g}} = 0.5794 \text{ mol}$; $\frac{0.5794}{0.194} = 2.99$

$Cr(NH_3)_5I_3$ is the empirical formula. If we assume an octahedral complex ion, then compound A is made of the octahedral $[Cr(NH_3)_5I]^{2+}$ complex ion and two I^- ions as counterions; the formula is $[Cr(NH_3)_5I]I_2$. Let's check this proposed formula using the freezing-point data.

iv. $\Delta T_f = iK_f m$; for $[Cr(NH_3)_5I]I_2$, i = 3.0 (assuming complete dissociation).

$\text{Molality} = m = \frac{0.601 \text{ g complex}}{1.000 \times 10^{-2} \text{ kg H}_2\text{O}} \times \frac{1 \text{ mol complex}}{517.9 \text{ g complex}} = 0.116 \text{ mol/kg}$

$\Delta T_f = 3.0 \times 1.86 \text{ °C kg/mol} \times 0.116 \text{ mol/kg} = 0.65\text{°C}$

Because ΔT_f is close to the measured value, this is consistent with the formula $[Cr(NH_3)_5I]I_2$. So our assumption of an octahedral complex ion is probably a good assumption.

CHAPTER 22

ORGANIC AND BIOLOGICAL MOLECULES

Questions

3. Carbon has the unusual ability to form bonds to itself, whether they be single, double, or triple bonds, as well as the ability to form long chains or rings of carbon atoms. Carbon also forms strong bonds to other nonmetals, such as hydrogen, oxygen, phosphorus, sulfur, and the halogens. Because of this bonding ability, carbon can form millions of compounds, including biomolecules which are the basis for the existence of life.

5. a. 1-sec-butylpropane

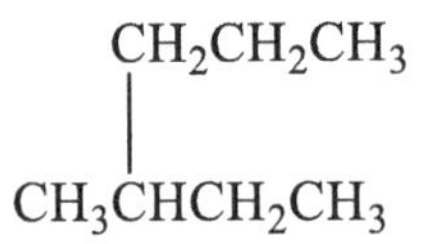

3-methylhexane is correct.

b. 4-methylhexane

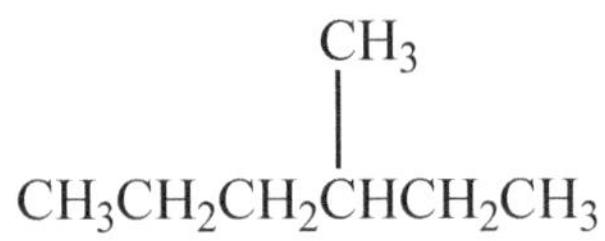

3-methylhexane is correct.

c. 2-ethylpentane

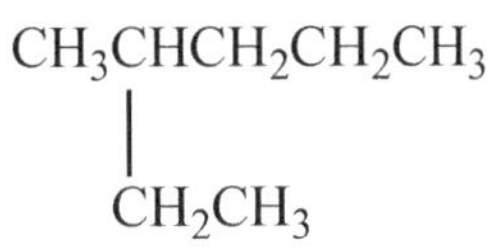

3-methylhexane is correct.

d. 1-ethyl-1-methylbutane

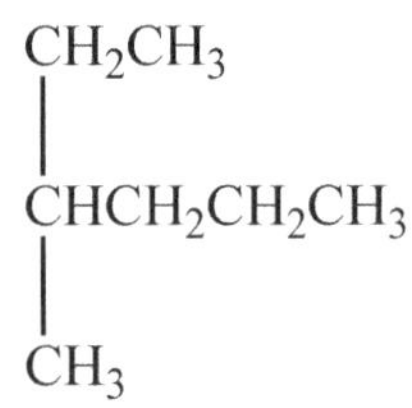

3-methylhexane is correct.

e. 3-methylhexane

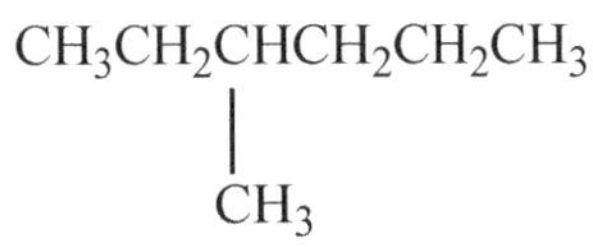

f. 4-ethylpentane

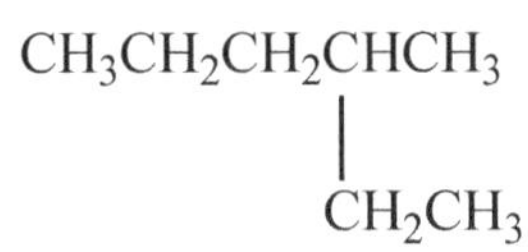

3-methylhexane is correct.

All six of these compounds are the same. They only differ from each other by rotations about one or more carbon-carbon single bonds. Only one isomer of C_7H_{16} is present in all these names, 3-methylhexane.

7. Hydrocarbons are nonpolar substances exhibiting only London dispersion forces. Size and shape are the two most important structural features relating to the strength of London dispersion forces. For size, the bigger the molecule (the larger the molar mass), the stronger

are the London dispersion forces, and the higher is the boiling point. For shape, the more branching present in a compound, the weaker are the London dispersion forces, and the lower is the boiling point.

9. Compounds with ring structures or compounds with double and/or triple bonds have restricted rotation about some of the bonds in the compounds. Only alkanes, compounds composed of C-C and C-H single bonds, have free rotation about all bonds in the compound. Ethane (answer b) is the only compound listed that has free rotation about every bond.

11. The general formula for cycloalkanes is C_nH_{2n} and the general formula for alkanes is C_nH_{2n+2}. Since the general formulas for alkanes and cycloalkanes are different, they are not structural isomers of each other. Answer b is the false statement. All the other statements are true. Cyclobutane is forced into 90° bond angles and is reactive because of this. Alkanes are composed of only single bonded atoms; atoms in these single bonds can rotate. All carbon atoms in alkanes exhibit tetrahedral bond angles, so all carbon atoms in alkanes are sp^3 hybridized. With the restricted rotation of the ring atoms is cycloalkanes, they can exhibit cis/trans isomerism.

13. The amide functional group is: 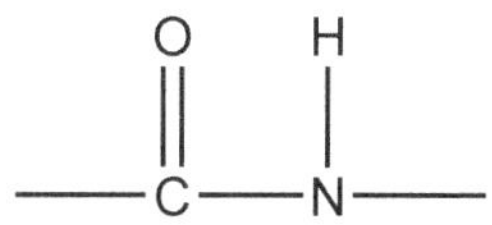

When the amine end of one amino acid reacts with the carboxylic acid end of another amino acid, the two amino acids link together by forming an amide functional group. A polypeptide has many amino acids linked together, with each linkage made by the formation of an amide functional group. Because all linkages result in the presence of the amide functional group, the resulting polymer is called a polyamide. For nylon, the monomers also link together by forming the amide functional group (the amine end of one monomer reacts with the carboxylic acid end of another monomer to give the amide functional group linkage). Hence nylon is also a polyamide.

The correct order of strength is:

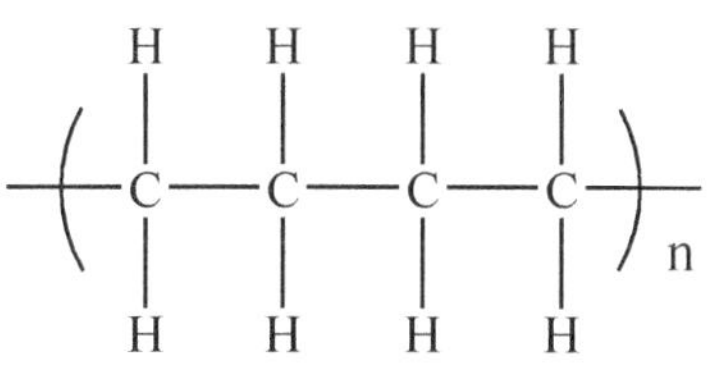

polyhydrocarbon
weakest fibers

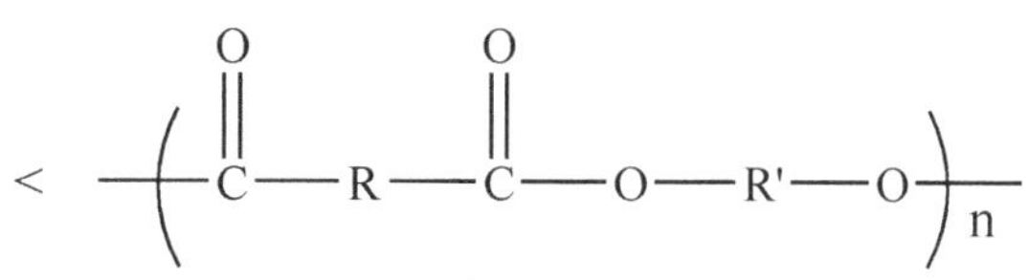

polyester

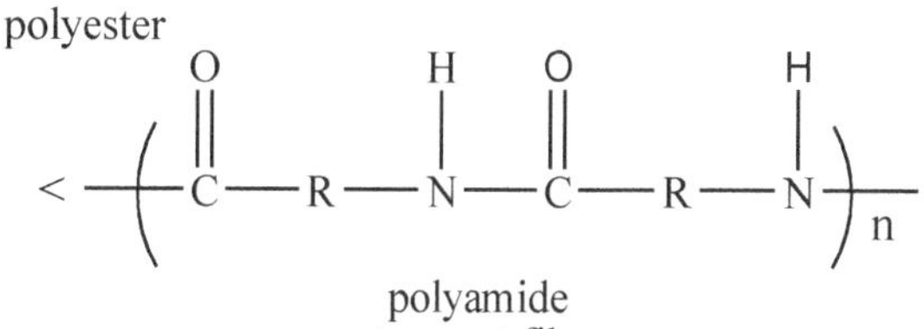

polyamide
strongest fibers

The difference in strength is related to the types of intermolecular forces present. All these types of polymers have London dispersion forces. However, the polar ester group in polyesters and the polar amide group in polyamides give rise to additional dipole forces. The polyamide can form relatively strong hydrogen-bonding interactions, hence why it would form the strongest fibers.

15. a. A polyester forms when an alcohol functional group reacts with a carboxylic acid functional group. The monomer for a homopolymer polyester must have an alcohol functional group and a carboxylic acid functional group present within the structure of the monomer.

b. A polyamide forms when an amine functional group reacts with a carboxylic acid functional group. For a copolymer polyamide, one monomer would have at least two amine functional groups present, and the other monomer would have at least two carboxylic acid functional groups present. For polymerization to occur, each monomer must have two reactive functional groups present.

c. To form an addition polymer, a carbon-carbon double bond must be present. Polyesters and polyamides are condensation polymers. To form a polyester, the monomer would need the alcohol and carboxylic acid functional groups present. To form a polyamide, the monomer would need the amine and carboxylic acid functional groups present. The two possibilities are for the monomer to have a carbon-carbon double bond, an alcohol functional group, and a carboxylic acid functional group present or to have a carbon-carbon double bond, an amine functional group, and a carboxylic acid functional group present.

17. Denaturation is the breakdown of the three-dimensional structure of the protein. Both the secondary and the tertiary structures of the protein are disrupted in denaturation. The primary structure (the order in which the amino acids link to each other) is not affected by denaturation.

Exercises

Hydrocarbons

19. i.

ii.

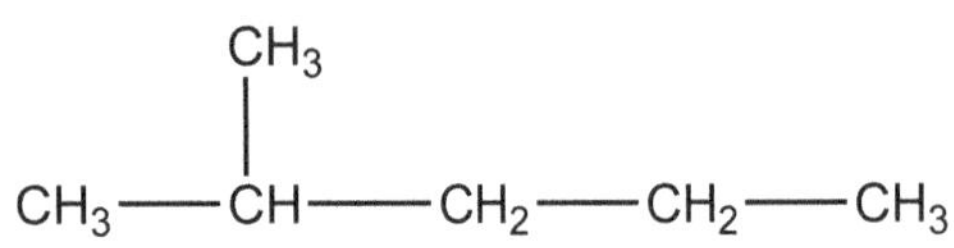

iii.

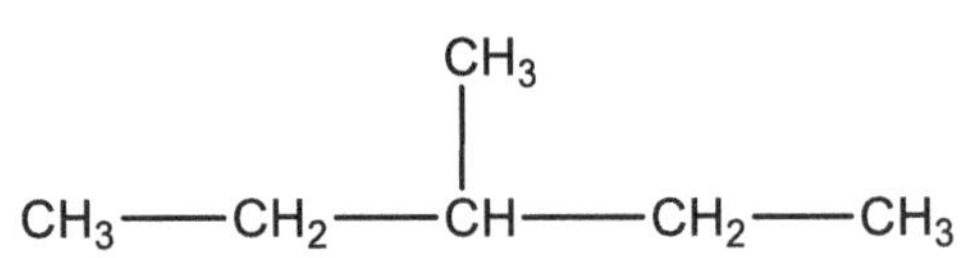

iv.

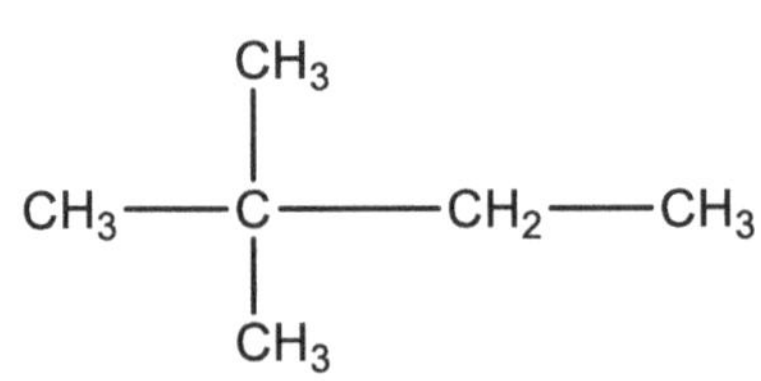

v.

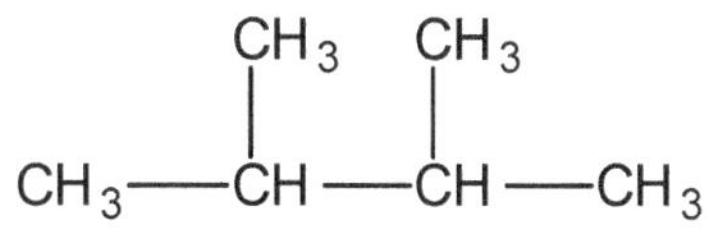

All other possibilities are identical to one of these five compounds.

21. A difficult task in this problem is recognizing different compounds from compounds that differ by rotations about one or more C–C bonds (called conformations). The best way to distinguish different compounds from conformations is to name them. Different name = different compound; same name = same compound, so it is not an isomer but instead is a conformation.

a.

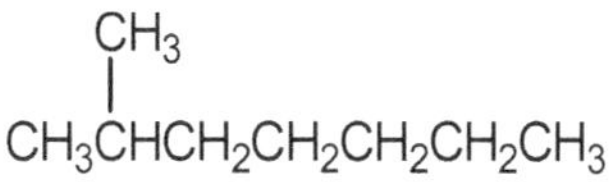

2-methylheptane

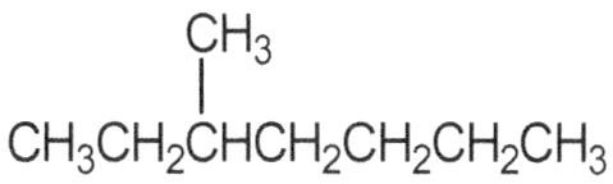

3-methylheptane

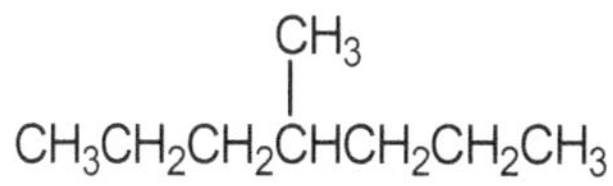

4-methylheptane

b.

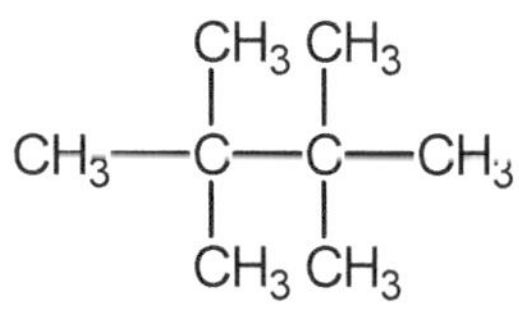

2,2,3,3-tetramethylbutane

23. a.

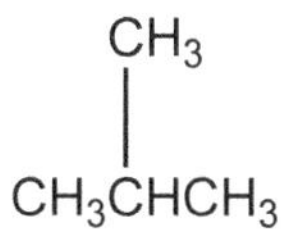

b.

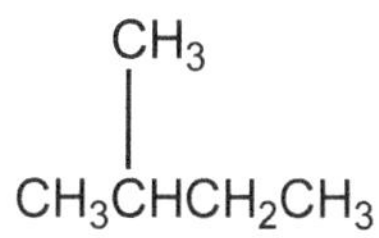

c.

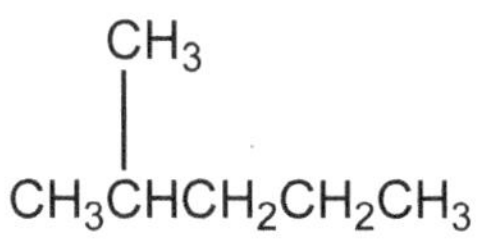

d.

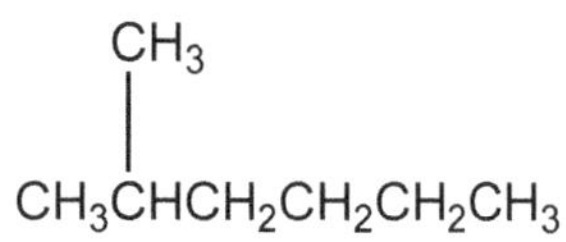

25. The structures of the compounds are:

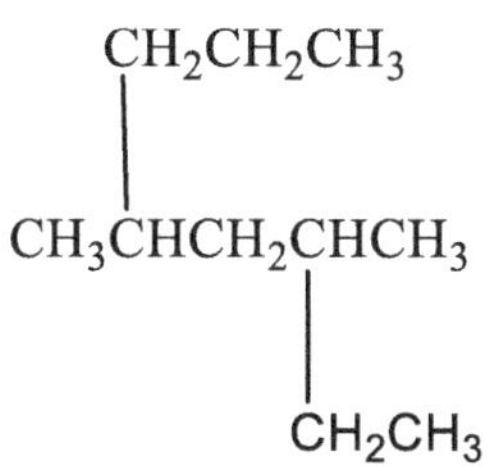

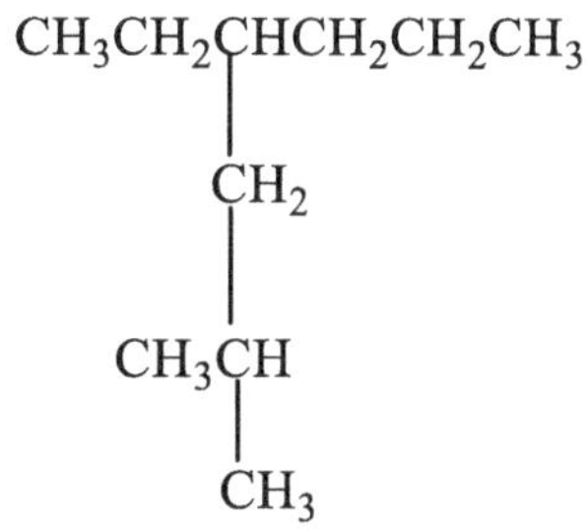

The correct name for the first structure is 3,5-dimethyloctane, and the correct name for the second structure is 4-ethyl-2-methylheptane.

27. a. 2,2,4-trimethylhexane b. 5-methylnonane c. 2,2,4,4-tetramethylpentane

d. 3-ethyl-3-methyloctane

Note: For alkanes, always identify the longest carbon chain for the base name first, then number the carbons to give the lowest overall numbers for the substituent groups.

29. The hydrogen atoms in ring compounds are commonly omitted. In organic compounds, carbon atoms satisfy the octet rule of electrons by forming four bonds to other atoms. Therefore, add C-H bonds to the carbon atoms in the ring in order to give each C atom four bonds. You can also determine the formula of these cycloalkanes by using the general formula C_nH_{2n}.

a. isopropylcyclobutane; C_7H_{14}

b. 1-tert-butyl-3-methylcyclopentane; $C_{10}H_{20}$

c. 1,3-dimethyl-2-propylcyclohexane; $C_{11}H_{22}$

31.

$CH_3—CH_2—CH_2—CH_3$

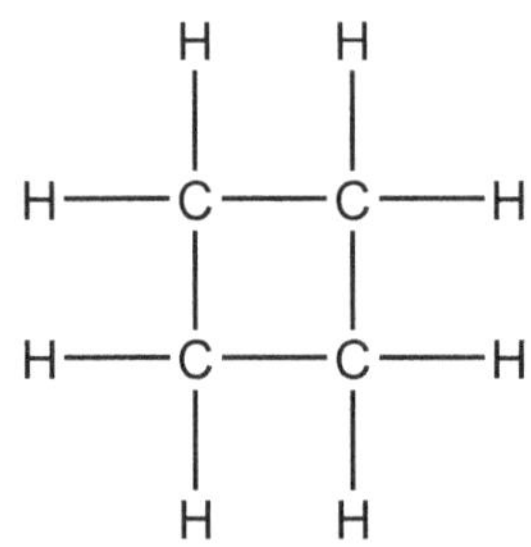

Each carbon is bonded to four other carbon and/or hydrogen atoms in a saturated hydrocarbon (only single bonds are present).

33. a. 1-butene b. 4-methyl-2-hexene c. 2,5-dimethyl-3-heptene

Note: The multiple bond is assigned the lowest number possible.

35. a. $CH_3–CH_2–CH=CH–CH_2–CH_3$ b. $CH_3–CH=CH–CH=CH–CH_2CH_3$

c.

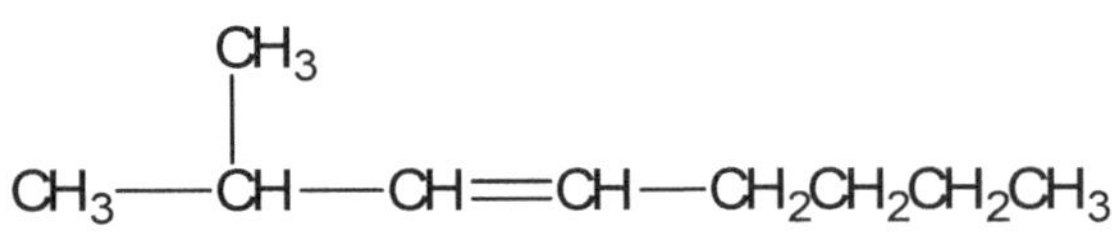

37. a.

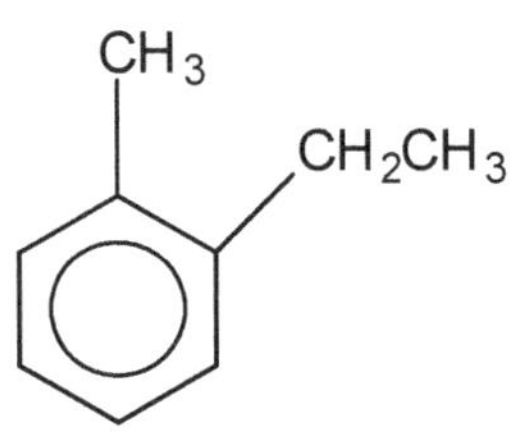

b.

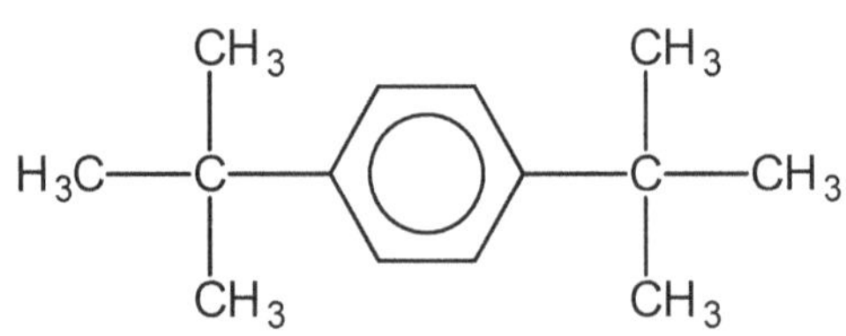

c.

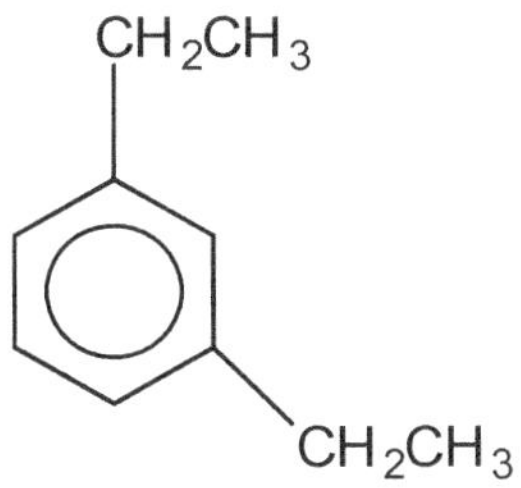

d.

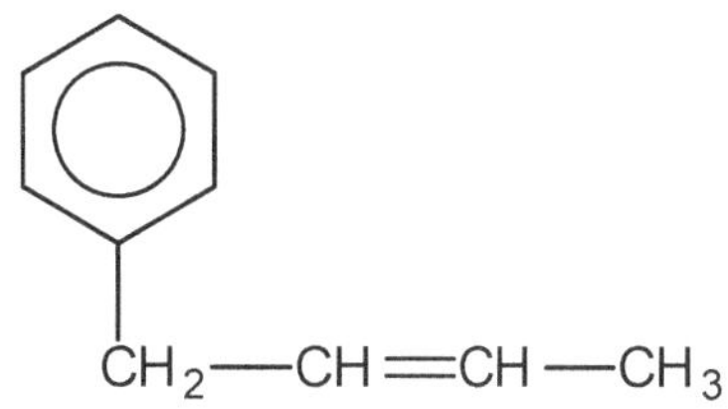

39. a. 1,3-dichlorobutane b. 1,1,1-trichlorobutane

c. 2,3-dichloro-2,4-dimethylhexane d. 1,2-difluoroethane

41. a. 3-chloro-l-butene b. 1-ethyl-3-methycyclopentene

c. 3-chloro-4-propylcyclopentene d. 1,2,4-trimethylcyclohexane

e. 2-bromotoluene (or o-bromotoluene or 1-bromo-2-methylbenzene)

f. 1-bromo-2-methylcyclohexane g. 4-bromo-3-methylcyclohexene

Note: If the location of the double bond is not given in the name, it is assumed to be located between C_1 and C_2. Also, when the base name can be numbered in equivalent ways, give the first substituent group the lowest number; e.g., for part f, 1-bromo-2-methylcyclohexane is preferred to 2-bromo-1-methycyclohexane.

Isomerism

43. $CH_2Cl–CH_2Cl$, 1,2-dichloroethane: In this compound, there is free rotation about the C–C single bond that doesn't lead to different compounds. CHCl=CHCl, 1,2-dichloroethene: This compound, however, has no free rotation about the C=C double bond. This creates the cis and trans isomers, which are different compounds.

45. All the compounds have a C_6H_{12} formula. The cis isomer has the bulkiest groups bonded to the carbons of the double on the same side of the double bond. The names of the various substances are:

I. trans-3-methyl-2-pentene; II. 2-methyl-1-pentene; III. trans-2-hexene; iv. cis-2-hexene

Only statement d is false. Structures I and IV are structural isomers of each other, by they are not geometric isomers of each other. Geometric isomers have the same bonds, but different spatial arrangement of some groups in the compound. Geometric isomers will always have the same base name. For example, structures III and IV have the same base name, but they have a different spatial arrangement of the groups bonded to the carbons with the double bond. Structures III and IV are geometric isomers of each other

47. To exhibit cis-trans isomerism, each carbon in the double bond must have two structurally different groups bonded to it. In Exercise 33, this occurs for compounds b and c. The cis isomer has the bulkiest groups on the same side of the double bond while the trans isomer has the bulkiest groups on opposite sides of the double bond. The cis and trans isomers for 33b and 33c are:

33b.

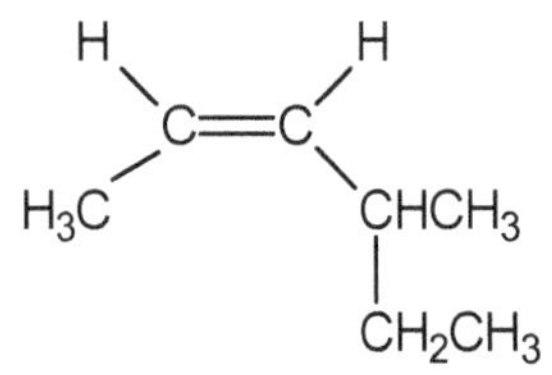

cis

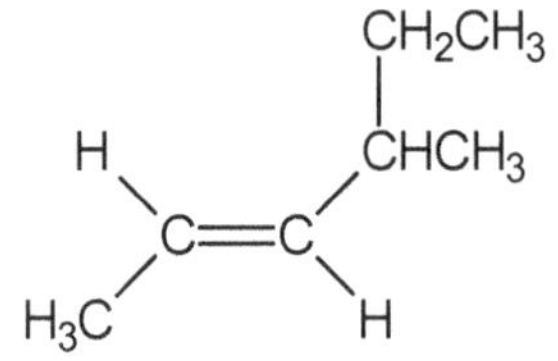

trans

33c.

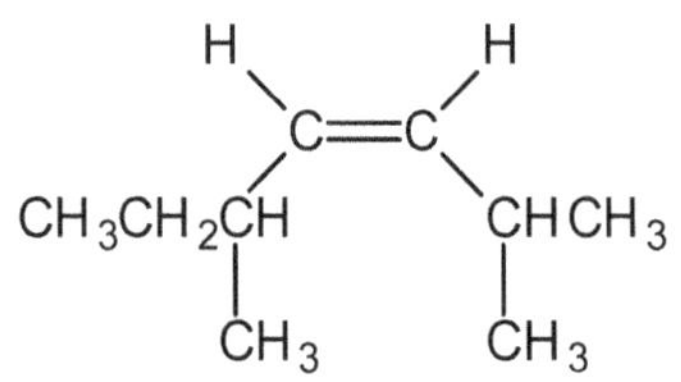

cis

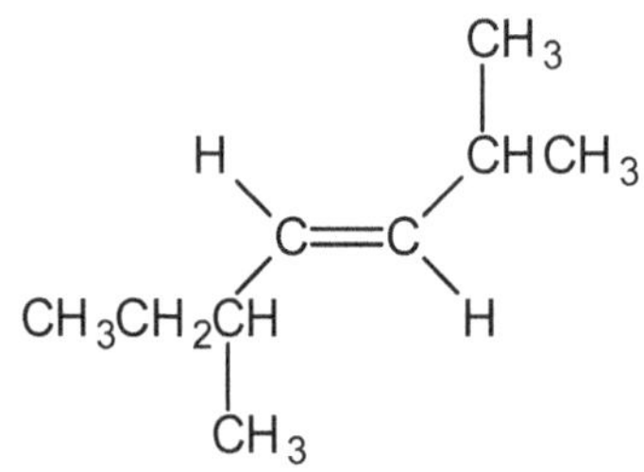

trans

Similarly, all the compounds in Exercise 35 exhibit *cis-trans* isomerism.

In compound a of Exercise 33, the first carbon in the double bond does not contain two different groups. The first carbon in the double bond contains two H atoms. To illustrate that this compound does not exhibit *cis-trans* isomerism, let's look at the potential c*is-trans* isomers.

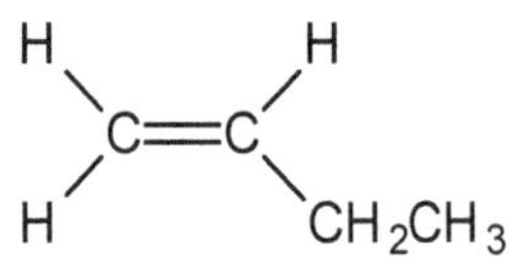

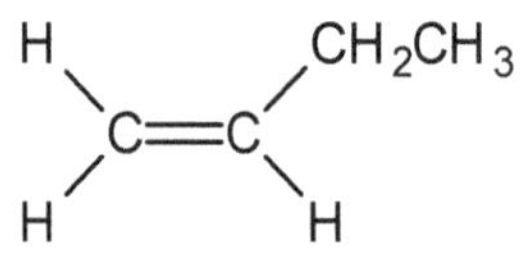

These are the same compounds; they only differ by a simple rotation of the molecule. Therefore, they are not isomers of each other but instead are the same compound.

49. To help distinguish the different isomers, we will name them.

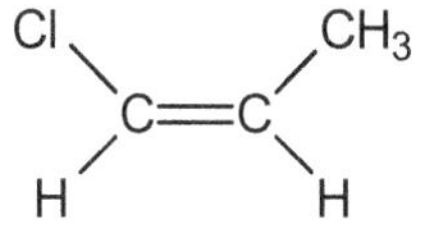

cis-1-chloro-1-propene

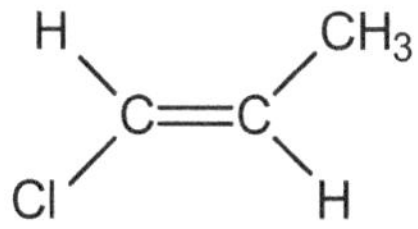

trans-1-chloro-1-propene

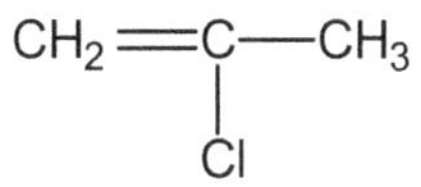

2-chloro-1-propene

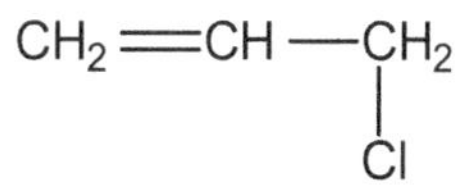

3-chloro-1-propene

chlorocyclopropane

51.

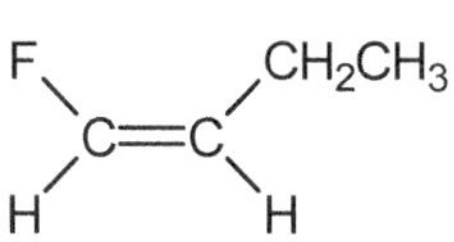

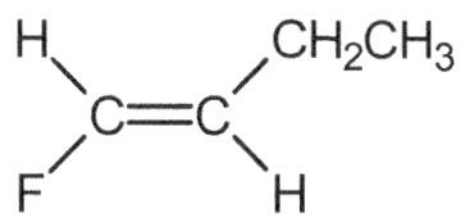

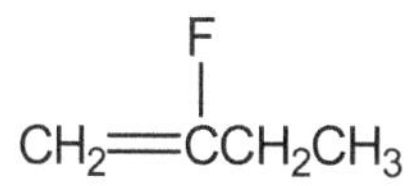

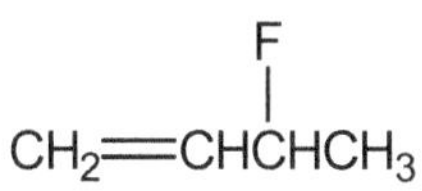

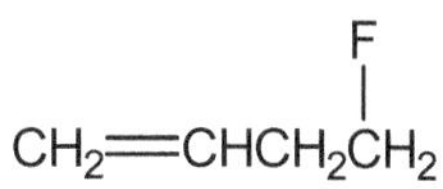

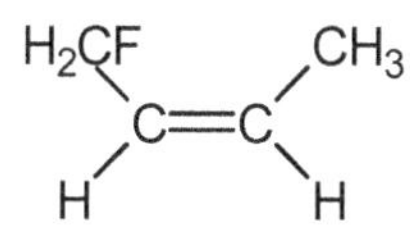

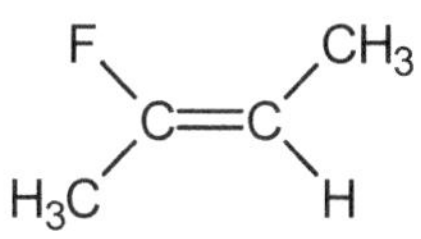

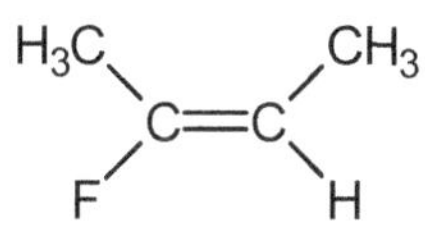

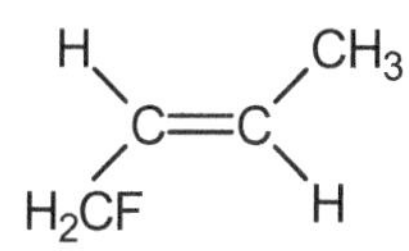

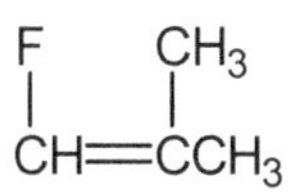

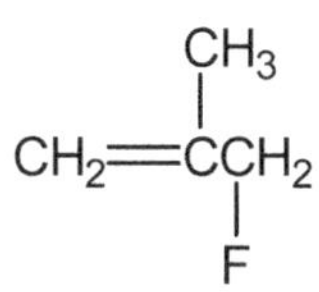

53. a. 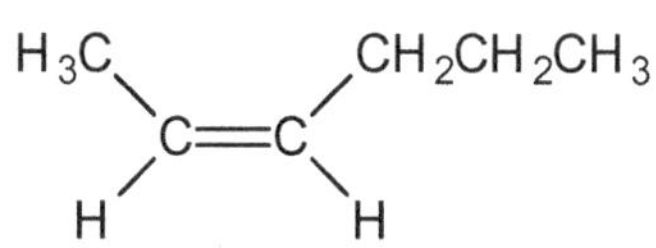

b. 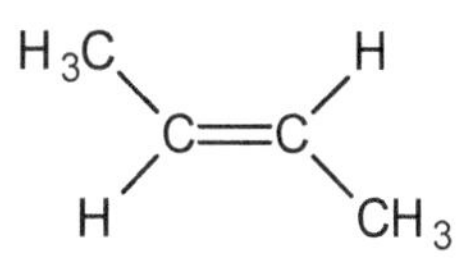

c. 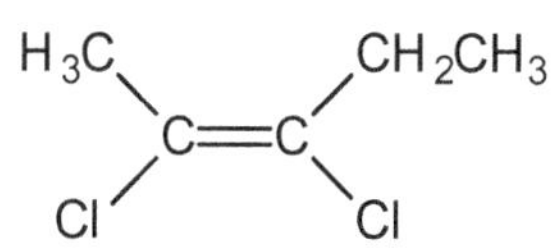

55. There are many possible structural isomers for the formula C_6H_{12}. Two structural isomers are:

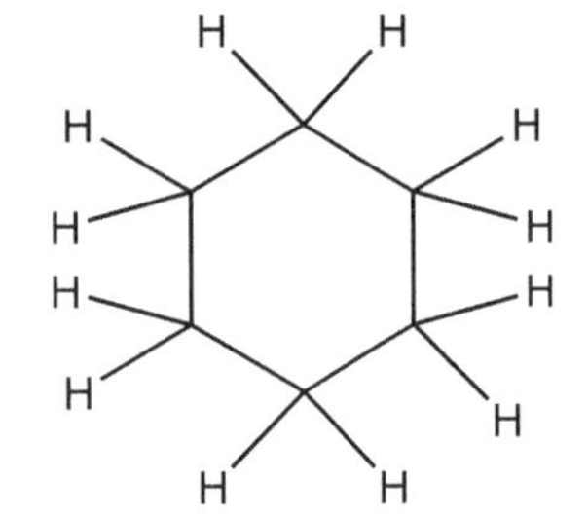

cyclohexane

$CH_2{=}CHCH_2CH_2CH_2CH_3$

1-hexene

The structural isomer 2-hexene (plus others) exhibits geometric isomerism.

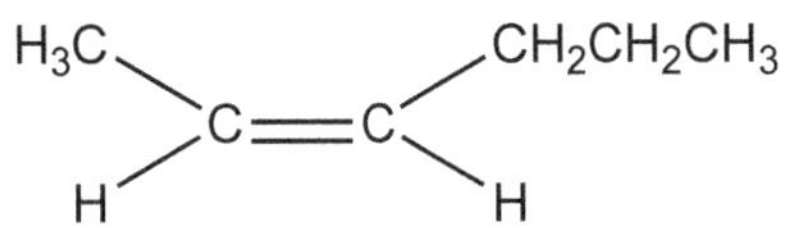

cis

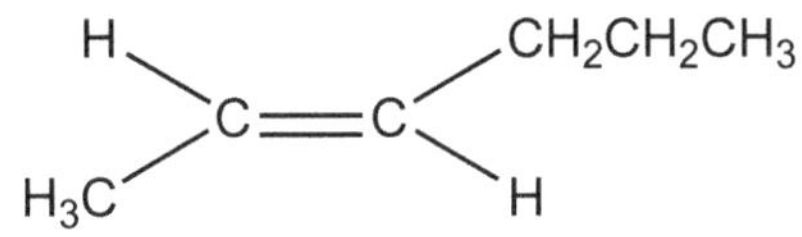

trans

The structural isomer 3-methyl-1-pentene exhibits optical isomerism (the asterisk marks the chiral carbon).

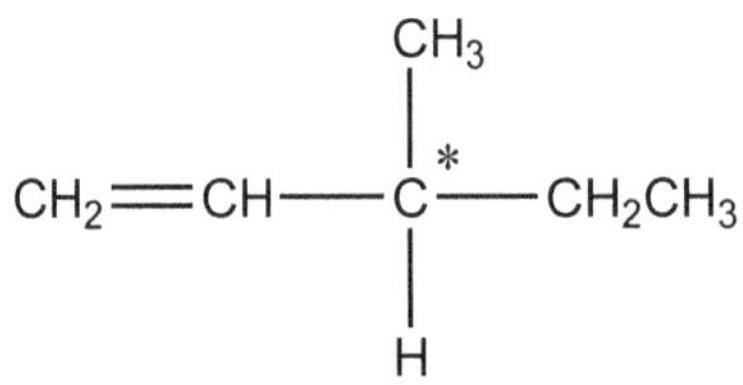

For this compound, the chiral carbon has four different groups bonded to it; the mirror image of this compound will be non-superimposable. Optical isomerism is also possible with some of the cyclobutane and cyclopropane structural isomers.

57. C_5H_{10} has the general formula for alkenes and cycloalkanes, C_nH_{2n}. The alkene isomers of C_5H_{10} are:

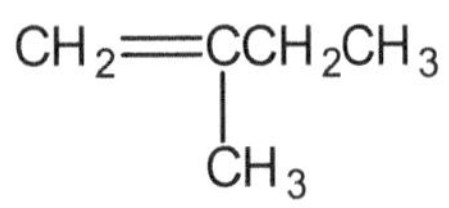

2-methyl-1-butene

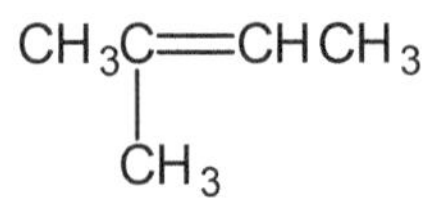

2-methyl-2-butene

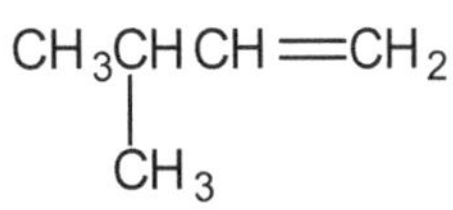

3-methyl-1-butene

$CH_2{=}CHCH_2CH_2CH_3$ 1-pentene $\qquad$ $CH_3CH{=}CHCH_2CH_3$ 2-pentene

The cycloalkane isomers of C_5H_{10} are:

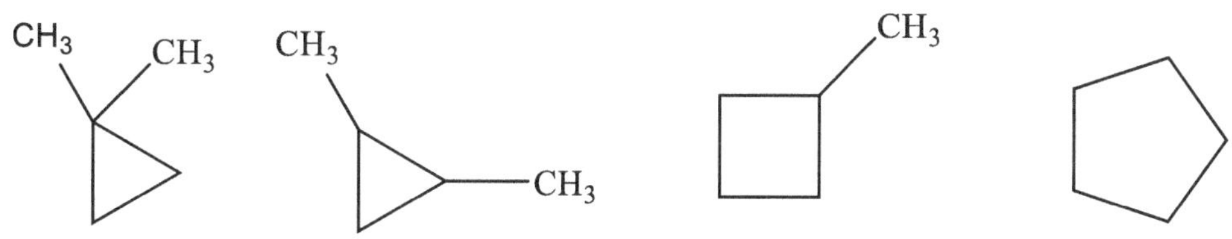

a. 2-pentene exhibits cis/trans isomerism.

b. 1-pentene, 2-methyl-1-butene, 2-methyl-2-butene, and 3-methyl-1-butene do not exhibit cis/trans isomerism.

c. 1,2-dimethylcyclopropane exhibits cis/trans isomerism.

d. 1,1-dimethylcyclopropane, methylcyclobutane, and cyclopentane do not exhibit cis/trans isomerism.

59. a. $\overset{*}{CH_3}-\overset{*}{CH_2}-\overset{*}{CH_2}-CH_2-CH_3$

There are three different types of hydrogens in n-pentane (see asterisks). Thus there are three monochloro isomers of n-pentane (1-chloropentane, 2-chloropentane, and 3-chloropentane).

b.

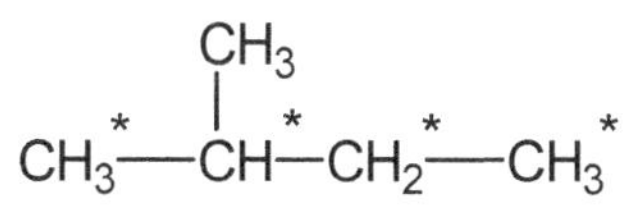

There are four different types of hydrogens in 2-methylbutane, so four monochloro isomers of 2-methylbutane are possible.

c.

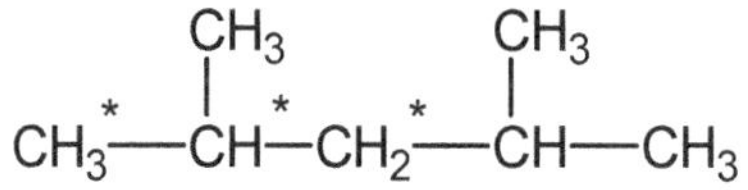

There are three different types of hydrogens, so three monochloro isomers are possible.

d.

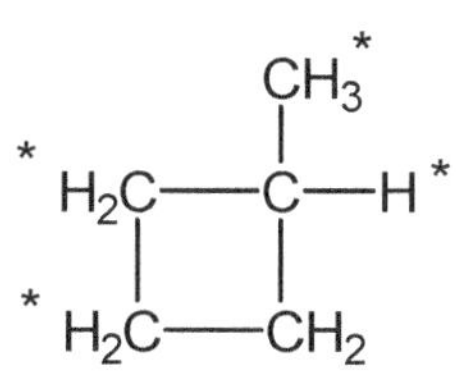

There are four different types of hydrogens, so four monochloro isomers are possible.

Functional Groups

61. Reference Table 22.4 for the common functional groups.

a. ketone b. aldehyde

c. carboxylic acid d. amine

63. a.

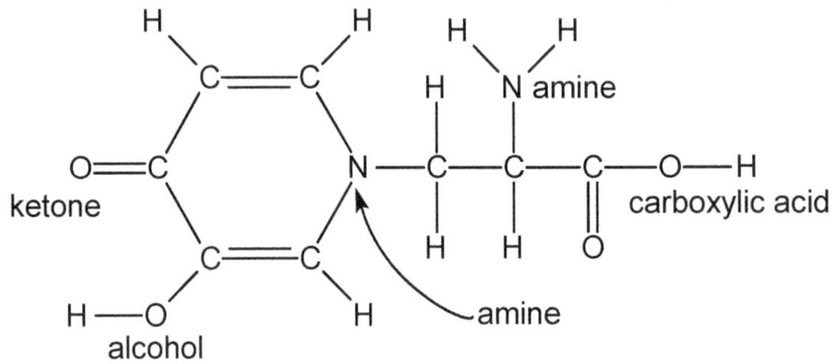

b. 5 carbons in the ring and the carbon in $-CO_2H$: sp^2; the other two carbons: sp^3

c. 24 sigma bonds; 4 pi bonds

65. a. 3-chloro-1-butanol; because the carbon containing the OH group is bonded to just 1 other carbon (1 R group), this is a primary alcohol.

b. 3-methyl-3-hexanol; because the carbon containing the OH group is bonded to three other carbons (3 R groups), this is a tertiary alcohol.

c. 2-methylcyclopentanol; secondary alcohol (2 R groups bonded to carbon containing the OH group). *Note*: In ring compounds, the alcohol group is assumed to be bonded to C_1, so the number designation is commonly omitted for the alcohol group.

67.

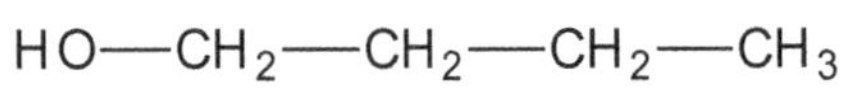

1-butanol

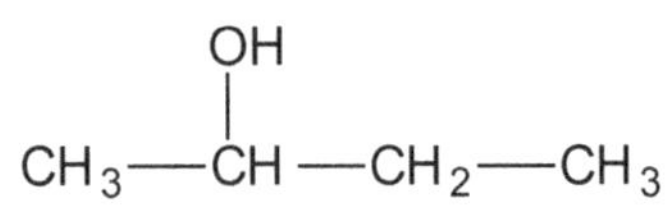

2-butanol

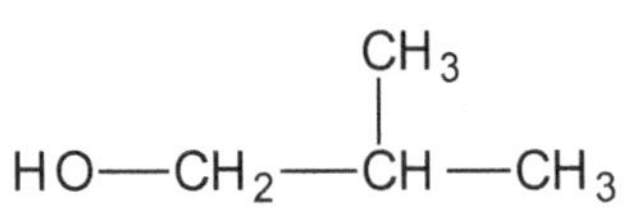

2-methyl-1-propanol

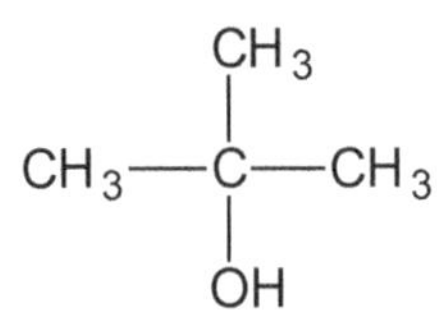

2-methyl-2-propanol

There are three possible ethers with the formula $C_4H_{10}O$. They are:

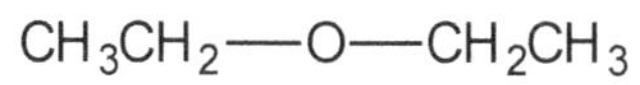

diethyl ether

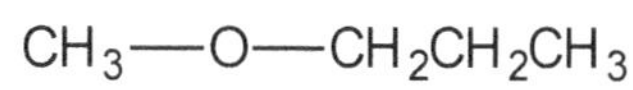

methylpropyl ether

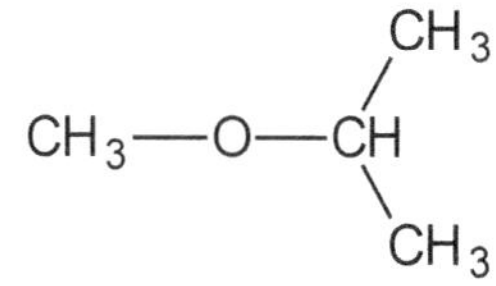

isopropylmethyl ether

69. Two possible aldehydes:

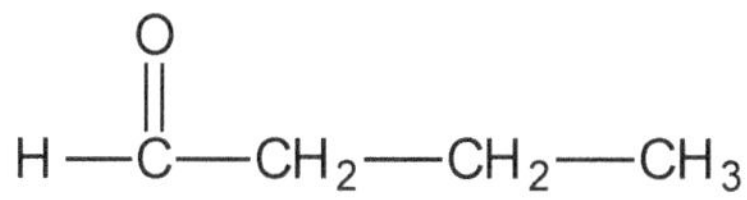

butanal

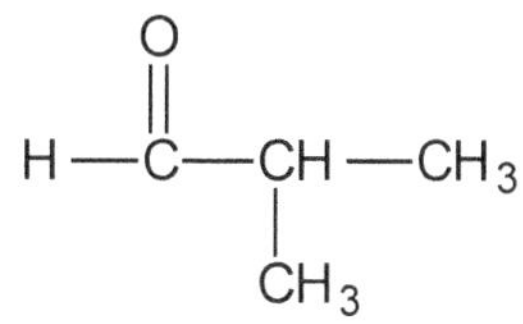

2-methylpropanal

One possible ketone:

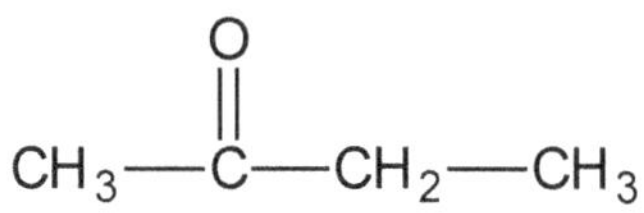

2-butanone

71. a. 4,5-dichloro-3-hexanone b. 2,3-dimethylpentanal

c. 3-methylbenzaldehyde or m-methylbenzaldehyde

73. a. 4-chlorobenzoic acid or p-chlorobenzoic acid

b. 3-ethyl-2-methylhexanoic acid

c. methanoic acid (common name = formic acid)

75. a. 1-pentanol; 2-methyl-2,4-pentadiol

b. trans-1,2-cyclohexadiol; 4-methyl-1-penten-3-ol (does not exhibit cis-trans isomerism)

77. Only statement d is false. The other statements refer to compounds having the same formula but different attachment of atoms; they are structural isomers.

a.

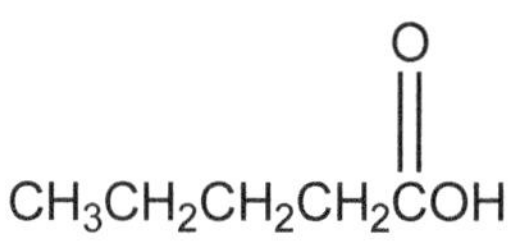

Both have a formula of $C_5H_{10}O_2$.

b.

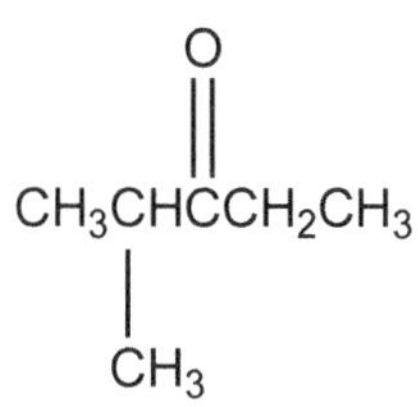

Both have a formula of $C_6H_{12}O$.

c.

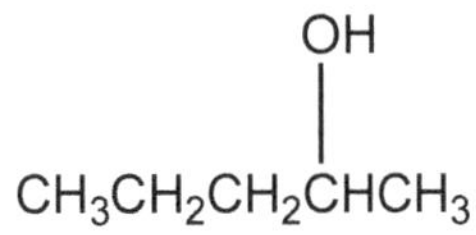

Both have a formula of $C_5H_{12}O$.

d.

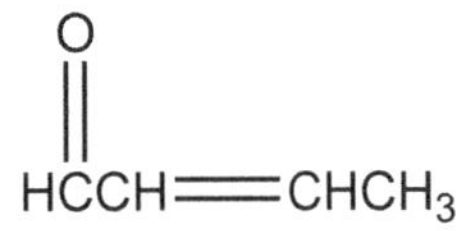

2-Butenal has a formula of C_4H_6O while the alcohol has a formula of C_4H_8O.

e.

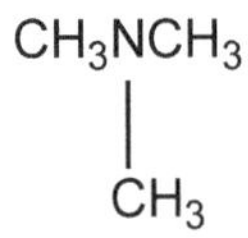

Both have a formula of C_3H_9N.

Reactions of Organic Compounds

79. The two monochlorination products are 1-chloro-2-methylpentane and 2-chloro-2-methylpentane. All the other possible products differ from one of these by a rotation.

81.

a. 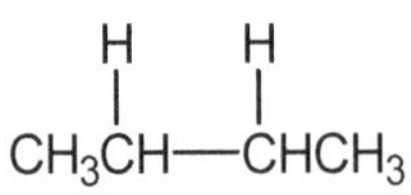

b. 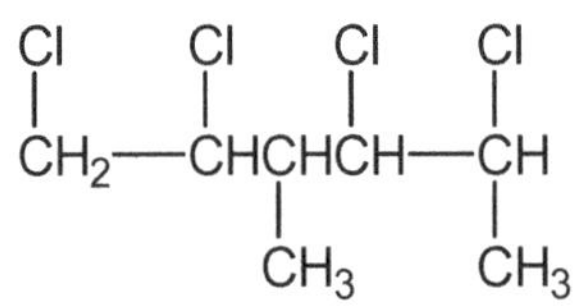

c. 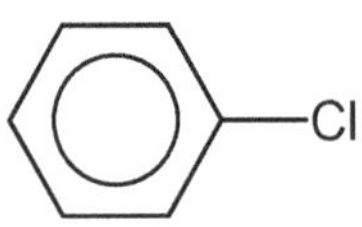 + HCl

d. $C_4H_8(g) + 6\ O_2(g) \rightarrow 4\ CO_2(g) + 4\ H_2O(g)$

83. The reaction is:

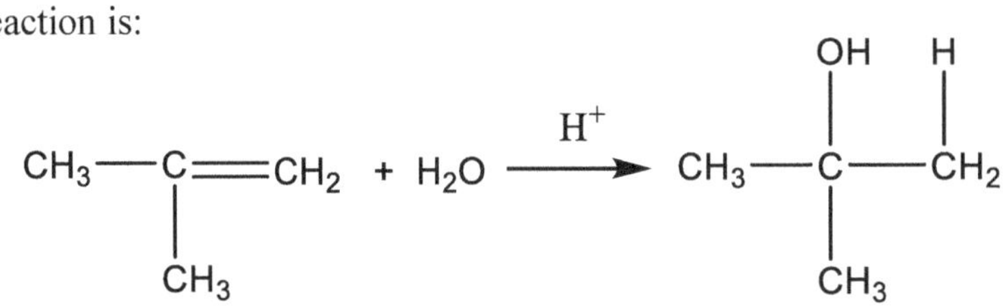

85.

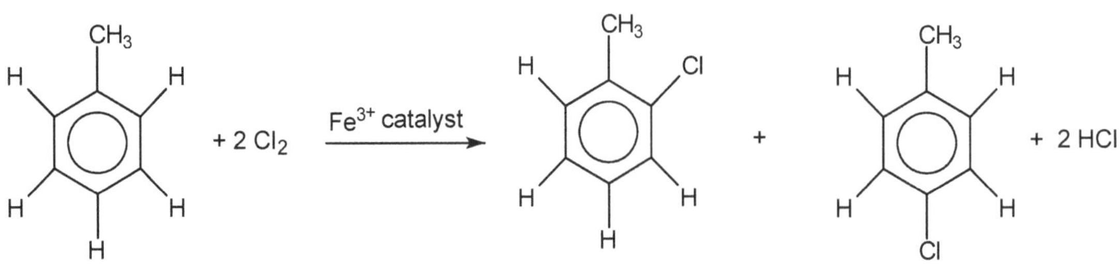

ortho para

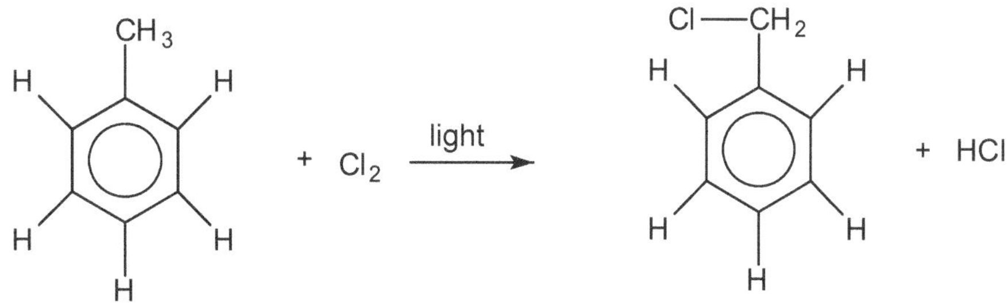

To substitute for the benzene ring hydrogens, an iron(III) catalyst must be present. Without this special iron catalyst, the benzene ring hydrogens are unreactive. To substitute for an alkane hydrogen, light must be present. For toluene, the light-catalyzed reaction substitutes a chlorine for a hydrogen in the methyl group attached to the benzene ring.

87. Primary alcohols (a, d, and f) are oxidized to aldehydes, which can be oxidized further to carboxylic acids. Secondary alcohols (b, e, and f) are oxidized to ketones, and tertiary alcohols (c and f) do not undergo this type of oxidation reaction. Note that compound f contains a primary, secondary, and tertiary alcohol. For the primary alcohols (a, d, and f), we listed both the aldehyde and the carboxylic acid as possible products.

a. 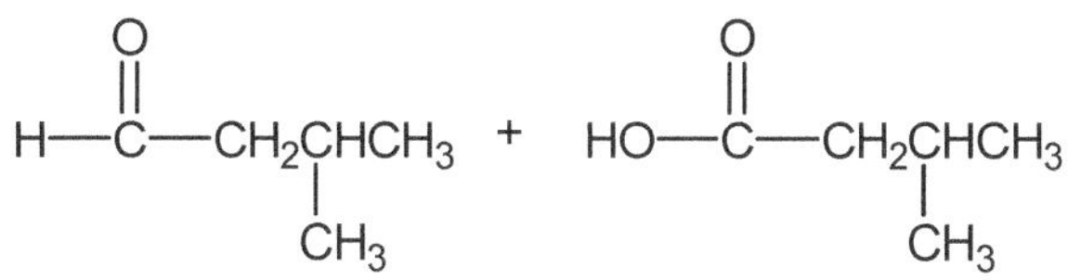

b. 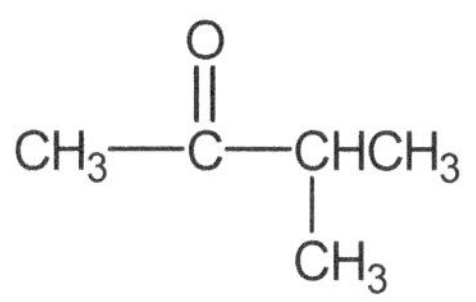

c. No reaction

d. 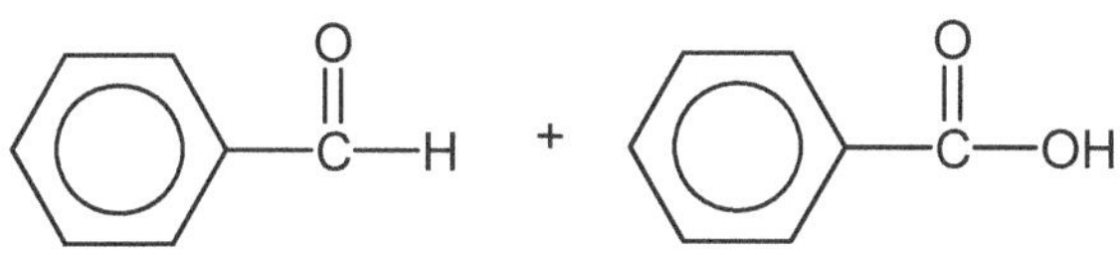

e. 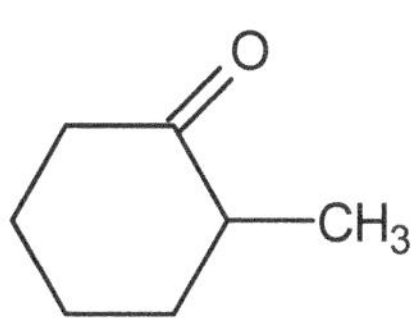

f. 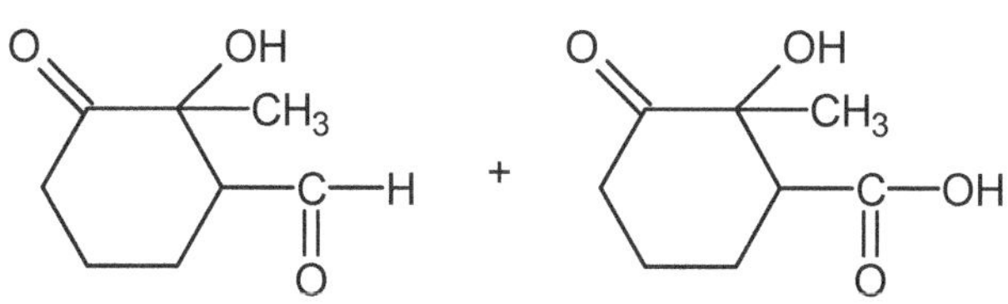

89. Primary alcohols and aldehydes are oxidized to carboxylic acids. Only 1-butanol is a primary alcohol and there are no aldehydes in the choices. So answer a (1-butanol) is correct.

91. a. $CH_3CH{=}CH_2 + Br_2 \rightarrow CH_3CHBrCH_2Br$ (addition reaction of Br_2 with propene)

b.

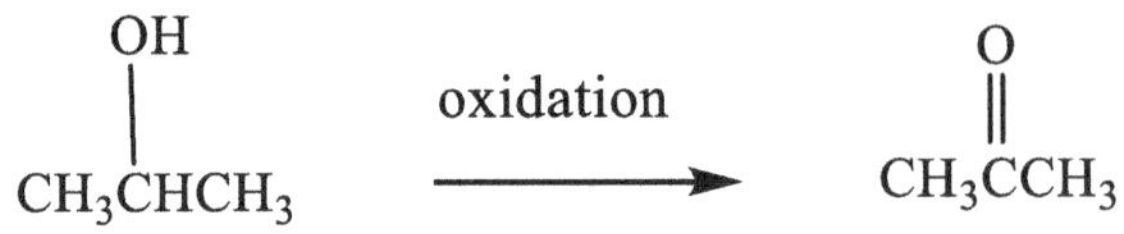

The oxidation of 2 propanol will yield acetone.

c.

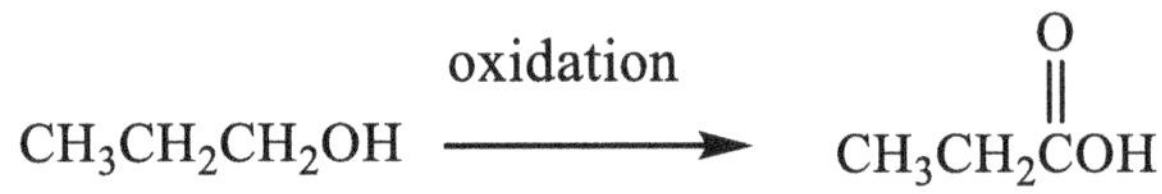

Oxidation of 1-propanol will eventually yield propanoic acid. The aldehyde propanal is an intermediate product in the reaction.

93. Reaction of a carboxylic acid with an alcohol can produce these esters.

Polymers

95. The backbone of the polymer contains only carbon atoms, which indicates that Kel-F is an addition polymer. The smallest repeating unit of the polymer and the monomer used to produce this polymer are:

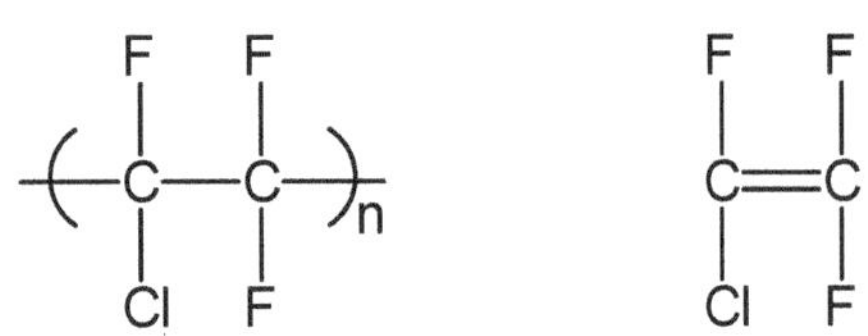

Note: Condensation polymers generally have O or N atoms in the backbone of the polymer.

97.

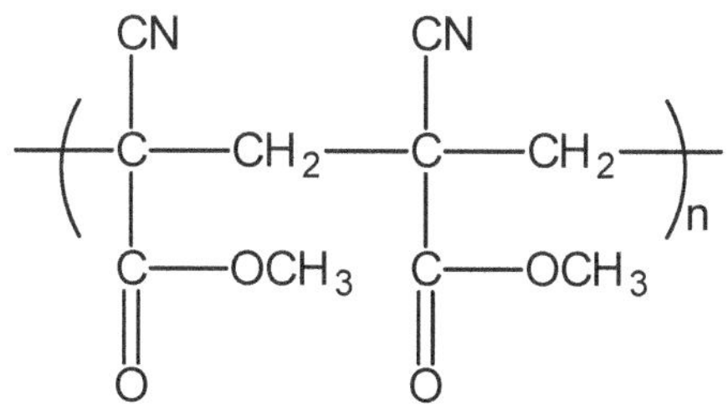

Super glue is an addition polymer formed by reaction of the C=C bond in methyl cyanoacrylate.

99. H_2O is eliminated when Kevlar forms. Two repeating units of Kevlar are:

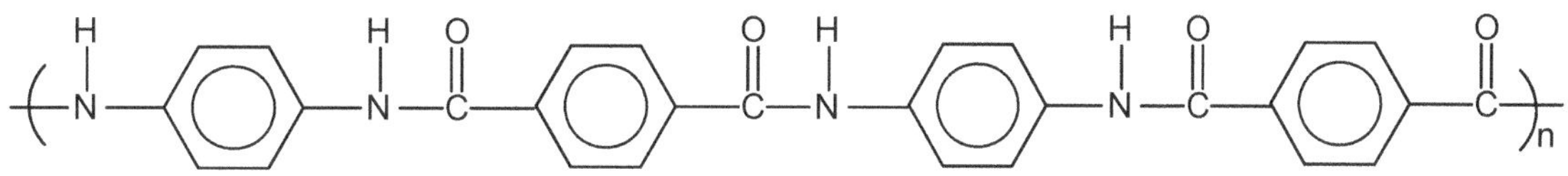

101. This is a condensation polymer, where two molecules of H_2O form when the monomers link together.

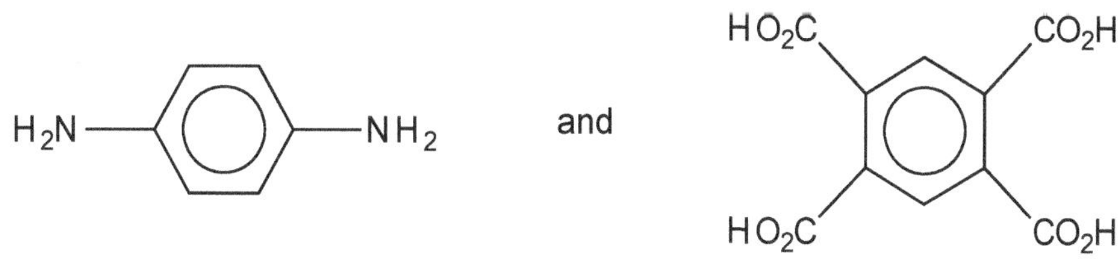

103. Divinylbenzene has two reactive double bonds that are used during formation of the polymer. The key is for the double bonds to insert themselves into two different polymer chains during the polymerization process. When this occurs, the two chains are bonded together (are cross-linked). The chains cannot move past each other because of the crosslinks, making the polymer more rigid.

105. a. The polymer formed using 1,2-diaminoethane will exhibit relatively strong hydrogen-bonding interactions between adjacent polymer chains. Hydrogen bonding is not present in the ethylene glycol polymer (a polyester polymer forms), so the 1,2-diaminoethane polymer will be stronger.

b. The presence of rigid groups (benzene rings or multiple bonds) makes the polymer stiffer. Hence the monomer with the benzene ring will produce the more rigid polymer.

c. Polyacetylene will have a double bond in the carbon backbone of the polymer.

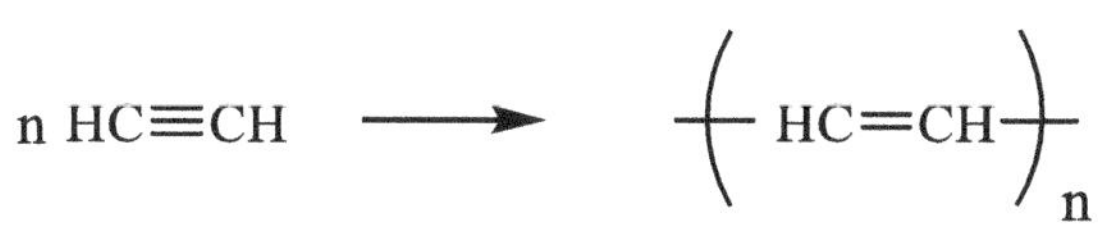

The presence of the double bond in polyacetylene will make polyacetylene a more rigid polymer than polyethylene. Polyethylene doesn't have C=C bonds in the backbone of the polymer (the double bonds in the monomers react to form the polymer).

Natural Polymers

107. a. Serine, tyrosine, and threonine contain the -OH functional group in the R group.

b. Aspartic acid and glutamic acid contain the -COOH functional group in the R group.

c. An amine group has a nitrogen bonded to other carbon and/or hydrogen atoms. Histidine, lysine, arginine, and tryptophan contain the amine functional group in the R group.

d. The amide functional group is:

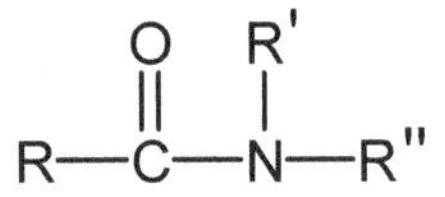

This functional group is formed when individual amino acids bond together to form the peptide linkage. Glutamine and asparagine have the amide functional group in the R group.

109. a. Aspartic acid and phenylalanine make up aspartame.

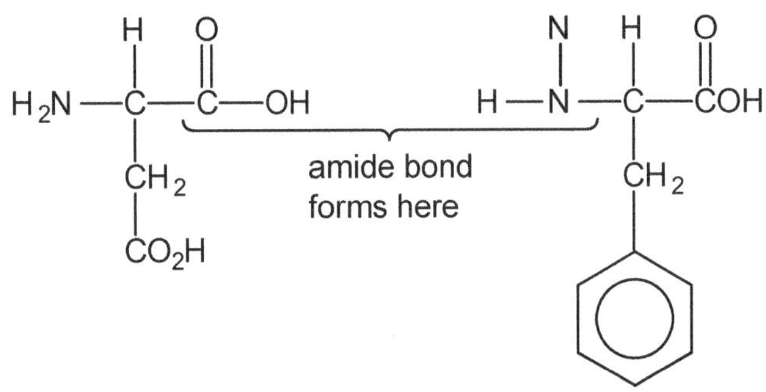

b. Aspartame contains the methyl ester of phenylalanine. This ester can hydrolyze to form methanol:

$$RCO_2CH_3 + H_2O \rightleftharpoons RCO_2H + HOCH_3$$

111.

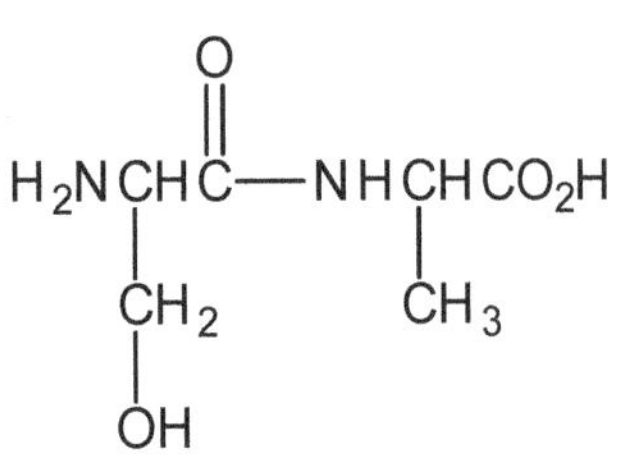

ser - ala

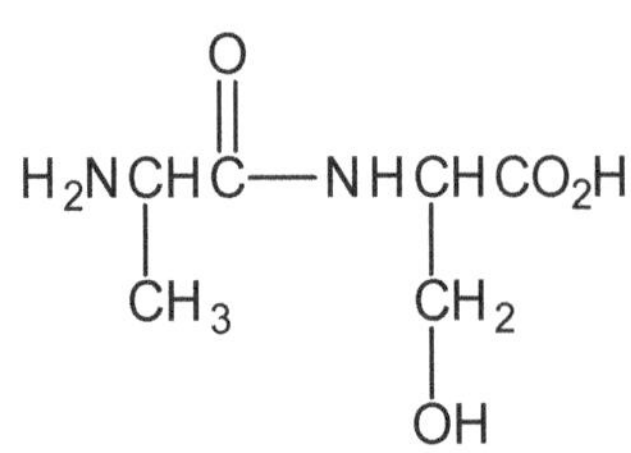

ala - ser

113. a. Six tetrapeptides are possible. From NH_2 to CO_2H end:

phe-phe-gly-gly, gly-gly-phe-phe, gly-phe-phe-gly,

phe-gly-gly-phe, phe-gly-phe-gly, gly-phe-gly-phe

b. Twelve tetrapeptides are possible. From NH_2 to CO_2H end:

phe-phe-gly-ala, phe-phe-ala-gly, phe-gly-phe-ala,

phe-gly-ala-phe, phe-ala-phe-gly, phe-ala-gly-phe,

gly-phe-phe-ala, gly-phe-ala-phe, gly-ala-phe-phe

ala-phe-phe-gly, ala-phe-gly-phe, ala-gly-phe-phe

115. The secondary structure of a protein describes the arrangement in space of the protein's polypeptide chain. The most common secondary structures are the α-helix and the pleated sheet. The α-helix secondary structure gives a protein elasticity; examples are wool, hair, and tendons. The pleated sheet secondary structure has hydrogen bonding interactions between protein chains. As several protein chains interact with each other through hydrogen bonding, fibers result that are very strong and are resistant to stretching. Examples of proteins having pleated sheet secondary structures are silk and muscle fibers.

117. a. Ionic: Need NH_2 on side chain of one amino acid with CO_2H on side chain of the other amino acid. The possibilities are:

NH_2 on side chain = His, Lys, or Arg; CO_2H on side chain = Asp or Glu

b. Hydrogen bonding: Need N–H or O–H bond present in side chain. The hydrogen bonding interaction occurs between the X– H bond and a carbonyl group from any amino acid.

X–H · · · · · · · O = C (carbonyl group)

X–H		O = C
Ser	Asn	Any amino acid
Glu	Thr	
Tyr	Asp	
His	Gln	
Arg	Lys	

c. Covalent: Cys–Cys (forms a disulfide linkage)

d. London dispersion: All amino acids with nonpolar R groups. They are:

Gly, Ala, Pro, Phe, Ile, Trp, Met, Leu, and Val

e. Dipole-dipole: Need side chain with OH group. Tyr, Thr and Ser all could form this specific dipole-dipole force with each other since all contain an OH group in the side chain.

119. Glutamic acid: R = $-CH_2CH_2CO_2H$; valine: R = $-CH(CH_3)_2$; a polar side chain is replaced by a nonpolar side chain. This could affect the tertiary structure of hemoglobin and the ability of hemoglobin to bind oxygen.

121. See Figures 22.29 and 22.30 of the text for examples of the cyclization process.

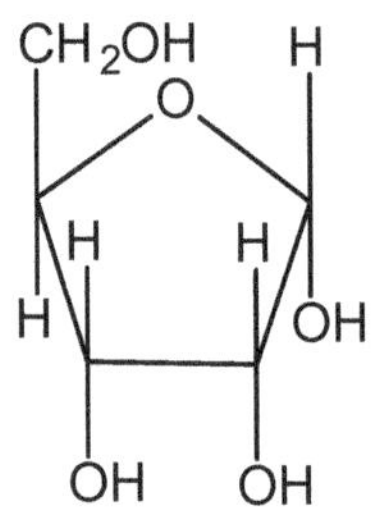

D-Ribose

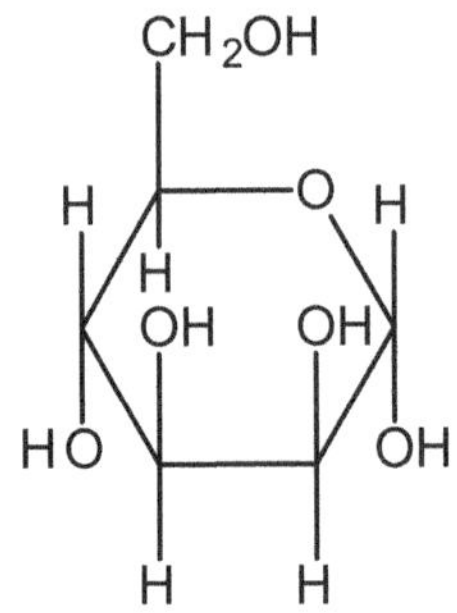

D-Mannose

123. The aldohexoses contain 6 carbons and the aldehyde functional group. Glucose, mannose, and galactose are aldohexoses. Ribose and arabinose are aldopentoses since they contain 5 carbons with the aldehyde functional group. The ketohexose (6 carbons + ketone functional group) is fructose, and the ketopentose (5 carbons + ketone functional group) is ribulose.

125. The α and β forms of glucose differ in the orientation of a hydroxy group on one specific carbon in the cyclic forms (see Figure 22.30 of the text). Starch is a polymer composed of only α-D-glucose, and cellulose is a polymer composed of only β-D-glucose.

127. A chiral carbon has four different groups attached to it. A compound with a chiral carbon is optically active. Isoleucine and threonine contain more than the one chiral carbon atom (see asterisks).

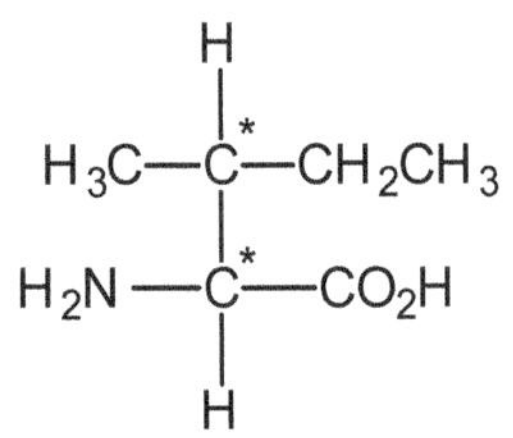

isoleucine

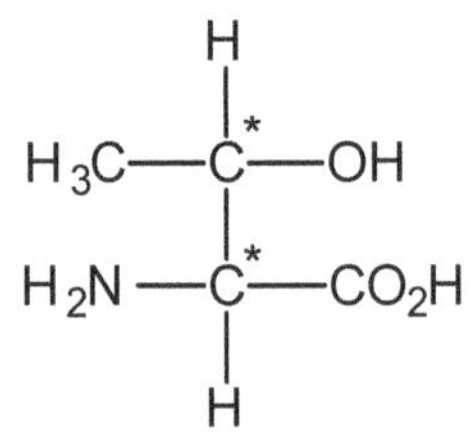

threonine

129. Only one of the isomers is optically active. The chiral carbon in this optically active isomer is marked with an asterisk.

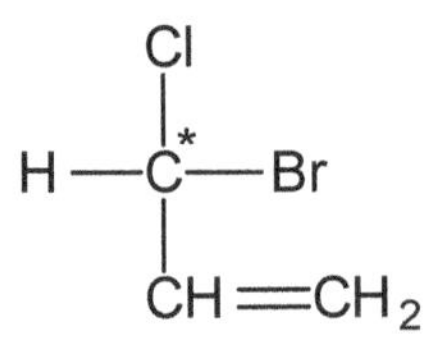

131. Aspartame has two chiral carbons (marked with an *). Only these two carbons have four different groups bonded to each of them.

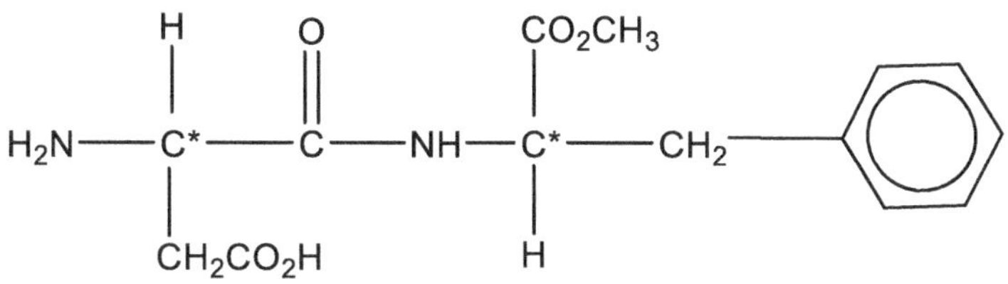

133. The chiral carbons are marked with an asterisk. A chiral carbon has four different groups bonded to it. The compound in answer a has only two chiral carbon atoms.

a.

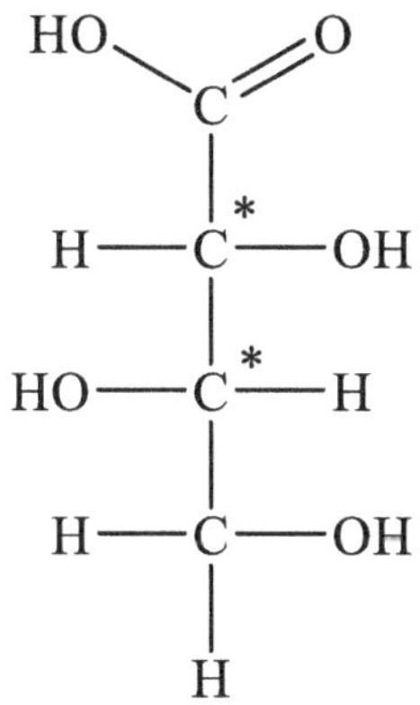

b.

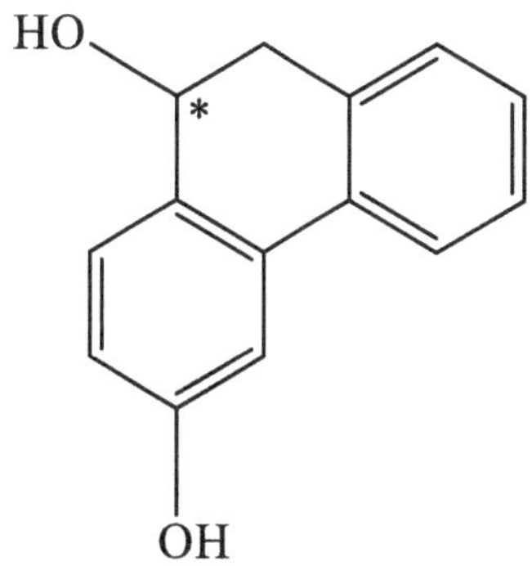

c.

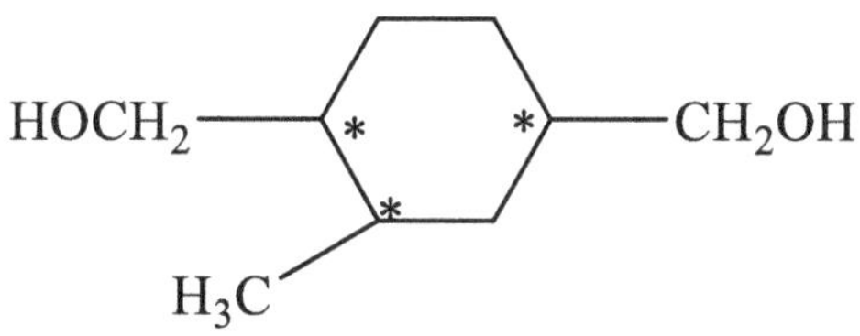

d.

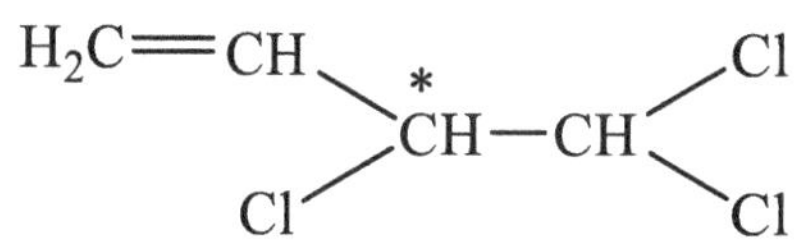

e.

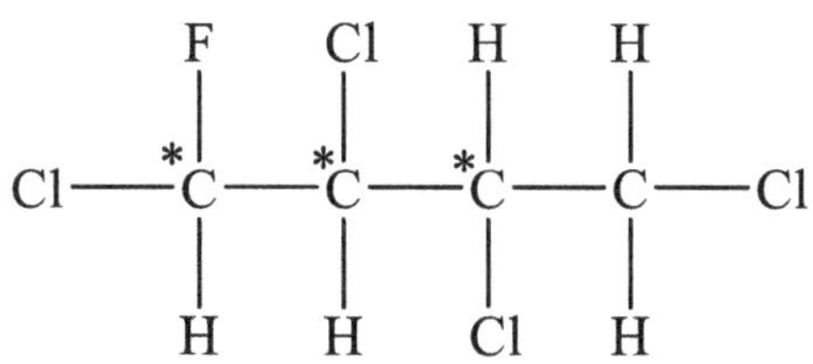

135. The complementary base pairs in DNA are cytosine (C) and guanine (G), and thymine (T) and adenine (A). The complementary sequence is C–C–A–G–A–T–A–T–G

137. Uracil will hydrogen bond to adenine. The dashed lines represent the H-bonding interactions.

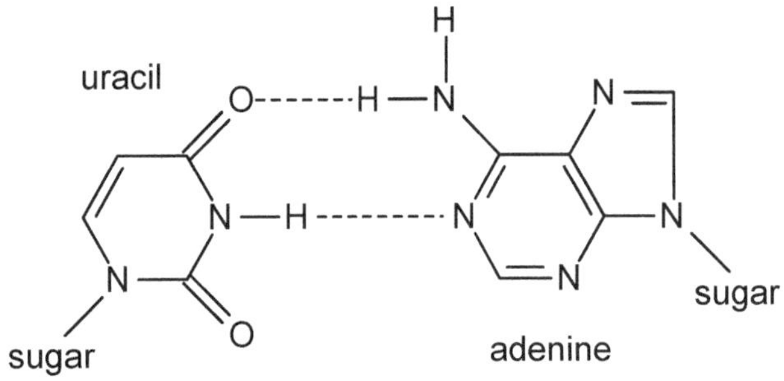

139. Base pair:

RNA	DNA
A	T
G	C
C	G
U	A

a. Glu: CTT, CTC — Val: CAA, CAG, CAT, CAC

Met: TAC — Trp: ACC

Phe: AAA, AAG — Asp: CTA, CTG

b. DNA sequence for trp-glu-phe-met:

ACC –CTT –AAA –TAC
or or
CTC AAG

c. Due to glu and phe, there is a possibility of four different DNA sequences. They are:

ACC–CTT–AAA–TAC or ACC–CTC–AAA–TAC or

ACC–CTT–AAG–TAC or ACC–CTC–AAG –TAC

d. T—A—C—C—T—G—A—A—G

met (TAC) asp (CTG) phe (AAG)

e. TAC–CTA–AAG; TAC–CTA–AAA; TAC–CTG–AAA

ChemWork Problems

141. We omitted the hydrogens for clarity. The number of hydrogens bonded to each carbon is the number necessary to form four bonds.

a.

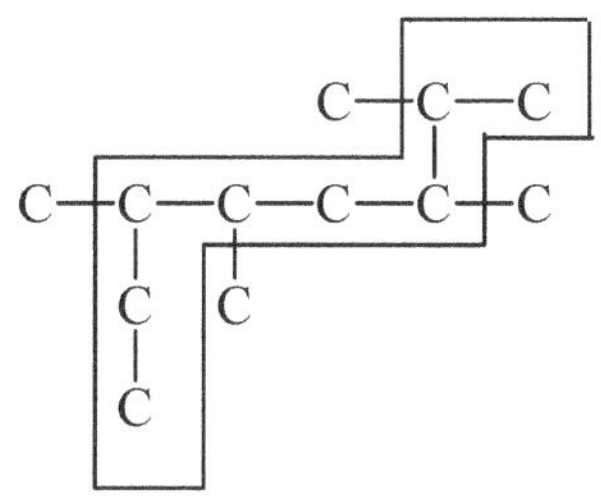

2,3,5,6-tetramethyloctane

b.

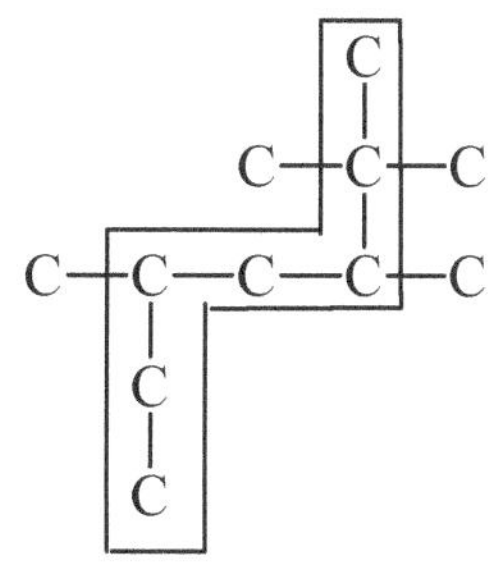

2,2,3,5-tetramethylheptane

c.

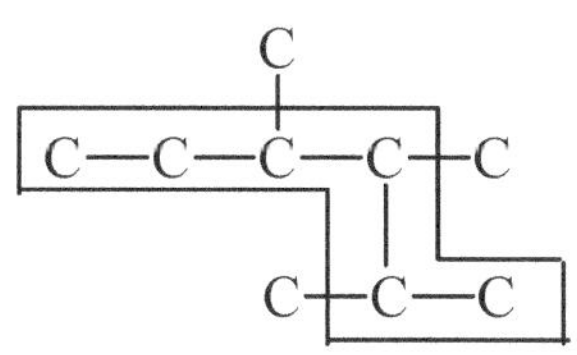

2,3,4-trimethylhexane

d.

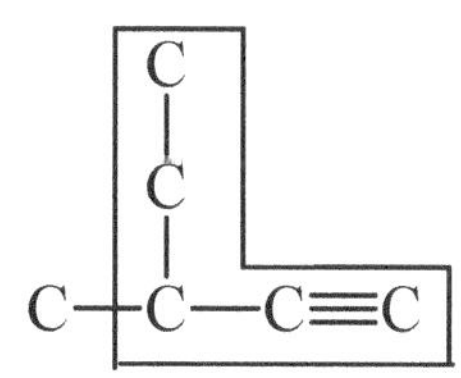

3-methyl-1-pentyne

143.

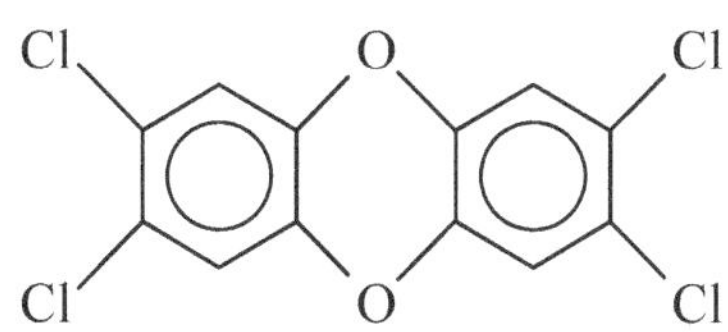

There are many possibilities for isomers. Any structure with four chlorines replacing four hydrogens in any four of the numbered positions would be an isomer; i.e., 1,2,3,4-tetrachloro-dibenzo-p-dioxin is a possible isomer.

145. The line notation for the five structural isomers of C_6H_{14} are:

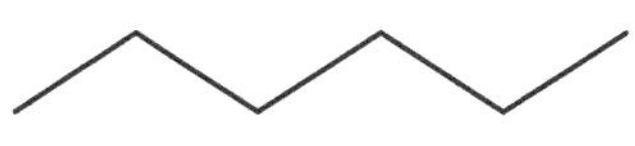

n-hexane

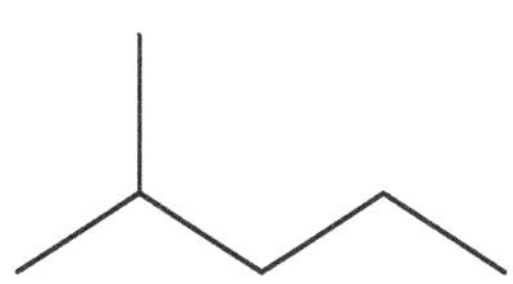

2-methylpentane

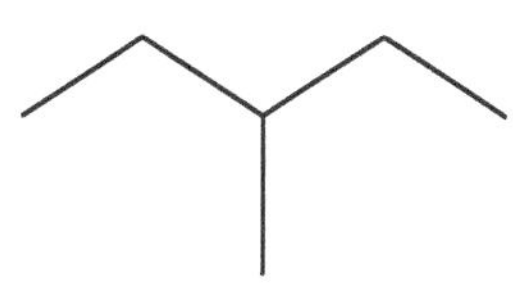

3-methylpentane

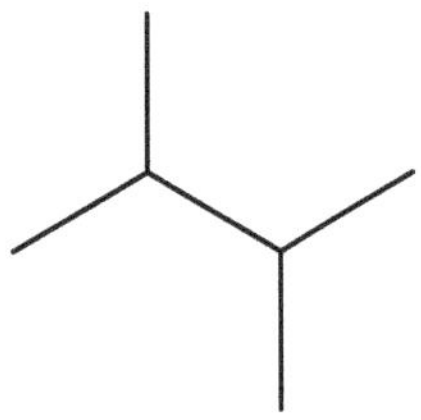

2,3-dimethylbutane

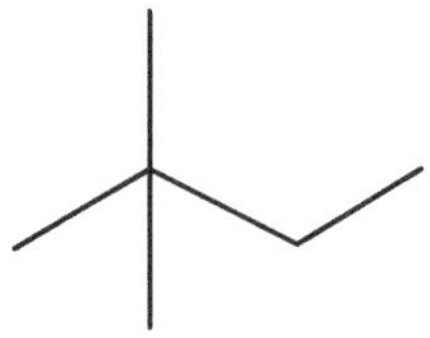

2,2-dimethylbutane

147. All the structures have a C_8H_{18} formula. To tell if a structure is a conformation or a different isomer, name it. If a structure has the same name, it is a conformation. If the compound has a different name, it is a different isomer. The names of the compounds are:

I. 2.4-dimethylhexane; II. 2,5-dimethylhexane; III. 2,3-dimethylhexane

IV. 2,4-dimethylhexane; V. 2,4-dimethylhexane

Structures I, IV, and V are all conformations of 2,4-dimethylhexane.

149. The isomers are:

$CH_3—O—CH_3$ CH_3CH_2OH

dimethyl ether, –23°C ethanol, 78.5°C

Ethanol, with its ability to form the relatively strong hydrogen-bonding interactions, boils at the higher temperature.

151. Alcohols consist of two parts, the polar OH group and the nonpolar hydrocarbon chain attached to the OH group. As the length of the nonpolar hydrocarbon chain increases, the solubility of the alcohol decreases in water, a very polar solvent. In methyl alcohol (methanol), the polar OH group overrides the effect of the nonpolar CH_3 group, and methyl alcohol is soluble in water. In stearyl alcohol, the molecule consists mostly of the long nonpolar hydrocarbon chain, so it is insoluble in water.

153. The structures, the types of intermolecular forces exerted, and the boiling points for the compounds are:

$CH_3CH_2CH_2C(=O)OH$	$CH_3CH_2CH_2CH_2CH_2OH$
butanoic acid, 164°C LD + dipole + H-bonding	1-pentanol, 137°C LD + H-bonding
$CH_3CH_2CH_2CH_2C(=O)H$	$CH_3CH_2CH_2CH_2CH_2CH_3$
pentanal, 103°C LD + dipole	n-hexane, 69°C LD only

All these compounds have about the same molar mass. Therefore, the London dispersion (LD) forces in each are about the same. The other types of forces determine the boiling-point order. Since butanoic acid and 1-pentanol both exhibit hydrogen bonding interactions, these two compounds will have the two highest boiling points. Butanoic acid has the highest boiling point since it exhibits H bonding along with dipole-dipole forces due to the polar C=O bond.

155. The necessary carboxylic acid is 2-chloropropanoic acid.

157. $KMnO_4$ will oxidize primary alcohols to aldehydes and then to carboxylic acids. Secondary alcohols are oxidized to ketones by $KMnO_4$. Tertiary alcohols and ethers are not oxidized by $KMnO_4$. The three isomers and their reactions with $KMnO_4$ are:

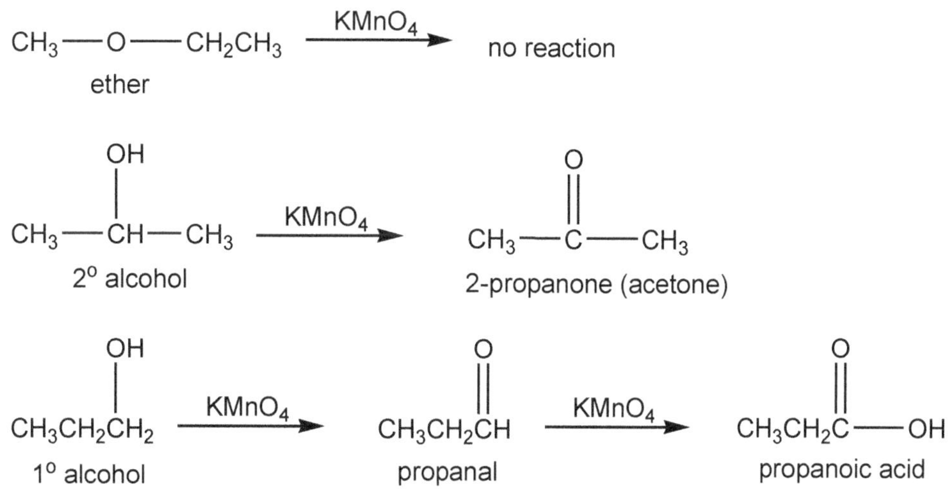

The products of the reactions with excess $KMnO_4$ are 2-propanone and propanoic acid.

159. In nylon, hydrogen-bonding interactions occur due to the presence of N–H bonds in the polymer. For a given polymer chain length, there are more N–H groups in Nylon-46 as

compared to Nylon-6. Hence Nylon-46 forms a stronger polymer compared to Nylon-6 due to the increased hydrogen-bonding interactions.

161. a.

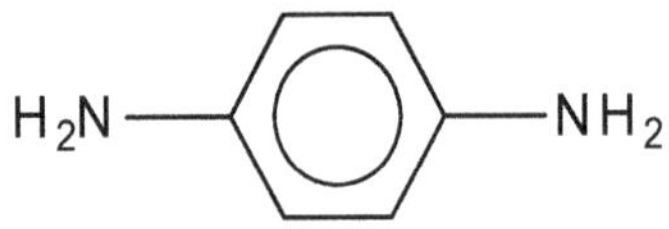

and

b. Repeating unit:

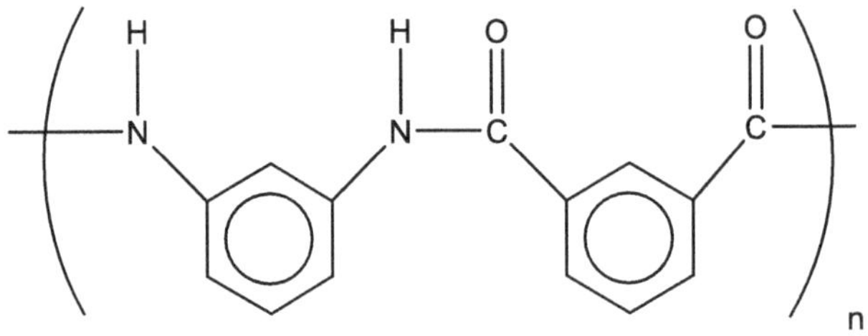

The two polymers differ in the substitution pattern on the benzene rings. The Kevlar chain is straighter, and there is more efficient hydrogen-bonding between Kevlar chains than between Nomex chains.

163. a. The bond angles in the ring are about 60°. VSEPR predicts bond angles close to 109°. The bonding electrons are closer together than they prefer, resulting in strong electron-electron repulsions. Thus ethylene oxide is unstable (reactive).

b. The ring opens up during polymerization; the monomers link together through the formation of O–C bonds.

$$\left(O-CH_2CH_2-O-CH_2CH_2-O-CH_2CH_2 \right)_n$$

165. Glutamic acid:

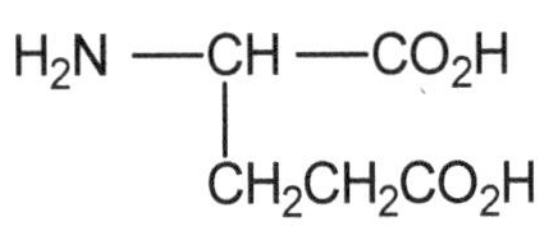

One of the two acidic protons in the carboxylic acid groups is lost to form MSG. Which proton is lost is impossible for you to predict.

Monosodium glutamate:

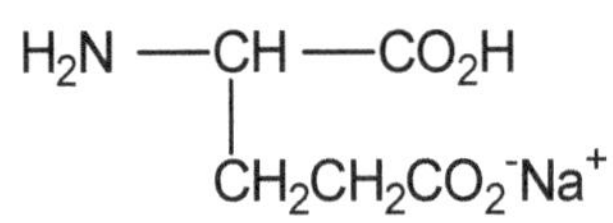

In MSG, the acidic proton from the carboxylic acid in the R group is lost, allowing formation of the ionic compound.

167. $\Delta G = \Delta H - T\Delta S$; for the reaction, we break a P–O and O–H bond and form a P–O and O–H bond, so $\Delta H \approx 0$ based on bond dissociation energies. ΔS for this process is negative (unfavorable) because positional probability decreases. Thus, $\Delta G > 0$ due to the unfavorable ΔS term, and the reaction is not expected to be spontaneous.

169. Alanine can be thought of as a diprotic acid. The first proton to leave comes from the carboxylic acid end with $K_a = 4.5 \times 10^{-3}$. The second proton to leave comes from the protonated amine end (K_a for $R{-}NH_3^+ = K_w/K_b = 1.0 \times 10^{-14}/7.4 \times 10^{-5} = 1.4 \times 10^{-10}$).

In 1.0 *M* H^+, both the carboxylic acid and the amine end will be protonated since H^+ is in excess. The protonated form of alanine is below. In 1.0 *M* OH^-, the dibasic form of alanine will be present because the excess OH^- will remove all acidic protons from alanine. The dibasic form of alanine follows.

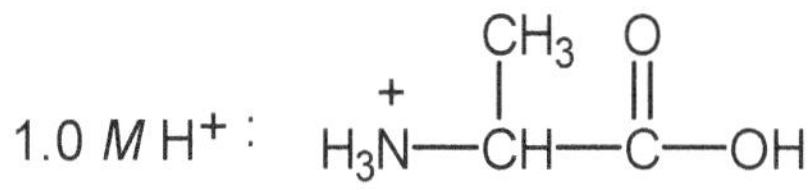

protonated form

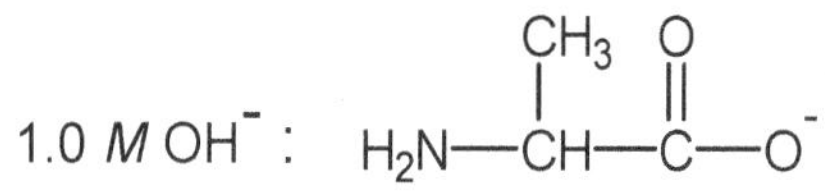

dibasic form

171. For denaturation, heat is added so it is an endothermic process. Because the highly ordered secondary structure is disrupted, positional probability increases, so entropy will increase. Thus ΔH and ΔS are both positive for protein denaturation.

173. a. Zn^{2+} has the $[Ar]3d^{10}$ electron configuration, and zinc does form 2+ charged ions.

$$\text{Mass \% Zn} = \frac{\text{mass of 1 mol Zn}}{\text{mass of 1 mol } CH_3CH_2ZnBr} \times 100 = \frac{65.38\text{ g}}{174.34\text{ g}} \times 100 = 37.50\%\text{ Zn}$$

b. The reaction is:

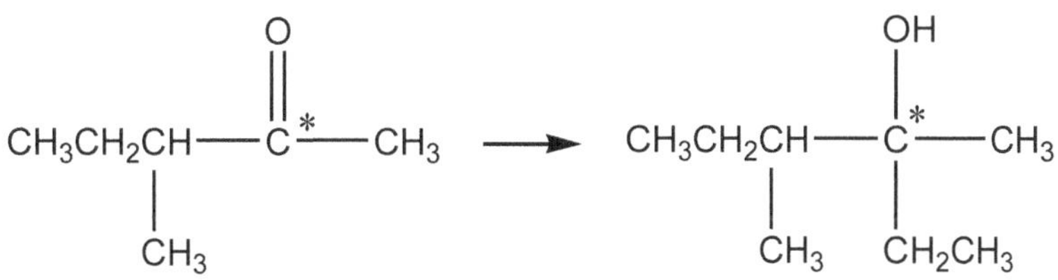

The hybridization changes from sp^2 to sp^3.

c. 3,4-dimethyl-3-hexanol

Challenge Problems

175. Out of 100.00 g:

$$71.89 \text{ g C} \times \frac{1 \text{ mol C}}{12.01 \text{ g C}} = 5.986 \text{ mol} \approx 6 \text{ mol C}$$

$$12.13 \text{ g H} \times \frac{1 \text{ mol H}}{1.008 \text{ g H}} = 12.03 \text{ mol} \approx 12 \text{ mol H}$$

$$15.98 \text{ g O} \times \frac{1 \text{ mol O}}{16.00 \text{ g O}} = 0.9988 \text{ mol} \approx 1 \text{ mol O}$$

The empirical formula is $C_6H_{12}O$.

The general reaction for this hydrolysis reaction is:

R_2 must be CH_3CH_2 because CH_3CH_2OH is one of the products. The molar mass of CO_2H is $\approx$ 45 g/mol, so the molar mass of R_1 is 172 – 45 = 127 g/mol. Because R_1 is a straight chain alkane, its general formula is $CH_3(CH_2)_n$. So 15 + n(14) = 127 and n = 112/14 = 8. Ethyl caprate is:

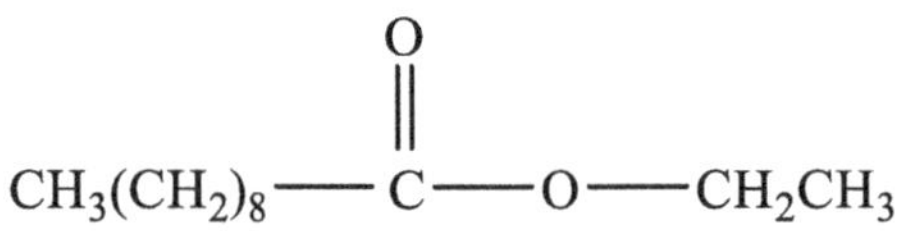

This compound has a molecular formula of $C_{12}H_{24}O_2$ with an empirical formula of $C_6H_{12}O$, which agrees with the mass percent data calculation above.

177. For the reaction:

$$^+H_3NCH_2CO_2H \rightleftharpoons 2\,H^+ + H_2NCH_2CO_2^- \quad K_{eq} = 7.3 \times 10^{-13} = K_a(-CO_2H) \times K_a(-NH_3^+)$$

$$7.3 \times 10^{-13} = \frac{[H^+]^2[H_2NCH_2CO_2^-]}{[^+H_3NCH_2CO_2H]} = [H^+]^2, \quad [H^+] = (7.3 \times 10^{-13})^{1/2}$$

$[H^+] = 8.5 \times 10^{-7}$ *M*; pH = –log[H^+] = 6.07 = isoelectric point

c. The chiral carbons are indicated with an asterisk.

179. a. Even though this form of tartaric acid contains 2 chiral carbon atoms (see asterisks in the following structure), the mirror image of this form of tartaric acid is superimposable. Therefore, it is not optically active. An easier way to identify optical activity in molecules with two or more chiral carbon atoms is to look for a plane of symmetry in the molecule. If a molecule has a plane of symmetry, then it is never optically active. A plane of symmetry is a plane that bisects the molecule where one side exactly reflects on the other side.

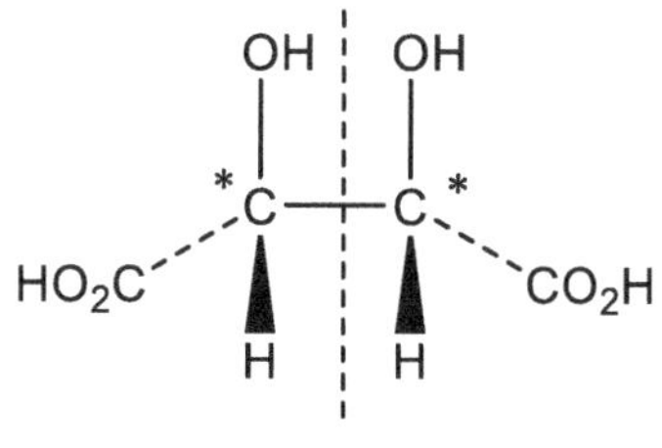

b. The optically active forms of tartaric acid have no plane of symmetry. The structures of the optically active forms of tartaric acid are:

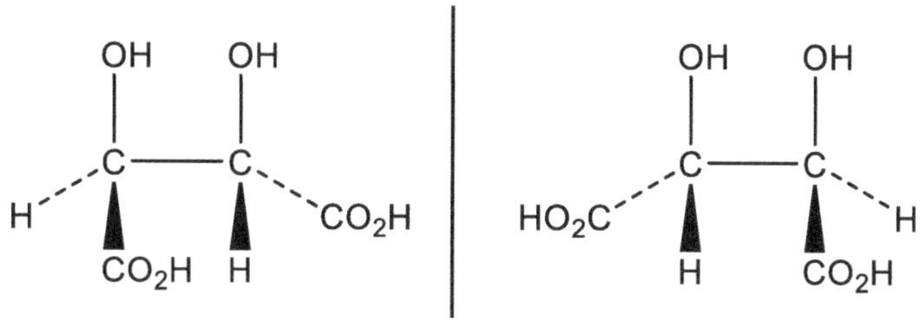

These two forms of tartaric acid are nonsuperimposable.

181.

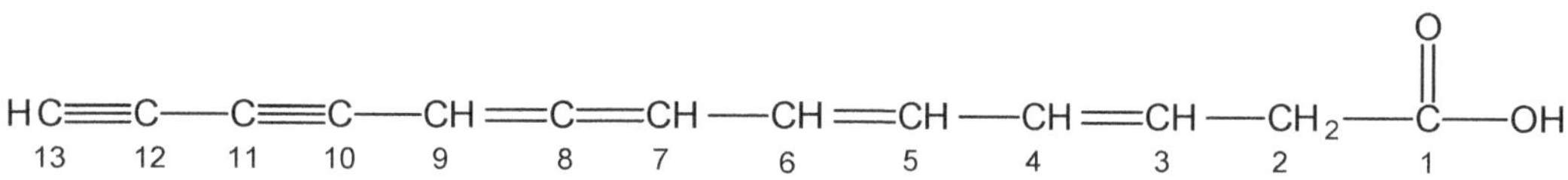

183. a. The three structural isomers of C_5H_{12} are:

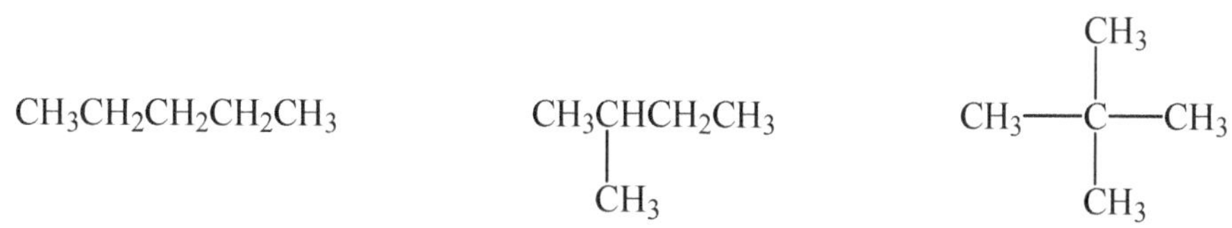

n-pentane 2-methylbutane 2,2-dimethylpropane

n-Pentane will form three different monochlorination products: 1-chloropentane, 2-chloropentane, and 3-chloropentane (the other possible monochlorination products differ by a simple rotation of the molecule; they are not different products from the ones listed). 2,2-Dimethylpropane will only form one monochlorination product: 1-chloro-2,2-dimethylpropane. 2-Methylbutane is the isomer of C_5H_{12} that forms four different monochlorination products: 1-chloro-2-methylbutane, 2-chloro-2-methyl-butane, 3-chloro-2-methylbutane (or we could name this compound 2-chloro-3-methylbutane), and 1-chloro-3-methylbutane.

b. The structure of 1-chloro-1-methylcyclohexane is:

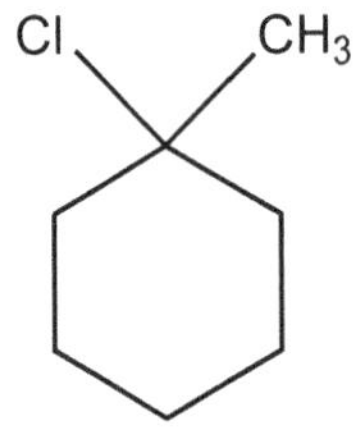

The addition reaction of HCl with an alkene is a likely choice for this reaction (see Exercise 82). The two isomers of C_7H_{12} that produce 1-chloro-1-methylcyclohexane as the major product are:

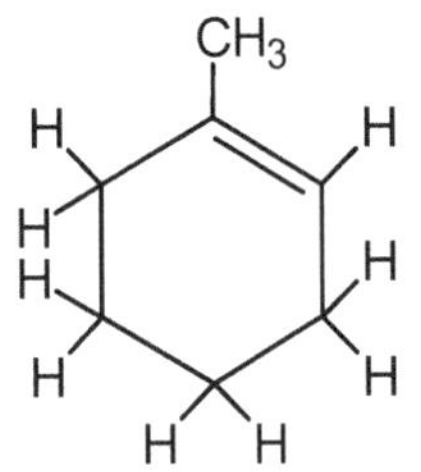

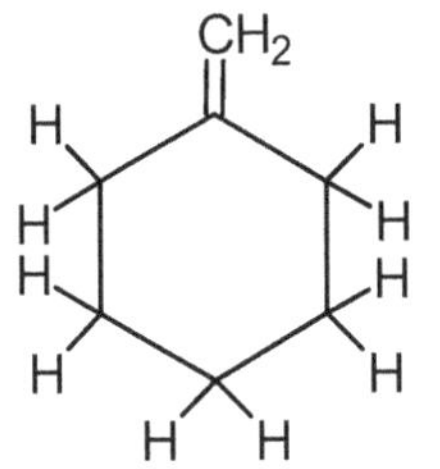

c. Working backwards, 2° alcohols produce ketones when they are oxidized (1° alcohols produce aldehydes, then carboxylic acids). The easiest way to produce the 2° alcohol from a hydrocarbon is to add H_2O to an alkene (with H^+ present). The alkene reacted is 1-propene (or propene).

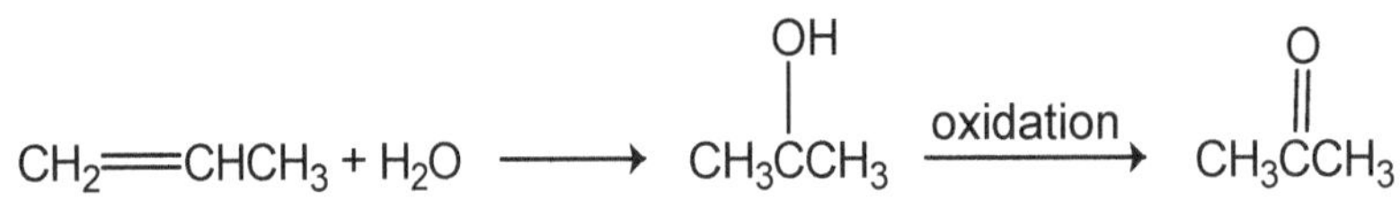

propene acetone

d. The $C_5H_{12}O$ formula has too many hydrogens to be anything other than an alcohol (or an unreactive ether). 1° Alcohols are first oxidized to aldehydes, then to carboxylic acids. Therefore, we want a 1° alcohol. The 1° alcohols with formula $C_5H_{12}O$ are:

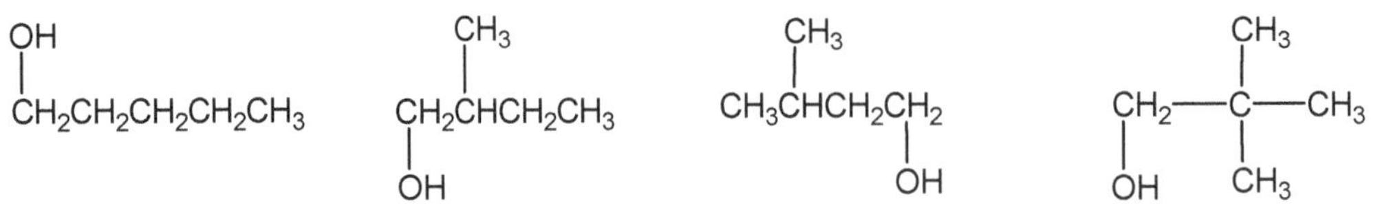

1-pentanol 2-methyl-1-butanol 3-methyl-1-butanol 2,2-dimethyl-1-propanol

There are other alcohols with formula $C_5H_{12}O$, but they are all 2° or 3° alcohols, which do not produce carboxylic acids when oxidized.

185.

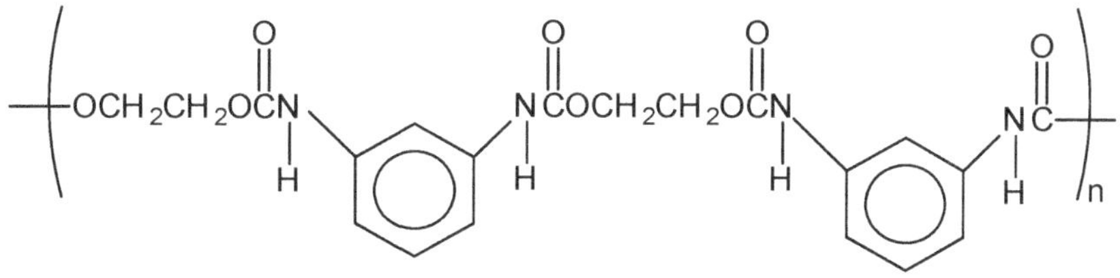

187. a. The temperature of the rubber band increases when it is stretched.

b. Exothermic because heat is released.

c. As the polymer chains that make up the rubber band are stretched, they line up more closely together, resulting in stronger London dispersion forces between the chains. Heat is released as the strength of the intermolecular forces increases.

d. Stretching is not spontaneous, so ΔG is positive. $\Delta G = \Delta H - T\Delta S$; since ΔH is negative, ΔS must be negative in order to give a positive ΔG.

e.

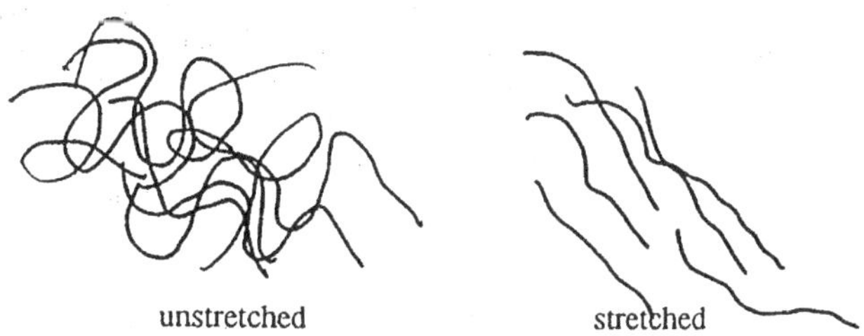

The structure of the stretched polymer chains is more ordered (has a smaller positional probability). Therefore, entropy decreases as the rubber band is stretched.

189. $$4.2 \times 10^{-3}\ \text{g}\ K_2CrO_7 \times \frac{1\ \text{mol}\ K_2Cr_2O_7}{294.20\ \text{g}} \times \frac{1\ \text{mol}\ Cr_2O_7^{2-}}{\text{mol}\ K_2Cr_2O_7} \times \frac{3\ \text{mol}\ C_2H_5OH}{2\ \text{mol}\ Cr_2O_7^{2-}}$$

$$= 2.1 \times 10^{-5}\ \text{mol}\ C_2H_5OH$$

$$n_{breath} = \frac{PV}{RT} = \frac{\left(750.\ \text{mm Hg} \times \frac{1\ \text{atm}}{760\ \text{mm Hg}}\right) \times 0.500\ \text{L}}{\frac{0.08206\ \text{L atm}}{\text{K mol}} \times 303\ \text{K}} = 0.0198\ \text{mol breath}$$

$$\text{Mole \%}\ C_2H_5OH = \frac{2.1 \times 10^{-5}\ \text{mol}\ C_2H_5OH}{0.0198\ \text{mol total}} \times 100 = 0.11\%\ \text{alcohol}$$